Dedication

This volume is dedicated to the memory of Evelyn Nelson who died this summer. Evelyn was a participant at the Workshop, and she chaired the first session. She will be missed both for her research in mathematics and computer science and for her lively interest in the work of others.

PREFACE

The Third Workshop on the Mathematical Foundations of Programming Language Semantics took place on the campus of Tulane University, New Orleans, Louisiana on April 8, 9 and 10, 1987. The First Workshop was at Kansas State University, Manhattan, Kansas in April, 1985 (see LNCS **239**), and the Second Workshop with a limited number of participants was at Kansas State in April, 1986. It was the intention of the organizers that the Third Workshop survey as many areas of the Mathematical Foundations of Programming Language Semantics as reasonably possible. Toward that end, following were the invited speakers:

Professor Stephen Brookes, Carnegie-Mellon University
Professor John Gray, University of Illinois
Professor Neil Jones, Copenhagen University
Professor Jimmie Lawson, Louisiana State University
Professor Gordon Plotkin, University of Edinburgh
Professor Dana Scott, Carnegie-Mellon University

The Workshop attracted 49 submitted papers, from which 28 papers were chosen for presentation. As the Table of Contents indicates, the papers ranged in subject from category theory and λ-calculus to the structure theory of domains and power domains, to implementation issues surrounding semantics. The papers appear in this volume in approximately the order in which they were presented at the Workshop.

In addition to the editors of this volume, the Program Committee for the Workshop consisted of David Benson, Boumediene Belkhouche, Tsutomu Kamimura, Ernie Manes, Jon Shultis, George Strecker, Adrian Tang and David Wise. In addition, David Black, Horst Herrlich, Karl Hofmann, Cordelia Hull and Melvin F. Janowitz also served as referees for the papers which were submitted for presentation. Boumediene Belkhouche also served as Local Arrangements Chairman for the Workshop.

This Workshop would not have taken place without the financial support of Tulane University, which bore most of the cost. Thanks go to Dr. Francis L. Lawrence, Provost and Vice President for Academic Affairs for his help in this regard. Thanks go also to the ACM SIGACT and SIGPLAN special interest groups for their sponsorship of the Workshop and to the IBM Corporation and Allan A. Kancher, the head of their New Orleans office, who also provided support for the Workshop. Thanks are also due to Dr. Mary Ann Maguire, Acting Dean of Newcomb College, for supporting the Workshop.

Finally, special thanks go to Mrs. Geralyn Caradona, Administrative Assistant of the Mathematics Department of Tulane University, and to Frank Elliott, a graduate student in the Mathematics Department, for their help in discharging all the "behind the scenes" details which are so important to the smooth running of a meeting such as this.

Michael Main   Austin Melton   Michael Mislove   David Schmidt

September, 1987

Table of Contents

I. Categorical and Algebraic Methods

II. Structure Theory of Continuous Posets and Related Objects

III. Domain Theory

IV. Domain Theory and Theoretical Computation

V. Implementation Issues

VI. New Directions

Part I
Categorical and Algebraic Methods

A Categorical Treatment of Polymorphic Operations

by John W. Gray
University of Illinois at Urbana-Champaign
Urbana, Illinois.

0. INTRODUCTION

Category theory is an algebraic tool for discussing mathematical structures precisely. It only becomes interesting when one uses it to clarify some part of mathematics or theoretical science. It began as an outgrowth of algebraic topology in the 1940's as part of an attempt to explain what "natural transformations" were. Today, in theoretical computer science, there are a number of things that need to be explained connected with the semantics of programming languages that lead to structures similar to those developed for dealing with natural transformations.

Historically, category theory played an important role in mathematics during the period approximately from 1945 to 1975. After the end of the second world war there was a tremendous growth in abstract mathematics. The set theoretic paradygm for mathematics began to be taken seriously and many kinds of structures which previously had only been considered for number systems were generalized to more abstract entities that were relevant in more general circumstances. As the complexity of the set theoretic descriptions rose, it became evident that some kind of simplification was essential. Category theory proposed that structures be described in terms of what they did rather than what they were. Two central driving forces behind this suggestion were generalized homology theories and structures like fiber bundles and sheaves that were described in terms of local conditions. Four achievements of this development can be singled out.

i) Concepts were clarified by phrasing them in categorical terms. E.g., the notion of a manifold grew to the general notion of an object modeled locally on simpler objects. One came to see how fiber spaces, sheaves, schemes, etc. were all special cases of a general idea. Less abstractly, the simple notions of products and coproducts in categories sufficied to codify the properties of a number of seeming different constructions.

ii) Uniform definitions of constructions were given in categorical terms. The specializations of these constructions in particular categories showed how seemingly ad hoc constructions were "really" the same as well understood constructions in other contexts. E.g., the Stone-Cech compactification was recognized as the precursor of and model for any number of left adjoint constructions.

iii) Abstraction led to simplification and generalization of concepts and proofs. By dissecting out the crucial ingredients of a proof and providing categorical descriptions of the constructions involved in the proof, one could often obtain a much more general categorical proof which could then be applied in quite different circumstances to yield new results.

iv) Complicated computations were guided by general categorical results. A general construction often makes transparent a lengthy computation in a particular category. For instance, general categorical formulas for adjoint constructions guided the computation of these entities in special cases.

Application of these results in new or unexpected categories led to increased understanding of particular areas of mathematics. The most interesting category theory was always that which particularized in the most interesting way.

Today the situation with regard to the mathematical role of category theory is quite different. Most mathematicians feel that the concepts with which they are working are clear and well understood. There is little need for the development of new concepts of a categorical nature, since the impression is that those concepts which are general enough to be relevant across a large spectrum of mathematics will no longer yield interesting new results. In particular, uniform definitions and abstraction are no longer interesting, and narrow specialization is the order of the day. It is felt that depth can only be achieved at the expense of breadth.

However, there are parts of theoretical computer science in which the situation now is similar to the situation in mathematics 20 or 30 years ago, in the sense that there is an exciting explosion in concept formation. Many notions which may be clear in the machine context lack precise theoretical descriptions. As was the case earlier in mathematics, many different formalisms may be proposed for what is essentially the same concept, and new ad hoc structures are erected to support these formalisms. The field is open for category theory to play the same role in these developments that it played earlier in mathematics; namely, clarify concepts, formulate uniform definitions, simplify through abstraction and guide computations. But this time, one doesn't have to create category theory at the same time, because category theory is already in place, replete with its whole arsenel of well understood, ready- made structures.

i) Cartesian closed categories. (Cf., [3], [8].) There is no doubt that the development of domain theory and models of the lambda calculus would have seriously suffered had the theory of cartesian closed categories not already been thoroughly developed, together with the relevant notions of limits and colimits that play a central role in fixed point semantics.

ii) The syntax and semantics of many-sorted algebraic theories have been thoroughly studied in category theory (cf., [1]) and this work provides the foundation for current work on algebraic semantics of data types in computer science. A central result here is the existence of a left adjoint for any algebraic functor, which allows one to construct any number of initial algebras essentially for free. (Pun intended.) Sketches have recently been proposed as a tool from category theory for the presentation of algebraic theories that has nicer formal properties than algebraic specifications.

iii) Toposes (cf. [6].) will be a significant topic in computer science since the current work on the effective topos turns out to have important consequences.

iv) Locally cartesian closed categories (cf., [11]) are intermediate between toposes and cartesian closed categories. They have been shown to be relevant in the discussion of models of Martin-Löf type theory.

v) It is the thesis of this paper that the theory of 2-categories (cf. [4], [7]) provides the proper expression for vague notions about operations which "behave the same way everywhere". Furthermore, 2-categorical notions are required to explain properties of models of the polymorphic lambda calculus.

vi) Fibred categories (cf., [5]) provide the correct setting for talking about "all models of all algebraic theories" and all morphisms between such models.

One fundamental, but unanswerable, question is: Is it easier to build an ad hoc structure than to find and adapt the appropriate structure from category theory, assuming that it exists there? The only response that I know is that if the structure is important enough and if it has sufficiently interesting properties, then people will try both approaches, each of them enriching the other. Another similar question is: Who should learn enough category theory to participate in the categorical approach? Here, I would say that every computer scientist working in any aspect of semantics of programming languages should be sufficiently aware of this approach to recognize when a discrete question or two directed to a category theorist might be appropriate. All graduate students in the field should have a brief course in the subject and the possibility of deeper studies if they are interested.

The notion we propose to explore here is that of "polymorphic" operations - operations that behave the same everywhere. Note that we are not concerned with polymorphic type theory here. The simplest example of a natural transformation is the map $\mu : V \to V^{**}$ from a vector space V to its double dual given by $\mu(v)(f) = f(v)$, which is obviously "natural" in that the construction is the same for all vector spaces and in fact can be seen to behave nicely with respect to linear transformations between vector spaces. Functors were invented as the things which natural transformations were transformations of. In the example, μ goes from the identity functor to the double dual functor. Finally, categories were invented for functors to operate on. In the example, the identity functor and the double dual functor are both functors from the category of vector spaces and linear transformations to itself. The general situation in category theory as derived from algebraic topology is illustrated in the diagram on the left. The current situation in theoretical computer science is illustrated on the right.

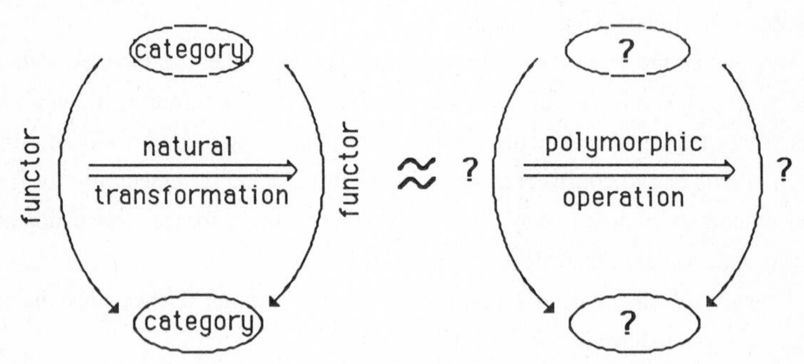

What the similarity sign is intended to suggest is that the two situations are the same; namely, that very many polymorphic operations in fact are natural transformations between functors which in turn are operations between categories.

1. Category Theory: Review of notation

The standard reference for category theory is MacLane [9]. Bold face letters **C, D**, etc. will denote categories. The class of objects of a category **C** will be denoted by ob(**C**). Elements of ob(**C**) will be denoted by A, B, X, Y, etc. The set of morphisms from X to Y will be denoted by **C**(X, Y), and called the *hom set* from X to Y. We write $f : X \to Y$ for $f \in \mathbf{C}(X, Y)$. X is called the *domain* and Y the *codomain* of f. Finally, composition of morphisms is an associative binary function

$$\bullet : \mathbf{C}(X, Y) \times \mathbf{C}(Y, Z) \to \mathbf{C}(X, Z).$$

$\bullet(f, g)$ is denoted by reverse infix notation $g \bullet f$. For any X, id_X denotes the identity morphism of X.

For instance, **SET** is the category whose objects are sets and whose morphisms are functions. If **C** and **D** are categories then $\mathbf{C} \times \mathbf{D}$ denotes the category whose objects are pairs (C, D) where $C \in \mathbf{C}$ and $D \in \mathbf{D}$, and whose morphisms are pairs (f, g) where $f \in \mathbf{C}$ and $g \in \mathbf{D}$. Also, $\mathbf{C}^{op}$ denotes the category with the same objects and morphisms as **C**, except that the direction of all morphisms is reversed.

A *functor* F from **C** to **D** consists of a function $F_{ob} : \text{ob}\,\mathbf{C} \to \text{ob}\,\mathbf{D}$ together with functions $F_{X,Y} : \mathbf{C}(X, Y) \to \mathbf{D}(F(X), F(Y))$ for all pairs of objects X, Y in **C**, such that

$$F_{X,X}(id_X) = id_{F(X)} \text{ and } F_{X,Z}(g \bullet f) = F_{Y,Z}(g) \bullet F_{X,Y}(f).$$

We drop the subscripts ob and X,Y from F from now on and use F ambiguously both for objects and morphisms.

For instance, if **C** is a category then the identity functor $Id_C : \mathbf{C} \to \mathbf{C}$ is defined by the rules: $Id_C(C) := C$ and $Id_C(f) := f$ for all objects C and morphisms f of **C**. Similarly, if $F : \mathbf{C} \to \mathbf{D}$ and $G : \mathbf{D} \to \mathbf{E}$ are functors then the composition of F and G is the functor $G \bullet F : \mathbf{C} \to \mathbf{E}$ given by the rules: $(G \bullet F)(C) := G(F(C))$ and $(G \bullet F)(f) := G(F(f))$ for all objects C and morphisms f of **C**. An important example of a functor is the *global hom functor*

$$\mathbf{C}(-, -) : \mathbf{C}^{op} \times \mathbf{C} \to \mathbf{SET}$$

given by the rules: $\mathbf{C}(-, -)(X, Y) := \mathbf{C}(X, Y)$ and if $f : X' \to X$, $g : Y \to Y'$, then

$$\mathbf{C}(f, g) = \lambda h.(g \bullet h \bullet f) : \mathbf{C}(X, Y) \to \mathbf{C}(X', Y')$$

If **C** and **D** are categories and F and G are functors from **C** to **D**, then a *natural transformation* $t : F \Rightarrow G$ is a family of morphisms in **D** of the form $t_C : F(C) \to G(C)$, indexed by the objects C in **C**. These morphisms satisfy the *naturality condition* which says that for any morphism $f : C \to C'$ in **C**, $t_{C'} \bullet F(f) = G(f) \bullet t_C$. The morphisms t_C are called the *components* of t. We frequently use the notation

$$\text{nat.trans.} : \text{functor} \Rightarrow \text{functor} : \text{category} \to \text{category};$$

e.g., $t : F \Rightarrow G : \mathbf{C} \to \mathbf{D}$.

For instance, if F is a functor, then the components of the identity natural transformation $id_F : F \to F$ are the identity morphisms $id_{F(C)}$ for each object; i.e., $(id_F)_C := id_{F(C)}$. Similarly, if F, G, and H are functors from **C** to **D** and if $t : F \Rightarrow G$ and $t' : G \Rightarrow H$ are natural transformations, then the composition of t and t' is the natural transformation $t' * t : F \Rightarrow H$ with components given by $(t' * t)_C := t'_C \bullet t_C$.

There is an important operation on functors and natural transformations as follows. Given functors and a natural transformation as illustrated,

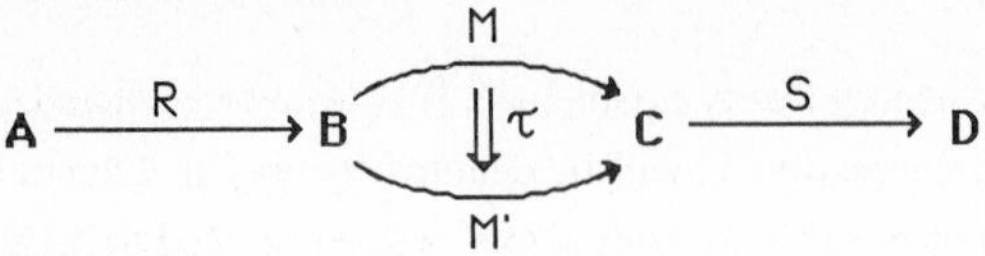

$$\mathbf{A} \xrightarrow{R} \mathbf{B} \underset{M'}{\overset{M}{\rightrightarrows}} \Downarrow \tau \; \mathbf{C} \xrightarrow{S} \mathbf{D}$$

then $S \cdot \tau : S \cdot M \Rightarrow S \cdot M'$ is the natural transformation with components

$$(S \cdot \tau)_B := S(\tau_B) : S(M(B)) \to S(M'(B))$$

and $\tau \cdot R : M \cdot R \Rightarrow M' \cdot R$ is the natural transformation with components

$$(\tau \cdot R_A) := \tau_{R(A)} : M(R(A)) \to M'(R(A)).$$

Two important examples of categories are the following:

a) $\mathbb{CAT}$ denotes the category whose objects are categories and whose morphisms are functors. Identity morphisms and composition are described above. $\mathbb{CAT}$ cannot literally consist of all categories and functors because of the usual paradoxes, but we can always assume that it contains all categories and functors of interest in any particular circumstance.

b) $[\mathbf{C} \to \mathbf{D}]$ denotes the category whose objects are functors from $\mathbf{C}$ to $\mathbf{D}$ and whose morphisms are natural transformations between such functors. Identity morphisms and composition are as described above. A category of the form $[\mathbf{C} \to \mathbf{D}]$ is called a *functor category* .

2. POLYMORPHISM

Natural transformations are *natural* candidates to represent polymorphic operations, by which we mean for the moment that they are families of morphisms which "act in the same way". For instance, a category $\mathbf{C}$ considered as an object of the category, **CAT**, of categories has an identity morphism; namely, the identity functor $Id_C : \mathbf{C} \to \mathbf{C}$. This functor, considered as an object of the functor category $[\mathbf{C} \to \mathbf{C}]$, has an identity morphism which is the natural transformation $id_{Id_C} : Id_C \Rightarrow Id_C$ whose components are given by $(id_{Id_C})_C := id_C$ (since $Id_C(C) = C$). In other words, the identity morphisms of the objects of $\mathbf{C}$ form the components of the identity natural transformation on the identity functor from $\mathbf{C}$ to $\mathbf{C}$. This is the sense in which they form a polymorphic operation.

Similarly, let 0 be an initial object in $\mathbf{C}$; i.e., 0 is an object such that there is a unique morphism $u_C : 0 \to C$ for every object C in $\mathbf{C}$. Then the morphisms $u_C : 0 \to C$ in $\mathbf{C}$ are the components of a natural transformation $u : K_C0 \Rightarrow Id_C$, where $K_C0 : \mathbf{C} \to \mathbf{C}$ is the constant functor taking every object to 0 and every morphism to the identity morphism of 0. Similarly, let 1 be a terminal object in $\mathbf{C}$; i.e., there is a unique morphism $!^C : C \to 1$ for every object C in **C**. Then the morphisms $!^C : C \to 1$ are the components of a natural transformation $! : Id_{\mathbf{C}} \Rightarrow K_{\mathbf{C}}1$. In each case, naturality follows from uniqueness.

Remark: We would like to identify natural transformations with polymorphic operations. This means we will think of u as a polymorphic morphism from 0 to any object in **C**, so we can write $u : 0 \to C$ without any subscript. Similarly, we can write $! : C \to 1$ without any subscript. However, this is not really the whole story. We will have to enlarge the scope of polymorphism to allow for two other aspects of naturality. First of all we will see that many of the natural transformations of interest in programming language semantics occur as parts of *natural families* of natural transformations. For instance, notice that the transformations id, u, and ! should have subscripts referring to the category **C** so that they belong to such families of natural transformations indexed by the collection of all categories. Second, it will turn out that naturality itself is too special a condition to cover all of the operations that we want to consider as polymorphic operations. This will necessitate the introduction of *generalized* natural transformations in 3.3.

First, let us describe what is actually going on with these families of natural transformations indexed by categories.

2.1 **Example.** Consider the identity morphism id_C of an object C in a category **C**. We have seen that these morphisms are the components of a natural transformation $id_{\mathbf{C}} : Id_{\mathbf{C}} \Rightarrow Id_{\mathbf{C}} : \mathbf{C} \to \mathbf{C}$ where $Id_{\mathbf{C}}$ is the identity functor on the category **C**. There is such a natural transformation for every category **C** and we want to show that these natural transformations are *natural* in the index **C**. That means that if $F : \mathbf{C} \to \mathbf{D}$ is a functor then the equation $F \bullet id_{\mathbf{C}} = id_{\mathbf{D}} \bullet F$ holds, or in terms of pictures that the diagram

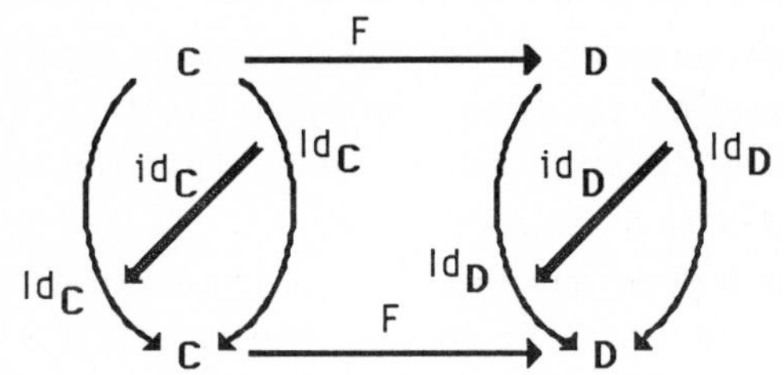

commutes. Here we are using the composition of functors and natural transformations described in section I. The equation makes sense because $F \bullet id_{\mathbf{C}}$ is a natural transformation from $F \bullet Id_{\mathbf{C}}$ to $F \bullet Id_{\mathbf{C}}$; i.e., from F to F. Similarly, $id_{\mathbf{D}} \bullet F$ is a natural transformation from $Id_{\mathbf{D}} \bullet F$ to $Id_{\mathbf{D}} \bullet F$; i.e., from F to F. To show that they are equal, we just have to check components. But $(F \bullet id_{\mathbf{C}})_C = F((id_{\mathbf{C}})_C) = F(id_C) = id_{F(C)}$ and $(id_{\mathbf{D}} \bullet F)_C = (id_{\mathbf{D}})_{F(C)} = id_{F(C)}$ so they are equal.

The preceeding is an example of a natural family $\{id_{\mathbf{C}} : Id_{\mathbf{C}} \Rightarrow Id_{\mathbf{C}}\}$ of natural transformations indexed by the collection of all categories, but it is too degenerate to illustrate the general notion. We will discuss the case of products and coproducts and all of the associated polymorphic operations in detail. First some results about binary products.

2.2 **Definition**: a) If X and Y are objects in a category **C**, then a (binary) *product* of X and Y is an object $X \times Y$ together with morphisms $pr_X : X \times Y \to X$, and $pr_Y : X \times Y \to Y$ such that given any pair of morphisms $f : Z \to X$ and $g : Z \to Y$ then there is a unique morphism, denoted by $<f, g> : Z \to X \times Y$ such that $pr_X \bullet <f, g> = f$ and $pr_Y \bullet <f, g> = g$, as illustrated:

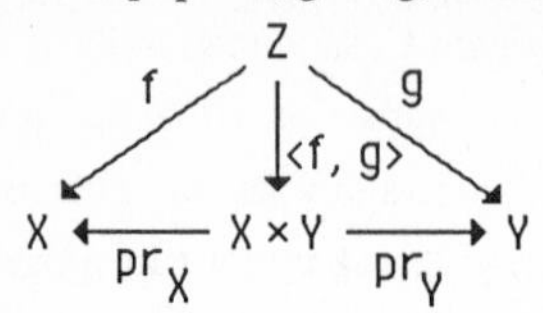

b) If $f : X \to X'$ and $g : Y \to Y'$, then $f \times g : X \times Y \to X' \times Y'$ denotes the unique morphism such that $pr_{X'} \bullet (f \times g) = f \bullet pr_X$ and $pr_{Y'} \bullet (f \times g) = g \bullet pr_Y$; i.e., $f \times g = <f \bullet pr_X, g \bullet pr_Y>$.

c) Coproducts are defined dually. Given $f : X \to Z$ and $g : Y \to Z$, the uniquely determined morphism from $X + Y$ to Z is denoted by [f, g]. Similarly, given $f : X \to X'$ and $g : Y \to Y'$, the uniquely determined morphism from $X + Y$ to $X' + Y'$ is denoted by $f + g$.

2.3 **Definition**: a) Let **C** and **D** be categories which have (binary) products and let F be a functor from **C** to **D**. F is said to *preserve (binary) products* if whenever

$$X \xleftarrow{pr_X} X \times Y \xrightarrow{pr_Y} Y$$

is a (binary) product in **C**, then

$$F(X) \xleftarrow{F(pr_X)} F(X \times Y) \xrightarrow{F(pr_Y)} F(Y)$$

is a (binary) product in **D**. Briefly, one writes: $F(X \times Y) \approx F(X) \times F(Y)$.

b) If explicit (binary) products are chosen (by the axiom of choice) for all pairs of objects in **C** and **D**, and if F applied to a chosen product in **C** *equals* the chosen product in **D**, then F is said to *preserve chosen products exactly*. $\mathbb{PRODCAT}$ denotes the category whose objects are categories with chosen binary products and functors which preserve chosen products exactly.

c) Preservation of coproducts and exact preservation of chosen coproducts is defined analogously.

2.4 **Proposition**: a) Let $F : \mathbf{C} \to \mathbf{D}$ preserve chosen products exactly. Then for any pair of morphisms $f : Z \to X$, $g : Z \to Y$, $F(<f, g>) = <F(f), F(g)>$ and for any pair of morphisms $f : X \to X'$, $g : Y \to Y'$, $F(f \times g) = F(f) \times F(g)$.

b) Dually, if F preserves chosen coproducts exactly, then for any pair of morphisms $f : X \to Z$, $g : Y \to Z$, $F([f, g]) = [F(f), F(g)]$ and for any pair of morphisms $f : X \to X'$, $g : Y \to Y'$, $F(f + g) = F(f) + F(g)$.

Proof. We prove the first part of a) for $f : Z \to X$ and $g : Z \to Y$. By definition, $<F(f), F(g)>$ is the unique morphism h such that $pr_{F(X)} \bullet h = F(g)$ and $pr_{F(Y)} \bullet h = F(g)$. But $pr_{F(X)} = F(pr_X)$ and $pr_{F(Y)} = F(pr_Y)$, so

$$pr_{F(X)} \bullet F(<f, g>) = F(pr_X) \bullet F(<f, g>) = F(pr_X \bullet <f, g>) = F(f)$$

$$pr_{F(Y)} \cdot F(<f, g>) = F(pr_Y) \cdot F(<f, g>) = F(pr_Y \cdot <f, g>) = F(g),$$

so $F(<f, g>) = <F(f), F(g)>$. The rest of the proofs are similar.

2.5 **Proposition:** Let **C** be a category with chosen binary products.

a) There is a functor $\times_{\mathbf{C}} : \mathbf{C} \times \mathbf{C} \to \mathbf{C}$ given by the rules

$$\times_{\mathbf{C}}(X, Y) := X \times Y \text{ and } \times_{\mathbf{C}}(f, g) := f \times g$$

b) There are functors

$$\lambda X.\lambda Y.\lambda Z.\ \mathbf{C}(X, Y) \times \mathbf{C}(X, Z) : \mathbf{C}^{\mathbf{op}} \times \mathbf{C} \times \mathbf{C} \to \mathbf{SET}$$

$$\lambda X.\lambda Y.\lambda Z.\ \mathbf{C}(X, Y \times Z) : \mathbf{C}^{\mathbf{op}} \times \mathbf{C} \times \mathbf{C} \to \mathbf{SET}$$

and a natural isomorphism (i.e., invertible natural transformation)

$$\mathrm{pair}_{\mathbf{C}} : \lambda X.\lambda Y.\lambda Z.\ \mathbf{C}(X, Y) \times \mathbf{C}(X, Z) \Rightarrow \lambda X.\lambda Y.\lambda Z.\ \mathbf{C}(X, Y \times Z)$$
$$: \mathbf{C}^{\mathbf{op}} \times \mathbf{C} \times \mathbf{C} \to \mathbf{SET}$$

with components: $(\mathrm{pair}_{\mathbf{C}})_{X,Y,Z} = \lambda f.\lambda g.<f, g>: \mathbf{C}(X, Y) \times \mathbf{C}(X, Z) \to \mathbf{C}(X, Y \times Z)$.

c) There are functors $(pr_1)_{\mathbf{C}} : \mathbf{C} \times \mathbf{C} \to \mathbf{C}$ and $(pr_2)_{\mathbf{C}} : \mathbf{C} \times \mathbf{C} \to \mathbf{C}$ given by $(pr_1)_{\mathbf{C}}(C, D) := C$, and $(pr_1)_{\mathbf{C}}(f, g) := f$, and similarly for $(pr_2)_{\mathbf{C}}$. Furthermore, there are natural transformations $\mathrm{fst}_{\mathbf{C}} : \times_{\mathbf{C}} \Rightarrow (pr_1)_{\mathbf{C}} : \mathbf{C} \times \mathbf{C} \to \mathbf{C}$ and $\mathrm{snd}_{\mathbf{C}} : \times_{\mathbf{C}} \Rightarrow (pr_2)_{\mathbf{C}} : \mathbf{C} \times \mathbf{C} \to \mathbf{C}$ with components given by $(\mathrm{fst}_{\mathbf{C}})_{(C, D)} := pr_C : C \times D \to C$ and $(\mathrm{snd}_{\mathbf{C}})_{(C, D)} := pr_D : C \times D \to D$.

Proof: a) Functoriality follows immediately from the definitions.

b) We leave as an exercise the description of the indicated functors as compositions of known functors. Naturality means that if $p : X' \to X$, $q : Y \to Y'$, and $r : Z \to Z'$, then the diagram

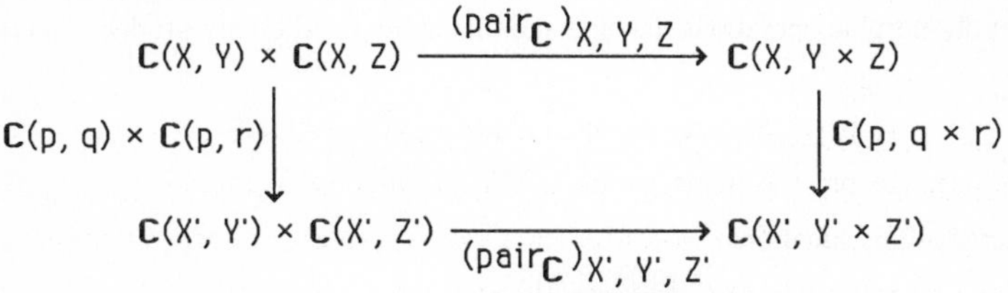

$$\begin{array}{ccc}
\mathbf{C}(X, Y) \times \mathbf{C}(X, Z) & \xrightarrow{(\mathrm{pair}_{\mathbf{C}})_{X, Y, Z}} & \mathbf{C}(X, Y \times Z) \\
{\scriptstyle \mathbf{C}(p, q) \times \mathbf{C}(p, r)} \downarrow & & \downarrow {\scriptstyle \mathbf{C}(p, q \times r)} \\
\mathbf{C}(X', Y') \times \mathbf{C}(X', Z') & \xrightarrow[(\mathrm{pair}_{\mathbf{C}})_{X', Y', Z'}]{} & \mathbf{C}(X', Y' \times Z')
\end{array}$$

commutes. To see this, start with (f, g) in the upper left hand corner. Then

$$(\mathbf{C}(p, q) \times \mathbf{C}(p, r))(f, g) = (q \cdot f \cdot p, r \cdot g \cdot p),$$

so $((\mathrm{pair}_{\mathbf{C}})_{X',Y',Z'} \cdot (\mathbf{C}(p, q) \times \mathbf{C}(p, r)))(f, g) = <q \cdot f \cdot p, r \cdot g \cdot p>$. On the other hand,

$$(\mathrm{pair}_{\mathbf{C}})_{X,Y,Z}(f, g) = <f, g>,$$

so $((\mathbf{C}(p, q \times r) \cdot (\mathrm{pair}_{\mathbf{C}})_{X,Y,Z})(f, g) = (q \times r) \cdot <f, g> \cdot p$. It is easily checked, using uniqueness, that

$$<q \cdot f \cdot p, r \cdot g \cdot p> = (q \times r) \cdot <f, g> \cdot p.$$

c) It is evident that $(pr_1)_{\mathbf{C}}$ and $(pr_2)_{\mathbf{C}}$ are functors and naturality of $\mathrm{fst}_{\mathbf{C}}$ and $\mathrm{snd}_{\mathbf{C}}$ is just the assertion that the diagram

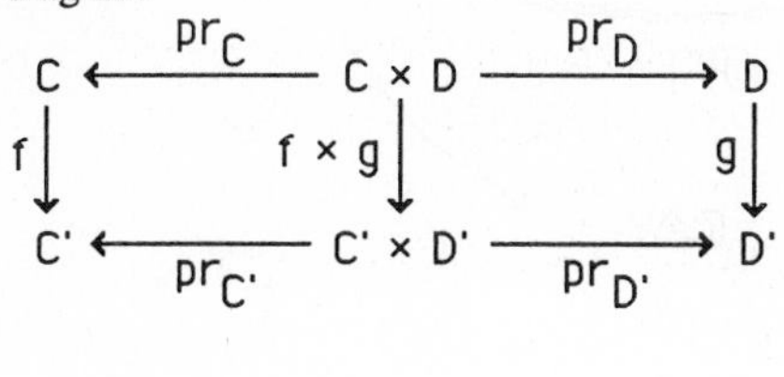

$$\begin{array}{ccccc}
C & \xleftarrow{pr_C} & C \times D & \xrightarrow{pr_D} & D \\
{\scriptstyle f} \downarrow & & {\scriptstyle f \times g} \downarrow & & \downarrow {\scriptstyle g} \\
C' & \xleftarrow[pr_{C'}]{} & C' \times D' & \xrightarrow[pr_{D'}]{} & D'
\end{array}$$

commutes, which is immediate from the definition of $f \times g$ as the unique morphism making this diagram commute.

2.6 **Proposition:** Dually, let **C** be a category with binary coproducts. There is a functor $+_{\mathbf{C}} : \mathbf{C} \times \mathbf{C} \to \mathbf{C}$ and a natural isomorphism with components: $(\mathrm{copair}_{\mathbf{C}})_{X,Y,Z} : \mathbf{C}(X, Z) \times \mathbf{C}(Y, Z) \to \mathbf{C}(X + Y, Z)$ given by $(\mathrm{copair}_{\mathbf{C}})_{Z,X,Y} = \lambda f.\lambda g.[f, g]$, as well as natural transformations

$$\mathrm{inl}_{\mathbf{C}} : (\mathrm{pr}_1)_{\mathbf{C}} \Rightarrow +_{\mathbf{C}} : \mathbf{C} \times \mathbf{C} \to \mathbf{C}$$
$$\mathrm{inr}_{\mathbf{C}} : (\mathrm{pr}_2)_{\mathbf{C}} \Rightarrow +_{\mathbf{C}} : \mathbf{C} \times \mathbf{C} \to \mathbf{C}$$

with components given by $(\mathrm{inl}_{\mathbf{C}})_{(C,\,D)} := \mathrm{in}_C : C \to C + D$ and $(\mathrm{inr}_{\mathbf{C}})_{(C,\,D)} := \mathrm{in}_D : D \to C + D$.

2.7 Consider the natural transformations $\mathrm{fst}_{\mathbf{C}}$ and $\mathrm{snd}_{\mathbf{C}}$ for all categories **C** with chosen binary products. To describe them as *natural families* of natural transformations, let $\mathrm{Square}_{\mathbb{PRODCAT}}$ be the functor from $\mathbb{PRODCAT}$ (cf., 2.3) to $\mathbb{CAT}$ that takes **C** to $\mathbf{C} \times \mathbf{C}$ and F to $F \times F$. (Square is defined for any category with chosen products.) Then there is a natural transformation

$$\mathrm{pr}_1 : \mathrm{Square}_{\mathbb{PRODCAT}} \Rightarrow \mathrm{Inc}_{\mathbb{PRODCAT}}$$

where $\mathrm{Inc}_{\mathbb{PRODCAT}}$ is the inclusion of $\mathbb{PRODCAT}$ in $\mathbb{CAT}$, whose components are the functors (=morphisms in $\mathbb{PRODCAT}$) $(\mathrm{pr}_1)_{\mathbf{C}}$ described above. Similarly for pr_2. The functors $\times_{\mathbf{C}}$ above are the components of a natural transformation from $\mathrm{Square}_{\mathbb{PRODCAT}}$ to $\mathrm{Inc}_{\mathbb{PRODCAT}}$. The next step up in abstraction is called a *modification* from one natural transformation to another, to be defined in a moment. The families of morphisms $\mathrm{fst} = \{\mathrm{fst}_{\mathbf{C}}\}$ and $\mathrm{snd} = \{\mathrm{snd}_{\mathbf{C}}\}$ are examples of such modifications. Thus there are higher order polymorphic operations fst and snd that apply to all binary products in all categories. They look like:

$$\mathrm{fst} \quad : \times \Rrightarrow \mathrm{pr}_1 : \mathrm{Square} \Rightarrow \mathrm{Inc} : \mathbb{PRODCAT} \to \mathbb{CAT}$$
$$\mathrm{snd} \quad : \times \Rrightarrow \mathrm{pr}_2 : \mathrm{Square} \Rightarrow \mathrm{Inc} : \mathbb{PRODCAT} \to \mathbb{CAT}$$

Here are a pair of pictures of this situation:

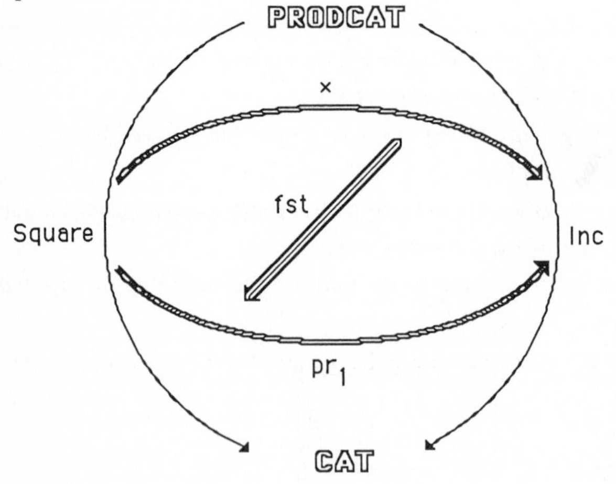

This first picture shows in three dimensions what is intended to be conveyed by the linear expression. The second picture shows a typical instance of what is going on in the target $\mathbb{CAT}$.

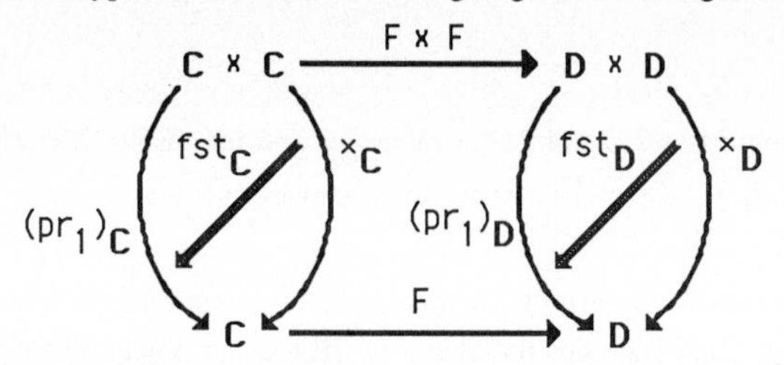

One sees that the modification fst consists of a family of natural transformations $\{fst_C\}$ indexed by the categories in $\mathbb{PRODCAT}$.

2.8 **Definition:** Let $\mathbb{C}$ and $\mathbb{C}'$ be categories of categories; i.e., the objects of $\mathbb{C}$ and $\mathbb{C}'$ are categories and the morphisms are functors. (We reserve outline letters for such categories.) Let Φ and Ψ be functors from $\mathbb{C}$ to $\mathbb{C}'$, and let σ and τ be natural transformations from Φ to Ψ. Thus, for each $\mathbf{C}$ in $\mathbb{C}$, $\Phi(\mathbf{C})$ and $\Psi(\mathbf{C})$ are categories and $\sigma_{\mathbf{C}}$ and $\tau_{\mathbf{C}}$ are functors from $\Phi(\mathbf{C})$ to $\Psi(\mathbf{C})$. A *modification* m from σ to τ, denoted by

$$m \ : \sigma \Rightarrow \tau : \Phi \Rightarrow \Psi : \mathbb{C} \to \mathbb{C}',$$

is a family of natural transformations $\{m_{\mathbf{C}}: \sigma_{\mathbf{C}} \Rightarrow \tau_{\mathbf{C}}\}$, indexed by the objects in $\mathbb{C}$, such that for any morphism $F : \mathbf{C} \to \mathbf{D}$ in $\mathbb{C}$ (i.e., functor in $\mathbb{C}$), the equation $\Phi(F) \bullet m_{\mathbf{C}} = m_{\mathbf{D}} \bullet \Psi(F)$ holds; i.e., the diagram

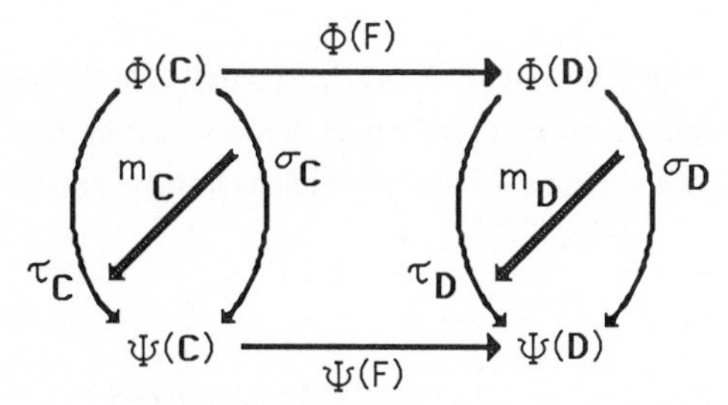

commutes. This means that for every object C in $\Phi(\mathbf{C})$, the equation $(\Phi(F) \bullet m_{\mathbf{C}})_C = (m_{\mathbf{D}} \bullet \Psi(F))_C$ holds; i.e., using the definition of composition of functors with natural transformations,

$$\Phi(F)((m_{\mathbf{C}})_C) = (m_{\mathbf{D}})_{(\Psi(F)(C))}.$$

2.9 **Proposition:** Fst and snd are modifications.

Proof. Let $F : \mathbf{C} \to \mathbf{D}$ be an exact product preserving functor between categories with binary products. Then

$$((fst_{\mathbf{D}}) \bullet (F \times F))(C, D) = pr_{F(C)} : F(C) \times F(D) \to F(C)$$

while $(F \cdot fst_C)(C, D) = F(pr_C) : F(C \times D) \to F(C)$. Exact preservation of products *means* that $F(C) \times F(D) = F(C \times D)$ and $pr_{F(C)} = F(pr_C)$ as well as $pr_{F(D)} = F(pr_D)$. Hence fst and snd are modifications.

2.10 **Proposition:** The analogues of 2.7 and 2.9 hold for inl and inr, where $\mathbb{PRODCAT}$ is replaced by $\mathbb{COPRODCAT}$.

Next we want to show that pair = $\{pair_C\}$ (and analogously, copair = $\{copair_C\}$) are modifications. This requires that we assume that $\mathbb{CAT}$ includes the category **SET** as an object. To avoid confusion, we always denote this object as **'SET**. We require the functor $Trip : \mathbb{PRODCAT} \to \mathbb{CAT}$ with values $Trip(\mathbf{C}) = \mathbf{C}^{op} \times \mathbf{C} \times \mathbf{C}$ and the constant functor $K'\mathbf{SET} : \mathbb{PRODCAT} \to \mathbb{CAT}$. Then, for every **C**, $\lambda X.\lambda Y.\lambda Z.\mathbf{C}(X, Y) \times \mathbf{C}(X, Z)$ and $\lambda X.\lambda Y.\lambda Z.\mathbf{C}(X, Y \times Z)$ are functors from $Trip(\mathbf{C})$ to $K'\mathbf{SET}(\mathbf{C})$, and $pair_C$ is a natural transformation between these functors.

2.11 **Proposition:** Pair is a modification.

Proof. Let $F : \mathbf{C} \to \mathbf{D}$ be a product preserving functor between categories with binary products. Then

$$((pair_D) \cdot (F^{op} \times F \times F))(X, Y, Z)(f, g) = \langle F(f), F(g) \rangle$$

while $(F \cdot pair_C)(X, Y, Z)(f, g) = F(\langle f, g \rangle)$. But, $\langle F(f), F(g) \rangle = F(\langle f, g \rangle)$ by 2.4, so pair is a modification.

2.12 **Proposition:** The analogue of 2.11 holds for copair.

Our view up to now of polymorphic operations is that they consist of such modifications; i.e., such natural families of natural transformations. It will turn out in the next section that this view has to be modified still further to allow for the generalized natural transformations that are introduced there.

3. POLYMORPHISM IN CARTESIAN CLOSED CATEGORIES

In order to discuss polymorphism for cartesian closed categories, we first have to describe the appropriate class of functors between such categories. First, recall the definition of a cartesian closed category.

3.1 **Definition:** Let **C** be a category with finite products. **C** is called *cartesian closed* if for every ordered pair of objects (Y, Z) of **C**, there are an object $[Y \to Z]_C$ in **C**, and a morphism

$$(app_C)_{Y,Z} : [Y \to Z]_C \times Y \to Z,$$

such that, given any $h : X \times Y \to Z$, there is a unique function $k : X \to [Y \to Z]_C$ making the diagram

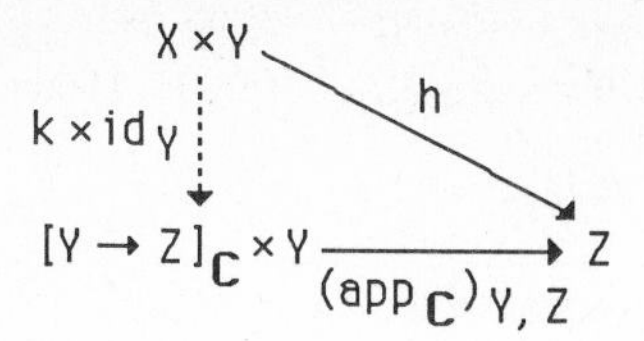

commute; i.e., such that $(app_C)_{Y,Z} \cdot (k \times id_Y) = h$. $[Y \to Z]_C$ is called the *function space object* or *exponential object* for Y and Z. The morphism k uniquely determined by h is often denoted by h# or curry (h). The subscripts **C** will be omitted whenever possible.

3.2 **Proposition:** $[- \to -] : \mathbf{C}^{op} \times \mathbf{C} \to \mathbf{C}$ is a functor of two variables, contravariant in the first and covariant in the second.

Proof: Let $f : Y' \to Y$ and $g : Z \to Z'$ be morphisms in **C**. We have to show how to construct a morphism $[f \to g] : [Y \to Z] \to [Y' \to Z']$ in **C** satisfying the appropriate properties. The definition is

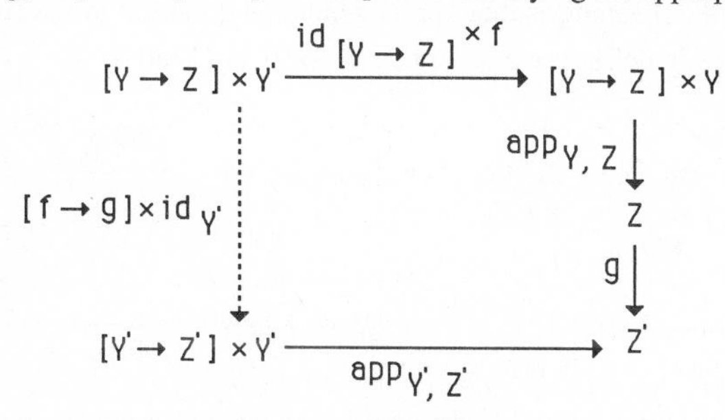

i.e., $[f \to g] = (g \cdot app_{Y,Z} \cdot (id_{[Y \to Z]} \times f))\#$. It is obvious from the diagram that $[id_Y \to id_Z] = id_{[Y \to Z]}$ and it follows from a standard uniqueness argument that if $f' : Y'' \to Y'$ and $g' : Z' \to Z''$ then $[f \cdot f' \to g' \cdot g] = [f' \to g'] \cdot [f \to g]$. Thus $[- \to -]$, or $\lambda Y.\lambda Z.\ [Y \to Z]$ is a functor. The special cases where one of the variables is an identity morphism are denoted by: $[f \to Z] : [Y' \to Z] \to [Y \to Z]$, and $[Y \to g] : [Y \to Z] \to [Y \to Z']$. For them, the composition rules read:

$$[f \cdot f' \to Z] = [f' \to Z] \cdot [f \to Z],$$

$$[Y \to g' \cdot g] = [Y \to g'] \cdot [Y \to g].$$

Unfortunately, (or perhaps fortunately), the morphisms $(app_C)_{Y,Z}$ are not the components of a natural transformation. However, they do have an important regularity property:

3.3 **Definition:** Let $T : \mathbf{C}^{op} \times \mathbf{D} \times \mathbf{C} \to \mathbf{E}$ and $S : \mathbf{D} \to \mathbf{E}$ be functors. A family of morphisms $\tau_{Y,Z} : T(Y, Z, Y) \to S(Z)$ in **E**, indexed by objects Y in **C** and Z in **D** is called a *generalized natural transformation* from T to $S \cdot pr_2$ if it is natural in Z in the usual sense, and if, for every morphism $f : Y' \to Y$ in **C**, the diagram

$$\begin{array}{ccc} T(Y,Z,Y') & \xrightarrow{T(Y,Z,f)} & T(Y,Z,Y) \\ {\scriptstyle T(f,Z,Y')}\downarrow & & \downarrow{\scriptstyle \tau_{Y,Z}} \\ T(Y',Z,Y') & \xrightarrow[\tau_{Y',Z}]{} & S(Z) \end{array}$$

commutes in **E.** (There are obviously other cases of generalized naturality, but this is the one we need here.)

3.4 **Proposition:** Let **C** be a cartesian closed category.
a) The morphisms $app_{Y,Z}$ are the components of a generalized natural transformation.
b) Currying establishes a natural isomorphism

$$(curry_C)_{X,Y,Z} : \mathbf{C}(X \times Y, Z) \approx \mathbf{C}(X, [Y \to Z]).$$

Proof: a) We will show that the family of morphisms $app_{Y,Z}$ is a generalized natural transformation from the functor

$$T_C = \lambda Y.\lambda Z.\lambda W.[Y \to Z] \times W \; : \; \mathbf{C^{op}} \times \mathbf{C} \times \mathbf{C} \to \mathbf{C}$$

to $Id_C \bullet pr_2$. Here **D** = **E** = **C**, but the meaning is that app is generalized natural in the first and third **C**'s and ordinary natural in the second as in definition 3.3. Let $f : Y' \to Y$ in **C** and $g : Z \to Z'$ in **C**. Then we must show that the diagrams

$$\begin{array}{ccc} [Y \to Z] \times Y' & \xrightarrow{[Y \to Z] \times f} & [Y \to Z] \times Y \\ {\scriptstyle [f \to Z] \times Y'}\downarrow & & \downarrow{\scriptstyle app_{Y,Z}} \\ [Y' \to Z] \times Y' & \xrightarrow[app_{Y',Z}]{} & Z \end{array} \qquad \begin{array}{ccc} [Y \to Z] \times Y & \xrightarrow{app_{Y,Z}} & Z \\ {\scriptstyle [Y \to g] \times Y}\downarrow & & \downarrow{\scriptstyle g} \\ [Y \to Z'] \times Y & \xrightarrow[app_{Y,Z'}]{} & Z' \end{array}$$

commute in **C.** (Here, for brevity, we often substitute the name of an object for its identity morphism.) But each of these is the appropriate specialization of the diagram in section 3.2 defining $[f \to g]$, and hence they both commute.

b) It is immediate that currying establishes a bijection $\mathbf{C}(X \times Y, Z) \approx \mathbf{C}(X, [Y \to Z])$. What has to be shown is that these bijections are the components of a natural transformation

$$\lambda X.\lambda Y.\lambda Z.\mathbf{C}(X \times Y, Z) \;\Rightarrow\; \lambda X.\lambda Y.\lambda Z.\mathbf{C}(X, [Y \to Z]),$$

these being functors from $\mathbf{C^{op}} \times \mathbf{C^{op}} \times \mathbf{C}$ to **SET.** It is easy to see that naturality in all three variables is equivalent to the validity of the three deduction rules

$$\frac{\begin{array}{ccc} X \times Y & \xrightarrow{h} & Z \\ {\scriptstyle f \times Y}\downarrow & \nearrow{\scriptstyle k} & \\ X' \times Y & & \end{array}}{\begin{array}{ccc} X & \xrightarrow{h^{*}} & [Y \to Z] \\ {\scriptstyle f}\downarrow & \nearrow{\scriptstyle k^{*}} & \\ X' & & \end{array}} \qquad \frac{\begin{array}{ccc} X \times Y & \xrightarrow{h} & Z \\ & \searrow{\scriptstyle k} & \downarrow{\scriptstyle g} \\ & & Z' \end{array}}{\begin{array}{ccc} X & \xrightarrow{h^{*}} & [Y \to Z] \\ & \searrow{\scriptstyle k^{*}} & \downarrow{\scriptstyle [Y \to g]} \\ & & [Y \to Z'] \end{array}} \qquad \frac{\begin{array}{ccc} X \times Y & \xrightarrow{h} & Z \\ {\scriptstyle X \times m}\downarrow & \nearrow{\scriptstyle k} & \\ X \times Y' & & \end{array}}{\begin{array}{ccc} X & \xrightarrow{h^{*}} & [Y \to Z] \\ & \searrow{\scriptstyle k^{*}} & \uparrow{\scriptstyle [m \to Z]} \\ & & [Y' \to Z] \end{array}}$$

where each deduction rule is read as asserting that the top triangle commutes if and only if the bottom one does. These rules follow easily from uniqueness arguments.

3.5 **Definition:** a) Let **C** and **D** be cartesian closed categories with chosen finite products and function space objects. Let $F: \mathbf{C} \to \mathbf{D}$ be a functor which preserves chosen finite products exactly. Then for each pair of objects (Y, Z) in **C**, there is a uniquely determined morphism $\psi_{Y,Z} : F([Y \to Z]) \to [F(Y) \to F(Z)]$ such that the diagram

$$\begin{array}{ccc} F([Y \to Z] \times F(Y) & \xrightarrow{F((app_{C})_{Y,Z})} & F(Z) \\ \downarrow{\scriptstyle \psi_{Y,Z} \times F(Y)} & \nearrow{\scriptstyle (app_{D})_{F(Y),F(Z)}} & \\ [F(Y) \to F(Z)] \times F(Y) & & \end{array}$$

commutes.

i) If $\psi_{Y,Z}$ is an isomorphism for all Y and Z, then F is said to *preserve function space objects* .

ii) If $\psi_{Y,Z}$ is an identity morphism for all Y and Z, then F is said to *preserve function space objects exactly* , or in this paper to be a *cartesian closed functor* .

iii) $\mathbb{CCCAT}$ denotes the category of cartesian closed categories with chosen finite products and chosen function space objects, and cartesian closed functors.

Because of app and similar examples we must expand our notion of polymorphism to include the notion of families of generalized natural transformations.

3.6 **Definition:** Let $\mathbb{C}$ and $\mathbb{C}'$ be categories of categories, let Φ and Ψ be functors from $\mathbb{C}$ to $\mathbb{C}'$, and let σ and τ be natural transformations from Φ to Ψ. A *generalized modification* m from σ to τ, denoted by

$$m \ : \sigma \Rightarrow \tau : \Phi \ \Rightarrow \tau : \ \mathbb{C} \to \mathbb{C}',$$

is a family of generalized natural transformations $\{m_C : \sigma_C \Rightarrow \tau_C\}$, indexed by the objects in $\mathbb{C}$, such that for any morphism $F : \mathbf{C} \to \mathbf{D}$ in $\mathbb{C}$, the equation $\Psi(F) \bullet m_C = m_D \bullet \Phi(F)$ holds; i.e., the diagram

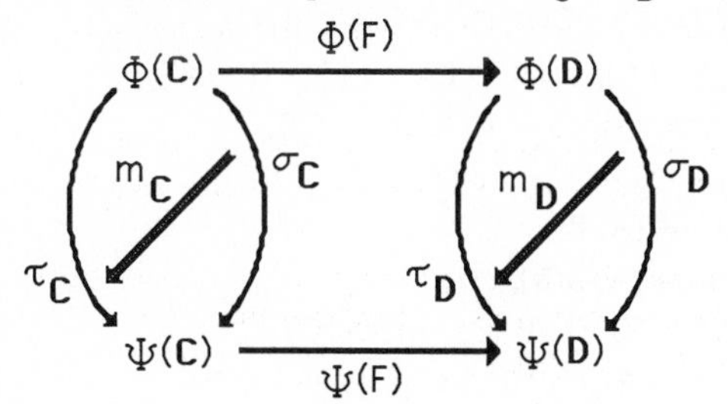

"commutes". This means that for every object C in $\Phi(\mathbf{C})$ for which m_C is defined, the equation

$$(\Psi(F) \bullet m_C)_C = (m_D \bullet \Phi(F))_C$$

holds; i.e., using the definition of composition of functors with natural transformations,

$$\Phi(F)((m_C)_C) = (m_D)_{(\Phi(F)(C))}.$$

3.7 **Proposition.** App is a generalized modification; i.e.,

$$\text{app: } [pr_1 \to pr_2] \times pr_3 \Rightarrow pr_2 : \lambda\mathbf{C}.(\mathbf{C}^{op} \times \mathbf{C} \times \mathbf{C}) \Rightarrow Id_{\mathbb{CCCAT}} : \mathbb{CCCAT} \to \mathbb{CCCAT}.$$

Proof: The picture here looks like:

$$\begin{array}{ccc} \mathbf{C}^{op} \times \mathbf{C} \times \mathbf{C} & \xrightarrow{F^{op} \times F \times F} & \mathbf{D}^{op} \times \mathbf{D} \times \mathbf{D} \\ pr_2 \downarrow \;\; \overset{app_{\mathbf{C}}}{\Leftarrow} \;\; \downarrow [pr_1 \to pr_2]_{\mathbf{C}} \times pr_3 & & pr_2 \downarrow \;\; \overset{app_{\mathbf{D}}}{\Leftarrow} \;\; \downarrow [pr_1 \to pr_2]_{\mathbf{D}} \times pr_3 \\ \mathbf{C} & \xrightarrow[F]{} & \mathbf{D} \end{array}$$

Suitable objects in $\mathbf{C}^{op} \times \mathbf{C} \times \mathbf{C}$ where app is defined are those of the form (Y, Z, Y) and commutativity says that for any such triple,

$$F((app_C)_{Y,Z}) : F([Y \to Z] \times Y) \to F(Z)$$
$$= (app_D)_{F(Y),F(Z)} : [F(Y) \to F(Z)] \times F(Y) \to F(Z).$$

But, in the definition of a cartesian closed functor, $\psi_{Y,Z}$ is an identity morphism, and hence the commutativity of the triangle in that definition is just the equality that is required here.

There are several more interesting modifications or generalized modifications for cartesian closed categories. We have to distinguish carefully between several pairs of families of morphisms.

i) $(name_C)_A : A \to \mathbf{C}(1, A)$,
$(\mathit{name}\ _C)_A : A \to [1 \to A]_C$.

ii) $(pair_C)_{C,A,B} : \mathbf{C}(C, A) \times \mathbf{C}(C, B) \to \mathbf{C}(C, A \times B)$,
$(\mathit{pair}\ _C)_{C,A,B} : [C \to A]_C \times [C \to B]_C \to [C \to A \times B]_C$.

iii) $(copair_C)_{A,B,C} : \mathbf{C}(A, C) \times \mathbf{C}(B, C) \to \mathbf{C}(A + B, C)$,
$(\mathit{copair}\ _C)_{A,B,C} : [A \to C]_C \times [B \to C]_C \to [A + B \to C]_C$.

iv) $(curry_C)_{A,B,C} : \mathbf{C}(A \times B, C) \to \mathbf{C}(A, [B \to C])$,
$(\mathit{curry}\ _C)_{A,B,C} : [A \times B \to C]_C \to [A \to [B \to C]_C]_C$.

v) $(const_C)_{B,A} : B \to \mathbf{C}(A, B)$,
$(\mathit{const}\ _C)_{B,A} : B \to [A \to B]_C$.

vi) $(comp_C)_{A,B,C} : \mathbf{C}(A, B) \times \mathbf{C}(B, C) \to \mathbf{C}(A, C)$,
$(\mathit{comp}\ _C)_{A,B,C} : [A \to B]_C \times [B \to C]_C \to [A \to C]_C$.

For instance, in i) $name_C$ only makes sense when $\mathbf{C} = \mathbf{SET}$, while $\mathit{name}\ _C$ is defined in any cartesian closed category as curry of $pr_A : A \times 1 \to A$. Thus, $(\mathit{name}\ _C)_A = curry(pr_A)$. Similarly, in v) $const_C$ only makes sense when $\mathbf{C} = \mathbf{SET}$, while $\mathit{const}\ _C$ is defined in any cartesian closed category; namely, it is curry of $pr_B : B \times A \to B$. Thus, $(\mathit{const}\ _C)_{B,A} = curry(pr_B)$. Clearly, $(\mathit{name}\ _C)_A = (\mathit{const}\ _C)_{A,1}$. In ii) $(pair_C)_{C,B,A}$ is defined in any category with finite products by the formula in terms of elements $(pair_C)_{C,B,A}(f, g) := \langle f,g \rangle$. We showed that it is a modification in section 2. However, $(\mathit{pair}\ _C)_{C,A,B}$ is only defined for a cartesian closed category. In terms of *generalized elements* $f : X \to [C \to A]_C$ and $g : X \to [C \to B]_C$, it is given by the formula $(\mathit{pair}\ _C)_{C,B,A} \bullet \langle f, g \rangle := \langle f^b, g^b \rangle^\#$. In this formula, $(-)^b$ denotes uncurrying; i.e., the inverse of currying. Here, $\langle f, g \rangle$ is a typical generalized element of $[C \to A]_C \times [C \to B]_C$. To see that the formula makes sense, we have

$$f^b : X \times C \to A$$
$$g^b : X \times C \to B,$$
$$\langle f^b, g^b \rangle : X \times C \to A \times B,$$
$$\langle f^b, g^b \rangle^\# : X \to [C \to A \times B].$$

We leave the proof that *pair* is a modification as an exercise.

We do want to treat curry and *curry* carefully. This requires an excursion into the slightly odd looking non-natural behavior of the global hom functor. For any functor $F : \mathbf{C} \to \mathbf{D}$ where $\mathbf{C}$ and $\mathbf{D}$ are arbitrary categories, there is a natural transformation, denoted by F^, in the following diagram:

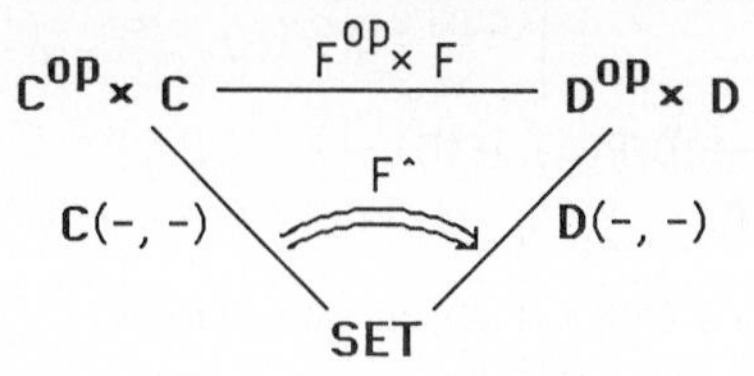

I.e., there is a natural transformation $F\^ : \mathbf{C}(-,-) \Rightarrow \mathbf{D}(-,-) \bullet (F^{op} \times F)$. Its components are given by the functions from hom sets in $\mathbf{C}$ to hom sets in $\mathbf{D}$ describing the behavior of F on morphisms; i.e.,

$$(F\^)_{A,B} := F_{A,B} : \mathbf{C}(A, B) \to \mathbf{D}(F(A), F(B)).$$

Naturality is an easy exercise. This can be viewed as saying the the global hom functor does not determine a natural transformation. (What it does determine is called a *lax natural transformation* See section 5.)

3.8 **Proposition**: The family of natural transformations $\{\text{curry}_{\mathbf{C}}\}$ does not form a modification. (It forms a *lax modification* .)

Proof: These morphisms are natural transformations between **SET**-valued functors. Our usual notation for a modification would require

i) The functor Trip : $\mathbb{CCCAT} \to \mathbb{CAT}$ taking $\mathbf{C}$ to $\mathbf{C}^{op} \times \mathbf{C}^{op} \times \mathbf{C}$.

ii) The functor K'**SET** : $\mathbb{CCCAT} \to \mathbb{CAT}$ taking $\mathbf{C}$ to '**SET**, where '**SET** is regarded as an object of $\mathbb{CAT}$.

iii) Two "natural transformations" σ and τ from Trip to K'**SET** whose components are the functors given by $\sigma_{\mathbf{C}} = \mathbf{C}(pr_1 \times_{\mathbf{C}} pr_2, pr_3)$ and $\tau_{\mathbf{C}} = \mathbf{C}(pr_1, [pr_2 \to pr_3]_{\mathbf{C}})$.

iv) Curry should then be a natural family of natural transformations between these natural families of functors, but it is not. The trouble is that since σ and τ involve the global hom functor $\mathbf{C}(-,-)$, they are not natural and neither is curry. Instead, the following diagram commutes (and that is what is meant by a modification between lax natural transformations):

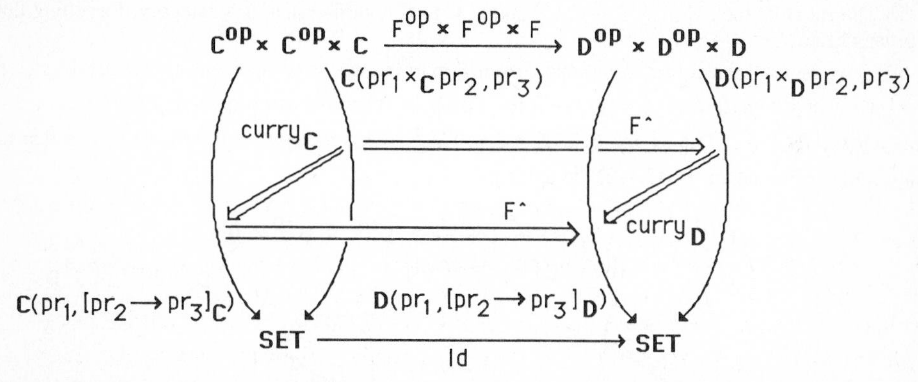

The commutativity here means with respect to composition of the natural transformations. Filling in components, it means that for any object $(A, B, C) \in \mathbf{C}^{op} \times \mathbf{C}^{op} \times \mathbf{C}$, the square

$$\begin{array}{ccc}
\mathbf{C}(A \times B, C) & \xrightarrow{F_{A \times B, C}} & \mathbf{D}(F(A) \times F(B), F(C)) \\
\downarrow {\scriptstyle (\mathrm{curry}_{\mathbf{C}})_{A, B, C}} & & \downarrow {\scriptstyle (\mathrm{curry}_{\mathbf{D}})_{F(A), F(B), F(C)}} \\
\mathbf{C}(A, [B \to C]_{\mathbf{C}}) & \xrightarrow[F_{A, [B \to C]_{\mathbf{C}}}]{} & \mathbf{D}(F(A), [F(B) \to F(C)]_{\mathbf{D}})
\end{array}$$

should commute. This in turn means that for any $h \in \mathbf{C}(A \times B, C)$, one should have

$$F((\mathrm{curry}_{\mathbf{C}})_{A,B,C}(h)) = (\mathrm{curry}_{\mathbf{D}})_{F(A),F(B),F(C)}(F(h)),$$

or, writing # for both instances of curry, $F(h\#) = F(h)\#$. But, if F is applied to the diagram on the left below, one gets the diagram on the right since F preserves products and function space objects exactly as well as app.

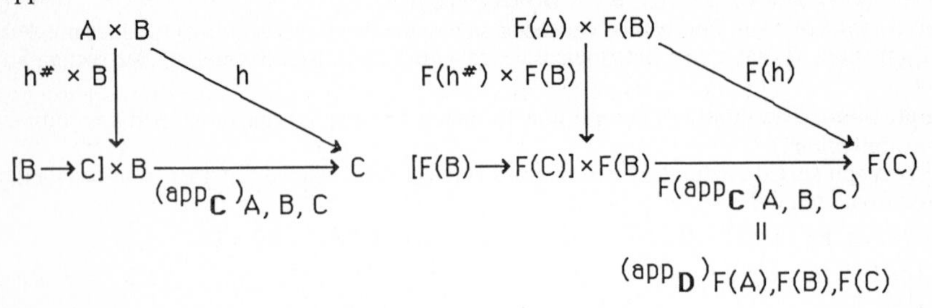

However, $F(h)\#$ is the unique morphism such that $F(h)\# \times F(B)$ makes the right hand triangle commute. Therefore, $F(h\#) = F(h)\#$ as desired.

3.9 **Proposition**: *Curry* is a modification.

Proof: The ingredients for this are similar to the preceeding proposition.

i) The functor $\mathrm{Trip} : \mathbb{CCCAT} \to \mathbb{CAT}$ taking $\mathbf{C}$ to $\mathbf{C}^{\mathrm{op}} \times \mathbf{C}^{\mathrm{op}} \times \mathbf{C}$.

ii) The functor $\mathrm{Inc} : \mathbb{CCCAT} \to \mathbb{CAT}$ taking $\mathbf{C}$ to $\mathbf{C}$ regarded just as a category, forgetting that it is cartesian closed.

iii) Two natural transformations σ and τ from Trip to Inc whose components are the functors given by $\sigma_{\mathbf{C}} = [pr_1 \times_{\mathbf{C}} pr_2 \to pr_3]_{\mathbf{C}}$ and $\tau_{\mathbf{C}} = [pr_1 \to [pr_2 \to pr_3]_{\mathbf{C}}]_{\mathbf{C}}$. Then the assertion is that there is a modi–fication: *curry* $: \sigma \Rrightarrow \tau : \mathrm{Trip} \Rightarrow \mathrm{Inc} : \mathbb{CCCAT} \to \mathbb{CAT}$. This means that it must be shown that for every cartesian closed functor $F : \mathbf{C} \to \mathbf{D}$ the diagram

$$\begin{array}{ccc}
\mathbf{C}^{\mathrm{op}} \times \mathbf{C}^{\mathrm{op}} \times \mathbf{C} & \xrightarrow{F^{\mathrm{op}} \times F^{\mathrm{op}} \times F} & \mathbf{D}^{\mathrm{op}} \times \mathbf{D}^{\mathrm{op}} \times \mathbf{D} \\
{\scriptstyle [pr_1 \to [pr_2 \to pr_3]_{\mathbf{C}}]_{\mathbf{C}}} \downarrow \; \overset{curry_{\mathbf{C}}}{\Leftarrow} \; \downarrow {\scriptstyle [pr_1 \times_{\mathbf{C}} pr_2 \to pr_3]_{\mathbf{C}}} & & {\scriptstyle [pr_1 \to [pr_2 \to pr_3]_{\mathbf{D}}]_{\mathbf{D}}} \downarrow \; \overset{curry_{\mathbf{D}}}{\Leftarrow} \; \downarrow {\scriptstyle [pr_1 \times_{\mathbf{D}} pr_2 \to pr_3]_{\mathbf{D}}} \\
\mathbf{C} & \xrightarrow[F]{} & \mathbf{D}
\end{array}$$

commutes. I.e., for any object (A, B, C) in $\mathbf{C^{op}} \times \mathbf{C^{op}} \times \mathbf{C}$, it must be shown that

$$F((\mathit{curry}\ \mathbf{C})_{A,B,C}) = (\mathit{curry}\ \mathbf{D})_{F(A),F(B),F(C)}.$$

This makes sense since we know from the preservation properties of F that

$$F([A \times_{\mathbf{C}} B \to C]_{\mathbf{C}}) = [F(A) \times_{\mathbf{D}} F(B) \to F(C)]_{\mathbf{D}}$$

$$F([A \to [B \to C]_{\mathbf{C}}]_{\mathbf{C}}) = [F(A) \to [F(B) \to F(C)]_{\mathbf{D}}]_{\mathbf{D}},$$

so the domains and codomains of these two morphisms are the same. To go farther, we need the following lemma.

3.10 **Lemma**: For any triple of objects, X, Y, Z $\in$ **D**,

$$(\mathit{curry}\ \mathbf{D})_{X,Y,Z} = (\mathrm{curry}_{\mathbf{D}})_{[X\times Y\to Z]\times X, Y, Z}\ ((\mathrm{curry}_{\mathbf{D}})_{[X\times Y\to Z], X, [Y\to Z]} \bullet (\mathrm{app}_{X\times Y, Z} \bullet a_{[X\times Y\to Z], X, Y}))$$

(pretty printed for legibility) or, omitting all subscripts and writing # for both curry's,

$$\mathit{curry} = ((\mathrm{app} \bullet a)\#)\#.$$

Proof: Here, for any W $\in$ **D**, $a_{W,X,Y} : (W \times X) \times Y \to W \times (X \times Y)$ is the associativity isomorphism. To derive this formula, note that *curry* is defined by means of the Yoneda Lemma using the sequence of natural transformations with components given by

$$\mathbf{D}(W, [X \times Y \to Z]) \xrightarrow{\mathrm{uncurry}_{W, X\times Y, Z}} \mathbf{D}(W \times (X \times Y), Z)$$
$$\xrightarrow{\mathbf{D}(a_{W,X,Y}, Z)} \mathbf{D}((W \times X) \times Y, Z)$$
$$\xrightarrow{\mathrm{curry}_{W\times X, Y, Z}} \mathbf{D}(W \times X, [Y \to Z])$$
$$\xrightarrow{\mathrm{curry}_{W, X, [Y\to Z]}} \mathbf{D}(W, [X \to [Y \to Z]])$$

One choses W = $[X \times Y \to Z]$ and then applies this sequence of morphisms to

$$\mathrm{id}_{[X\times Y\to Z]} \in \mathbf{D}([X \times Y \to Z], [X \times Y \to Z])$$

to get $(\mathit{curry}\ \mathbf{D})_{X,Y,Z} \in \mathbf{D}([X \times Y \to Z], [X \to [Y \to Z]])$. But

$$\mathrm{uncurry}_{[X\times Y\to Z], X\times Y, Z}(\mathrm{id}_{[X\times Y\to Z]}) = \mathrm{app}_{X\times Y, Z}$$

(since, obviously, curry(app) = id), while for any f, **D**(a, Z)(f) = f $\bullet$ a, giving the required formula.

End of proof of 3.9. The result is now immediate since

$$(\mathit{curry}\ \mathbf{D})_{F(A),F(B),F(C)}$$
$$= ((\mathrm{app}_{F(A)\times F(B), F(C)} \bullet a_{[F(A)\times F(B)\to F(C)], F(A), F(B)})\#)\#$$
$$= ((F(\mathrm{app}_{A,B,C}) \bullet F(a_{[A\times B\to C], A, B}))\#)\#$$
$$= ((F(\mathrm{app}_{A,B,C} \bullet a_{[A\times B\to C], A, B}))\#)\#$$
$$= F(((\mathrm{app}_{A,B,C} \bullet a_{[A\times B\to C], A, B})\#)\#)$$
$$= F((\mathit{curry}\ \mathbf{C})_{A,B,C}).$$

3.11 **Corollary**: Let $F : \mathbf{C} \to \mathbf{D}$ be a cartesian closed functor between cartesian closed categories.

a) If $h : A \times B \to C$ in **C**, then $F(\mathrm{curry}_{\mathbf{C}}(h)) = \mathrm{curry}_{\mathbf{D}}(F(h))$.

b) For all A, B, C in **C**, $F((\mathit{curry}\ \mathbf{C})_{A,B,C}) = (\mathit{curry}\ \mathbf{D})_{F(A),F(B),F(C)}$.

Proof: These equations just summarize the results of 3.8 and 3.9.

The conclusion is that in order to describe the kinds of polymorphic operations that arise in semantics of programming languages, one has to progress through an increasing more general sequence of concepts:

- natural transformations

- natural families of natural transformations; i.e., modifications
- generalized natural transformations
- natural families of generalized natural transformations
- lax natural transformations
- modifications of lax natural transformations.

What is missing is the precise description of these last two concepts. They are illustrated in this section . In the appendix, we give their definitions and verify the appropriate properties for our examples.

4. Afterword

The discussion of polymorphic operations has involved talking about all categories of an appropriate kind and all suitable functors between them. It would be nice to cut this down to a smaller structure by finding a *generic* category which reflects all of these constructions in itself. That means a category $\mathbf{C}_0$ whose morphisms correspond to polymorphic operations. In order to have any hope of finding such a $\mathbf{C}_0$, we have to restrict attention to natural families of natural transformations. In the case of cartesian closed categories, there is such a category; namely, the free cartesian closed category generated by countable many objects. (The category $\mathbb{CCCAT}$ is "sketchable" so such a free object exists. Cf. [1].) Any natural family of natural transformations in n variables has a component for the first n objects of $\mathbf{C}_0$ and this component determines all other components of the natural family uniquely. The properties of this $\mathbf{C}_0$ will be discussed elsewhere.

5. APPENDIX

5.1 **Definition:** Let $\mathbb{C}$ and $\mathbb{C}'$ be categories of categories, and let Φ and Ψ be functors from $\mathbb{C}$ to $\mathbb{C}'$. A *lax natural transformation* σ from Φ to Ψ consists of:

i) a family of functors $\sigma_C : \Phi(C) \to \Psi(C)$ in $\mathbb{C}'$, indexed by the objects $\mathbf{C} \in \mathbb{C}$,

ii) a family of natural transformations $\sigma_F : \Psi(F) \bullet \sigma_C \Rightarrow \sigma_{C'} \bullet \Phi(F)$

indexed by the functors $F : \mathbf{C} \to \mathbf{C}'$ in $\mathbb{C}$. These natural transformation are required to satisfy two conditions:

a) $\sigma_{idC} = id_{\sigma C}$

b) If $F : \mathbf{C} \to \mathbf{C}'$ and $G : \mathbf{C}' \to \mathbf{C}''$ in $\mathbb{C}$, then

$$\sigma_{G \bullet F} = (\sigma_G \bullet \Phi(F)) * (\Psi(G) \bullet \sigma_F)$$

Here "*" is composition of natural transformations. The following picture illustrates this situation.

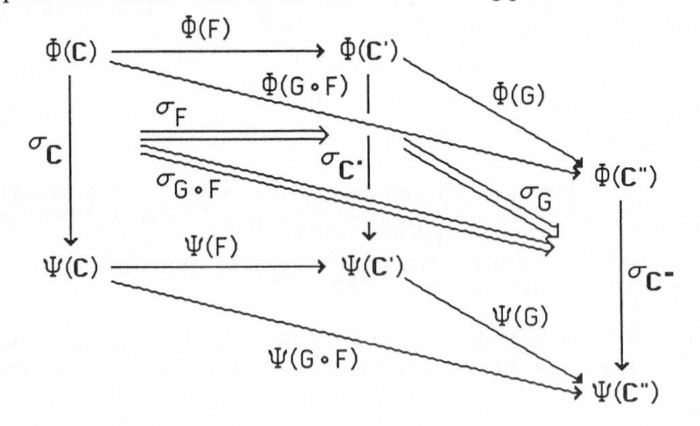

Note that a natural transformation is a special case of a lax natural transformation in which all natural transformations σ_F are identities.

5.2 **Definition**: Let $\mathbb{C}$ and $\mathbb{C}'$ be categories of categories, let Φ and Ψ be functors from $\mathbb{C}$ to $\mathbb{C}'$, and let σ and ψ be lax natural transformations from Φ to Ψ. A *modification* m, from σ to τ, denoted by

$$m : \sigma \Rightarrow \tau : \Phi \Rightarrow \Psi : \mathbb{C} \to \mathbb{C}',$$

is a family of natural transformations $\{m_C : \sigma_C \Rightarrow \tau_C\}$, indexed by the objects $\mathbf{C}$ in $\mathbb{C}$, such that for any morphism $F : \mathbf{C} \to \mathbf{C}'$ in $\mathbb{C}$, the equation

$$\tau_F * (\Psi(F) \bullet m_C) = (m_{C'} \bullet \Phi(F)) * \sigma_F$$

holds; i.e., the diagram

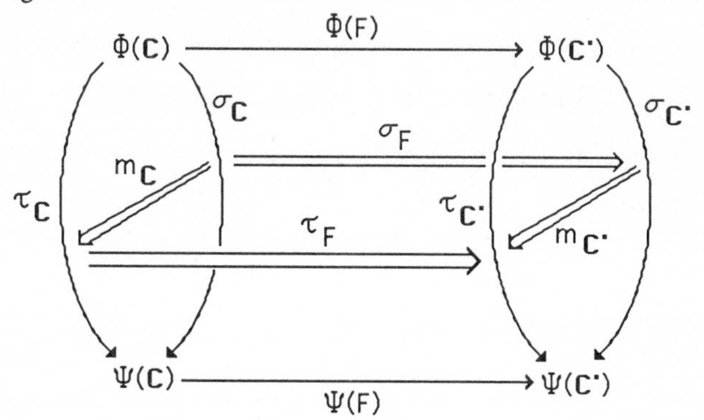

commutes.

5.3 **Proposition**: a) The global hom functor determines a lax natural transformation from the functor $\Phi(C) = \mathbf{C^{op}} \times \mathbf{C}$ to Ψ = K'SET.

b) Curry determines a modification between the lax natural transformations σ and τ with components

$$\sigma_C = C(pr_1 \times_C pr_2, pr_3)$$
$$\tau_C = C(pr_1, [pr_2 \to pr_3]_C).$$

Proof. a) The diagram for the global hom functor is just before 3.8. The only thing to be verified is 4.1, ii), b). Since Ψ is a constant functor, this reduces to the obvious equation

$$(G \bullet F)^{\wedge}_{X,Y} = G^{\wedge}_{F(X),F(Y)} \bullet F^{\wedge}_{X,Y}.$$

b) It is easily seen that σ and τ are lax natural transformations. The equation required in 5.2 has already been verified in 3.8.

References

[1] : M. Barr and C. Wells, *Toposes, Triples and Theories*, Springer-Verlag, New York, 1985.

[2] : J. Benabou, Introduction to Bicategories, *Reports Midwest Category Seminar I*, Lecture Notes in Mathematics 47, Springer-Verlag, New York, 1967.

[3] : S. Eilenberg and G. M. Kelly, Closed categories, in *Proceedings of the Conference on Categorical Algebra, La Jolla, 1965*, Springer-Verlag, New York, 1966.

[4] : J. W. Gray, *Formal Category Theory: Adjointness for 2-Categories*, Lecture Notes in Mathematics 391, Springer Verlag, New York, 1974.

[5] : J. W. Gray, Categorical aspects of data type constructors, *Theoretical Computer Science* , 141 (to appear).

[6] : P. Johnstone, *Topos Theory* , Academic Press, New York, 1977.

[7] : G. M. Kelly and R. Street, Review of the elements of 2-categories, in *Category Seminar, Sydney 1972/73* , Lecture Notes in Mathematics 420, Springer-Verlag, New York, 1974.

[8] : J. Lambek and P. J. Scott, *Introduction to Higher Order Categorical Logic* . Cambridge studies in advanced mathematics 7, Cambridge University Press, New York, 1986

[9] : S. Mac Lane, *Categories for the Working Mathematician* , Springer Verlag, New York, 1972.

[10] : B. Pareigis, *Kategorien und Funktoren* , B. G. Teubner, Stuttgart, 1969.

[11] : R. Seely, Locally cartesian closed categories and type theory, *Math. Proc. Camb. Phil. Soc* . 95 (1984), 33 - 48.

A CATEGORICAL APPROACH TO REALIZABILITY AND POLYMORPHIC TYPES

AURELIO CARBONI, PETER J. FREYD, and ANDRE SCEDROV

ABSTRACT. A categorical calculus of relations is used to derive a unified setting for higher order logic and polymorphic lambda calculus.

INTRODUCTION

There has been considerable interest recently in the development of a mathematical framework for polymorphic types, which have become commonplace in powerful programming languages such as Ada, ML, HOPE, and CLU [Bruce & al. 87, Cardelli & Wegner 85, Cousineau & al. 86]. The current syntactic paradigm for polymorphic typing is higher-order lambda calculus [Girard 71, Reynolds 74], for which there are no nontrivial set-theoretic interpretations [Reynolds 84]. In a recent work of Moggi and Hyland, however, a recursive interpretation of polymorphic types obtained in [Girard 72] is inferred within the setting for an intuitionistic theory of sets given by the Realizability Universe. In a related work discussed at this Workshop, D. Scott has proposed the Realizability Universe as a setting for making the theory of computational domains [Scott 82] more closely knit with the ambient higher-order logic. In short, the Realizability Universe is emerging as an important framework for higher-order logic, domain theory, and polymorphism.

The Realizability Universe has been studied in [Hyland 82] for other purposes (related work is discussed e. g. in [Beeson 85] and [McCarty 86]). The main idea behind its construction (due to D. Scott and related to earlier work of W. Powell) was to consider sets of natural numbers as "truth-values" in the Kleene-Kreisel-Troelstra recursive realizability interpretation of a syntax of intuitionistic higher-order logic.

We consider a different approach based on a calculus of relations that allows us to show that the recursive functions and realizability are in fact forced by the required higher-order logical structure of the category in question. Various new interpretations of polymorphism (as well as Moggi's result mentioned above) are then easily obtained by using reflections into several small full subcategories. The method itself and the basic results are explained here. More involved results in this vein are discussed in [Freyd & Scedrov 87] , [Bainbridge & al. 88], and [Scedrov 88].

1. A CONSTRUCTION OF THE REALIZABILITY UNIVERSE: STEP ONE

Let $\mathbb{K}$ be a collection of partial endofunctions on the set of natural numbers $\mathbb{N}$. We make the following assumptions on $\mathbb{K}$:

(i) $\mathbb{K}$ contains the identity,

(ii) $\mathbb{K}$ is closed under composition,

(iii) $\mathbb{K}$ contains two total functions $\mathbb{l}$ and $\mathbb{r}$ such that for any φ, ψ in $\mathbb{K}$, there exists θ in $\mathbb{K}$ defined on the common domain of φ and ψ , such that φ contains $\theta\mathbb{l}$ and ψ contains $\theta\mathbb{r}$:

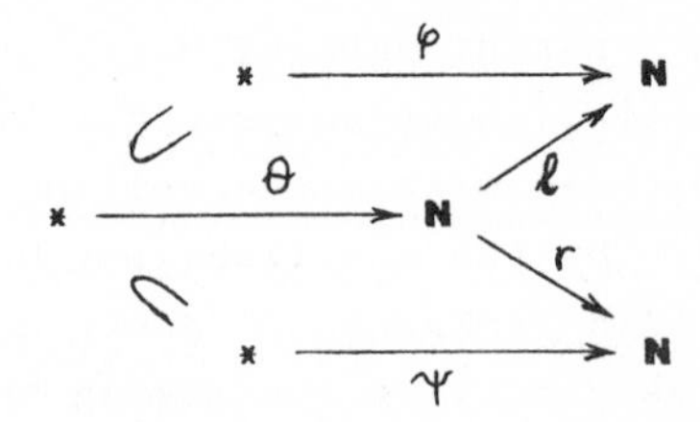

Convention: Here and throughout the paper, composition is written in the diagrammatic order and all displayed diagrams commute.

If $(\_ , \_) : \mathbb{N}\times\mathbb{N} \longrightarrow \mathbb{N}$ is a coding of pairs and $(n, k)\mathbb{l} = n$, $(n, k)\mathbb{r} = k$ for all numbers n , k , then in the condition (iii) one may let $\theta = (\varphi, \psi)$.

We first consider an auxiliary category $\mathbb{A}$ of ASSEMBLIES $A = (X, \{Y_n\}_{n\geq 0})$, where X is a set and $\{Y_n\}_{n\geq 0}$ a sequence of subsets $Y_n \subset X$ called CAUCUSES and written $A|_n$. We note that different caucuses are not necessarily disjoint. The CARRIER of an assembly A is the set $|A| = \cup_n A|_n$. A morphism

$f\colon A \longrightarrow B$ is an ordinary map $f\colon |A| \longrightarrow |B|$ for which there exists φ in $\mathbb{K}$ (a <u>MODULUS</u> of f) so that for every $n \in \mathbb{N}$, $x \in |A|$:

$$\text{if } x \in A|_n \text{, then } \varphi(n) \text{ is defined and } f(x) \in B|_{\varphi(n)} .$$

We recall some categorical terminology:

A <u>SUBOBJECT</u> of an object A in any category is named by a monomorphism (monic) $B \longrightarrow A$. Given monomorphisms $f\colon B \longrightarrow A$ and $f'\colon B' \longrightarrow A$, we say that B is included in B' and write $B \subset B'$ iff there is (necessarily monic) $g\colon B \longrightarrow B'$ so that $f = gf'$. f and f' name the same subobject iff $B \subset B'$ and $B' \subset B$.

An <u>IMAGE</u> of a morphism $f\colon A \longrightarrow B$ is the minimal subobject $B' \subset B$ for which f factors into $A \longrightarrow B' \subset B$.

A morphism $f\colon A \longrightarrow B$ is a <u>COVER</u> iff its image is B . If equalizers exist, every cover is an epimorphism.

A <u>REGULAR</u> <u>CATEGORY</u> is a category with finite products, equalizers, and images, in which pullbacks transfer covers.

PROPOSITION 1. Let $\mathbb{K}$ be a set of partial endofunctions on the natural numbers that satisfies the conditions (i) - (iii) above. Then the category $\mathbb{A}$ is a regular category with an initial object.

PROOF. Initial object 0 is given by the empty set. Given a pair of morphisms $f, g : A \longrightarrow B$, their equalizer is obtained as $|E| = \{x \in |X| : f(x) = g(x)\}$ (that is, the ordinary equalizer of the maps f and g in the category of sets), and $E|_n$ is the ordinary intersection of $|E|$ and $A|_n$. Terminal object is given by a one-element set 1 so that $1|_n = 1$. As for binary products, let $(A \times B)|_n = A|_{nl} \times B|_{nr}$. We recall that equalizers and binary products yield pullbacks. Regarding the images, given a morphism $f\colon A \longrightarrow B$, we wish to obtain the minimal subobject B' of the codomain such that f factors through B . Let $|B'|$ be the ordinary image of the map f and let $B'|_n = f(A|_n)$. B' is a subobject of B named by inclusion of the same modulus as f . (Note that we will have therefore shown that any subobject may be named by a monic given by inclusion with some modulus.) The required factor $A \longrightarrow B'$ is obtained as f with the identity as its modulus. We leave the minimality and the stability under pullbacks to the reader. ∎

We observe that the partial ordering of the subobjects of an object (given by inclusion) has finite infima: without loss of generality, we may assume that two given subobjects B , C of A are named by inclusions, so we let $(B \cap C)|_n$ be the ordinary intersection of $B|_n$ and $C|_n$. This is, of course, the pullback of $B \subset A$ and $C \subset A$.

Note the functor ∇: **Sets** $\longrightarrow \mathbf{A}$ given by $|\nabla X| = (\nabla X)|_n = X$, all n; for $f: X \longrightarrow Y$ let ∇f be f with the identity as modulus. ∇ preserves initial object, equalizers, and finite products.

Two special kinds of assemblies deserve attention. A <u>MODEST ASSEMBLY</u> is one in which every caucus has at most one member. An assembly is <u>PARTITIONED</u> iff its caucuses are disjoint.

The full subcategory of modest assemblies is a <u>REFLECTIVE SUBCATEGORY</u> of $\mathbf{A}$:

PROPOSITION 2. Under the assumptions of Proposition 1, for every assembly A there exists a modest assembly A_{mod} and a cover $q: A \longrightarrow A_{mod}$ such that for any morphism f from A to a modest M there exists a unique:

$$\begin{array}{ccc} A & \xrightarrow{q} & A_{mod} \\ & {}_{f}\searrow & \downarrow \\ & & M \end{array}$$

If A is modest, then q is an isomorphism. A_{mod} is called the <u>MODEST REFLECTION</u> of an assembly A .

PROOF. $q: |A| \longrightarrow |A_{mod}|$ must collapse each caucus to a single point and may defined as an onto map that does that and no more collapsing than is necessary. ∎

PROPOSITION 3. Under the assumptions of Proposition 1 , any assembly may be covered by a partition.

PROOF. Let $|A|$ be the disjoint union of the caucuses $B|_n$ and let $A|_n$ be a copy of $B|_n$. There is an obvious cover $B \longrightarrow A$. ∎

The category of assemblies does not have quotients of all equivalence relations. In section 2 we will freely adjoin these quotients by one of the basic methods of the general calculus of relations introduced in [Freyd 74] and developed in [Freyd & Scedrov 88] and [Carboni & Walters 88].

More precisely, in any category we may consider <u>RELATIONS</u> R from an object A to an object B as named by pairs of morphisms:

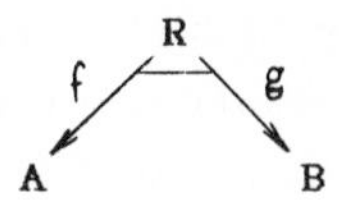

with the joint cancellation property: given any $v, w: C \longrightarrow R$, if $vf = wf$ and $vg = wg$, then $v = w$. (This property is indicated by the horizontal line in the diagram.) Inclusion of relations $R \subset S$ from A to B is given by the existence of a (necessarily monic) morphism $R \longrightarrow S$ such that:

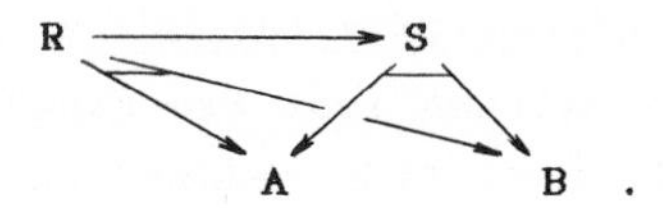

R and S name the same relation from A to B iff each is included in the other. (Note that in the presence of binary products, relations from A to B is isomorphic to subobjects of $A \times B$. In our view, however, it is important to give a definition without referring to products.) In **A**, R may be named by a sequence $\{R_n\}$ of ordinary relations from $|A|$ to $|B|$ for which there exist φ, ψ in $\mathbb{K}$ such that for every $x \in |A|$, $y \in |B|$, and $n \in \mathbf{N}$: if $x \; R_n \; y$, then $\varphi(n)$ and $\psi(n)$ are defined, $x \in A|_{\varphi(n)}$, and $y \in B|_{\psi(n)}$. Definition of inclusion bears repeating: one of such sequences $\{R_n\}$ is included into another $\{S_n\}$ iff there exists φ in $\mathbb{K}$ such that for each $n \in \mathbf{N}$, $x \in |A|$, $y \in |B|$, if $x \; R_n \; y$, then $\varphi(n)$ is defined and $x \; S_{\varphi(n)} \; y$. Two sequences name the same relation iff each is included in the other. We will refer to relations by their names.

The <u>COMPOSITION</u> <u>OF</u> <u>RELATIONS</u> R (from A to B) and S (from B to C) is the relation RS from A to C defined by: $x \; (RS)_n \; z$ iff there exists $y \in |B|$ such that $x \; R_{n_l} \; y$ and $y \; S_{n_r} \; z$. (General definition uses finite products, equalizers, and images. Associativity of composition of relations requires that pullbacks transfer covers and hence images.) We thus obtain the category $Rel(\mathbf{A})$ whose objects are assemblies and whose morphisms from A to B are relations from A to B. We shall capitalize morphisms of $Rel(\mathbf{A})$ and use small letters for the morphisms of $\mathbf{A}$.

For any object A there is a special relation IDENTITY named by: $x\,(id_A)_n\,x'$ iff $x = x'$ and $x \in A|_n$. Moreover, given a relation $R: A \longrightarrow B$ named by $\{R_n\}$, its RECIPROCAL $R^o: B \longrightarrow A$ may be named by the sequence of ordinary reciprocals $\{(R_n)^o\}$.

Notice that the morphisms in $\mathbf{A}$ from A to B are precisely those relations $R: A \longrightarrow B$ such that $id_A \subset RR^o$ (that is: R is total on A) and $R^oR \subset id_B$ (that is: R is single-valued). In other words, $\mathbf{A}$ is SUBCATEGORY OF MAPS of $Rel(\mathbf{A})$.

A relation $R: A \longrightarrow A$ is an EQUIVALENCE RELATION iff $id_A \subset R$, $R^o = R$, and $RR \subset R$. Moduli of these inclusions are called, respectively, MODULI OF REFLEXIVITY, SYMMETRY, AND TRANSITIVITY. The reader should check that the last two conditions imply that R is an idempotent, $RR = R$. (The reader should note that ordinary equivalence relations are precisely those ordinary relations that satisfy the three conditions listed above, applied in the context of ordinary reciprocation, ordinary composition of relations, and with id as the ordinary equality.)

2. A CONSTRUCTION OF THE REALIZABILITY UNIVERSE: STEP TWO

We recall categorical folklore on "splitting a class of idempotents". First recall that in any category, one says that a pair of morphisms $v: A \longrightarrow B$, $w: B \longrightarrow A$ splits an idempotent $e: A \longrightarrow A$ of that category iff

$$\begin{array}{ccc} A & \xrightarrow{e} & A \\ {\scriptstyle v}\downarrow & \nearrow{\scriptstyle w} & \downarrow{\scriptstyle v} \\ B & \xrightarrow[id]{} & B \end{array}\quad .$$

In the category $Rel(\mathbf{Sets})$ of ordinary sets and relations, for example, an equivalence relation $E: A \longrightarrow A$ is split by the mapping $P: A \longrightarrow A/E$ of an element of A to its equivalence class in the quotient A/E and the relation $R: A/E \longrightarrow A$ of an equivalence class to its representatives. Still in the same category, consider a <u>COREFLEXIVE</u> relation $E: A \longrightarrow A$ (that is: E is contained in the identity). Such a relation is clearly symmetric and transitive, and hence idempotent. Let $\mathrm{Dom}(E)$ be the domain of E (in this case, the domain and the range are the same). Let $i: \mathrm{Dom}(E) \longrightarrow A$ be the inclusion. E is split by (i^o, i).

In $\mathrm{Rel}(\mathbb{A})$, a coreflexive relation $E: A \longrightarrow A$ is one for which $E \subset id_A$. E is symmetric and idempotent. Let $\mathrm{Dom}(E)$ be the subobject of A given by $\mathrm{Dom}(E)|_n = \{ x \in |A| : x \, E_n \, x \}$. Let $i: \mathrm{Dom}(E) \longrightarrow A$ be the map given by inclusion of modulus $\mathbb{I}$. (i^o, i) splits E. Thus all coreflexive morphisms in $\mathrm{Rel}(\mathbb{A})$ split.

Let $f: A \longrightarrow B$ be a cover in $\mathbb{A}$. In $\mathrm{Rel}(\mathbb{A})$, the pair (f, f^o) splits the equivalence relation $ff^o: A \longrightarrow A$. In fact, for any regular category $\mathbb{C}$, an equivalence relation $E: A \longrightarrow A$ is split as an idempotent in $\mathrm{Rel}(\mathbb{C})$ iff there exists a cover $f: A \longrightarrow B$ in $\mathbb{C}$ such that $E = ff^o$ and $f^o f = id_A$. We call such equivalence relations <u>EFFECTIVE</u>. An <u>EFFECTIVE REGULAR CATEGORY</u> is a regular category all of whose equivalence relations are effective. $\mathbb{A}$ is not effective. This can be remedied by formally splitting all equivalence relations (as idempotents):

Consider the new category $\mathbb{E}$ whose objects are pairs $\langle A, I \rangle$ (we will write A/I to suggest the formal quotient), where A is an assembly and $I: A \longrightarrow A$ in $\mathrm{Rel}(\mathbb{A})$ is an equivalence relation.

A relation $R: A \longrightarrow B$ in $\mathrm{Rel}(\mathbb{A})$ is considered a morphism from A/I to B/J in $\mathbb{E}$ iff $IR = R = RJ$ in $\mathrm{Rel}(\mathbb{A})$:

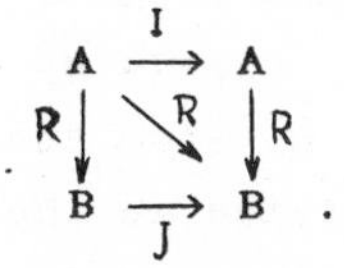

$$\begin{array}{ccc} A & \xrightarrow{I} & A \\ {\scriptstyle R}\downarrow & \searrow^{R} & \downarrow{\scriptstyle R} \\ B & \xrightarrow[J]{} & B \end{array} .$$

Finally, let $\mathbb{R}$ be the subcategory of maps of $\mathbb{E}$, that is: the subcategory of $\mathbb{E}$ with the same objects, but whose morphisms are those morphisms $R: A/I \longrightarrow B/J$ of $\mathbb{E}$ for which $I \subset RR^o$ and $R^o R \subset J$.

We now show that $\mathbb{R}$ is in fact the effective reflection of $\mathbb{A}$ (in the large category of regular categories and representations thereof, that is: functors that preserve finite products, equalizers, and images):

PROPOSITION 4. Assume the hypotheses of Proposition 1. Then $\mathbb{E}$ is isomorphic to $Rel(\mathbb{R})$. $\mathbb{R}$ is an effective regular category with an initial object. Furthermore, there is a full and faithful representation $F: \mathbb{A} \longrightarrow \mathbb{R}$ such that for any effective regular category $\mathbb{B}$ and any representation G of $\mathbb{A}$ in $\mathbb{B}$ there is a representation:

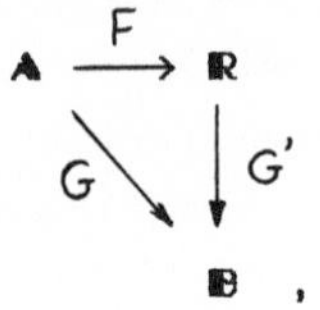

$$\begin{array}{ccc} \mathbb{A} & \xrightarrow{F} & \mathbb{R} \\ & {\scriptstyle G}\searrow & \downarrow{\scriptstyle G'} \\ & & \mathbb{B} \end{array},$$

unique up to natural equivalence.

PROOF. If a morphism $E: A/I \longrightarrow A/I$ in $\mathbb{E}$ is an equivalence relation, then $E: A \longrightarrow A$ in $Rel(\mathbb{A})$ is an equivalence relation. E is split in $\mathbb{E}$ by the pair $E: A/I \longrightarrow A/E$, $E: A/E \longrightarrow A/I$. Note that $E: A/I \longrightarrow A/E$ is a map and hence a morphism in $\mathbb{R}$. F sends $f: A \longrightarrow B$ to $f: A/id_A \longrightarrow B/id_B$. F is a full and faithful representation. The rest is left to the reader. Also note that F preserves 0 and that G preserves 0 iff G' does. ∎

REMARK. The category $\mathbb{E}$ may also be presented as the result of splitting all symmetric idempotents of the full subcategory of $Rel(\mathbb{A})$ of those assemblies that have all caucuses equal (in other words, those assemblies that are simply sets). In fact, $Rel(\mathbb{A})$ is itself the result of splitting all coreflexive morphisms of this full subcategory. $\mathbb{R}$ is obtained from $\mathbb{E}$ in the same way that $\mathbb{A}$ may be obtained from $Rel(\mathbb{A})$, namely by taking the subcategory of maps. The construction of $\mathbb{R}$ from the full subcategory of $Rel(\mathbb{A})$ described above is related to the construction of "tagging the types by partial equivalence relations" studied in [Breazu-Tannen & Coquand 87].

3. THE REALIZABILITY UNIVERSE AS A TOPOS

In this section we determine when the category $\mathbb{R}$ has a higher-order logical structure. A TOPOS is a category with finite products and equalizers such that for every object A there is a POWER-OBJECT $\mathbb{P}A$ and a relation:

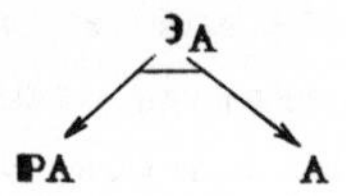

(UNIVERSAL RELATION targeted at A) such that any relation R with codomain A may be uniquely represented as a pullback:

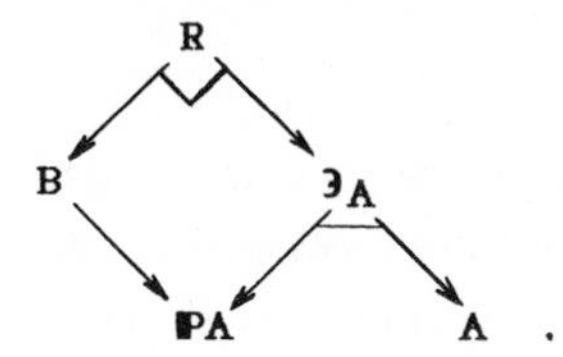

Note that for each A , $\mathbb{P}A$ and $\ni_A$ are determined uniquely up to isomorphism. (In the category of sets, $\mathbb{P}A$ is given by the set of all subsets of A , $S \ni_A a$ iff S contains a as an element, and $f(b) = \{a: b R a\}$. Clearly $b R a$ iff $f(b) \ni_A a$.)

Topoi are also often defined as cartesian closed categories with equalizers, in which there exists an object Ω and a morphism $t: 1 \longrightarrow \Omega$ (1 a terminal object) such that for any object A , any subobject $A' \subset A$ may be uniquely represented as a pullback:

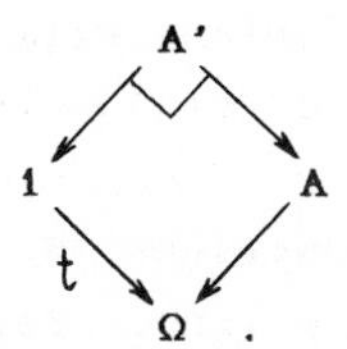

Observe that $\Omega = \mathbb{P}1$, $\mathbb{P}A = [A \to \Omega]$. The notion of topos was introduced by Lawvere in the late 1960's as the categorical framework for higher-order (intuitionistic) logic. In addition, a topos with "an object of natural numbers" (see section 4) may be viewed as (a model of) a collection of assertions in an intuitionistic theory of sets. The reader is referred to [Lambek & Scott 86] and [Boileau & Joyal 81] for a detailed exposition.

THEOREM 1. Let $\mathbb{K}$ be a collection of partial endofunctions on the natural numbers that satisfies the conditions (i) - (iii) and that contains the identity, constant zero, the successor, and the conditionals $\delta(n) = k_0$ if $n = 0$ else k_1 (for each k_0, k_1). Suppose that the functions $\mathfrak{l}$ and $\mathfrak{r}$ are given explicitly in terms of these functions (say by finite summation and multiplication). Then whenever $\mathbb{R}$ is a topos, every partial recursive function is a restriction of a partial endofunction in $\mathbb{K}$ to a recursively enumerable set. Conversely, if $\mathbb{K}$ is the collection of all partial recursive functions, then the conditions (i) - (iii) are met and the category $\mathbb{R}$ is a topos, called the <u>REALIZABILITY UNIVERSE</u> .

REMARK. If $\mathbb{K}$ is such that $\mathbb{R}$ is a topos, $\mathbb{R}$ is not necessarily the Realizability Universe. For example, $\mathbb{K}$ may be the collection of partial functions that are recursive relatively to a highly complex oracle (say, relatively to the halting problem).

PROOF. Replacing $\mathbb{K}$ by the collection of all restrictions of functions in $\mathbb{K}$ to arbitrary subsets does not affect the category $\mathbb{R}$ and the conditions (i) - (iii) . Thus we may assume that $\mathbb{K}$ is closed under restriction. Throughout the proof, the symbol $\simeq$ will mean "one side is defined iff the other is and the two sides are equal when defined". The symbol $\succsim$ will mean "if the left-hand side is defined, then so is the right-hand side and they are equal". The symbol $\downarrow$ will be used as a shorthand for "is defined". Unary partial endofunctions on $\mathbb{N}$ will be written in small letters; binary partial functions will be capitalized. We write $n_{\mathfrak{l}}$ and $n_{\mathfrak{r}}$ for $\mathfrak{l}(n)$ and $\mathfrak{r}(n)$, respectively.

Let $\mathbb{K}_2$ be the set of partial functions $\Psi\colon \mathbb{N}\times\mathbb{N} \longrightarrow \mathbb{N}$ such that for some ψ in $\mathbb{K}$, $\Psi(n_{\mathfrak{l}}, n_{\mathfrak{r}}) \simeq \psi(n)$ for all n . Observe that both projections are in $\mathbb{K}_2$. We first show that if $\mathbb{R}$ is a topos, there exists a special Φ in $\mathbb{K}_2$ such that for every Ψ in $\mathbb{K}_2$:

$$\Psi(n_{\mathfrak{l}}, n_{\mathfrak{r}}) \succsim \Phi(\varepsilon(n_{\mathfrak{l}}), n_{\mathfrak{r}}) \quad \text{for some } \varepsilon \text{ in } \mathbb{K} .$$

If $\mathbb{R}$ is a topos, then the lower semilattice of subobjects of any object is cartesian closed (see e. g. [Lambek & Scott 86]). In particular, consider the assembly $\nabla\mathbb{N}$ and its subobjects V (vertical) and H (horizontal) given by: $V|_k = \{n : n_{\mathfrak{l}} = k\}$, $H|_k =$

$\{n : n_r = k\}$. Then there must exist a subobject $V \Rightarrow H \subset \nabla\mathbf{N}$ maximal among all $A \subset \nabla\mathbf{N}$ such that $A \cap V \subset H$. In particular, we must have $V \Rightarrow H \cap V \subset H$. Notice that if φ is a modulus of an inclusion $A \cap V \subset H$, then φ restricted to $\{n : (A \cap V)|_n$ inhabited$\}$ is still a modulus of that inclusion. Let $\Phi(n_l , n_r) = \varphi(n)$. Obviously, Φ is well-defined and hence in $\mathbb{K}_2$. Thus we may assume that an inclusion $A \cap V \subset H$ has a binary modulus. In particular, let Φ be such a modulus of inclusion $V \Rightarrow H \cap V \subset H$. Let $m \in (V \Rightarrow H)|_{n_l}$. Then if $m \in V|_{n_r}$, then $\varphi(n)\downarrow$ and $m \in H|_{\varphi(n)}$. Thus, in any event, $(V \Rightarrow H)|_{n_l}$ must be included in:

$$\bigcap_{k:\ k_l = n_r} \{m: \text{ if } m \in V|_{k_r} \text{ , then } \varphi(n)\downarrow \text{ and } m \in H_{\varphi(n)}\} ,$$

i.e.,

$$\bigcap_{k:\ k_l = n_r} \{m: \text{ if } m_l = k_r \text{ , then } \varphi(n)\downarrow \text{ and } \varphi(n) = m_r\} .$$

in other words, the "graph" of Φ . Let Ψ be any binary function in $\mathbb{K}_2$. Let $\psi(n) \simeq \Psi(n_l , n_r)$. Define an assembly $A \subset \nabla\mathbf{N}$ by $A|_n$ = the above intersection with ψ instead of φ . Clearly $A \cap V \subset H$ (with modulus Ψ) , hence $A \subset (V \Rightarrow H)$. Let ξ be a modulus of $A \subset V \Rightarrow H$: if $m \in A|_i$, then $\xi(i)\downarrow$ and $m \in (V \Rightarrow H)|_{\xi(i)}$. Hence for any m , $\Psi(m_l , m_r) \succeq \Phi(\xi(m_l) , m_r)$.

We now restrict Φ so that in the previous sentence $\succeq$ may be replaced by $\simeq$ for a suitable subcollection in $\mathbb{K}_2$. The desired subset is defined inductively, e.g., mimicking the inductive definition of partial recursive functions in [Kleene 59] , with numerical arguments only and with $\Phi(e, \_)$ playing the role of Kleene's $\{e\}(\_)$. Let Φ' be the obtained restriction of Φ . The collection of functions of the form $\Phi'(e, \_)$ for some number e (and the corresponding collection of functions of several arguments) inherits all of the closure properties listed in the conditions of the theorem. We have gained the enumeration property (by construction). The s-m-n property is obtained by a modulus for the tautology $A \cap (A \cap B \Rightarrow C) \subset B \Rightarrow C$ in any subobject lattice. (In the inductive definition above, we ensure that such a modulus belongs to the subcollection being defined.) Now it is a recursion-theoretic routine to establish closure under μ-operator and hence under primitive recursion, say as in [Kleene 59: XVI] . Therefore the subcollection of functions thus constructed contains the least collection of functions satisfying the latter closure properties, that is: the collection of partial recursive functions. The reader may find related results and techniques in [Wagner 69] .

Given an object X/I , we construct its power object $[X]/E$. Let the assembly X' be such that $X'|_n = \{x \in |X| : x\ I_n\ x\}$. Let $[X]|_n$ consist of all assemblies $S \subset X'$ of modulus $\Phi(n_r, \_)$ such that $\Phi(n_l, \_)$ is a modulus of $IS \subset S$. Let $S\ E_n\ S'$ iff $S \in [X]|_{n_{ll}}$, $S' \in [X]|_{n_{lr}}$, and $\Phi(n_r, \_)$ is a modulus of $S = S'$. The universal relation $\ni$ with target $[X]/E$: $S \ni_n x$ iff $S \in [X]|_{n_l}$ and $x \in S|_{n_r}$. ∎

In [Hyland 82] the Realizability Universe is called the Effective Topos.

4. TOWARD EXPLORATION OF THE REALIZABILITY UNIVERSE

We conclude by pointing out some special kinds of objects in the Realizability Universe $\mathbb{R}$ (that is, in the topos $\mathbb{R}$ when $\mathbb{K}$ is the collection of all partial recursive functions). In fact, most of our discussion could take place in any topos.

We consider those objects D for which the property "Every map from Ω to D is constant" holds internally in $\mathbb{R}$. More precisely, we say that an object D in a topos is <u>DISCRETE</u> iff the morphism $D \longrightarrow [\Omega \to D]$ induced by the projection $\Omega \times D \longrightarrow D$ is an isomorphism. In the case of $\mathbb{R}$, in this definition Ω may be replaced by $\nabla 2$, where 2 is a two-element set. This comes down to requiring that for $D = X/I$ there exists ψ in $\mathbb{K}$ such that for all $x, x' \in |X|$, $n \in \mathbb{N}$: if $x\ I_n\ x$ and $x'\ I_n\ x'$ then $\psi(n)$ is defined and $x\ I_{\psi(n)}\ x'$. In other words, whenever x, x' exist in D for the same reason, they are certifiably equal.

The notion of a discrete object and Proposition 7 are related to the issues discussed in [Freyd & Kelly 72] .

PROPOSITION 5. In any topos, the full subcategory of discrete objects is a cartesian closed subcategory. In fact, if D is discrete, so is $[A \to D]$ for any object A .

PROOF. $[\Omega \to D \times D'] = [\Omega \to D] \times [\Omega \to D'] = D \times D'$.
$[\Omega \to [A \to D]] = [\Omega \times A \to D] = [A \to [\Omega \to D]] = [A \to D]$. ∎

In $\mathbb{R}$, an assembly is discrete iff it is modest, that is: every caucus has at most one member. Particular attention should be paid to the discrete partitioned assembly N whose n-th caucus has n itself as the sole member. N is a <u>proper</u> subobject of $\nabla\mathbb{N}$. If E is an equivalence relation on a subobject $X \subset N$, then X/E is discrete.

Since constant 0 and the successor function s are in $\mathbb{K}$, one has the morphisms $1 \longrightarrow N$ with modulus 0 and $N \longrightarrow N$ with modulus s . In fact, one may show that $1 \longrightarrow N \longrightarrow N$ is initial among all diagrams $1 \longrightarrow A \longrightarrow A$ in $\mathbb{R}$, that is: N is a <u>NATURAL NUMBER OBJECT</u> in $\mathbb{R}$:

PROPOSITION 6. Let $\mathbb{R}$ be the Realizability Universe. For any diagram $1 \longrightarrow A \longrightarrow A$ in $\mathbb{R}$, there exists a unique morphism $N \longrightarrow A$ such that:

$$\begin{array}{ccccc} 1 & \longrightarrow & N & \longrightarrow & N \\ & \searrow & \downarrow & & \downarrow \\ & & A & \longrightarrow & A \end{array} .$$

PROOF. Recursion in $\mathbb{K}$. ∎

In any topos, the full subcategory of discrete objects is a reflective subcategory:

PROPOSITION 7. In any topos, for any object A there is a discrete object A_{disc} and a cover $r: A \longrightarrow A_{disc}$ such that for any map from A to a discrete D there exists a unique:

$$\begin{array}{ccc} A & \longrightarrow & A_{disc} \\ & \searrow & \downarrow \\ & & D \end{array} .$$

If A is discrete, r is an isomorphism. A_{disc} is the <u>DISCRETE REFLECTION</u> of A .

PROOF. In intuitionistic higher-order logic, let E be the intersection of all equivalence relations on A whose quotient is discrete. Then let $A_{disc} = A/E$. Because this argument is completely intuitionistic, it can be repeated in any topos (see remarks in section 3). ∎

We should point out that by using classical logic in the topos of ordinary sets, a discrete set has at most one element and hence A/E is empty iff A is empty, and A/E is a one-element set iff A is nonempty.

In the case of $\mathbb{R}$, this argument may be redescribed as follows. First, it is easily shown independently that the discrete reflection of a partitioned assembly P is the partitioned assembly P_{disc} such that $(P_{disc})|_n$ has the sole element n iff $P|_n$ is inhabited (else $(P_{disc})|_n$ is empty). Given an arbitrary object X/I, let φ and ψ in $\mathbb{K}$ be the moduli associated with I. Define the partitioned assembly P as the disjoint union of caucuses $P|_n = \{\langle n, x\rangle : x \; I_n \; x\}$. Let R be the relation on P_{disc} given as $\varphi(k) \; R_k \; \psi(k)$ iff $x \; I_k \; x'$ for some x, x'. In general, R is neither reflexive, nor symmetric, nor transitive. The least symmetric relation S on P_{disc} containing R may be presented as $S_{2n} = R_n$, $S_{2n+1} = R_n$. We write $n_1, \ldots, n_5$ for the five projections of n obtained as suitable mixed iterations of $\mathbb{l}$ and $\mathbb{r}$. Φ refers to a universal Turing machine (see proof of Theorem 1). Define the least equivalence relation E on P_{disc} containing S by: $i \; E_k \; j$ iff for all sequences $\{Q_m\}_{m\geq 0}$ of ordinary relations on natural numbers, if $\Phi(n_1, \_), \ldots, \Phi(n_5, \_)$ are moduli of $Q \subset N{\cdot}N$, of $R \subset Q$, of reflexivity, symmetry, and transitivity of Q, respectively, then $m = \Phi(k, n)$ is defined and $i \; Q_m \; j$. (E may also be described in terms of a recursive enumeration of finite words of natural numbers: if $w = j_1 \ldots j_m$, $m \geq 0$, then let E_w be the m-fold composition $S_{j_1} S_{j_2} \ldots S_{j_m}$.) Finally, let A_{disc} be the object P_{disc}/E. The reader may easily verify the required property of A_{disc}. If A is already discrete, then the identity morphism on A must factor through the cover r. We have shown:

PROPOSITION 8. In the Realizability Universe, for every discrete object D there exists a subobject X of the natural number object N and an equivalence relation E on X such that D is isomorphic to X/E. In the Realizability Universe, furthermore, any quotient of a discrete object is discrete. ∎

Because of this representation of discrete objects, there exists an object $\mathbb{D} \subset \mathbb{P}\mathbb{P}N$ in the Realizability Universe such that discrete objects corresponds to morphisms $1 \longrightarrow \mathbb{D}$. Also note that all discrete objects in the Realizability Universe are subobjects of a single object, e.g. $\mathbb{P}N$.

REMARK. The discrete reflection of an assembly is not necessarily the modest reflection discussed in Proposition 2 .

In order to elucidate the remark just made, we again recall a few basic facts from topos theory. In the proof of Theorem 1, we have already used the fact that in a topos, the poset of subobjects of any object A is cartesian closed. (It also has finite unions, but we do not need the additional structure here.) Because the initial object of a topos is a subobject of any object, for any subobject $A' \subset A$ there exists a subobject $\neg A' = A' \Rightarrow 0$ of A (called the negation or complement of A') maximal among all $B \subset A$ such that $A \cap B = 0$. A subobject $A' \subset A$ will be called <u>CLOSED</u> iff $\neg\neg A' = A'$. A subobject $B \subset A$ will be called <u>DENSE</u> iff $\neg\neg B = A$. We will say that an object A is <u>SEPARATED</u> iff any two morphisms with source A must be equal whenever they are equal on a dense subobject of A . This requirement is equivalent to the condition that the relation:

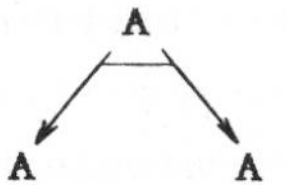

is closed (as a subobject of $A \cdot A$) . In the Realizability Universe, every assembly is separated and an object A is separated iff it is a subobject of ∇S for some set S .

REMARK. In any topos, $[A \to S]$ is separated whenever S is, and a finite product of separated objects is separated. This is related to the fact that in intuitionistic logic, double negation may be transfered inward across universal quantification, implication, and conjunction.

PROPOSITION 9. In any topos, the full subcategory of separated objects is a reflective subcategory. In the Realizability Universe, every separated object is isomorphic to an assembly.

PROOF. In general, work in intuitionistic higher-order logic. Given A , let E be the intersection of all stable equivalence relations on A . The separated reflection A_{sep} is the quotient A/E. In $\mathbb{R}$ this argument may be redescribed as follows. Given an object $A = X/I$, the union $\bigcup_{n\geq 0} I_n$ is an equivalence relation on $|X|$. Let A_{sep} be the assembly for which $(A_{sep})|_n$ consists of the equivalence classes of elements of $X|_n$. The rest is left to the reader. ▮

We say that an object in a topos is a <u>MODEST OBJECT</u> iff it is separated and discrete. We obtain:

PROPOSITION 10. In any topos, the full subcategory of modest objects is reflective. In the Realizability Universe, every modest object is isomorphic to a quotient of a closed subobject of the natural number object N by a closed equivalence relation. In the Realizability Universe, a quotient of a modest object by a closed equivalence relation is modest.

PROOF. In $\mathbb{R}$, the modest reflection of A is the separated reflection of the discrete reflection of A. ▮

Beacuse of the representation of modest objects given by Proposition 10, there is an object $\mathbb{M} \subset \mathbb{D} \subset \mathbb{PP}N$ in the Realizability Universe such that modest objects correspond to morphisms $1 \longrightarrow \mathbb{M}$. We also note, as above, that all modest objects of the Realizability Universe are subobjects of a single object, e.g. $\mathbb{P}N$.

REMARK. Modest objects of the Realizability Universe were called effective in [Hyland 82]. The present terminology is due to D. Scott. Moggi's interpretation of polymorphism mentioned in the introduction was given in the full subcategory of modest objects [Rosolini 86]. Discrete and modest objects in topoi are studied also in [Hyland & al. 87].

PROPOSITION 11. In any topos, the full subcategory of modest objects is a cartesian closed subcategory. Furthermore, $[A \to M]$ is modest whenever M is.

PROOF. In general, use Proposition 10 and the remark before Proposition 9. The calculation in the Realizability Universe is left to the reader. ▮

We recall some trivial categorical folklore about products in reflective subcategories:

PROPOSITION 12. Let $\mathbb{C}'$ be a full reflective subcategory of a category $\mathbb{C}$ closed under isomorphism type. Then an equalizer in $\mathbb{C}$ of a pair of maps between objects of $\mathbb{C}'$ is an equalizer in $\mathbb{C}'$. Furthermore, whenever a collection of objects of $\mathbb{C}'$ has a product in $\mathbb{C}$, that product is the product in $\mathbb{C}'$. ∎

We illustrate the argument on the case of binary products. Let A, B be objects of $\mathbb{C}'$ and let $A \times B$ be their product in $\mathbb{C}$. Let $(A \times B)'$ be its reflection. Then:

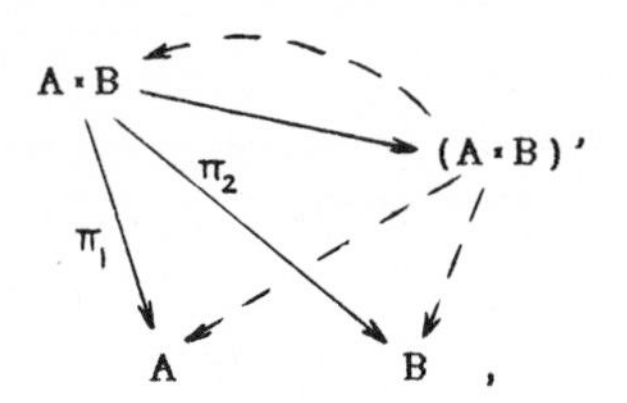

where $(A \times B)' \dashrightarrow A \times B$ is obtained because of the universal property of the product. Now use the universal properties of products and of reflection to show that this morphism is an isomorphism.

Now in any topos, any internally defined collection of objects (say by a morphism $A \longrightarrow \mathbb{P}B$) has a product. Thus the preceding discussion may be summarized by:

THEOREM 2. In any topos, the full subcategory of discrete, respectively modest, objects is reflective, cartesian closed, internally complete subcategory. In the Realizability Universe, neither subcategory is a poset and furthermore, all discrete (and hence all modest) objects are subobjects of a single object. ∎

Either subcategory, therefore, may be used as a setting for polymorphism by considering objects as types, $[A \to B]$ as a function type constructor, arbitrary products as type abstraction, and morphisms as terms.

REMARK. It follows from the work of Powell and Pitts (described in section 17 of [Hyland 82]) that the set of Turing degrees may be embedded in the collection of intermediate full subcategories of the Realizability Universe between the modest and the discrete cases that enjoy the properties given by Theorem 2. Such intermediate subcategories are obtained by modifying the notion of a separated object.

ACKNOWLEDGEMENTS. We wish to thank D. Scott, A. Nerode, and P. J. Scott for helpful discussions and suggestions. Scedrov is partially supported by N. S. F.

REFERENCES

Bainbridge E. S., Freyd P. J., Scedrov A., Scott P. J. [88], *Functorial polymorphism*. In: "Logical Foundations of Functional Programming" (G. Huet, ed.), Addison-Wesley, to appear.

Beeson M. [85], *"Foundations of constructive mathematics"*. Ergeb. der Math., Springer-Verlag, New York.

Boileau A., Joyal A. [81], *La logique des topos*. J. Symb. Logic 46, 6-16.

Breazu-Tannen V., Coquand T. [87], *Extensional models for polymorphism*. Proc. TAPSOFT '87 - CFLP, Pisa, Theor. Comp. Sci., to appear.

Bruce K. B., Meyer A. R., Mitchell J. C. [87], *The semantics of second-order lambda calculus*. Information & Computation, to appear.

Carboni A., Walters R. F. C. [88], *Cartesian bicategories I*. J. Pure Appl. Algebra, to appear.

Cardelli L., Wegner P. [85], *On understanding types, data abstraction, and polymorphism*. ACM Comp. Surveys 17 (4) 471-522.

Cousineau G., Curien P. L., Robinet B. (eds.) [86], *"Combinators and functional programming languages"*. Springer LNCS 242 .

Freyd P. J. [74], *On functorializing model theory, or: on canonizing category theory*. Manuscript.

Freyd P. J., Kelly G. M. [72], *Categories of continuous functors I*. J. Pure Appl. Alg. 2 , 169-191; Erratum, ibid. 4 (1974), 121.

Freyd P. J., Scedrov, A. [87], *Some semantic aspects of polymorphic lambda calculus*. Proc. 2nd IEEE Symp. Logic in Computer Science, Ithaca, N. Y., pp. 315-319.

Freyd P. J., Scedrov A. [88], *"Geometric logic"*. Math. Library Ser., North-Holland, Amsterdam, to appear.

Girard J. Y. [71], *Une extension de l'interprétation de Gödel* ... Proc. 2nd Scandinavian Logic Symp. (J. Fenstad, ed.), North-Holland, Amsterdam, 63-92.

Girard J. Y. [72], *"Interprétation fonctionelle et élimination des coupures* ... *"*. Thèse de Doctorat d'Etat. Université Paris VII.

Hyland J. M. E. [82], *The effective topos*. Proc. Brouwer Centenary Symposium (A. S. Troelstra, D. van Dalen, eds.), North-Holland, Amsterdam, 165-216.

Hyland J. M. E., Robinson, E., Rosolini, G. [87], *Discrete objects in a topos*. Preprint.

Kleene S. C. [59], *Recursive functionals and quantifiers of finite type I*. Trans. Amer. Math. Soc. 91, 1-52.

Lambek J., Scott P. [86], *"Introduction to higher-order categorical logic"*. Cambridge Univ. Press.

McCarty D. C. [86], *Realizability and recursive set theory*. Ann. Pure Appl. Logic 32, 153-183.

Mitchell J. C. [86], *A type-inference approach to reduction properties and semantics of polymorphic expressions*. Symp. on Lisp and Functional Programming.

Reynolds J. C. [74], *Towards a theory of type structure*. Springer LNCS 19, 408-425.

Reynolds J. C. [84], *Polymorphism is not set-theoretic*. Springer LNCS 173.

Rosolini G. [86], *About modest sets*. Preprint

Scedrov A. [88], *Recursive realizability and the calculus of constructions*. In: "Logical Foundations of Functional Programming" (G. Huet, ed.), Addison-Wesley, to appear.

Scott D. S. [82], *Domains for denotational semantics*. ICALP 82. Springer LNCS 140 .

Seely R. A. G. [87], *Categorical semanticsfor higher-order polymorphic lambda calculus*. J. Symb. Logic, to appear.

Wagner E. [69], *Uniformly reflective structures: on the nature of gödelizations and relative computability*. Trans. Amer. Math. Soc. 144 , 1-41.

AUTHORS' ADDRESSES:

For A. Carboni:

Dipartimento di Matematica
Università di Milano
Via Saldini 50
20133 Milano, Italy

mcvax!i2unix!cca_umi!carboni

For P. Freyd and A. Scedrov:

Department of Mathematics
University of Pennsylvania
Philadelphia, PA 19104-6395
U. S. A.

ARPANET: Andre@cis.upenn.edu

RULE-BASED SEMANTICS FOR AN EXTENDED LAMBDA-CALCULUS

György E. Révész
IBM T. J. Watson Research Center
Yorktown Heights, N.Y.

Abstract

Many implementation techniques proposed for functional languages in the literature are based on lambda-calculus and the theory of combinators [5, 12, 14]. The main advantage of functional languages over the more conventional programming languages is their lack of side-effects which makes their semantics much simpler [2].

This paper presents a λ-calculus dialect with list-handling extensions that can be used for defining an operational semantics for functional programs. This semantics is based on a set of elementary α-rules and β-rules which collectively implement the substitution operation without explicitly using it [10]. Two extra reduction rules, called γ-rules, have been added to the system for list manipulations [11]. Combining λ-calculus with list-handling capabilities makes an efficient vectorization of λ-calculus possible. This way we obtain a very elegant treatment of mutual recursion in our system.

The reduction rules described in this paper represent, in fact, an operational semantics for our extended λ-notation. An interpreter program for this extended λ-calculus has been developed by a direct implementation of the reduction rules which makes the correctness proof of the interpreter very easy. The notion of *controlled reduction* is introduced here to guarantee the existence of a normal form for λ-expressions representing recursively defined functions using the **Y** combinator.

1. A direct extension of the λ-notation

It is well known that list structures and list manipulating functions can be encoded in pure λ-notation, which means that list manipulation can be implemented in the standard λ-calculus without introducing new reduction rules. The theoretical implications of such an approach are very important but its practical value is minimal. A direct representation of list structures and a direct implementation of their elementary operations appears to be much more efficient. The same is true for many other higher order functions or combinators. A case in point is the **Y** combinator which can be expressed as a λ-term and thus, it can be implemented via the standard β-reduction without any extra rule. A direct implementation, however, based on its characteristic property is clearly more efficient.

The extension of the λ-notation implies, of course, the extension of its semantic definition, as well. The problem is to define the semantics of the extended λ-notation in such a way that is both practical and theoretically sound. We claim that our approach to be presented in Section 4 of this paper satisfies these requirements.

*The semantics of a language **L** can be defined in general as a mapping μ from **L** to **M**, where **M** is the set of "meanings"*. For the sake of simplicity, the elements of ***L*** will be called ***syntactic objects*** while the elements of ***M*** will be called ***semantic objects***. The syntactic objects considered here will be the λ-expressions which form a context-free language denoted by Λ. Actually, we use an extended λ-notation described by the following syntax.

THE SYNTAX OF λ-EXPRESSIONS

<λ-expression>::=<variable> | <abstraction> |
<application> | <list> | <constant>

<variable>::=<identifier>

<abstraction>::=λ<variable>.<λ-expression>

<application>::=(<λ-expression>)<λ-expression>

<list>::=[<λ-expression><list-tail> | []

<list-tail>::=,<λ-expression><list-tail> |]

<constant>::=<number> | <operator> | <combinator>

<operator>::=<arithmetic operator> | <relational operator> |

<predicate> | <boolean operator> | <list operator>

<arithmetic operator>::=+ | - | * | / | **succ** | **pred** | **mod**

<relational operator>::=< | ≤ | = | ≥ | > | ≠

<predicate>::=**zero** | **null**

<boolean operator>::=**and** | **or** | **not**

<list operator>::=**head** | **tail** | **cons** | **map** | **append**

<combinator>::=**true** | **false** | **Y**

The lexical details of the identifiers and numbers are not specified here since they are not important for our discussion. Note, however, that in an application we always put the operator, rather than the operand, between parentheses. Hence, we use *(f)x*, instead of *f(x)*, to denote the application of a function *f* to the argument *x*.

The empty list is denoted by []. The primitive functions **pred** and **succ** represent the predecessor and the successor functions on integers. The predicate **zero** represents the test for integer type zero while **null** represents the test for the empty list. The constant symbols **head**, **tail**, and **cons** denote the usual list operations. All functions are *Curry-ed*, which means that they take only one argument at a time. Two or more arguments are supplied to a function by repeated applications. Thus, for example, the *Curry-ed* version of the binary addition is written as ((+)a)b.

The use of combinators is, of course, redundant because they can also be represented by pure λ-expressions. Nevertheless, we use them as special symbols in order to make their implementation more efficient.

For the semantics of λ-expressions, we follow Church, who regarded λ-expressions without normal form as meaningless. The main difficulty with this view is that recursive functions are usually defined by λ-expressions without normal form. They use the **Y** combinator in such a way that gives rise to an infinite reduction sequence. Thus, we are faced with the awkward situation of having meaningful λ-expressions containing meaningless subexpressions. An attempt at fixing this problem will be presented in Section 4.

If we choose $\boldsymbol{M} \subset \Lambda$ to denote *the set of λ-expressions being in normal form,* and define the mapping μ to map a λ-expression to its normal form (if any), then we cannot give an independent syntactic description of $\boldsymbol{M}$ without referring to the reduction rules. But that is only a minor problem. The definition of μ will describe precisely how the program works even if it does not terminate. This approach is, in fact, closely related to the so called operational semantics where the emphasis is placed on the computational procedures represented by the program.

2. The reduction rules and their implementation

The reduction rules represent "meaning preserving" transformations on λ-expressions. The execution of a program (algorithm) represented by a λ-expression can be described as a sequence of reduction steps, each of which is a single application of some reduction rule.

Here we shall use the notation {z/x}E to represent the operation of renaming all free occurrences of x in E by z. Our reduction rules are the following:

ALPHA RULES

(α1) $\{z/x\}x \to z$

(α2) $\{z/x\}E \to E$ if x does not occur free in E

(α3) $\{z/x\}\lambda y.E \to \lambda y.\{z/x\}E$ for every λ-expression E, if $x \not\equiv y \not\equiv z$.

(α4) $\{z/x\}(E_1)E_2 \rightarrow (\{z/x\}E_1)\{z/x\}E_2$

(α5) $\{z/x\}[E_1, ..., E_n] \rightarrow [\{z/x\}E_1, ..., \{z/x\}E_n]$ for $n \geq 0$

BETA RULES

(β1) $(\lambda x.x)Q \rightarrow Q$

(β2) $(\lambda x.E)Q \rightarrow E$ if x does not occur free in E

(β3) $(\lambda x.\lambda y.E)Q \rightarrow \lambda z.(\lambda x.\{z/y\}E)Q$ if $x \not\equiv y$, and z is neither free nor bound in (E)Q.

(β4) $(\lambda x.(E_1)E_2)Q \rightarrow ((\lambda x.E_1)Q)(\lambda x.E_2)Q$

GAMMA RULES

(γ1) $([f_1, ..., f_n])E \rightarrow [(f_1)E, ..., (f_n)E]$ for $n \geq 0$

(γ2) $\lambda x.[E_1, ..., E_n] \rightarrow [\lambda x.E_1, ..., \lambda x.E_n]$ for $n \geq 0$

LIST OPERATORS

(**head**)[] → [], (**head**)$[E_1, E_2 ..., E_n] \rightarrow E_1$

(**tail**)[] → [], (**tail**)$[E_1, E_2, ..., E_n] \rightarrow [E_2, ..., E_n]$

(**null**)[] → **true**, (**null**)$[E_1, ..., E_n] \rightarrow$ **false** for $n \geq 1$

((**cons**)A)$[E_1, ..., E_n] \rightarrow [A, E_1, ..., E_n]$ for $n \geq 0$

((**map**)F)$[E_1, ..., E_n] \rightarrow [(F)E_1, ..., (F)E_n]$ for $n \geq 0$

((**append**)$[A_1, ..., A_m]$)$[E_1, ..., E_n] \rightarrow [A_1, ..., A_m, E_1, ..., E_n]$ for $n \geq 0$

PROJECTIONS

$(1)[E_1, ..., E_n] \rightarrow E_1$ for $n \geq 1$

$(i)[E_1, ..., E_n] \rightarrow (i-1)[E_2, ..., E_n]$ for $n \geq i \geq 2$

COMBINATORS

((**true**)A)B → A, ((**false**)A)B → B

(Y)E → (E)(Y)E

The reduction rules for the remaining operators are straightforward. As we have stated before, *the evaluation of λ-expressions will be performed by reducing them to normal form using the above rules.* This means that the execution of a program can be defined in terms of program transformations performed directly on its source representation. Note that the substitution prefix does not occur in our reduction rules since we need not define the operation of substitution at all. Instead, we have four different β-rules which is clearly a departure from the traditional approach [3, 7]. Also, it should be noted that we have eliminated the overhead of bracket abstraction (see e.g. in [4, 9, 13]) in favor of a run-time routine to check if a variable occurs free in a given subexpression.

We claim (without proof) that the Church-Rosser property is preserved in our extended λ-calculus. This may appear to be obvious, but it is not really so, because it is very easy to destroy that property. If, for instance, we introduce **is-application** as a new predicate symbol with the reduction rules

(is-application)(P)Q → **true**

(is-application)λx.E → **false**

then the resulting system is no longer Church-Rosser. This must be clear from the example

(is-application)(λy.λx.y)z

which reduces to **true** or **false** depending on whether or not the argument is reduced first to λx.z.

The correctness proof of our interpreter is very simple, because we implement the reduction rules in a direct manner. We use a graph-reduction technique, which seems to be more efficient than string-reduction when working on a conventional computer with random access storage. (See, e. g. [1, 12, 14].) The outlines of a graph-reduction technique for implementing our extended λ-calculus have been described in [11].

3. Dealing with mutual recursion

Nonrecursive function definitions can be treated in lambda-calculus in a fairly simple way. Assume, that we have a sequence of definitions of the form

$$f_1 = d_1$$

$$\vdots$$

$$f_n = d_n$$

where f_i are variables (function names) and d_i are λ-expressions. Assume further, that E is a λ-expession containing free occurrences of the variables f_i (i=1,...,n). Now, the value of E with respect to the given definitions can be computed by evaluating the combined expression

$$(\lambda f_1. \ldots (\lambda f_n.E)d_n \ldots)d_1$$

Thus, the set of equations can be treated as mere syntactic sugar having a trivial translation into pure lambda-notation. This simple minded approach, however, does not always work. As can be seen from the given translation, the form of the combined expression reflects the order of the equations. Therefore, a free occurrence of f_i in d_j will be replaced by d_i if and only if $i<j$.

In other words, *previously defined function names can be used on the right hand side of the equations, but no forward reference can be made to a function name defined only later in the sequence.* This clearly excludes mutual recursion, which always involves some forward reference. If, for instance, f_1 is defined in terms of f_2 (i.e., f_2 occurs in d_1) and vice versa then the forward reference cannot be eliminated simply by changing the order of the equations.

It should be clear that in the absence of mutual recursion the equations can be rearranged in such a way that no forward reference occurs. Immediate recursion is not a problem, since it can be resolved with the aid of the **Y** combinator. Mutual recursion can be resolved in a similar fashion in our extended lambda-calculus without lambda-lifting [8] or any other complicated preprocessing. Namely, the list manipulating power of our calculus makes it possible to represent a list of variables by a single one just as it is done in vector algebra. Hence, *the solution of a set of simultaneous equations can be expressed in a compact form with a single occurrence of the* **Y** *combinator.*

Consider namely the following set of simultaneous recursion equations

$$f_1 = d_1$$

$$\vdots$$

$$f_n = d_n$$

First we rewrite each d_k by substituting (i)F for the free occurrences of f_i ($1 \le i \le n$). The resulting expressions will be denoted by D_k which form an ordered n-tuple.

So, we get the defining equation

$$F = [D_1, \ldots, D_n]$$

in place of the original ones. Abstracting with respect to F on the right hand side yields

$$F = (\lambda F.[D_1, \ldots, D_n])F$$

whose solution is expressed by

$$F = (Y)\lambda F.[D_1, \ldots, D_n]$$

Now, the value of any λ-expression E with respect to the given recursive definitions can be computed as the value of the combined expression

$$(\lambda F.H)(Y)\lambda F.[D_1, \ldots, D_n]$$

where

$$H = (\lambda f_1. \ldots(\lambda f_n.E)(n)F \ldots)(1)F$$

This formula is correct also for nonrecursive definitions but the simple minded approach described at the beginning of this section is clearly more efficient. Therefore, an optimizing compiler should treat recursive and nonrecursive definitions separately. For that purpose, one can compute the dependency relation between the given definitions and check if it forms a partial order on the set of functions f_i (i = 1,...,n). If so, then no mutual recursion is present and the definitions can be arranged in a sequence suitable for the simple minded solution. Otherwise, one should try to isolate the minimal sets of mutually recursive definitions and solve them separately, before putting them back to their proper place in the sequence.

4. Adjusting the reduction rules

As we have mentioned in Section 2, the mapping μ cannot be identified with the "reduction to normal form" without that loss of generality which rules out recursive definitions. On the other hand, it seems desirable to make use of the reduction rules for the definition of μ as much as possible, because the meaning must be invariant under the equivalence generated by reduction. Moreover, when a λ-expression does have a simple normal form then this must be the same as its meaning.

So, as a compromise, we shall define ***controlled reduction*** by modifying the reduction rules as follows:

The rule for the **Y** *combinator will be changed to*

(Y')E → (E)(Y)E

where **Y'** *is a newly introduced combinator. Furthermore, the* $\beta 2$*-rule will be modified as*

($\beta 2$')   (λx.E)Q → E' *if x does not occur free in* E

where E' *is the same as* E, *except that each (if any) occurrence of* **Y** *is replaced by* **Y'**.

For the new system to work, all recursive definitions should be written with the aid of the new **Y'** combinator in place of **Y**. The latter is namely disabled in the new system until it gets changed to **Y'** in a $\beta 2$-reduction step.

To see how this system works on a simple example consider the following recursive definition of the factorial function.

(fact)n = *if* n = 0 *then* 1 *else* ((*)n)(fact)(**pred**)n

which will be written in our λ-notation as

(fact)n = (((**zero**)n)1)((*)n)(fact)(**pred**)n

that is

fact = λn.(((**zero**)n)1)((*)n)(fact)(**pred**)n

The solution of this recursion equation will be expressed by

fact = (**Y'**)λf.λn.(((**zero**)n)1)((*)n)(f)(**pred**)n

where the right hand side reduces to

(λf.λn.(((**zero**)n)1)((*)n)(f)(**pred**)n)(**Y**)λf.λn.(((**zero**)n)1) ((*)n)(f)(**pred**)n

which further reduces to the normal form

λn.(((**zero**)n)1)((*)n)((**Y**)λf.λn.(((**zero**)n)1)((*)n)(f)(**pred**)n) (**pred**)n

If, however, we supply an argument, say 5, to the function "fact" then we get the correct result 120 as can be easily checked by the reader. The trick of *controlled reduction* lies in the fact that the **Y'** combinator can fire only once, but each time an argument A is presented to an expression of the form

(F)(**Y**)F

an attempt at evaluating the expression

((F)(**Y**)F)A

is made, which in turn involves an attempt at substituting the argument A for some variable x in **Y**. But that would change **Y** to **Y'**, since no variable occurs free in **Y**.

It is assumed, however, that the expression F has at least one more abstraction besides the abstraction on the function name which is being defined by the recursion. This is the case in the above example where we have λn besides the λf. Indeed, any well-founded recursion must have a condition which would terminate the recursion after a finite number of steps. That condition must depend on the argument(s), otherwise there could be no change in its truth value throughout the recursion.

It is interesting to note that this modified system is extensionally incomplete, since we have

$(\lambda x.\mathbf{Y'})Q = (\lambda x.\mathbf{Y})Q$ for any Q,

which would imply $\lambda x.\mathbf{Y'} = \lambda x.\mathbf{Y}$ in an extensionally complete system. (Actually, we may assume that this is the case even though $\mathbf{Y'} \neq \mathbf{Y}$.)

In any case, *the meaning function μ can be defined as reducing to normal form in the new system.* If the normal form of an expression has a subexpression of the form (Y)F then we may say that this subexpression denotes the fixed-point of F.

Note that the **Y** combinator can also be used for defining infinite lists. For instance, an infinite list of zeroes can be defined recursively as

zeroes = ((**cons**)0)zeroes

hence,

zeroes = (**Y'**)λz.((**cons**)0)z

where the right hand side reduces to

(λz.((**cons**)0)z)(**Y**)λz.((**cons**)0)z

which has the normal form

((**cons**)0)(**Y**)λz.((**cons**)0)z

Now, the problem is that a finite projection of this infinite list is not computable in the new system, because the infinite list is not the operator but the operand of the projection. In order to be able to compute the *i*-th element of an infinite list we have to define projections for arbitrary λ-expressions, P and Q, this way

(1)((**cons**)P)Q → P

(i)((**cons**)P)Q → (i-1)Q for $i \geq 2$

In other words, operations on lists will be defined *lazily* so that we can use infinite lists as arguments. Thus, in our example we get

$(i)((\mathbf{cons})0)(\mathbf{Y})\lambda z.((\mathbf{cons})0)z \rightarrow (i\text{-}1)(\mathbf{Y})\lambda z.((\mathbf{cons})0)z$

Now, in order to continue the evaluation we have to change the **Y** combinator back to **Y'**. This requires further modifications of the reduction rules, because the $\beta 2'$-rule is not applicable in this case. All list manipulating functions, which are the projections (with an integer constant), the **head tail, cons, null, map,** and **append**, will be defined lazily. The last two need not be defined separately since they can be defined recursively with the aid of the others. Namely,

$((\mathbf{map})f)E = \textit{if}\ E = []\ \textit{then}\ []\ \textit{else}\ ((\mathbf{cons})(f)(\mathbf{head})E)((\mathbf{map})f)(\mathbf{tail})E$

$((\mathbf{append})E)F = \textit{if}\ E = []\ \textit{then}\ F\ \textit{else}\ ((\mathbf{cons})(\mathbf{head})E)((\mathbf{append})(\mathbf{tail})E)F$

Furthermore, **cons** does not evaluate its arguments [6], so its reduction rule need not be changed. For the remaining list operations we introduce the following rules:

$(i)(\mathbf{Y})E \rightarrow (i)(\mathbf{Y'})E$ for $i \geq 1$

$(\mathbf{head})(\mathbf{Y})E \rightarrow (\mathbf{head})(\mathbf{Y'})E$

$(\mathbf{tail})(\mathbf{Y})E \rightarrow (\mathbf{tail})(\mathbf{Y'})E$

$(\mathbf{null})(\mathbf{Y})E \rightarrow (\mathbf{null})(\mathbf{Y'})E$

Recursively defined infinite lists can thus be used as arguments in these operations and each time the operator bounces into an occurrence of the **Y** combinator it will change it back to **Y'**.

Our controlled reduction is clearly related to the idea of suspension as discussed in many papers following a suggestion by Friedman and Wise [6]. The distinguishing feature of our approach can be characterized by the fact that we implement suspension via the reduction rules themselves rather than by some extraneous control mechanism which would prevent somehow the application of the wrong rules.

5. Conclusion and future research

The system described in this paper forms the basis of a software interpreter developed for the IBM PC and ported to the IBM 3081. Various experiments have been performed with the interpreter whose performance is encouraging. Currently a parallel implementation is being developed using a shared memory multiprocessor system.

We feel that our approach represents an interesting application of formal semantics to practical computing problems. Building a bridge between a formal theory of semantics and a practical implementation is a challenging problem. Apart from that, our modified reduction rules seem to suggest some theoretical questions. The model theoretical consequences of the new system are not clear at this point. The relationship between the **Y** combinator and its primed version appears to be rather unusual in combinatory logic. Similar features, however, are quite common in other types of rewriting systems.

References

1. Arvind, Kathail, V., and Pingali, K., Sharing Computation in Functional Language Implementations, *Proc. Internat. Workshop on High-level Computer Architecture,* Los Angeles, May 21-25, 1985, pp. 5.1-5.12 .
2. Backus, J., The algebra of functional programs: Function level reasoning, linear equations, and extended definitions, *Proc. International Colloquium on the Formalization of Programming Concepts,* Lecture Notes in Computer Science, No. 107, Springer-Verlag 1981, pp. 1-43.
3. Barendregt, H. P., *The Lambda Calculus - Its Sytax and Semantics* (revised edition), Studies in Logic and the Foundation of Mathematics, Vol.103, North-Holland 1984.

4. Burton, F. W., A linear space translation of functional programs to Turner combinators, *Information Processing Letters,* Vol.14, No.5, 1982, pp. 201-204.
5. Clarke, T. J. W., Gladstone, P. J. S, MacLean, C. D., and Norman, A. C., SKIM - The S, K, I reduction macine, *LISP Conference Records,* Stanford Univ., Stanford, CA 1980, pp. 128-135.
6. Friedman, D. P. and Wise, D. S., Cons should not evaluate its arguments, in *Automata, Languages, and Programming* (Michaelson S. and Milner, R. eds) pp. 257-284. Edinburgh University Press, Edinburgh (1976).
7. Hindley, J. R., and Seldin, J. P., *Introduction to Combinators and λ-calculus,* London Mathematical Society Student Text 1, Cambridge Univerity Press 1986.
8. Johnsson, T., Lambda lifting: Transforming programs to recursive equations, *IFIP Conf. on Functional Programming Languages and Computer Architecture,* Sept. 16-19, 1985, Nancy, France, Lecture Notes in Computer Science, Vol.201, Springer Verlag 1985, pp. 190-203.
9. Noshita, K., and Hikita, T., The BC-chain method for representing combinators in linear space, *New Generation Computing,* Vol.3, No.2, (1985), pp. 131-144.
10. Revesz, G., Axioms for the theory of lambda-conversion, *SIAM Journal on Computing,* Vol.14, No.2, (1985) pp. 373-382.
11. Revesz, G., An extension of lambda-calculus for functional programming, *The Journal of Logic Programming,* Vol.1, No.3, (1984) pp. 241-251.
12. Scheevel, M., NORMA: A graph reduction processor, *Proc. of the 1986 ACM Conference on LISP and Functional Programming,* Cambridge, Mass. (Aug. 1986), pp. 212-219.
13. Turner, D. A., Another algorithm for bracket abstraction, *Journal of Symbolic Logic,* Vol.44, No.2, (June 1979), pp. 267-270.
14. Turner, D. A., A new implementation technique for applicative languages, *Software - Practice and Experience,* Vol.9, No.1, (1979) pp. 31-49.

Semantics of Block Structured Languages with Pointers

Eric G. Wagner
Mathematical Sciences Department
IBM T.J. Watson Research Center,
Yorktown Heights, N.Y. 10598 / U.S.A.

Abstract

This paper presents an algebraic and categorical approach to the mathematical modeling of imperative programming languages. In particular we model languages with block structure, records and variants, user definable recursive types, and pointers, etc., and with "control constructs" such as primitive recursion (generalized to recursive types), while-do, if-then-else, and assignment. In our earlier papers on this subject ([4], [5], [6]) we showed how data types and operations can be defined in an algebraic framework. In this paper we present a more mathematically sophisticated version of that framework, and we show how it can be used to provide a new approach to languages that have block structure together with objects, such as pointers, which are dynamically declared and may persist outside the block in which they are declared. The main new mathematical concept, and the key to the development, is the concept of an EDHT-category which is an extension of the DHT-symmetric categories introduced by Hoehnke [13] as a categorical framework for partial algebras.

1. Introduction

We are interested in languages where a program starts with a collection of given TYPEs (perhaps none!) and additional TYPEs are defined (declared) throughout the course of the program. We treat this abstractly by defining TYPEs either axiomatically (as in [2]) or as tensor products and categorical coproducts of existing TYPEs (building on the ideas in [11]). The later approach yields RECORDs, VARIANTs, and recursively defined TYPEs (e.g., STRINGs, STACKs, TREEs, etc.). The basic operations on these TYPEs are defined in terms of the product projections, the coproduct injections, and the mediating morphisms arising from the universal properties of the coproducts. We also employ fixpoint constructions on coproduct mediators to get iteration forms such as WHILE-DO and primitive recursion.

We view the STORE (set of memory states) the product of designated TYPEs (the VALUES) indexed by other designated TYPEs (the VARIABLES). A concurrent assignment statement on the STORE corresponds directly to a mediating morphism for the product. Whether or not a language contains POINTERs depends on choice of VALUES and VARIABLES. To a first approximation, a language has POINTERs if some objects serve as both VALUEs and VARIABLEs.

We put these ideas together mathematically by extending Hoehnke's DHT symmetric categories, [13], [15], to include coproducts, and then working with certain of the algebras corresponding to these categories. Hoehnke's original motivation in developing the DHT-symmetric categories was to develop a categorical framework for partial algebras, analogous to the framework provided by Lawvere, [17], for algebras over **Set**. We need the partiality in order to deal with non-termination (as in WHILE-DO), we need the coproducts in order to be able to handle the familiar recursive data TYPEs, and to get the basic control operators such as IF-THEN-ELSE and WHILE-DO. The general concept is captured by what we call an *extended DHT-symmetric category* (generally abbreviated to "EDHT-category"). The EDHT-categories provide a clean treatment of the desired algebras, and presentations of algebras, in which some carriers are restricted to being products or coproducts of designated carriers, and in which the notion of a "derived operation" is extended beyond operations derived using composition and tupling to include operations derived using the universal properties (the mediating morphisms) of these products and coproducts together with certain fixpoint operations on the coproduct mediators. We further specialize this to the concept of an S-sorted EDHT-category, in which S is a set (think of it as the set of declared TYPEs) and the objects of the category are the set $(S^*)^*$ of strings of strings on S.

The S-sorted EDHT-categories are used as the basis for the definition of pointer-presentations and pointer-algebras (Ptr-presentations and Ptr-algebras) which intuitively, are EDHT-categories with an added store or memory. While these Ptr-algebras (and Ptr-presentations) are the ones used to model imperative programming languages, all the important mathematical constructions take place at the level of the underlying EDHT-categories.

Our earlier papers ([4], [5], [6]) show how an less powerful version this framework can be used to define the operations found in languages without block structure. In this paper we extend these ideas to languages with block structure and POINTERs.

In the next section of the paper we present an example of a program fragment in order to motivate and explicate our approach. Sections 3 through 7 present the underlying mathematics: the basic concept of an EDHT-category is introduced in Section 3; the S-sorted EDHT-categories, which provide our syntactic framework, are defined in Section 4; the principle categories of EDHT-categories are introduced in section 5, EDHT-algebras are defined in Section 6, and it is shown how they can be represented by morphisms between EDHT-categories; finally Pointer-algebras are introduced in Section 7, and are used to defined the ASSIGNMENT operation. In Sections 8 through 11 we show how these concepts may be employed to provide semantics for the example in Section 2: Section 8 discusses the semantics of RECORDs, VARIANTs, and recursive TYPEs; Section 9 examines the declaration of TYPEs and VARIABLEs; Section 10 considers FUNCTION declarations, and Section 11 gives a treatment of BLOCKs, LOCAL VARIABLEs, and POINTERs. Section 12 provides a brief summary of what has been done.

Some remarks on notation. Composition of morphisms (including functions) is almost always written in diagrammatic order, i.e., if $f:A \to B$, and $g:B \to C$ then $f \bullet g:A \to C$. In consequence, if $f:A \to B$, and $a \in A$ we generally write af for the application of f to a. Given a morphism $f:A \to B$ in a category we refer to A as its *source* and B as its *target*, we reserve the word *domain* to refer to the subset of the source of a partial function on which it is defined. Given a category **C** we write $|\mathbf{C}|$ for its class of objects, and **C** for its class of morphisms. In order to avoid confusion in the use of words that have somewhat different computer science and mathematical meanings (e.g., "variable" and "function") we generally capitalize words corresponding to programming constructs.

2. A Motivating Example

In this section we present a program written in a PASCAL-like language, we discuss its meaning from a non-mathematical computer science viewpoint and then from a (rather informal) mathematical viewpoint. The remaining sections are where we make the mathematics precise. The "programming language" used in the example is not been spelled out in detail. As a rule it follows the conventions of PASCAL, however, unlike PASCAL, it allows recursive TYPE definitions, and the declaration of local VARIABLEs. In addition, the kind of definition-by-cases used in the FUNCTION declarations differs from that in PASCAL. These differences allow us to build up the needed TYPEs and FUNCTIONs "starting from scratch" rather than from as set of given "primitive" TYPEs. On the other hand, the (syntactic) treatment of POINTERs is exactly as found in the PASCAL manual [3], but our semantics is considerably more precise and avoids the implementation dependencies found in PASCAL. While we do not present a complete language, we contend that the fragment we do present is both more powerful and more elegant than the corresponding part of PASCAL, and that the improvement would be a direct reflection of the prescriptive effects of formulating concepts categorically.

Consider the program fragment shown in Figure 1. The fragment starts with the declarations of RECORD, VARIANT and POINTER TYPEs, FUNCTIONs, and VARIABLEs. This is followed by a portion of the body of the program. The portion consists of a single block which includes the declaration of a VARIABLE `L` local to the block, and of a WHILE-DO loop in which POINTERs "are created" (using a PASCAL-like `NEW` statement) and employed to construct a "singly linked list of length 5".

Now let us review the intended informal semantics in a more leisurely fashion. The program begins with five TYPE declarations for TYPEs named NULL, BOOL, NAT, LINK, and ITEM. All the TYPEs are either RECORDs or VARIANTs.

RECORDs are a very familiar programming concept -- a RECORD TYPE (with n fields) has a declaration of form

```
R = RECORD
     field1 : Type1
      ...
     fieldn : Typen
    end;
```

An object of this TYPE may be regarded as an { $\mathtt{field}_1$,..., $\mathtt{field}_n$ }-indexed set of objects where the object with index $\mathtt{field}_i$
is of TYPE $\mathtt{Type}_i$.

VARIANTs are not quite as familiar a programming construct as are RECORDs. The idea of VARIANT we have in mind is roughly similar to the VARIANT-RECORD of PASCAL, or the union TYPE of ALGOL68. A VARIANT (with n tags) has a declaration of the form

```
TYPE
  NULL = Record
  end;
  BOOL = Variant
           True:  Null;
           False: Null;
         end;
  NAT = Variant
          Zero: Null;
          Su:   Nat;
        end;
  LINK = ↑ITEM;
  ITEM = Record
           Value: NAT;
           Next:  LINK;
         end;
FUNCTION Pred( N:NAT): NAT;
  Case N
      tag(N) = Zero then Zero;
      tag(N) = Su then val(N);
  end;
FUNCTION Eq( N, P: NAT): BOOL;
  Case N,P
      (tag(N)=Zero and tag(P)= Zero) then True;
      (tag(N)=Zero and tag(P)= Su) then False;
      (tag(N)=Su and tag(P)=Zero) then False;
      (tag(N)=Su and tag(P)=Su) then Eq(val(N), val(P));
  end;
VAR
   N, Q: NAT;
   B1, B2: BOOL;
   A, B: LINK;
BEGIN
  BEGIN
     Var L:LINK;
     A:= Nil;
     N:= 5;
     WHILE Eq(N, Zero) = False DO
       BEGIN
          NEW(L);
          L↑.Next:= A;
          L↑.Value:= N;
          A:= L;
          N:=Pred(N);
       end;
  end;
   ...
end.
```

Figure 1. A Sample Program

```
V = VARIANT
     tag_1 : Type_1
      ...
     tag_n : Type_n
   end;
```

Speaking informally, an object of this TYPE can be viewed a pair <t, v> where t is a tag, $t \in \{$ tag_1 ,..., tag_n $\}$, and v is an object of the corresponding TYPE (i.e., if t = tag_i then v is an object of TYPE $Type_i$). If $i \neq j$, then $tag_i \neq tag_j$, however $Type_i = Type_j$ is allowed and, indeed, very useful (see below). We think of a VARIANT as coming equipped with operation for each tag_i which, given an object v of TYPE $Type_i$, will form the appropriate tagged-object corresponding to the pair <tag_i , v>. We will write "tag_i" for this operation as well as for the tag itself. To ensure TYPE safety we always use CASE statements (with a case for each possible tag) to operate on elements of VARIANT TYPEs. (PASCAL does not have this restriction, and so the semantics of PASCAL Variant-records is ill-defined.) In the sample program the case statements make use of two auxiliary operations, tag and val, where tag(<t, v>) = t, val(<t, v>)=v.

Each TYPE declaration defines both a sort corresponding to the TYPE (e.g., BOOL) and a sort corresponding to VARIABLEs of that TYPE (e.g., BOOL-VAR).

The first declaration, that of NULL, defines it to be the TYPE consisting of (the unique!) RECORD with no fields, i.e., it is a singleton set. In the second declaration, BOOL is defined as a VARIANT. Note that the information is in the tags rather than the TYPE NULL. In the third declaration we define NAT as a recursive VARIANT Type, i.e., the definition is self-referential. NAT is intended to be the TYPE of the Natural numbers. Letting * denote the unique object of TYPE NULL, the correspondence is as follows

0 ≡ <Zero, *> = Zero

1 ≡ <Su, <Zero, *>> = <Su, Zero> = (Zero)Su

2 ≡ <Su, <Su, <Zero, *>>> = <Su, 1> = ((Zero)Su)Su

etc.

An object of TYPE LINK is "a POINTER to an object of TYPE ITEM". Speaking informally, a POINTER is either a "location" (in which one can put objects of TYPE ITEM and extract them as desired) or it is "Nil" and does not actually "point at anything". We make this precise by defining a LINK as a VARIANT which is either a ITEM-VAR with tag Pt, or a NULL object with tag Nil. Finally, object of TYPE ITEM is a RECORD with two fields: The first field "Value" being of TYPE NAT, the second field, "Next" being of TYPE LINK. So one can think of an ITEM as consisting of a natural number n, together with a LINK object "pointing at the Next ITEM" if there is one, and "pointing at nothing" if it is "Nil". With the help of such POINTERs we can build up a "chain" of ITEMs each one pointing at its successor.

The next two declarations are FUNCTION declarations. The FUNCTIONs here are defined by cases, exploiting the fact that the objects in question are VARIANTs. While it is not all that obvious from the programming syntax, the form of these FUNCTION definitions is quite close to the form of primitive recursive definitions. Note that while the definition of Pred is a straight-forward definition by cases, the definition of Eq is recursive (self-referential).

At this point we come to the program body. All that is shown there is a single BLOCK. Within that BLOCK a LOCAL VARIABLE L, of TYPE LINK, is declared. Then there is a WHILE-DO loop in which the statement NEW(L) occurs. In PASCAL this NEW(L) statement "creates an object of TYPE LINK (i.e., a POINTER to an object of TYPE ITEM) and assigns it to L". The intuition is that the created object will be a "location" capable of holding an object of TYPE ITEM. Recall that an ITEM is a RECORD with fields Value and Next, then the next two instructions are assignments of appropriate values to these fields. Next, we have the assignment A:= L which assigns the POINTER in L to A, and then an assignment that reduces the value in N by 1. The end effect of the WHILE-DO loop will be to create a "singly-linked-list" of length five. Pictorially this result is often represented by a picture such as,

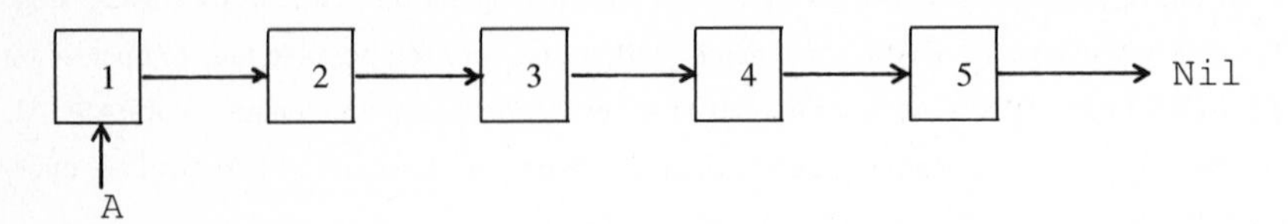

where each box corresponds to an ITEM (RECORD), the contents of the box is the contents of the Value field, while the outgoing arrow corresponds to the Next field (and shows what the POINTER is pointing at), while the POINTER in the LINK VARIABLE A points at the "head of the list", and the "tail of the list" has the value Nil.

The supposition is that when we leave the BLOCK, that the VARIABLE L will "cease to exist" but the POINTERs, and thus the singly-linked-list, will persist. Indeed, it is expected that any "link" of the singly-linked-list can be accessed via the LINK-VARIABLE A .

How then do we take all the above computer-science-eze and translate it into something more mathematical?

Our claim is that the declarations before the body of the program should be viewed as constructing a partial algebra (an algebra of sets and partial functions). The desired partial algebra, A, can be pictured informally as in Figure 2, where each circle denotes a carrier of the algebra, and the (multi-) arrows denote operations on the carriers. Note that, to avoid clutter, the TYPE NULL does not appear explicitly -- but all arrows without sources should be viewed as coming from NULL. A key aspect of the algebra A, that is not indicated in the diagram, is that we assume that (where + denotes coproduct (see definition below) and $\otimes$ denotes Cartesian product) that

$A_{NULL} = \{*\}$ ("the singleton set")

$A_{BOOL} = A_{NULL} + A_{NULL}$ with coproduct injections True and False.

$A_{NAT} = A_{NULL} + A_{NAT}$ with coproduct injections Zero and Su.

$A_{ITEM} = A_{NAT} \otimes A_{LINK}$ with projections Value and Next.

$A_{LINK} = A_{ITEM\text{-}VAR} + A_{NULL}$ with coproduct injections Pt and Nil.

The carriers of A corresponding to the sorts BOOL-VAR, NAT-VAR, ITEM-VAR, and LINK-VAR, are just sets containing the indicated constants b1, b2, N, Q, A and B . Finally the M-carrier, A_M, is taken to be the categorical product

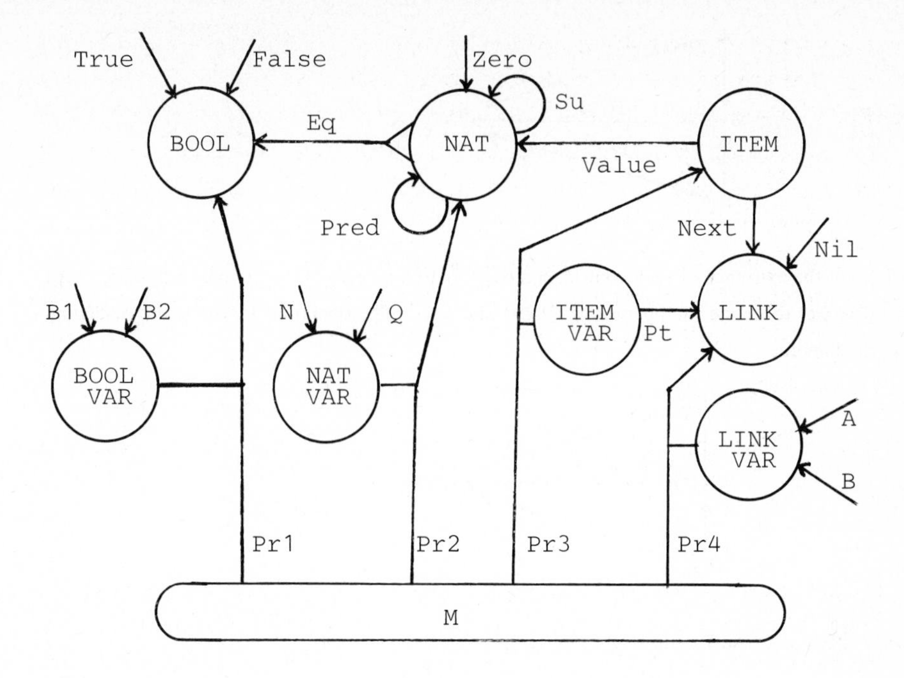

Figure 2. The Declarations Yield a (Partial) Algebra

$$A_M = (A_{BOOL})^{A_{BOOL\text{-}VAR}} \times (A_{NAT})^{A_{NAT\text{-}VAR}} \times (A_{ITEM})^{A_{ITEM\text{-}VAR}} \times (A_{LINK})^{A_{LINK\text{-}VAR}}$$

where, A^B denotes the set of all partial mappings of B into A.

The definition of a VARIANT that we gave above (as a set of value-tag pairs) could be taken here as a concrete definition of the coproduct -- i.e., a coproduct of A+B with injections I_A:A→A+B and I_B:B→A+B, can be viewed as the set of pairs

$$A+B = \{ \langle a, I_A \rangle \mid a \in A \} \cup \{ \langle b, I_B \rangle \mid b \in B \}$$

with

$$I_A: a \mapsto \langle a, I_A \rangle, \text{ and } I_B: b \mapsto \langle b, I_B \rangle.$$

However, it is really better for our purposes to view it abstractly that is, a coproduct for objects A and B in a category **C** is an object of **C**, denoted A+B, together with a pair of morphism $\langle \kappa:A \to A+B, \lambda:B \to A+B \rangle$ such that for any object X, and any pair of morphisms $\langle f:A \to X, g:B \to X \rangle$ there exists a unique morphism from A+B to X, denoted [f, g] and called *the coproduct mediator*, such that $\kappa \bullet [f, g] = f$, and $\lambda \bullet [f, g] = g$.

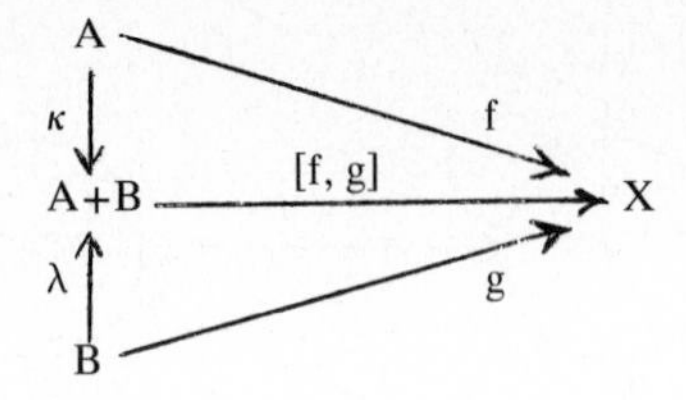

We can now exploit the existence of such coproduct mediators to take care of the definition-by-cases used in our sample program to define the operations Pred and Eq . The operation Pred is the mediating morphism for the diagram

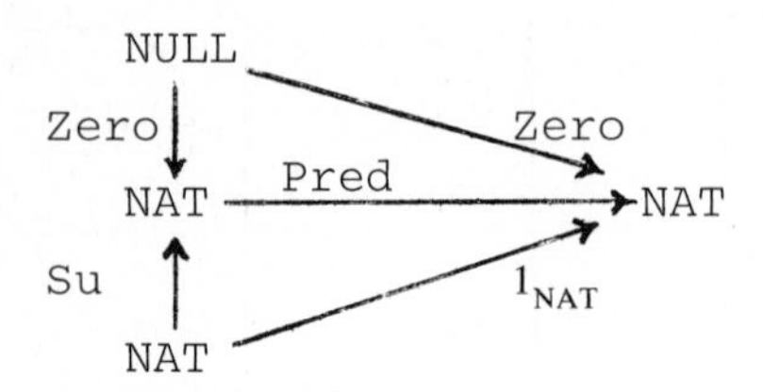

where Zero = κ, Su = λ, and Pred = [Zero, 1_{NAT}].

By similarly exploiting the properties of tensor products (Cartesian products) we can define the basic operations on RECORDs. Then, by exploiting the properties of the categorical product in the category, **Pfn**, of sets and partial functions, we are able to define the fetch and assignment statements of imperative programming.

The semantics of the body of the program given Figure 1 are more complicated because the body involves a combination of VARIABLE declarations, POINTER creations, and BLOCK exits, which modify the underlying algebra, and operations, such as ASSIGNMENTs and CASE STATEMENTs that are naturally viewed as derived operations within the underlying algebra. The key idea is that a declaration is modeled by a pushout in the underlying category of presentations of Ptr-algebras. A BLOCK, for the purposes of this paper, consists of a declaration part followed by a body part, followed by the exit from the BLOCK. The semantics of the declaration part is given by a pushout which produces a new algebra of a different presentation (including, for example, the local VARIABLEs and TYPEs). The semantics body part of the BLOCK is a transformation of this algebra to another one with the same presentation. The semantics of the exit is given by a reduct which transforms the algebra into on of the original signature. This construction has the desired property of "forgetting the LOCAL VARIABLEs upon exiting the block" while not "deallocating" objects, such as POINTERs, that we want to persist. Using these ideas we can provide a semantics for operations such as the NEW operation in PASCAL (which creates a new POINTER and assigns it to a VARIABLE) which completely avoids an *ad hoc* treatment of such constructs. In particular, unlike [7] and [8], we do not have to posit an initial infinite set of POINTERs (or locations) and a mysterious mechanism for picking the new POINTER (or location) from this set. The category theoretic approach, using pushouts, not only produces a "new POINTER" (or new location) in an abstract manner, it also provides the needed operations and precludes the non-explicit introduction of extraneous operations. Speaking informally, the POINTERs persist because, while their names are forgotten, they are not deallocated (more formally, they disappear from the presentation but not from the carriers of the algebra). A related approach is employed in [7]. However, our new framework, in contrast to that in [7], also provides a model of the handling of LOCAL VARIABLEs where they disappear from the algebra as well as from the presentation. Finally, by using these purely algebraic and categorical means we avoid the use of auxiliary mechanisms to keep track of the states, such as the stacks employed in [10], or [9].

3. EDHT-Categories, Basic Definitions and Concepts

Our development is built upon the concept of EDHT-categories. As noted in the introduction, EDHT-categories are an extension of the DHT-categories of Hoehnke, [13], [15]. The extension consists of adding coproducts to the categories. The formulation of DHT-category we use is based on that given by Robinson and Rosolini [12], rather than on Hoehnke's original definition [13], or Schreckenberger's later version [15]. However, we have gone even further than Robinson and Rosolini in the direction of presenting the category equationally rather than in terms of categorical concepts such as monoidal categories. It is our hope that the equational approach makes the material more accessible to those with a limited background in category theory.

In reading the following definition it is helpful to have in mind as an example, the category, **Pfn**, of sets and partial functions. Briefly, in **Pfn**: the functor $\otimes$ is the Cartesian product, ($X\otimes Y = \{ <x,y> \mid x\in X, y\in Y \}$ -- recall that the categorical product in **Pfn** is not the Cartesian product); $p_{X,Y}$, and $q_{X,Y}$ are the projection functions for the Cartesian product; $\delta_X:X\to X\otimes X$ is the diagonal function ($x\delta_X = <x,x>$); $\alpha_{X,Y,Z}$ is "the" isomorphism between $X\otimes(Y\otimes Z)$ and $(X\otimes Y)\otimes Z$; $\tau_{X,Y}$ is "the" isomorphism between $X\otimes Y$ and $Y\otimes X$; I is "the" one-element set; 0 is the empty set; + is coproduct ($X+Y = \{ <0, x> \mid x\in X \}\cup\{<1,y> \mid y\in Y \}$); $\kappa_{X,Y}$ and $\lambda_{X,Y}$ are the coproduct injections ($x\kappa = <0, x>$, $y\lambda = <1, y>$); $\theta_X:X+X\to X$ where $<i, x>\theta = x$; and $\mu_{X,Y,Z}$ is "the" isomorphism between $(X+Y)\otimes Z$ and $(X\otimes Z)+(Y\otimes Z)$.

Definition 3.1. An *Extended DHT-category* (*EDHT-Category*) is a category **C** equipped with:
$\otimes:\mathbf{C}\times\mathbf{C}\to\mathbf{C}$, a bifunctor, i.e.,

F.1 $(f_1\otimes g_1)\bullet(f_2\otimes g_2) = ((f_1\bullet f_2)\otimes(g_1\bullet g_2))$

F.2 $(1_X\otimes 1_Y) = 1_{X\otimes Y}$

$p_{X,Y}:(X\otimes Y)\to X$, a family of morphisms natural in X, i.e.,

P.1. $p_{X,Z}\bullet f = (f\otimes 1_Z)\bullet p_{Y,Z}$, for all $f:X\to Y$, and all $Z\in|\mathbf{C}|$.

$q_{X,Y}:(X\otimes Y)\to Y$, a family of morphisms natural in Y, i.e.,

Q.1 $q_{X,Y}\bullet f = (1_X\otimes f)\bullet q_{X,Z}$, for all $f:Y\to Z$, and $X\in|\mathbf{C}|$.

$\delta_X:X\to(X\otimes X)$, a family of morphisms natural in X, i.e.,

D.1 $f\bullet\delta_Y = \delta_X\bullet(f\otimes f)$ for all $f:X\to Y$.

Plus the following identities on $\otimes$, p, and q:

B.1 $\delta_X\bullet p_{X,X} = 1_X = \delta_X\bullet q_{X,X}$ B.2. $\delta_{X\otimes Y}\bullet(p_{X,Y}\otimes q_{X,Y}) = 1_{X\otimes Y}$

B.3 $(1_X\otimes p_{Y,Z}\bullet p_{X,Y}) = p_{X,Y\otimes Z}$ B.4 $(1_X\otimes q_{Y,Z})\bullet p_{X,Z} = p_{X,Y\otimes Z}$

B.5 $(p_{X,Y}\otimes 1_Z)\bullet q_{Y,Z} = q_{X\otimes Y,Z}$ B.6 $(q_{X,Y}\otimes 1_Z)\bullet q_{Y,Z} = q_{X\otimes Y,Z}$.

The operations

$$\alpha_{X,Y,Z} = \delta_{X\otimes(Y\otimes Z)}\bullet((1_X\otimes p_{Y,Z})\otimes q_{X,Y\otimes Z}\bullet q_{Y,Z}):X\otimes(Y\otimes Z)\rightarrow(X\otimes Y)\otimes Z$$

and

$$\tau_{X,Y} = \delta_{X\times Y}\bullet(q_{X,Y} \times p_{X,Y}):X\times Y\rightarrow Y\times X$$

are natural in all variables, i.e.,

A.1 $\alpha_{X,Y,Z}\bullet((f\otimes g)\otimes h) = (f\otimes(g\otimes h))\bullet\alpha_{X',Y',Z'}$.

T.1 $\tau_{X,Y}\bullet(g\otimes f) = (f\otimes g)\bullet\tau_{X',Y'}$.

$I\in|\mathbf{C}|$, and $\{ t_X:X\rightarrow I \mid X\in|\mathbf{C}| \}$

I.1 $p_{X,I}\bullet\delta_X\bullet(1_X\otimes t_X) = 1_{X\otimes I}$

I.2 $\delta_X\bullet(1_X\otimes t_X)\bullet p_{X,I} = 1_X$

$0\in|\mathbf{C}|$, $\{ 0_{X,Y}\in\mathbf{C}(X,Y) \mid X,Y\in|\mathbf{C}| \}$

Let $X,Y,Z\in|\mathbf{C}|$, $\beta:Y\rightarrow Z$, $f:X\rightarrow 0$, $g:0\rightarrow X$, and $\gamma:W\rightarrow Z$, then

O.1 $0_{X,Y}\bullet\beta = 0_{X,Z}$ O.2 $\beta\bullet 0_{Z,X} = 0_{Y,X}$

O.3 $f = 0_{X,0}$ O.4 $g = 0_{0,X}$

O.5 $(\gamma\otimes 0_{X,Y}) = 0_{W\otimes X,Z\otimes Y}$ O.6 $(0_{X,Y}\otimes\gamma) = 0_{X\otimes W,Y\otimes Z}$

$+:\mathbf{C}\times\mathbf{C}\rightarrow\mathbf{C}$, a bifunctor, i.e.,

G.1 $(f_1+g_1)\bullet(f_2+g_2) = ((f_1\bullet f_2)+(g_1\bullet g_2))$

G.2 $(1_X+1_Y) = 1_{X+Y}$

$\kappa_{X,Y}:X\rightarrow X+Y$ natural in X and Y, i.e.,

K.1 $\kappa_{X,Y}\bullet(f+g) = f\bullet\kappa_{X',Y'}$ for all $f:X\rightarrow X'$, and $g:Y\rightarrow Y'$.

$\lambda_{X,Y}:Y\rightarrow X+Y$ natural in X and Y, i.e.,

L.1 $\lambda_{X,Y}\bullet(f+g) = g\bullet\lambda_{X',Y'}$, for all $f:X\rightarrow X'$, and $g:Y\rightarrow Y'$

$\theta_X:(X+X)\rightarrow X$, a family of morphisms natural in X, i.e.,

E.1 $(f+f)\bullet\theta_Y = \theta_X\bullet f$, for all $f:X\rightarrow Y$.

Plus the following identities on $+$, κ, and λ, and for $f:X\rightarrow Z$, $g:Y\rightarrow Z$, and $h:X+Y\rightarrow Z$,

C.1 $\kappa_{X,Y}\bullet(f+g)\bullet\theta_Z = f$

C.2 $\lambda_{X,Y}\bullet(f+g)\bullet\theta_Z = g$

C.3 $(\kappa_{X,Y}\bullet h + \lambda_{X,Y}\bullet h)\bullet\theta_Z = h$.

$\mu_{X,Y,Z}:(A+B)\otimes C \rightarrow(A\otimes C)+(B\otimes C)$ natural in all variables, i.e.,

H.1 $\mu_{X,Y,Z}\bullet((f\otimes h)+(g\otimes h)) = ((f+g)\otimes h)\bullet\mu_{X',Y',Z'}$

and subject to the identities.

M.1 $((\kappa_{X,Y}\otimes 1_Z)+(\lambda_{X,Y}\otimes 1_Z))\bullet\theta_{(X+Y)\otimes Z}\bullet\mu_{X,Y,X} = 1_{(X\otimes Z)+(Y\otimes Z)}$

M.2 $\mu_{X,Y,Z,}\bullet((\kappa_{X,Y}\otimes 1_Z)+(\lambda_{X,Y}\otimes 1_Z))\bullet\theta_{(X+Y)\otimes Z} = 1_{(X+Y)\otimes Z}$. □

The above axioms are redundant, but there seems to be little advantage in removing axioms at the cost of adding lemmas. The most obvious example of an EDHT-category is the category, **Pfn**, of partial functions, for additional examples of computer science interest see Section 4.

Definition 3.2. Let $\mathbf{C}_1$ and and $\mathbf{C}_2$ be EDHT-categories, then by an *EDHT-morphism:* $H:\mathbf{C}_1\rightarrow\mathbf{C}_2$ we mean a functor between $\mathbf{C}_1$ and $\mathbf{C}_2$ which preserves all the structure, i.e.,

$(f\otimes g)H = fH\otimes gH$, $(\delta_X)H = \delta_{XH}$,

$(p_{X,Y})H = p_{XH,YH}$, $(q_{X,Y})H = q_{XH,YH}$,

$(0_{X,Y})H = 0_{XH,YH}$, $(t_X)H = t_{XH}$,

$(f+g)H = fH+gH,$ $(\kappa_{X,Y})H = \kappa_{XH,YH},$

$(\lambda_{X,Y})H = \lambda_{XH,YH},$ $(\theta_X)H = \theta_{XH},$

$(\mu_{X,Y,Z})H = \mu_{XH,YH,ZH},$ □

Two important aspects of partial functions are that we can define their domain (if $f:A \rightarrow B$, then the domain(f) = $\{ a \in A \mid f(a) \text{ is defined } \}$), and we can give a partial order on functions with the same source and target (if $f,g:A \rightarrow B$, then $f \leq g$ iff for every $a \in \text{domain}(f)$, we have $a \in \text{domain}(g)$ and $f(a)=g(a)$). Both these concepts carry over to arbitrary DHT-categories (and also to arbitrary EDHT-categories) by means of the following definitions:

Definition 3.3. Let **C** be an EDHT-category and let $f:A \rightarrow B$ in **C**. We define

$$\text{dom}(f) = \delta_A \bullet (1_A \otimes f) \bullet p_{A,B}.$$

We say that $f:A \rightarrow B$ is a *total morphism* iff $\text{dom}(f) = 1_A$. □

Definition 3.4. Let $f,g:A \rightarrow B$ in **C**, then we say $f \leq g$ iff

$$\delta_A \bullet (g \otimes f) \bullet p_{B,B} = f.$$ □

We leave it to the reader to show that these definitions work in **Pfn**. The following result shows that $\otimes$ does indeed operate like the Cartesian product in the precise sense that in an EDHT-category **C** it is the categorical product with respect to the subcategory of **C** consisting of the total morphisms.

Lemma 3.5. Let **C** be an EHDT-category, and let X,Y, and $Z \in |\mathbf{C}|$, and let $f:Z \rightarrow X$ and $g:Z \rightarrow Y$ be total (i.e., such that $\text{dom}(f) = 1_Z = \text{dom}(g)$), then there exists a unique $h:Z \rightarrow X \otimes Y$ such that $h \bullet p_{X,Y} = f$ and $h \bullet q_{X,Y} = g$. Indeed, $h = \delta_Z \bullet (f \otimes g)$. □

Definition 3.6. Let **C** be an EDHT-category. Given total morphisms $f_1:Z \rightarrow X$ and $f_2:Z \rightarrow Y$, we write $<f_1, f_2>$ for the morphism $\delta_Z \bullet (f_1+f_2):Z \rightarrow X \otimes Y$, which, by the above Proposition, has the property of being the unique morphism h from Z to $X \otimes Y$ such that $h \bullet p_{X,Y} = f_1$ and $h \bullet q_{X,Y} = f_2$, i.e., in the subcategory of **C** consisting of all total morphisms, it is the *product mediator* for f and g, while $p_{X,Y}:X \otimes Y \rightarrow X$, and $q_{X,Y}:X \otimes Y \rightarrow Y$, are the *product projections* . This definition is easily generalized to give n-ary mediators $<f_1, f_2,...,f_n>:Z \rightarrow (X_1 \otimes (X_2 \otimes ...X_n)...)$. □

Definition 3.7. Let **C** be an EDHT-category. Given $f_1:A_1 \rightarrow B$ and $f_2:A_2 \rightarrow B$, we write $[f_1, f_2]$ for the morphism $(f_1+f_2) \bullet \theta_B:A_1+A_2 \rightarrow B$, which, directly from the axioms, has the property of being the unique morphism h from A_1+A_2 to B such that $\kappa_{A_1,A_2} \bullet h = f$ and $\lambda_{A_1,A_2} \bullet h = g$, i.e., it is *coproduct mediator* for f and g, while κ_{A_1,A_2} and λ_{A_1,A_2} are the *coproduct injections* . Again, this notation is easily generalized to n-ary mediators for coproducts $[f_1, f_2,...,f_n]:(A_1+(A_2+ ...A_n)...) \rightarrow B$. □

The following concept plays a central role in our development.

Definition 3.8. We say that an EHDT-category **C** has a *strong singleton* if **C**(I, I) has exactly two elements, and, for all $f,g:I \rightarrow A+B$,

$$f \bullet [1_A, 0_{B,A}] = g \bullet [1_A, 0_{B,A}] \wedge f \bullet [0_{A,B}, 1_B] = g \bullet [0_{A,B}, 1_B] \text{ implies } f = g.$$ □

The key result concerning strong singletons is the following.

Proposition 3.9. Let **C** be a EDHT-category with a strong singleton. Then for $A,B \in |\mathbf{C}|$ and $f:I \to A+B$ exactly one of the following holds,

i) $f = 0_{I,A+B}$

ii) $f \neq 0_{I,A+B}$, but $f = f \bullet [1_A, 0_{B,A}] \bullet \kappa_{A,B} = f \bullet [\kappa_{A,B}, 0_{B,A+B}]$

iii) $f \neq 0_{I,A+B}$, but $f = f \bullet [0_{A,B}, 1_B] \bullet \lambda_{A,B} = f \bullet [0_{A,A+B}, \lambda_{A,B}]$ □

4. S-sorted Strict EDHT-categories

Definition 4.1. Let ω denote the set of natural numbers, $\omega = \{ 0, 1, 2, ...\}$. For any $n \in \omega$ let $[n] = \{ 1,...,n \}$, so $[0] = \emptyset$. Given a set S, let S*, the *set of strings on S*, be the set of all mappings $w:[n] \to S$, $n \in \omega$. The unique string $\lambda:[0] \to S$, is called *the empty string on S*. We form a category $\mathbf{Str}_S$ with objects $|\mathbf{Str}_S| = S^*$, where for $w,u \in |\mathbf{Str}_S|$, with $w:[n] \to S$, and $u:[p] \to S$, a *string morphism* $f:w \to u$ is a mapping $f:[n] \to [p]$ such that $f \bullet u = w$, i.e., such that the diagram

$$\begin{array}{ccc} [n] & \xrightarrow{f} & [p] \\ & {\scriptstyle w} \searrow \quad \swarrow {\scriptstyle u} & \\ & S & \end{array}$$

commutes, i.e., $f \bullet u = w$. □

Definition 4.2. Given a set S, let (S*)* be the set of *strings of strings of S*. That is, an element of (S*)* is a mapping $w:[n] \to S^*$, for some $n \in \omega$. Let $w_1,...,w_n \in S^*$, where $w_i = w_{i,1}w_{i,2}...w_{i,p_i}$, and let $w:[n] \to S^*$, such that $w(i) = w_i$ for each $i \in [n]$, then we shall write w as

$$w = (w_1)(w_2)...(w_n)$$

$$= (w_{1,1}w_{1,2}...w_{1,n_1})(w_{2,1}w_{2,2}...w_{2,n_2})...(w_{n,1}w_{n,2}...w_{n,n_n}).$$

Given $u = (u_1)(u_2)...(u_n)$ and $v = (v_1)(v_2)...(v_p) \in (S^*)^*$, we define

$$u+v = (u_1)(u_2)...(u_n)(v_1)(v_2)...(v_n)$$

and,

$$u \otimes v = (u_1v_1)(u_1v_2)...(u_1v_p)(u_2v_1)...(u_nv_p)$$

or, more formally, letting $\langle \_,\_ \rangle:[n] \times [p] \to [np]$ where $\langle i,j \rangle = (i-1)p + j$,

$$(u+v):[n+p] \to S^*$$

$$i \mapsto \text{IF } i \leq n \text{ THEN } u_i \text{ ELSE } v_{i-n}$$

and,

$$(u \otimes v):[np] \to S^*$$

$$\langle i, j \rangle \mapsto u_i v_j$$ □

Definition 4.3. Let S be a set (generally finite), then by a *strict S-sorted EDHT-category* we mean an $((S^*)^* \times (S^*)^*)$-sorted algebra **T** where, with $+, \otimes:(S^*)^* \times (S^*)^* \to (S^*)^*$ defined as in Definition 4.2, and with **T**(u,v) denoting the carrier with indices u and v, we have, for all $s,t,u,v \in (S^*)^*$, operations:

$\bullet_{u,v,w}:\mathbf{T}(u,v) \times \mathbf{T}(v,w) \to \mathbf{T}(u,w)$, $1_u \in \mathbf{T}(u,u)$,
$\otimes:\mathbf{T}(s,u) \times \mathbf{T}(t,v) \to \mathbf{T}(s \otimes t, u \otimes v)$, $\delta_u \in \mathbf{T}(u, u \otimes u)$,
$p_{s,t} \in \mathbf{T}(s \otimes t, s)$, $q_{s,t} \in \mathbf{T}(s \otimes t, t)$,
$\alpha_{s,t,u,v}:(s \otimes (u \otimes v)) \to (s \otimes u) \otimes v$. $\tau_{u,v}:(u \otimes v) \to (v \otimes u)$,
$I = (\lambda)$ (where λ is the empty string on S $(\lambda[0] \to S)$),

$0 = ()$ (the empty string on S^* ($():[0] \to S^*$)),

$0_{0,u} \in \mathbf{T}(0,u)$, $\quad 0_{u,0} \in \mathbf{T}(u,0)$,

$0_{u,v} = 0_{u,0} 0_{0,v}$, $\quad +:\mathbf{T}(s, u) \times \mathbf{T}(t, v) \to \mathbf{T}(s+t, u+v)$,

$\theta_u \in \mathbf{T}(u+u, u)$, $\quad \kappa_{s,t} \in \mathbf{T}(s, s+t)$,

$\lambda_{s,t} \in \mathbf{T}(t, s+t)$ $\quad \mu_{s,u,v}:(s \otimes u)+(t \otimes u) \to ((s+t) \otimes u)$.

satisfying all the axioms for an EDHT-category (including the category axioms for $\bullet$ and 1) and *strict* in that α and μ are both identity morphisms. Note that when +, and $\otimes$ are defined on $(S^*)^*$ as in Definition 4.2 we have, $(s \otimes (u \otimes v)) = ((s \otimes u) \otimes v)$, and, $(s \otimes u)+(t \otimes u) = ((s+t) \otimes u)$, so that strictness makes sense. However, the strictness of μ is dependent on just how we defined the coproduct, +, since, as is easily seen, $(u \otimes s)+(u \otimes t) \neq (u \otimes (s+t))$. □

Example 4.4. The simplest example of an S-sorted EDHT-category is *the category,* $\mathbf{SSt}_S$, *of strings of strings on* S , which is the category with set of objects $|\mathbf{SSt}_S| = (S^*)^*$, and, letting $u = (u_1)(u_2)...(u_n)$, $v = (v_1)(v_2)...(v_m)$, and $w = (w_1)(w_2)...(w_p)$, be three elements in $(S^*)^*$, then a $\mathbf{SSt}_S$- *morphism* $g:u \to v$ is given by the data

$$<g_0:[n] \multimap [m], < g_i:v_{ig_0} \to u_i \mid i \in [n] \text{ and } ig_0 \text{ is defined}>>$$

where $g_0:[n] \multimap [m]$ is a partial mapping, and each g_i is a string morphism in $\mathbf{Str}_S$.

Given $f:u \to v$ and $g:v \to w$, then $f \bullet g = h$ where $h_0 = f_0 \bullet g_0$, and, if $i \in [n]$, such that ih_0 is defined, then $h_i = g_{if_0} \bullet f_i$, so h is given by the data

$$< f_0 \bullet g_0, < g_{if_0} \bullet f_i \mid i \in [n] \text{ and } if_0 \text{ and } (if_0)g_0 \text{ are defined}>>.$$ □

5. Categories of strict EDHT Categories

Definition 5.1. Given a finite set S (of sorts) define **S-SEDHT** to be the *category of all S-sorted strict EDHT categories*, where the objects in **S-SEDHT** are the the S-sorted, strict, EHDT-categories, and, given S-sorted EDHT-categories $\mathbf{P}_1$ and $\mathbf{P}_2$, a *S-sorted EDHT-morphism* $H:\mathbf{P}_1 \to \mathbf{P}_2$ is just a homomorphism from $\mathbf{P}_1$ to $\mathbf{P}_2$ viewed, as in Definition 4.3, as algebras. That is, H consists of a family of mappings

$$< H_{u,v}:\mathbf{P}_1(u, v) \to \mathbf{P}_2(u, v) \mid u,v \in (S_1{}^*)^* >$$

such that the expected structure is preserved, i.e,

$(f \bullet g)H = fH \bullet gH$, $\quad (1_u)H = 1_u$,

$(f \otimes g)H = fH \otimes gH$, $\quad (\delta_u)H = \delta_u$,

$(p_{u,v})H = p_{u,v}$, $\quad (q_{u,v})H = q_{u,v}$,

$(0_{u,v})H = 0_{u,v}$, $\quad (t_u)H = t_u$,

$(f+g)H = fH+gH$, $\quad (\kappa_{u,v})H = \kappa_{u,v}$,

$(\lambda_{u,v})H = \lambda_{u,v}$, $\quad (\theta_u)H = \theta_u$,

$(\mu_{u,v,w})H = \mu_{u,v,w}$. □

Theorem 5.2. The category **S-SEDHT** is complete and cocomplete. □

Definition 5.3. Let S_1 and S_2 be finite sets, and let $h:S_1 \to S_2$, then we shall write h^* for the mapping $h^*:S_1^* \to S_2^*$, such that, for all $u \in S^*$, $uh^* = u \bullet h$. Similarly, we shall we shall write h^{**} for the mapping $h:(S_1^*)^* \to (S_2^*)^*$ such that $wh^{**} = w \bullet h^*$ for all $w \in (S_1^*)^*$. □

Definition 5.4. Define **SEDHT** to be the *category of sorted strict EDHT-Categories* where the objects in **SEDHT** are the union, as S ranges over all sets, of the S-sorted, strict, EHDT-categories, and, given an S_1-sorted strict EDHT-category $\mathbf{P}_1$ and an S_2-sorted strict EDHT-category $\mathbf{P}_2$, a *SEDHT-morphism* $<h, H>:\mathbf{P}_1 \to \mathbf{P}_2$ consists of a mapping $h:S_1 \to S_2$ together with a family of mappings

$$< H_{u,v}:\mathbf{P}_1(u, v) \to \mathbf{P}_2(uh^{**}, vh^{**}) \mid u,v \in (S_1{}^*)^* >$$

(using the above extension h^{**} of h to $(S_1{}^*)^*$) such that the expected structure is preserved, i.e,

$(f \bullet g)H = fH \bullet gH$, $\quad (1_u)H = 1_{uh^{**}}$,

$(f \otimes g)H = fH \otimes gH$, $\quad (\delta_u)H = \delta_{uh^{**}}$,

$(p_{u,v})H = p_{uh^{**},vh^{**}}$, $\quad (q_{u,v})H = q_{uh^{**},vh^{**}}$,

$(0_{u,v})H = 0_{uh^{**},vh^{**}}$, $\quad (t_u)H = t_{uh^{**}}$,

$(f+g)H = fH+gH$, $\quad (\kappa_{u,v})H = \kappa_{uh^{**},vh^{**}}$,

$(\lambda_{u,v})H = \lambda_{uh^{**},vh^{**}}$, $\quad (\theta_u)H = \theta_{uh^{**}}$,

$(\mu_{u,v,w})H = \mu_{uh^{**},vh^{**},wh^{**}}$, □

Proposition 5.5. The category **SEDHT** is cocomplete. □

6. EDHT-algebras

Definition 6.1. By a *flat poset* we mean a poset A, with ordering $\leq$, and a with a distinguished element $\perp_A$ such that, for all $a,b \in A$, $b \leq a$ iff $b = \perp_A$ or $b = a$. Let **Flt** be *the category of flat posets* where the class of objects is the class of all flat posets and the morphisms are the strict mappings, i.e., if $f:A \to B$ in **Flt** then $\perp_A f = \perp_B$. The category **Flt** is closely related to the category **Pfn** of partial functions, but is more convenient for our purposes because "the undefined" appears explicitly as $\perp$, and the ordering is built-in. □

Definition 6.2. Let **P** be an S-sorted strict EDHT-category. Then by a **P**-*partial-algebra* (or, for short, a **P**-algebra) we mean an EDHT-morphism $A:\mathbf{P} \to \mathbf{Flt}$. □

The following result establishes the existence of **P**-algebras for **P** with strong singletons, we leave it to the reader to show that they exist more generally. The next result, and its corollary, provide the means for viewing **P**-algebras as morphisms from **P** in **SEDHT** which we employ later as the basis for our treatment of declarations and BLOCKs.

Theorem 6.3. Let **P** be an S-sorted strict EDHT-category with and with a strong singleton. Then the Hom-functor $\mathbf{P}(I, \_\_):\mathbf{P} \to \mathbf{Flt}$ is a **P**-algebra. □

Corollary 6.4. Let $<h, H>:\mathbf{P} \to \mathbf{P}_1$ where $\mathbf{P}_1$ has a strong singleton object. Then the hom-functor

$$(H)A_P = \mathbf{P}_1(I, \_\_H):\mathbf{P} \to \mathbf{Flt}$$

is a **P**-partial-algebra. Furthermore, if $<f, F>:\mathbf{P} \to \mathbf{P}_2$ where $\mathrm{Card}(\mathbf{P}_2(I, I)) = 2$ and $\mathbf{P}_2$ has a strong singleton object, and $<g, G>:\mathbf{P}_1 \to \mathbf{P}_2$ such that $<h, H> \bullet <g, G> = <f, F>$ the family

$$(G)A_P = <G_{I,sf}:\mathbf{P}_1(I, sh) \to \mathbf{P}_2(I, sf) \mid s \in S>$$

is a homomorphism from $(H)A_P$ to $(F)A_P$. Indeed, let **GOOD** be the full subcategory of **SEDHT** determined by the categories with strong singleton objects, and let (**P**↓**GOOD**) be the comma category [1] with, as objects, EDHT-morphisms from **P** into objects in **GOOD**. Then we may regard A_P as a functor

$$A_P:(\mathbf{P}\downarrow\mathbf{GOOD}) \to \mathrm{Alg}_P$$ □

Theorem 6.5. A_P has a left adjoint H_P, and for any **P**-algebra A, $((A)H_P)A_P = A$. Furthermore, letting A be a **P**-algebra with $AH_P:\mathbf{P}\to\mathbf{P}_A$, and letting $J:\mathbf{P}\to Y$ and letting $<L:\mathbf{P}_A\to X,\ K:Y\to X>$ be the pushout of $<AH_P, J>$ in **SEDHT**, then $X\in|\mathbf{GOOD}|$ implies $K\cong KA_YH_Y$. □

7. Pointer-Presentations and Pointer-Algebras

In terms of the informal treatment of Section 2, S-sorted **EDHT**-categories and their algebras provide the underlying algebra minus the memory or store. To add the memory we introduce the concepts of pointer-presentations and pointer-algebras.

Definition 7.1. An *S-sorted pointer-presentation* (Ptr-presentation) is given by a pair $<\mathbf{P}, \rho>$ where P is an S-sorted **EDHT**-category and ρ is a relation on S. View ρ as a set of ordered pairs, $\rho\subseteq S\times S$, then the intuition is that $<s, s'>\in\rho$ means that s is (the name of) a set of VARIABLEs for the set of objects of TYPE (named) s′. □

Definition 7.2. PTRP, *the category of pointer-presentations*, is the category with all pointer-presentations as objects, and where a morphisms from $<\mathbf{P}_1, \rho_1>$ to $<\mathbf{P}_2, \rho_2>$ (with $\mathbf{P}_i$ S_i-sorted) are all the **SEDHT**-morphisms $<f, F>$ such that $<s, s'>\in\rho_1$ implies $<sf, s'f>\in\rho_2$. □

Definition 7.3. Given a pointer-presentation $<\mathbf{P}, \rho>$ let $\mathbf{P}_\rho$ denote the $(S\cup\{M\})$-sorted **EDHT**-category that results from freely adjoining to **P** a new sort M ($M\notin S$) together with an operation $Pr_{<s.s'>}:M\otimes s\to s'$ for each $<s, s'>\in\rho$.

Given $B = <B_i \mid i\in I>$ a family of flat posets from **Flt**, let $XB = X<B_i \mid i\in I>$ denote their *categorical* product in **Flt** and let

$$x_i:(XB)\to B_i$$

denote the ith projection. (Note that the categorical product of flat posets $A_\perp$ and $B_\perp$ is $A_\perp\times B_\perp \cong A_\perp+(A_\perp\otimes B_\perp)+B_\perp \cong$ their Cartesian product as sets $A_\perp = A\cup\{\perp_A\}$, and $B_\perp=B\cup\{\perp_B\}$.)

Definition 7.4. Let $<\mathbf{P}, \rho>$ be a Ptr-presentation, then a $<\mathbf{P}, \rho>$*-algebra* consists of a **P**-algebra $A:P\to\mathbf{Flt}$ extended to a $\mathbf{P}_\rho$-algebra, also denoted A, such that $A_M = X<A_{s'}{}^{A_s} \mid <s, s'>\in\rho>$ and such that for each $<s, s'>$

$$(Pr_{<s,s'>})_A:A_M\otimes A_s\to A_{s'}$$
$$<a_m, a_s>\mapsto a_s(a_m x_{<s,s'>})$$

Note that $a_m x_{<s,s'>}:A_s\to A_{s'}$ in **Flt**. □

Fact 7.5. Let $A_\perp = A\cup\{\perp_A\}$ and $B_\perp = B\cup\{\perp_B\}$ be flat posets, then $(A_\perp)^{B_\perp}$ ($=\mathbf{Flt}(B_\perp, A_\perp)$) is isomorphic to the B-indexed categorical product of $A_\perp$ with itself in **Flt**. □

Speaking informally, this means that A_M is the VARIABLE-indexed product of the respective TYPEs, and this suggests that we can exploit the universal properties of the the product (the mediating morphisms) to define new operations. In particular, we can define operations corresponding to assignment statements. To make this more precise, let A be a $<\mathbf{P}, \rho>$-algebra (where **P** is S-sorted) and define *the set of typed VARIABLEs from A* to be

$$VAR_A = \{ <v, s, s'> \mid s,s' \in S, v \in A_{s'}, <s, s'> \in \rho \} \cong \Sigma <A_s \mid \exists s', <s, s'> \in \rho >$$

(where Σ is coproduct in **Flt**).

Fact 7.6. If $t \in (S \cup \{M\})$, and we have a VAR_A-indexed family

$$< \gamma_{<v,s,s'>} : A_t \to A_{s'} \mid <v, s, s'> \in VAR_A >$$

of **Flt**-morphisms, then there exists a unique **Flt**-morphism $\gamma : A_t \to A_M$ such that

$$(\gamma, v)(Pr_{s,s'})_A = \gamma_{<v,s,s'>}$$

for every $<v,s,s'> \in VAR_A$. □

A special case of the above that is of particular interest from a programming point of view is the following abstract version of the ASSIGNMENT STATEMENT.

Corollary 7.7. Let $<s_1, s_2> \in \rho$ then for all morphisms $f: A_M \to A_{s_1}$ and $g: A_M \to A_{s_2}$ there exists a unique morphism $h_{f,g}: A_M \to A_M$ such that, for all $m \in A_M$, if $mf = \bot$ then $mh_{f,g} = \bot$, but otherwise

$$(mh_{f,g}, mf)Pr_{s_1,s_2} = mg$$

and, for all $b \in A_{s_1}$, $b \neq mf$ implies

$$(mh_{f,g}, b)Pr_{s_1,s_2} = (m, b)Pr_{s_1,s_2}$$

and for all $<s'_1, s'_2> \neq <s_1, s_2>$, and $c \in A_{s'_1}$,

$$(mh_{f,g}, c)Pr_{s'_1,s'_2} = (m, c)Pr_{s'_1,s'_2}.$$

Slightly abusing the usual programming notation, we shall write (f:=g) for $h_{f,g}$.

Proof: The desired morphism, (f:=g), is given by the VAR_A-indexed family

$$< \gamma_{<v,s,s'>} : A_M \to A_{s'} \mid <v, s, s'> \in VAR_A >$$

such that, for each $m \in A_M$,

$$m\gamma_{<v,s\,s'>} = \begin{cases} mg \text{ if } s=s_1, s'=s_2, \text{ and } v = mf \\ (m, v)Pr_{s,s'} \text{ otherwise.} \end{cases}$$

□

Note, in writing assignments we shall adopt the usual programming convention of writing "N:=N+1" for what we should formally write as "(N:=(__,N)$Pr_{NAT\text{-}VAR,NAT}$+1)". Thus, all VARIABLEs A to the right of the ":=" should be viewed "shorthand for the corresponding "fetch operation", (__,A)Pr."

8. RECORDs, VARIANTs, and Recursive Types

Having presented the underlying mathematical concepts we now show how to apply them to the programming syntax and semantics issues raised in the introduction and Section 2. We start by looking at TYPE definitions for RECORDs, VARIANTs, and recursive TYPEs. We will want to be able to define any recursive TYPE that can be defined using tensor product, $\otimes$, and coproduct, +. For convenience we will take a syntactically restricted approach suggested by our motivating example where, in effect, we use only RECORDs and VARIANTs, but, in contrast to, say, PASCAL, we allow recursive definitions. As is easily shown, this syntactic restriction does not result in a semantic restriction. As a first step we introduce some convenient notation for the finite tensor-products and coproducts that will be needed.

Definition 8.1. Let $u = s_1 s_2 ... s_n \in S^*$, then define

$$\prod(u) = (u)$$

i.e., the mapping $\prod(u):[1]\rightarrow S^*$, $(1)\prod(u) = u$, and define

$$\coprod(u) = (s_1)(s_2)...(s_n)$$

i.e., the mapping $\coprod(u):[n]\rightarrow S^*$, $(i)\coprod(u) = s_i$ (where $s_i:[1]\rightarrow S$, $(1)s_i = s_i$). □

Fact 8.2. $\prod(\lambda) = (\lambda) = I$, while $\coprod(\lambda) = \Lambda = 0$. □

We want to show how the algebra, informally presented in Figure 2, can be formally presented using an S-sorted EDHT-category. In this Section we show how the algebra (minus the FUNCTIONs Pred and Eq) can be defined by means of a Ptr-presentation. In subsequent sections we will look in greater detail at the TYPE and VARIABLE declarations that build up this algebra, how it can be extended to include the declared FUNCTIONs, and last of all we will look into the dynamic declarations corresponding to declaration of local VARIABLEs and the "creation of POINTERs".

The intuition given in Section 2 is that a declaration of the form

```
s = RECORD
      field_1 : s_1
      . . .
      field_n : s_n
    end;
```

should mean that, in the corresponding algebra, A, the s-carrier, A_s, should be a tensor product, $A_{s_1}\otimes...\otimes A_{s_n}$, of the s_1- through s_n-carriers. We claim then that the desired counterpart of this at the presentation level is that $s\cong\prod(s_1...s_n)$ Correspondingly, the counterpart of the variant

```
s = VARIANT
      tag_1 : s_1
      . . .
      tag_n : s_n
    end;
```

is that $s \cong \coprod(s_1...s_n)$. Let Γ represent either $\prod$ or $\coprod$. We can force $s\cong\Gamma(s_1...s_n)$ by appending operations

$$\iota_s:\Gamma(s_1...s_n)\rightarrow s \text{ and } \omega_s:s\rightarrow\Gamma(s_1...s_n)$$

and axioms

$$\omega_s\bullet\iota_s = 1_s \text{ and } \iota_s\bullet\omega_s = 1_{\Gamma(s_1...s_n)}$$

to those for an S-sorted EDHT-category.

When dealing with programming examples we shall identify the field-name, $field_i$, of a RECORD $s\cong\prod(s_1...s_n)$ (the tag-name, tag_i, of a VARIANT $s\cong\coprod(s_1...s_n)$) with the corresponding projection morphism (injection morphism). That is, we shall write $field_i$ for $\omega_s\bullet p_i:s\rightarrow s_i$, and, tag_i for $\kappa_i\bullet\iota_s:s_i\rightarrow s$.

Informally speaking, $field_i$ is then a function which reads out the ith field of a RECORD, while tag_i is a function for constructing a VARIANT with tag tag_i from elements of s_i. On the other hand, a tensor-product mediator, $\langle f_1,...,f_n\rangle:Z\rightarrow s$, may be viewed as a constructor for the corresponding RECORDs, while a coproduct mediator, $[f_1,...,f_n]:s\rightarrow Z$, corresponds to a case-statement (or, more generally, a case-expression) for operating on a VARIANT to get an element of Z. Before illustrating the use of such mediators, we will show how to use the above notation (i.e. $s\cong\prod(s_1...s_n)$ and $s\cong\coprod(s_1...s_n)$) to write a specification for the presentation of the algebra of the motivating example.

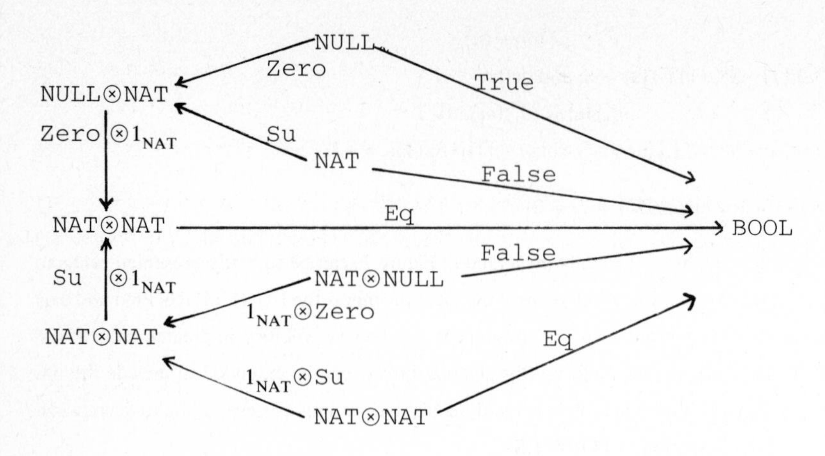

Figure 3. The Coproduct diagram for the function Eq

Example 8.3.

sorts:

S = { BOOL-VAR, NAT-VAR, ITEM-VAR, LINK-VAR,
NULL, ITEM, BOOL, NAT, LINK }

NULL ≅ ∏(λ) ITEM ≅ ∏(NAT.LINK)

BOOL ≅ ∐(NULL.NULL) NAT ≅ ∐(NULL.NAT)

LINK ≅ ∐(ITEM-VAR.NULL)

Additional "Constants"

B1, B2:I→BOOL-VAR, N,Q:I→NAT-VAR, and, A,B:I→LINK-VAR

where, for each "constant" X we have an axiom dom(X) = 1_I.

Naming of injections and projections:

Value ≡ $\omega \bullet p$ Next ≡ $\omega \bullet q$ Pt ≡ $\kappa \bullet \iota$ Nil ≡ $\lambda \bullet \iota$

This gives us an S-sorted EDHT-category, to get the desired Ptr-presentation we need only take

ρ = { <BOOL-VAR, BOOL>, <NAT-VAR, NAT>,
<ITEM-VAR, ITEM>, <LINK-VAR, LINK> } □

It can be shown that the resulting S-sorted EDHT-category has a strong singleton object.

We have already shown, in Section 2, how the function Pred:NAT→NAT can be defined as a mediating morphism for a coproduct. The definition of Eq is somewhat more complex. First of all, in order to handle two variables, we must take advantage of the fact that, in the category **Pfn** of sets and partial functions we have

$$(X+Y) \otimes Z \cong (X \otimes Z)+(Y \otimes Z)$$

Second, in order to capture the iterative (recursive) part of the definition, we must employ a fixpoint. The relevant diagram is shown in Figure 3. The claim is that the desired equality function is the unique fixpoint of this diagram, i.e., the unique function $\mathtt{Eq}:A_{NAT} \otimes A_{NAT} \rightarrow A_{BOOL}$ making the diagram commute.

Another example of the use of the coproduct mediator is the definition of the "L↑" operation found in the body of our working example. Informally, the value of L↑ is the "location pointed at by the POINTER in L" -- however L may have the value `Nil`, in which case L↑ is undefined. Thus, we can view L↑ as a mapping $L\uparrow : A_M \rightarrow A_{ITEM\text{-}VAR}$, such that for any $m \in A_M$,

$$m(L\uparrow) = (m, L)Pr_{LINK\text{-}VAR,LINK} \bullet [1_{ITEM\text{-}VAR}, 0_{I,ITEM\text{-}VAR}]$$

Recall that, in our example L is a VARIABLE of TYPE LINK (i.e., it is a constant of TYPE LINK-VAR) where LINK ≅ ⊔⊔(ITEM-VAR.NULL), thus the general operation is

$$\_\uparrow : A_{LINK\text{-}VAR} \otimes A_M \rightarrow A_{ITEM\text{-}VAR},$$

and corresponds to the diagram

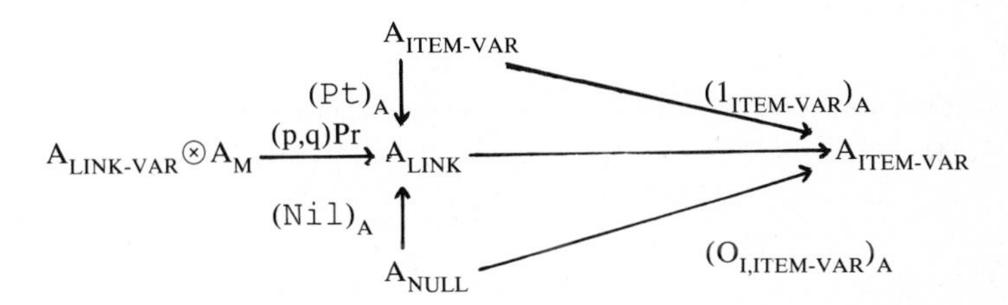

The tensor product mediators play a lesser role in our development than that played by the coproduct mediators. But they are of use in assignments to RECORDs. For example the instruction

```
L↑.Next := A;
```

in the body of our working example should be interpreted as the morphism

$$(L\uparrow := \langle(\_, L\uparrow)Pr_{ITEM\text{-}VAR,ITEM} \bullet \mathtt{Value}, (\_, A)Pr_{NAT\text{-}VAR,NAT}\rangle) : A_M \rightarrow A_M$$

where the the expression on the right-hand side of the ":=" is just the tensor product mediator, μ, of the diagram

$$\begin{array}{ccl} & \xrightarrow{(\_,A)Pr} & A_{LINK} \\ & & \uparrow \mathtt{Next} \\ A_M & \xrightarrow{\mu} & A_{ITEM} = A_{LINK} \otimes A_{NAT} \\ & & \downarrow \mathtt{Value} \\ & \xrightarrow{(\_,L\uparrow)Pr \bullet \mathtt{Value}} & A_{NAT} \end{array}$$

9. Declarations of TYPEs and VARIABLEs

In Section 8 we showed that the algebra desired as the result of the TYPE and VARIABLE declarations of the example given in Section 2, corresponds to a suitable **Ptr**-presentation and its initial algebra. Our aim now is to show how that presentation can be built up by a sequence of declarations, carried out by means of pushouts in **PTRP**.

Recall, to begin with, that given a category **C**, and two morphisms $f:X \rightarrow Y$ and $g:X \rightarrow Z$ with a common source X, that a *pushout* for $\langle f, g\rangle$ is a pair $\langle f':Y \rightarrow W, g':Z \rightarrow W\rangle$ such that $f \bullet f' = g \bullet g'$ and, for any pair

$<f'':Y \to U, g'':Z \to U>$ such that $f \bullet f'' = g \bullet g''$, there exists a unique $h:W \to U$ such that $f' \bullet h = f''$ and $g' \bullet h = g''$.

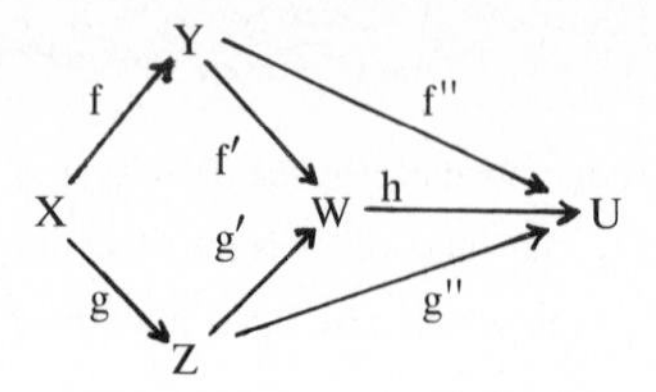

How declarations work in our abstract setting is perhaps best seen by looking at a simple example of the declaration of a VARIABLE. Let **P** be the S-sorted EDHT-category, with

S = { BOOL-VAR, NAT-VAR, ITEM-VAR,
NULL, ITEM, BOOL, NAT, LINK }

given in Section 8. Now say we wished to declare an additional VARIABLE, say R, of TYPE NAT, that is, we wish, in effect, to give give the counterpart of the PASCAL statement

```
VAR  R : NAT;
```

We claim that , viewed abstractly, this declaration corresponds to a diagram

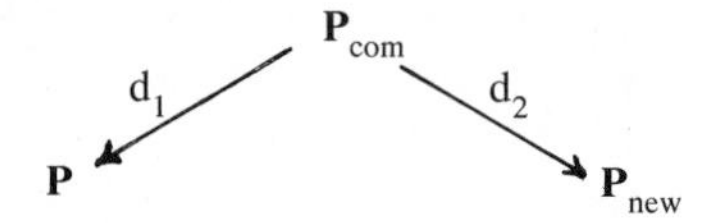

where, the EDHT-category $\mathbf{P}_{new}$ corresponds to the data

S_{new} = { NV }

$R:I \to NV$

Axiom: $dom(R) = 1_I$

So $\mathbf{P}_{new}$ is a 1-sorted EDHT-category with a single "constant" R. While P_{com} corresponds to the the 1-sorted EDHT-category given by

S_{com} = { N } (with no operations, and no axioms)

and d_1, and d_2 are the **SEDHT**-morphisms

$d_1:P_{com} \to \mathbf{P}$	$d_2:P_{com} \to P_{new}$
$N \mapsto$ NAT-VAR	$N \mapsto NV$

Now we claim that "the pushing out" of d_1 and d_2 identifies the sort NV in $\mathbf{P}_{new}$ with the sort NAT-VAR in **P** and so, in effect, adds the "constant" R to NAT-VAR. More precisely, if $<d_1':\mathbf{P} \to \mathbf{P}_{res}$, $d_2':\mathbf{P}_{new} \to \mathbf{P}_{res}>$ is the pushout for $<d_1, d_2>$, then the pushout object $\mathbf{P}_{res}$ is the desired **SEDHT**-category, where the usual (implicit) programming convention is to write: R for Rd_2', N for Nd_1', and Q for Qd_1' (recall that N and Q are constants of TYPE NAT-VAR in **P**) and, of course, to write X for Xd_1' for each $X \in S$ = { BOOL-VAR, NAT-VAR, ITEM-VAR, NULL, ITEM, BOOL, NAT, LINK }.

These implicit conventions have the consequence that a repeated declaration such as

```
VAR
    R : NAT;
    R : NAT;
```

is regarded as a syntax error in many programming languages. However, in the formal context provided here such a repeated declaration has a straightforward meaning. Putting it in general terms, given a declaration $< d_1:\mathbf{P}_{com} \to \mathbf{P}, d_2:\mathbf{P}_{com} \to \mathbf{P}_{new}>$, the result of applying it once is the pushout, call it $<e_1, e_2>$, of $<d_1, d_2>$, and the result of applying it twice is the pushout, call it $<f_1, f_2>$, of $<d_1 \bullet e_1, d_2>$. In the final pushout object the "first R" appears as $Re_2 \bullet f_1$, and the "second R" as Rf_2.

The declaration of TYPEs works in exactly the same way as the declaration of VARIABLEs, except that, in general, $\mathbf{P}_{com}$ and $\mathbf{P}_{new}$ have more structure, i.e., are more complex EDHT-categories. The EDHT-categories playing the $\mathbf{P}_{new}$ role for the sequence of declarations in our working program (Figure 1) are the five EDHT-categories $\mathbf{P}_0$, ..., $\mathbf{P}_4$ given below. The construction starts with $\mathbf{P}$ the ø-sorted EDHT-category, and at each step the requisite "common" category $\mathbf{P}_{com}$ is the evident one needed to make the requisite identifications of sorts. The following list of five Ptr-presentations $\mathbf{P}_0$,...,$\mathbf{P}_4$ correspond to the five TYPE declarations in Figure 1.

$\mathbf{P}_0$ (the declaration of NULL)

$S_0 = \{ NULL_0 \}$

$NULL_0 \cong \prod(\lambda)$

$\rho_0 = \emptyset$

P_1 (the declaration of BOOL)

$S_1 = \{ NULL_1, BOOL_1, BOOL\text{-}VAR_1 \}$

$BOOL_1 = \coprod(NULL_1.NULL_1) = NULL_1+NULL_1$

$\rho_1 = \{< BOOL\text{-}VAR_1, BOOL_1 > \}$

$True_1 \equiv \kappa \bullet \iota$, $False_1 = \lambda \bullet \iota$

P_2 (the declaration of NAT)

$S_2 = \{ NULL_2, NAT_2, NAT\text{-}VAR_2 \}$

$NAT_2 \cong \coprod(NULL_2+NAT_2)$

$\rho_2 = \{ <NAT\text{-}VAR_2, NAT_2> \}$

$Zero \equiv \kappa \bullet \iota$, $Su \equiv \lambda \bullet \iota$

P_3 (the declaration of LINK)

$S_3 = \{ NULL_3, ITEM_3, LINK_3, ITEM\text{-}VAR_3 \}$

$LINK_3 \cong \coprod(ITEM\text{-}VAR_3.NULL_3) = ITEM\text{-}VAR_3+NULL_3$

$\rho_3 = \{ <ITEM\text{-}VAR_3, ITEM_3 > \}$

$Pt \equiv \kappa \bullet \iota$, $Nil \equiv \lambda \bullet \iota$

P_4 (the declaration of ITEM)

$S_4 = \{ ITEM_4, LINK_4, NAT_4, LINK\text{-}VAR_4 \}$

$ITEM_4 \cong \prod(NAT_4.LINK_4) = NAT_4 \otimes LINK_4$

$\rho_4 = \{ < LINK\text{-}VAR_4, LINK_4 > \}$

$Value = \omega \bullet p$, $Next = \omega \bullet q$.

Given an EDHT-category **P** with **P**-algebra $A:\mathbf{P}\rightarrow\mathbf{Flt}$, we know, by Theorem 6.5, that we can represent A by a morphism $AH_P:\mathbf{P}\rightarrow\mathbf{P}_A$. Furthermore, we know that if we have a declaration $<d_1:\mathbf{P}_{com}\rightarrow\mathbf{P}, d_2:\mathbf{P}_{com}\rightarrow\mathbf{P}_{new}>$ with a pushout $<e_1:\mathbf{P}\rightarrow\mathbf{P}_{res}, e_2:\mathbf{P}_{new}\rightarrow\mathbf{P}_{res}>$, and the pushout of $<AH_P, e_1>$ is $<f_1:\mathbf{P}_A\rightarrow\mathbf{P}_B, f_2:\mathbf{P}_{res}\rightarrow\mathbf{P}_B>$ then $f_2 \cong f_1 A_{P_{res}} H_{P_{res}}$, and the pair $<e_2, f_1>$ induces a homomorphism from A to $B = f_2 A_{P_{res}}$.

Part of this theorem goes over directly to Ptr-algebras, namely the part dealing with the presentations, including the requisite ρ's. However there is a problem with the algebras. For example, if the declaration identifies VARIABLEs that were previously distinct, then there is no way to extend the homomorphism of EHDT-algebras given by $<e_1, f_1>$ to the corresponding Ptr-algebras A and B. The problem is that there is no mapping $h_M:A_M\rightarrow B_M$ with the the requisite homomorphism property, i.e., such that

$$(m,v)(Pr_{s,s'})_A h_{s'} = (mh_M, vh_s)(Pr_{se_2,s'e_2})_B$$

for all $v\epsilon A_s$ and $m\epsilon A_M$.

However a short examination of all the declarations that we have employed in our exposition shows that they do extend (non-uniquely) from the EDHT-algebras to the **Ptr**-algebras. Furthermore, while the extension is not unique there is always a least extension, namely one in which "all the new VARIABLEs have value $\perp$". We say that a declaration is *nice* if, for every **P** declaration $<d_1, d_2>$, and every **P**-algebra, A, the EDHT-morphism given by the pushouts has such a minimal extension to the corresponding Ptr-algebras.

Nice declarations are important in Section 11 where we deal with the semantics of LOCAL VARIABLEs and POINTER creation. In a language without LOCAL VARIABLEs and/or POINTERs the whole computation can be viewed as being carried out within a single algebra, namely the algebra, A, corresponding to the result of executing the declarations preceding the program body. In such a language the meaning of the program is a **Flt**-morphism $A_M\rightarrow A_M$. However, when we declare LOCAL VARIABLEs or create POINTERs then the meaning of a program will be a partial function on the set of full-states, where a *full-state* is a triple

$$<\mathbf{P}, A:\mathbf{P}\rightarrow\mathbf{Flt}, \sigma\epsilon A_M>$$

consisting of an Ptr-presentation, **P**, a **P**-algebra, A, and an element $\sigma\epsilon A_M$. (Of course, to get a set of full-states we must restrict the category of Ptr-presentations to be small -- but that is no problem.) We shall refer to σ as the *memory-state* . Obviously, declarations change the full-state since they change the underlying Ptr-presentation and its algebra. The significance of a declaration being nice is that this provides a natural way to determine the new memory-state corresponding to the change in the other components of the full-state. Intuitively, extending the homomorphism of EDHT-presentations to the Ptr-algebras by making the new components have value $\perp$ corresponds to "maintaining the values of the old n VARIABLEs and leaving the new VARIABLEs uninitialized". One could also consider other extensions of the homomorphism, and they would correspond to other initialization strategies such as having designated default values.

10. Iteration and FUNCTION declarations

Let **P** be an S-sorted strict EHDT-category. Let $u,v \in (S^*)^*$, and define **P**[u,v] to be the A-sorted strict EHDT-category that results from freely adjoining an additional morphism $\zeta : u \to v$ to **P**. Then we have an injection $\iota : \mathbf{P} \to \mathbf{P}[u,v]$, and, given any EDHT-category **C**, any EDHT-morphism $H:\mathbf{P} \to \mathbf{C}$, and any $f \in \mathbf{C}(u,v)$, there exists a unique EHDT-Morphism, $H_f : \mathbf{P}[u,v] \to \mathbf{C}$ such that $\iota \bullet H_f = H$, and $\zeta H_f = f$.

Given **P** and $u,v \in (S^*)^*$, where $u = (u_1)(u_2)...(u_n)$, with coproduct injections $\iota_i : (u_i) \to u$, $i=1,...,n$. Let **C** and $H:\mathbf{P} \to \mathbf{C}$ be as above. Then for any $g_1,...,g_n \in \mathbf{P}[u,v]$, where $g_i : (u_i) \to v$, $i=1,...,n$, define

$$\mathcal{I}[g_1,...,g_n] : \mathbf{C}(u, v) \to \mathbf{C}(u, v)$$
$$f \mapsto [\, g_1 H_f,\, g_2 H_f,\, ...,\, g_n H_f \,]$$

Proposition 10.1. Let everything be as above, but with $\mathbf{C} = \mathbf{Flt}$ and $H = A:\mathbf{P} \to \mathbf{Flt}$ a **P**-algebra. Then for every choice of $g_1,...,g_n$, $\mathcal{I}[g_1,g_2,...,g_n]$ has a least fix point in **Flt**(u,v), i.e., there exists $f \in \mathbf{Flt}(u, v)$ such that $f\mathcal{I}[g_1,g_2,...,g_n] = f$, and, for every $f' \in \mathbf{Flt}(u,v)$ $f'\mathcal{I}[g_1,g_2,...,g_n] = f'$ implies $f \leq f'$.

Assuming the algebra A is known, we shall generally write $\mathcal{I}[g_1,...,g_n]^\dagger$ to denote the least fix point of $\mathcal{I}[g_1,...,g_n]$ with respect to A. □

This result provides the basis for an approach to FUNCTION declaration which covers the cases shown in our working example and many others as well. Given and S-sorted EDHT-category **P**, a **P**-algebra $A:\mathbf{P} \to \mathbf{Flt}$, and given a desired function representable as the least fix point

$$\sigma = \mathcal{I}[g_1,...,g_n]^\dagger : uA \to vA.$$

the result of the declaration is **P**[u, v] and **P**[u,v]-algebra A_σ. Furthermore, where $A_\sigma H_{P[u,v]} : \mathbf{P}[u,v] \to \mathbf{P}_\sigma$ there will exist **SEDHT**-morphisms I_1 and I_2 such that the diagram

$$\begin{array}{ccc} \mathbf{P} & \xrightarrow{I_1} & \mathbf{P}[u,v] \\ {\scriptstyle AH_P}\downarrow & & \downarrow{\scriptstyle A_\sigma H_{P[u,v]}} \\ \mathbf{P}_A & \xrightarrow[I_2]{} & \mathbf{P}_\sigma \end{array}$$

commutes.

The examples given in the program in Figure 1 are of the form of definition-by-cases with, or without, recursion. That is really a special case. We can also use essentially the same approach to define FUNCTIONs via WHILE-DO or even GOTOs.

11. BLOCKs, Local VARIABLEs and POINTERs

In this Section we look at BLOCK structure in programs with particular emphasis on the question of how to handle local VARIABLEs which disappear at the end of the BLOCK, together with the creation of POINTERs which persist after the end of the BLOCK. We are able to provide a simple model for these programming phenomena within the framework already developed.

Informally, a BLOCK begins with one or more declarations, followed by a "body of code" (which may contain further BLOCKs), and ends with the "forgetting of the VARIABLEs" declared at the beginning of the BLOCK. In the light of Section 9 we can model beginning of the BLOCK by a pushout corresponding to the declarations. More precisely, let us assume that upon entering the BLOCK, the declared presentation is given by an EDHT-category **P**, and is interpreted by a **P**-algebra A, which, of course, we can represent as a morphism $AH_P:\mathbf{P}\to\mathbf{P}_A$. Then, the declaration corresponds to a pair of morphisms $d_1:\mathbf{P}_{com}\to\mathbf{P}$, $d_2:\mathbf{P}_{com}\to\mathbf{P}_{new}$, where $\mathbf{P}_{new}$ is "the declaration itself" and the morphisms $d_1:\mathbf{P}_{com}:\to\mathbf{P}$ and $d_2:\mathbf{P}_{com}:\to\mathbf{P}_{new}$, serve to identify the appropriate objects and morphisms in **P** and $\mathbf{P}_{new}$. Pushing out $<d_1, d_2>$, to get $<e_1, e_2>$, and then pushing out $<AH_P, e_1>$ to get $<f_1, f_2>$ yields the pushout diagram

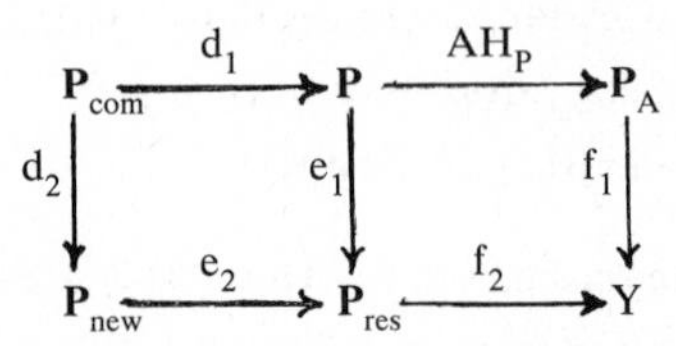

The Ptr-presentation $\mathbf{P}_{res}$ is the new presentation resulting from the declaration. Furthermore, if Y has a strong singleton object, then $f_2:\mathbf{P}_{res}\to Y$ (or, rather, the corresponding $\mathbf{P}_{res}$-algebra $A_{res} = f_2A_{\mathbf{P}_{res}}$) provides the desired interpretation. This, in terms of the discussion at the end of Section 9, is that, if we enter the block in full-state $<\mathbf{P}, A, \sigma>$, then the result of the declaration $<d_1, d_2>$ (assuming it is a nice declaration) is the full-state $<\mathbf{P}_{res}, A_{res}, \sigma_{res}>$ where σ_{res} is the extension of σ in which "all new VARIABLEs have the value $\perp$".

Our thesis is that the meaning of the body of a block will be a partial function from full-states to full-states. Further, we assume that if the full-state on leaving the body of the block is defined, then it will be of the form $<\mathbf{P}_{res}, B_{res}, \sigma>$ and there will exist morphisms g and G such that the diagram

$$\begin{array}{ccc} \mathbf{P} & \xrightarrow{\;g\;} & \mathbf{P}_{res} \\ {\scriptstyle AH_P}\downarrow & & \downarrow{\scriptstyle BH_{\mathbf{P}_{res}}} \\ \mathbf{P}_A & \xrightarrow[\;G\;]{} & \mathbf{P}_B \end{array}$$

commutes. Then our claim is that the desired result upon exiting the BLOCK is the **P**-algebra $(g\bullet(BH_{\mathbf{P}_{res.}}))A_P$. That is, this simple construction provides the necessary mechanism for "forgetting" LOCAL VARIABLES upon leaving the BLOCK. The validity of this thesis depends, of course, on what one allows as program STATEMENTs and BLOCKs and their semantics. However, it does hold in the "programming language" used for our working example, although to prove this we would first have to complete the definition of the language. However the reader should have no real difficulty convincing himself that the thesis holds in the sample program.

What we want to show now is that the above mechanism for declarations and forgetting, simple though it is, is enough to handle both LOCAL VARIABLEs and "PERSISTENT VARIABLEs" such as POINTERs. The basic idea is that if we want a VARIABLE declaration

```
VAR A1,...,An : T (QUALITY)
```

where `QUALITY` equals `LOCAL` or `PERSISTENT`, then this can be captured in a declaration of form $<d_1:\mathbf{P}_{com}\to\mathbf{P}, d_2:\mathbf{P}_{com}\to\mathbf{P}_{new}>$ where, in either case, we take $\mathbf{P}_{new}$ to be of the form

$S_{new} = \{ T_1, V_1 \}$

$A_1,\ldots,A_n{:}I \rightarrow V_1$

Axioms: $dom(A_i) = 1_I$, $i=1,\ldots,n$

$\rho = \{ <V_1, T_1> \}$

and for QUALITY = LOCAL we take $\mathbf{P}_{com}$ to be

$S_{com} = \{ T' \}$

with

$$d_1{:}T' \mapsto T,\ d_2{:}T' \mapsto T_1,$$

while for QUALITY = PERSISTENT we take $\mathbf{P}_{com}$ to be

$S_{com} = \{ T', V' \}$

with

$$d_1{:}T' \mapsto T, \quad d_2{:}T' \mapsto T_1,$$
$$d_1{:}V' \mapsto V_1, \quad d_2{:}V' \mapsto V$$

for some V such that $<V, T> \epsilon \rho$ in **P**. In the LOCAL case the new VARIABLEs appear, in the algebra, in a new carrier which is forgotten upon leaving the block, while in the PERSISTENT case the new VARIABLEs remain in the V-carrier of the algebra (even though they do not appear in **P**).

It is now time to show, in some detail, how this all relates to POINTERs, and, in particular, to give an explanation of the operation NEW(_) in PASCAL and our working example. The intuitive ideas are possibly best grasped by rewriting the BLOCK from the body of the working example as follows:

```
BEGIN
   Var L:LINK (LOCAL);
   A:= Nil;
   N:= 5;
   WHILE Eq(N, Zero) = False DO
     BEGIN

                      ( BEGIN
                      |   VAR Z:ITEM (PERSISTENT);
          NEW(L)  =   {   Z:= <Zero, Nil>
                      |   L:= Pt(Z);
                      ( END;

          L↑.Next:= A;
          L↑.Value:= N;
          A:= L;
          N:=Pred(N);
     end;
end;
```

All we have done is expand the statement NEW(L) into a small BLOCK and indicate whether declarations of VARIABLEs are LOCAL or PERSISTENT .

Let $<\mathbf{P}_0, A_0, \sigma_0>$ denote the full-state resulting from the declarations preceding the body of the program, where

$$A_0 = \mathbf{P}_0(I, \_\_)$$

and

$$\sigma = \bot_M$$

(the completely uninitialized memory state). Then the declaration of L changes the full-state to <$\mathbf{P}_1$, A_1, σ_1> where $\mathbf{P}_1$ is the result of the corresponding local declaration, $A_1 = \mathbf{P}_1(I, \_\_)$, and $\sigma_1 = \bot$ (uninitialized memory-state for the memory with the additional VARIABLE L). Note that the declaration, because it is local, introduces both a new sort, call it V, and a new constant L:I→V in $\mathbf{P}_1$, and thus also has an element of the new carrier $(A_1)_V = \mathbf{P}_1(I, V)$. The pushout will also give us a morphism $g:\mathbf{P}_0 \to \mathbf{P}_1$, with the following obviously commuting diagram

$$\begin{array}{ccccc} & \mathbf{P}_0 & \xrightarrow{g} & \mathbf{P}_1 & \\ (\mathbf{P}_0(I,\_\_))H_{P_0} = 1_{P_0} & \downarrow & & \downarrow & 1_{P1} = (\mathbf{P}_1(I,\_\_))H_{P_1} \\ & \mathbf{P}_0 & \xrightarrow[g]{} & \mathbf{P}_1 & \end{array}$$

In accordance with the semantics of ASSIGNMENT given in Section 7, the next two statements change the memory-state to

$$\sigma_2 = (\sigma_1(A{:=}\mathtt{Nil}))(N{:=}5)$$

(where, of course, `5` ≡ `(((((Zero)Su)Su)Su)Su)Su)`.

We now enter the WHILE-DO BLOCK. The semantics of this BLOCK (in keeping with our general thesis on the semantics of BLOCKs) is to be a partial function on an appropriate set, F_{P_1}, of full-states with first component $\mathbf{P}_1$. (F_{P_1} will be appropriate if it contains all the $\mathbf{P}_1$-algebras, A, that we can construct by means of finite sequences of nice declarations, and in which we have $A_{BOOL} \cong$ { `True`, `False` }, $A_{NAT} \cong$ { 0,1,2,... }, etc.) Clearly then, the expression `Eq(N, Zero)` induces a (possibly partial) function

$$(\mathtt{Eq(N,\ Zero)}):F_{P_1} \to \{\mathtt{True, False}\},$$

and if we can show that the body of the WHILE-DO corresponds to a function $\beta:F_{P_1} \to F_{P_1}$, then, as usual, the meaning of the WHILE-DO applied to <$\mathbf{P}_1$, A_1, σ_1> will be <$\mathbf{P}_1$, A_1, σ_1>β^n where n is the least natural number, $n \geq 0$, if any, such that (<$\mathbf{P}_1$, A_1, σ_1>β^n)(`Eq(N, Zero)`) = `True`, and is undefined if no such n exists.

It remains then to define the function β corresponding to the body of the BLOCK in the working example. The only part that is not an elementary matter of composing ASSIGNMENTs is the initial subBLOCK corresponding to the `NEW(L)` operation. Given that this subBLOCK is entered in a full-state <$\mathbf{P}_1$, A_i, σ_i> where A_i corresponds, by Theorem 6.5, to the EDHT-morphism, $A_iH_P = H_i:\mathbf{P}_1 \to \mathbf{P}_{A_i}$, then the persistent declaration at the beginning of the subBLOCK is given by the evident pushout <d_1, d_2> which yields the diagram

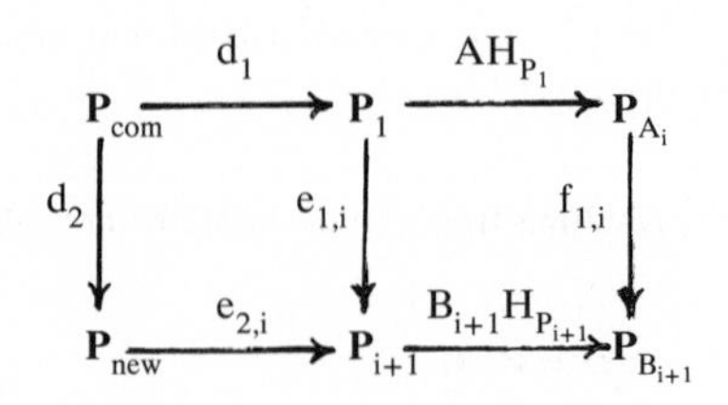

The resulting full-state is then $<\mathbf{P}_{i+1}\ B_{i+1}, \sigma_{i+1}>$ where σ_{i+1} is the minimal extension of σ_i. The effect of this declaration is to add a new constant "Z" to the ITEM-VAR-sort in $\mathbf{P}_i$ getting $\mathbf{P}_{i+1}$, and to make a similar change to $\mathbf{P}_{A_i}$ to get $\mathbf{P}_{B_{i+1}}$. The two ASSIGNMENTs then take σ_{i+1} to

$$\sigma'_{i+1} = (\sigma_{i+1}(\texttt{Z:=<zero, Nil>}))(\texttt{L:=Pt(Z)})$$

But then, exiting the subBLOCK yields full-state

$$<\mathbf{P}_1, A_{i+1}, \sigma'_{i+1}>.$$

where $A_{i+1} = (e_{1,i}\bullet(B_{i+1}H_{P_{i+1}}))A_{P1}$. Note that the newly declared constant "Z", while it is not in $\mathbf{P}_1$, will still be in $(A_{i+1})_{\text{ITEM-VAR}}$. The remaining ASSIGNMENTs only serve to change the memory-state to

$$\sigma''_{i+1} = (((\sigma'_{i+1}(\texttt{L}\uparrow\texttt{Next:=A}))(\texttt{L}\uparrow\texttt{.Value:=N}))(\texttt{A:=L}))(\texttt{N:=Pred(N)})$$

and thus the resulting function, β, has the required form for the WHILE-DO.

It is easy to see that

$$<\mathbf{P}_1, A_6, \sigma''_6> = <\mathbf{P}_1, A_1, \sigma_1>\beta^5$$

satisfies the predicate of the WHILE-DO. Thus the WHILE-DO terminates, then we exit from the block as a whole, and get the full-state

$$<\mathbf{P}_0, (g\bullet(A_6H_{P6})A_{P_0}, \sigma_6>$$

where, because L was declared local, this reduct forgets the V-sort, and thus the VARIABLE L, and forms the final memory-state by means of the evident projection.

12. Conclusions

In this paper we have given a mathematical framework, the EDHT-categories and their algebras, and then sketched how it can be used to understand various aspects of PASCAL-like imperative programming languages. Some of the key ideas are:

★ RECORDs are Cartesian products (tensor products)

★ VARIANTs are coproducts

★ Recursive TYPEs via tensor products and coproducts.

- Build up a "typical language" from "nothing".
- The injections and projections as basic operations.

★ The STORE (Memory) is a categorical product

★ Closure of derived operators under mediating morphisms

- Case statements for VARIANTs correspond to coproduct mediators.
- RECORD constructors are tensor product mediators.
- Assignment statements are categorical product mediators from the STORE to itself.

★ Iterators (Primitive Recursion, WHILE-DO, etc.) via fix points.

★ VARIABLEs and POINTERs as sorts.

- Nil POINTERs via coproducts.

★ TYPE and VARIABLE Declarations via pushouts.

★ BLOCKs via pushouts and reducts

- persistent declarations (as for POINTERs)
- transient declarations (as for local VARIABLEs)

• VARIABLE initializations via extension of homomorphisms.

★ etc.

This (still incomplete) categorical formulation of the basic concepts of imperative programming languages was carried out with an eye to the prescriptive nature of the categorical framework. That is, we have been guided by the rule, "a good programming concept should have a natural categorical counterpart".

13. Bibliography

[1] MacLane, S., *Categories for the Working Mathematician*, Springer-Verlag, (1971).

[2] Goguen, J. A., Thatcher, J. W., Wagner, E. G., "An Initial Algebra Approach to the Specification, Correctness, and Implementation of Abstract Data Types," *Current Trends in Programming Methodology IV: Data Structuring*, (R. Yeh, Ed.) Prentice-Hall (1978).

[3] Jensen, K., and Wirth, N., *PASCAL User Manual and Report,* 2nd Ed., Springer-Verlag, New York, Heidelberg, and Berlin, (1978)

[4] Wagner, E. G., "Categorical Semantics, or Extending Data Types to Include Memory," (Invited Paper) *Recent Trends in Data Type Specification: 3rd Workshop on Theory and Applications of Abstract Data Types, Selected Papers* (H.-J. Kreowski, Ed.) Informatik-Fachberichte 116, Springer-Verlag, (1985) pp. 1-21.

[5] Wagner, E. G., "Algebraic Theories, Data Types, and Control Constructs," *Fundamenta Informaticae* IX (1986) 343-370.

[6] Wagner, E. G., "Categories, Data Types, and Imperative Languages," (invited lecture) *Category Theory and Computer Programming,* (Tutorial and Workshop, Guildford, U.K., September 1985, Proceedings) (Pitt, D., Abramsky, S., Poigne, A., Rydeheard, D., Eds.) LNCS 240, Springer-Verlag (1986), pp 143-162.

[7] Milne, R. E., and Strachey, C., *A Theory of Programming Language Semantics* Chapman and Hall, London, and John Wiley, New York, (1976)

[8] Stoy, J. E., *Denotational Semantics: The Scott-Strachey Approach to Programming Language Theory,* MIT Press, Cambridge, Mass. and London, England, (1977)

[9] Oles, F. J., *A Category-Theoretic Approach to the Semantics of Programming Languages,* PhD Dissertation, School of Computer Science, Syracuse University, August 1982.

[10] Landin, P. J., "The Mechanical Evaluation of Expressions," *Computer Jour.*, 6, (1964) pp.308-320.

[11] Lehmann, D. J., and Smyth, M. B., "Data Types:, *Proceedings of the 18th IEEE Symposium on the Foundations of Computing,* Providence, R.I., Nov. 1977, pp.7-12.

[12] Robinson, E., and Rosolini, G., *Categories of Partial Maps,* Manuscript, submitted for publication.

[13] Hoehnke, H-J., "On Partial Algebras," *Colloquia Mathematica Societatis Janos Bolyai*, 29. Universal Algebra, Esztergom (Hungary), 1977. pp 373-412.

[14] Hoehnke, H-J., "On Yoneda-Schreckenberger's Embedding of a class of Monoidal Categories," 1985, submitted for publication.

[15] Schreckenberger, J., "Über die Einbettung von dht-symmetrischen Kategorien in die Kategorie der partiellen Abbildungen zwischen Mengen," Preprint P-12/80, Akad. Wiss. der DDR, ZI Math. und Mech., Berlin 1980.

[16] Bloom, S. L., and Wagner, E. G., "Many-sorted theories and their Algebras with some Applications to Data Types," *Algebraic Methods in Semantics,* (M. Nivat, and J. C. Reynolds, Eds.), Cambridge University Press, 1985, pp 133-168.

[17] Lawvere, W., Functorial Semantics of Algebraic Theories, *Proc. Nat. Acad. Sci., 21* (1963) pp 1-23.

ASSERTIONAL CATEGORIES

Ernie Manes
Department of Mathematics and Statistics
Lederle Research Center Tower
University of Massachusetts
Amherst, MA 01003

Abstract

Assertional categories provide a general algebraic framework for the denotation of programs. While the axioms deal exclusively with the abstract structure of coproducts, it is possible to express Boolean structure, loop-free constructs and predicate transformers and to deduce basic properties associated with propositional dynamic logic. At the foundational level, new "quarks" to build the atomic constructions of programs are espoused, leading to a new categorical duality principle for predicate transformers (based on a semilattice completion by ideals) and to the "grand unification principle" that "composition determines semantics". The first-order theory of assertional categories is more general than dynamic logic in that nondeterminism may include repetition count, that is, $f + f$ need not be f. On the other hand, an adaptation of a theorem of Kozen shows that, at least for iteration-free sentences about predicate transformers, semantics is standard. Several algebraic characterizations of Dijkstra's definition of determinism are offered and one leads to a technique to reduce loop-free expressions to guarded commands. Axioms for iteration include a "uniformity principle" that "related programs have related iterates" and the Segerburg induction axiom follows.

0. Introduction and preliminaries

The assertional categories introduced in this paper provide an algebraic model of propositional dynamic logic [Fischer and Ladner '79] in which the coproduct semantics of loop-free programs of [Elgot '75] and the partially-additive sum of [Arbib and Manes '80] play a central role. We begin with more detail on background.

The set of multi-valued functions (= relations) from a set to itself represents a universe for the denotation of programs. An algebraic model for the "tests" appearing in programs is found in the concept of a Boolean algebra and the representation theory of such structures is well established [Stone '36]. A number of authors considered algebraic models for the algebra of relations and its representation theory [McKinsey '40, Tarski '41, Everett and Ulam '46, Jonsson and Tarski '48, Lyndon '50]. The notion of "dynamic algebra" introduced (independently) by [Kozen '80, Pratt '79] (axioms are given in Section 14 below) is a two-sorted algebra in which a Boolean algebra B and a semiring K (abstracting the relations on a set D) are married by a function $K \times B \longrightarrow B$ $(f,Q) \longmapsto \langle f \rangle Q$ with standard meaning "a state in D satisfies <f>Q just in case f started in that state can terminate in a state satisfying Q". ("Dynamic logic", a form of modal logic in which there is a "diamond operator" <f> for each program f, had already been studied -- see [Harel '79, Parikh '81] for an expository treatment -- and dynamic algebras give an algebraic model for the propositional case). [Kozen '80] proved a Stone-style representation theorem for dynamic algebras which specialized both to a representation theorem for the relation algebras mentioned above and to the Stone theorem in the Boolean case. This largely reduced the first-order theory of dynamic algebras (with iteration deleted) to the standard relational case.

When regarding a dynamic algebra as a model of program semantics (there are other interpretations too) the fundamental paradigm is that programs denote endomorphisms of an object D of "states" in the category Mfn (see 2.1 below) of sets and relations. In contrast to this "endomorphism model", [Elgot '75] emphasized a finer "category model" which incorporated the fact that a single-input, single-output flowscheme is synthesized from multi-input, multi-output schemes. Elgot modeled an m-input, n-output scheme as a morphism from $m \cdot D$ to $n \cdot D$ where $k \cdot D$ denotes the k-fold coproduct (= disjoint union in many cases) of D with itself. Elgot showed how to provide the semantics of loop-free schemes in any category with finite coproducts.

[Arbib and Manes '80] introduced "partially-additive categories" which incorporated those ideas of Elgot mentioned above but which gave center stage to a "sum" operation $\sum$ on the set of morphisms between two objects which in Mfn is just union but which in the subcategory Pfn of sets and partial functions is defined again as union but only for those countable families (f_i) whose domains of definition are pairwise disjoint. Elgot's loop-free semantic constructs take a matrix-theoretic form here (based on 3.15-16) and least-fixed-point recursion is an infinite sum [Arbib and Manes '82]. The linear expressions that arise by manipulating composition and sum simplify in the "high-school way" (cf. [Backus '78, '81]); see the example of 3.10.

These remarks on background aside, what do assertional categories have to offer?

To appreciate the foundational issues which motivate this paper, we consider an analogy. Many treatments follow program syntax, viewing composition, conditional and iteration as indivisibles just as protons and neutrons are indivisibles in elementary atomic theory. Physicists introduced smaller building blocks (quarks) in a quest for a unifying framework in which new symmetries appear. The morphisms in an assertional category act as the quark from which the basic constructions can be built, and at least one new symmetry appears, namely category-theoretic duality. (For example, for $f:X \longrightarrow Y$ in Mfn, the inverse predicate transformer <f>Q is categorically dual to the direct predicate transformer <P>f which is satisfied by a state y in Y just in case every initial state which can terminate with y under execution of f necessarily satisfies P; thus these operators have the same abstract theory).

There is also a "grand unification principle" namely that "composition determines semantics". Composition determines a category and one can do much in a category by exploiting universal mapping properties. Because the first-order description of a category is as a set of morphisms (representing "denotations"), such universal properties are first-order sentences. Such category models appear in a number of mathematical contexts. We owe a special debt to Freyd's axioms for an abelian category [Freyd '64]. Many constructions in an assertional category have correspondents in an abelian category and a brief comparison between the two is given at the close of this section. The closest we come to an assertional version of the full embedding theorem for abelian categories [Freyd '64, Mitchell '65] is a direct adaptation of Kozen's theorem which appears as 9.2 below.

Despite our emphasis of a category model it would not be apt to characterize the work reported here as part of category theory. For one thing, very little such is used (preliminaries are provided below). More to the point, one could avoid categories altogether by axiomatizing the structure of the endomorphisms of an object as a "bio-ring" and we do this in Section 14 to allow an explicit comparison with dynamic algebras.

Apart from foundational concerns, assertional categories are intended to be more general than dynamic algebras. Thus even though the set of morphisms between two objects is always a poset, $f + g$ need not always be supremum. Indeed, it is possible to have $f + f \neq f$ so that models such as Mset (2.1) in which multiplicity is counted are permitted. Although the general theory is denotational in that aspects of the intermediate state are not expressible in general, this may be circumvented by tailoring the choice of model (cf. 2.5).

In [Dijkstra '76], deterministic programs are defined as those whose weakest precondition operators preserve unions. We give several characterizations of determinism in assertional categories including one which would surprise Dijkstra: f is deterministic if and only if $wp(f,wp(g,R)) = wp(fg,R)$ for all g. Dijkstra stated this property as being true for all f, deterministic or not. The discrepancy is explained by noting that he intended a composition of multi-valued functions different from that of Mfn. In Mfn, given multi-valued functions $f:X \longrightarrow Y$ and $g:Y \longrightarrow Z$, xfg is the set of all $z \in yg$ for some $y \in xf$ whereas Dijkstra collapses xfg to the empty set if any $y \in xf$ has empty yg. We feel that this composition is unacceptable because it is incompatible with Scott-Strachey semantics [Milne and Strachey '76, Stoy '77] since it is not a monotone (let alone a continuous) operation. In all, many characterizations of determinism are given (11.1-2) and it is shown in Section 12 that a straight-through expression whose components are boundedly nondeterministic can be reduced to a guarded command.

Assertional categories are single-sorted first-order models. To make this point clear we shall shortly write down an explicit such "morphisms-only" presentation for categories (even though we shall use the standard two-sorted objects and morphisms notation and terminology throughout the paper). More specific environments can be expressed by first-order extensions of the theory of assertional categories. For example, to model "while programs" which map states to states where states are identifier-indexed families of values would require constants V, S for "values" and "states" together with appropriate axioms on how these constants are to be interpreted (such as "V is an object" and "S is the product of I compies of V" where I is the (externally given) set of identifiers. A bit of this is done in Section 10 but a detailed treatment of imperative features must be given elsewhere. The subcategory of "atomic morphisms" developed in Section 10 can be used to construct many data types. The hope, overall, is that assertional categories are sufficiently robust to tailor to the needs of specific program environments without altering the basic properties developed in this paper.

We must make two apologies to the reader. Firstly, limited space necessitates that proofs be deferred to a later publication for the most part. Secondly, our treatment of iteration is very preliminary. It is clear (e.g. from [Conway '71]) that there are many possible first-order properties an iteration can satisfy, but no one set seems to dominate all others. Our choice of axioms in Section 4 emphasizes the "axiom of uniformity" 4.2 which stems from [Arbib and Manes '78].

The sum operation in an assertional category leads to a matrix-theoretic technique to reduce vector iterative program expressions to scalar ones. The discussion outlined in Section 13 is but the tip of the iceberg and a more complete study will appear elsewhere.

Others have done related work. Category models of semantics have been given using the algebraic theories of [Lawvere '63]; see [Elgot '75, Wagner, Bloom and Thatcher '85] and the references cited there. In general, algebraic theories cannot express propositional dynamic logic for lack of the kernel-domain decomposition of Axiom 1.3. The assertional Lawvere theories are characterized as matrix theories in Theorem 3.19. In addition to dynamic algebras, there are some first-order (usually many-sorted, equational) structures for what we have called the "endomorphism model" of semantics above. See [Bloom and Esik '85, Elgot and Shepherdson '82, Nelson '83 and Tiuryn '81]. The Boolean vector spaces of [Subrahmanyam '64, '65, '67] have some points in common with dynamic algebras, although no connections with program semantics were studied in the papers just cited.

Too many colleagues to list individually have provided useful discussion which led to improvements in the work reported here. I am particularly grateful to my collaborators Michael Arbib and David Benson, to my student Martha Steenstrup and to Calvin Elgot who greatly inspired me prior to his much regretted untimely death. I also wish to thank the referee for Example 4.6 and N. V. Subrahmanyam who pointed out that two of the equations for if-then-else in [Manes '85a] were redundant. The updated version appears in 6.8 below.

Some background in category theory is necessary to read this paper. We recommend older texts such as [Freyd '64, Mitchell '65] which most emphasize the ideas needed here. A very small fraction of these books need be mastered, however, and we shall summarize the concepts needed a bit later in this section.

We view categories as models over a single-sorted first-order theory with equality with one ternary relation. The elements of the carrier set are the morphisms and the ternary relation is composition, i.e. "c = fg". Far from being novel, this viewpoint is clearly stated in the founding paper of category theory [Eilenberg and Mac Lane '45, page 237] who imposed four axioms:

(i) If $c = fg$ and $d = fg$ then $c = d$. (We hence write fg).

(ii) $(fg)h$ is defined if and only if $f(gh)$ is defined and then they are equal.

(iii) If fg and gh are defined then fgh is defined.

Say that e is an <u>identity</u> if whenever eg is defined $eg = g$ and if whenever fe is defined then $fe = f$. The fourth axiom is

(iv) For all f there exist identities e_1, e_2 with e_1f and fe_2 defined.

This describes the <u>first-order theory of categories</u> whose models are categories. As promised earlier, we henceforth use the more conventional two-sorted objects and morphisms description of category theory obtained by defining the objects to be the identities. Thus, if $\underline{C}$ is a category we will use capital letters $W, X, Y, Z,\ldots$ for objects, denote the set of $\underline{C}$-morphisms from X to Y by $\underline{C}(X,Y)$ and write $f:X \longrightarrow Y$ for such morphisms. (For display purposes, f may also be written above or below the arrow, etc.) Composition of $f:X \longrightarrow Y$, $g:Y \longrightarrow Z$ will be written in "algebra order" $fg:X \longrightarrow Z$ unless otherwise noted.

An object in a category is <u>terminal</u> (respectively, <u>initial</u>) if it admits a unique morphism from (to) each object. Such is unique up to a unique isomorphism. A <u>zero object</u>, denoted 0, is both initial and terminal; the unique morphism $X \longrightarrow 0 \longrightarrow Y$ is also written $0:X \longrightarrow Y$ and $0f = 0 = f0$ for all f. We denote the object part of a finite coproduct by infix +, e.g. $X + Y + Z$. (In many of the important examples of this paper, objects are sets and coproduct is disjoint union. For many other categories, coproducts, if they exist, are constructed very differently. Loosely speaking, our axioms in Section 1 are designed to make coproducts behave like disjoint unions). For I a set (or for n a natural number) we shall use the notation $I\cdot Y$ (or $n\cdot Y$) for the coproduct of I (of n) copies of Y, the empty case being the initial object (= zero object if such exists). We shall use various notations for coproduct injections, the most common being $in_X:X \longrightarrow X + Y$ and variations thereof. Pointy brackets denote morphisms induced by the universal mapping property of the coproduct, so that given $f_i:X_i \longrightarrow Y$, $<f_i>:X_1+\ldots+X_n \longrightarrow Y$ or $<f_1,\ldots,f_n>$ denotes the unique h with $in_i h = f_i$ for all i.

The following definitions are possible where there is a zero object. A morphism $f:X \longrightarrow Y$ is <u>total</u> if given $z:Z \longrightarrow X$ with $zf = 0$, necessarily $z = 0$. Dually, f is <u>cototal</u> if $fz = 0$ implies $z = 0$. The <u>i-projection</u> $\rho_i:X_1+\ldots+X_n \longrightarrow X_i$ is defined as $<0,\ldots,0,id_{X_i},0,\ldots,0>$. The symbols $\rho,\theta,\ldots$ will be used to denote such projections. The morphism $<id_Y,\ldots,id_Y>:n\cdot Y \longrightarrow Y$ will be denoted σ. A diagram

$$P \underset{\rho}{\overset{i}{\rightleftarrows}} X \underset{\theta}{\overset{j}{\leftrightarrows}} Q$$

is called a coproduct system if the top row is a coproduct with ρ, θ the projections as above; if, further, the bottom row is a product, it is called a <u>biproduct system</u>. In general we will use infix $\times$ for products, denoting product projections (<u>not</u> to be confused with the coproduct projections above) as $\pi_i:Y_1\times\ldots\times Y_n \longrightarrow Y_i$ and the morphism $X \longrightarrow Y_1\times\ldots\times Y_n$ induced by the universal mapping property of the product by $f_i:X \longrightarrow Y_i$ as $[f_1,\ldots,f_n]$. The product of I copies of Y will be written using exponent notation as Y^I. Products not arising from biproducts play only an auxiliary role in our theory. (See Section 10 where we find use for these as well as for equalizers).

Let X be an object in a category with zero object. Say that $i:P \longrightarrow X$ is a <u>presummand</u> of X if there exists a coproduct diagram

$$P \xrightarrow{\ i\ } X \xleftarrow{\ j\ } Q\ .$$

As $i\rho_P = id_P$, i is a monomorphism. The subobject class represented by i is called a <u>summand</u> of X and the class of all summands of X will be denoted $Summ(X)$. Under the usual subobject ordering (namely $P \subset Q$ if $P \longrightarrow X$ factors through $Q \longrightarrow X$), $(Summ(X),\subset)$ is a poset with least element $0 \longrightarrow X$ and greatest element $id_X:X \longrightarrow X$ (we rarely distinguish between summands and presummands in our notation because constructions of interest are only unique up to isomorphism anyway).

In a category with zero object, a <u>kernel</u> of $f:X \longrightarrow Y$ is a morphism $i:K \longrightarrow X$ with $if = 0$ and such that for all $w:W \longrightarrow X$, if $wf = 0$ there exists unique $v:W \longrightarrow K$ with $vi = w$. Dually, $d:Y \longrightarrow C$ is a <u>cokernel</u> of f if $fd = 0$ subject to the property that if $fw = 0$ then $w = dv$ for unique v. In any biproduct system as above, i is a kernel of θ and θ is a cokernel of i.

For the reader wishing to compare assertional categories and abelian categories, a couple of brief remarks should prove helpful. The monomorphism hierarchies differ as follows ("higher" means "subclass of"):

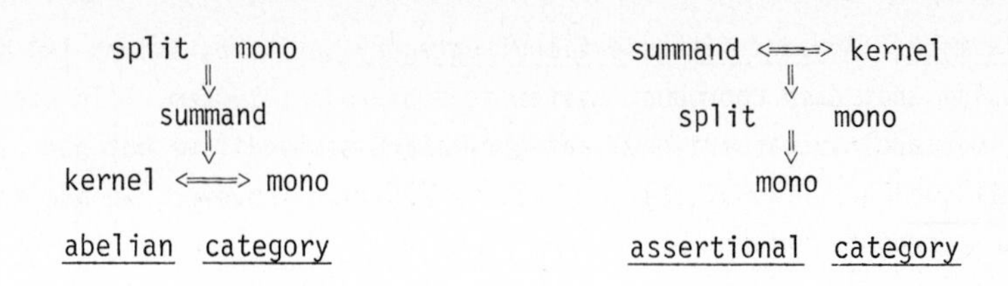

In an assertional category, Summ(X) is a Boolean algebra (infimum is pullback) whereas in an abelian category (such as real vector spaces) a summand may have distinct complements.

1. The axioms

All categories we study satisfy the following four sentences in the first-order theory of categories.

1. Axiom of zero. There is a zero object, 0.

2. Axiom of coproduct. Each pair X, Y of objects has a coproduct.

3. Axiom of bikernels. Each morphism $f:X \longrightarrow Y$ has a kernel-domain decomposition (K,i,D,j) as well as a cokernel-range decomposition (C,t,R,u). The former is a coproduct diagram

$$K \xrightarrow{i} X \xleftarrow{j} D$$

such that $if = 0$ whereas jf is total. The latter is a coproduct system

$$C \underset{\rho}{\overset{t}{\rightleftarrows}} Y \underset{\theta}{\overset{u}{\leftrightarrows}} R$$

such that $f\rho = 0$ and f_θ is cototal.

4. Axiom of antisymmetry. For $f,g:X \longrightarrow Y$, the relation $f \leq g$ defined by the existence of a fill-in

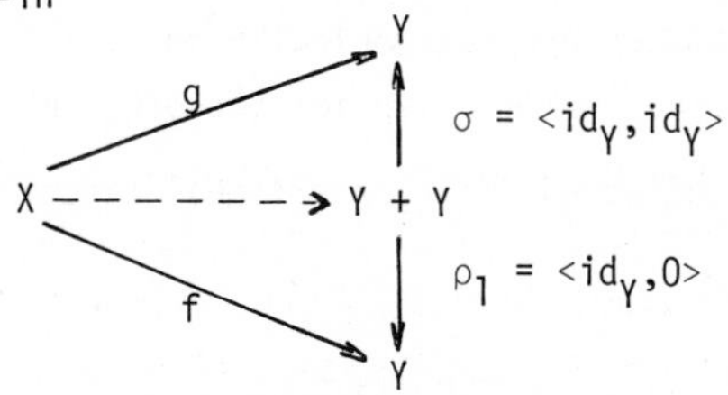

is antisymmetric.

5. Definition. A self-dual assertional category satisfies Axioms 1-4 and possesses an involution $f:X \longrightarrow Y \longmapsto f^-:Y \longrightarrow X$ subject to the axioms

$(fg)^- = g^-f^-$

$f^{--} = f$

If $P \xrightarrow{i} X \xleftarrow{j} Q$ is a coproduct, $ii^- = id_P$, $ji^- = 0$.

6. Definition. An additive-assertional category satisfies Axioms 1-4 and the additional axiom that each coproduct system is a biproduct system. (In standard terminology, our additive-assertional categories are semiadditive but not additive. Since no nontrivial assertional category can be additive, however, we use the hypen instead of a redundant semi).

7. Definition. An assertional category satisfies Axioms 1-4 as well as the following axiom scheme: For each integer $n \geq 1$, given the commutative diagram on the left there exists a unique fill-in as indicated on the right:

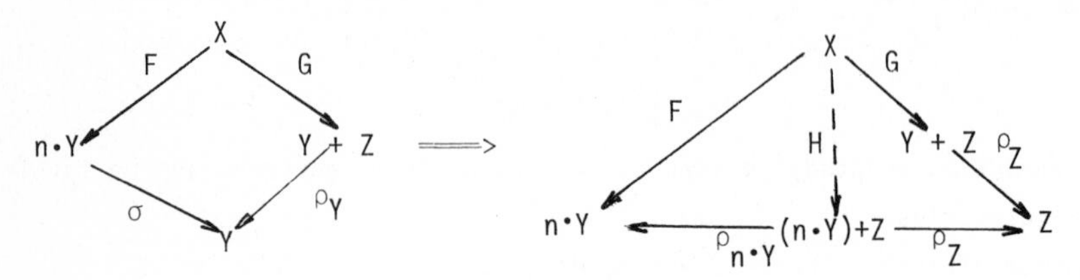

We note that in 5 the involution is an additional unary operation. Conceivably, a category may be self-dual in more than one way.

8. Proposition. A self-dual assertional category is additive-assertional, whereas an additive-assertional category is assertional.

Proof. Axioms 5 imply that

$$P \underset{i^-}{\overset{i}{\rightleftarrows}} X \underset{j^-}{\overset{j}{\leftrightarrows}} Q$$

is a byproduct system if the top row is a coproduct. For the second part, H in 7 exists uniquely because the bottom row is a product.

9. Remark. In an additive-assertional category, kernel-domain decomposition and cokernel-range decomposition are categorically dual. It follows that if an involution exists, either half of Axiom 3 implies the other half.

10. Remark. The dual of an additive-assertional category is additive-assertional.

2. Basic examples

1. Matrix categories. Matrix categories over positive semirings induce a broad class of self-dual assertional categories.

A positive semiring is a pair (R,τ) with τ an infinite cardinal and R a τ-complete unital semiring, that is, families indexed by sets of cardinal $< \tau$ have sums. (Write ∞ for τ to indicate that there is no cardinality constraint). By "positive" we mean that R is antisymmetric (i.e. that the relation $f \leq g$ defined by $g = f + h$ for some h is antisymmetric) and has no zero-divisors. This definition is a mild strengthening of that of [Eilenberg '74, page 125].

The <u>matrix category</u>. $Mat_{(R,\tau)}$ of (R,τ) has sets of cardinal $<\tau$ as objects and X-by-Y matrices with entries in R (X indexing rows and Y indexing columns) as morphisms with the usual matrix multiplication and identity matrices for composition and identities.

This category is self-dual assertional providing R has no zero-divisors. The empty set 0 provides the zero object. (For example, if X, Y are non-empty, the X-by-Y matrix obtained as the composition $X \longrightarrow 0 \longrightarrow Y$ has all entries 0 because the empty sum in R is 0). Each relation (in particular, each partial function or total function) $\subset X \times Y$ may be represented by its incidence matrix. So representing the inclusion functions of a disjoint union provides coproducts. For Axiom <u>1.3</u>, given X-by-Y (f_{xy}),

$$K = \{x \in X : f_{xy} = 0 \text{ for all } y\} \qquad C = \{y \in Y : f_{xy} = 0 \text{ for all } x\}$$

with D the complement of K and R the complement of C provide kernel-domain and cokernel-range decompositions (but the proof requires the absence of zero-divisors). (Indeed, (f_{xy}) is total if and only if for all x there exists y with $f_{xy} \neq 0$ and is cototal if its transpose is total). Further, $(f_{xy}) \leq (g_{xy})$ if and only if $f_{xy} \leq g_{xy}$ in R for all x, y so antisymmetry in the sense of Axiom <u>1.4</u> follows from the antisymmetry of R.

The category <u>Mfn</u> of sets and multi-valued functions arises as the matrix category over the Boolean semiring $(\{0,1\},\infty)$.

The category <u>Mset</u> of sets and multiset-valued functions is the matrix category over $(\{0,1,2,\ldots\} \cup \{\infty\},\infty)$ with the usual addition and multiplication.

2. <u>Definition</u>. A subcategory <u>D</u> of an assertional category <u>C</u> is <u>exact</u> if

(i) The zero object and all zero morphisms of <u>C</u> are in <u>D</u>.

(ii) If X_1, X_2 are objects of <u>D</u> then for each <u>C</u>-coproduct $in_i : X_i \longrightarrow X_1 + X_2$, in_1, in_2 are in <u>D</u> and, moreover, if $f_i : X_i \longrightarrow Y$ are in <u>D</u> then so is $\langle f_1, f_2 \rangle$.

(iii) If $f : X \longrightarrow Y$ is in <u>D</u> then so are all components of any kernel-domain or cokernel-range decomposition in <u>C</u>.

At this point, <u>D</u> satisfies Axioms <u>1.1-4</u>. We also require that <u>D</u> be assertional, that is, that Axiom <u>1.7</u> be satisfied by <u>D</u>.

3. <u>Partial functions</u>. The category <u>Pfn</u> of sets and partial functions is an exact subcategory of <u>Mfn</u>. To prove Axiom <u>1.7</u> define

$$xH = \begin{cases} xF & \text{if} \quad xF \text{ is defined} \\ xG\rho_Z & \text{if} \quad xG\rho_Z \text{ is defined} \\ \text{undefined otherwise} . \end{cases}$$

To this end it must be ensured that xF and $xG\rho_Z$ are never both defined for the same x. Indeed, ρ_Y, ρ_Z are defined only on Y, Z respectively so if $xG\rho_Z$ is defined, $xF\sigma = xG\rho_Y$ is not. As σ is total, xF is not defined either.

4. Example. The subcategory of one-to-finite multi-valued functions is exact in Mfn. This category is additive-assertional but admits no involution because it has a characteristic which its dual does not (e.g. infinite coproducts).

5. Example. Let $\Omega = (\Omega_n : n \geq 0)$ be a ranked alphabet and let $\bot$ be a distinguished nullary label in Ω_0. For each set Y define $\Omega(Y)$ to be the set of all (possibly infinite) trees t with leaves in $\Omega_0 + Y$ and each node of outdegree $n \geq 1$ labelled by an element of Ω_n such that either $t = \dot{\bot}$ or t has only finitely-many nonisomorphic subtrees and no subtree has all its leaves in Ω_0.

Let $\underline{\Omega}$ be the category whose objects are sets with $\underline{\Omega}(X,Y)$ the set of all total functions from X to $\Omega(Y)$. Given $f:X \longrightarrow \Omega(Y)$, $g:Y \longrightarrow \Omega(Z)$, $x(fg)$ is defined by substituting yg for each leaf y in xf and reducing all subtrees with all leaves in Ω_0 to $\bot$ until no such subtrees remain. The identity morphism maps x to the tree $\dot{x}$. Composition is well-defined and $\underline{\Omega}$ is an assertional category. The empty set is the zero object and zero morphisms are constantly $\dot{\bot}$. Each partial function $f:X \longrightarrow Y$ may be represented in $\underline{\Omega}(X,Y)$ by $x \longmapsto \dot{y}$ if $xf = y$, $x \longmapsto \dot{\bot}$ else. With this representation, the inclusions of a disjoint union provide coproducts. Total morphisms satisfy $xf \neq \dot{\bot}$ for all x. The constructions for kernel-domain and cokernel-range decomposition are

$$K = \{x : xf \neq \dot{\bot}\} \qquad R = \{y : y \text{ is a leaf in } xf \text{ for some } x\}$$

with D the complement of K and C the complement of R. The relation $f \leq g$ of Axiom 1.4 is "xg has more information than xf" in the sense that for all x, xf can be obtained from xg by replacing some of the leaves with $\bot$ and reducing as in the definition of composition. The axiom of 1.7 is somewhat delicate; xH is derived from xG by replacing each leaf y with a leaf of form (y,i) in $n \cdot Y$. To discover the correct i for y, observe that y remains as a leaf of $xG\rho_Y$ which has the same tree shape as xF and (y,i) with the desired i is the corresponding leaf in xF.

Notice that if $\Omega_0 = \{\bot\}$ and all other Ω_n are empty then $\underline{\Omega}$ is just Pfn.

6. Construction. Any product of {self-dual} {additive-} assertional categories is again such.

7. Construction. Any diagram category [Mitchell '65] over a {self-dual} {additive-} assertional category is again such.

3. Sums

1. Definition. A finite family $f_i : X \longrightarrow Y$ $(i \in I)$ in an assertional category is said to be summable if there exists $H : X \longrightarrow I \cdot Y$ such that $H\rho_i = f_i$ for all $i \in I$. As a special case of the uniqueness of the fill-in in Axiom 1.7, such H is unique if it exists. We can therefore define the sum $\sum f_i$ by

2. $\sum f_i = X \xrightarrow{H} I\cdot Y \xrightarrow{\sigma} Y$.

Infix + will also be used for sum (though this notation is ambiguous because the coproduct functor, also written +, acts on morphisms; this is rarely a problem, however).

3. Definition. In an additive-assertional category, for each object X and finite set I there is the "diagonal morphism":

$$X \xrightarrow{\Delta} I\cdot X$$

defined by the product property by $\Delta\rho_i = id_X$ $(i \in I)$. It is well-known [Freyd '64, Mitchell '65] that the sum of 2 coincides with the "cosum"

4. $\sum f_i = X \xrightarrow{\Delta} I\cdot X \xrightarrow{\langle f_i \rangle} Y$.

Examples are given in Table 5.

category	conditions that $\sum f_i$ exists	value of $\sum f_i$
$Mat_{(R,\tau)}$	none	entry-wise matrix sum
Pfn	pairwise disjoint domains	union
Ω	$\leq$ -consistent, $\leq$ as in 2.5	$\leq$ -supremum

Table 5: Examples of sum

6. Proposition. Sum has the following properties.

(i) If two families are isomorphic, one is summable if and only if the other is and then the sums are equal.

(ii) The empty family is summable with sum 0. Finitely many 0 may be added or deleted without affecting summability or sum.

(iii) A unary family sums to itself.

(iv) Given $f:W \to X$, $g_i:X \to Y$, $h:Y \to Z$, if (g_i) is summable then so is (fg_ih) and $\sum fg_ih = f(\sum g_i)h$.

(v) (Right ideal property). If $f_i:X \to Y$ is summable and $g_i:Y \to Z$ is arbitrary then (f_ig_i) is summable.

(vi) (Partition-associativity). If I, J are finite sets and if $(I_j : j \in J)$ partitions I (with no restriction on how often I_j is empty) then

$$\sum_{i\in I} f_i = \sum_{j\in J} \sum_{i\in I_j} f_i$$

in the strong sense that one side is defined if and only if the other is.

(vii) The relation $\leq$ of 1.4 is given by $f \leq g$ if and only if $g = f + h$ for some h. Hence $\leq$ is a partial order.

(viii) If $f_1+\ldots+f_n = 0$ then each $f_i = 0$.

The axiom of bikernels is not needed in the proof, most of which is routine. The hardest part is to show that the left side in (vi) exists if the right side does. This requires Axiom 1.7 and is adapted from an idea of Adámek (personal communication). The axiom of antisymmetry is used to show (viii).

7. Observation. The right ideal property 6(v) makes it clear that an assertional category is additive-assertional if and only if $id_X + id_X$ exists for all objects X.

8. Observation. If $f,f_1:X \longrightarrow Y$, $g,g_1:Y \longrightarrow Z$ in an assertional category, then if $f \leq f_1$ and $g \leq g_1$, also $fg \leq f_1g_1$.

9. Observation. In a self-dual assertional category, $(\sum f_i)^- = \sum f_i^-$. In particular, $f \leq g$ if and only if $f^- \leq g^-$.

We demonstrate the use of sums by analyzing the flowscheme of Fig. 10, in an arbitrary assertional category. To follow convention, we compose and evaluate in "analysis order", $g(f(x)$ rather than xfg. Conventional analysis would break the flowscheme into three natural cases as follows:

Case 1. We reach (*), so that $p(a)$ is false. If $q(f(a))$ is true we reach (**) which must output $f(a)$.

Case 2. We reach (*), but now $q(f(a))$ is false. Hence $p(a)$ must be false and the output is $g(f(a))$.

Case 3. Otherwise. Thus we get to (**) with $p(a)$ true and the output is $h(a)$.

In an assertional category, the interpretation of Fig. 10 is as follows. For a fixed object X, $f,g,h:X \longrightarrow X$ are morphisms. We think of $p,q:X \longrightarrow X$ as "guard morphisms" which formally means that there exists $p':X \longrightarrow X$ such that

11. $$p + p' = id_X, \quad pp' = 0 = p'p$$

and q, q' similarly. (In Pfn $xp = x$ (we now return to algebraic order of evaluation and composition) when p is true for x and is undefined otherwise). The assignment $x: = a$ will be modelled as a "constant function" $\tilde{a}:X \longrightarrow X$. A reasonable set of first-order properties to define this could be

12. (i) $\tilde{a}$ is total.

(ii) If p is a guard as in 11 and if $t, u:X \longrightarrow X$ are arbitrary with t deterministic (as defined in 11.1 below -- in Mfn this just means t is a partial function), then $\tilde{a}tpu\tilde{a}tp' = 0$.

(iii) For all $t:X \longrightarrow X$, $\tilde{a}t\tilde{a}t = \tilde{a}t$.

The semantics of 10 is then defined to be the sum of its paths:

$$\tilde{a}((p + p'fq\tilde{a})(p'f + ph) + p'fq'g)$$

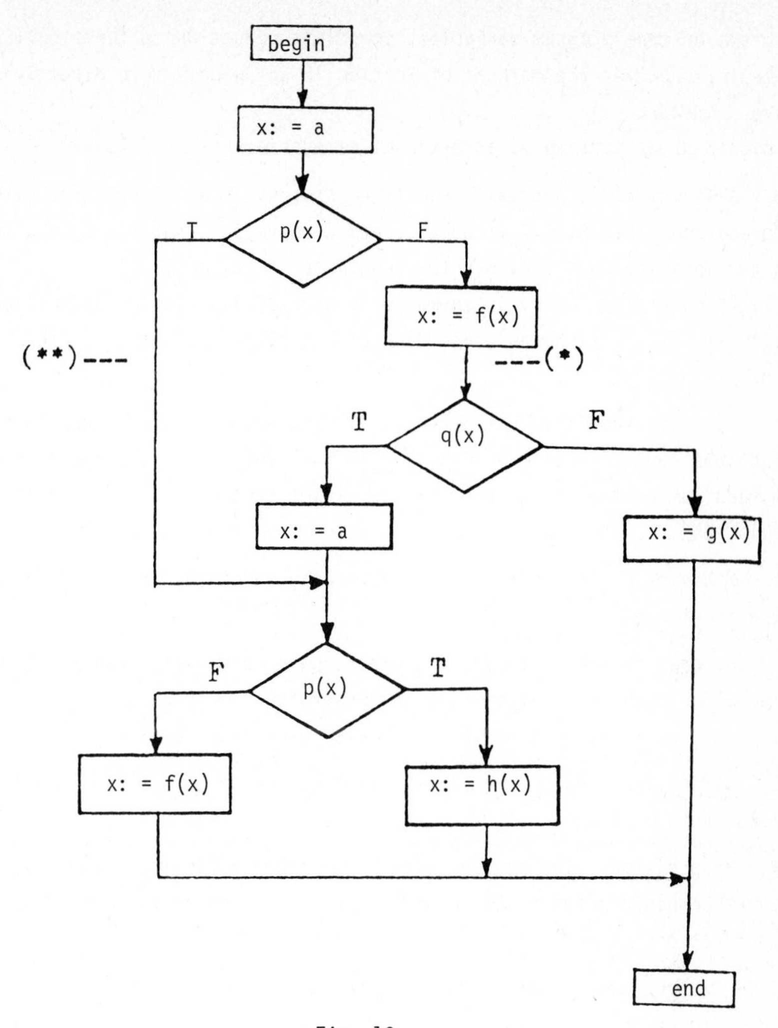

Fig. <u>10</u>

a sum which necessarily exists using the right ideal property. Expanding and using <u>11</u>, <u>12</u> (and noting that $pp = p$ -- see <u>5.1</u> below) yields

$$\tilde{a}(pp'f + pp' + p'fq\tilde{a}p'f + p'fq\tilde{a}ph + p'fq'g)$$

$$= \tilde{a}0 + \tilde{a}ph + \tilde{a}p'fq\tilde{a}p'f + \tilde{a}p'fq\tilde{a}ph + \tilde{a}p'fq'g \ .$$

Now $\tilde{a}p'fq\tilde{a}ph = (\tilde{a}tp'u\tilde{a}tp)h$ if $t = id_X$, $u = fq$, so this is 0. Additionally, $\tilde{a}p'fq\tilde{a}p'f = (\tilde{a}p'fq)(\tilde{a}p'fq) + \tilde{a}p'fq\tilde{a}p'fq' = \tilde{a}p'fq$ since the second term is 0 as above. Hence the semantics reduces to

$$\tilde{a}ph + \tilde{a}p'fq + \tilde{a}p'fq'g$$

which is just Case 3 + Case 1 + Case 2.

For two or more program variables, immediate reduction to functional form is problematic. However the methods of Section 10 below deal more directly with an imperative framework.

A higher-order version of assertional categories is as follows.

13. Definition. A partially-additive category is an assertional category in which every countable family of objects has a coproduct, subject to the following axioms for infinite sums (defined for countable I as in 1, 2).

(i) A countable family is summable if each of its finite subfamilies is.

(ii) If (f_i) is summable and each of its finite subsums is $\leq h$, then $\sum f_i \leq h$.

In a partially-additive category $\underline{C}$, all the properties of 6 continue to hold for countable sums. Further, $(\underline{C}(X,Y) \leq)$ is a domain (= ω-complete poset with bottom) and composition on either side is strict continuous. See [Manes '85, Manes and Arbib '86].

14. Examples. Pfn and Mfn are partially-additive. $\mathrm{Mat}_{(R,\tau)}$ is partially-additive if $\tau > \aleph_0$.

As mentioned in the introduction, others have considered Lawvere theory models of computation. We now characterize the assertional such categories. We start with a definition and a proposition which are also useful in Section 13:

15. Definition. An $m \times n$ array with entries in $\underline{C}(D,D)$ is a matrix over D if all row sums exist.

16. Proposition. The passage $f:m \cdot D \longrightarrow n \cdot D \longmapsto [in_i f \rho_j]$ establishes a bijective correspondence between $\underline{C}(m \cdot D, n \cdot D)$ and $m \times n$ matrices over D. Composition of such functions corresponds to matrix multiplication.

If A is an $m \times n$ matrix, we use the same symbol $A:m \cdot D \longrightarrow n \cdot D$ for the corresponding morphism.

We have found that discussions of Lawvere theories in the computer science literature are often unnecessarily technical and wish to offer the following definition, equivalent to that of [Lawvere '63], as a pedagogical improvement:

17. Definition. Let $\underline{E}$ be any category with finite coproducts (including the empty case 0, an initial object). Given an object D of $\underline{E}$ and a specific copower $n \cdot D$ for each natural number n, define the category $\underline{L}$ with natural numbers as objects and with morphisms $\underline{L}(m,n) = \underline{E}(m \cdot D, n \cdot D)$ and composition and identities as in $\underline{E}$. A category of form $\underline{L}$ is a Lawvere theory.

If a Lawvere theory as in 17 is also assertional then it follows easily from 15 and 16 that $\underline{L}$-morphisms correspond to matrices over $R = \underline{L}(1,1)$ and that composition in $\underline{L}$ is matrix multiplication. Since $\mathrm{Summ}(n) \cong 2^n$ in $\underline{L}$, $\mathrm{Summ}(1) \cong \{0,1\}$ and thus each nonzero r in R is total. In particular, if $r,s \in R$ and $rs = 0$ with $s \neq 0$ then $r = 0$ so R has no nontrivial zero-divisors. By now, $\underline{L}$ looks very much like the matrix category of 2.1. The following definition and theorem formalizes the desired result:

18. Definition. A partial semiring (see [Steenstrup '85, Manes and Arbib '86] for a more careful treatment) is $(R,\sum,\cdot,1)$ as follows.

(i) $\sum$ is a partial operation mapping finite families of R to elements of R and satisfying the obvious analogs of 6(vi) (which implies 6(i,ii)) and 6(iii), and which is antisymmetric in that $r \leq s$ if $s = r + t$ for some t describes an antisymmetric relation. (Such $\leq$ is then a partial order).

(ii) $(R,\cdot,1)$ is a monoid.

(iii) The analogs of 6(iv,v) hold.

The matrix category $\underline{Mat}_R$ is then defined as in 2.1 save for our purposes we restrict the sets X, Y there to natural numbers, and 15 is used to define "matrix". The right ideal property guarantees that matrix multiplication is well-defined.

19. Theorem. The Lawvere theories which are assertional categories are precisely the categories $\underline{Mat}_R$ for R a partial semiring with no zero-divisors.

We conclude this section with an observation and a definition for assertional categories in which sum is always defined.

20. Observation. An assertional category is additive-assertional (1.6) if and only if for all $f,g:X \longrightarrow Y$, $f + g$ is defined.

21. Definition. A semilattice-assertional category is an additive-assertional category in which $f + g$ is the supremum of f, g for all $f,g:X \longrightarrow Y$.

4. Iteration

In an additive-assertional partially-additive category, a suitable Kleene iterate $f^*:X \longrightarrow X$ for $f:X \longrightarrow X$ is given by

1. $$f^* = id_X + f + f^2 + f^3 + \ldots$$

While 1 is not a first-order property we should like to find first-order axioms on the construction $f \longmapsto f^*$ which guarantee that 1 holds when the right hand side is defined, at least for partially-additive categories. We begin by considering the

2. Axiom of uniformity. If $fh \leq h\bar{f}$ then $f^*h \leq h\bar{f}^*$. If $fh \geq h\bar{f}$ then $f^*h \geq h\bar{f}^*$. This axiom guarantees that $f \longmapsto f^*$ is not done "differently" for different f in that if h relates f to $\bar{f}$ then h also relates f^* to $\bar{f}^*$. In particular, note that if h is an isomorphism with $\bar{f} = h^{-1}fh$ then $\bar{f}^* = h^{-1}f^*h$.

Axiom 2 is not enough because $f^* = 0$ also satisfies it. An additional axiom which prevents this case is the

3. Fixed point axiom. $f^* = ff^* + id_X$.

4. Proposition. For any additive-assertional category $\underline{C}$ there is at most one operation $f \mapsto f^*$ satisfying Axioms $\underline{2}$, $\underline{3}$ as well as $id_X^* = id_X$ for all X. If such exists then

(i) $\underline{C}$ is a semilattice-assertional category.

(ii) For all $f:X \longrightarrow X$, f^* is the least h with $id_X \leq h$, $fh \leq h$, and so is, in particular, the least h with $h = fh + id_X$.

(iii) For all $f:X \longrightarrow X$, f^* is the reflexive, transitive closure of f, that is, the least h with $id_X \leq h$, $f \leq h$, $hh \leq h$.

(iv) In the self-dual case, $f^{-*} = f^{*-}$.

Proof. (i) If $id_X^* = id_X$ then by $\underline{3}$ $id_X = id_X + id_X$ so $f = f + f$ and $+$ is a join-semilattice. That $f \leq g$ if and only if $f + g = g$ is easily seen.

(ii) By $\underline{3}$, $id_X \leq f^*$ and $ff^* \leq f^*$. If $id_X \leq h$ and $fh \leq h$ then $fh \leq h\, id_X$ so $f^*h \leq h\, id_X^* = h$ and $f^* id_X \leq f^*h \leq h$.

(iii) From $id_X \leq f^*$, $f \leq ff^* \leq f^*$. As $ff = ff$, $ff^* = f^*f$ so $f^*f \leq f^* = id_X f^*$ whence $f^*f^* \leq id_X^* f^* = f^*$. Now let $id_X \leq h$, $f \leq h$, $hh \leq h$. As $hh \leq h\, id_X$, $hh^* = h^*h \leq h\, id_X^* = h$. As $f\, id_X \leq id_X h$, $f^* \leq h^*$. Hence $f^* \leq h^* = id_X h^* \leq hh^* \leq h$.

(iv) As $f^{*-*-} = (f^{**})^- = f^{*-}$ it follows from (iii) that $f^{-*} \leq f^{*-}$. Substituting f^- for f and taking $(-)^-$ yields the reverse inequality.

5. Definition. A Kleene iterate on an additive-assertional category is an operation $f \mapsto f^*$ subject to the axiom of uniformity $\underline{2}$ as well as the fixed point axiom $\underline{3}$.

As noted above, $ff^* = f^*f$. It follows at once that the same f^* provides an iterate for the dual category.

6. Example. In $\underline{Mfn}$, $\underline{1}$ provides the unique iterate of Proposition $\underline{4}$ but there is another iterate namely f^* the greatest fixed point of $h = fh + id_X$. Here $x\, id_X^* = X$.

For assertional categories generally the right hand side of $\underline{3}$ is not always defined, even in the partially-additive case. To fix this, first say that $f:X \longrightarrow X$, $g:X \longrightarrow Y$ are compatible if there exists a fill-in

7. 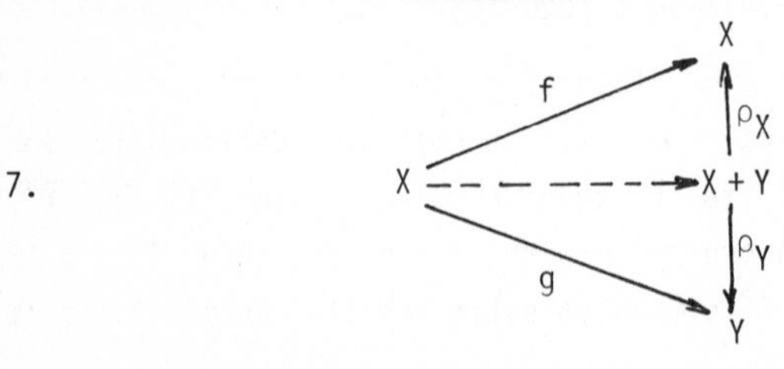

It is not hard to show that if f, g are compatible then $fh + g$ is defined for all $h:X \longrightarrow Y$ and then $g + fg + \ldots + f^n g$ exists for all n. The intended iteration is then the operation for compatible f, g expressed by

8. $\qquad (f,g)^{\dagger} = g + fg + f^2g + f^3g + \ldots \quad .$

9. Definition. (cf. [Arbib and Manes '78]). An iterate on an assertional category is a binary operation $(f,g)^{\dagger}$ defined for all compatible $f:X \longrightarrow X$, $g:X \longrightarrow Y$ subject to the uniformity axioms

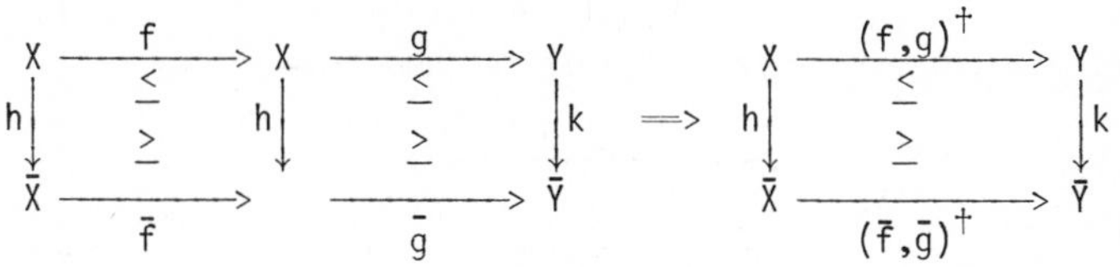

$$
\begin{array}{ccccc}
X & \xrightarrow{\;f\;} & X & \xrightarrow{\;g\;} & Y \\
h\downarrow & \begin{smallmatrix}\le\\ \ge\end{smallmatrix} & h\downarrow & \begin{smallmatrix}\le\\ \ge\end{smallmatrix} & \downarrow k \\
\bar{X} & \xrightarrow[\;\bar{f}\;]{} & & \xrightarrow[\;\bar{g}\;]{} & \bar{Y}
\end{array}
\quad \Longrightarrow \quad
\begin{array}{ccc}
X & \xrightarrow{\;(f,g)^{\dagger}\;} & Y \\
h\downarrow & \begin{smallmatrix}\le\\ \ge\end{smallmatrix} & \downarrow k \\
\bar{X} & \xrightarrow[\;(\bar{f},\bar{g})^{\dagger}\;]{} & \bar{Y}
\end{array}
$$

(the inequalities in the boxes have two readings, top and bottom; the top reading on the left side asserts that $h\bar{f} \le fh$ and $h\bar{g} \le gk$), and also the fixed point equation

$$h = (f,g)^{\dagger} \quad \text{satisfies} \quad h = fh + g.$$

10. Observation. For additive-assertional categories, $(f,g)^{\dagger}$ and f^* are coextensive because $f^* = (f,id_X)^{\dagger}$, $(f,g)^{\dagger} = f^*g$ maps one to the other. The latter is seen from $id_X f = f\, id_X$ and $id_X g = id_X g$ implies $f^*g = id_X(f,g)^{\dagger}$.

11. Remark. It is shown in [Manes and Arbib '86] that a partially-additive category has iterate given by

$$(f,g)^{\dagger} = \sum f^n g.$$

It is constructed by the Kleene fixed point theorem.

5. Boolean structure

Let $\underline{C}$ be an assertional category. Boolean structure comes for free. We begin with

1. Theorem. [Manes and Benson '85]. For each object X of $\underline{C}$, the set

$$\text{Guard}(X) = \{X \xrightarrow{\;p\;} X: \text{ there exists } p' \text{ as in } \underline{3.11}\}$$

of guard morphisms of X is a Boolean algebra under the operations

least element: 0	greatest element: id_X	supremum: $p + p'q$
infimum: pq	complement: p'	

Further, if $p \in \text{Guard}(X)$ and $q \le id_X$ then $pq = qp$.

2. Theorem. The kernel-domain decomposition of Axiom 1.3 is unique up to isomorphism and we denote it $(\text{Ker}(f),i,\text{Dom}(f),j)$. $\text{Ker}(f)$ is the kernel of f

and Dom(f) is the largest summand restricted to which f is total. The passage $p \mapsto \mathrm{Dom}(p)$ (with inverse passage displayed in 4) establishes a Boolean isomorphism of Guard(X) with the poset $(\mathrm{Summ}(X),\subset)$. The summands in a binary coproduct are Boolean complements. Guard(X) is countably complete in the partially-additive case.

3. Notation. Small letters p, q, r, ... denote guards and the corresponding capital letters P, Q, R, ... denote the corresponding summands.

4. Observation. Dom(p) = P whereas p is defined from P by

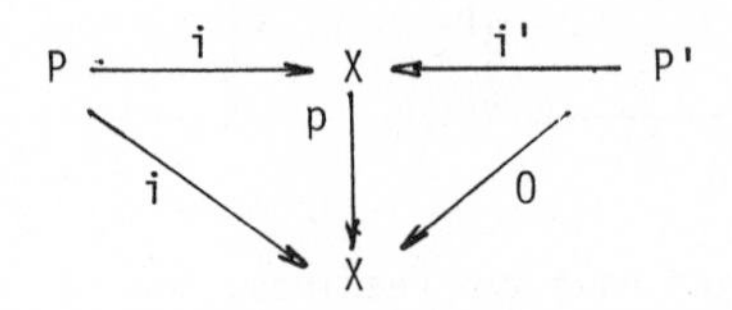

5. Proposition. The cokernel-range decomposition of Axiom 1.3 is unique up to isomorphism and we denote it (Cok(f),t,Ran(f),u). Cok(f) is the cokernel of f whereas Ran(f) is the smallest summand of Y through which f factors. This factorization of f decouples f as the composition f = ci with c cototal and i a (pre)summand and such cototal-summand factorizations are unique up to isomorphism.

In all of the examples of Section 2, Summ(X) is the Boolean algebra of all subsets of X and Guard(X) consists of the usual guard (partial) functions, suitably represented in the category as discussed earlier. In all these examples, cototal means "surjective" in the sense that for all y there exists x such that y is nontrivial in xc (e.g. a leaf, a nonzero matrix entry, etc.)

6. Definition. The extension ordering $f \sqsubseteq g$ for $f,g:X \to Y$ is defined by "$\mathrm{Dom}(f) \subset \mathrm{Dom}(g)$ and f, g agree on Dom(f)" which is easily formalized in an assertional category.

7. Proposition. For $f,g:X \to Y$, $f \sqsubseteq g$ if and only if f = pg for some $p \in \mathrm{Guard}(X)$. If $f \sqsubseteq g$ then $f \leq g$; indeed, $f \sqsubseteq g$ if and only if g = f + h with $\mathrm{Dom}(f) \cap \mathrm{Dom}(h) = 0$.

The analog of 3.8 fails for the extension ordering. The best that can be achieved is that $fg \sqsubseteq f_1 g$ if $f \sqsubseteq f_1$.

6. Control structures

The Boolean structure of the preceding section is useful in describing a number of control structures in an arbitrary assertional category.

1. Definition. For $P \in \mathrm{Summ}(X)$, $f,g:X \to Y$, if P then f else g is the morphism $X \to Y$ defined as pf + p'g. This sum exists by the right ideal property 3.6(v).

2. Definition. A guarded command is a morphism $X \longrightarrow Y$ of form $\sum p_i f_i$ for $p_i \in \text{Guard}(X)$ and $f_i : X \longrightarrow Y$. This is Dijkstra's

$$\underline{\text{if}}\ p_1 \to f_1\ \square \dots \square\ p_n \to f_n\ \underline{\text{fi}}$$

[Dijkstra '76] in Mfn.

3. Proposition. [Manes and Benson '85]. If $p_1, \dots, p_n \in \text{Guard}(X)$ with $p_i p_j = 0$ for $i \neq j$ then $p_1 + \dots + p_n$ exists and is $p_1 \vee \dots \vee p_n$. Thus the guarded command $\sum p_i f_i$ exists for any $f_i : X \longrightarrow Y$ by the right ideal property.

4. Definition. For an assertional category with iterate, define

$$\underline{\text{while}}\ \ P\ \ \underline{\text{do}}\ \ f = (pf, p')^{\dagger}$$

for $P \in \text{Summ}(X)$, $f : X \longrightarrow X$. The right ideal property implies that pf and p' are compatible.

5. Definition. Given $f, f_1 : X \longrightarrow Y$, $Q \in \text{Summ}(X)$, write $f \sqsubseteq_Q f_1$ to mean that $f \sqsubseteq f_1$ (as in 5.6) with $\text{Dom}(f_1) \cap Q = \text{Dom}(f)$.

6. Proposition. In any assertional category, if P then f else g is the unique operation I, $(P,f,g) \longmapsto I_P(f,g)$ satisfying

(i) $I_X(f,g) = f$ for all $f,g : X \longrightarrow Y$.

(ii) $I_0(f,g) = g$ for all $f,g : X \longrightarrow Y$.

(iii) If $f, f_1, g, g_1 : X \longrightarrow Y$, $P, P_1, Q \in \text{Summ}(X)$ with $P \cap Q = P_1 \cap Q$, $f \sqsubseteq_Q f_1$, $g \sqsubseteq_Q g_1$ then $I_P(f,g) \sqsubseteq_Q I_{P_1}(f_1, g_1)$.

For a proof see [Manes, to appear].

7. Definition. Let B be a Boolean algebra and let $(F,+,0)$ be an abelian monoid. A representation of a total function $I : B \times F \times F \longrightarrow F$, $(b,f,g) \longrightarrow I_b(f,g)$ is given by an additive-assertional category $\underline{C}$, objects X, Y of $\underline{C}$, a Boolean embedding $(-)^{\sim} : B \longrightarrow \text{Summ}(X)$ and a monoid embedding $(-)^{\sim} : F \longrightarrow \underline{C}(X,Y)$ such that $(I_b(f,g))^{\sim} = \underline{\text{if}}\ b^{\sim}\ \underline{\text{then}}\ f^{\sim}\ \underline{\text{else}}\ g^{\sim}$.

8. Theorem. In the context of 7, I admits a representation if and only if the following equations hold:

$$I_1(f,g) = f \qquad\qquad I_{b \wedge c}(f,g) = I_b(I_c(f,g),g)$$

$$I_{b'}(f,g) = I_b(g,f) \qquad\qquad I_b(f+f_1, g+g_1) = I_b(f,g) + I_b(f_1,g_1).$$

In that case, a representation exists in Mset. Representations in Mfn are characterized by adjoining the equation $f + f = f$ to the above, that is, let F be a join semilattice in 7.

That all representations satisfy the equations is routine. The remaining details are proved in [Manes '85a].

We note that the equations of 8 are single-sorted, regarding I as a

B-indexed family of binary operations.

9. Open question. What theorem analogous to 8 is possible for while-do?

7. Assertions and PDL

1. Hoare assertions. [Hoare '69]. For $f:X \longrightarrow Y$, $P \in \mathrm{Summ}(X)$, $Q \in \mathrm{Summ}(Y)$, say that $\{P\}\ f\ \{Q\}$ if there exists a (necessarily unique) fill-in

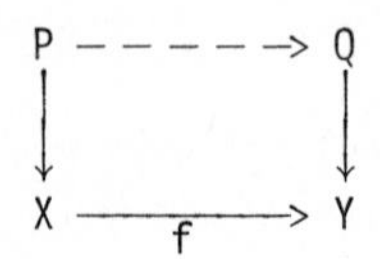

2. Inverse predicate transformers. (cf. e.g. [Harel '79, Pratt '79, Kozen '80]). For $f:X \longrightarrow Y$ and $Q \in \mathrm{Summ}(Y)$ define

$$[f]Q = \mathrm{Ker}(fq') \in \mathrm{Summ}(x)$$

$$\langle f\rangle Q = ([f]Q')' = \mathrm{Dom}(fq) \in \mathrm{Summ}(X)$$

$$wp(f,q) = \mathrm{Dom}(f) \cap [f]Q \in \mathrm{Summ}(X).$$

In Mfn, $[f]Q = \{x \in X : xf \subset Q\}$ and $wp(f,q) = \{x \in X : \emptyset \neq xf \subset Q\}$ are the weakest liberal precondition and weakest precondition operators as in [Dijkstra '76] (save that, as noted in the introduction, Dijkstra was actually in a different category in the nondeterministic case) whereas $\langle f\rangle Q = \{x \in X : xf \cap Q \neq \emptyset\}$ is the diamond operator of dynamic logic.

3. Direct predicate transformers. For $f:X \longrightarrow Y$, $P \in \mathrm{Summ}(X)$, define

$$P[f] = \mathrm{Ran}(pf) \in \mathrm{Summ}(Y)$$

$$P\langle f\rangle = \mathrm{Cok}(p'f) \in \mathrm{Summ}(Y)$$

$$sp(P,f) = \mathrm{Ran}(f) \cap P\langle f\rangle \in \mathrm{Summ}(Y) \quad \text{(strongest postcondition).}$$

In Mfn, $P[f] = \{y \in Y:$ there exists $x \in P$ with $y \in xf\}$, $P\langle f\rangle =$ $\{y \in Y:$ if $y \in xf$ then $x \in P\}$ and $sp(P,f) = \{y \in Y:$ some x exists with $y \in xf$ and each such x is in $P\}$.

4. Adjointness relationships.

$$P \subset [f]Q \iff \{P\}\ f\ \{Q\} \iff P[f] \subset Q$$

$$P\langle f\rangle \supset Q \iff \{P'\}\ f\ \{Q'\} \iff P \supset \langle f\rangle Q\ .$$

5. Faithfulness. Any of the six predicate transformers of 2, 3 can be expressed in terms of any of the others. Thus

$$[f]Q = \mathrm{Ker}(f) \cup wp(f,q)$$

so that the three inverse operators are interexpressible and similarly for the direct ones because

$$P\langle f\rangle = \mathrm{Cok}(f) \cup \mathrm{sp}(P,f).$$

The cross-connection is made by adjointness, e.g.

$$P[f] = \mathrm{Min}(Q : P \subset [f]Q).$$

Say that an assertional category is faithful ([Kozen '80] in an analogous situation says "separable") if morphisms are determined by their predicate transformers. Mfn and Pfn are faithful whereas Mset is not.

6. Definition. Let $\underline{C}$, $\underline{D}$ be assertional categories and let $F:\underline{C} \longrightarrow \underline{D}$ be a functor. F is additive if it preserves the zero object and binary coproducts (equivalently, F preserves all sums). F is exact if it is additive, if it preserves total and cototal morphisms (equivalently, F is additive and preserves kernels and cokernels) and if, additionally, F is injective on guards, that is, if $p,q \in \mathrm{Guard}(X)$ satisfy $pF = qF$, then $p = q$. F is Boolean if it is exact and also surjective on guards, that is, every guard in $\mathrm{Guard}(XF)$ has form pF for some $p \in \mathrm{Guard}(X)$.

The inclusion functor of an exact subcategory as in 2.2 is exact. If F is Boolean it establishes a Boolean isomorphism $\mathrm{Guard}(X) \longrightarrow \mathrm{Guard}(XF)$ so that f and fF have identical predicate transformers.

The assertional category $\underline{\Omega}$ of 2.5 admits a Boolean functor to Mfn -- the multi-valued function assigned to $f:X \longrightarrow \Omega(Y)$ maps x to the Y-leaves of xf. Hence predicate transformers in $\underline{\Omega}$ are at the level Mfn and, since different trees can have the same set of leaves, $\underline{\Omega}$ is not faithful.

In a similar way, $\mathrm{Mat}_{(R,\tau)}$ admits a Boolean functor to Mfn, namely map x to the set of y with $f_{xy} \neq 0$ for a given (f_{xy}). Hence matrix categories are rarely faithful.

7. Proposition. If $\underline{C}$ is assertional and if $\underline{D}(X,Y)$ is defined as $\underline{C}(X,Y)/\sim$ where $f \sim g$ means that f, g have the same predicate transformers, then the quotient category $\underline{D}$ is faithful and assertional and the canonical projection functor $\underline{C} \longrightarrow \underline{D}$ is Boolean. Further, $\underline{D}$ is additive-assertional, semilattice-assertional or self-dual accordingly as $\underline{C}$ is.

8. Properties of Hoare assertions.

(i) Three conditions equivalent to $\{P\}\ f\ \{Q\}$ are $pfq' = 0$, $pf = pfq$ and $pf \leq fq$.

(ii) $\{P \cap Q\}\ f\ \{R\}$ and $\{P' \cap Q\}\ g\ \{R\}$ if and only if $\{Q\}$ if P then f else g $\{R\}$. (Proof: Using 3.6(viii), $qpfr' = 0 = qp'gr'$ if and only if $q(pf + p'g)r' = 0$).

(iii) $\{P\}\ f\ \{Q \cap R\}$ if and only if $\{P\}\ f\ \{Q\}$ and $\{P\}\ f\ \{R\}$.

(iv) $\{P \cup Q\}\ f\ \{R\}$ if and only if $\{P\}\ f\ \{R\}$ and $\{Q\}\ f\ \{R\}$.

(v) If $\{P\}\ f\ \{Q\}$ and $\{Q\}\ g\ \{R\}$ then $\{P\}\ fg\ \{R\}$.

(vi) $\{P\}\ \sum f_i\ \{Q\}$ if and only if $\{P\}\ f_i\ \{Q\}$ for all i.

(vii) For a guarded command $\sum r_i f_i$, $\{P\}\ \sum r_i f_i\ \{Q\}$ if and only if $\{P \cap R_i\}\ f_i\ \{Q\}$ for all i.

(viii) $\{P \cap Q\}\ f\ \{Q\}$ then $\{Q\}$ $\underline{\text{while}}$ P $\underline{\text{do}}$ $f\ \{Q \cap P'\}$.

<u>Proof</u>. For any iterate, we have the diagrams

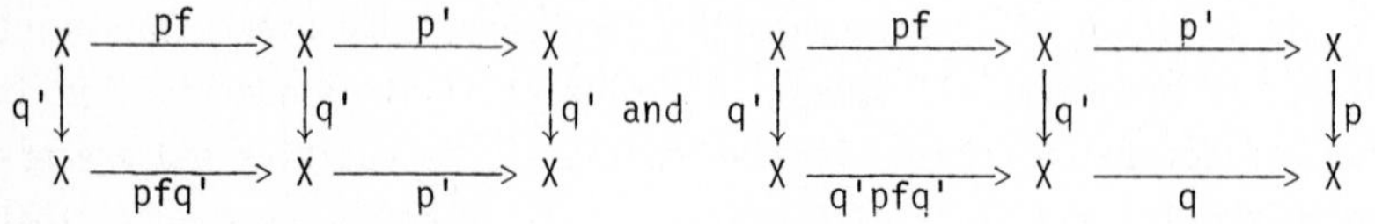

$$\begin{array}{ccccc} X & \xrightarrow{pf} & X & \xrightarrow{p'} & X \\ {\scriptstyle q'}\downarrow & & \downarrow{\scriptstyle q'} & & \downarrow{\scriptstyle q'} \\ X & \xrightarrow[pfq']{} & X & \xrightarrow[p']{} & X \end{array} \quad \text{and} \quad \begin{array}{ccccc} X & \xrightarrow{pf} & X & \xrightarrow{p'} & X \\ {\scriptstyle q'}\downarrow & & \downarrow{\scriptstyle q'} & & \downarrow{\scriptstyle p} \\ X & \xrightarrow[q'pfq']{} & X & \xrightarrow[q]{} & X \end{array}$$

because $qpfq' = 0$ using (i) so that $pfq' = (q + q')pfq' = q'pfq'$. By 4.9 it follows that $q\ (\underline{\text{while}}\ P\ \underline{\text{do}}\ f)\ q' = q\ (q'\ (\underline{\text{while}}\ P\ \underline{\text{do}}\ fq')) = 0$ and $q\ (\underline{\text{while}}\ P\ \underline{\text{do}}\ f)\ p = q\ q'\ (\text{while}\ Q'\ \text{do}\ pfq') = 0$. Thus $q\ (\underline{\text{while}}\ P\ \underline{\text{do}}\ f)\ (qp')' = q\ (\underline{\text{while}}\ P\ \underline{\text{do}}\ f)(q' + pq) = 0$ as desired.

(ix) If $\{P\}\ f\ \{P\}$ then $\underline{\text{while}}\ P\ \underline{\text{do}}\ f = p'$.

9. <u>Properties of $[f]Q$ for $f{:}X \longrightarrow Y$.</u>

(i) $[f]Y = X$, $[f]0 = \mathrm{Ker}(f)$.

(ii) $[id_X]Q = Q$, $[0]Q = X$.

(iii) If $g{:}Y \longrightarrow Z$, $[fg]R = [f]([g]R)$.

(iv) $[f](Q \cap R) = ([f]Q) \cap ([f]R)$.

(v) $[\sum f_i]Q = \cap [f_i]Q$.

(vi) $P \cup Q = [p']Q$.

(vii) $[\underline{\text{if}}\ P\ \underline{\text{then}}\ f\ \underline{\text{else}}\ g]Q = (P' \cup [f]Q) \cap (P \cup [g]Q)$.

All iterates have the same predicate transformers:

(viii) In an additive-assertional category with iterate $(-)^*$, $[f^*]Q$ is the greatest solution of the fixed point equation $S = Q \cap [f]S$, and

(ix) (the induction axiom of [Segerburg '77])

$$P \cap [f^*](P \Rightarrow [f]P) \leq [f^*]P\ .$$

The proof uses [Pratt '80, Lemma 1].

(x) $[\underline{\text{while}}\ P\ \underline{\text{do}}\ f]Q$ is the greatest solution of the fixed point equation $S = (P' \cap Q) \cup (P \cap [f]S)$.

10. <u>Properties of $\langle f\rangle Q$ for $f{:}X \longrightarrow Y$.</u>

(i) $\langle f\rangle Y = \mathrm{Dom}(f)$, $\langle f\rangle 0 = 0$.

(ii) $<id_X>Q = Q, \quad <0>Q = 0.$

(iii) If $g:Y \longrightarrow Z, \quad <fg>R = <f>(<g>R).$

(iv) $<f>(Q \cup R) = (<f>Q) \cup (<f>R).$

(v) $<\sum f_i>Q = \cup <f_i>Q$.

(vi) $P \cap Q = <p>Q.$

(vii) $<$<u>if</u> P <u>then</u> f <u>else</u> $g>Q = (P \cap <f>Q) \cup (P' \cap <g>Q).$

(viii) In an additive-assertional category with iterate, $<f^*>Q$ is the least solution of the fixed point equation $S = Q \cup <f>S$.

(ix) $<$<u>while</u> P <u>do</u> $<f>Q$ is the least solution of the fixed point equation $S = (P' \cap Q) \cup (P \cap <f>S)$.

11. <u>Properties of wp(f,Q) for f:X ⟶ Y</u>

(i) $wp(f,Y) = Dom(f), \quad wp(f,0) = 0.$

(ii) $wp(id_X,Q = Q, \quad wp(0,Q) = 0.$

(iii) $wp(f,Q \cap R) = wp(f,Q) \cap wp(f,R).$

(iv) $wp(\sum f_i,Q) = (\cup Dom(f_i)) \cap \cap [f_i]Q.$

(v) $wp(p,Q) = P \cap Q.$

(vi) wp(<u>if</u> P <u>then</u> f <u>else</u> g, $Q) = (P \cap wp(f,Q)) \cup (P' \cap wp(g,Q)).$

12. <u>Further properties</u>

(i) If r is a guard, $\{P\}\ rf\ \{Q\}$ if and only if $P \subset R' \cup [f]Q$.

(ii) There exists a unique fill-in such that the square is a pullback:

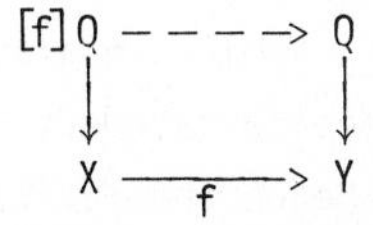

$$\begin{array}{ccc} [f]Q & \dashrightarrow & Q \\ \downarrow & & \downarrow \\ X & \xrightarrow[f]{} & Y \end{array}$$

Thus binary intersections in Summ(Y) are pullbacks. More generally, <u>any</u> infima that happen to exist in Summ(Y) are collective pullbacks (hint: consider the range). Hence $[f](-)$ preserves all intersections that exist.

(iii) There exists a unique fill-in such that the square is a pushout:

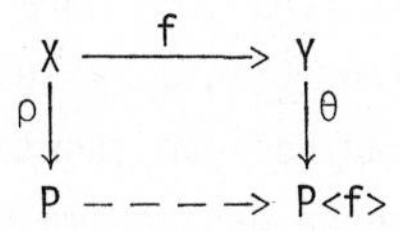

$$\begin{array}{ccc} X & \xrightarrow{f} & Y \\ \rho\downarrow & & \downarrow\theta \\ P & \dashrightarrow & P<f> \end{array}$$

All suprema that happen to exist in Summ(X) are collective pushouts and $(-)<f>$ preserves all suprema that exists. (Cf. Kozen's proof in [Parikh '81, p. 131]).

An immediate consequence of (ii, iii) is

(iv) In a self-dual assertional category, $[f]Q = Q\langle f^{-}\rangle$. (E.g., in Mfn this says $xf \subset Q$ if and only if $x \in yf^{-}$ implies $y \in Q$).

(v) $\langle f\rangle Q = Q[f^{-}]$.

8. Duality and the semilattice completion

1. Comments on duality. If $\underline{C}$ is additive-assertional, its dual $\underline{C}^{op}$ also is. Guards are self-dual. Since $P \subset Q$ in $\underline{C} \iff pq = p$ in $\underline{C} \iff qp = p$ in $\underline{C}^{op} \iff P \subset Q$ in $\underline{C}^{op}$, the posets Summ(X) are exactly the same in $\underline{C}$ and $\underline{C}^{op}$. Let $f:X \longrightarrow Y$ in $\underline{C}$ also be written as $f^{op}:Y \longrightarrow X$ in $\underline{C}^{op}$. Then as kernels and cokernels are dual we have

$$[f]Q = Q\langle f^{op}\rangle, \quad \langle f\rangle Q = Q[f^{op}], \quad \text{and} \quad wp(f,Q) = sp(Q,f^{op}).$$

Comparing with 7.12(iv,v) shows that the predicate transformers of f^{-} are available in the general additive-assertional case via f^{op}. This also shows that even if an assertional category supports more than one involution, the predicate transformers of f^{-} are independent of the choice of involution.

2. "De Morgan's law". In an additive-assertional category, $\{P\}\ f\ \{Q\}$ if and only if $\{Q'\}\ f^{op}\ \{P'\}$.

3. Theorem. Every additive-assertional category admits a Boolean functor to an additive-assertional category.

4. Metatheorem. The dual of an iteration-free statement about predicate transformers (such as those of 7.8-11 -- but no mention of iteration is permitted) which is true for all assertional categories is itself true.

To illustrate this metatheorem, consider the dual of 7.9(iv). This asserts that $(Q \cap R)\langle f\rangle = Q\langle f\rangle \cap R\langle f\rangle$. If $\underline{C}$ admits an exact functor to an additive assertional category $\underline{D}$ then in the opposite category of $\underline{D}$ the desired statement is just that of 7.9(iv), so is true. Hence the desired statement holds in $\underline{D}$ and then, by exactness, also holds in $\underline{C}$.

Theorem 3 is established by constructing the "ideal completion" of the given assertional category $\underline{C}$. The approach is as follows. We seek suitable axioms for a subset J of $\underline{C}(X,Y)$ to be an "ideal". One required property is to be that every subset A generates an ideal $I(A)$. We hope to construct a category $\hat{\underline{C}}$ with the same objects as $\underline{C}$ but with $\hat{\underline{C}}(X,Y)$ a suitable set of ideals in $\underline{C}(X,Y)$ such that $f \longmapsto I(f)$ is the desired Boolean functor, and such that for $J,K \in \hat{\underline{C}}(X,Y)$, $J + K = I(J \cup K)$. These desiderata motivate two axioms on ideals:

5. Axiom. If $f,g \in J$ and if $f + g$ exists then $f + g \in J$.

6. Axiom. If $f \leq g$ with $g \in J$ then $f \in J$.

To explain these, if $f,g \in J$ then $f + g \in I(f,g) \subset J$ whereas if $f \leq g$ so that $g = f + h$ and $g \in J$ then $f \in I(f,h) = I(g) \subset J$.

To formalize, define an ideal in $C(X,Y)$ to be a subset of $\underline{C}(X,Y)$ which satisfies axioms 5-6. Then for $A \subset \underline{C}(X,Y)$

7. $\quad I(A) = \cap(J:J$ is an ideal, $A \subset J)$

is an ideal ($= \{0\}$ if A is empty) and hence is the smallest ideal containing A. While no convenient construction for $I(A)$ is currently known, 7 suffices to show that the following formulas are well-defined and have the properties claimed.

8. $\hat{\underline{C}}(X,Y)$, defined as the set of finitely-generated ideals in $C(X,Y)$ (that is, $I(A)$ with A finite) makes $\hat{C}$ a category with composition

$$I(A)\, I(B) = I(fg:f \in A, g \in B)$$

and with $I(id_X)$ for identity morphisms.

9. If $P \underset{\rho}{\overset{i}{\rightleftarrows}} X \underset{\theta}{\overset{j}{\leftrightarrows}} Q$ is a coproduct system in C, then

$P \underset{I(\rho)}{\overset{I(i)}{\rightleftarrows}} X \underset{I(\theta)}{\overset{I(j)}{\leftrightarrows}} Q$ is a biproduct system in $\hat{\underline{C}}$. Here,

$$\langle I(A),I(B)\rangle = I(\langle f,g\rangle:f \in A, g \in B)$$

$$[I(A),I(B)] = I(\{fi:f \in A\} \cup \{gj:g \in B\})\ .$$

10. $\underline{C}$ is semilattice-assertional with

$$Ker(I(A)) = \cap\ (Ker(f):f \in A)$$

$$Dom(I(A)) = \cup\ (Dom(f):f \in A)$$

$$Cok(I(A)) = \cap\ (Cok(f):f \in A)$$

$$Ran(I(A)) = \cup\ (Ran(f):f \in A)$$

$$I(A)+I(B) = I(A \cup B).$$

11. The functor $f:\underline{C} \longrightarrow \hat{C}$, $f \longmapsto I(f)$, establishes $\hat{\underline{C}}$ as the reflection of $\underline{C}$ in the category of semilattice-assertional categories and additive functors. Specifically, F is additive and, in 12, if $\underline{D}$ is semilattice-assertional and

12. 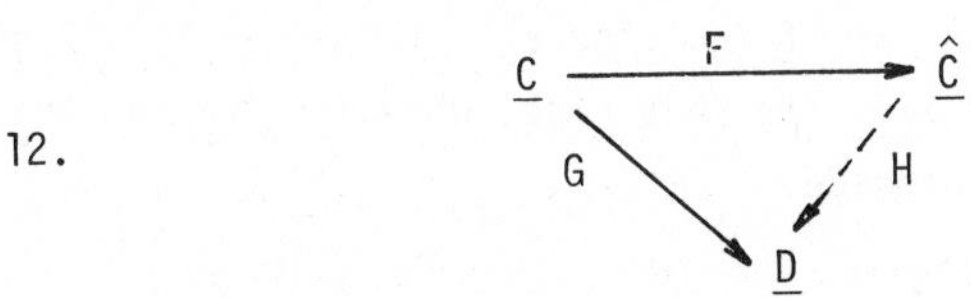

G is additive then there exists unique additive H with $FH = G$, namely

$(X \xrightarrow{I(A)} Y)H = Sup(fG:f \in A).$

13. In fact, F is Boolean and, in 12, if G is exact or Boolean so, respectively, is H.

9. Iteration-free Stone representation

The classic theorem of [Stone '36] represents an arbitrary Boolean algebra B as an algebra of subsets of the set $B\beta$ of ultrafilters of B. Specifically,

1. $B \longrightarrow 2^{B\beta}, \quad p \longmapsto \{U \in B\beta : p \in U\}$

is an injective Boolean homomorphism. [Kozen '80] extended this result to represent dynamic algebras, and the following (including the proof details) amounts to a straightforward adaptation of his work to our setting.

2. Theorem. Every assertional category C admits an exact functor to the category Mfn of sets and multi-valued functions.

Such exact $\beta : \underline{C} \longrightarrow \underline{Mfn}$ is constructed as follows:

3. $X\beta$ = set of ultrafilters on Summ(X) .

For $X \xrightarrow{f} Y$, $U(f\beta) = \{V : Q \in V \Longrightarrow \langle f\rangle Q \in U\}$.

It is easy to see that for $i:P \longrightarrow X \in \mathrm{Summ}(X)$, $\langle i\rangle Q = P \cap Q$ so that 3 reduces to the Stone representation 1 of Summ(X) when f is restricted to range over such i.

4. Metatheorem. An iteration-free statement about predicate transformers involving no existential quantification which is true in Mfn is true in all assertional categories.

Thus, for example, the intuitive reason supporting the equivalence of (i) and (vi) in Theorem 11.1 below, namely that this is so in Mfn, amounts to a formal proof for all assertional categories. This line of attack fails in attempting to show the equivalence of (i) and (iii) in that theorem, however, because p is existentially quantified in (iii) and β in 3 is not surjective on guards.

We conclude with some additional comments about the functor β.

5. Proposition. If $f,g:X \longrightarrow Y$ with $\mathrm{Dom}(f) \neq \mathrm{Dom}(g)$ then $f\beta \neq g\beta$. Further β is faithful (that is, $f\beta \neq g\beta$ if $f \neq g$ for all $f,g:X \longrightarrow Y$) if and only if C is faithful in the sense of 7.5. In that case, whenever the sum and the supremum of a family both exist, they coincide.

10. Atomicity and states

Space allows only a sketchy treatment of imperative features here, but the basic idea is as follows. The first task is to isolate within the given assertional category C a subcategory E which, in most examples, is a subcategory of the category of sets and total functions which is closed under finite limits. Finite products in $\underline{E}$ construct states as "identifier-indexed tuples of values". Assignment is easy to define and the expected types of predicates can be written down using the predicate transformers of Section 7 and equalizers.

We begin with necessary definitions, facts and axioms needed to define $\underline{E}$.

1. Definition. Say that $a:A \longrightarrow X$ is an atom of X if it represents an atom in Summ(X). A morphism $f:X \longrightarrow Y$ is atomic if af is an atom whenever a is. The subcategory of atomic morphisms will be denoted $\underline{C}_{atomic}$ and is our candidate for the subcategory $\underline{E}$ above.

We impose two axioms:

2. Axiom of extent. There exists an object G in $\underline{C}$ such that whenever $f,g:X \longrightarrow Y$ with $f \neq g$ then there exists an atom $x:G \longrightarrow X$ with $xf \neq xg$.

3. Axiom $\underline{C}_{atomic}$ has finite limits. If $E \longrightarrow X$ is an equalizer of $f,g:X \longrightarrow Y$ in $\underline{C}_{atomic}$ then $E \in \mathrm{Summ}(X)$ in $\underline{C}$.

4. Fact. G is unique up to isomorphism and $\mathrm{Summ}(G) \cong \{0,1\}$. Further, G is the terminal object of $\underline{C}_{atomic}$.

Now we define states. For values, let N be an algebra or first-order structure in $\underline{C}_{atomic}$, so that there are "functions" or "relations" R such as

$$N^k \xrightarrow{f} N \qquad R \subset N^k .$$

Here the product N^k is in $\underline{C}_{atomic}$, f is atomic and R is a $\underline{C}$-summand. The case $k = 0$ is not excluded; indeed, $N^0 = G$. This additional structure on N can be adjoined to the first-order theory of categories, one constant for each f and for each R as well as a constant for N itself.

Let I be a finite set of "identifiers" $A,B,\ldots$. The object of states is then the product

$$S = N^I$$

in $\underline{C}_{atomic}$.

If $A \in I$ and $v:G \longrightarrow N$ is a value, $A := v$ is the morphism $S \longrightarrow S$ defined by

$$(A := v)\pi_B = \begin{cases} \pi_B & \text{if } B \neq A \\ N^I \longrightarrow G \xrightarrow{v} N & \text{if } B = A \end{cases}$$

(Here, $\pi_B:N^I \longrightarrow N$ denotes the product projection).

A "term" such as $f(v_1,\ldots,v_k)$ for $v_i:G \longrightarrow N$ values and $f:N^k \longrightarrow N$ built in the usual way from the given "functions" has semantics

$$G \xrightarrow{[v_1,\dots,v_k]} N^k \xrightarrow{f} N .$$

A summand corresponding to, e.g., "$A < g(B)$" with $g:N \longrightarrow N$ and $<$ a summand of N^2 is obtained in $\underline{C}$ as $[[\pi_A,\pi_B g]] <$ (the outer box is the predicate transformer and the inner one is the map induced into the product). "$A = g(B)$" is similarly expressed since the equality relation in $N \times N$ is the equalizer of the two product projections in $\underline{C}_{atomic}$ and so is a summand in $\underline{C}$.

All constructs discussed earlier in the paper are available to construct morphisms from S to itself since, after all, S is an object of $\underline{C}$ as well as $\underline{C}_{atomic}$.

If $\underline{C}$ is $\underline{Mfn}$, $\underline{Pfn}$, $\underline{Mset}$ or $\underline{\Omega}$, $\underline{C}_{atomic}$ is the category of sets and total functions so that the above constructions are the usual ones.

We conclude this section with a comment on the effect of atomicity on Theorem 9.2.

5. Proposition. If the assertional category $\underline{C}$ satisfies the axiom of extent $\underline{2}$ then the functor β may be replaced by

$$\underline{C} \xrightarrow{F} \underline{Mfn}, \qquad XF = \{G \xrightarrow{x} X : x \text{ is an atom}\}$$

$$x(X \xrightarrow{f} Y)F = \{y \in YF : x \subset \langle f \rangle y\}$$

for such F is an exact functor.

11. Determinism

According to [Dijkstra '76], f is deterministic if $wp(f,-)$ preserves union. We adopt this definition. The first result establishes several equivalent forms.

1. Theorem. In any assertional category, the following conditions on $f:X \longrightarrow Y$ are equivalent and define when f is deterministic.

(i) $wp(f,Q \cup R) = wp(f,Q) \cup wp(f,R)$ for all $Q,R \in Summ(Y)$.

(ii) $[f]Q \cup [f]Q' = X$ for all $Q \in Summ(Y)$.

(iii) For all $q \in Guard(Y)$ there exists $p \in Guard(X)$ with $pf = fq$.

(iv) For all $g:Y \longrightarrow Z$, $Dom(fg) = wp(f,Dom(g))$.

(v) For all $g:Y \longrightarrow Z$, $wp(fg,R) = wp(f,wp(g,R))$ for all $R \in Summ(Z)$.

(vi) $\langle f \rangle Q \subset [f]Q$ for all $Q \in Summ(Y)$.

2. Proposition. A necessary and sufficient condition that all morphisms in an assertional category are deterministic is that for every $f:X \longrightarrow Y$, each $t:T \longrightarrow X$ such that tf is total factors through $Dom(f)$.

3. Remark. Every nontrivial additive-assertional category contains a nondeter-

ministic morphism. For let $X \neq 0$ and form the coproduct

$$X \xrightarrow{\ i\ } X + X \xleftarrow{\ j\ } X .$$

As all sums exist, $f = i + j$ exists $X \longrightarrow X + X$. This is a diagonal morphism as in 3.3; indeed, in Mfn, it maps x to the doubleton, one x in each copy of X, a prime candidate for a nondeterministic morphism. Coproduct projections are deterministic (use 7.12(ii) and the distributivity of the Boolean lattice $\mathrm{Summ}(X + X)$). By 7.9(v) and 7.10(v), $\langle f\rangle Q = \langle i\rangle Q \cup \langle j\rangle Q$ and $[f]Q = [i]Q \cap [j]Q$. Using 1(vi) above, if Q exists with $[i]Q \cap [j]Q = 0$ whereas $\langle i\rangle Q \cup \langle j\rangle Q \neq 0$, f is not deterministic. Indeed, choose Q to be the summand represented by i for then $[j]Q = 0$ (by 7.12(ii) whereas $\langle i\rangle Q = ([i]Q')' = 0' = X$.

4. Definition. For arbitrary $F:X \longrightarrow X + X$, $f,g:X \longrightarrow Y$ in an assertional category, define the "Elgot conditional"

$$\mathrm{if}_F(f,g) = X \xrightarrow{\ F\ } X + X \xrightarrow{\ \langle f,g\rangle\ } Y .$$

5. Proposition. Given $P \in \mathrm{Summ}(X)$, $f,g:X \longrightarrow Y$, there exists total deterministic F (namely $F = p\ in_1 + p'\ in_2$) with *if* P *then* f *else* $g = \mathrm{if}_F(f,g)$. Conversely, if $F:X \longrightarrow X + X$ is deterministic then there exist deterministic endomorphisms $t,u:X \longrightarrow X$ such that

$$\mathrm{if}_F(f,g) = \underline{\mathrm{if}}\ ([F]X \xrightarrow{\ in_1\ } X + X)\ \underline{\mathrm{then}}\ tf\ \underline{\mathrm{else}}\ ug.$$

6. Proposition. In an assertional category with iterate, let $f:X \longrightarrow X$, $g:X \longrightarrow Y$ be compatible so that $h = f\ in_X + g\ in_Y:X \longrightarrow X + Y$ exists. Then if h is deterministic,

$$(f,g)^{\dagger} = (\underline{\mathrm{while}}\ wp(h,X)\ \underline{\mathrm{do}}\ f)g .$$

7. Proposition. The subcategory $\underline{C}_{det}$ of deterministic morphisms in the assertional category $\underline{C}$ is an exact subcategory of $\underline{C}$ and is an assertional category in particular.

We conclude this section by presenting a strengthened form of semilattice completion for deterministic categories.

8. Theorem. Every assertional category $\underline{C}$ in which all morphisms are deterministic is an exact subcategory of a semilattice-assertional category $\underline{C}^{\bullet}$ with the properties:

(i) Each morphism in $\underline{C}^{\bullet}$ is a finite supremum of deterministic $\underline{C}$-morphisms.

(ii) Every deterministic morphism in $\underline{C}$ is deterministic in $\underline{C}^{\bullet}$.

The construction is quite similar to that of Theorem 8.3. Here, ideals are replaced by guarded ideals $I \subset \underline{C}(X,Y)$, where such satisfy Axiom 8.6 and are closed under disjoint guarded sum which means that $\sum p_i f_i \in I$ whenever $f_1,\dots,f_n \in I$ and $p_1,\dots,p_n \in \mathrm{Guard}(X)$ are pairwise disjoint -- the sum necessarily exists under these conditions.

The details of the proof construction for Theorem 8 are too similar to those of 8.3 to bear repeating, but we point out that there is a useful formula for the guarded ideal generated by A, namely the set of all guarded sums $\sum p_i f_i$ with each $f_i \le a_i$ for some $a_i \in A$.

9. Example. If $\underline{C}$ is $\underline{Pfn}$, $\underline{C}^{\bullet}$ is the subcategory of $\underline{Mfn}$ of finite-valued multi-valued functions.

12. Reduction to guarded commands

Guarded commands $\sum p_i f_i$ were defined in 6.2.

Let $\underline{C}$ be an assertional category. By a straight-through expression in $\underline{C}$ we mean an expression such as

$$p(f + (gq)f)$$

in morphisms $f,g,\ldots$ and guards $p,q,\ldots$ built using sum and composition in such a way that all sums are defined.

Theorem. Each straight-through expression each of whose morphisms is a finite sum of deterministic morphisms (i.e. is "boundedly nondeterministic") can be canonically reduced to a guarded command which evaluates to the same morphism.

Proof. "Sums of deterministic morphisms" is a subcategory closed under sum so the only problem is to convert $(f_1+\ldots+f_k)q$ to a guarded command when each f_i is deterministic. This is immediate from 11.1(iii).

13. Steenstrup compression

Let $\underline{C}$ be an assertional category.

1. Definition. Iterative expressions are defined as in Section 12 save that we now add the binary operation $(f,g)^{\dagger}$.

We now propose an approach to program expression transformation which has two points in common with the use of Jordan form for the matrix linear differential system $\dot{X} = AX$, $X(0) = B$. Such a system has solution

2. $\qquad X = Ue^{Jt}U^{-1}B$, $J = U^{-1}AU$ the Jordan form of A.

This method does not trivialize the solution of linear differential equations nor is it even an effective procedure (eigenvalues!) but it does lead to a theoretical understanding of the qualitative behavior of such systems.

To begin, consider the program expression $((p + g(f,g)^{\dagger}),f)^{\dagger}$. If the value of this expression in $\underline{C}(D,D)$ is t and if $u = (f,g)^{\dagger}$ then the fixed point equations for t and u give rise to the system

$$t = (p + gu)t + f$$
$$u = fu + g$$

which is "non-linear" because of the gut term. A different method leads to linear equations as follows. Each program expression induces a regular set in the alphabet consisting of the function and guard variables in an obvious way. For the example at hand, the corresponding regular set is $(p + gf^*g)^*f$ (where $+$ denotes union and set braces have been omitted). An automaton realization of this regular set is

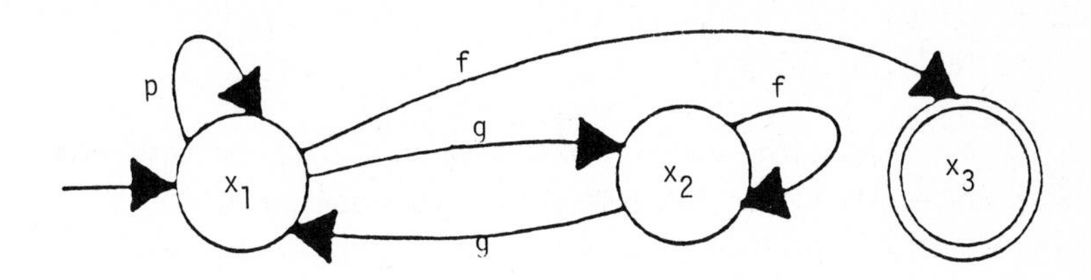

with linear Arden equations

3. $$x_1 = px_1 + gx_2 + fx_3$$
$$x_2 = gx_1 + fx_2$$
$$x_3 = 1 \quad .$$

To what extend does this work for program expressions interpreted in $\underline{C}(D,D)$?

To make this question precise, we need the following

4. <u>Definition</u>. Recall <u>3.15-16</u>. A <u>vector iterative program with</u> n <u>loops</u> is (A,B,n) where A is $n \times n$ and B is $n \times 1$ such that $A{:}B$ is a matrix over D. The semantics of (A,B,n) is defined to be

$$D \xrightarrow{in_1} n \cdot D \xrightarrow{(A,B)^\dagger} D \; .$$

At the time of this writing we are not able to prove that every program expression may be effectively transformed to a set of Arden equations whose vector iterative program in every $\underline{C}(C,D)$ has the same semantics. The construction outlined above works in standard $\underline{C}$. Thus <u>3</u> leads to $(A,B,2)$ with

$$A = \begin{bmatrix} p & g & f \\ g & f & 0 \\ 0 & 0 & 0 \end{bmatrix} \qquad B = \begin{bmatrix} 0 \\ 0 \\ id_D \end{bmatrix} .$$

Even though such A need not be a matrix in <u>Pfn</u>, <u>Pfn</u> embeds in <u>Mfn</u> (where all arrays are matrices) and the <u>Mfn</u> semantics is a partial function. But we conjecture that a general result exists (cf. [Bloom, Elgot and Wright, 80, 80a] in the context of iterative theories).

Based on an observation of [Steenstrup '85], matrix similarity transform may be used in some cases to effectively transform an n-loop vector program to one with a single loop. Let $D = \{d_0, d_1, d_2, \ldots\}$ be a countable set and let $u_i : D \longrightarrow D$ be the total function $d_k u_i = d_{i+kn}$. Let u_i^{-1} be the usual partial function inverse. Then $U = [u_1 \ldots u_n]^T : n \cdot D \longrightarrow D$ is an isomorphism in Pfn with inverse $[u_1^{-1} \ldots u_n^{-1}] : D \longrightarrow n \cdot D$. Define $J = U^{-1}AU$ so that J is a single morphism $D \longrightarrow D$. By 4.9, $(A,B)^\dagger = U(J, U^{-1}B)^\dagger$, which is analogous to 2. If one checks the details, the semantics of the original (A,B,n) is

$$5. \qquad \left(\sum_{i,j} u_i^{-1} a_{ij} u_j , \sum u_i^{-1} b_i\right)^\dagger$$

if $A = [a_{ij}]$, $B = [b_i]$. The necessary axioms on $u_1, \ldots, u_n$ to make this construction work are easily stated for general D and assertional $\underline{C}$.

14. Dynamic algebras and bio-rings

There is some variation in the definition of "dynamic algebra" found in the literature. One version is the following.

1. Definition. [Pratt '79, Kozen '80]. A *dynamic algebra* is $(K, +, 0, ;, 1, (-)^-, (-)^*, B, \langle\ \rangle)$ such that $(K, +, 0, ;, 1)$ is a unital semiring (we write multiplication as fg instead of $f;g$), B is a Boolean algebra and $\langle\ \rangle : K \times B \longrightarrow B$ is a total function written $(f,Q) \longmapsto \langle f \rangle Q$, all subject to the equations

(A) $\quad f + f = f \qquad$ (i.e., $(K,+,0)$ is a join-semilattice).

(B) $\quad (fg)^- = g^- f^-$

$\qquad (f + g)^- = f^- + g^-$

$\qquad f^{--} = f$

(C) $\quad \langle f + g \rangle Q = \langle f \rangle Q \cup \langle g \rangle Q$

$\qquad \langle f \rangle (Q \cup R) = \langle f \rangle Q \cup \langle f \rangle R$

$\qquad \langle fg \rangle Q = \langle f \rangle (\langle g \rangle Q)$

$\qquad \langle f \rangle 0 = \langle 0 \rangle Q = 0$

$\qquad \langle 1 \rangle Q = Q$

as well as the first-order axiom

(D) $\quad P \le [f]\langle f^- \rangle P$

and the higher-order property

(sup-star) $\quad fg^*h = \mathrm{Sup}\ fg^n h.$

[Kozen '81] also weakened (sup-star) to the first-order "induction axiom" of

[Segerburg '77] which is

$$P \cap [f^*](P \Longrightarrow [f]P) \leq [f^*]P \quad .$$

As noted in 7.9(ix), this axiom holds for any iterate in an additive-assertional category.

2. Proposition. If D is an object in a self-dual assertional category $\underline{C}$ then $K = \underline{C}(D,D)$, $B = \mathrm{Summ}(D)$ satisfies (B,C,D).

Proof. (B) is the definition of an involution and 3.9. (C) is 7.10. To prove (D),

$$\begin{aligned} P \leq [f]\langle f^{-}\rangle P &\Longleftrightarrow \{P\}\ f\ \{\langle f^{-}\rangle P\} && \text{(by 7.4)} \\ &\Longleftrightarrow \{P\}\ f\ \{P[f]\} && \text{(by 7.12(v))} \\ &\Longleftrightarrow P[f] \subset P[f] && \text{(by 7.4).} \end{aligned}$$

Property (A) does not seem to be important for proofs. Similarly, (B) can be done without if $\underline{C}$ is additive-assertional. For example, the proof of (D) above, using 8.1 instead of 7.12(v), proves

$$P \subset [f]\langle f^{op}\rangle P.$$

From 4.11 we then have

3. Proposition. In the context of 2, if $\underline{C}$ is partially-additive then (sup-star) holds.

We conclude by presenting an intrinsic axiomatization of $\underline{C}(D,D)$, building upon [Manes and Benson '85, Steenstrup '85].

4. Definition. A bo-ring (bi-kerneled ordered semiring) is $R = (R,+,0,;1)$ subject to the axioms in (E,F) below:

(E) $(R,+,0,;,1)$ is an ordered semiring, that is, a unital semiring for which the relation $f \leq g$ if $g = f + h$ for some h is antisymmetric, and hence a partial order. We write $f;g$ as fg.

A guard is $p \in R$ for which there exists (necessarily unique) p' with $p + p' = 1$, $pp' = 0 = p'p$.

(F) Every $f \in R$ has a kernel K_f, that is,

(i) K_f is a guard

(ii) $K_f f = 0$

(iii) If $gf = 0$ then $gK_f = g$.

Also, every $f \in R$ has a cokernel C_f, which means

(i) C_f is a guard

(ii) $fC_f = 0$

(iii) If $fg = 0$ then $C_f g = g$.

A bio-ring (bikerneled iterative ordered semiring) is a bo-ring equipped with an iterate $f \mapsto f^*$ satisfying

(G) (i) $h = f^*$ satisfies $h = fh + 1$

(ii) If $hf \leq gh$ then $hf^* \leq g^*h$

(iii) If $fh \leq hg$ then $f^*h \leq hg^*$.

A bio-set is a bio-ring for which $1^* = 1$.

Note that the dual of a bo-ring, bio-ring or bio-set (reverse the order of composition) is again one.

5. Theorem. Every faithful bo-ring can be represented in Mfn in the style of 2. Here faithful has the same meaning as in 7.5, it being clear that predicate transformers are definable with the same formulas used for assertional categories. The proof construction is the same as for Theorem 9.2.

References

M. A. Arbib and E. G. Manes, Functorial iteration, Notices Amer. Math. Soc. 25, 1978, A-381.

M. A. Arbib and E. G. Manes, Partially-additive categories and the semantics of flow diagrams, J. Algebra 62, 1980, 203-227.

M. A. Arbib and E. G. Manes, The pattern-of-calls expansion is the canonical fixpoint for recursive definitions, J. Assoc. Comput. Mach. 29, 1982, 557-602.

J. Backus, Can programming be liberated from the von Neumann Style? A functional style and its algebra of programs, Commun. Assoc. Comput. Mach. 21, 1978, 613-641.

J. Backus, The algebra of functional programs: function level reasoning, linear equations and extended definitions, in J. Diaz and T. Ramos (eds.), Formalization of programming concepts, Lecture Notes in Computer Science 107, Springer-Verlag, 1981, 1-43.

S. L. Bloom, C. C. Elgot and J. B. Wright, Solutions of the iteration equation and extensions of the scalar iteration operation, SIAM J. Comput. 9, 1980, 25-45.

S. L. Bloom, C. C. Elgot and J. B. Wright, Vector iteration in pointed iterative theories, SIAM J. Comput. 9, 1980a, 525-540.

S. L. Bloom and Z. Esik, Axiomatizing schemes and their behaviors, J. Comput. Sys. Sci. 31, 1985, 375-393.

J. H. Conway, Regular Algebra and Finite Machines, Chapman and Hall, London, 1971.

E. W. Dijkstra, A Discipline of Programming, Prentice-Hall, 1976.

S. Eilenberg, Automata, Languages, and Machines, Vol. A, Academic Press, 1974.

S. Eilenberg and S. Mac Lane, General theory of natural equivalences, Trans. Amer. Math. Soc. 58, 1945, 231-294.

C. C. Elgot, Monadic computation and iterative algebraic theories, in H. E. Rose and J. C. Shepherdson (eds.), Proc. Logic Colloq. '73, North-Holland, 1975, 175-230.

C. C. Elgot and J. C. Shepherdson, An equational axiomatization of reducible flow-chart schemes, IBM Research Report RC-8221, April 1980; in S. L. Bloom (ed.), Calvin C. Elgot, Selected Papers, Springer-Verlag, 1982, 361-409.

C. J. Everett and S. Ulam, Projective algebra I, Amer. J. Math. 68, 1946, 77-88.

M. J. Fischer and R. E. Ladner, Propositional dynamic logic of regular programs, J. Comput. Sys. Sci. 18, 1979, 194-211.

P. Freyd, Abelian Categories, Harper and Row, 1964.

D. Harel, First-order Dynamic Logic, Lecture Notes in Computer Science 68, Springer-Verlag, 1979.

C. A. R. Hoare, An axiomatic basis for computer programming, Comm. Assoc. Comput. Mach. 12, 1969, 576-580, 583.

B. Jonsson and A. Tarski, Representation problems for relation algebras, abstract 89t, Bull. Amer. Math. Soc. 54, 1948, 80.

D. Kozen, A representation theorem for models of *-free PDL, in J. W. de Bakker and J. Van Leeuwen (eds.), Automata, Languages and Programming, ICALP '80, Lecture Notes in Computer Science 85, Springer-Verlag, 1980, 351-362.

D. Kozen, On induction vs. *-continuity, in E. Engeler (ed.), Logics of Programs, Lecture Notes in Computer Science 131, 1981, 167-176.

F. W. Lawvere, Functorial semantics of algebraic theories, Ph.D. Dissertation, Columbia University, 1963.

R. C. Lyndon, The representation of relation algebras, Ann. Math. 51, 1950, 707-729.

E. G. Manes, Additive domains, in A. Melton (ed.), Mathematical Foundations of Programming Semantics, Lecture Notes in Computer Science 239, Springer-Verlag, 1985, 184-195.

E. G. Manes, Guard modules, Algebra Universalis 21, 1985a, 103-110.

E. G. Manes, A transformational characterization of if-then-else, to appear.

E. G. Manes and M. A. Arbib, Algebraic Approaches to Program Semantics, Springer-Verlag, 1986.

E. G. Manes and D. B. Benson, The inverse semigroup of a sum-ordered semiring, Semigroup Forum 31, 1985, 129-152.

J. C. C. McKinsey, Postulates for the calculus of binary relations, J. Symbolic Logic 5, 1940, 85-97.

R. Milne and C. Strachey, A Theory of Programming Language Semantics, Parts a and b, Chapman and Hall, London, 1976.

B. Mitchell, Theory of Categories, Academic Press, 1965.

E. Nelson, Iterative algebras, Theoret. Comp. Sci. 25, 1983, 67-94.

R. Parikh, Propositional dynamic logics of programs: a survey, in E. Engeler (ed.), Logic of Programs, Lecture Notes in Computer Science 125, 1981, 102-144.

V. R. Pratt, Models of program logics, Proc. 20th IEEE Symp. Found. Comp. Sci., IEEE 79CH1471-2C, 1979, 115-122.

V. R. Pratt, Dynamic algebras and the nature of induction, Proc. 12th ACM Symposium on Theory of Computing, May 1980, 22-28.

J. Reiterman and V. Trnková, Dynamic algebras which are not Kripke structures, Proc. 9th Symposium on Mathematical Foundations of Computer Science, Aug. 1980, 528-538.

K. Segerburg, A completeness theorem in the modal logic of programs, Notices Amer. Math. Soc. 24, 1977, A-522.

M. E. Steenstrup, Sum-ordered partial semirings, Ph.D. Dissertation, University of Massachusetts at Amherst, 1985.

M. H. Stone, The theory of representations for Boolean algebras, Trans. Amer. Math. Soc. 40, 1936, 37-111.

J. E. Stoy, Denotational Semantics: The Scott-Strachey Approach to Programming Language Theory, M. I. T. Press, 1977.

N. V. Subrahmanyam, Boolean vector spaces I, Math. Zeit. 83, 1964, 422-433.

N. V. Subrahmanyam, Boolean vector spaces II, Math. Zeit. 87, 1965, 401-419.

N. V. Subrahmanyam, Boolean vector spaces, III, Math. Zeit. 100, 1967, 295-313.

A. Tarski, On the calculus of relations, J. Symbolic Logic 6, 1941, 73-89.

J. Tiuryn, Unique fixed points vs. least fixed points, Theoret. Comput. Sci. 13, 1981, 229-254.

E. G. Wagner, S. L. Bloom and J. W. Thatcher, Why algebraic theories?, in M. Nivat and J. C. Reynolds (eds.), Algebraic Methods in Semantics, Cambridge Univ. Press, 1985, 607-634.

KAN EXTENSIONS IN EFFECTIVE SEMANTICS

Philip S. Mulry
Colgate University
Hamilton, New York 13346

ABSTRACT

An extension property for maps between domains is generalized to a categorical setting where the notions of adjoint and Kan extension are utilized to prove an extension property for functors. The results are used in an effective setting to provide a new characterization for certain computable mappings.

Introduction

The use of ordered sets and continuous lattices and domains by Scott and others has provided an elegant approach to the theory of computation and provided a conceptual framework for the semantics of higher type programming languages [S1,S2]. Also important is the pioneering work of Ersov who has given categorical and topological constructs, through the theory of f-spaces, similar to those of Scott [E].

The question was raised in [M2] whether the topological and lattice theoretic constructs used in the above-mentioned work might not fall under a common categorical umbrella which would generalize and unify components of the continuous and computable cases. In this paper we introduce an approach through the use of adjoints, and particularly Kan extensions, which we hope will prove useful in answering this question. In particular, in Section 3 we prove a general extension property for functors which generalizes similar results of Scott and Ersov. We utilize this approach in Section 4 to provide a new characterization for certain computable mappings. In Section 4 the topos of recursive sets [M1] is used to motivate and clarify the discussion. In fact, its use is appropriate to any discussion of computable maps on non-effectively generated domains. What is not discussed is the full extent of its role in the results of this paper, but rather will appear in a forthcoming paper.

Section 1

In this section we will provide some background and motivation for what is to follow. It is assumed the reader is familiar with the basic definitions of continuous lattice, domain and continuous poset and is urged to consult [J, S2] for details.

Theorem 1.1 (Scott) Every continuous lattice is an injective space under the induced topology.

Proof. See [S1].

Remarks: If $X \xrightarrow{e} Y$ is a subspace embedding of two T_0 spaces and f is a continuous function from X to a continuous lattice Z then Scott showed f has a unique maximal continuous extension $\overline{f}$. The requirement that Z be a continuous lattice can be weakened if we strengthen the embedding property as the following result of Ersov shows. Let Z be a complete f_0 space (see [E] for definitions and details) or a domain.

Theorem 1.2. (Ersov) If $X \longrightarrow Y$ is a dense subspace inclusion then any continuous function $X \xrightarrow{f} Z$ has a maximal continuous extension $\overline{f}$ to all of Y.

The two theorems utilize the same key construction of $\overline{f}$. Its characterization as a maximal extension is reminiscent of an algebraic description. This view is strengthened by the observation that the role of the continuity of f is simply to ensure that $\overline{f}$ is an actual extension. Thus it seems useful to consider a more algebraic viewpoint by considering posets and monotonic maps. We provide some preliminaries.

Consider the categories **Pos** and $\mathbf{T}_0$. The objects of **Pos** are posets P, Q with arrows order preserving maps. Recall every poset P can be considered a category where for $p_1, p_2 \in P$, $p_1 \longrightarrow p_2$ iff $p_1 \leq p_2$. In this interpretation order preserving maps are simply functors between categories. $\mathbf{T}_0$ is the category of T_0 spaces with continuous functions. We have functors between **Pos** and $\mathbf{T}_0$ which we denote $\mathbf{T}_0 \underset{\Omega}{\overset{S}{\rightleftarrows}} \mathbf{Pos}$. For P a poset, $\Omega(P)$ is the topological space P with opens all upper sets, i.e. $U \subset P$ is open iff $(\forall p,q)[p \leq q \ \& \ p \in U \Longrightarrow q \in U]$. For X a topological space S(X) is the poset X with ordering $x \leq y$ iff for all open U, $x \in U$ implies $y \in U$. The ordering induced by the topology is called the specialization ordering. The following result is well known.

Theorem 1.3 $\Omega \dashv S$.

For X a T_0 space and P a poset, the theorem states there exists a natural isomorphism $\mathrm{Hom}_{\mathbf{T}_0}(\Omega P, X) \simeq \mathrm{Hom}_{\mathbf{Pos}}(P, SX)$. This isomorphism follows easily from the above definitions. In general the Scott topology on a poset P is weaker than Ω (P). When composed with S, however, the consequent order, for both topologies, is the original poset ordering. A great deal more could be said about the functorial properties of these and other categories such as the category of locales. Details can be found in [J].

We now return to the point raised earlier concerning the continuity of f.

Theorem 1.4. In theorems 1.1 and 1.2, if $f:X \longrightarrow Z$ is only assumed to be monotonic then $\overline{f}:Y \longrightarrow Z$ exists and is continuous. $\overline{f}$ is an actual extension of f iff f is continuous.

Proof. A careful examination of either proof shows that only the monotonicity of f is used to prove $\overline{f}$ is continuous. That $\overline{f}$ is an actual extension follows from the statement of the theorems.

There is of course a price paid for requiring only monotonicity. In general $\overline{f}$ only extends f in the following weak sense: $\overline{f}$ is the largest continuous map $Y \longrightarrow Z$ whose restriction to X is less than or equal to f. This last characterization is suggestive of the categorical notion of Kan extension. We take up this notion in the next section and elucidate its connection with the above examples in the succeeding section.

Section 2.

In this section we introduce briefly the notion of Kan extensions. The reader can find a fuller account in [ML]. Consider the categories **D, E** and **A** and functors K, F

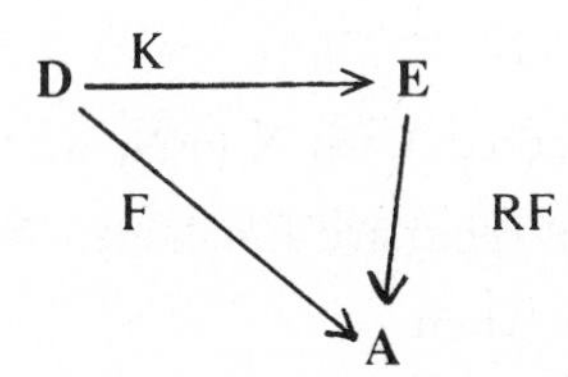

Definition 2.1. The right Kan extension of F along K is a pair RF, ϵ where $\mathbf{E} \xrightarrow{RF} \mathbf{A}$ is a functor and ϵ is a natural transformation $RF \circ K \overset{\cdot}{\longrightarrow} F$ which is universal as an arrow from $\mathbf{A}^{\mathbf{E}} \xrightarrow{\mathbf{A}^K} \mathbf{A}^{\mathbf{D}}$ to F. Restated, there exists a natural isomorphism between natural transformations $G \circ K \overset{\cdot}{\longrightarrow} F$ and natural transformations $G \overset{\cdot}{\longrightarrow} RF$ where G is any functor $\mathbf{E} \xrightarrow{G} \mathbf{A}$.

Note that $\mathbf{A}^K$ is just the functor that restricts to K, so when right Kan extensions exist for every $F \in \mathbf{A}^{\mathbf{D}}$ then $\mathbf{A}^K$ has a right adjoint. A similar definition exists for left Kan extensions, which we will denote LF.

When do Kan extensions exist and how do we compute them? RF, when it exists, is defined as follows. For $e \in \mathbf{E}$, $RF(e) = \text{limit } [e \downarrow K \xrightarrow{Q} \mathbf{D} \xrightarrow{F} \mathbf{A}]$ where Q is the projection functor. Thus the existence of such limits in **A** for each $e \in \mathbf{E}$ implies the existence of RF.

Example 2.2. If **A** is the category Set, then **A** has all limits and colimits so Set^K has both a left and right adjoint and both LF and RF exist for all $F \in \text{Set}^{\mathbf{D}}$.

Kan extensions are extremely useful constructions since, as is pointed out in [ML], the notion of Kan extension "subsumes all the other fundamental concepts of category theory." We have, for instance,

Example 2.3. Every left adjoint is a suitably described right Kan extension. Let G be a functor between two categories **B** and **A**, $\mathbf{B} \xleftarrow{G} \mathbf{A}$. If G had a left adjoint it is just the right Kan extension of $id_{\mathbf{A}}$ along G.

Example 2.4. In the case where **D**, **E** and **A** are posets with **A**, for example, a complete lattice, the descriptions of RF and LF are particularly easy.

$$RF(e) = \bigwedge_{e \leqslant Kd} Fd. \qquad LF(e) = \bigvee_{Kd \leqslant e} Fd.$$

Kan extensions are not in general true extensions. If $\mathbf{D} \xrightarrow{K} \mathbf{E}$ is full and faithful, however, then left and right Kan extensions are real extensions in the sense that $RF \circ K \simeq F$, $LF \circ K \simeq F$.

Can either the left or right Kan extension represent directly the bar construction $(\overline{})$ discussed earlier? The following example answers this question negatively.

Example 2.5. Let **P** denote all partial functions from **N** to **N**, where **N** denotes the natural numbers, and let **R** denote all total recursive functions. $\mathbf{N}^*$ is defined to be $\mathbf{N} \cup \{\bot, \top\}$ where for all $x, y, z \in \mathbf{N}^*$ $x \leq y$ when

$x = \bot$ or $x = y$ or $y = \top$.

R is a subspace of **P**, a T_0 space, and $\mathbf{N}^*$ is a continuous lattice. Let $f: \mathbf{R} \to \mathbf{N}^*$ be the constant 0 function which is continuous on **R**. Both left and right Kan extensions of f along the subspace inclusion exist. For $\phi \in \mathbf{P}$

$$Lf(\phi) = \begin{cases} 0 & \text{if } \phi \in \mathbf{R} \\ \bot & \text{otherwise} \end{cases}$$

and

$$Rf(\phi) = \begin{cases} 0 & \text{if } \phi \text{ has a total recursive extension} \\ \top & \text{otherwise} \end{cases}$$

Neither Lf nor Rf is continuous nor corresponds to $\overline{f}$ which is just the constant 0 functional on **P**.

The last example does not mean Kan extensions play no role in our discussion. What is missing in our discussion is the consideration of the role that the topological structure of X and Y played in the examples of section 1. Further, our analysis can take place in the more general arena where category replaces poset and functor replaces monotonic map.

Section 3

In this section we begin by composing Kan extensions to provide us with an extension property. This construction is then used to unify several examples from this and the next section under one common framework. For the remainder of this section we fix categories **C**, **D**, **E** and **A** and functors K, i and F as follows:

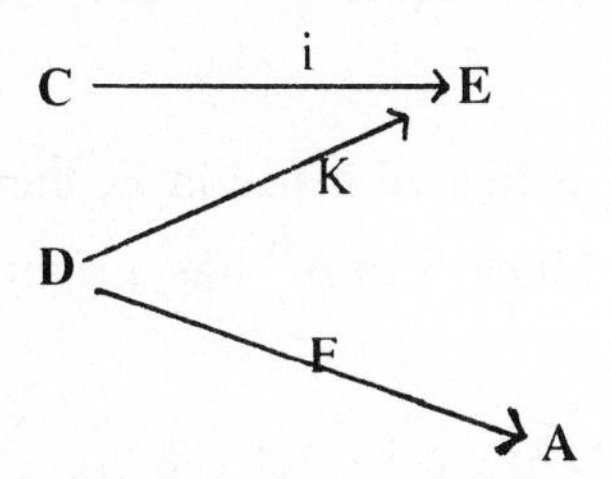

Lemma 3.1 Suppose **A** has all limits and colimits. There exists a functor $\overline{F}$: **E** → **A** as well as a natural transformation $\overline{F} \circ K \to F$ where $\overline{F}$ is in general distinct from the left and right Kan extensions of F along K.

Proof

Since **A** has all limits and colimits, left and right Kan extensions exist along both K and i. In the following diagram

$$A^D \underset{R}{\overset{A^K}{\rightleftarrows}} A^E \underset{A^i}{\overset{L}{\rightleftarrows}} A^C$$

R denotes the right adjoint of A^K and L the left adjoint of A^i. Let F^* denote $A^i \circ R(F)$ and $\overline{F}$ denote $L(F^*)$. The natural transformation $\overline{F} \circ K \rightarrow F$ is just the counit of the adjunction $A^K \circ L \dashv A^i \circ R$ (since composition preserves adjoints). It is easily seen that $\overline{F}$ differs in general from either Kan extension of F along K. For instance, in example 2.3, if **C**, **D**, **E**, and **A** are **F**, **R**, **P** and $\mathbf{N}^*$ respectively (where **F** denotes the finite partial functions) and i and K are just inclusions, then the left and right Kan extensions are not continuous and so cannot equal $\overline{F}$ which is.

Lemma 3.1 is useful for establishing the basic $(\overline{\quad})$ construction. We must be more specific before we are ready to utilize this construction in particular examples.

Definition 3.2. Given functor i, **C** → **E**, we say i is *dense* if for every e ∈ **E**, $e = \text{colim}\,(\, i \downarrow e \xrightarrow{Q} \mathbf{C} \xrightarrow{i} \mathbf{E})$ where Q is the first projection of the comma category. If the colimit diagram is filtered, we say i is *filtered dense.*

Definition 3.3. Suppose **E** has filtered colimits. An element e in **E** is called *finitely presentable* if the functor $\text{Hom}_{\mathbf{E}}(e, \;)$ preserves filtered colimits. Denote by $\mathbf{E}_{fp}$ the full subcategory of finitely presentable objects in **E**.

Example 3.4. If **E** is a poset then $\mathbf{E}_{fp}$ consists of the finite or compact elements of **E**.

Theorem 3.5. Suppose in Lemma 3.1, $\mathbf{C} = \mathbf{E}_{fp}$ is filtered dense in **E**, then the restriction of A^K to filtered colimit preserving (fcp) functors in $A^{\mathbf{E}}$ has a right adjoint.

Proof

From the diagram in the proof of Theorem 3.1 $A^K \circ L \dashv A^i \circ R$ since composition preserves adjoints. If i is filtered dense and $H \in A^{\mathbf{E}}$ preserves all filtered colimits we have $L \circ A^i(H) \simeq H$ and so $A^{\mathbf{D}}(H \circ K, F) \simeq A^{\mathbf{D}}(A^K \circ L \circ A^i(H), F) \simeq A^{\mathbf{C}}(A^i(H), F^*) \simeq A^{\mathbf{E}}(H, \overline{F})$ where the last isomorphism follows in one direction by taking filtered colimits and in the other direction since i is full and faithful. Finally $\overline{F}$ itself preserves filtered colimits since i is a dense full inclusion.

Remarks. The condition of completeness and cocompleteness of **A** in Theorem 3.5 is considerably stronger than what is actually needed. In fact, **A** need only be closed under those limits and colimits necessary to ensure the existence of the Kan extensions. The following is immediate.

Corollary 3.6. Theorem 3.5 holds if **A** is closed under limits and filtered colimits.

Definition 3.7. K is codense in i when ic $\simeq$ limit (ic $\downarrow$ Kd $\xrightarrow{Q}$ D $\xrightarrow{K}$ E) for every c in C.

Corollary 3.8. If K is codense in i then Theorem 3.5 holds when **A** is closed under non-empty limits and filtered colimits.

Proof. If K is codense in i then **A** need not have a terminal object for the right adjoint to exist.

Example 3.9. Theorem 1.1 now follows from Theorem 3.5. Let Y* denote the neighborhood system of open neighborhoods of Y thought of as a category ($U \longrightarrow V$ when $V \leq U$) and $\overline{Y}$ denote the domain determined by Y*. It is easy to check that $\overline{Y}_{fp} \simeq Y^*$, that the inclusion $Y^* \to \overline{Y}$ is filtered dense, and there exists an inclusion $Y \longrightarrow \overline{Y}$. Since Z a continuous lattice has limits and colimits, by Theorem 3.5 there exists a filtered colimit preserving functor $\overline{f}$: $\overline{Y} \to Z$ and its restriction to Y is the desired result.

Note: Theorem 3.5 does not guarantee that $\overline{f}$ in the last example is an actual extension of f but only that there exists an isomorphism, fcp $Z^{\overline{Y}}(H,\overline{f}) \simeq Z^X$ (H o K, f), for $H:\overline{Y} \longrightarrow Z$. Since f is continuous, however, Theorem 1.4 provides the desired result. More on this point shortly.

Example 3.10. In a similar fashion Theorem 1.2 follows from Corollary 3.8. Here $X \xrightarrow{e} Y$ a dense subspace inclusion (topologically) guarantees that e will be codense (categorically) in the inclusion $Y^* \longrightarrow \overline{Y}$.

Example 3.11. In Theorem 3.5 let **D** and **E** denote the poset of partial functions **P** and let **A** be $\mathbf{N}^*$ defined in Example 2.5. Fix $\phi \in \mathbf{P}$ a partial function with infinite domain. Define $F:\mathbf{P} \to \mathbf{N}^*$ by

$$F(\tau) = \begin{cases} 0 & \text{if } \phi \wedge \tau \text{ is infinite} \\ \perp & \text{otherwise.} \end{cases}$$

F is not continuous; however $\overline{F}$, which is just the constant $\perp$ functional, is.

The last example indicates that $F \in \mathbf{A}^{\mathbf{D}}$ need not satisfy $F \simeq \overline{F}$ o K. The next result clarifies this situation.

Corollary 3.12. The restriction of the adjunction of Theorem 3.5 to filtered colimit preserving functors in A^{D} forms an equivalence of categories.

Proof.

The result follows immediately from the definition of $(\overline{})$ where F is filtered colimit preserving in A^{D} if whenever d $\in$ **D** can be expressed as a filtered colimit e_j in **E**, then $F\,d \simeq \operatorname{colim} \overline{F} e_j$.

Example 3.13. It is well known that a map g: M $\rightarrow$ N between two continuous posets (lattices) is Scott continuous iff g preserves filtered colimits. Thus if L is a subposet (sublattice) of M and f:L $\rightarrow$ N is monotonic, then $\overline{f}$ exists and $\overline{f}|_L$ is f iff f preserves filtered colimits which in turn occurs iff f is continuous with the subspace Scott topology. This generalizes the remark after Example 3.9.

Section 4

In this section we relate the considerations of the last section to effective spaces with effective mappings. We will find that the approach of the previous section allows us to analyze why certain extension properties hold (and fail) and to characterize an important class of effective mappings in a new way.

Because we are dealing with effective spaces our considerations can no longer include general categories and must be limited to some categorical setting where an intrinsic internal effective structure is present. To be as general as possible, we will not fix any such category but will allow the reader to assume the objects of discussion come from some suitable setting such as the constructive f_0 spaces of Ersov or the effective domains of Scott or from the topos of recursive sets *Rec*. The obvious advantage of the latter is the freedom to construct higher type objects and to have a r.e. partial map classifier. So for example the object of total computable functions from **N** to itself exists in *Rec* but not in the category of constructive f_0 spaces.

We give a brief introduction to *Rec*. Consider the monoid of recursive functions with functional composition as the binary operation. Thought of as a category with one object, denoted N, arrows N $\longrightarrow$ N correspond to total recursive functions. *Rec* is defined to be the Grothendieck topos of canonical sheaves on the monoid.

Using Yoneda's lemma we can give a more intuitive description of *Rec*. An object X of *Rec* can be thought of as a set of paths satisfying a closure condition (closed under the recursive action). For example, if $N \xrightarrow{x} X$ is a path in X and f is a recursive function then x o f is a path in X. These paths satisfy a further glueing condition (sheaf condition). For example, if E and O denote the even and odd natural numbers and x and y are paths in X defined on E and O respectively then there exists a unique path on N which extends x and y.

Example 4.1. Following the usual convention we have denoted the representable functor Hom (, N) by N. Thus the set of paths for N consists of the total recursive functions. We can think of N as a non-standard natural number object and note further that the global or fixed elements of N concide exactly with the set of natural numbers **N**.

Example 4.2. Let X be an arbitrary set. The set of sequences in X, $X^{\mathbf{N}}$ is an object of *Rec*. The paths in $X^{\mathbf{N}}$ are just sequences $\{x_n\}$ of elements in X. Closure follows easily since $\{x_n\}$ a sequence in X and f recursive implies $\{x_{fn}\}$ is a sequence in X. Likewise if $\{x_n\}$ and $\{y_n\}$ are sequences in X then so is the sequence $x_0, y_0, x_1, y_1, \ldots$ showing glueing holds.

Arrows between objects of *Rec*, $X \xrightarrow{F} Y$, are natural transformations. Thus F takes paths in X to paths in Y and in so doing preserves the recursive action. For example, maps from N^N to N in *Rec* correspond to classical Banach-Mazur functionals. Thus arrows in *Rec* can be thought of as generalized Banach-Mazur functionals.

Any object X in *Rec* generates a r.e. partial map classifier, denoted $\tilde{X}_{re}$, which satisfies the following: any partial map $A \xrightarrow{\phi} X$ where A is an r-e subobject of N has a unique extension ϕ making the diagram

$$\begin{array}{ccc} A & \xrightarrow{\phi} & X \\ \downarrow & & \downarrow \\ N & \xrightarrow{\phi} & \tilde{X}_{re} \end{array}$$

a pullback. Note that $\tilde{X}_{re}$ is r.e. injective in *Rec*.

Example 4.3. $\tilde{N}_{re}$ along with a universal surjective path $N \longrightarrow \tilde{N}_{re}$ replaces $\tilde{N} = N \cup \{\bot\}$ in the continuous case. Paths in $\tilde{N}_{re}$ correspond to the partial recursive functions. We can form $\tilde{N}_{re}^{N}$, the object of partial recursive functions. Arrows $\tilde{N}_{re}^{N} \longrightarrow \tilde{N}_{re}$ correspond to the classical effective operations. We could continue inductively for higher types.

Example 4.4. Scott's category of effective domains and Ersov's category of enumerated sets are both fully embedded inside *Rec*. In each case N acts like the standard natural number object in these categories.

We can interpret in *Rec* the spaces (or categories) *R*, *PR*, *F* of total recursive, partial recursive and finite functions which, along with $\tilde{N}_{re}$, provide a nice environment for the analysis of effective functional properties. See [M1] for a more detailed description of *Rec* and its properties. Our arguments, however, remain external.

We begin by recalling a definition of classical recursion theory.

Definition 4.5. A functional F on *R* is an *effective operation* if there exists a partial recursive function ψ so that for all $x \in \mathbf{N}$, $\phi_x \in R$ implies

(a) $F(\phi_x)$ converges iff $\psi(x)$ converges
(b) $F(\phi_x)$ converges implies $F(\phi_x) = \psi(x)$.

F is called a *total effective operation* on *R* if in addition $\phi_x \in R$ implies $\psi(x)$ converges.

An effective operation on *PR* is defined in a similar manner.

Theorem 4.6 All the above mentioned effective operations as well as the standard recursive functionals and enumeration operators are representable in *Rec*.

Proof. See [M2] or [M3].

We apply the above to a concrete example. Consider the spaces *R*, *PR*, *F*, and $\tilde{N}_{re}$ as posets (with the usual effective relations between the objects) and therefore categories. (Note *PR* and $\tilde{N}_{re}$ are also constructive f_o spaces and effective domains.) We ask whether a right adjoint to the restriction functor $\tilde{N}_{re}^{R} \longleftarrow \tilde{N}_{re}^{PR}$ exists in a similar fashion to section 3. We emphasize these functor categories consist of effective mappings between the spaces, i.e. natural transformations in the sense defined earlier. In short, the answer is no, but let us examine an example carefully to see why.

Theorem 4.7 (Friedberg). There exists an effective operation $R \longrightarrow \tilde{N}_{re}$ which does not extend to a recursive functional on *PR*.

A quick glance at the proof in Rogers [R] shows that the effective operation defined by the partial recursive ψ is not continuous. What we must be careful about is not to assume that Friedberg's result provides a negative answer to our above question. It doesn't! Theorem 4.3 states there exists no *extension* of F. The lesson gleaned from section 3 is that our canonical construction of $\overline{F}$ gave a true extension only when F itself was continuous. In fact, a continuous functional $\overline{F}$ from *PR* to $\tilde{N}$ exists by Theorem 3.5 and $\overline{F}(\lambda x [0]) = \bot$ (so the contradiction that results in Rogers' proof no longer holds). $\overline{F}$, however, is not a morphism of *Rec* (and therefore not partial recursive) nor does there exist a largest effective mapping on *PR* whose restriction to *R* is less than F.

The above example provides a negative answer to the existence of a right adjoint because $\tilde{N}_{re}$ (or more generally, any constructively complete f_0 space or effective domain) is closed under r.e. sups but not under r.e. infs. Thus a functor $R \xrightarrow{F} \tilde{N}_{re}$ may induce a non-effective $F \xrightarrow{F^*} \tilde{N}$. This leads naturally to the following definition.

Definition 4.8. An effective functional $R \xrightarrow{F} \tilde{N}_{re}$ in *Rec* is *inf-effective* iff F* is an effective mapping, i.e. a morphism of *Rec*.

We are almost in a position to state when the restriction functor has a right adjoint. We need a lemma first.

Lemma 4.9. Functionals in *Rec* of the form $PR \longrightarrow \tilde{N}_{re}$ preserve r.e. filtered sups.

Proof. See [M2].

Theorem 4.10. The restriction functor $\tilde{N}_{re}^{R} \longleftarrow \tilde{N}_{re}^{PR}$ has a right adjoint when we restrict to the subcategory of $\tilde{N}_{re}^{R}$ consisting of inf-effective functors.

Proof.

The statement of the theorem is just the effective analogue of Theorem 3.5 applied to our particular example. *F* is r.e. filtered dense in *PR* and consists of the r.e. finitely presentable objects of *PR*. Thus if a functional $F: R \longrightarrow \tilde{N}_{re}$ is inf-effective, by Lemma 4.9 and the above discussion both F* and $\overline{F}$ exist and are effective.

Example 4.11. The subcategory of N^{R} consisting of inf-effective functors corresponds to the total effective operations on *R*.

The above example obviously holds for higher type effective operations as well. Also, the above results can be generalized to apply to other effective spaces where the notion of inf-effective would have to be suitably modified to the spaces at hand.

The author is presently working on such a generalization by internalizing results inside the topos *Rec*. We end with an example.

Example 4.12. All partial recursive functions are inf-effective. Restating, every functor in $\tilde{N}_{re}^{N}$ is inf-effective relative to $\tilde{N}$ (i.e. $\tilde{N} = N \cup \{\bot\}$ replaces *F* in this example). $\tilde{N}$ is r.e. filtered dense in $\tilde{N}_{re}$ and consequently the right adjoint to the restriction functor $\tilde{N}_{re}^{N} \longleftarrow \tilde{N}_{re}^{\tilde{N}_{re}}$ is defined as follows. For any non-constant partial recursive function ψ, $\overline{\psi}$ is just the effective left Kan extension of ψ. For ψ constant, the obvious constant extension does the trick.

References

[E] Ersov, Ju. Model C of Partial Continuous Functions, in *Logic Colloquium 76.* Amsterdam: North Holland, 1977.

[J] Johnstone, P. T. *Stone Spaces.* Cambridge: Cambridge University Press, 1982.

[ML] MacLane, S. *Categories for the Working Mathematician.* New York: Springer-Verlag, 1971.

[M1] Mulry, P. S. Generalized Banach-Mazur Functionals in the Topos of Recursive Sets, *Journal of Pure and Applied Algebra,* 26 (1982), 71-83.

[M2] Mulry, P. S. Adjointness in Recursion, *Annals of Pure and Applied Logic,* 32 (1986).

[M3] Mulry, P. S. A Categorical Approach to the Theory of Computation. Preprint, 1986.

[R] Rogers, H. *Theory of Recursive Functions and Effective Computability.* New York: McGraw-Hill, 1967.

[S1] Scott, D. Continuous Lattices, in *Toposes, Algebraic Geometry and Logic.* New York: Springer-Verlag, 1972.

[S2] Scott, D. *Lectures on a Mathematical Theory of Computation.* Technical Monograph PRG-19. Oxford University, 1981.

Part II

Structure Theory of Continuous Posets and Related Objects

THE VERSATILE CONTINUOUS ORDER

Jimmie D. Lawson
Department of Mathematics
Louisiana State University
Baton Rouge, Lousiana 70803

Abstract

In this paper we survey some of the basic properties of continuously ordered sets, especially those properties that have led to their employment as the underlying structures for constructions in denotational semantics. The earlier sections concentrate on the order-theoretic aspects of continuously ordered sets and then specifically of domains. The last two sections are concerned with two natural topologies for sets with continuous orders, the Scott and Lawson topologies.

The purpose of this article is to survey some of the principal aspects of the theory of domains and continuous orders. A significant portion of the theory has arisen in an attempt to find suitable mathematical structures to serve as a framework for modeling concepts and constructions from theoretical computer science. In the course of our presentation, we try to include (in a rather sketchy way) some of this background.

There are certain distinctive and novel mathematical ideas that have grown out of the theory of continuous orders that appear to be worthwhile mathematical contributions independent of whatever future role the theory may play in the field of theoretical computer science. A basic aim of this paper is to point out explicitly some of these unique features of the theory. The knowledgable reader will not find much new herein (although here and there we include some improvements on earlier results); rather we seek to present some of the basic ideas for the non-expert in an illustrative and comprehensible manner. A significant departure from the treatment given in [**COMP**] is our emphasis on continuous partially ordered sets instead of continuous lattices. This is consistent with the way that the theory has developed since the publication of that book.

I. Complete Partial Orders

Plato envisioned an "ideal" world where everything existed in perfection; in contrast, the objects of our universe were viewed as approximations to the ideal. Similarly in mathematics we postulate ideal abstractions (e.g. the points, lines, etc., of geometry) of physical objects or phenomena. An idea of ordering comes into play when we consider whether one approximation to an ideal is better or worse than (or incomparable to) another. It is often the case that the approximations determine the ideal in the sense that the ideal is the limiting case of increasingly finer approximations (we could consider, for example, algorithms that compute the number π to increasingly better accuracy). It would, of course, be pretentious to claim that a certain mathematical model captured these philosophical and metamathematical ideas better than any other or that one specific approach was the most useful for treating them. These ideas are helpful, however, for gaining intuitive insight into many of the mathematical structures and constructions that appear in the following.

We first introduce the mathematical concepts that model the idea of increasingly better approximations or stages of a computation.

Let $(U, \leq)$ be a partially ordered set. An ω-**chain** is an increasing sequence, a **chain** is a totally ordered subset, and a **directed** set is one for which any two elements have an upper bound.

An ω-chain: $x_1 \leq x_2 \leq x_3 \leq \ldots$
A chain C: $x, y \in C \Rightarrow x \leq y$ or $y \leq x$
A directed set D: $x, y \in D \Rightarrow \exists z \in D$ such that $x, y \leq z$.

Completeness in the current context means the existence of "ideal" objects or the limits of computations. From this viewpoint, a distinctive feature of the following systems is that they incorporate *both* the ideal objects and the approximations. It is a standard (though non-trivial) result that completeness with respect to chains is equivalent to completeness with respect to directed sets (see e.g., [**Ma76**]). Of course, ω-completeness is a slightly weaker concept, but is more suitable for recursive considerations.

A **chain-complete partially ordered set** (CPO) is a partially ordered set U such that every directed subset (equivalently, every chain) of U has a least upper bound in U. Frequently one postulates a least element $\perp$ also.

We can view the suprema of directed sets as "ideal" elements in our universe U, and the elements of the directed sets as increasingly better approximations or as stages in a "computation".

A **continuous** morphism (or CPO-morphism) is an order preserving function which also preserves suprema of directed sets (equivalently, suprema of chains). These may be viewed as the "computation-preserving" functions.

Ideal Completions

Let $(P, \leq)$ be a partially ordered set. A subset I of P is an **ideal** if

(i) $I = \downarrow I$, where $\downarrow I = \{y \in I \colon \exists x \in I$ such that $y \leq x\}$,
(ii) I is directed.

Let $Id(P)$ denote the set of ideals of P ordered by inclusion. Then $Id(P)$ is a CPO and the mapping $i: P \to Id(P)$ defined by $x \mapsto \downarrow x$ is an order embedding. The pair $(Id(P), i)$ is called the **ideal completion** of P.

The ideal completion can be characterized alternately as arising from the adjoint functor to the forgetful functor from the category of CPO's and continuous morphisms to the category of partially ordered sets and order preserving functions. Or it may be characterized in terms of the universal property that any order preserving mapping α from P to a CPO Q extends uniquely to a CPO mapping $\hat{\alpha}$ from $Id(P)$ to Q, which sends an ideal I to $\sup \alpha(I)$ (see the paper of Markowsky and Rosen [**MR76**] for this and other results on the ideal completion).

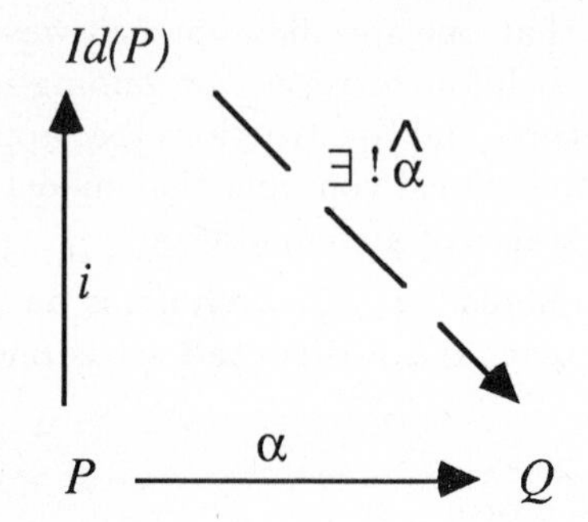

Example. *The Cantor Tree.*

Consider the set P of all finite strings of $\{0,1\}$ (including the empty string). We take the prefix order and form the ideal completion $Id(P)$. $Id(P)$ can be identified with all finite and infinite strings of $\{0,1\}$. The longer finite strings give increasingly better approximations to the infinite strings. Alternately $Id(P)$ can be identified with the finite and infinite points in the free dyadic tree (with P being embedded as the finite points).

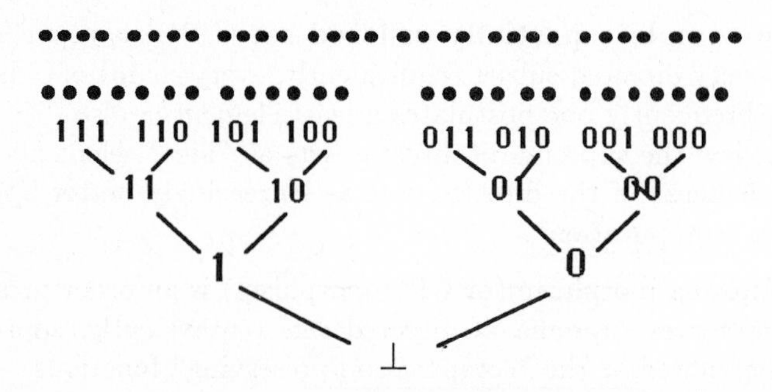

Figure 1. The Cantor Tree

Such constructions as those arising in the previous example also have other interesting interpretations. Suppose that $(X_n, f_n^m: X_n \to X_m)$ is an inverse system of finite sets indexed by the integers. For $m < n$, we define $x < y$ for $x \in X_m$

and $y \in X_n$ if $f_n^m(y) = x$. (In the previous example the sets X_m are the strings of length m and f_n^m sends a string of length n to the prefix of length m.) The ideal points then arise at the top and may be viewed as the (inverse) limit of the system.

Special classes of partially ordered sets often have important alternate characterizations in terms of their ideal completions. If P itself is a CPO and we consider the identity mapping on P, then this extends to a continuous morphism SUP: $Id(P) \to P$ which sends an ideal to its supremum. Observe that the composition $\mathrm{SUP} \circ i = 1_P$ and $i \circ \mathrm{SUP} \geq 1_{Id(P)}$. The existence of an order preserving function from $Id(P)$ to P with these composition properties characterizes CPO's (we leave the straightforward verification of this assertion as an exercise for the reader). Note, however, in this case that the inclusion mapping $i: P \to Id(P)$ is not continuous.

An important class of CPO's are what have been called domains. There is some variation in the way they are defined. We define a **domain** to be a partially ordered set which is order-isomorphic to the ideal completion of a partially ordered set. Frequently one restricts oneself to **countably based domains**, which are order-isomorphic to the ideal completion of a countable partially ordered set. The countability restriction lends itself to recursive considerations.

Fixed Points

The Banach Contraction Theorem states that a contraction in a complete metric space has a unique fixed point and is used (among other things) to guarantee the existence of solutions of differential equations. A somewhat analogous role is played by the Tarski Fixed Point Theorem for CPO's and shows why chain-completeness is an appropriate form of completeness in the current context.

Proposition. *Let P be a CPO with $\bot$, and let $f: P \to P$ be a continuous morphism. Then f has a least fixed point.*

Proof. $\bot \leq f(\bot) \leq f^2(\bot) \ldots \leq f^n(\bot) \ldots$. The sup of this sequence is the least fixed point. ∎

This proposition is the basis for inductive or recursive constructions and solutions of equations in CPO's. The construct to be defined or solution to some equation is given by the least fixed point of an appropriate continuous function.

Example. Suppose we wish to solve the equation $W = 01W$ on the set of finite and infinite words with alphabet $\{0, 1\}$ (which may be identified with the Cantor tree, the ideal completion of the free dyadic tree). To find a solution, we consider the continuous transformation $T(W) = 01W$. The solution is the least fixed point of the transformation T, and consists of the infinite word consisting of repeated 01's.

More generally one can order a (universal) set of countably based domains by inclusion to form a domain of domains and use the Tarski theorem for finding solutions to domain equations (see [**WL84**] for a nice presentation of this approach).

II. A New Kind of Order

A significant contribution of the theory of continuous orders has been the explicit definition and use of a new order relation, one that sharpens the traditional notion of order.

Let P be a CPO, x,y in P. We say x is **essentially below** y (traditionally, "way-below"), written $x \ll y$, if given a directed set $D \subseteq P$ such that $y \leq \sup D$, then $x \leq d$ for some $d \in D$.

If we think of D as a computation of y, then D yields x at some "finite" stage. In the Cantor tree, $x \ll y$ if and only if x is a finite prefix of y. (Hence the alternate terminology x is "finite" in y.) In this context the idea of essentially below is that any method of "approximating" or "computing" y must give x at some stage along the way.

A partially ordered set P is a **continuous CPO** if it is a CPO (complete with respect to directed sups) and satifies

$$y \in P \Rightarrow y = \sup\{x : x \ll y\} = \sup \Downarrow y,$$

and the set on the right is directed.

Every element of P can be "approximated" by or "computed" from the elements essentially below it.

We remark that if $\Downarrow y$ contains a directed set with supremum y, then from the definition of $\ll$ it follows that this set must be cofinal in $\Downarrow y$, and hence that $\Downarrow y$ is itself directed.

The following example of D. Scott [**Sc76**] nicely illustrates these ideas.

Example. *A Data Type Structure.*

Let $\mathbb{R}^* = \mathbb{R} \cup \{-\infty, \infty\}$, and let D^* consist of all closed intervals $[\underline{x}, \overline{x}]$ where $\underline{x} \leq \overline{x}$ for $\underline{x}, \overline{x} \in \mathbb{R}^*$. We partial order D^* by reverse inclusion. Then D^* is a complete semilattice containing the least element $\perp = \mathbb{R}^*$, the "perfect" reals–the one-point intervals–$x = [x, x]$, and the "approximate" reals $[\underline{x}, \overline{x}]$ with $\underline{x} < \overline{x}$.

The interval $[\underline{y}, \overline{y}]$ is a "better approximation" to $r \in R^*$ or "contains more information" than $[\underline{x}, \overline{x}]$ if $\underline{y} \leq r \leq \overline{y}$ and $[\underline{x}, \overline{x}] \leq [\underline{y}, \overline{y}]$, i.e., $[\underline{y}, \overline{y}] \subseteq [\underline{x}, \overline{x}]$.

If $[\underline{x}, \overline{x}] \in D^*$ and $\underline{x} < r < \overline{x}$, then any algorithm for computing r will eventually give an upper bound $\leq \overline{x}$ and a lower bound $\geq \underline{x}$. Hence the interval of error will be greater than $[\underline{x}, \overline{x}]$ in the order on D^*. Thus $[\underline{x}, \overline{x}] \ll [r, r] = r$.

Other Completeness Properties

The requirement that the set $\Downarrow y = \{x : x \ll y\}$ be directed is sometimes difficult to check and is often not preserved by standard constructions on CPO's. Thus it is frequently convenient or necessary to work with partially ordered sets with stronger completeness properties so that the directedness happens automatically.

It is immediate from the definition that if $a \ll y$, $b \ll y$, and $c = \sup\{a, b\}$, then $c \ll y$. Thus if any two elements below y have a supremum, it follows that the set of elements essentially below y *must* be directed. We consider some important classes where this holds.

A CPO is said to be **bounded complete** if any two elements that are bounded above have a least upper bound. Note that since a CPO is already chain-complete, it follows that any non-empty subset which has an upper bound must have a least upper bound. It follows that any interval $[a, b] = \{x: a \leq x \leq b\}$ is a complete lattice in the restricted order, and indeed this property is an alternate characterization of bounded complete CPO's. Thus the class of **bounded complete continuous CPO's** is a convenient class of continuous CPO's.

A semilattice is said to be **complete** if it is chain complete (i.e., a CPO) and every non-empty subset has an infimum. Another important subclass is the class of **continuous complete semilattices**, continuous CPO's which are also complete semilattices. It is an elementary exercise to show that complete semilattices are bounded complete and conversely that a bounded complete CPO with bottom $\perp$ is a complete semilattice. (Similarly a CPO P is bounded complete iff $P_\perp$, the set P with a bottom element adjoined, is a complete semilattice.) Hence the complete semilattices are precisely the bounded complete CPO's with $\perp$. These are sometimes alternately referred to as being **consistently complete**. It is sometimes required as part of the definition of a domain that it be consistently complete.

Another important class is the class of **continuous lattices**, complete lattices with a continuous order. These are the bounded complete continuous CPO's that have both a top and a bottom. A CPO is a bounded complete continuous CPO iff when a top and bottom are adjoined to it, a continuous lattice results.

Rounded Ideals

We consider some of the basic properties of the relation $\ll$ in a continuous CPO.

(1) $a \ll b \Rightarrow a \leq b$

(2) $a \ll d,\ b \ll d \Rightarrow \exists c$ such that $a, b \leq c$ and $c \ll d$

(3) $a \leq b \ll c \leq d \Rightarrow a \ll d$

(4) $a \ll c \Rightarrow \exists b$ such that $a \ll b \ll c$

(5) $\perp \ll a$

The fourth property plays a crucial role in the theory and is referred to as the "interpolation" property.

An arbitrary relation on a poset P that satisfies the preceding axioms is called an **auxiliary** relation. If P is a poset equipped with an auxiliary relation, then we can consider the set of "rounded" ideals, i.e., those ideals I satisfying $x \in I \Rightarrow \exists y \in I$ such that $x \ll y$. These ideals ordered by inclusion form a completion, called the **rounded ideal completion**. This completion is a continuous CPO and P maps into the completion by sending x to $\{y: y \ll x\}$. This completion generalizes the ideal completion. (Indeed the ideal completion arises if $\ll$ is chosen as $\leq$.)

Example. The data structure example is the rounded ideal completion of the approximate rational data structure consisting of those intervals $[p, q]$ with p, q rational and $p < q$ with the inherited order and essentially below relation.

If one wishes to consider the notion of a recursively defined continuous CPO, then one can consider those continuous CPO's which arise as the rounded ideal completion of a countable set with a recursively defined partial order and auxiliary relation. More directly, we say that a continuous CPO P is **countably based** if there exists a countable subset B of P such that $p \ll q$ in P implies there exists $b \in B$ with $p \ll b \ll q$.

Locally Compact Spaces

We consider another illustrative example of naturally occuring continuous orders. The results of the next two sections are mainly drawn from [**HL78**] or [**COMP**, Chapter V].

Let X be a topological space, let $O(X)$ denote the lattice of open sets, and let $U, V \in O(X)$. Then $U \ll V$ iff for every open cover of V, there is a finite subcollection that covers U. In this context it seems appropriate to say that U is **compact** in V.

We say that X is **core compact** if given $x \in V \in O(X)$, there exists U open, $x \in U \subseteq V$, such that U is compact in V.

Theorem. *X is core compact $\Leftrightarrow$ $O(X)$ is a continuous lattice.*

For Hausdorff spaces, these are precisely the locally compact spaces. They appear to be the appropriate generalization of local compactness to the non-Hausdorff setting (in the sense that many basic mapping properties of locally compact spaces are retained in this setting). For example, X is core compact iff $1_X \times f: X \times Y \to X \times Z$ is a quotient mapping whenever $f: Y \to Z$ is a quotient mapping [**DK70**]. Also appropriate modifications of the compact-open topology for function spaces exist so that one gets an equivalence between $[X \times Y, Z]$ and $[X, [Y, Z]]$ if Y is core compact (see [**COMP**, Chapter II] and for later developments, [**SW85**] or [**LP85**]). Of course this equivalence is closely related to cartesian closedness, a topic to which we return at a later point.

Spectral Theory

Spectral theory seeks to represent a lattice as the lattice of open sets of a topological space. However, the constructions are more intuitive if one works with the lattice of closed sets. We take this approach initially, and set everything on its head at a later stage.

Suppose that X is a T_1-space, and let L be the lattice of closed sets. We let $\hat{X}$ denote the set of atoms in L (which correspond to the singleton subsets of X) and topologize $\hat{X}$ by defining a closed set to be all the atoms below a fixed member of the lattice L, i.e., $\{\,\{x\}: \{x\} \subseteq A\}$ where A is a closed subset of X. Then the mapping from X to $\hat{X}$ which sends an element to the corresponding singleton set is a homeomorphism. Thus X may be recovered (up to homeomorphism) from the lattice of closed sets.

The situation becomes more complex (and more interesting) for a T_0-space X. In this case we let an element of X correspond to the closure of the corresponding singleton set in the lattice L of closed sets. The fact that X is T_0 is precisely the condition needed for this correspondence to be one-to-one. But how does one distinguish in a lattice-theoretic way the closed sets that arise in this fashion? One easily verifies that sets that are closures of points are **irreducible**, i.e., not the union of two strictly smaller closed sets. We are thus led to define the cospectrum, Cospec(L), to be the set of coprime elements (p is **coprime** if $p \leq \sup\{x, y\}$ implies $p \leq x$ or $p \leq y$) equipped with the **hull-kernel** topology with closed sets of the form $hk(a) = \{p \in L\colon p \text{ is coprime}, p \leq a\}$.

A space is **sober** if every irreducible closed set is the closure of a unique point. In precisely this case the embedding of X into the cospectrum of the closed sets is a homeomorphism. For any topological space X, there is a largest T_0-space $\hat{X}$ having the same lattice of closed (or open) sets, called the **sobrification** of X. The sobrification of X can be obtained by taking $\hat{X}$ to be the cospectrum of the closed sets; X maps to the sobrification by sending a point to its closure. It can be shown that a space is core compact iff its sobrification is locally compact. (A space is **compact** if every open cover has a finite subcover, and **locally compact** if every (not necessarily open) neighborhood of a point contains a compact neighborhood of that point.)

We now dualize the preceding notions to the lattice of open sets. An element $p \in L$, $p \neq 1$ is **prime** (resp. **irreducible**) if $x \wedge y \leq p \Rightarrow x \leq p$ or $y \leq p$ (resp. $x \wedge y = p \Rightarrow x = p$ or $y = p$). It can be shown that the irreducible elements of a continuous lattice order generate (i.e., every element is an infimum of such elements) and that the prime elements of a distributive continuous lattice order generate.

The collection of sets of the form PRIME $L \cap \uparrow x$ for $x \in L$ forms the closed sets for a topology on PRIME L, called the **hull-kernel** topology. PRIME L equipped with the hull-kernel topology is called the **spectrum** of L, and denoted Spec L. The following theorem results by showing that the spectrum is sober (which is always the case) and locally compact when L is continuous.

Theorem. *Given any continuous distributive lattice L, there exists a locally compact sober space X (namely the spectrum) such that L is order-isomorphic to $O(X)$.*

As a consequence of the preceding considerations there results a duality between distributive continuous lattices and locally compact sober spaces.

III. Domains

In this section we turn our attention to domains, which have been the most important structures in the theory of continuous CPO's from the viewpoint of computer science.

In earlier sections we have seen how to obtain a distributive continuous lattice as the lattice of open sets of a locally compact sober space and conversely how to obtain a locally compact sober space as the spectrum of such a lattice. Such inverse constructions (and the dualities to which they often lead) are a pervasive feature of the theory of continuous orders. We turn now to an inverse construction for the ideal completion.

An element $k \in P$ is **compact** if $k \ll k$, i.e., if $\sup D \geq k$ for D directed, then $k \leq d$ for some $d \in D$. A CPO P is **algebraic** if every element is a directed sup of compact elements.

Note that algebraic CPO's are a special subclass of the class of continuous CPO's. In an algebraic CPO the relation $\ll$ is characterized by $x \ll y$ iff there exists a compact element k such that $x \leq k \leq y$.

For any partially ordered set P, the ideal completion $Id(P)$ is an algebraic poset, and P embeds in $Id(P)$ as the compact elements. Conversely, if Q is an algebraic poset, one may consider the subset of compact elements $K(Q)$ with the inherited order. The operators Id and K are inverse operators between the class of all posets and the class of algebraic CPO's (in the sense that if one is applied and then the other, one obtains a poset order isomorphic to the original.) It follows that the class of algebraic CPO's is precisely the class of domains. Indeed it is customary to define a domain to be an algebraic CPO instead of the way we have done it.

When one is working in the context of algebraic CPO's, properties of continuous CPO's can generally be given alternate characterizations in terms of the partially ordered set of compact elements. For example, an algebraic CPO is countably based iff the set of compact elements is countable.

This "object" correspondence can be extended to a functor for various categories. If continuous morphisms $f: P \to Q$ are considered in the class of domains, then for the corresponding morphisms in the poset category one considers the restrictions of their subgraphs (i.e., the relation $\{(x, y) \in K(P) \times K(Q): y \leq f(x)\}$).

If a domain (i.e., algebraic CPO) is a complete semilattice, respectively, complete lattice, then it is called an **algebraic semilattice**, respectively, **algebraic lattice**. Algebraic semilattices are the consistently complete domains. If P is an algebraic semilattice, then (since the sup of two compact elements is again compact) $K(P)$ is also consistently complete. Hence $K(P)$ can be characterized as a partially ordered set with $\perp$ such that $\downarrow x$ is a sup semilattice for each x. Such sets have the property that their ideal completions are algebraic semilattices. Further restricting these inverse constructions, one obtains that the set of compact elements of an algebraic lattice forms a sup semilattice with bottom, and the ideal completion of a sup semilattice with bottom is an algebraic lattice.

Information Systems and Convexity

One is interested in trying to present some version of domain theory in as intuitive and usable form as possible. One recent approach has been in terms of informations systems.

An **information system** is a triple $\mathbf{A} = (A, Con, \Rightarrow)$ for which

1) A is a non-empty set (whose members are called "tokens" and thought of as statements or items of information);
2) Con is a family of *finite* subsets of A (the finite "consistent" subsets) satisfying
   i) $Y \subseteq X \in Con$ implies $Y \in Con$,
   ii) $\{a\} \in Con$ for all $a \in A$;
3) $\Rightarrow$ is a relation (a subset of $Con \times A$) satisfying
   iii) If $X \in Con$ and $a \in A$, then $X \cup \{a\} \in Con$ if $X \Rightarrow a$,

iv) If $X, Y \in Con$ and $c \in A$, and if $X \Rightarrow b$ for all $b \in Y$, and if $Y \Rightarrow c$, then $X \Rightarrow c$.

An arbitrary subset $Y \subseteq A$ is **consistent** if $X \in Con$ for every finite subset $X \subseteq Y$. The set Y is **deductively closed** if $X \subseteq Y$, X finite, and $X \Rightarrow a$ together imply $a \in Y$. The **elements** of **A**, denoted $|\mathbf{A}|$, are defined by

$$|\mathbf{A}| = \{Y \subseteq A : Y \text{ is consistent and deductively closed}\}.$$

The elements form a consistently complete domain, i.e., an algebraic semilattice. The compact elements are the smallest deductively closed subsets containing a finite set, which is just the set of tokens implied by the finite set. In general, there is no way of recovering the information system from the domain of elements (indeed distinct information systems may lead to isomorphic domains of elements). However, there is associated with each consistently complete domain (=algebraic semilattice) in a canonical way an information system which gives rise to the domain; indeed the construction is an alternate formulation of the inverse constructions between domains and partially ordered sets given in the preceding section. Given an algebraic semilattice D, one defines the tokens to be the compact elements, the sets in Con to be those finite subsets that have an upper bound in D, and the relation $\Rightarrow$ by $F \Rightarrow a$ if $a \leq \sup F$. One verifies that this is an information system whose set of elements is order isomorphic to the original domain.

One may alternately view information systems as abstract convexities. If A is a (not necessarily convex) subset of a convex set X, then one defines Con to be all finite subsets of A whose convex hull is a subset of A; one defines $F \Rightarrow a$ to mean that a is in the convex hull of F. The structure so obtained is an information system; the set of elements consists of all convex sets contained in A. An **abstract convexity** on a set X consists of a collection of subsets that are closed under arbitrary intersections and directed unions. These are all the properties one needs to show that any subset A gives rise to an information system as just defined, and it is not hard to see that the notions of information system and abstract convexity are equivalent notions via this correspondence. Abstract convexities have been investigated in the dissertation of Robert Jamison [**J**]; numerous papers of Marcel van de Vel deal with various aspects of abstract convexities in topological spaces, see e.g. [**VDV85**].

Retracts and Projections

Retracts play an important role in the theory of continuous CPO's. We consider some of their most basic properties.

Let P be a CPO. An **(internal) retraction** is a continuous morphism $r: P \to P$ such that $r \circ r = r$. It was Scott's observation that a continuous retract of a continuous lattice is again a continuous lattice [**Sc72**], and the proof carries over to continuous CPO's.

Proposition. *Let P be a continuous CPO and let $r: P \to P$ be a retraction. Then $r(P)$ is a continuous CPO, and the inclusion $j: r(P) \to P$ is continuous.*

Proof. Let $A = r(P)$. If D is a directed set in A, let $w = \sup D$ in P. Then $r(D) = D$, so by continuity $r(w) = \sup r(D) = \sup D$. Hence $w = r(w)$. Thus A is closed in P with respect to directed sups. This in turn implies that A is a CPO and the inclusion $j: A \to P$ is continuous.

Let $y \in A$. If $x \ll y$ in P, we claim $r(x) \ll y$ in A. Let D be a directed set in A with $\sup D \geq y$. Then $x \leq d$ for some $d \in D$. Hence $r(x) \leq r(d) = d$, and thus $r(x) \ll y$ in A. Clearly the set of all $r(x)$ for $x \ll y$ is directed and has supremum $r(y) = y$. ∎

A CPO A is a **retract** of a CPO P if there exist continuous morphisms $r: P \to A$ and $j: A \to P$ such that $r \circ j = 1_A$. In this case the function r is called an **(external) retraction**. Note that $j \circ r$ is an internal retraction on P and that $j: A \to j(A)$ is an order isomorphism. Thus the previous proposition yields

Corollary. *A retract of a continuous CPO is a continuous CPO.*

A special type of (external) retraction is the **projection**, where in addition to the preceding conditions we require that $j \circ r \leq 1_P$. In this case we write $P \overset{r}{\rightleftharpoons} Q$. If r is a projection, then j is unique, is automatically continuous, and is given by $j(y) = \inf\{x: r(x) \geq y\}$.

Continuous CPO's have an alternate characterization in terms of their ideal completions, namely a CPO P is continuously ordered if and only if the mapping SUP: $Id(P) \to P$ is a projection. The continuous embedding $j: P \to Id(P)$ is given by $j(x) = \Downarrow x$, which is the smallest ideal with supremum greater than or equal to x.

It follows that every continuously ordered set is the retract of a domain and that the class of continuously ordered sets is the smallest class of CPO's that contains the domains and is closed with respect to taking retracts.

IV. The Scott Topology

A distinctive feature of the theory of continuous orders is that many of the considerations are closely interlinked with topological and categorical ideas. The result is that topological considerations and techniques are basic to significant portions of the theory.

The Scott topology is the topology arising from the convergence structure given by $D \to x$ if D is a directed set with $x \leq \sup D$. Thus a set A is **Scott closed** if $A = \downarrow A$ and if $D \subseteq A$ is directed, then $\sup D \in A$. Similarly U is **Scott open** if $U = \uparrow U$ and $\sup D \in U$ for a directed set D implies $d \in U$ for some $d \in D$.

By means of the Scott topology one can pass back and forth between an order-theoretic viewpoint and a topological viewpoint in the study of CPO's. Generally order-theoretic properties have corresponding topological properties and vice-versa. For example, continuous morphisms between CPO's are precisely those functions which are continuous with respect to the Scott topologies.

Example. The Scott-open sets in $\mathbb{R}^*$ consist of open right rays. For a topological space X, the set of Scott-continuous functions $[X, \mathbb{R}^*]$ consists of the lower semicontinuous functions.

Suppose that a CPO P is equipped with the Scott topology, so that it is now a topological space. Then the original order may be recovered from the topological space as the **order of specialization**, which is defined by $x \leq y$ iff $x \in \overline{\{y\}}$. Note that any topological space has an order of specialization, and that this order is a partial order precisely when the space is T_0.

There are useful alternate descriptions of the Scott topology for special classes of CPO's. For a continuous CPO P, let $\Uparrow z = \{x : z \ll x\}$. It follows from the interpolation property that $\Uparrow z$ is a Scott open set. That these form a basis for the Scott topology follows from the fact that each $x \in P$ is the directed supremum of $\Downarrow x$. It follows that a continuous CPO is countably based iff the Scott topology has a countable base. Alternately the Scott open filters also form a basis for the Scott topology in a continuous CPO.

For domains, a basis for the Scott open sets is given by all sets of the form $\uparrow z$, where z is a compact element. The argument is analogous to the continuously ordered case.

Topological Properties

Given a partially ordered set P, there are a host of topologies on P for which the order of specification agrees with the given order. The finest of these is the **Alexandroff discrete** topology, in which every upper set is an open set, and the coarsest of these is the **weak** topology, in which $\{\downarrow x : x \in P\}$ forms a subbasis for the closed sets. The Scott topology is the finest topology giving back the original order with the additional property that directed sets converge to their suprema. It is this wealth of topologies that makes the study of CPO's from a topological viewpoint (as opposed to an order-theoretic viewpoint) both richer and more complex.

What spaces arise by equipping continuous CPO's with the Scott topology? A result of Scott's [**Sc72**] asserts that continuous lattices equipped with the Scott topology are precisely the injective T_0-spaces (a continuous function from a subspace A of a T_0-space X into L extends to a continuous function on all of X). In general, a continuous CPO equipped with the Scott topology gives rise to a locally compact, sober (T_0-)space. (A base of compact neighborhoods of x in this case is given by $\uparrow z$ for all $z \ll x$.) Indeed, the lattice of Scott-open sets in this case is a completely distributive lattice (a lattice is completely distributive if arbitrary joins distribute over arbitrary meets and vice-versa; these are a special class of distributive continuous lattices). Conversely the spectrum of a completely distributive lattice turns out to be a continuous CPO (with respect to the order of specification) equipped with the Scott topology. Hence another characterization of continuous CPO's equipped with their Scott topologies is that they are the spectra of completely distributive lattices (see [**La79**] or [**Ho81a**]).

There is an inclusion functor from category of sober spaces into the category of T_0-spaces and there is a functor from the category of T_0-spaces into the category of partially ordered sets which sends a space to the order of specialization. Both of these functors have adjoints. The adjoint functor for the order of specialization functor equips a partially ordered set with the Alexandroff discrete topology. The functor sending a space to its sobrification is the adjoint of the inclusion of sober spaces into T_0-spaces. The composition of these two functors sends a partially ordered set to the sobrification of the Alexandroff discrete topology, which turns out to be the ideal completion equipped with the Scott topology. Thus the sobrification of the Alexandroff discrete topology gives a topological analog of the ideal completion. The topological retracts of these sobrified Alexandroff discrete spaces are the retracts in our earlier sense and, as we have seen previously, are the continous CPO's. (These results appear in [**Ho81b**].)

Function Spaces

A crucial and characteristic property of countably based continuous CPO's is that they are closed under a wide variety of set-theoretic operations. This allows one to carry along a recursive theory. Such constructions break down in the category of sets because one obtains sets of larger cardinality. Also one can employ these stability features of continuous CPO's to obtain examples which reproduce isomorphic copies of themselves under a variety of set-theoretic operations. (This is essentially the idea of solving domain equations.) It is these features that provide strong motivation for moving from the category of sets to some suitable category of domains or continuous CPO's.

One of the most basic constructs is that of a function space. If X and Y are CPO's, then $[X \to Y]$ denotes the set of continuous morphisms (the order preserving functions which preserve suprema of directed sets) from X to Y. For a directed family of continuous morphisms, the pointwise supremum is again continuous. So the set $[X \to Y]$ with the pointwise order is again a CPO.

For topological spaces X and Y let $[X \to Y]$ denote the set of continuous functions from X to Y. If X or Y is a CPO, then we identify it with the topological space arising from the Scott topology. If Y is a CPO, then $[X \to Y]$ is also a CPO with respect to the pointwise order on functions. One verifies that the supremum of a directed family of continuous functions is again continuous, so directed suprema are computed pointwise in $[X \to Y]$. If X and Y are both CPO's equipped with the Scott topology, then the function space $[X \to Y]$ is just the set of continuous morphisms of the previous paragraph.

Suppose additionally that X is a continuous CPO. Let $f: X \to Y$ be a (not necessarily continuous) order preserving function. Then there exists a largest continuous morphism $\underline{f}: X \to Y$ which satisfies $\underline{f} \leq f$; $\underline{f}$ is given by $\underline{f}(x) = \sup\{f(z): z \ll x\}$. Thus if Y^X denotes the set of all order-preserving functions from X to Y, the mapping $f \to \underline{f}$ from Y^X to $[X \to Y]$ is a projection. If X is an algebraic CPO, then $\underline{f}$ is the unique continuous extension of the restriction of f to the set of compact elements $K(X)$.

Under what conditions will $[X \to Y]$ be a continuous CPO? Let us first consider the case that $Y = 2$, where $2 = \{0,1\}$ denotes the two-element chain with $0 < 1$ equipped with the Scott topology (sometimes called the Sierpinski space). Then $f: X \to 2$ is continuous iff f is the characteristic function of an open set of X. Hence there is a natural order isomorphism between $O(X)$, the lattice of open sets, and $[X \to 2]$. Since $O(X)$ is a continuous lattice iff X is core compact, we conclude that the same is true for $[X \to 2]$.

More generally, let us suppose that X is core compact and that Y is a continuous CPO with $\bot$. Let $f \in [X \to Y]$, $a \in X$, and $f(a) = b$. Let $z \ll b$. Pick U open in X containing a such that $f(U) \subseteq \Uparrow z$ (which we can do since f is Scott continuous). Pick V open with $a \in V$ such that $V \ll U$. Define $g \in [X \to Y]$ by $g(x) = z$ if $x \in V$ and $g(x) = \bot$ otherwise. It is straightforward to verify that $g \ll f$ in $[X \to Y]$ (see [**COMP**, Exercise II.4.20]) and that f is the supremum of such functions. However, one needs additional hypotheses on X and/or Y to be able to get a directed set of such functions. If S is a continuous complete semilattice, then one can take finite suprema of such functions g and obtain the principal implication of

Theorem. *Let S be a non-trivial CPO equipped with the Scott topology. Then $[X \to S]$ is a continuous complete semilattice iff X is core compact and S is a continuous complete semilattice.*

The proof of the reverse implication follows from the fact that $O(X) \cong [X \to 2]$ and S are both retracts of $[X \to S]$ (see [**COMP**, Section II.4]).

If additionally X is compact, then one only needs that S is bounded complete. (Apply the preceding theorem to $[X \to S_\bot]$ and note that if $f(X) \subseteq S$, then by compactness the sup of finitely many of the g's constructed earlier must also not take on the value $\bot$, and hence give an element essentially below f in $[X \to S]$.)

Problem. Suppose P is a continuous CPO and that $[X \to P]$ is a continuous CPO for all core compact spaces X. Is P a continuous complete semilattice?

If P is a CPO without a bottom element, then it is frequently more appropriate to consider the function space $[X \rightharpoonup P]$ of all continuous partial functions with domain some open subset of X. This function space corresponds to $[X \to P_\bot]$ (by extending a partial function to be $\bot$ where not defined), and by this device it can be treated as a full function space. Hence the preceding theorem yields that if X is core compact and P is a bounded complete continuous CPO, then $[X \rightharpoonup P]$ is a continuous complete semilattice.

It is frequently desirable to model the notion of self-application (we may think of programs that act on other programs, including themselves, or programming languages that incorporate the λ-calculus, where objects are also functions and vice-versa). This involves building spaces X homeomorphic to $[X \to X]$. These can be constructed in suitable subcategories of continuously ordered sets by using projective limit constructions, where the bonding maps are projections. This was the original approach of Scott in [**Sc72**]. In these constructions one needs to know that the function space $[X \to X]$ is back in the category that one is considering, that one has natural projections from $[X \to X]$ to X, and that taking function space is preserved by inverse limits. Similar remarks apply for trying to solve other types of domain equations by the technique of projective limits (see [**COMP**, Chapter IV]). The preceding theorem shows that continuous complete semilattices form a good category in this regard (since $[X \to X]$ is another such).

Cartesian Closedness

Let X, Y, Z be sets and let $\alpha: X \times Y \to Z$, and define $\hat{\alpha}: X \to [Y \to Z]$ by $\hat{\alpha}(x)(y) = \alpha(x, y)$. This induces the exponential (or currying) function $E_{XYZ} = E: [X \times Y \to Z] \to [X \to [Y \to Z]]$ sending $\alpha: X \times Y \to Z$ to the associated function $\hat{\alpha}: X \to [Y \to Z]$, and E_{XYZ} is a bijection (a type of exponential law). In general, we call a category **cartesian closed** if products and function spaces are again in the category and the exponential function is always a bijection. This is a convenient property for constructions such as in the preceding section and for other purposes.

Note that E restricted to the category of CPO's and continuous morphisms is still a bijection, for if X, Y, Z are all CPO's, then one verifies directly that α preserves directed sups if and only if $\hat{\alpha}$ does (where $[Y \to Z]$ is given the pointwise order). Hence the category of CPO's and continuous morphisms is also cartesian closed.

Again things rapidly become more complicated when one moves to a topological viewpoint. First of all, one has to have a means of topologizing the function spaces $[Y \to Z]$. In this regard we recall certain basic notions from topology (see e.g. [**Du**, Chapter XII]).

A topology τ on $[Y \to Z]$ is **splitting** if for every space X, the continuity of $\alpha: X \times Y \to Z$ implies that of the associated function $\hat{\alpha}: X \to [Y \to Z]_\tau$ (where $\hat{\alpha}(x)(y) = \alpha(x, y)$). A topology τ on $[Y \to Z]$ is called **admissible** (or **conjoining**) if for every space X, the continuity of $\hat{\alpha}: X \to [Y \to Z]_\tau$ implies that of $\alpha: X \times Y \to Z$. Thus for fixed Y, Z we have that E_{XYZ} is a bijection for all X if and only if the topology τ on $[Y \to Z]$ is both splitting and admissible.

We list some basic facts about splitting and admissible topologies. A topology τ is admissible iff the evaluation mapping $\epsilon: [X \to Y]_\tau \times X \to Y$ defined by $\epsilon(f, x) = f(x)$ is continuous. A topology larger than an admissible topology is again admissible, and a topology smaller than a splitting topology is again splitting. Any admissible topology is larger than any splitting topology, and there is always a unique largest splitting topology. Thus a function space can have at most one topology that is both admissible and splitting, and such a topology is the largest splitting topology and the smallest admissible topology.

A standard function space topology is the compact-open topology. We need a slight modification of this that is suitable for core compact spaces. Let X and Y be spaces, let H be a Scott open set in the lattice $O(X)$ of open sets on X, and let V be an open subset of Y. We define the **Isbell topology** on $[X \to Y]$ by taking as a subbase for the open sets all sets of the form

$$N(H, V) = \{f \in [X \to Y]: f^{-1}(V) \in H\}.$$

If X is locally compact, then the Isbell topology is just the compact-open topology. The next theorem asserts that the core compact spaces are the exponentiable spaces (see [**Is75**], [**SW85**], or [**LP85**]).

Theorem. *Let Y be a core compact space. Then for any space Z the space $[Y \to Z]$ admits an (unique) admissible, splitting topology, the Isbell topology, and with respect to this topology the exponential function E_{XYZ} is a bijection for all X.*

What happens if Y is not core compact? Then results of Day and Kelly [**DK70**] show that the Scott topology on $[Y \to 2]$ is not admissible, but it is the inf of admissible topologies. Thus there is no smallest admissible topology on $[Y \to 2]$, hence no topology that is both admissible and splitting. In this case there is no topology on $[Y \to Z]$ such that E_{XYZ} is a bijection for all X. Thus any category of topological spaces which contains 2, is closed with respect to taking function spaces with respect to some appropriate topology, and is cartesian closed must be some subcategory of core compact spaces. These considerations reduce the search for a largest cartesian closed category in Top to the following problem:

Problem. Is there a largest collection of core compact spaces containing 2 which is closed with respect to taking finite products and function spaces equipped with the Isbell topology (since this is the one that yields that the exponential function is a bijection)?

Suppose now that Z is a CPO equipped with the Scott topology. Then $[Y \to Z]$ is again a CPO, and one can investigate how the Scott and Isbell topology compare on $[Y, Z]$. A direct argument from the definition of the Isbell topology yields that a directed set of functions converges to its pointwise supremum in the Isbell topology, and hence the Isbell topology is coarser than the Scott topology. Since we have seen that the Isbell topology is an admissible topology if Y is core compact, it follows that the Scott topology is also admissible. Gierz and Keimel [**KG82**] have shown that if Y is locally compact and Z is a continuous lattice, then the compact-open and Scott topology agree on $[Y \to Z]$. Analogously Schwarz and Weck [**SW85**] have shown that if Y is core compact and Z is a continuous lattice, then the Isbell topology agrees with the Scott topology on $[Y \to Z]$. In the later section on supersober and compact ordered spaces we generalize these results.

If Y is core compact and second countable (i.e., the topology has a countable base) and if Z is also second countable, then $[Y \to Z]$ equipped with the Isbell topology is second countable (see [**LP 85**, Proposition 2.17]). Hence if Y is core compact and second countable (e.g., Y is a countably based continuous CPO), Z is a countably based continuous CPO, and $[Y \to Z]$ is a continuous CPO on which the Scott and Isbell topologies agree, then $[Y \to Z]$ is a countably based continuous CPO (since being countably based is equivalent to the second countability of the Scott topology).

Strongly algebraic and finitely continuous CPO's

The category of finite partially ordered sets and order preserving functions is cartesian closed. The full subcategories with objects lattices or (meet) semilattices are also cartesian closed. One can extend these categories by taking projective limits where the bonding mappings are projections. For the finite lattices (resp. semilattices), one gets the algebraic lattices (resp. the algebraic semilattices). For all finite partially ordered sets one obtains objects which are called **strongly algebraic** CPO's. They form a larger cartesian closed category than the algebraic semilattices and were introduced by Plotkin [**Pl76**] to have a cartesian closed category available where one could carry out certain power domain constructions and remain in the category. The morphisms in these categories (as earlier) are the Scott continuous morphisms, and the function spaces are the CPO's arising from the pointwise order of functions. In the section on supersober and compact ordered spaces we will relate these function spaces to the topological considerations of the previous section.

One can consider all retracts of strongly algebraic CPO's and obtain an even larger cartesian closed category. These objects have been called **finitely continuous** CPO's by Kamimura and Tang and studied in several of their papers (see in particular [**KT86**]). A CPO P is a finitely continuous CPO iff there exists a directed family D of continuous functions from P into P with supremum the identity function on P such that the $f(P)$ is finite for each $f \in D$. The strongly algebraic CPO's are characterized by requiring in addition that each member of D be a projection. We take these characterizations for our working definition of these concepts. Frequently one's attention is restricted to the countably based case. Here the the directed family of functions, respectively, projections with finite range may be replaced by an increasing *sequence* of functions.

Let S be a continuous complete semilattice and let F be a finite subset of S containing $\bot$. Enlarge F to G by adjoining the supremum of each subset of F that is bounded above. Then G is still finite and has the property that $\Downarrow x \cap G$ has a largest element for each $x \in S$. The mapping that sends x to the largest element of $\Downarrow x \cap G$ is continuous and below the identity mapping. Furthermore, the family of all such mappings for all finite sets is a directed family whose supremum is the identity. Hence S is a finitely continuous CPO.

We list some basic properties of finitely continuous CPO's.

Proposition. *A retract of a finitely continuous CPO is again a finitely continuous CPO.*

Proof. Let $r : P \to Q$ be a retract with inclusion $i : Q \to P$. Let D be the directed family of functions in $[P \to P]$ with finite range and with supremum 1_P. Then $\{r \circ f \circ i : f \in D\}$ gives the desired directed family on Q. ∎

Proposition. *Let P and Q be finitely continuous CPO's. Then $[P \to Q]$ is a finitely continuous CPO.*

Proof. Let D be the directed family for P and D' for Q. Then $\{h \mapsto f' \circ h \circ f : f \in D, f' \in D'\}$ gives the desired directed family on $[P \to Q]$. ∎

It follows directly from the last proposition that the finitely continuous CPO's form a cartesian closed subcategory of the CPO category.

Plotkin [**Pl76**] gave an alternate characterization of strongly algebraic CPO's in terms of the partially ordered set of compact elements, which we do not pursue here. Smyth [**Sm83a**] used these to derive the following significant result:

Theorem. *Let P be a countably based algebraic CPO with $\bot$. If $[P \to P]$ is also an algebraic CPO, then P is a strongly algebraic CPO.*

This theorem shows that the largest cartesian closed full subcategory of countably based algebraic CPO's consists of the strongly algebraic CPO's.

Problem. Do the finitely continuous CPO's form the largest cartesian closed full subcategory contained in the category of continuous CPO's?

Problem. Give an internal description of a finitely continuous CPO that one can apply directly to determine whether a given continuous CPO is finitely continuous.

Problem. Find a topological description of the spaces obtained by endowing a finitely continuous CPO with the Scott topology.

V. Dual and Patch Topologies

Given a T_0-topology, each open set is an upper set and each closed set is a lower set with respect to the order of specification $x \leq y \Leftrightarrow x \in \overline{\{y\}}$. There are methods for creating "dual" topologies from the given topology in which open sets in the new dual topology are now lower sets (with respect to the original order of specification). "Patch" topologies then arise as the join of a topology and its dual.

Suppose $d: X \times X \to \mathbb{R}^+$ satisfies the triangular inequality. We use d to generate a topology on X by declaring a set U open if for each $x \in U$, there exists a *positive* number r such that $N(x;r) \subseteq U$, where $N(x;r) = \{y: d(x,y) \leq r\}$. (This is slightly at variance with the usual approach, but allows us momentarily a useful generalization.) Then $d^*(x,y) = d(y,x)$ gives rise to a **dual** topology.

Example. Define $d: \mathbb{R}\times\mathbb{R} \to \mathbb{R}^+$ by $d(x,y) = \max\{0, x-y\}$. Then d generates the Scott topology on $\mathbb{R}$, d^* gives the reverse of the Scott topology (the Scott topology on the order dual), and the join of the two topologies is the usual topology.

The situation can be considerably generalized by considering functions satisfying the triangular inequality into much more general semigroups than the positive reals $\mathbb{R}^+$ (see e.g. [**Ko87**]). In this case we need to specify an ordered semigroup S and a subset of positive elements S^+ for the codomain of the "distance" function. Suppose that P is a continuous CPO. We set S equal to the power set of P with addition being the operation of union. We let S^+, the set of positive elements, be the cofinite subsets. We define the metric d by $d(x,y) = \Downarrow x \setminus \downarrow y$, and then define the open sets precisely as in the earlier paragraph for real metrics. This metric is called the **canonical generalized metric** for a continuous CPO.

Proposition. *The topology generated by d is the Scott topology.*

Proof. Consider the set $N(x;A)$, where A is a cofinite subset. Let F be the complement of A. Then one verifies directly that

$$N(x;A) = \{y: F \cap \Downarrow x \subseteq \downarrow y\} = \bigcap\{\uparrow z: z \in F \cap \Downarrow x\}.$$

Such sets contain x in their interior in the Scott topology if P is a continuously ordered set (since $x \in \Uparrow z \subseteq \uparrow z$); hence the metric open sets are Scott open. Conversely if $x \in U$ where U is Scott open, pick $z \in U$ such that $z \ll x$ (this is possible since $\Downarrow x$ is directed). Let $A = P \setminus \{z\}$. Then $N(x;A) = \uparrow z \subseteq U$. Hence U is metric open. ∎

An approach that has received more attention has been the following (see [**Sm83b**]). Let X be a T_0-topological space. A set is said to be **saturated** if it is the intersection of open sets (this is equivalent to being an upper set in the order of specification). One defines the **dual** topology by taking as a subbasis for the closed sets all saturated compact sets. The join of these two topologies is called the **patch** topology.

For a partially ordered set P, the **weak** topology is defined by taking as a subbase for the closed sets all principal lower sets $\downarrow x$ for $x \in P$. The **weak**^{d} topology is defined to be the weak topology on the dual of P, the set P with the order reversed. All sets of the form $\uparrow x$ form a subbasis for the closed sets for the weak^d topology.

Proposition. *Let P be a continuous CPO. Then the dual topology for the canonical generalized metric and the dual topology for the Scott topology both agree and both yield the weak^d topology.*

Proof. Let $d(x,y) = \Downarrow x \setminus \downarrow y$. Let A be cofinite in P and let F be its complement. Then

$$x \in P \setminus \bigcup \{\uparrow z : z \in F \setminus \downarrow x\} \subseteq P \setminus \bigcup \{\Uparrow z : z \in F \setminus \downarrow x\}$$
$$= \{y : \Downarrow y \cap F \subseteq \downarrow x\} = \{y : \Downarrow y \setminus \downarrow x \subseteq A\} = N_{d^*}(x, A).$$

Since the first set is open in the weak^d topology, it follows that dual open sets are open in the weak^d topology.

Conversely we show that a subbasic open set $U = P \setminus \uparrow w$ is open in the dual topology. Let $x \in U$. Then there exists $z \ll w$ such that $z \not\leq x$. Let $F = \{z\}$ and let $A = P \setminus F$. Then as before $N_{d^*}(x, A) = \{y : \Downarrow y \setminus \downarrow x \subseteq A\} = P \setminus \Uparrow z \subseteq P \setminus \uparrow w$. Thus U is open in the dual topology.

We turn now to the second case. Since $\uparrow x$ is trivially compact in the Scott topology, it follows that every weak^d open set is open in the dual topology. Conversely let A be an upper set which is compact in the Scott topology and pick $y \notin A$. For each $x \in A$, pick $z_x \ll x$ such that $z_x \not\leq y$. Since $\{\Uparrow z_x\}$ is an open cover of A, there exist finitely many such that $A \subseteq \bigcup\{\uparrow z_i : 1 \leq i \leq n\}$. Note that the righthand set is closed in the weak^d topology and misses y. Since y was arbitrary, it follows that A is the intersection of sets closed in the weak^d topology, and hence is itself closed in the weak^d topology. ∎

The Lawson Topology

The **Lawson** topology on a CPO is obtained by taking the join of the Scott topology and the weak^d topology. It follows from the last proposition of the preceding section that if P is a continuous CPO, then the Lawson topology is the patch topology defined from the canonical generalized metric and it is also the patch topology arising from the Scott topology.

The Lawson topology on a continuous CPO P is Hausdorff, for if $x \not\leq y$, then there exists $z \ll x$ such that $z \not\leq y$, and $\Uparrow z$ and $P \setminus \uparrow z$ are disjoint neighborhoods of x and y resp. Indeed the set $\Uparrow z \times P \setminus \uparrow z$ misses the graph of the order relation

$\leq$, so that the order relation is closed in $P \times P$. Such spaces (in which the order is closed) are called **partially ordered spaces**.

If P is an algebraic CPO, then the Lawson topology is generated by taking all sets $\uparrow x$ for compact elements x to be *both* open and closed. It follows that P with the Lawson topology is a 0-dimensional space. Hence it is the continuous (as opposed to the algebraic) CPO's that can give rise to continuum-like properties with respect to the Lawson topology.

If S is a complete semilattice, then one can take all complete subsemilattices which are upper sets or lower sets as a subbase for the closed sets and again obtain the Lawson topology. If S is a continuous complete semilattice, then the Lawson topology is compact and Hausdorff, the operation $(x, y) \mapsto x \wedge y$ is continuous, and S has a basis of neighborhoods at each point which are subsemilattices. Conversely, if a semilattice admits a topology with these properties, then the semilattice is a continuous complete semilattice and the topology is the Lawson topology (see [**COMP**, VI.3]).

Example. Let X be a compact Hausdorff space and let L be the semilattice of closed non-empty subsets ordered by reverse inclusion and with the binary operation of union. Then X is a continuous complete semilattice, the traditional Vietoris topology on L agrees with the Lawson topology, and this is the unique compact Hausdorff topology on L for which the binary operation of union is continuous.

Supersober and Compact Ordered Spaces

A **compact supersober** topological space X is one in which the set of limit points of an ultrafilter is the closure of a unique point. These spaces are in particular sober and also turn out to be locally compact (and hence the lattice of open sets is a continuous lattice). The patch topology on such a space is compact and Hausdorff, and the order of specification is closed in $(X, \text{patch}) \times (X, \text{patch})$. Hence in a natural way a compact ordered space results.

Conversely, if X is a compact ordered space, consider the space $(X, \mathcal{U})$, where $\mathcal{U}$ consists of all open *upper* sets. Then $(X, \mathcal{U})$ is a compact supersober space (with the set of limit points of an ultrafilter being the lower set of the point to which the ultrafilter converged in the original topology). The dual topology consists of all open lower sets, the patch topology is the original topology, and the order of specification is the original order (see [**COMP**, VII.1 Exercises] for the preceding results). Specializing to CPO's and the Scott topology, we obtain

Theorem. *A CPO P is compact supersober with respect to the Scott topology iff the Lawson topology is compact. In this case P is a compact ordered space with respect to the Lawson topology.*

We note that the order dual of a compact partially ordered space is another such. Hence the topology consisting of the open lower sets is also a compact supersober space with dual topology the open upper sets.

The preceding theorem quickly yields

Proposition. *If the Lawson topology is compact for a CPO P, then the same is true for any retract.*

Proof. Let $r: P \to Q$ be a retract with inclusion $i: Q \to P$. We show that Q is compact supersober. Let $\mathcal{U}$ be an ultrafilter in Q. Then the ultrafilter $i(\mathcal{U})$ has a largest limit point $p \in P$. Then $r(p)$ is a limit point of the original ultrafilter in Q, and if q is another limit point, then $i(q) \le p$, so $q = r(i(q)) \le r(p)$. ∎

Note that the preceding result is really a topological result, namely that the retract of any compact supersober space is again compact supersober.

It was shown in [**COMP**] that a continuous lattice or continuous complete semilattice is compact in the Lawson topology. This result extents to finitely continuous CPO's.

Proposition. *A finitely continuous CPO is compact in the Lawson topology.*

Proof. Let $\mathcal{F}$ be the directed family of Scott continuous functions with finite range that approximate the identity in the finitely continuous CPO P. Let $\mathcal{U}$ be an ultrafilter in P. Then $f(\mathcal{U})$ is an ultrafilter in the finite set $f(P)$ for each $f \in \mathcal{F}$, and hence contains a singleton set $\{p_f\}$. Since $\mathcal{U}$ is a filter and the family $\mathcal{F}$ is directed, it follows that the family $\{p_f: f \in \mathcal{F}\}$ is directed, and hence has a supremum p. We claim that the ultrafilter converges to p, which will establish the compactness of P.

Let U be a Scott open set containing p. Then $p_f \in U$ for some $f \in \mathcal{F}$. Since $f \le 1_P$, it follows that p_f is a lower bound for each member of the ultrafilter whose image under f is $\{p_f\}$. Since U is an upper set, U contains each of these sets. Hence the ultrafilter converges to p in the Scott topology. Suppose now that $p \notin \uparrow q$. Pick $f \in \mathcal{F}$ such that $f(q) \not\le p$. Then $f(q) \not\le p_f$, which in turn implies $\uparrow q \notin \mathcal{U}$. Since $\mathcal{U}$ is an ultrafilter, $P \setminus \uparrow q \in \mathcal{U}$. Thus the ultrafilter also converges to p in the weak^d topology, and hence in the Lawson topology. ∎

We consider function spaces for the compact supersober continuous CPO's. First we need a lemma.

Lemma. *Let P be a continuous CPO for which the Lawson topology is compact. Suppose that $A \subseteq \Downarrow x$ and $x = \sup A$. If U is Scott open and $x \in U$, then there exists a finite subset F of A such that $a \le b$ for all $a \in F$ implies $b \in U$.*

Proof. The sets $\uparrow a_1 \cap \uparrow a_2 \cap \cdots \cap \uparrow a_n$ for $a_1, \ldots, a_n \in A$ form a descending family of Lawson closed sets with intersection $\uparrow x \subseteq U$. It follows from compactness that $\uparrow a_1 \cap \uparrow a_2 \cap \cdots \cap \uparrow a_n \subseteq U$ for some finite subset. If $a_i \le b$ for $1 \le i \le n$, then b is in this intersection and hence in U. ∎

Theorem. *Let X be a core compact space, and let P be a continuous CPO with $\perp$. If $[X \to P]$ is a continuous CPO for which the Lawson topology is compact, then the Scott topology on $[X \to P]$ is the Isbell topology (which is the compact-open topology if X is locally compact).*

Proof. We have seen in the section on function spaces that the Scott topology is admissible and that the Isbell topology is coarsest of the admissible topologies. If we show that the Scott topology is also the coarsest of the admissible topologies, then we are done. So suppose that τ is a topology on $[X \to P]$ such that the evaluation mapping $\epsilon: [X \to P] \times X \to P$ is continuous.

Let $f \in [X \to Y]$, $a \in X$, and $f(a) = b$. Let $z \ll b$. By joint continuity of the evaluation mapping, there exists a τ-open set W containing f and an open set U in X containing a such that $h(x) \in \Uparrow z$ for all $h \in W$ and $x \in U$. Pick V open with $a \in V$ such that $V \ll U$. Define $g \in [X \to Y]$ by $g(x) = z$ if $x \in V$ and $g(x) = \perp$ otherwise. As we saw in the section on function spaces, $g \ll f$ in $[X \to Y]$ and f is the supremum of such functions. Note also that $W \subseteq \uparrow g$ since $z \leq h(x)$ for $x \in V$, $h \in W$, and $\perp \leq h(x)$ otherwise. By the preceding lemma, for any Scott open set Q containing f, there exists finitely many such g_i such that $\uparrow g_1 \cap \cdots \cap \uparrow g_n \subseteq Q$. Since each $\uparrow g_i$ is a neighborhood of f in the τ-topology (as we have just established), it follows that Q is also. It follows that the τ-topology is finer than the Scott topology. ∎

Adjunctions

Let $f^+: P \to Q$ and $f^-: Q \to P$ be order-preserving functions between the partially ordered sets P and Q. The pair (f^+, f^-) is called an **adjunction** if $y \leq f^+(x) \Leftrightarrow f^-(y) \leq x$. (Such pairs are also sometimes referred to as Galois connections, but many authors prefer to define Galois connections in terms of antitone functions.) Adjunctions can be alternately characterized by the property that $1_Q \leq f^+ \circ f^-$ and $1_P \geq f^- \circ f^+$. Hence f^+ is called the **upper adjoint** and f^- the **lower adjoint**. The mapping f^- is sometimes referred to as a **residuated** mapping.

The upper adjoint f^+ has the property that the inverse of a principal filter $\uparrow q$ in Q is again a principal filter in P (indeed this property characterizes mappings that arise as upper adjoints). Hence if P and Q are CPO's, then f^+ is Scott continuous iff it is Lawson continuous. If Q is a continuous CPO, then f^+ is Scott continuous iff f^- preserves the relation $\ll$ (see [**COMP**, Exercise IV.1.29]). Note that projections are upper adjoints (with the lower adjoint being the inclusion mapping), and hence are continuous in the Lawson topology.

The preceding remarks show that the Scott continuous upper adjoints form a good class of morphisms to consider if one is working with the Lawson topology. If P and Q are both continuous lattices, then these mappings are precisely the Lawson continuous $\wedge$-homomorphisms, which in turn are the mappings that preserve infima of non-empty sets and suprema of directed sets. As we have seen in the previous paragraph, there results a dual category consisting of the same objects with morphisms the lower adjoints which preserve the relation $\ll$. If one restricts to algebraic lattices, then the lower adjoint must preserve the compact elements. Its restriction to the compact elements is a $\vee$-preserving and $\perp$-preserving mapping. In this way one obtains the Hofmann-Mislove-Stralka duality [**HMS**] between the category of algebraic lattices with morphisms the Scott continuous upper adjoints and the category of sup-semilattices with $\perp$ and morphisms preserving $\perp$ and the $\vee$-operation.

Powerdomains

A powerdomain is a CPO together with extra algebraic structure for handling nondeterministic values. Their consideration is motivated by the desire to find semantic models for nondeterministic phenomena. Examples are frequently obtained by taking some appropriate subset of the power set of a given CPO P (hence the terminology "powerdomain"). We think of the subsets as keeping track of the possible outcomes of a nondeterministic computation. Again one is motivated to find categories where powerdomain constructions remain in the category.

We quickly overview some of the standard powerdomain constructions. If P is a CPO with $\perp$, then one can construct the Hoare powerdomain as all non-empty Scott closed subsets. If P is a continuous CPO, then this set is anti-isomorphic to the lattice of open sets, and hence forms a continuous (indeed completely distributive) lattice. The Smyth powerdomain is obtained by taking all the upper sets which are compact in the Scott topology. (We refer to [**Sm73b**] for a nice topological development of these ideas in a general setting.) In the case of a continuous CPO for which the Lawson topology is compact, these are just the closed sets in the weak^d topology, which is again anti-isomorphic to the lattice of weak^d open sets. We have seen previously that in the case that the Lawson topology is compact, this topology is compact supersober, hence locally compact, and hence the lattice of open sets is continuous.

One of the most interesting of the powerdomain constructions is the so-called Plotkin powerdomain. This again lends itself to nice description in the case that D is a continuous CPO for which the Lawson topology is compact (which we assume henceforth). It will also be convenient to assume certain basic facts about compact partially ordered spaces (see [**COMP**, VI.1]). Let $P(D)$ denote the set of all non-empty Lawson closed order-convex subsets. If $A \in P(D)$, then A is compact, and hence $\downarrow A$ and $\uparrow A$ are closed. Since A is order convex, $A = \downarrow A \cap \uparrow A$. Hence $A \in P(D)$ iff it is the intersection of a closed upper and closed lower set. We order $P(D)$ with what is commonly referred to as the Egli-Milner ordering: $A \leq B \Leftrightarrow A \subseteq \downarrow B$ and $B \subseteq \uparrow A$.

Theorem. *$(P(D), \leq)$ is a continuous CPO for which the Lawson topology is compact.*

Proof. Let A_α be a directed family in $P(D)$ (i.e., $\alpha \leq \beta$ implies $A_\beta \subseteq \uparrow A_\alpha$ and $A_\alpha \subseteq \downarrow A_\beta$). Then one verifies directly that the supremum of this family is given by

$$\sup A_\alpha = \overline{\bigcup \downarrow A_\alpha} \cap \bigcap \uparrow A_\alpha = \bigcap (\overline{\bigcup \downarrow A_\alpha} \cap \uparrow A_\alpha).$$

From the first equality it follows that this set is closed and order convex and from the second that it is non-empty (since it is the intersection of a descending family of non-empty compact sets). Thus $P(D)$ is a CPO.

Let A be closed and order-convex. Suppose that F is a finite set such that $A \subseteq \bigcup\{\Uparrow z : z \in F\}$ and $F \subseteq \bigcup\{\Downarrow a : a \in A\}$. We claim the order-convex hull $h(F) = \downarrow F \cap \uparrow F$ satisfies $h(F) \ll A$ in $P(D)$. Suppose that A_α is a directed family in $P(D)$ with $A \leq \sup A_\alpha$. Then

$$\sup A_\alpha = \bigcap (\overline{\bigcup \downarrow A_\alpha} \cap \uparrow A_\alpha) \subseteq \uparrow A \subseteq \bigcup\{\Uparrow z : z \in F\}.$$

Since the latter set is open in the Scott and hence Lawson topology, it follows that $A_\alpha \subseteq \overline{\bigcup \downarrow A_\alpha} \cap \uparrow A_\alpha \subseteq \bigcup\{\Uparrow z : z \in F\}$ for all α sufficiently large.

Conversely, for $z \in F$ there exists $a \in A$ with $z \ll a$. Since $A \leq \sup A_\alpha$, we conclude there exists $b \in \sup A_\alpha$ with $a \leq b$. Then $\Uparrow z$ is a Scott open neighborhood of b, so there exists $c \in \bigcup \downarrow A_\alpha$ such that $z \leq c$. Then $c \in \downarrow A_\gamma$ for some γ. It follows that $z \in \downarrow A_\beta$ for all indices $\beta \geq \gamma$ since $z \leq c \in \downarrow A_\gamma \subseteq \downarrow A_\beta$. Carrying this out for each $z \in F$, we conclude that $F \subset A_\alpha$ for all indices large enough, and hence the same obtains for $h(F)$. We conclude that $h(F) \leq A_\alpha$ for all indices large enough, and hence $h(F) \ll A$.

We next show that the collection of all $h(F)$ as constructed in the preceding is directed. Suppose that we are given $h(F_1)$ and $h(F_2)$. Then $A \subseteq (\bigcup\{\Uparrow y : y \in F_1\}) \cap \bigcup\{\Uparrow z : z \in F_2\}$, and the latter is a Scott open set U. Hence for each $a \in A$ there exists $z_a \ll a$ such that $z_a \in U$. Since A is compact, finitely many of the $\Uparrow z_a$ cover A. Let G be this finite set of z_α. Now given $f \in F_1$, there exists $a \in A$ such that $f \ll a$. Then there exists $g \in F_2$ such that $g \ll a$. Since $\Downarrow a$ is directed, there exists $h \ll a$ with $f, g \leq h$. For each $f \in F_1 \cup F_2$, pick such an h_f. Then the finite set F consisting of G and all the h_f satisfies the earlier conditions and $h(F)$ is above $h(F_1)$ and $h(F_2)$ in the Egli-Milner ordering. The fact that A is the suprema of the $h(F)$ follows fairly directly from the formula in the first paragraph for the calculation of suprema.

We show finally that $P(D)$ is compact. Indeed, more is true. The mapping from the space of non-empty closed subsets $Cl(D)$ with the Vietoris (=Lawson) topology to $P(D)$ which sends A to $h(A)$ is continuous. Since the former is compact, so is the latter. To show continuity, let A_α be a net in $Cl(D)$ converging to some closed set A in the Vietoris topology. Let $h(F) \ll h(A)$ for F finite. Then $\Uparrow h(F)$ is a basic open set containing $h(A)$ in the Scott topology and $A \subseteq h(A) \subseteq \operatorname{int} \uparrow h(F)$. Since the set of all closed sets contained in $\operatorname{int} \uparrow h(F)$ is a neighborhood of A in the Vietoris topology, we conclude that $A_\alpha \subseteq \uparrow h(F)$ for all indices sufficiently large; hence the same holds for $h(A_\alpha)$. For each $z \in F$, there exists $a \in h(A)$ with $z \ll a$. By definition of $h(A)$, there exists $b \in A$ with $a \leq b$. Then $\Uparrow z$ is an open set in D meeting A, and hence $A_\alpha \cap \Uparrow z \neq \emptyset$ for all indices large enough. We repeat this for all $z \in F$, and conclude that $h(F) \subseteq \downarrow h(A_\alpha)$ for all indices sufficiently large. It follows that $h(A_\alpha)$ converges to $h(A)$ in the Scott topology on $P(D)$.

Now let $B \in P(D)$ such that $B \not\leq h(A)$. Then the complement of the upper set of the singleton B in $P(D)$ is a subbasic open set around $h(A)$ in the weak^d topology on $P(D)$. If $h(A) \not\subseteq \uparrow B$, then there exists $a \in A$ with $a \notin \uparrow B$. Since the complement of $\uparrow B$ is an open set meeting A, this complement also meets A_α for all indices large enough. Then $h(A_\alpha) \not\subseteq \uparrow B$. On the other hand, if $B \not\subseteq \downarrow h(A) = \downarrow A$, then there exists $b \in B$ with $\uparrow b \cap A = \emptyset$. Then the complement of $\uparrow b$ is an open set containing A and hence A_α for all α large enough. The complement then contains $h(A_\alpha)$. Thus in either case $B \not\leq h(A_\alpha)$ for large indices. Hence $h(A_\alpha)$ converges to $h(A)$ in both the Scott and weak^d topologies, and hence in the Lawson topology. ∎

We remark that Plotkin introduced the strongly algebraic (countably based) CPO's because the Plotkin powerdomain is another such [**PL76**]. The same is true for finitely continuous CPO's, as has been shown by Kamimura and Tang [**KT87**]. To get the directed family of functions which approximate the identity and have finite range on $P(D)$ from those on D, simply consider $A \mapsto h(f(A))$ for each f in the approximating family on D. The same technique works to obtain projections if D is strongly algebraic, and in the countably based case one obtains a sequence of functions.

REFERENCES

[DK70] B. J. Day and G. M. Kelly, "On topological quotient maps preserved by pullbacks or products," *Proc. of the Cambridge Phil. Soc.* **67** (1970), 553–558.

[Du] J. Dugundji, **Topology**, Allyn and Bacon, Boston, 1964.

[COMP] G. Gierz, K. H. Hofmann, K. Keimel, J. D. Lawson, M. Mislove, and D. Scott, **A Compendium of Continuous Lattices**, Springer-Verlag, Heidelberg, 1980.

[Ho81a] R.-E. Hoffmann, "Continuous posets, prime spectra of completely distributive complete lattices, and Hausdorff compactifications," **Continuous Lattices, Lecture Notes in Mathematics 871**, edited by B. Banaschewski and R.-E. Hoffmann, Springer-Verlag, 1981, 125–158.

[Ho81b] R.-E. Hoffmann, "Projective sober spaces," **Continuous Lattices, Lecture Notes in Mathematics 871**, edited by B. Banaschewski and R.-E. Hoffmann, Springer-Verlag, 1981, 125–158.

[HL78] K. H. Hofmann and J. D. Lawson, "The spectral theory of distributive continuous lattices," *Trans. Amer. Math. Soc.* **246** (1978), 285–310.

[HMS] K. H. Hofmann, M. Mislove, and A. Stralka, **The Pontryagin Duality of Compact 0-Dimensional Semilattices and its Applications, Lecture Notes in Mathematics 396**, Springer-Verlag, 1974.

[Is75] J. R. Isbell, "Function spaces and adjoints," *Symposia Math.* **36** (1975), 317–339.

[Ja74] R. E. Jamison, **A general theory of convexity**, Doctoral Dissertation, University of Washington, 1974.

[KT86] T. Kamimura and A. Tang, "Retracts of SFP objects," Lecture Notes in Computer Science **239**, Springer-Verlag, 1986, 135–148.

[KT87] T. Kamimura and A. Tang, "Domains as finitely continuous CPO's," preprint.

[KG82] K. Keimel and G. Gierz, "Halbstetige Funktionen und stetige Verbände," **Continuous Lattices and Related Topics**, Proceedings of the Conference on Topological and Categorical Aspects of Continuous Lattices (Workshop V), edited by R.-E. Hoffmann, Mathematik-Arbeitspapiere, Universität Bremen **27**, 1982, 59–67.

[Ko87] R. Kopperman, "All topologies come from generalized metrics," *Amer. Math. Monthly*, (to appear).

[LP85] P. Th. Lambrinos and B. Papadopoulos, "The (strong) Isbell topology and (weakly) continuous lattices," **Continuous Lattices and Their Applications**, edited by R.-E. Hoffmann and K. H. Hofmann, Marcel Dekker, 1985, 191–211.

[La79] J. D. Lawson, "The duality of continuous posets," *Houston J. Math.* **5** (1979), 357–386.

[Ma76] G. Markowsky, "Chain-complete posets and directed sets with applications," *Algebra Univ.* **6** (1976), 53–68.

[MR76] G. Markowsky and B. K. Rosen, "Bases for chain-complete posets," *IBM Journal of Research and Development* **20** (1976), 138–147.

[Pl76] G. D. Plotkin, "A powerdomain construction," *SIAM J. Comp.* **5** (1976), 452–487.

[Sc70] D. Scott, "Outline of a mathematical theory of computation," *Proc. 4th Ann. Princeton Conf. on Inform. Sci. and Systems*, 1970, 169–176.

[Sc72] D. Scott, "Continuous lattices," **Toposes, Algebraic Geometry, and Logic**, Springer LNM **274**, Springer-Verlag, Berlin, 1972.

[Sc76] D. Scott, "Data types as lattices," *SIAM Journal on Computing* **5** (1976), 522–587.

[SW85] F. Schwarz and S. Weck, "Scott topology, Isbell topology and continuous convergence," **Continuous Lattices and Their Applications**, edited by R.-E. Hoffmann and K. H. Hofmann, Marcel Dekker, 1985, 251–273.

[Sm83a] M. Smyth, "The largest cartesian closed category of domains," *Theoretical Computer Sci.* **27** (1983), 109–119.

[Sm83b] M. Smyth, "Powerdomains and predicate transformers: a topologica view," **ICALP 83, Lecture Notes in Computer Science 154**, edited by J. Diaz, Springer-Verlag, 1983, 662–676.

[VDV85] M. Van de Vel, "Lattices and semilattices: a convex point of view," **Continuous Lattices and Their Applications**, edited by R.-E. Hoffmann and K. H. Hofmann, Marcel Dekker, 1985, 279–302.

[WL84] G. Winskel and K. Larsen, "Using information systems to solve recursive domain equations effectively," **Semantics of Data Types, Lecture Notes in Computer Science 173** edited by G. Kahn and G. D. Plotkin, Springer-Verlag, 1984, 109–130.

On the Smyth Power Domain

Michael Mislove
Department of Mathematics
Tulane University
New Orleans, LA 70118

Abstract

This paper explores the connection between the Smyth power domain $\mathbf{PS}(D)$ of a domain D and the domain D itself. The Smyth power domain is the most prevalent of the three power domain constructions commonly used to model nondeterminism in the denotational semantics of high-level programming languages. One definition of the Smyth power domain $\mathbf{PS}(D)$ is as the set of all Lawson-closed upper sets X from the domain D, so there is the natural inclusion $x \mapsto \uparrow x : D \to \mathbf{PS}(D)$. On the other hand, the inf map $X \mapsto \bigwedge X : \mathbf{PS}(D) \to D$ is an upper adjoint to this inclusion, and we use this adjunction to obtain information about $\mathbf{PS}(D)$ from the domain D. If D is distributive, spectral theory implies that each element X of $\mathbf{PS}(D)$ satisfies $\bigwedge X$ is the infimum of a unique set of primes minimal with respect to being contained in X. Results which characterize when a domain D does not contain a copy of $2^{\mathbf{N}}$ are invoked to show that the set of such primes is finite in certain cases. We indicate how these results can be generalized to the case that D is *locally distributive* or *semiprime*. Our results are motivated by an interest in understanding the Smyth power domain $\mathbf{PS}(D)$ in terms of the domain D, and we feel they should have application to the semantics of high-level programming languages. An indication of some possible applications of these results is given at the end of the paper.

0. Introduction. Suppose we are given a nondeterministic computation f defined on a domain D. Then the output of f may well be a *set* of elements from D, rather than a single element of D. Using the Smyth power domain to model this situation means we are identifying the output of f with the Lawson-closed upper set X from D which that output generates. This makes sense as long as we realize that we must then be willing to accept any element of X as a possible outcome of the computation. In particular, any of the minimal elements of X are possible outcomes of the computation, and these represent the minimal possible information we might receive. If the output is one element of D, then that element is the only minimal element of the upper set it generates, but if there is more than one possible outcome, and they are incomparable, then any one of them may be the element we must accept. So we may well ask what is the most information we can *guarantee* to be true of the computation f? If f outputs the set X, then this is precisely the element $\bigwedge X$ of D, since this element represents the largest amount of information which is common to all the elements of X, and hence to all the possible outputs of f. In this paper we investigate this map $X \mapsto \bigwedge X : \mathbf{PS}(D) \to D$ and what it tells us about the Smyth power domain construction in general. We also point out some special circumstances where there are more precise results available.

I. Some Background. Throughout this paper, D will denote a *domain*, by which we mean a consistently complete ω-algebraic complete partial order. As has been noted numerous times, such objects can be obtained by taking an algebraic lattice whose identity element is compact and deleting the identity element. That is, if D is a domain, then $D^{\top}$, the domain D with an isolated identity adjoined, is an algebraic lattice. We use D^0 to denote the set of *compact* elements of D; i.e., $k \in D^0$ if, for every directed subset X of D, $k \leq \bigvee X$, then there is some $x \in X$ with $k \leq x$. These elements are also called *finite* or *isolated*, but we prefer the term compact since it reduces the potential for confusion. Since D is "almost" an algebraic lattice, there are a number of things which are true for algebraic lattices which apply to D as well. The standard reference for the results about algebraic lattices is [Hof74]; the duality which is described there is commonly refered to as *HMS-duality*, an acronym derived from the initials of the authors of [Hof74]. The basic result of that duality is that there is a one-to-one correspondence between discrete inf-semilattices with identity, on the one hand, and algebraic lattices on the other. The duality is implemented on the object level by showing that an algebraic lattice D is isomorphic to the lattice of ideals $\mathbf{Id}(D^0)$ of the inf-semilattice D^0 of compact elements of D, and, conversely, that any inf-semilattice S with identity is isomorphic to the inf-semilattice of compact elements of the ideal lattice $\mathbf{Id}(S)$.

The *Scott topology* for a domain D is the topology generated by all sets of the form $\uparrow k = \{d \in D | k \leq d\}$, where $k \in D^0$. This topology is sober, but not generally T_1. From a mathematical viewpoint, a second and equally important topology for D is the *Lawson topology*. It is gotten by taking the common refinement of the Scott topology and the *lower topology*; the latter has for a basis of open sets all sets of the form $D \backslash \uparrow x = \{d \in D | d \notin \uparrow x\}$, for $x \in D$. A basis for the Lawson open sets consists of all sets of the form $\uparrow k \backslash (\bigcup \{\uparrow k_i | i = 1, ..., n\})$, where $k \in D^0$ and $k_i \in D^0$ are finitely many elements from D^0. A now standard argument (which dates back to the work of Marshall Stone on the duality of Boolean algebras, cf. [Sto36]) shows that the Lawson topology on an algebraic lattice is compact and Hausdorff, from which it readily follows that the same is true for a domain D. The sets $\uparrow k$, for $k \in D^0$, are compact and open, and this implies that D is a compact totally disconnected space in the Lawson topology. Moreover, the map $(x, y) \mapsto x \wedge y \colon D \times D \to D$ is continuous if we endow D with the Lawson topology and $D \times D$ with the product topology. This same result also holds for the Scott topology as well. The following theorem summarizing this discussion generalizes the corresponding theorem for algebraic lattices with countably many compact elements which is Theorem 3.13 of Chapter VI in [Gie80].

Theorem 1.1. *Let D be a domain. Then D is a compact zero-dimensional semilattice in the Lawson topology and D has a countable basis for its Lawson topology. Conversely, if S is any second countable compact zero-dimensional semilattice, then S is a domain in the natural order, and the topology of S is the Lawson topology.*

An element p of an inf-semilattice S is *irreducible* if, whenever $p = x \wedge y$ for elements x, y in S, then $p = x$ or $p = y$. If S has a largest element 1, then 1 is not regarded as being irreducible. In the absence of an identity, any maximal element of S is an irreducible element. For domains, irreducible elements abound because of the following:

Lemma 1.2 ([Gie80], Proposition 3.6, p.69) *Let D be a domain and let k be a compact element of D. If M denotes the set of maximal elements of $D\backslash\uparrow k$, then $D\backslash\uparrow k = \downarrow M$, and each element of M is irreducible.*

Proof. Let $d \in D\backslash\uparrow k$, and choose a maximal chain $C \subset D\backslash\uparrow k$ containing d. Since C is clearly directed and D is a domain, the element $\bigvee C$ exists in D. Since k is compact and $C \subset D\backslash\uparrow k$, it follows that $p = \bigvee C \in D\backslash\uparrow k$. Suppose that $x, y \in D$ with $x \wedge y = p$. If $x, y \in \uparrow k$, then $x \wedge y \in \uparrow k$, which contradicts $x \wedge y = p \in D\backslash\uparrow k$. So, at least one of x and y is in $D\backslash\uparrow k$. Assuming that one is x, then $p = x \wedge y \leq x$ implies that $C \cup \{x\}$ is a chain in $D\backslash\uparrow k$ which contains d. Since C is a maximal such chain, $C = C \cup \{x\}$, which implies that $x \leq \bigvee C = p$. Thus $x = p$. This argument shows that $p \in M$ is irreducible and that $\downarrow M = D\backslash\uparrow k$. The remaining claim that each element of M is irreducible is also easily established from this argument.

We denote the set of irreducible elements of the domain D by $IRR(D)$. Another way to state Lemma 1.2 is to say that every element of D is the infimum of those irreducibles which lie above the element; i.e., for every $d \in D$, $d = \bigwedge(\uparrow d \bigcap IRR(D))$; we refer to this by saying D is *order generated* by irreducibles.

There is a more special notion than irreducible. An element p in the domain D is *prime* if whenever $x \wedge y \leq p$, then either $x \leq p$ or $y \leq p$. We denote the set of primes of D by *Spec* D (as with irreducibles, the identity element is not prime). Any prime is an irreducible element of D: if $x \wedge y = p$ and p is prime, then $x \leq p$ or $y \leq p$, and so $x = p$ or $y = p$ since $p = x \wedge y \leq x, y$.

Definition 1.3. The domain D is **distributive** if $D^\top$, the domain D with a largest element adjoined, is a distributive lattice.

Our interest in distributive domains stems from the following:

Theorem 1.4. *For a domain D the following are equivalent:*
1) D is distributive,
2) Spec D order-generates D,
3) Spec $D = IRR(D)$.

The proof relies on the fact that the corresponding result holds for algebraic (indeed, continuous) lattices, and the fact that, by definition, a domain D is distributive if and only if $D^\top$ is distributive in the usual lattice-theoretical sense. See Section I.3 of [Gie80] for some of the relevant details for continuous lattices.

There are several definitions of distributivity for inf-semilattices in the literature, but none of them appear to be equivalent to ours (cf., e.g., [Hof74], Section III-1). For example, the most commonly used definition for distributivity was discovered by Grätzer (cf. [Grä68]), and it states that the domain D is *G-distributive* if, for every x, y and z in D, if $x \wedge y \leq z$, then there are $x' \geq x$ and $y' \geq y$ with $x' \wedge y' = z$. However, the domain $T = \{\perp, a, b\}$ with a and b incomparable

If D is a domain of compactly finite breadth, is each element of D the infimum of only finitely many irreducible elements from D?

and $\perp$ the bottom element does not satisfy Grätzer's definition, but it does satisfy ours. Similar examples show that the other definitions given in ([Hof74], Section III-1) are inequivalent to ours; it would be interesting to have a purely algebraic characterization of our definition for distributivity. We can prove the following:

Lemma 1.5. *If the domain D is G-distributive, then D is distributive.*

Proof. Suppose that D is G-distributive. We show that each irreducible element of D is prime; Theorem 1.3 then implies that D is distributive. So, let p be an irreducible element of D, and suppose that $x, y \in D$ satisfy $x \wedge y \leq p$. Since D is G-distributive, there are $x' \geq x$ and $y' \geq y$ with $x' \wedge y' = p$. Since p is irreducible, either $x' = p$ or $y' = p$. Assuming $x' = p$, since $x \leq x'$, we have $x \leq p$. Thus p is prime.

We will also use the following notion:

Definition 1.6. *Let D be a domain. The finite subset $A \subset D$ is* **meet irredundant** *if $\bigwedge A < \bigwedge B$ for every proper subset B of A. D has* **locally finite meet breadth** *if $\wedge\!-br(x)$ is finite for every $x \in D$, where*

$$\wedge\!-br(x) = SUP\{|A| \mid x = \wedge A \text{ and } A \text{ is meet irredundant}\}.$$

D has **compactly finite breadth** *if, for every non-empty compact subset K of D, there is a finite subset $F \subset K$ with $\bigwedge F = \bigwedge K$.*

Theorem 1.7. *Let D be a domain. The following conditions are equivalent:*
1) D has compactly finite breadth.
2) D admits no injection of $2^{\mathbf{N}}$ preserving all infs and all directed sups.
Moreover, if D is distributive, then the above conditions are equivalent to
3) D has locally finite meet breadth.
4) Each element x of D is the infimum of finitely many primes of D.

Proof. See [Liu83] for a proof that 1) and 2) are equivalent; a proof that 3) and 4) are equivalent for distributive domains may be found in [Gie85].

In any domain, 3) implies 2) since $2^{\mathbf{N}}$ does not have locally finite meet breadth (indeed, the least element can be expressed as the infimum of finite sets of arbitrarily large cardinality), and so the same holds for any domain containing a copy of $2^{\mathbf{N}}$.

Conversely, suppose that D is a distributive domain of compactly finite breadth. If x is any element of D, then $x = \bigwedge(\overline{\uparrow x \bigcap Spec\ D})$, and since this set is compact, it has a finite subset F with $x = \bigwedge F$. Since F is finite, it follows that for each prime $p \geq x$, there is some $q_p \in F$ with $q_p \leq p$, and so $\uparrow x \bigcap Spec\ D \subset \uparrow F$. Hence $\overline{\uparrow x \bigcap Spec\ D} = \uparrow F$ since $\uparrow x \bigcap Spec\ D$ is dense in $\overline{\uparrow x \bigcap Spec\ D}$ and $F \subset \overline{\uparrow x \bigcap Spec\ D}$. Any element which is the limit of primes in its upper set must also be prime, from which it follows easily that F consists of prime elements. Since F is finite, we have shown that 4) holds.

An unsettled question in this regard is the following:

This question is important since, if it were true, then it could lead to a representation of a compactly finite breadth domain D in terms of the finite subsets of $IRR(D)$. We comment further on this in our closing remarks.

II. Hyperspaces and the Smyth Power Domain. For a topological space X, a *hyperspace* is a family of subspaces of X endowed with a topology usually arising from the topology on X. For example, there is the hyperspace of closed subsets of a space X, endowed with Vietoris topology, and there is the hyperspace of subcontinua of X endowed with the relative Vietoris topology. Our interest is in hyperspaces of a domain D, where we endow D with the Lawson topology and use this topology to generate topologies on the relevant hyperspaces. We denote these hyperspaces as $\mathbf{U}(D)$, $\mathbf{D}(D)$, and $\mathbf{C}(D)$, and they are defined as follows:

Definition 2.1. *Let D be a domain.*

We define the space $\mathbf{U}(D) = \{A \subset D | A = \uparrow A = \overline{A} \neq \emptyset\}$, the space of all non-empty Lawson-closed upper sets from D, in the opposite order; i.e., $A \leq B$ if and only if $B \subset A$.

We define $\mathbf{D}(D) = \{A \subset D | A = \downarrow A = \overline{A} \neq \emptyset\}$, the space of all non-empty Lawson-closed lower sets from D, in the usual order.

Finally, we define $\mathbf{C}(D) = \{A \subset D | A = \downarrow A \cap \uparrow A = \overline{A} \neq \emptyset\}$, the family of all non-empty Lawson-closed order convex subsets of D, in the opposite order.

Proposition 2.2. *For a domain D, each of the hyperspaces $\mathbf{U}(D)$, $\mathbf{D}(D)$ and $\mathbf{C}(D)$ is a domain in the indicated order. Moreover, if P is any one of these hyperspaces, the map $X \mapsto \bigwedge X : P \to D$ is a continuous map.*

Proof. Although this result is not new, we sketch a proof. That $\mathbf{D}(D)$ is a domain can be most easily seen by realizing that $\mathbf{D}(D)$ is anti-isomorphic to the lattice $\Sigma(D)$ of Scott-open subsets of D. The compact elements of $\mathbf{D}(D)$ are easily seen to be the elements of the form $D \backslash \uparrow F$, where F is a finite subset of D^0; since D^0 is countable, so is $\mathbf{D}(D)^0$.

If we let $\Gamma(D)$ denote the lattice of Lawson-closed subsets of D in the opposite order, then $\Gamma(D)$ is an algebraic lattice, and the map $A \mapsto \uparrow A : \Gamma(D) \to \mathbf{U}(D)$ is a kernel operator preserving directed sups (i.e., descending intersections), as is easily verified. This implies that $\mathbf{U}(D)$ is a continuous semilattice. Since the elements of the form $\uparrow F$, for F a finite subset of D^0 are easily seen to be compact elements in $\mathbf{U}(D)$ and every element of $\mathbf{U}(D)$ is the intersection of such elements, it follows that $\mathbf{U}(D)$ is a domain. Finally, the map $A \mapsto \uparrow A \bigcap \downarrow A : \Gamma(D) \to \mathbf{C}(D)$ is a projection, so similar arguments show that $\mathbf{C}(D)$ is also a domain. (See Corollary 1.7, p.181 of [Gie80] for the relevant facts about kernel operators and projections).

The map $X \mapsto \bigwedge X : \mathbf{D}(D) \to D$ is continuous since it is constant. For a compact semilattice S, one characterization that S is a domain is that the map $X \mapsto \bigwedge X : \Gamma(S) \to S$ is continuous (see [Liu83] for a proof); in particular the map $X \to \bigwedge X : \Gamma(D) \to D$ is continuous for any domain D. The continuity of $X \mapsto \bigwedge X : \mathbf{U}(D) \to D$ follows from the fact that this map can be induced on $\mathbf{U}(D)$ from the map $X \mapsto \bigwedge X : \Gamma(D) \to D$, since for any set X, $\bigwedge X = \bigwedge \uparrow X$. The same argument applies to $X \mapsto \bigwedge X : \mathbf{C}(D) \to D$.

The following theorem summarizes the relations between these hyperspaces and the commonly used power domains for denotational semantics. A proof may be found in [Smy82].

Theorem 2.3. *Let D be a domain.*
1) The Hoare power domain for D is isomorphic to the hyperspace $\mathbf{D}(D)$.
2) The Smyth power domain $\mathbf{PS}(D)$ for D is isomorphic to the hyperspace $\mathbf{U}(D)$.
3) The Plotkin power domain for D is isomorphic to the hyperspace $\mathbf{C}(D)$.

Thus, each of the power domains P is a domain over the domain D, and the map $X \mapsto \bigwedge X: P \to D$ is continuous for each domain P. With the Hoare power domain, this map tells us nothing since it is constant. For a set A in the Plotkin power domain it tells us only as much about A as it tells us about the set $\uparrow A$, which is an element of the Smyth power domain. We therefore investigate what the function $X \mapsto \bigwedge X: \mathbf{U}(D) \to D$ tells us about $\mathbf{U}(D)$, the Smyth power domain of D.

In the proof of 1) implies 4) of Theorem 1.7 we showed that for distributive domains the set $\uparrow \overline{(\uparrow x \bigcap Spec\ D)}$ is a closed upper set (i.e., an element of $\mathbf{U}(D)$) whose infimum is x. The following result shows that $\uparrow \overline{(\uparrow x \bigcap Spec\ D)}$ is the smallest such set in $\mathbf{U}(D)$:

Lemma 2.4. ([Gie77]). *Let D be a domain, and let p be a prime in D. If X is a closed upper set in D with $\bigwedge X \leq p$, then $p \in X$.*

Proof. We include the proof since it is easily given. Suppose that $\bigwedge X \leq p$ with p prime and X a closed upper set from D. If $p \notin X$, then since X is an upper set, for each $x \in X$ we can find a compact element $k_x \leq x$ with $p \notin \uparrow k_x$. Since X is compact, finitely many of the sets $\uparrow k_x$ cover X, say $\uparrow k_1, \ldots, \uparrow k_n$. Then $k_1 \wedge k_2 \wedge \cdots \wedge k_n \leq \bigwedge X \leq p$, and since p is prime, this implies that $k_i \leq p$ for some index i. This contradiction to $p \notin \uparrow k_x$ implies that $p \in X$.

Lemma 2.4 implies that $\uparrow \overline{(\uparrow x \bigcap Spec\ D)}$ is the smallest closed upper set X with $\bigwedge X \leq x$ for each x in D; it is easy to show that this property characterizes distributive domains:

Proposition 2.5. *Let D be a domain. The following are equivalent:*
1) D is distributive.
2) For each $x \in D$, there is a smallest closed upper set X with $\bigwedge X \leq x$.

Proof. Lemma 2.4 shows that 1) implies 2), where $X = \uparrow \overline{(\uparrow x \bigcap Spec\ D)}$ for each $x \in D$. Conversely, suppose that 2) holds. Now, the map $A \mapsto \bigwedge A: \mathbf{C}(D) \to D$ preserves all infs and all directed sups since it is continuous, and so this map has a lower adjoint which preserves compact elements (cf. [Gie80], Theorem 3.4, p.19 and Theorem 1.4, p.180). Hypothesis 2) amounts to assuming that this map also has an upper adjoint, which implies that $A \mapsto \bigwedge A: \mathbf{C}(D) \to D$ preserves all sups (cf. op cit, Theorem 3.2, p.19). This means the map $A \mapsto \bigwedge A: \mathbf{C}(D) \to D$ preserves all infs and sups, and so its image, D, must be distributive since its domain, $\mathbf{C}(D)$, is distributive.

There is a notion which lies between distributive domains and arbitrary domains.

Definition 2.6. *The domain D is* **semiprime** *if each element x of D is the infimum of primes in the upper set of x. That is, if $Spec(\uparrow x)$ denotes those elements of D which are prime in $\uparrow x$ (but not ncessarily prime in all of D), then D is semiprime iff $x = \bigwedge(Spec(\uparrow x))$ for every $x \in D$.*

An example where such domains arise is the following. Let K be a compact convex subset of a locally convex vector space E. Then $Con(K)$, the family of all compact convex subsets of K, is a continuous lattice in the opposite order (i.e., $A \leq B$ iff $B \subset A$). If $ext(K)$ denotes the set of extreme points of K, then $Spec(Con(K)) = \{\{x\} | \, x \in ext(K)\}$; since E is locally convex, every element A of $Con(K)$ is the closed convex hull of its extreme points, so $Con(K)$ is semiprime. $Con(K)$ is not distributive, however, since, for any x in K, the set $\{x\}$ is irreducible in $Con(K)$. (For details, see Example 1.22, p.50 and Exercise 3.39, p.80 of [Gie80].)

One might ask whether Proposition 2.5 generalizes to the case of semiprime domains. We believe that this more general result should hold, but we can only verify it in the two cases indicated below. Recall that a domain is *arithmetic* if the infimum of compact elements is again compact; equivalently, D is arithmetic if D^0 is an inf-subsemilattice of D.

Proposition 2.7. *Let D be a domain, and consider the following conditions:*
1) D is a semiprime domain.
2) For each $x \in D$ there is a smallest element X in $\mathbf{U}(D)$ *with $x = \bigwedge X$.*
Then 1) implies 2) always holds. Conversely, if D is arithmetic or if D has compactly finite breadth, then 2) implies 1).

Proof. 1) implies 2): If D is semiprime, then $x = \bigwedge(Spec(\uparrow x))$ for each $x \in D$. If Y is any closed upper set in D with $\bigwedge Y = x$, then $Y \subset \uparrow x$, and so Lemma 2.4 applied to the domain $\uparrow x$ implies that $Spec(\uparrow x) \subset Y$. Clearly $\uparrow(\overline{Spec(\uparrow x)})$ is then the smallest closed upper set whose infimum is x.

2) implies 1): Suppose first that D is algebraic. Fix $x \in D$, and let X denote the smallest set in $\mathbf{U}(D)$ with $\bigwedge X = x$. Since X is closed, the set M of minimal elements of X satisfies $X = \uparrow M$. We show that each element of M is a prime in $\uparrow x$. Indeed, suppose to the contrary that $m \in M \backslash Spec(\uparrow x)$. Then there are $a, b \in \uparrow x \backslash \uparrow m$ with $a \wedge b \leq m$. Then there are elements k_a, $k_b \in D^0$ with $k_a \leq a$, $k_b \leq b$, but $m \notin \uparrow k_a \bigcup \uparrow k_b$. Since D is arithmetic, $k_a \wedge k_b \in D^0$ and clearly $k_a \wedge k_b \leq m$. Thus $\uparrow(k_a \wedge k_b)$ is an open set containing $\uparrow m$, and so $\uparrow(X \backslash \uparrow(k_a \wedge k_b))$ is a closed subset of $\uparrow x$. So also is the set $\uparrow(X \backslash \uparrow(k_a \wedge k_b)) \bigcup(\uparrow k_a \bigcup \uparrow k_b)$, and this upper set satisfies $\bigwedge(\uparrow(X \backslash \uparrow(k_a \wedge k_b)) \bigcup(\uparrow k_a \bigcup \uparrow k_b)) = \bigwedge \uparrow(X \backslash \uparrow(k_a \wedge k_b)) \wedge \bigwedge(\uparrow k_a \bigcup \uparrow k_b) = \bigwedge X = x$. But $\uparrow(X \backslash \uparrow(k_a \wedge k_b)) \bigcup(\uparrow k_a \bigcup \uparrow k_b)$ is a proper closed upper set of X, which contradicts the fact that X is the smallest closed upper set whose infimum is x. Thus m must be prime in $\uparrow x$, and so $M \subset Spec(\uparrow x)$, which proves that D is semiprime.

Now, suppose that D has compactly finite breadth, and, again, let X be the smallest closed upper set in D whose infimum is the element $x \in D$. Since D has compactly finite breadth, there is a finite subset F of X with $\bigwedge X = \bigwedge F$; the properties of X imply that $\uparrow F = X$. Suppose that $f \in F \setminus Spec(\uparrow x)$. Then there are elements a, b in D with $a \wedge b \leq f$, but $f \notin \uparrow a \bigcup \uparrow b$. It is routine to show that $\uparrow(F \setminus \{f\}) \bigcup (\uparrow a \bigcup \uparrow b)$ is a closed upper set whose infimum is still x, but that this is a proper subset of $X = \uparrow F$. This contradiction shows that $F \subset Spec(\uparrow x)$, so once again, 1) is satisfied.

Note: Proposition 2.7 was discovered in collaboration with J. D. Lawson.

III. Some Special Cases. In this section we generalize what we have found about the inf-map $X \mapsto \bigwedge X: \mathbf{U}(D) \to D$ to more general domains than those which are distributive or semiprime. In the most restrictive case, if D is a distributive domain of locally finite breadth, then the results from Section 1 imply that each element of D is the infimum of finitely many primes from D. So, if X is an element of $\mathbf{U}(D)$, then $\bigwedge X = p_1 \wedge p_2 \wedge \cdots \wedge p_n$ for finitely many primes $p_1, \ldots, p_n$. Lemma 2.4 implies that $p_i \in X$ for each index i, and so $\uparrow\{p_i | \, i = 1, \ldots, n\} \subset X$ is the smallest element of $\mathbf{U}(D)$ whose infimum is $\bigwedge X$. In fact the domain D is isomorphic to the domain $\mathbf{I}(Spec\ D)^0$ of all non-empty finitely generated upper sets of $Spec\ D$ in this case, where the order on $\mathbf{I}(Spec\ D)^0$ is the containment order. For finite subsets F, $G \subset Spec\ D$, this order translates to

$$F \leq G \; iff \; \forall g \in G, \; \exists f \in F \; with \; f \leq g \; iff \; F \subset \downarrow G.$$

The lattice $D^\top$ is then isomorphic to the lattice of all finite subsets of $Spec\ D$ under this same order.

But even when D is only distributive, Lemma 2.4 still implies that the element $\bigwedge X$ is the infimum of (perhaps infinitely many) primes, all of which lie within X for each X in $\mathbf{U}(D)$. In this case, however, one must resort to spectral theory to gain a complete description of the domain D in terms of the subsets of $Spec\ D$; see [Gie80], Chapter V for details. We summarize this in the following:

Proposition 3.1. *Let D be a distributive domain.*

1) *For each X in $\mathbf{U}(D)$, $\bigwedge X = \bigwedge(X \bigcap Spec\ D)$ is the infimum of primes in X, and $\uparrow \overline{(\uparrow(\bigwedge X) \bigcap Spec\ D)}$ is the smallest element of $\mathbf{U}(D)$ whose infimum is $\bigwedge X$.*
2) *If D has locally finite breadth, then there is a finite subset $F \subset X \bigcap Spec\ D$ with $\bigwedge X = \bigwedge F$ for each X in $\mathbf{U}(D)$, so each $x \in D$ is the infimum of finitely many primes, all of which lie in X.*

For distributive domains D, describing the primes of D gives information about the elements of the Smyth power domain of D. However, it is unusual that a domain is distributive; e.g., flat domains are not distributive. In order to make this theory applicable to more general domains, we introduce the following notion.

Definition 3.2. *A domain D is* **locally distributive** *if $\downarrow x$ is distributive for each $x \in D$. For a locally distributive domain D, we denote the primes of $\downarrow x$ by $Spec(\downarrow x)$.*

Flat domains are locally distributive, as are the dI-domains of Berry [Ber78] (see also [Co88] in this volume). This notion allows us to localize the theory we have presented so far. In a locally distributive domain D, if $X \in \mathbf{U}(D)$ and if $x \in X$, then $\bigwedge X = \bigwedge(\uparrow(x \wedge X) \bigcap \mathit{Spec}(\downarrow x))$ since $\bigwedge X \in \downarrow x$ and $\downarrow x$ is a distributive domain. Moreover, we are free to choose whichever $x \in X$ we want to calculate $\bigwedge X$. In other words, if there is some $x \in X$ for which we know the elements of $\mathit{Spec}(\downarrow x)$, then we know that $\bigwedge X = \bigwedge(\uparrow(x \wedge X) \bigcap \mathit{Spec}(\downarrow x))$ for this particular x. In addition, if the subdomain $\downarrow x$ of D is also of locally finite meet breadth, then there is a finite subset F of $\uparrow(x \wedge X) \bigcap \mathit{Spec}(\downarrow x)$ with $\bigwedge X = \bigwedge F$.

From a computational viewpoint, it is not only important to realize $\bigwedge X$ as the infimum of some special set of elements of the domain with which one is dealing, it is also important to know those elements represent computable elements of the domain, or at the least, that there is a uniform way to approximate them with computable elements. We therefore introduce the following notion.

Definition 3.3. *The domain D is* **powerlike** *if D^0 is an ideal of D; i.e., if $\downarrow D^0 = D^0$.*

An example of such a domain is the lattice $P\omega \simeq 2^{\mathbf{N}}$, from which powerlike domains get their name. These domains are intuitively appealing for computational purposes for the following reason. If we view a domain D as modeling computation so that the compact elements of D represent the information which we could receive in finite time, then it reasonable to assume that anything in D which lies below a compact element of D should also be compact; this is exactly the assumption that D^0 be an ideal of D.

Locally distributive powerlike domains are relevant to our discussion of $\mathbf{U}(D)$ and D because of the following. If we have a subset X of D which is in $\mathbf{U}(D)$, and if $x_0 \in X \bigcap D^0$, then, as we have seen, $\bigwedge X = \bigwedge(\uparrow(x_0 \wedge X) \bigcap \mathit{Spec}(\downarrow x_0))$, and since D is powerlike, the set $\uparrow(x_0 \wedge X) \bigcap \mathit{Spec}(\downarrow x_0)$ consists of compact elements of D. If, in addition, D is of locally finite meet breadth, then there is a finite subset $F \subset \uparrow(x_0 \wedge X) \bigcap \mathit{Spec}(\downarrow x_0)$ with $\bigwedge X = \bigwedge F$, and, once again, the set F consists of compact elements of D. Finally, even if X contains no compact elements, the fact that the inf-map $X \mapsto \bigwedge X : \mathbf{U}(D) \to D$ is continuous implies that for any finite subset F of D^0 with $X \subset \uparrow F$, $\bigwedge F$ approximates $\bigwedge X$. If D is powerlike (or even just arithmetic), then $\bigwedge F$ is compact since each element of F is, and so the compact elements of $\mathbf{U}(D)$ which approximate X naturally give rise to compact elements of D which approximate $\bigwedge X$. We summarize all of this in the following.

Theorem 3.4. *Let D be a domain.*

I. *If D is locally distributive, then for each element X of $\mathbf{U}(D)$ and each $x \in X$, we have $\bigwedge X = \bigwedge(\uparrow(x \wedge X) \bigcap \mathit{Spec}(\downarrow x))$. Moreover, if D is of locally finite meet breadth and if $x_0 \in X$, then there is some finite subset F of $\uparrow(x_0 \wedge X) \bigcap \mathit{Spec}(\downarrow x)$ with $\bigwedge X = \bigwedge F$.*

II. *If D is also a powerlike domain and $x_0 \in X \bigcap D^0$, then $\bigwedge X = \bigwedge(\uparrow(x_0 \wedge X) \bigcap \mathit{Spec}(\downarrow x_0))$ expresses $\bigwedge X$ as the infimum of compact elements which are primes in $\downarrow x_0$. If D is also of locally finite meet breadth, then $\bigwedge X = \bigwedge F$ expresses $\bigwedge X$ as the infimum of finitely many such elements.*

III. *If D is arithmetic and F is a finite subset of D^0 with $X \subset \uparrow F$, then $\bigwedge F$ approximates $\bigwedge X$ and $\bigwedge F$ is a compact element of D.*

Local distributivity - as well as distributivity itself - allow for a "top-down" analysis of realizing the infimum of an element of the Smyth power domain as an infimum of primes. On the other hand, semiprime domains combine this with a "bottom-up" approach. Namely, in a semiprime domain D, each element X of the Smyth power domain $\mathbf{U}(D)$ is an infimum of primes, but the primes whose infimum is $\bigwedge X$ are only prime in $\uparrow(\bigwedge X)$, and not generally prime in the entire domain. Even so, for any element Y of $\mathbf{PS}(D)$ with $\bigwedge Y = \bigwedge X$, we have $Spec(\uparrow\bigwedge X) \subset Y$. Moreover, if D has compactly finite breadth, then each element X of the Smyth power domain has its infimum, $\bigwedge X$, given as the infimum of finitely many "$\bigwedge X$-primes", all of which lie in the set X. Since the elements of $Spec(\uparrow(\bigwedge X))$ are irreducible in the domain D, we have the following result: for a semiprime domain D of compactly finite breadth, if X is an element of the Smyth power domain over D, then the element $\bigwedge X$ is given as the infimum of finitely many irreducible elements. Moreover, in this case, those irreducibles must lie in the set X.

If the domain D is arithmetic or is of compactly finite breadth, then Proposition 2.6 says D is semiprime if and only if every element is the infimum of a unique smallest element of the Smyth power domain, and this gives an alternative test for the domain to be semiprime.

Even in the complete absence of distributivity, some things can be said. If it were true that a domain D had compactly finite breadth if and only if each element were the infimum of finitely many irreducible elements, then the infimum $\bigwedge X$ would be given as the infimum of finitely many irreducible elements of D for each set X in the Smyth power domain over D. But, we can no longer assert that those irreducible elements lie in X. Understanding what irreducible elements and what prime elements are would mean understanding somewhat better the relationship between the Smyth power domain and the domain upon which it is built. After all, every domain is order-generated by irreducible elements, and although only the distributive domains are order-generated by primes, certainly primes should exist to some extent in every domain. Note that since any powerlike domain is arithmetic, the semiprime powerlike domains are characterized by the property that each element is the infimum of a unique smallest element in the Smyth power domain. So, the results from Section 2 imply that in such domains, each element X of the Smyth power domain is the infimum of finitely many $\bigwedge X$-primes, all of which again lie in X.

In [BHR] the *failures model* for CSP is developed, and an improved model is presented in [BR]. In these models, communicating sequential processes are modeled by their *failures*, which contain possible traces of actions the processes may have performed, along with the set of actions the process could then refuse to perform. It is easy to check that the models constructed are locally distributive powerlike domains, and so the results we have outlined above apply. In fact, these models are *Smyth-like domains*, in that the order reflects how non-deterministic a process is; the higher it is in the order, the more deterministic a process is (this definition is from [Old86]). The maximal processes are thus the most deterministic processes. As commented in [BR], there have been attempts to use these ideas to model non-deterministic processes in terms of deterministic ones; from our comments above, this clearly amounts to writing any process P as the infimum of the deterministic processes which lie above it in the order. The condition that every process is an infimum of *finitely many* deterministic processes implies that the system under study has bounded nondeterminism, but what condition actually corresponds to bounded nondeterminism is unclear. An investigation of what the *prime* processes are and what role they play in such models is also a topic for further research.

References

[Ber78] Berry, G., *Stable models of the typed λ-Calculi*, Lecture Notes in Computer Science **62** (1978), Springer-Verlag, 72 - 89.

[BHR] Brookes, S. D., C. A. R. Hoare and A. W. Roscoe, *A theory of communicating sequential processes*, Journal of the ACM **31** (1984), 560 - 599.

[BR] Brookes, S. D. and A. W. Roscoe, *An improved failures model for communicating processes*, Lecrture Notes in Computer Science **197** (1985), Springer-Verlag, 281 - 305.

[Co88] Coquand, T., C. Gunter and G. Winskel, *DI-domains as models for polymorphism*, This Volume.

[Gie77] Gierz, G. and K. Keimel, *A lemma on primes appearing in algebra and analysis*, Houston Journal of Mathematics **3** (1977), 207 - 224.

[Gie80] Gierz, G., K. H. Hofmann, K. Keimel, J. D. Lawson, M. Mislove and D. Scott, **A Compendium of Continuous Lattices**, Springer-Verlag, Berlin, Heidelberg and New York (1980), 371 pp.

[Gie85] Gierz, G., J. D. Lawson and A. R. Stralka, *Intrinsic topologies for semilattices of finite breadth*, Semigroup Forum **31** (1985), 1 - 18.

[Grä68] Grätzer, G., **Universal Algebra**, D. van Nostrand, Toronto (1968), 368 pp.

[Hof74] Hofmann, K. H., M. Mislove and A. R. Stralka, **The Pontryagin Duality of Compact 0-Dimensional Semilattices and Its Applications**, Lecture Notes in Mathematics **396** (1974), Springer-Verlag, 122 pp.

[Liu83] Liukkonen, J. R. and M. Mislove, *Measure algebras of locally compact semilattices*, Lecture Notes in Mathematics **998** (1983), Springer-Verlag, 202 - 214.

[Old86] Olderog, E.-R. and C. A. R. Hoare, *Specification-oriented semantics for communicating sequential processes*, Acta Informatica **23** (1986), 9 - 66.

[Smy82] Smyth, M. B., *Power domains and predicate transformers: a topological view*, Lecture Notes in Computer Science **154** (1982), Springer-Verlag, 662 - 675.

[Sto36] Stone, M. H., *The theory of representations for Boolean algebras*, Transactions of the American Mathematical Society **40** (1936), 37-111.

The Metric Closure Powerspace Construction

Robert E. Kent
Department of Electrical Engineering and Computer Science
University of Illinois at Chicago
Chicago, IL., 60680

Abstract

In this paper we develop a natural *powerobject* construction in the context of enriched categories, a context which generalizes the traditional order-theoretic and metric space contexts. This powerobject construction is a *subobject transformer* involving the *dialectical flow* of closed subobjects of enriched categories. It is defined via factorization of a *comprehension schema* over metrical predicates, followed by the fibrational inverse image of metrical predicates along *character*, the left adjoint in the comprehension schema. A fundamental continuity property of this metrical powerobject construction vis-a`-vis greatest fixpoints is established by showing that it preserves the limit of any *Cauchy* ω^{op}*-diagram.* Using this powerobject construction we unify two well-known fixpoint semantics for concurrent interacting processes.

Introduction

In order to define the semantics of interacting processes, concurrent programs, and programming languages with nondeterministic or parallel features, we would like to solve recursive equations of the form $X = F(X)$ where F is a composite operator containing the powerset operator P as one component. An attempt to find the greatest fixpoint solution of the nonlinear recursive domain equation $X \cong 1+A\times P(X)$ in the category of sets and functions, ends in failure, because of cardinality; but in such an attempt, the naturally induced prefix metric suggests changing from the relatively unstructured context of sets and functions to the enriched context of metric spaces and suitable metric space morphisms (in this case, contractions), and using the closure operator in conjunction with the powerset operator. We call this conjunction the *closure powerspace functor*, and we denote it by $\overline{P}$. If this functor is used in place of P in the above equation, then the greatest fixpoint solution exists. This greatest fixpoint solution consists of the metric space of finite or infinite node–labelled trees, where tree branching is unordered and without duplicates. The closure powerspace functor and its continuity properties are the main concern of this paper. Application of the closure powerspace functor to the solution of the above equation gives the metric space of closed forests of node–labelled trees. We show that the subspace of closed forests of full node–labelled trees is isomorphic to the greatest fixpoint solution of the recursive domain equation $X \cong \overline{P}(A \times X)$, the metric space of finite or infinite edge–labelled trees.

So in this paper we follow the metric space approach. However, we use the more general context of "arbitrary, and not necessarily complete, asymmetric quasimetric spaces" for the following reasons:

- It has a very natural categorical definition (it is a normed enriched category of enriched V–categories, Lawvere [1973]).

- The asymmetry in the definition allows us to subsume the order–theoretic context of preorders.

- Some applications in computer science (Golson and Rounds [1983], Kent [1984,1985]) need quasimetrics, not metrics.

- We prefer to let the completeness property arise naturally out of the greatest fixpoint construction, since, as we have observed elsewhere (Kent [1984]), the greatest fixpoint solution, in the categorical theory of abstract data types, is (often) the metric space completion, using the naturally induced prefix metric, of the least fixpoint solution.

We make extensive use of the *Hausdorff metric,* which we believe is very natural and conceptually very simple. To demonstrate this simplicity and buttress our belief, we develop the metric space specializations of *existential Kan quantification* and *comprehension schema*, the adjunction between the *character* of subobjects of enriched categories (Kuratowski [1966]) and the *extension* of predicates over enriched categories, and show how the Hausdorff metric is "naturally" induced from these. Our approach was originally advocated by Lawvere [1973], but was not developed in much detail. We give a detailed development in this paper.

Our main result is a categorical analogue to Hahn's theorem (DeBakker and Zucker p.111 [1982], Nivat p.28 [1980]). In this analogy, ω^{op}–diagrams correspond to Cauchy sequences of closed sets, and categorical limits of ω^{op}–diagrams correspond to metrical limits of Cauchy sequences of closed sets using the Hausdorff metric (and in more detail, elements in the categorical limit of an ω^{op}–diagram correspond to appropriate Cauchy sequences of elements; namely, to elements in the above metrical limit). In the analogy, limits of ω^{op}–diagrams also correspond to the completion of metric spaces.

Our main application is the development of Milner's strong congruence on nondeterministic transition systems, a categorical analogue to the classical Nerode congruence on deterministic transition systems, in terms of the categorical greatest fixpoint solution of the recursive domain equation $X \cong \overline{P}(A \times X)$. Milner's congruence forms the basis for his semantics of synchronous communicating systems called SCCS (Milner [1983]). Our application builds a strong link between two important and well–known theories of concurrency: the complete metric space approach of DeBakker and Zucker [1982] and Nivat [1979], and the algebraic approach of Milner [1980,1983].

The metric closure powerspace construction is a canonical *subobject transformer* for enriched categories, which is related to the *predicate transformer* of existential Kan quantification through a comprehension schema (character–extension adjunction). Both subobject tranformers and predicate transformers are special cases of the notion of *dialectical flow*: the motion of structured entities in dialectical systems. This theory of dialectical motion has already (Kent [1986,1987]) been applied to four important areas of computer science:

- concurrent systems, where it distinguishes the notions of observational equivalence and dialectical motion of transition systems;
- OBJ–like functional programming (Kent [1986]), where it generalizes the notion of *institutions*, and is based upon the doctrinal diagram associated with algebraic theories;
- generalized Petri net theory, where it unites notions of *nets* with the notion of *predicate transformers* in Domain theory, and is based upon the notion of Kan quantification and normed categories (this paper); and
- first order logic (Kent [1987]), where it unifies the semantics of Horn clause logic with that of relational databases, and is based upon the notion of model doctrines.

Dialectical systems (Bernow and Raskin [1976]) contain the following essential aspects:

- based upon contradictions or "opposing tendencies";
- interacting objects or entities;
- movement, motion or development; and
- reproduction or renewal of entities.

All of these aspects are present in subobject and predicate transformers and the parallel composition of concurrent interacting transition systems (processes). F.W. Lawvere gave the theory of dialectical systems its most succinct expression: "Category Theory ≡ Objective Dialectics". Indeed dialectics invests the *dynamical view* of systems theory with the fundamental ideas of *category theory*, such as adjunctions, fibrations, limits, tensors and Kan extensions (the components of *dialectical flow*); but in turn, it gives these categorical notions that dynamical view. In short, the theory of dialectics studies both the "structure of motion (development)" and the "motion of structure".

1. Quasimetric Spaces and Preorders

A pair $X=<X,d>$ consisting of a set X and a function $d: X \times X \longrightarrow R$, where $R=[0,\infty]$ is the set of nonnegative real numbers with infinity, is a *quasimetric*, or *gauge, space* when $d(x_1,x_2)+d(x_2,x_3) \geq d(x_1,x_3)$, for all $x_1,x_2,x_3 \in X$; and $0=d(x,x)$, for all $x \in X$. A quasimetric space is precisely an **R**-*enriched category* for the closed poset $\mathbf{R}=<R,\geq,+,0,\dot{-}>$, where subtraction $\dot{-}$ is defined below. A pair $X=<X,d>$ consisting of a set X and a function $d: X \times X \longrightarrow I$, where $I=[0,1]$ is the *interval*, that is, the bounded set of real numbers $I=\{r|0 \leq r \leq 1\}$, is an *ultraquasimetric*, or *ultragauge, space* when $\max\{d(x_1,x_2),d(x_2,x_3)\} \geq d(x_1,x_3)$, for all $x_1,x_2,x_3 \in X$; and $0 = d(x,x)$, for all $x \in X$. The quasimetric (ultraquasimetric) space $<X,d>$ is a *metric (ultrametric) space* when it satisfies the additional condition: if $d(x_1,x_2) = 0$ and $d(x_2,x_1) = 0$ then $x_1 = x_2$. In our metric spaces we allow the possibility of zero distance $d(x_1,x_2) = 0$ (in one direction) between two distinct points $x_1 \neq x_2$. The function d is called a *metric, gauge* or *distance*. Clearly an ultra(quasi)metric space is a (quasi)metric space, since $r_1+r_2 \geq \max(r_1,r_2)$ for all $r_1,r_2 \in R$. Notice that in general our metrics are asymmetrical: $d(x_1,x_2) \neq d(x_2,x_1)$. A fundamental example of a quasimetric space is the metric space $\mathbf{R}=<R,d>$ where the distance $d: R \times R \longrightarrow R$ is defined by $d(r_1,r_2) = r_2 \dot{-} r_1 = r_2 - r_1$, if $r_2 \geq r_1$; $= 0$, otherwise. This is not a classical metric space, but the symmetrization $d^{sym}(r_1,r_2) = (r_2 \dot{-} r_1)+(r_1 \dot{-} r_2) = |r_1-r_2|$ is the usual metric space on the reals R. Similarly, a fundamental example of an ultraquasimetric space is the ultrametric space $\mathbf{I}=<I,d>$, where the distance is defined by $d(r_1,r_2) = r_2 \boxminus r_1 = r_2$, if $r_2 > r_1$; $= 0$, otherwise. Symmetrization $d^{sym}(r_1,r_2) = \max\{r_2 \boxminus r_1, r_1 \boxminus r_2\} = \max\{r_1,r_2\}$ if $r_1 \neq r_2$, $= 0$ if $r_1=r_2$, is the usual ultrametric on the interval I.

Associated with every quasimetric space $X=<X,d>$ is a preorder $X=<X,\leq>$ where $x \leq y$ when $d(x,y)=0$, and x and y are unrelated when $d(x,y)>0$. For the nonnegative reals $\mathbf{R}=<R,d>$, $d(r_1,r_2)=0$ iff $r_1 \geq r_2$. So $U_{\square}(<R,d>) = <R,\geq>$ is the usual order on the reals: the order used in the definition of quasimetric spaces. In the other direction, associated with every preorder $X=<X,\leq>$ is a quasimetric space $X=<X,d>$ where $d(x,y) = 0$, if $x \leq y$; $= \infty$, otherwise. Let us denote this construction by $<X,d> = F_{\square}(<X,\leq>)$.

Let $X=<X,d_X>$ and $Y=<Y,d_Y>$ be two (ultra)quasi metric spaces. A function $f: X \longrightarrow Y$ is a *contraction* from X to Y when $d_X(x_1,x_2) \geq d_Y(f(x_1),f(x_2))$ for all $x_1,x_2 \in X$. We denote this by $f: <X,d_X> \longrightarrow <Y,d_Y>$.

Let **QMet** denote the category whose objects are quasimetric spaces and whose morphisms are contractions. Let **UQMet** denote the full subcategory of **QMet** whose objects are ultraquasimetric spaces. There is an obvious *underlying set functor* U_{QMet}: **QMet** $\longrightarrow$ **Set**, where $U_{QMet}(<X,d>) = X$ and $U_{QMet}(f) = f$. Any function $f: X \longrightarrow Y$ from a quasimetric space determines a direct image quasimetric space $f^*(<X,d_X>) = <Y,d^f>$ over the set Y, where the metric d^f is defined by $d^f(y,y') =$ $\inf\{d_X(x,x_1)+d_X(x_1',x_2)+\dots+d_X(x_{n-1}',x_n)+d_X(x_n',x') \mid \text{some } x,x_1,x_1',\dots,x_n,x_n',x' \in X \text{ with } y=f(x), f(x_1)=f(x_1'),\dots,f(x_n)=f(x_n'), f(x')=y' \text{ and } 0 \leq n\}$ where the inside expression for $n=0$ is $d_X(x,x')$, if y and y´ are in the ordinary set–theoretic direct image of f; = 0, if y = y´; and = ∞, otherwise. Then $f:\langle X,d_X\rangle \longrightarrow f^*(\langle X,d_X\rangle)$ is a contraction. If $f:\langle X,d_X\rangle \longrightarrow \langle Y,d_Y\rangle$ is a contraction for some metric d_Y on Y, then $Id_Y : f^*(\langle X,d_X\rangle) \longrightarrow \langle Y,d_Y\rangle$ is a contraction: that is, $d^f(y,y') \geq d_Y(y,y')$ for all $y,y' \in Y$. Any function $f: X \longrightarrow Y$ into a quasimetric space determines an inverse image quasimetric space $f_*(\langle Y,d_Y\rangle) = \langle X,d_f\rangle$ over the set X, where the metric d_f, called the *kernel metric* of f, is defined by $d_f = d_Y.(f{\times}f) : X{\times}X \longrightarrow Y{\times}Y \longrightarrow R$; that is, $d_f(x,x') = d_Y(f(x),f(x'))$ for all $x,x' \in X$. Then $f: f_*(\langle Y,d_Y\rangle) \longrightarrow \langle Y,d_Y\rangle$ is a contraction; actually an isometry. If $f:\langle X,d_X\rangle \longrightarrow \langle Y,d_Y\rangle$ is a contraction for some metric d_X on X, then $Id_X : \langle X,d_X\rangle \longrightarrow f_*(\langle Y,d_Y\rangle)$ is a contraction; that is, $d_X(x',x) \geq d_f(x',x)$ for all $x',x \in X$. In this case, the factorization $f = f.Id_X : \langle X,d_X\rangle \longrightarrow \langle X,d_f\rangle \longrightarrow \langle Y,d_Y\rangle$, is called the *full image factorization* of f, with Id_X giving f's action on R–values and the isometry f giving f's action on elements of X.

Fact. *Quasimetric spaces (* **R**–*categories) form a* 01*-fibration over sets; that is, the underlying set functor* U_{QMet} *is a* 01*-fibration.*

Let **Met** denote the full subcategory of **QMet** whose objects are metric spaces, and let $Inc_{Met} : \mathbf{Met} \longrightarrow \mathbf{QMet}$ be the associated inclusion functor. Since $\langle X,d^{sym}\rangle$ is a symmetric quasimetric space, the preorder associated with it is an equivalence relation $\leq^{sym} = \equiv$ defined by $x \equiv y$ iff $x \leq y$ and $y \leq x$. For every quasimetric space ⟨X,d⟩ the associated equivalence relation ≡, defines a quotient metric space $\langle X/{\equiv},d_{\equiv}\rangle$, where $X/{\equiv} = \{[x] | x \in X\}$ is the set of equivalence classes with the metric $d_{\equiv}$ defined by $d_{\equiv}([x],[y]) = d(x,y)$. Denote this quotient space by $F_{Met}(\langle X,d\rangle)$. Then $F_{Met}(\langle X,d\rangle)$ is the metric space induced by ⟨X,d⟩, and $U_{\square}(F_{Met}(\langle X,d\rangle))$ is the partial order induced by $U_{\square}(\langle X,d\rangle) = \langle X,\leq\rangle$. For any contraction $f:\langle X,d_X\rangle \longrightarrow \langle Y,d_Y\rangle$ define $F_{Met}(f) : \langle X/{\equiv_X},d_{X,\equiv}\rangle \longrightarrow \langle Y/{\equiv_Y},d_{Y,\equiv}\rangle$ by $F_{Met}(f)([x]) = [f(x)]$. It is easy to see that $F_{Met}(f)$ is well–defined and that $F_{Met} : \mathbf{QMet} \longrightarrow \mathbf{Met}$ is a functor. Obviously there is a canonical isometry $\eta_{Met}(\langle X,d\rangle) = [\,] : \langle X,d\rangle \longrightarrow \langle X/{\equiv},d_{\equiv}\rangle$.

Proposition. $F_{Met} : \mathbf{QMet} \longrightarrow \mathbf{Met}$ *is left adjoint to the inclusion functor* $Inc_{Met} : \mathbf{Met} \longrightarrow \mathbf{QMet}$ *with unit* $\eta_{Met} : Id_{QMet} \Longrightarrow Inc_{Met}.F_{Met}$ *and counit* $\varepsilon_{Met} = Id : F_{Met}.Inc_{Met} \Longrightarrow Id_{Met}$.

This is a reflection: F_{Met} reflects **QMet** into its subcategory **Met**. Since Inc_{Met} is a right adjoint it preserves all limits.

A *metric strecher* for quasimetric spaces is a function $\lambda : R \longrightarrow R$ which is order–preserving: (i) if $r_1 \geq r_2$ then $\lambda(r_1) \geq \lambda(r_2)$; subadditive: (ii) $\lambda(r_1)+\lambda(r_2) \geq \lambda(r_1+r_2)$; and zero–preserving: (iii) $0 = \lambda(0)$. The following functions are metric strechers for quasimetric spaces: 1. $\alpha(r) = \alpha.r$, for some $\alpha \in R$, the "constant strecher"; 2. $sat(r) = r$, if $r \leq 1$, the "saturation strecher"; = 1, otherwise. A metric strecher for ultraquasimetric spaces is a function $\lambda: I \longrightarrow I$ which is order–preserving and satisfies the inequalities: $\max\{\lambda(r_1),\lambda(r_2)\} \geq \lambda(\max\{r_1,r_2\})$; and $0 \geq \lambda(0)$. For any constant $\alpha \in R$ the constant saturation strecher $sat(\alpha) = sat.\alpha : I \longrightarrow R \longrightarrow I$ defined by $sat(\alpha)(r) = \alpha.r$, if $r \leq 1/\alpha$, $= 1$, otherwise, is a metric strecher for ultraquasimetric spaces. A metric strecher λ defines for each (ultra)quasimetric space X=<X,d> another "streched" (ultra)quasimetric space $\lambda X = \langle X, \lambda d\rangle$ where $\lambda d = \lambda.d = X \times X \longrightarrow R \longrightarrow R$. This defines a functor λ: **QMet** ⟶ **QMet**. For any constant $\alpha \in R$, $\alpha \neq 0,\infty$, α: **QMet** ⟶ **QMet** is an isomorphism with inverse functor $1/\alpha$. $\infty = F_{\Box}.U_{\Box}$: **QMet** ⟶ **PO** ⟶ **QMet** and 0 : **QMet** ⟶ **QMet** is the constant functor giving the trivial contiguous space. For $\alpha \in I = [0,1]$ α extends a functor on **UQMet.** Obviously there is a canonical contraction $\eta_{\alpha}(\langle X,d\rangle) = Id_X : \langle X,d\rangle \longrightarrow \alpha.sat(1/\alpha)(\langle X,d\rangle)$. For every constant $\alpha \in I$, $\alpha \neq 0$, $sat(1/\alpha)$: **UQMet** ⟶ **UQMet** is left adjoint to α : **UQMet** ⟶ **UQMet** with unit $\eta_{\alpha} : Id_{UQMet} \Longrightarrow \alpha.sat(1/\alpha)$ and counit $\varepsilon_{\alpha} = Id : sat(1/\alpha).\alpha \Longrightarrow Id_{UQMet}$. This is a reflection: $sat(1/\alpha)$ reflects **UQMet** into itself; that is, "constant strechers define reflectors". Since α is a right adjoint it preserves all limits.

In **QMet** every contraction $f : \langle X,d_X\rangle \longrightarrow \langle Y,d_Y\rangle$ is a monotonic function $f : \langle X,\leq_X\rangle \longrightarrow \langle Y,\leq_Y\rangle$, where $\leq_X$ are $\leq_Y$ are the associated preorders. So the construction $U_{\Box}$ is actually a functor $U_{\Box}$: **QMet** ⟶ **PO**, the *underlying preorder functor*, where **PO** is the category of preorders and monotonic functions. Associated with every contraction $f : \langle X,d_X\rangle \longrightarrow \langle Y,d_Y\rangle$ is a binary relation $\leq_f$ on X, called the *kernel preorder* of f, and defined by: $x \leq_f x'$ when $f(x) \leq_Y f(x')$. Clearly $\leq_f$ is a preorder containing the preorder $\leq_X$, $U_{\Box}(\langle X,d_f\rangle) = \langle X,\leq_f\rangle$, and $U_{\Box}$ maps the factorization $f = f.Id_X : \langle X,d_X\rangle \longrightarrow \langle X,d_f\rangle \longrightarrow \langle Y,d_Y\rangle$ to the factorization $f = f.Id_X : \langle X,\leq_X\rangle \longrightarrow \langle X,\leq_f\rangle \longrightarrow \langle Y,\leq_Y\rangle$.

Fact. *Quasimetric spaces (***R***–categories) form a 01–fibration over preorders (***2***–categories); that is, the underlying preorder functor* $U_{\Box}$ *is a 01–fibration.*

Every quasimetric space <X,d> has associated with it a canonical contraction $\varepsilon_{\Box}(\langle X,d\rangle) = Id_X : U_{\Box}.F_{\Box}(\langle X,d\rangle) \longrightarrow \langle X,d\rangle$.

Proposition. $F_{\Box}$: **PO** ⟶ **QMet** *is left adjoint to* $U_{\Box}$: **QMet** ⟶ **PO** *with unit* $\eta_{\Box} = Id : Id_{PO} \Longrightarrow U_{\Box}.F_{\Box}$ *and counit* $\varepsilon_{\Box} : F_{\Box}.U_{\Box} \Longrightarrow Id_{QMet}$. *This type of adjunction is called a coreflection:* $U_{\Box}$ *coreflects* **QMet** *into its subcategory* **PO.**

In this sense the order–theoretic context of **PO** is extended and generalized to the metrical context of **QMet.** For any quasimetric space <X,d> we may think of the distance d(x,y) as a measure of how close we are to having $x \leq y$ hold. So this point of view regards a quasimetric space to be an *approximation* or *fuzzy preorder*, and regards a metric space to be an *approximation* or *fuzzy partial order.*

Let U_{PO} : **PO** ⟶ **Set** be the obvious underlying set functor. Associated with each set X is the trivial identity preorder: $x \leq y$ iff $x=y$. This defines a functor F_{PO} : **Set** ⟶ **PO**. Every preorder $\langle X,\leq\rangle$ has associated with it a canonical monotonic function $\varepsilon_{PO}(\langle X,\leq\rangle)=Id_X : F_{PO}.U_{PO}(\langle X,\leq\rangle) \longrightarrow \langle X,\leq\rangle$.

Proposition. F_{PO} : **Set** ⟶ **PO** *is left adjoint to the underlying set functor* U_{PO} : **PO** ⟶ **Set** *with unit* $\eta_{PO}=Id : Id_{Set} \Longrightarrow U_{PO}.F_{PO}$ *and counit* $\varepsilon_{PO} : F_{PO}.U_{PO} \Longrightarrow Id_{PO}$. *This is a coreflection:* U_{PO} *coreflects* **PO** *into its subcategory* **Set.**

Define the functor $F_{QMet} = F_{\square}.F_{PO}$: **Set** ⟶ **PO** ⟶ **QMet.**

Proposition. F_{QMet} : **Set** ⟶ **QMet** *is left adjoint to the underlying set functor* U_{QMet} : **QMet** ⟶ **Set,** *and this adjunction (coreflection of* **QMet** *into its subcategory* **Set***) is the adjunction composition* $F_{QMet} \dashv U_{QMet} = (F_{\square} \dashv U_{\square}).(F_{PO} \dashv U_{PO})$ *of the coreflection of* **QMet** *into its subcategory* **PO** *with the coreflection of* **PO** *into its subcategory* **Set.**

2. Existential Kan Quantification

The trivial one element space $1=\langle\{0\},d_1\rangle$ is a terminal object in the category **QMet.** Let $\langle X,d_X\rangle$ and $\langle Y,d_Y\rangle$ be any two quasimetric spaces. The cartesian product set $X\times Y$ forms a quasimetric space $\langle X\times Y,d\rangle$ with the *product metric* $d : (X\times Y)\times(X\times Y) \longrightarrow R$ defined by $d(\langle x,y\rangle,\langle x',y'\rangle) = d_X(x,x') + d_Y(y,y')$ (for ultraquasimetric spaces use "max" instead of "+"). We call $X\times Y$ the *product quasimetric space* of X and Y. This is a categorical product with obvious projection contractions $pr_X : X\times Y \longrightarrow X$ and $pr_Y : X\times Y \longrightarrow Y$. If we define $f\times g$ by $f\times g(\langle x,y\rangle) = \langle f(x),g(y)\rangle$, the operation "$\times$" becomes a bifunctor $\times$: **QMet**$\times$**QMet** ⟶ **QMet.** In particular, for any fixed space $X=\langle X,d\rangle$ we have a functor $X\times(\,)$ on quasimetric spaces. If $X=\langle X,d_X\rangle$ and $Y=\langle Y,d_Y\rangle$ are two quasimetric spaces, let Y^X be the set of all contractions $Y^X = \{f : X \longrightarrow Y \mid f \text{ is a contraction}\}$. Y^X forms a quasimetric space $Y^X=\langle Y^X,d\rangle$ with the *sup metric* $d : Y^X\times Y^X \longrightarrow R$ defined by $d(f,g) = \sup\{d_Y(f(x),g(x)) \mid x\in X\}$. If Y is a metric space, then so is Y^X. We call Y^X the *exponential (morphism) quasimetric space* from X to Y. If we define g^f by $g^f(h) = g.h.f$, the exponentiation operation becomes a bifunctor $(\,)^{(\,)}$: **QMet**$^{op}\times$**QMet** ⟶ **QMet.** In particular, for any fixed space $X=\langle X,d\rangle$ we have a functor $(\,)^X$ on quasimetric spaces. The product functor $X\times(\,)$: **QMet** ⟶ **QMet** is left adjoint to the exponential functor $(\,)^X$: **QMet** ⟶ **QMet.**

Fact. **QMet** *is a (cartesian) closed category.*

With the above definition of a metric on the hom–set Y^X, **QMet** becomes an enriched **QMet**–category. A functor F: **QMet** $\longrightarrow$ **QMet** on quasimetric spaces is a *contracting (isometric) functor* when the maps $F_{X,Y}: Y^X \longrightarrow FY^{FX}: f \mapsto F(f)$ are contractions (isometries); that is, when $\sup\{d_Y(f(x),g(x)) \mid x \in X\} \geq (=) \sup\{d_{FY}(Ff(x),Fg(x)) \mid x \in FX\}$ for all contractions f and g. $U_{\Box}$ preserves products: $U_{\Box}(X \times Y) = U_{\Box}(X) \times U_{\Box}(Y)$ and $U_{\Box}$ commutes with $()^{op}$: $U_{\Box}.()^{op} = ()^{op}.U_{\Box}$. In particular, $U_{\Box}(Y^{op} \times X) = U_{\Box}(Y)^{op} \times U_{\Box}(X)$. The preorder $U_{\Box}(Y^X)$ is described by: $f \leq g$ iff $d(f,g) = 0$ iff $f(x) \leq_Y g(x)$ for all $x \in X$ iff $f \leq_{ptwise} g$. So that $U_{\Box}$ preserves exponents: $U_{\Box}(Y^X) = U_{\Box}(Y)^{U_{\Box}(X)}$. A functor F: **QMet** $\longrightarrow$ **QMet** on quasimetric spaces is a *monotonic functor* when the maps $F_{X,Y}: U_{\Box}(Y^X) \longrightarrow U_{\Box}(FY^{FX})$ are monotonic functions; that is, when for all $x \in X$, $f(x) \leq g(x)$ implies for all $x \in FX$, $Ff(x) \leq Fg(x)$ for all contractions f and g. Clearly, contracting functors are monotonic functors. Since both contracting functors and monotonic functors are closed under composition, they each form a (2–)category.

Two contractions $f: \langle X,d_X \rangle \longrightarrow \langle Y,d_Y \rangle$ and $g: \langle Y,d_Y \rangle \longrightarrow \langle X,d_X \rangle$ are called a *reflection pair* (or a *projection–embedding pair*), and symbolized by $\langle f,g \rangle : \langle X,d_X \rangle \longrightarrow \langle Y,d_Y \rangle$ (this arrow notation is opposite to that for an adjoint pair of functors), when $Id_X \geq g.f$ and $f.g = Id_Y$ (in the exponential preorder $U_{\Box}(X^X)$). Here f is called the *projection* and g the *embedding*. Two contractions $f: \langle X,d_X \rangle \longrightarrow \langle Y,d_Y \rangle$ and $g: \langle Y,d_Y \rangle \longrightarrow \langle X,d_X \rangle$ are called an *adjoint pair* (or a *Galois connection*), and symbolized by $\langle f \dashv g \rangle : \langle X,d_X \rangle \longrightarrow \langle Y,d_Y \rangle$, when $Id_X \geq g.f$ and $f.g \geq Id_Y$. Equivalently, f and g is an adjoint pair $\langle f \dashv g \rangle$ when they satisfy the condition $f(x) \geq y$ iff $x \geq g(y)$ for all $x \in X$ and $y \in Y$. Here f is called the *upper adjoint* and g the *lower adjoint*. Clearly a reflection pair is an adjoint pair. Reflection pairs are closed under composition. So they form a subcategory $\mathbf{QMet}_{refl}$ of **QMet**, since the identity Id_X is a reflection pair $\langle Id_X,Id_X \rangle$ for each space $\langle X,d \rangle$. Similarly, adjoint pairs form a subcategory $\mathbf{QMet}_{adj}$ of **QMet** which contains $\mathbf{QMet}_{refl}$. Monotonic functors preserve adjoint pairs, and hence also reflection pairs.

A particularly interesting exponential space is the metric space $E(X) = R^{X^{op}}$, the contractions from X^{op} to the reals R. In this paper we take the point of view that a contraction $\phi: X^{op} \longrightarrow R$ is a *metrical predicate* on X which represents an ordinary (closed) subset of the quasimetric space X. We say that "x is approximately a member of ϕ" when $\phi(x) < \varepsilon$ for some small approximation tolerance ε, and we say that "x is exactly a member of ϕ" when $\phi(x) = 0$. So ϕ represents the ordinary subset $\{x \mid \phi(x) = 0\}$. The partial order $U_{\Box}(E(\langle X,d \rangle))$ is described by: $\phi \leq_{EX} \psi$ iff $\phi(x) \geq \psi(x)$ for all $x \in X$.

Let $f: X \longrightarrow Y$ be a contraction. Composition on the right with $f^{op}: X^{op} \longrightarrow Y^{op}$ defines a contraction $R^{f^{op}}: R^{Y^{op}} \longrightarrow R^{X^{op}}$, which we interpret as the *metrical inverse image function* of metrical predicates along contraction f. As f varies this is clearly functorial: there is a contravariant functor $R^{()^{op}}$: $\mathbf{QMet}^{op} \longrightarrow$ **QMet** which we call *right composition.*

Let $f: X \longrightarrow Y$ be a contraction, where $X = \langle X,d_X \rangle$ and $Y = \langle Y,d_Y \rangle$. For any element $\phi \in R^{X^{op}}$, that is, for any contraction $\phi: X^{op} \longrightarrow R$, define the function $Ef(\phi): Y^{op} \longrightarrow R$, by $Ef(\phi)(y) =$

$\inf\{\phi(x)+d_Y(y,f(x)) \mid x \in X\}$. We interpret Ef to be the *metrical direct image function* of f, in the following sense: y is approximately a member of $Ef(\phi)$ if and only if there is an $x \in X$ such that x is approximately a member of ϕ and y is approximately equal to f(x), the image of x under f. There exists a functor $E:\mathbf{QMet} \longrightarrow \mathbf{QMet}$, called *existential Kan quantification*, which is defined as follows: on objects $E(X) = R^{X^{op}}$; and on morphisms $E(f:X \longrightarrow Y) = Ef:R^{X^{op}} \longrightarrow R^{Y^{op}}$ as above. For each contraction $f:X \longrightarrow Y$ the direct image function $Ef:E(X) \longrightarrow E(Y)$ is left adjoint $Ef \dashv R^{f^{op}}$ to the inverse image function $R^{f^{op}}:E(Y) \longrightarrow E(X)$. The embedding $\mathbf{QMet} \longrightarrow \mathbf{Cat}$ allows us to describe <existential Kan quantification – right composition> as a functor $E:\mathbf{QMet} \longrightarrow \mathbf{Adj}$ into the category of adjunctions. Associated with this functor E (Gray [1965]) is a category $\mathbf{Sub}_E$, called the *category of metrical predicates*, and a 01–fibration $P_E:\mathbf{Sub}_E \longrightarrow \mathbf{QMet}$. Objects in $\mathbf{Sub}_E$ are pairs $\langle X,\phi\rangle$, where $X=\langle X,d_X\rangle$ is a quasimetric space and $\phi \in E(X)=R^{X^{op}}$ is a metrical predicate $\phi:X^{op} \longrightarrow R$ on X. Morphisms $f:\langle X,\phi\rangle \longrightarrow \langle Y,\psi\rangle$ in $\mathbf{Sub}_E$ are contractions $f:X \longrightarrow Y$ satisfying: $Ef(\phi) \leq_{EY} \psi$; that is, $Ef(\phi)(y) \geq \psi(y)$ for all $y \in Y$; or equivalently, $\phi \leq_{EX} R^{f^{op}}(\psi)=\psi.f^{op}$; that is, $\phi(x) \geq \psi(f(x))$ for all $x \in X$. The functor $P_E:\mathbf{Sub}_E \longrightarrow \mathbf{QMet}$ is called the *underlying quasimetric space functor*, and is defined by $P_E(\langle X,\phi\rangle) = X$ and $P_E(f) = f$.

The point of view that regards category theory as objective dialectics is centered on the identification of adjoint pairs with dialectical contradictions (also called, opposing tendencies). Dialectical systems are based upon systems of contradictions or opposing tendencies. Since an adjoint pair is a morphism in **Adj** the category of adjunctions, the categorical approach represents such a dialectical base, or system of contradictions, as a diagram $\mathbf{B} \longrightarrow \mathbf{Adj}$ in the category of contradictions **Adj**. Equivalently (Gray [1965]), a *dialectical base* (Kent [1986]) $P:\mathbf{E} \longrightarrow \mathbf{B}$ is a 01–fibration which is *fiber complete*: each fiber $\mathbf{E}_B = P^{-1}(B)$ is a bicomplete category.

Proposition. *Metrical predicates form a dialectical base over quasimetric spaces; that is, the underlying quasimetric space functor* $P_E:\mathbf{Sub}_E \longrightarrow \mathbf{QMet}$ *is a dialectical base.*

So existential Kan quantification and right composition form a base which allows for the specification of metrical dialectical systems. These metrical dialectical systems define the *dialectical flow* of metrical predicates. The dialectical flow of predicates is our meaning of the term *predicate transformer*. Other dialectical contexts define the dialectical flow of: bits (truth values), natural numbers, real numbers, subsets (see below), transition systems (see below), databases with constraints, and functional programming modules (see Kent [1986,1987]).

A *bimodule* to $\langle Y,d_Y\rangle$ from $\langle X,d_X\rangle$ is a contraction $\phi:E(\langle Y,d_Y\rangle) \longleftarrow \langle X,d_X\rangle$; or equivalently, a contraction $\phi:Y^{op}{\times}X \longrightarrow R$. We denote this by $\phi:\langle Y,d_Y\rangle \leftarrow\!\!\shortmid\!\!-\!\!- \langle X,d_X\rangle$. We considered bimodules to be *approximation*, or *fuzzy relations*. Quasimetric spaces as objects and bimodules as morphisms form an enriched category **BIMet** with d_X the identity at $\langle X,d_X\rangle$ and composition defined by $\psi \circ \phi(z,x) =$

$\inf\{\psi(z,y)+\phi(y,x) \mid y\in Y\}$ for all composable pairs $\psi : \langle Z,d_Z\rangle \leftarrow\!\!\!\!\mid \langle Y,d_Y\rangle$ and $\phi : \langle Y,d_Y\rangle \leftarrow\!\!\!\!\mid \langle X,d_X\rangle$.

BIMet incorporates **QMet** as a subcategory: a contraction $f : \langle Y,d_Y\rangle \longleftarrow \langle X,d_X\rangle$ is regarded as the bimodule $f_* : \langle Y,d_Y\rangle \leftarrow\!\!\!\!\mid \langle X,d_X\rangle$ defined by $f_*(y,x) = d_Y(y,f(x))$. This construction is a functor $(\,)_* : \mathbf{QMet} \longrightarrow \mathbf{BIMet}$. The isomorphism $(\,)^{op} : \mathbf{QMet} \longrightarrow \mathbf{QMet}$ and the symmetry of the product space operation "$\times$" give us an alternate way of representing **QMet** in **BIMet**: since the contraction $f : \langle Y,d_Y\rangle \longleftarrow \langle X,d_X\rangle$ is also a contraction $f : \langle Y,{d_Y}^{op}\rangle \longleftarrow \langle X,{d_X}^{op}\rangle$, it can be regarded as the bimodule $f^* : \langle X,d_X\rangle \leftarrow\!\!\!\!\mid \langle Y,d_Y\rangle$ defined by $f^*(x,y) = d_Y(f(x),y)$. This construction is a functor $(\,)^* : \mathbf{QMet}^{op} \longrightarrow \mathbf{BIMet}$. The notion of bimodules and the embedding of contractions as bimodules by means of the functor $(\,)_*$, allows for a simple definition of the functor E. For any contraction $f : \langle Y,d_Y\rangle \longleftarrow \langle X,d_X\rangle$ we can define $Ef(\phi) = f_* \circ \phi$.

We can define a **BIMet** exponential space Y^X by defining the metric $d(\phi,\psi) = \sup\{d_R(\phi(y,x),\psi(y,x)) \mid y,x\} = \sup\{\psi(y,x) \dot{-} \phi(y,x) \mid y,x\}$ for any two bimodules $\phi,\psi : \langle Y,d_Y\rangle \leftarrow\!\!\!\!\mid \langle X,d_X\rangle$. For any two contractions $f,g : \langle Y,d_Y\rangle \longleftarrow \langle X,d_X\rangle$, $d(f,g) \geq d(f_*,g_*)$. So $(\,)_* : \mathbf{QMet} \longrightarrow \mathbf{BIMet}$ is a contracting, and hence monotonic, functor. Similarly $d(f,g) \geq d(g^*,f^*)$. So $(\,)^* : \mathbf{QMet}^{op} \longrightarrow \mathbf{BIMet}$ is a contracting functor, and monotonic in the sense that $f \leq g$ implies $f^* \geq g^*$.

Let $f : \langle X,d_X\rangle \longrightarrow \langle Y,d_Y\rangle$ be a contraction. The kernel metric d_f is a bimodule $d_f : X^{op}\times X \longrightarrow R$, called the *kernel bimodule,* which can be expressed as a composition, $d_f = f^* \circ f_*$, of the bimodules associated with f. The *norm* $|f|$ of a contraction f is defined to be the distance from the kernel bimodule of f to the identity bimodule in the **BIMET**–exponential space X^X: $|f| = d(d_f,d_X) = d(f^* \circ f_*,d_X)$ $= \sup\{d_X(x',x) \dot{-} d_Y(f(x'),f(x)) \mid x',x\}$. The norm of f measures how close f is to being an isometry. For monotonic functions $f : \langle X,\leq_X\rangle \longrightarrow \langle Y,\leq_Y\rangle$ we have $|f| = 0$, if $\leq_X = \leq_f$; and $= \infty$, if $\leq_X \subset \leq_f$.

Proposition. *The norm* $|\;|$ *satisfies the following properties:* 1. $|g| + |f| \geq |g.f|$, *for composable pairs* f *and* g; 2. $0 = |g|$ *iff* g *is an isometry; or, more generally* $|f| = |g.f|$ *for all composable* f, *if* g *is an isometry; and* 3. $|\alpha f| = \alpha|f|$, *where* $\alpha f = f : \alpha\langle X,d_X\rangle \longrightarrow \alpha\langle Y,d_Y\rangle$.

An adjoint pair $\langle f \dashv g\rangle : \langle X,d_X\rangle \longrightarrow \langle Y,d_Y\rangle$ has a special norm defined on it. Equivalent conditions that $\langle f \dashv g\rangle$ be an adjoint pair are: 1. $Id_X \geq g.f$ and $f.g \geq Id_Y$; 2. $d_X \geq g_* \circ f_*$ and $g^* \circ f^* \leq d_Y$; or 3. $d(g_* \circ f_*,d_X) = 0$ and $d(g^* \circ f^*,d_Y) = 0$. So define the *adjoint norm* $|f|_{adj}$ by $|f|_{adj} = d(d_X,g_* \circ f_*) + d(d_Y,g^* \circ f^*)$. It is straightforward to show that an equivalent definition for this norm is $|f|_{adj} =$ $\sup\{d_X(x,g(f(x))) + d_Y(f(g(y)),y) \mid x,y\} = \sup\{g_* \circ f_*(x,x) \mid x\} + \sup\{g^* \circ f^*(y,y) \mid y\} = \int_X g_* \circ f_* + \int_y g^* \circ f^*$, where $\int$ denotes an *end.* The adjoint norm measures how close f is to being an isomorphism.

Proposition. *The norm* $|\;|_{adj}$ *satisfies the following properties:* 1. $|f'|_{adj} + |f|_{adj} \geq |f'.f|_{adj}$, *for composable adjoint pairs* $\langle f \dashv g\rangle$ and $\langle f' \dashv g'\rangle$; 2. $0 = |f|_{adj}$ *iff the adjoint pair* $\langle f \dashv g\rangle$ *is an isomorphism pair with* $f^{-1} =$

g; 3. $|\alpha f|_{adj} = \alpha |f|_{adj}$*; and 4.* $|f|_{adj} \geq |f|$ *for any adjoint pair* $<f\dashv g>$.

Since a reflection pair $<f,g>$ is an adjoint pair which satisfies the further condition that $f.g \leq Id_Y$ equivalently $g^* \circ f^* \geq d_Y$ equivalently $d(d_Y, g^* \circ f^*) = 0$, we can define the *reflection norm* $|f|_{refl}$ by $|f|_{refl} = |f|_{adj} = d(d_X, g_* \circ f_*) = \sup\{d_X(x, g(f(x))) \mid x\} = \sup\{g_* \circ f_*(x,x) \mid x\} = \int_X g_* \circ f_* = \int_X (g.f)_*$.

A *normed category* $<\mathbf{K}, |\ |_{\mathbf{K}}>$ is a category $\mathbf{K}$ with norms on hom–sets $|\ |_{X,Y} : [X \longrightarrow Y] \longrightarrow R$, and a collection of functors $\alpha : \mathbf{K} \longrightarrow \mathbf{K}$, one for each constant $\alpha \in R$, which satisfy: 1. $|g| + |f| \geq |g.f|$ for composable pairs f and g; 2. $0 = |Id_X|$ for objects X; and 3. $|\alpha(f)| = \alpha|f|$. Any quasimetric space $X = <X, d_X>$ is trivially a normed category with homsets being singletons $[x \longrightarrow y] = \{xy\}$ and with norms given by $|xy|_X = d_X(x,y)$. Denote this trivial construction by Inc(X). Collecting together the norms $\mathbf{K}(X,Y) = \{|f|_{\mathbf{K}} \mid f : X \longrightarrow Y\}$ allows us to view a normed category $\mathbf{K}$ as a category valued in the closed preorder L(R) $= <PR, \leq_{PR}, +_{ptwise}, \{0\}, \dot{-}_{ptwise}>$, the preorder of subsets of reals with the *lower* (or *Hoare*) order (see the next section for the definition of powerorder L). A *norm–decreasing functor* $F : <\mathbf{K}, |\ |_{\mathbf{K}}> \longrightarrow <\mathbf{L}, |\ |_{\mathbf{L}}>$ between two normed categories is a functor which satisfies $|f|_{\mathbf{K}} \geq |F(f)|_{\mathbf{L}}$ for all morphisms f. Any contraction $f : X \longrightarrow Y$ is trivially a norm–decreasing functor $f : Inc(X) \longrightarrow Inc(Y)$. The product functor $A\times(\ ) : \mathbf{QMet} \longrightarrow \mathbf{QMet}$ is norm–decreasing; in fact, $A\times(\ )$ is norm–preserving since $|f| = |A\times(f)|$ for all contractions f. Normed categories and norm–decreasing functors form the category **NormCat**, a subcategory of L(R)–**Cat**, the L(R)–valued categories. The construction Inc is the *inclusion functor* $Inc : \mathbf{QMet} \longrightarrow \mathbf{NormCat}$ of quasimetric spaces as normed categories. The infimum operation on the reals is a closed functor (monotonic function) $\inf : L(R) \longrightarrow R$ from subsets of reals to reals. So any normed category $\mathbf{K} = <\mathbf{K}, |\ |_{\mathbf{K}}>$ becomes a quasimetric space $Inf(\mathbf{K}) = <|\mathbf{K}|, d_{\mathbf{K}}>$ by defining $d_{\mathbf{K}}(X,Y) = \inf\{|f|_{\mathbf{K}} \mid f : X \longrightarrow Y\}$. With the above definition of norm on contractions, the enriched category **QMet** becomes a normed category. The number $d_{\mathbf{QMet}}(X,Y) = \inf\{|f| \mid f : X \longrightarrow Y\}$ is a measure of how "geometrically similar" X is to a subspace of Y. Similarly the above norms on adjoint pairs and reflection pairs make $\mathbf{QMet}_{adj}$ and $\mathbf{QMet}_{refl}$ into normed categories. For any normed category $\mathbf{K} = <\mathbf{K}, |\ |_{\mathbf{K}}>$ the identity functor $Id_{\mathbf{K}} : \mathbf{K} \longrightarrow \mathbf{K}$ is a canonical norm–decreasing functor $\eta_{\mathbf{K}} = Id_{\mathbf{K}} : \mathbf{K} \longrightarrow Inc(Inf(\mathbf{K}))$. Also, any quasimetric space X satisfies $Inf(Inc(X)) = X$.

Theorem. *The infimum functor is left adjoint* $Inf \dashv Inc$ *to the inclusion functor. In fact, this is a reflection with infimum* Inf *reflecting the category* **NormCat** *of normed categories into its subcategory* **QMet** *of quasimetric spaces.*

3. Comprehension Schema and Closure Powerspace

In the unstructured context of sets and functions (**Set**) the notions of *subobject* (subset) and *predicate* (characteristic function) are equivalent since $P(X) \cong 2^X$. The same is true in the structured context of preorders and monotonic functions (**PO**), where closed suborders and order–theoretic predicates are equivalent, since $\overline{L}(X) \cong 2^{X^{op}}$, where $\overline{L}$ is the lower (or Hoare) closure powerorder construction (see comments below). However, in the context of quasimetric spaces and contractions (**QMet**) the notions of subobject (closed subspace) and predicate (metrical predicate) are related through a fundamental adjunction called a *comprehension schema.* A comprehension schema incorporates the opposing tendencies of *character* of subobjects and *extension* of predicates.

Let $X=\langle X,d\rangle$ be a quasimetric space. Define the *diameter* $\delta(A)$ of a subset of pairs $A \subseteq X{\times}X$ by $\delta(A) = \sup\{d(a_1,a_2) | (a_1,a_2) \in A\}$. Define the diameter $\delta(f)$ of a contraction $f: \langle A,d_A\rangle \longrightarrow \langle X,d\rangle$ by $\delta(f) = \delta(f(A){\times}f(A))$. Define the *quasidistance* $\rho(f,g)$ between any two contractions $f: \langle A,d_A\rangle \longrightarrow \langle X,d\rangle$ and $g: \langle B,d_B\rangle \longrightarrow \langle X,d\rangle$ with common target $\langle X,d\rangle$, by $\rho(f,g) = \inf\{d(f(a),g(b)) | a \in A \text{ and } b \in B\}$. Note that: $\rho(f,g) = \infty$ if $A=\emptyset$ or $B=\emptyset$; $\rho(x,y) = d(x,y)$ for all points $x,y \in X$; $\rho(f,h) \le \rho(f,g)+\rho(g,h)+\delta(g)$ for contractions f,g,h; and the function $\rho(f): X \longrightarrow R : x \mapsto \rho(x,f)$, for contraction $f: \langle A,d_A\rangle \longrightarrow \langle X,d\rangle$, is a contraction; that is, $\rho(f)$ is a metrical predicate $\rho(f) \in E(X) = R^{X^{op}}$. So $\rho_X : \mathbf{QMet}{\downarrow}X \longrightarrow E(X)$ is a functor, from the comma category of quasimetric spaces over $X=\langle X,d\rangle$ to $E(X) = \langle EX, \le_{EX}\rangle$ the complete poset of metrical predicates of X. We call this functor the metrical *character* functor. So quasidistance from points to contractions defines character.

Associated with any metrical predicate $\phi: X^{op} \longrightarrow R$ of a quasimetric space $X=\langle X,d\rangle$ is an ordinary subset $\{X|\phi\} \subseteq X$ of X, called the *extension* of ϕ, and defined by $\{X|\phi\} = \{x \in X | \phi(x) = 0\}$. Any subset A $\subseteq$ X of a quasimetric space $\langle X,d\rangle$ is a subspace $A=\langle A,d\rangle$ of $X=\langle X,d\rangle$, and can be represented by the inclusion map $Inc_A : \langle A,d\rangle \longrightarrow \langle X,d\rangle$, which is an injection and an isometry. So as ϕ varies, the extension of predicates defines a functor $\{X|-\} : E(X) \longrightarrow \mathbf{QMet}{\downarrow}X$.

Fact *(categorical comprehension schema). For every quasimetric space* $X=\langle X,d\rangle$ *character* $\rho_X : \mathbf{QMet}{\downarrow}X \longrightarrow E(X)$ *is left adjoint* $\rho_X \dashv \{X|-\}$ *to extension* $\{X|-\} : E(X) \longrightarrow \mathbf{QMet}{\downarrow}X$.

In categories with suitable image factorization *objects_over* and *subobjects_of* are related by a morphism of doctrines (or 01–fibrations). **QMet** is such a category.

Fact *(doctrinal diagram). For every quasimetric space* $X=\langle X,d\rangle$ *image factorization* Factor : $\mathbf{QMet}{\downarrow}X \longrightarrow P(X)$ *of contractions into* X *is left adjoint* Factor $\dashv$ Inc *to inclusion* Inc : $P(X) \longrightarrow \mathbf{QMet}{\downarrow}X$ *of subobjects as objects over* X, *where* $P(X)=\langle PX,\subseteq\rangle$ *is the poset of subspaces of*

X *under inclusion. This is a reflection.*

The categorical comprehension schema factors through the doctrinal diagram, and hence "cuts down" from categories to preorders.

Fact *(order–theoretic comprehension schema). For every quasimetric space* X=<X,d> *character of subspaces* $\rho_X : P(X) \longrightarrow E(X)$ *is left adjoint* $\rho_X \dashv \{X|-\}$ *to extension of metrical predicates* $\{X|-\} : E(X) \longrightarrow P(X)$. *Moreover, the categorical comprehension schema is the adjunction composition* $(\rho_X \dashv \{X|-\}) = (\mathrm{Factor} \dashv \mathrm{Inc}) . (\rho_X \dashv \{X|-\})$ *of doctrinal diagram with order–theoretic comprehension schema.*

When X is a set, and hence the trivial discrete metric space on X ($d(x,y)=\infty$ when $x \neq y$), then $\rho_X(A)$ the character of A reduces to $\rho_X(A) = \kappa_A$, the (negation of the) usual characteristic function of A. When X=<X,≤> is a preorder then $\rho_X(A)$ the character of A reduces to $\rho_X(A) = \kappa_{\downarrow A} : X \longrightarrow \{0,\infty\}$, the (negation of the) usual characteristic function of $\downarrow A = \{x | x \leq a \text{ for some } a \in A\}$ the *closure–below* of A. Moreover, $\rho_X(\downarrow A) = \rho_X(A)$ for every subset $A \subseteq X$.

Each exponential space Y^X is a normed quasimetric space with the norm $|\ |$. However, the special exponential spaces $E(X) = R^{X^{op}}$ have a special norm defined on them. Compare each predicate $\phi \in E(X)$ with its associated predicate $\rho_X(\{X|\phi\})$. The counit in the above adjunction implies that $d_{EX}(\rho_X(\{X|\phi\}),\phi) = 0$. Define the E–*norm* $|\phi|_E$ on E(X) by $|\phi|_E = d_{EX}(\phi,\rho_X(\{X|\phi\})) = \sup\{d_R(\phi(x),\rho(x,\{X|\phi\})) | x\} = \sup\{[\rho(x,\{X|\phi\}) \dot{-} \phi(x)] | x\}$. The E–norm measures how close ϕ is to being the characteristic function $\rho(A)$ for some closed set A. The norm $|\ |_E$ satisfies the following norm properties: 1. $|\phi|_E + |\phi'|_E = |\phi + \phi'|_E$ for $\phi,\phi' \in E(X)$; 2. $0 = |\phi|_E$ iff $\phi = \rho(A)$ for some closed set A (namely $A = \{X|\phi\}$); 3. $|\alpha\phi|_E = \alpha|\phi|_E$; and 4. $|\phi| \geq |\phi|_E$ for all $\phi \in E(X)$. The two norms $|\phi|$ and $|\phi|_E$ define two metrics on E(X), $d(\phi,\psi) = |\psi \dot{-} \phi|$ and $d_E(\phi,\psi) = |\psi \dot{-} \phi|_E$, where $d \geq d_E$. These can be compared to the standard sup metric d_{EX} on E(X).

Facts. *Character* ρ *is a natural transformation* $\rho : P.U_{QMet} \Longrightarrow U_{QMet}.E : \mathbf{QMet} \longrightarrow \mathbf{Set}$ *"from" the powerset functor* $P : \mathbf{Set} \longrightarrow \mathbf{Set}$ *"to" the quasimetric existential Kan quantification functor* $E : \mathbf{QMet} \longrightarrow \mathbf{QMet}$. *Character* ρ *is also a natural transformation* $\rho : \overline{L} . U_{\square} \Longrightarrow U_{\square} . E : \mathbf{QMet} \longrightarrow \mathbf{PO}$ *"from" the preorder existential Kan quantification (the closure powerorder construction, see below)* $\overline{L} : \mathbf{PO} \longrightarrow \mathbf{PO}$ *"to" the quasimetric existential Kan quantification functor* $E : \mathbf{QMet} \longrightarrow \mathbf{QMet}$.

We use the character natural transformation ρ to "induce" a quasimetric space functor $P : \mathbf{QMet} \longrightarrow \mathbf{QMet}$, which "lifts" both the powerset functor $P : \mathbf{Set} \longrightarrow \mathbf{Set}$ and the closure powerorder functor $\overline{L}: \mathbf{PO} \longrightarrow \mathbf{PO}$ (see DeBakker and Zucker, p.113, Corollary B4 [1982], where P is implicitly shown to be a functor by other less general methods). This "induction" of a quasimetric space functor works in a fully general setting.

Proposition. *For any category* **I** *both* $(U_{QMet})^{\mathbf{I}}: \mathbf{QMet}^{\mathbf{I}} \longrightarrow \mathbf{Set}^{\mathbf{I}}$ *and* $(U_{\square})^{\mathbf{I}}: \mathbf{QMet}^{\mathbf{I}} \longrightarrow \mathbf{PO}^{\mathbf{I}}$ *are 01–fibrations.*

Proof. Both U_{QMet} and $U_{\square}$ are 01–fibrations (see Gray [1965]).

The above proposition was motivated by the character natural transformation, both $\rho: P.U_{QMet} \Longrightarrow U_{QMet}.E: \mathbf{QMet} \longrightarrow \mathbf{Set}$ and $\rho: \overline{L}.U_{\square} \Longrightarrow U_{\square}.E: \mathbf{QMet} \longrightarrow \mathbf{PO}$. For this special case character induces a *powerspace functor* $P: \mathbf{QMet} \longrightarrow \mathbf{QMet}$ which lifts $P: \mathbf{Set} \longrightarrow \mathbf{Set}$, or more precisely $P.U_{QMet}: \mathbf{QMet} \longrightarrow \mathbf{Set}$, through $U_{QMet}: \mathbf{QMet} \longrightarrow \mathbf{Set}$ (and lifts $\overline{L}: \mathbf{PO} \longrightarrow \mathbf{PO}$, or more precisely $\overline{L}.U_{\square}: \mathbf{QMet} \longrightarrow \mathbf{PO}$, through $U_{\square}: \mathbf{QMet} \longrightarrow \mathbf{PO}$), and an isometric natural transformation $\rho: P \Longrightarrow E: \mathbf{QMet} \longrightarrow \mathbf{QMet}$, again called character, which lifts $\rho: P.U_{QMet} \Longrightarrow U_{QMet}.E$ through $U_{QMet}: \mathbf{QMet} \longrightarrow \mathbf{Set}$. The embedding $\mathbf{QMet} \longrightarrow \mathbf{Cat}$ allows us to describe powerspace as a covariant functor $P: \mathbf{QMet} \longrightarrow \mathbf{Cat}$.

For any space $X=\langle X,d \rangle$, the powerspace $P(X)$ is defined by $P(X)=\langle PX,d_{PX} \rangle$ where PX is the powerset of X and $d_{PX} = d_{\rho_X}$ is the kernel of the map ρ_X which sends subsets to their associated metrical predicates: $d_{PX}(A,B) = d_{EX}(\rho_X(A),\rho_X(B)) = \sup\{d_R(\rho(x,A),\rho(x,B)) \mid x \in X\} = \sup\{\rho(x,B) \dot{-} \rho(x,A) \mid x \in X\}$. This is the nonsymmetric *Hausdorff metric* on the powerset of X. For any contraction $f: X \longrightarrow Y$ the direct image function $Pf: PX \longrightarrow PY$ is also a contraction. In short, we say that "powerspace is the inverse image of existential Kan quantification along character".

Fact. $d_{PX}(A,B) = \sup\{\rho(a,B) \mid a \in A\} = \sup\{\inf\{d(a,b) \mid a \in A, b \in B\}\}$.

The composite monotonic function $(\overline{\ })_X = \{X|-\}.\rho_X: P(X) \longrightarrow P(X)$ is called the *closure function* on X. For each subset $A \subseteq X$ the *closure* of A, denoted by $\overline{A}$, is defined by $\overline{A} = \{X|\rho_X(A)\} = \{x|\rho(x,A)=0\}$. Note that this is an asymmetric closure in the sense that $\overline{A} \neq \{x|\rho(A,x)=0\}$, in general. For a preorder $\langle X,\leq \rangle$ with associated metric d, the closure of a subset $A \subseteq X$ is $\overline{A} = \{x|\rho(x,A)=0\} = \downarrow A$, the elements below A. Closure obeys the Kuratowski closure axioms.

Fact. *Let* $\langle X,d \rangle$ *be a quasimetric space. Then for any subsets* $A,B \subseteq X$, *we have:* 1. $\overline{A \cup B} = \overline{A} \cup \overline{B}$; 2. $A \subseteq \overline{A}$; 3. $\overline{\phi} = \phi$; *and* 4. $\overline{\overline{A}} = \overline{A}$. *If* X *is a metric space, then* 5. $\overline{\{x\}}=\{x\}$, *for each point* $x \in X$.

A *topological space* $X=\langle X,(\overline{\ }) \rangle$ is a set X and a function $(\overline{\ })$ called closure satisfying axioms 1–4 in the above proposition. Thus, any quasimetric space X=<X,d> induces a unique topological space $X=\langle X,\overline{(\ )}\rangle$ by the above definition for closure. We say that a subset $A \subseteq X$ is *closed* when $\overline{A} = A$ (a subset $A \subseteq X$ of a preorder is said to be *closed below* when $\downarrow A = \overline{A} = A$). The *specialization preorder* of a topological space $X=\langle X,\overline{(\ )}\rangle$ is the preorder $X=\langle X,\leq\rangle$, where $x \leq x'$ when $x' \in \overline{A}$ implies $x \in \overline{A}$ for all subsets $A \in PX$. The specialization preorder of the topological space induced by a quasimetric space X=<X,d> is just the underlying preorder $U_{\Box}(X) = \langle X,\leq\rangle$: [one way] $\rho(x,A) \leq \rho(x,y) + \rho(y,A) + \delta(y) = d(x,y) + \rho(y,A)$; so $x \leq y$ iff $d(x,y) = 0$ implies ($\rho(y,A) = 0$ implies $\rho(x,A) = 0$); [other way] Let A be the singleton $A = \{y\}$; $\rho(y,\overline{A}) = \rho(y,A) = \rho(y,\{y\}) = 0$; so $0 = \rho(x,\{y\}) = d(x,y)$. A contraction $f: X \longrightarrow Y$ is called a *closed map* when the image of each set closed in X is closed in Y. A function $f: X \longrightarrow Y$ is said to be *continuous* when $f(\overline{A}) \subseteq \overline{f(A)}$ for every subset $A \subseteq X$. Equivalently, f is continuous when the inverse image $f^{-1}(B)$ of each set B closed in Y is closed in X.

Corollary. $Id_{PX} \leq \overline{(\ )}_X = \{X|-\}.\rho_X$ *the unit law, or equivalently,* $A \subseteq \overline{A}$ *for all subsets* $A \in PX$; $\rho_X.\{X|-\} \leq Id_{EX}$ *the counit law, or equivalently,* $\phi \geq_{EX} \rho_X(\{X|\phi\})$, *or equivalently,* $\phi(x) \leq \rho(x,\{y|\phi(y)=0\})$ *for all* $x \in X$. $\rho_X.\{X|-\}.\rho_X = \rho_X$, *or equivalently,* $\rho_X(\overline{A}) = \rho_X(A)$ *for all subsets* $A \in PX$; $\{X|-\}.\rho_X.\{X|-\} = \{X|-\}$, *or equivalently, the extension* $\{X|\phi\}$ *is a closed subset of* X *for all metrical predicates* ϕ.

Corollary. $d_{PX}(\overline{A},\overline{B}) = d_{PX}(A,B)$ *for any subsets* $A,B \subseteq X$. *If* $\overline{A}_1 = \overline{A}_2$ and $\overline{B}_1 = \overline{B}_2$ then $d_{PX}(A_1,B_1) = d_{PX}(A_2,B_2)$.

Now that we have defined closure we can describe the specialization poset $\langle PX,\leq_{PX}\rangle = U_{\Box}(P(X))$. The order $\leq_{PX}$ for the asymmetric Hausdorf metric is $A \leq_{PX} B$ iff $d_{PX}(A,B) = 0$ iff $\sup\{\rho(a,B) | a \in A\} = 0$ iff $\rho(a,B) = 0$ for all $a \in A$ iff $A \subseteq \overline{B}$. So $\leq_{PX} = L(\langle X,\leq\rangle)$, the preorder of subsets of $\langle X,\leq\rangle = U_{\Box}(X)$ with the *lower* (or *Hoare*) order. Symbolically, $U_{\Box}(P(X)) = L(U_{\Box}(X))$. In words, the specialization preorder of the powerspace construction is the powerorder construction of the specialization preorder. The relationships between the metric powerspace construction and the Hoare, Smyth and Plotkin orders, are discussed in an extended version of this paper (Kent [1984]).

Let $f: X \longrightarrow Y$ be a contraction. The composite monotonic function $f_{\bullet} = \{X|-\}.R^{f^{op}}.\rho_Y: P(Y) \longrightarrow E(Y) \longrightarrow E(X) \longrightarrow P(X)$ is called the *inverse image function* of subsets along the contraction f. For any subset $B \in PY$ we have $f_{\bullet}(B) = \{x | \rho(B)(f(x))=0\} = \{x | \rho(f(x),B)=0\} = \{x | f(x) \in \overline{B}\} = f^{-1}(\overline{B})$, a closed subset of X. As f varies this is functorial, since $f_{\bullet}(g_{\bullet}(C)) = (g.f)_{\bullet}(C)$. So this is a contravariant functor $()_{\bullet}: \mathbf{QMet}^{op} \longrightarrow \mathbf{PO}$ called the *inverse image functor*. The embedding $\mathbf{PO} \longrightarrow \mathbf{Cat}$ allows us to describe inverse image as a contravariant functor $()_{\bullet}: \mathbf{QMet}^{op} \longrightarrow \mathbf{Cat}$.

Fact. *For each contraction* $f: X \longrightarrow Y$ *the direct image contraction* $Pf: P(X) \longrightarrow P(Y)$ *is left adjoint* $Pf \dashv f_{\bullet}$

to the inverse image monotonic function $f_\bullet : P(Y) \longrightarrow P(X)$.

The embeddings $\mathbf{PO} \longrightarrow \mathbf{Cat}$ and $\mathbf{QMet} \longrightarrow \mathbf{Cat}$ allow us to describe <powerspace – inverse image> as a functor $P : \mathbf{QMet} \longrightarrow \mathbf{Adj}$. Associated with this powerspace functor P is a category $\mathbf{Sub}_P$, called the *category of subspaces*, and a 01–fibration $P_P : \mathbf{Sub}_P \longrightarrow \mathbf{QMet}$. Objects in $\mathbf{Sub}_P$ are pairs $\langle X,A\rangle$, where $X=\langle X,d_X\rangle$ is a quasimetric space and $A \in PX$ is a subspace of X. Morphisms $f : \langle X,A\rangle \longrightarrow \langle Y,B\rangle$ in $\mathbf{Sub}_P$ are contractions $f : X \longrightarrow Y$ satisfying: $Pf(A) \subseteq \overline{B}$; or equivalently, $A \subseteq f_\bullet(B) = f^{-1}(\overline{B})$. The functor $P_P : \mathbf{Sub}_P \longrightarrow \mathbf{QMet}$ is called the *underlying quasimetric space functor*, and is defined by $P_P(\langle X,A\rangle) = X$ and $P_P(f) = f$.

Proposition. *Subspaces form a dialectical base over quasimetric spaces; that is, the underlying quasimetric space functor* $P_P : \mathbf{Sub}_P \longrightarrow \mathbf{QMet}$ *is a dialectical base.*

Now character $\rho : P \Longrightarrow E$, and in particular $\rho_X : P(X) \longrightarrow E(X)$ for each quasimetric space X, represents ordinary subsets as metrical predicates. But this is not a faithful representation since ρ_X is not injective. We can make it injective by factorization. Consider the symmetric (and hence ordinary) kernel of $\rho_X : P(X) \longrightarrow E(X)$: $A \equiv_{PX} B$ iff $\rho_X(A) = \rho_X(B)$. Clearly, if $A \equiv_{PX} B$ then $\overline{A} = \{x \mid \rho(x,A)=0\} = \{x \mid \rho(x,B)=0\} = \overline{B}$. Thus, the above proposition implies that there is only one closed subset in the equivalence class of A. So, the function $\rho_X : P(X) \longrightarrow E(X)$ factors $\rho_X = \overline{\rho}_X . (\overline{\ \ })_X : P(X) \longrightarrow \overline{P}(X) \longrightarrow E(X)$ where $\overline{P}(X) = \{\overline{A} \mid A \in P(X)\}$ is the set of all closed subsets of X, $(\overline{\ \ })_X : A \longrightarrow \overline{A}$ is a surjection, and $\overline{\rho}_X : \overline{A} \longrightarrow \rho_X(\overline{A}) = \rho_X(A)$, the restriction of ρ_X to closed subsets of X, is an injection. If we let $\overline{P}(X) = \langle \overline{P}X, d_{\overline{P}X}\rangle$ denote the inverse image of the space E(X) along the function $\overline{\rho}_X$, then $d_{\overline{P}X}(A,B) = \sup\{\rho(x,B) \dot{-} \rho(x,A) \mid x \in X\} = \sup\{\rho(a,B) \mid a \in A\}$ for $A,B \in \overline{P}(X)$. Of course, $\overline{P}(X)$ is just the subspace of P(X) of closed subsets. Then $\overline{\rho}_X$ is an isometry, and hence an embedding of $\overline{P}(X)$ into E(X). Obviously, using the above corollary, $(\overline{\ \ })_X$ is also an isometry.

Can the construction $\overline{P}$ be extended to a functor? If so, we would like $\overline{P}$ to satisfy the following equalities: 1. $\overline{P}f.(\overline{\ \ })_X = (\overline{\ \ })_Y.Pf$; and 2. $Ef.\overline{\rho}_X = \overline{\rho}_Y.\overline{P}f$; for every contraction $f : X \longrightarrow Y$ So, if $f : X \longrightarrow Y$ is a contraction, then define $\overline{P}f : \overline{P}X \longrightarrow \overline{P}Y$ by $\overline{P}f(A) = \overline{f(A)}$, the closure of the direct image of A under f. Now f is continuous iff $f(\overline{A}) = \overline{f(A)}$ for every subset $A \subseteq X$; that is, f is continuous precisely when f satisfies equality 1 above. It is easy to show that contractions are continuous functions. Also, if $f : X \longrightarrow \langle Y,d_Y\rangle$ is a function, and $g : \langle Y,d_Y\rangle \longrightarrow \langle Z,d_Z\rangle$ is a continuous function, then for any subset $A \subseteq X$ we have $\overline{g(\overline{f(A)})} = \overline{g.f(A)}$. There exists a *closure powerspace functor* $\overline{P} : \mathbf{QMet} \longrightarrow \mathbf{QMet}$ defined as follows: on objects $\overline{P}[\langle X,d_X\rangle] = \langle \overline{P}X, d_{\overline{P}X}\rangle$; and on morphisms $\overline{P}[f : \langle X,d_X\rangle \longrightarrow \langle Y,d_Y\rangle] = \overline{P}f : \langle \overline{P}X, d_{\overline{P}X}\rangle \longrightarrow \langle \overline{P}Y, d_{\overline{P}Y}\rangle$, where for $A \in \overline{P}X$, $\overline{P}f(A)$ is the closure of f(A) in $\langle Y,d_Y\rangle$.

It is easy to see that P and $\overline{P}$ extend functors on **UQMet**; that is, if $\langle X,d_X\rangle$ is an ultraquasimetric space, then $P(\langle X,d_X\rangle) = \langle PX,d_{PX}\rangle$ and $\overline{P}(\langle X,d_X\rangle) = \langle \overline{P}X,d_{\overline{P}X}\rangle$ are ultraquasimetric spaces also. Note that $\overline{P}$ and P agree on closed maps, in the sense that, $f:X \longrightarrow Y$ is a closed map if and only if $\overline{P}f$ and Pf (restricted to $\overline{P}X$) are equal. A useful property of reflection pairs can be expressed in terms of $\overline{P}$: if $\langle f,g\rangle : \langle X,d_X\rangle \longrightarrow \langle Y,d_Y\rangle$ is a reflection pair and A is a closed subspace of $\langle X,d_X\rangle$, then the restriction $\langle f,g\rangle : \langle A,d_X\rangle \longrightarrow \langle \overline{P}f(A),d_Y\rangle$ of f to A and g to $\overline{P}f(A)$ is also a reflection pair. We have defined the functor $\overline{P}$ in terms of the functor P and the closure operator $(\overline{\ \ })$. How are these related categorically? Closure $(\overline{\ \ })$ is an objectwise surjective natural transformation $(\overline{\ \ }) : P \Longrightarrow \overline{P} : \mathbf{QMet} \longrightarrow \mathbf{QMet}$ from powerspace P to closure powerspace $\overline{P}$. Now character factors as $\rho_X = \overline{\rho}_X \cdot (\overline{\ \ })_X$. Both character ρ and closure $(\overline{\ \ })$ are natural transformations. Is this true for the operator $\overline{\rho}$? Closure character $\overline{\rho}$ is an objectwise injective natural transformation $\overline{\rho} : \overline{P} \Longrightarrow E : \mathbf{QMet} \longrightarrow \mathbf{QMet}$ from closure powerspace to existential Kan quantification.

The above development says that "closure is the image factorization of character", and that closure character is an "embedding of closure powerspace into existential Kan quantification".

Fact. *For each contraction* $f:X \longrightarrow Y$ *the closure direct image contraction* $\overline{P}f : \overline{P}(X) \longrightarrow \overline{P}(Y)$ *is left adjoint* $\overline{P}f \dashv f^{-1}$ *to the ordinary inverse image monotonic function* $f^{-1} : \overline{P}(Y) \longrightarrow \overline{P}(X)$ *restricted to closed subspaces.*

The embeddings $\mathbf{PO} \longrightarrow \mathbf{Cat}$ and $\mathbf{QMet} \longrightarrow \mathbf{Cat}$ allow us to describe <closure powerspace – inverse image> as a functor $\overline{P} : \mathbf{QMet} \longrightarrow \mathbf{Adj}$. Associated with this closure powerspace functor $\overline{P}$ is a category $\mathbf{Sub}_{\overline{P}}$, called the *category of closed subspaces*, and a 01–fibration $P_{\overline{P}} : \mathbf{Sub}_{\overline{P}} \longrightarrow \mathbf{QMet}$. Objects in $\mathbf{Sub}_{\overline{P}}$ are pairs $\langle X,A\rangle$, where $X = \langle X,d_X\rangle$ is a quasimetric space and $A \in \overline{P}X$ is a closed subspace of X. Morphisms $f : \langle X,A\rangle \longrightarrow \langle Y,B\rangle$ in $\mathbf{Sub}_{\overline{P}}$ are contractions $f:X \longrightarrow Y$ satisfying: $\overline{P}f(A) \subseteq B$; or equivalently, $A \subseteq f^{-1}(B)$. The functor $P_{\overline{P}} : \mathbf{Sub}_{\overline{P}} \longrightarrow \mathbf{QMet}$ is called the *underlying quasimetric space functor*, and is defined by $P_{\overline{P}}(\langle X,A\rangle) = X$ and $P_{\overline{P}}(f) = f$.

Proposition. *Closed subspaces form a dialectical base over quasimetric spaces; that is, the underlying quasimetric space functor* $P_{\overline{P}} : \mathbf{Sub}_{\overline{P}} \longrightarrow \mathbf{QMet}$ *is a dialectical base.*

In the last section of this paper we will use this dialectical base of closed subspaces over spaces to describe the dialectical structure of metrical transition systems.

Let **K** be a cartesian closed category. A *(semi) power object* in **K** is a pair $\langle P,\in\rangle$ defining for each **K**–object X a **K**–object P(X) and a **K**–morphism $\in_X : X^{op} \times P(X) \longrightarrow \Omega$, where $\Omega = P(1)$, such that the map

$$\frac{f: Y \longrightarrow P(X)}{\in(f): X^{op} \times Y \longrightarrow \Omega}$$

defined by $\in(f) = \in_X.(Id_{X^{op}} \times f)$ (mapping extended **K**–morphisms to **K**–relations) is a (injection) bijection. The (X^{op},R)–th component $\varepsilon_X : X^{op} \times R^{X^{op}} \longrightarrow R : \langle x,\phi \rangle \longmapsto \phi(x)$ of the counit $\varepsilon : X \times (\,).(\,)^X \Longrightarrow Id_{QMet}$ of the product–exponential adjunction $X^{op} \times (\,) \dashv (\,)^{X^{op}}$ is a bimodule $\varepsilon_X : \langle X,d_X \rangle \leftarrow \langle EX,d_{EX} \rangle$, called *evaluation.* The existential Kan quantification and evaluation pair $\langle E,\varepsilon \rangle$ is a power object in the cartesian closed category **QMet**. Any finite ordinal n is isomorphic in **PO**, and hence in **QMet**, to $\overline{P}^n(\emptyset) = \overline{P}^{n-1}(1)$. Let N_α be the greatest fixpoint solution of the recursive domain equation $X \cong \overline{P}(\alpha X)$. Since $\overline{P}(\alpha X) \cong \overline{P}(1 \times \alpha X)$, this solution exists by the development below. Then $N = U_\square(N_\alpha)$ is a natural numbers object in **QMet**. Each component $\overline{\rho}_X$ of the natural transformation $\overline{\rho} : \overline{P} \Longrightarrow E$ is a contraction $\overline{\rho}_X : \overline{P}(X) \longrightarrow R^{X^{op}}$, and hence a bimodule $\overline{\rho}_X : \langle X,d_X \rangle \leftarrow \langle \overline{P}X,d_{\overline{P}X} \rangle$. Applying $U_\square$ to $\overline{\rho}_X$ gives us a "bimodule" $U_\square(\overline{\rho}_X) : X^{op} \times \overline{P}(X) \longrightarrow R \longrightarrow 2 = \overline{P}(1)$, which we also denote by $\overline{\rho}_X$, and call *closure membership.* The closure powerspace and closure membership pair $\langle \overline{P},\overline{\rho} \rangle$ is a semipower object in the cartesian closed category **QMet**.

Let Met denote the *metrization monad* on **QMet** which is generated by the adjunction (reflection) $F_{Met} \dashv Inc_{Met} : \mathbf{Met} \longrightarrow \mathbf{QMet}$; that is, Met = $\langle Met,\eta_{Met},\mu_{Met} \rangle$ where (i) Met = $Inc_{Met}.F_{Met} : \mathbf{QMet} \longrightarrow \mathbf{QMet}$, so that $Met(\langle X,d \rangle) = \langle X/\equiv,d_\equiv \rangle$; (ii) η_{Met} is the unit of the adjunction $F_{Met} \dashv Inc_{Met}$, $\eta_{Met}(\langle X,d \rangle) = [\,] : \langle X,d \rangle \longrightarrow \langle X/\equiv,d_\equiv \rangle$; and (iii) $\mu_{Met} = Inc_{Met}\varepsilon_{Met}F_{Met} = Id_{Met} : Met.Met \Longrightarrow Met$ (since $F_{Met} \dashv Inc_{Met}$ is a reflection, Met is idempotent). For every quasimetric space $\langle X,d \rangle$ there is a fundamental canonical isometry $\lambda(\langle X,d \rangle) : P.Met(\langle X,d \rangle) \longrightarrow Met.P(\langle X,d \rangle)$ defined by $\lambda(\langle X,d \rangle)(\{E_i | i \in I\}) = [\{x_i | i \in I\} | x_i \in E_i \text{ for all } i \in I]$, where $[\,]$ denotes the metrization equivalence class with respect to d_{PX}: $[A] = [B]$ iff $A \equiv_{PX} B$ iff $\overline{A} = \overline{B}$.

Metrization Monad Theorem. 1. $\lambda : P.Met \Longrightarrow Met.P$ *is an objectwise isometric distributive law, which distributes powerspace* P *over metrization* Met*;* 2. $\overline{P}$ = Met.P *the composite monad defined by* λ*; that is, closure powerspace is the same as metrization powerspace; and* 3. $\overline{\rho}$ = Met.ρ*; that is, closure character is the same as metrization character.* (The two identities $\rho = \overline{\rho}.(\overline{\ \ })$ and $\overline{\rho}$ = Met.ρ show that character ρ and closure character $\overline{\rho}$ are equivalent concepts.) 4. *Closure* $(\overline{\ \ }) : P \Longrightarrow \overline{P} = Met.P$ *is a monad map.*

4. Cauchy Diagrams

We now make some necessary categorical definitions. Let the symbol "ω^{op}" denote the graph $\ldots \longrightarrow n+1 \longrightarrow n \longrightarrow \ldots \longrightarrow 0$ of natural numbers, where edges "$n+1 \longrightarrow n$" replace order "$n+1 \geq n$". Let **K** be any category. An ω^{op}–diagram D in **K** is a diagram $\ldots \longrightarrow D_{n+1} \xrightarrow{d_n} D_n \longrightarrow \ldots \longrightarrow D_0$

of **K**–objects and **K**–morphisms of shape ω^{op}. An ω^{op}*–cone* (C,γ) from a **K**–object C to an ω^{op}–diagram D is a collection of **K**–morphisms $\gamma_n : C \longrightarrow D_n$ satisfying the commutativity conditions $d_n.\gamma_{n+1} = \gamma_n$ for each $n\in\omega^{op}$. An ω^{op}*–limit* of an ω^{op}–diagram D is an ω^{op}–cone $(LimD,\alpha)$, where $\alpha_n : LimD \longrightarrow D_n$ is called the *n–th projection*, with the property that given any other ω^{op}–cone (C,γ) to D, there is a unique **K**–morphism $g : C \longrightarrow LimD$ satisfying the commutativity conditions $\alpha_n.g = \gamma_n$ for all $n\in\omega^{op}$. We call g the ω^{op}*–limit tupling* of the ω^{op}–cone $\gamma = \{\gamma_n | n\in\omega^{op}\}$. An ω^{op}–limit of D is unique up to isomorphism.

Let $F : \mathbf{K} \longrightarrow \mathbf{K}$ be a functor on the category **K**, and suppose $\{\alpha_n : LimD \longrightarrow D_n | n\in\omega^{op}\}$ is "the" ω^{op}–limit of an ω^{op}–diagram $D = \{d_n : D_{n+1} \longrightarrow D_n | n\in\omega^{op}\}$ in **K**. Then F is said to *preserve* this limit (or to be D*–continuous* at LimD) when $F(LimD) = \{F(\alpha_n) : F(LimD) \longrightarrow F(D_n) | n\in\omega^{op}\}$ is "the" ω^{op}–limit of the ω^{op}–diagram $FD = \{F(d_n) : F(D_{n+1}) \longrightarrow F(D_n) | n\in\omega^{op}\}$; that is, $F(LimD) = Lim(FD)$. F is said to be ω^{op}*–continuous* when it is D–continuous at LimD for all ω^{op}–diagrams D.

Constant functors and the identity functor on **K** are ω^{op}–continuous. The pointwise product of two ω^{op}–continuous functors is ω^{op}–continuous. For the category **K** = **QMet** , the pointwise coproduct of ω^{op}–continuous functors is ω^{op}–continuous. If $F : \mathbf{K} \longrightarrow \mathbf{K}$ is D–continuous at LimD and $G : \mathbf{K} \longrightarrow \mathbf{K}$ is FD–continuous at F(LimD)=Lim(FD), then GF is D–continuous at LimD.

For any ω^{op}–diagram $D = \{f_n : \langle A_{n+1},d_{n+1}\rangle \longrightarrow \langle A_n,d_n\rangle | n\in\omega^{op}\}$ in the category **QMet** there is a canonical ω^{op}–limit LimD called the *cartesian* ω^{op}*–limit* of D. The underlying set of points of LimD consists of the D*–compatible sequences* $\{\langle a_n | n\in\omega^{op}\rangle | a_n\in A_n$ and $f_n(a_{n+1})=a_n$ for all $n\in\omega^{op}\}$. These D–sequences are analogous to Cauchy sequences. The metric on LimD, $d : LimD\times LimD \longrightarrow LimD$, is defined by $d(\langle a_n\rangle,\langle b_n\rangle) = \sup\{d_n(a_n,b_n) | n\in\omega^{op}\}$. Since the functions f_n are contractions $d_{n+1}(a_{n+1},b_{n+1}) \geq d_n(f_n(a_{n+1}),f_n(b_{n+1})) = d_n(a_n,b_n)$, and hence the metric d has the simpler definition $d(\langle a_n\rangle,\langle b_n\rangle) = \lim\{d_n(a_n,b_n) | n \longrightarrow \infty\}$.

For any ω^{op}–diagram D, as above, and any quasimetric space $\langle X,d_X\rangle$, consider the exponential space $(LimD)^X$. Now $(LimD)^X \cong Lim(D^X)$, the cartesian ω^{op}–limit of the ω^{op}–diagram $D^X = \{f_n{}^X : A_{n+1}{}^X \longrightarrow A_n{}^X | n\in\omega^{op}\}$, since the exponential functor $()^X : \mathbf{QMet} \longrightarrow \mathbf{QMet}$ is right adjoint to the product functor $X\times() : \mathbf{QMet} \longrightarrow \mathbf{QMet}$, and hence continuous. So if $\phi = \{\phi_n : \langle X,d_X\rangle \longrightarrow \langle A_n,d_n\rangle | n\in\omega^{op}\}$ and $\gamma = \{\gamma_n : \langle X,d_X\rangle \longrightarrow \langle A_n,d_n\rangle | n\in\omega^{op}\}$ are any two cones from $\langle X,d_X\rangle$ to D with corresponding ω^{op}–limit tuplings $f : \langle X,d_X\rangle \longrightarrow \langle LimD,d\rangle$ and $g : \langle X,d_X\rangle \longrightarrow \langle LimD,d\rangle$, respectively, then d(f,g) (in $(LimD)^X$) = $d(\phi,\gamma)$ (in $Lim(D^X)$). In particular, the tuplings $f \leq g$ iff $\phi \leq \gamma$ iff the components in the cones $\phi_n \leq \gamma_n$ for all $n\in\omega^{op}$.

In order to utilize the powerspace functor $P : \mathbf{QMet} \longrightarrow \mathbf{QMet}$ in the fixpoint solution of recursive domain equations, this functor must preserve the limits of appropriate ω^{op}–diagrams (see the greatest fixpoint theorem below). Unfortunately, it does not preserve all ω^{op}–limits. An attempt to prove that it does shows the need to restrict P. A very natural restriction on P is the restriction to the closed subsets of quasimetric spaces; that is, we use the closure powerspace functor $\overline{P}$ in place of P in recursive domain equations. We will show that $\overline{P}$ preserves the limits of a useful class of ω^{op}–diagrams D; that is, that $\overline{P}(\mathrm{Lim}D) \cong \mathrm{Lim}(\overline{P}D)$ for each of these diagrams.

Proposition. *For any* ω^{op}*–diagram* $D = \{f_n : \langle A_{n+1}, d_{n+1}\rangle \longrightarrow \langle A_n, d_n\rangle \mid n \in \omega^{op}\}$ *in* **QMet**, *there exists a canonical contraction* $\mathrm{seq} : \overline{P}(\mathrm{Lim}D) \longrightarrow \mathrm{Lim}(\overline{P}D)$.

Proposition. *For any* ω^{op}*–diagram* D *as above, there is a canonical monotonic function* $\lim : \mathrm{Lim}(\overline{P}D) \longrightarrow \overline{P}(\mathrm{Lim}D)$ *such that* $\mathrm{Id}_{\mathrm{Lim}(\overline{P}D)} \geq \mathrm{seq}.\lim$, *and* $\lim.\mathrm{seq} \geq \mathrm{Id}_{\overline{P}(\mathrm{Lim}D)}$; *that is, such that* $\langle \lim \dashv \mathrm{seq}\rangle : \mathrm{Lim}(\overline{P}D) \longrightarrow \overline{P}(\mathrm{Lim}D)$ *is a Galois connection (an adjoint pair of monotonic functions).*

An ω^{op}–diagram $D = \{f_n : \langle A_{n+1}, d_{n+1}\rangle \longrightarrow \langle A_n, d_n\rangle \mid n \in \omega^{op}\}$ in **QMet** is called a *reflection* ω^{op}*–diagram* when f_n is the projection part of a reflection pair $\langle f_n, g_n\rangle : \langle A_{n+1}, d_{n+1}\rangle \longrightarrow \langle A_n, d_n\rangle$ for all $n \in \omega^{op}$. Clearly every reflection ω^{op}–diagram D has an associated ω–diagram $D^{op} = \{g_n : \langle A_n, d_n\rangle \longrightarrow \langle A_{n+1}, d_{n+1}\rangle \mid n \in \omega\}$. Define $f_n^{\ m} = f_m \cdot \ldots \cdot f_{n-1} : \langle A_n, d_n\rangle \longrightarrow \langle A_m, d_m\rangle$ and $g_m^{\ n} = g_{n-1} \cdot \ldots \cdot g_m : \langle A_m, d_m\rangle \longrightarrow \langle A_n, d_n\rangle$ for all $n > m$, and $f_n^{\ n} = g_n^{\ n} = \mathrm{Id}_{A_n}$. Then $\langle f_n^{\ m}, g_m^{\ n}\rangle : \langle A_n, d_n\rangle \longrightarrow \langle A_m, d_m\rangle$ is a reflection pair for all $n \geq m$. Define the function $i_n : A_n \longrightarrow \mathrm{Lim}D$ by: $pr_m(i_n(a_n)) = g_n^{\ m}(a_n)$, if $m > n$; $= a_n$, if $m = n$; and $= f_n^{\ m}(a_n)$, if $m < n$. Clearly i_n is well-defined and is a contraction $i_n : \langle A_n, d_n\rangle \longrightarrow \langle \mathrm{Lim}D, d\rangle$ since $g_n^{\ m}$ and $f_n^{\ m}$ are contractions for all $m \in \omega^{op}$. It is easy to show that $i_{n+1} \cdot g_n = i_n$ for all $n \in \omega$, so that $\iota = \{i_n : \langle A_n, d_n\rangle \longrightarrow \langle \mathrm{Lim}D, d\rangle \mid n \in \omega\}$ is a D^{op}–cocone. Hence, there is a unique contraction $i : \mathrm{Colim}(D^{op}) \longrightarrow \mathrm{Lim}D$, the colimit cotupling of cocone ι, satisfying $i.in_n = i_n$ for all $n \in \omega$. It is easy to show that $\langle pr_n, i_n\rangle : \langle \mathrm{Lim}D, d\rangle \longrightarrow \langle A_n, d_n\rangle$ is a reflection pair for each $n \in \omega^{op}$. Any monotonic functor (in particular, a contracting functor) preserves reflection ω^{op}–diagrams. Since P and $\overline{P}$ are monotonic they preserve reflection ω^{op}–diagrams.

Proposition. *If* D *is a reflection* ω^{op}*–diagram, then* seq *is a surjection with right inverse* lim; *that is,* $\mathrm{Id}_{\mathrm{Lim}(\overline{P}D)} = \mathrm{seq}.\lim$.

We next show that the closure powerspace functor $\overline{P}$ preserves the limits for a class of ω^{op}–diagrams in **QMet** that includes many useful approximation diagrams used in the categorical greatest fixpoint theorem. A reflection ω^{op}–diagram $D = \{f_n : \langle A_{n+1}, d_{n+1}\rangle \longrightarrow \langle A_n, d_n\rangle \mid n \in \omega^{op}\}$ in the category **QMet** is a *Cauchy* ω^{op}*–diagram* when $|f_n^{\ m}| \longrightarrow 0$ as $n,m \longrightarrow \infty$; that is, when the kernel bimodules of D converge uniformly to identity bimodules. The objects in a Cauchy ω^{op}–diagram beyond n become increasingly more "geometrically similar" as n gets large. Associated with any ω^{op}–diagram D is a metric d_D defined by: $d_D(A_n, A_m) = |f_n^{\ m}|$, for $n > m$; and $= 0$, for $n \leq m$. Obviously, $d_D(A_n, A_m) \geq d_{|\,|}(A_n, A_m)$. and when

D is Cauchy we have $d_D(A_n,A_m) \geq d_{||}(A_n,A_m) \longrightarrow 0$ as $n,m \longrightarrow \infty$. So if D is a Cauchy ω^{op}–diagram, then $<\ldots, A_n, \ldots, A_0>$ is a Cauchy sequence in the quasimetric space $<$**QMet**$,d_{||}>$. Any contracting norm–decreasing functor on **QMet** preserves Cauchy ω^{op}–diagrams.

The Cauchy axiom implies that for any $\varepsilon>0$ there is an $N \in \omega^{op}$ such that for all $n>N$ if $b_n,c_n \in A_n$ are any two elements of A_n, then $d_n(b_n,c_n) < d_N(f_n^{\ N}(b_n),f_n^{\ N}(c_n)) + \varepsilon$. In particular, the axiom implies that for any given $\varepsilon>0$ there is an N such that for any two elements $b,c \in \text{LimD}$ we have $d(b,c) \leq d_N(pr_N(b),pr_N(c)) + \varepsilon$. Of course, for any $\varepsilon>0$ and any two elements $b,c \in \text{LimD}$ we can find an N such that the above inequality holds, even when D is not Cauchy. With the Cauchy axiom we can choose N uniformly and independent of b and c. We use this property below to show that seq is an isometry and an injection.

The notion of Cauchy ω^{op}–diagrams in **PO** (considered to be the full subcategory $F_\square$(**PO**) of **QMet**) does not seem to be directly useful. If D is a Cauchy diagram in **PO**, then there is an N such that for all $n \geq N$ we have $\leq_{f_{n-1}} = \leq_n$; that is, the kernel bimodules of D, not only *converge to* the identity bimodules, but *eventually equal* the identity bimodules. The prefix orders $<A_n,d_n^{\ 1}>$ for the stacks and trees approximation diagrams do *not* satisfy this condition, and hence are *not* Cauchy. Whereas, the asymmetrical prefix metrics $<A_n,d_n^{\ \alpha}>$ for $\alpha \neq 1$ for the stacks and trees approximation diagrams *are* Cauchy.

We can define the notion of Cauchy ω^{op}–diagrams in **UQMet** in the same way. A Cauchy diagram in **UQMet** is also Cauchy in **QMet**. If $D = \{f_n : <A_{n+1},\leq_{n+1}> \longrightarrow <A_n,\leq_n> | n \in \omega^{op}\}$ is a reflection ω^{op}–diagram in the category **PO**, then for each constant $\alpha \in I=[0,1]$, $\alpha \neq 1$, there is a sequence of metrics $\{d_n | n \in \omega^{op}\}$ making this a Cauchy ω^{op}–diagram in the category **UQMet**. Let d_0 be the metric $d_0(a,b)=1$ when $a \not\leq_0 b$ and $d_0(a,b)=0$ when $a \leq b$. Define d_{n+1} recursively as follows: $d_{n+1}(a,b) = 0$, when $a \leq_{n+1} b$; $d_{n+1}(a,b) = \alpha^n$, when $a \leq_{f_n} b$ and $a \not\leq_{n+1} b$; and $d_{n+1}(a,b) = d_n(f_n(a),f_n(b))$, when $a \not\leq_{f_n} b$. Then $<A_n,d_n>$ is an ultraquasimetric space, and $f_n : <A_{n+1},d_{n+1}> \longrightarrow <A_n,d_n>$ and $g_n : <A_n,d_n> \longrightarrow <A_{n+1},d_{n+1}>$ are contractions for all $n \in \omega^{op}$. Denote this reflection ω^{op}–diagram in **UQMet** by $D_\alpha =$ $\{f_n : <A_{n+1},d_{n+1}> \longrightarrow <A_n,d_n> | n \in \omega^{op}\}$. Obviously, for $\alpha \neq 1$, D_α is Cauchy. Also $U_\square(<A_n,d_n>) = <A_n,\leq_n>$ and the closure of any subset $A \subseteq A_n$ is $\overline{A} = \downarrow A$ the elements below A for all $n \in \omega^{op}$. In fact, for each $\alpha \in I=[0,1]$, the above construction defines a functor $F_\alpha : \mathbf{PO}^{\omega^{op}} \longrightarrow \mathbf{UQMet}^{\omega^{op}}$ to the category of Cauchy diagrams, a subcategory of $\mathbf{UQMet}^{\omega^{op}}$, such that $U_\square^{\ \omega^{op}} . F_\alpha = Id_{\mathbf{PO}^{\omega^{op}}}$. For the special case $\alpha=1$, we have $F_1 = F_\square^{\ \omega^{op}}$. So when functor $\hat{K}$: **UQMet** $\longrightarrow$ **UQMet** extends functor K: **PO** $\longrightarrow$ **PO**, we have $F_1(S(K)) = F_\square^{\ \omega^{op}}(S(K)) = S(\hat{K})$ (where S maps an endofunctor to its approximation diagram; see below); that is, the approximation diagram $S(\hat{K})$ of $\hat{K}$ consists only of preorders and monotonic functions, and hence is *not* a Cauchy ω^{op}–diagram. Certain functors $\hat{K}$ also satisfy the condition that $F_\alpha(S(K)) = S(\hat{K}.\alpha)$ for $\alpha \in I=[0,1]$, $\alpha \neq 1$; so that $S(\hat{K}.\alpha)$ *is* a Cauchy ω^{op}–diagram. The trees–constructing functor E_α, discussed in section 5, is such a functor, and so its approximation diagram $S(E_\alpha)$ is Cauchy.

Proposition. *If* D *is a Cauchy* ω^{op}*–diagram, then* $\lim : Lim(\overline{P}D) \longrightarrow \overline{P}(LimD)$ *is a contraction.*

Proposition. *If* D *is a Cauchy* ω^{op}*–diagram, then* seq *is an injection with left inverse* lim*; that is,* lim.seq $= Id_{\overline{P}(LimD)}$.

The main technical theorem of this paper now follows.

Continuity Theorem. *The closure powerspace functor* $\overline{P}$: **QMet** $\longrightarrow$ **QMet** *preserves the limit of any Cauchy* ω^{op}*–diagram.*

Proof. Let $D = \{f_n : \langle A_{n+1}, d_{n+1}\rangle \longrightarrow \langle A_n, d_n\rangle \mid n \in \omega^{op}\}$ be a Cauchy ω^{op}–diagram. By the above propositions the canonical contraction $seq : \overline{P}(LimD) \longrightarrow Lim(\overline{P}D)$ is a **QMet**–isomorphism. We have effectively proven that any *closed subdiagram (subobject)* of a Cauchy ω^{op}–diagram D is itself a Cauchy ω^{op}–diagram, and is isomorphic via lim to a *closed subobject* of LimD.

Corollary. *The closure powerspace functor* $\overline{P}$: **UQMet** $\longrightarrow$ **UQMet** *preserves the limit of any Cauchy* ω^{op}*–diagram.*

5. Synchronization Trees and Milner's Strong Congruence

If $F : \mathbf{K} \longrightarrow \mathbf{K}$ is any functor on a category **K**, then an F*-algebra* is a pair $\langle A,\alpha\rangle$ where A is a **K**–object and $\alpha : F(A) \longrightarrow A$ is a **K**–morphism, and an F*–coalgebra* is a pair $\langle C,\gamma\rangle$ where C is a **K**–object and $\gamma : C \longrightarrow F(C)$ is a **K**–morphism. An F*–coalgebra morphism* $h : \langle A,\alpha\rangle \longrightarrow \langle B,\beta\rangle$ is a **K**–morphism $h : A \longrightarrow B$ satisfying $F(h).\alpha = \beta.h$. F–coalgebras and F–coalgebra morphisms form a category $\mathbf{K}_F$. A *terminal* F*–coalgebra* is a terminal object in $\mathbf{K}_F$; that is, an F–coalgebra $\langle C_\infty,\gamma_\infty\rangle$ satisfying the property that if $\langle C,\gamma\rangle$ is any other F–coalgebra then there is a unique F–coalgebra morphism $h : \langle C,\gamma\rangle \longrightarrow \langle C_\infty,\gamma_\infty\rangle$. Terminal F–coalgebras are unique up to F–coalgebra isomorphism. In fact, terminal F–coalgebras *are* isomorphisms $\gamma_\infty : C_\infty \cong F(C_\infty)$, and hence, F–fixpoints (Arbib and Manes [1982]).

Let **K** be a category, let $F : \mathbf{K} \longrightarrow \mathbf{K}$ be a functor on **K**, and let $\langle A,\alpha\rangle$ be an F–algebra. The ω^{op}*-approximation diagram (sequence) of* F *starting from* $\langle A,\alpha\rangle$ is the ω^{op}–diagram $S(F,A,\alpha) = \{F^n(\alpha) : F^{n+1}(A) \longrightarrow F^n(A) \mid n \in \omega^{op}\}$. Suppose that **K** has a terminal object **1**. Then the ω^{op}*-approximation diagram (sequence) of* F is $S(F) = S(F,\mathbf{1},\Delta)$, where Δ is the unique **K**–morphism $\Delta : F(\mathbf{1}) \longrightarrow \mathbf{1}$.

We are now ready to give the categorical greatest fixpoint theorem. Our version of the categorical greatest fixpoint theorem below is a generalization of the version of Arbib and Manes [1982] in that our functor F is only required to be continuous at one **K**–object, the limit of its own ω^{op}–approximation diagram. The greatest fixpoint theorem is the categorical dual of the least fixpoint theorem (Adamek and Koubek [1979], Lehmann and Smyth [1981]). Although categorically it is just the dual, the details of the greatest fixpoint solutions are much more interesting, since they often have various "completeness" properties.

Greatest Fixpoint Theorem. *Let* **K** *be a category with a terminal object* **1**, *and let* $F: \mathbf{K} \longrightarrow \mathbf{K}$ *be a functor on* **K**. *If* $C_\infty = \mathrm{Lim}(S(F))$ *exists in* **K**, *and if* F *preserves this limit* (F *is* S(F)*–continuous at* C_∞) *then there is a* **K**–*morphism* $\gamma_\infty : C_\infty \longrightarrow F(C_\infty)$ *making* $\langle C_\infty, \gamma_\infty \rangle$ *the terminal* F*–coalgebra (the greatest fixpoint of* F*).*

We now give a very important application of the closure powerspace functor: an extension of the DeBakker–Zucker–Nivat metric space approach to concurrency which allows us to define Milner's algebraic approach. We identify Milnerian processes (Milner [1980,1983]) with states of nondeterministic transition systems. A nondeterministic transition system on state set Q over alphabet A is a ternary relation $R \subseteq Q \times A \times Q$. R is considered to be a set of A–labelled transitions (edges) over Q: we interpret any edge $(p,a,q) \in R$ to be a transition from current state p to next state q with label a. Equivalently an A–transition system is a transition function $\gamma: Q \longrightarrow P(A \times Q)$. We enrich and extend the notion of a transition system by defining a distance between states which is respected by transitioning (transitioning is contracting). More precisely, a metrical A–transition system is a pair $\langle Q,R \rangle$, where $Q = \langle Q,d \rangle$ is an underlying ultraquasimetric (state) space and R is a ternary relation $R \subseteq Q \times A \times Q$. A morphism $\langle f,h \rangle : \langle A,Q,R \rangle \longrightarrow \langle B,Q',R' \rangle$ of metrical transition systems is a function $f: A \longrightarrow B$ and a contraction $h: Q \longrightarrow Q'$ which satisfy $\overline{P}(h \times f \times h) \subseteq R'$; or equivalently, $R \subseteq (h \times f \times h)^{-1}(R')$. Metrical transition systems and their morphisms form the category **Sys**. There is an *underlying set functor* $P_{\mathbf{Sys}} : \mathbf{Sys} \longrightarrow \mathbf{Set}$, where $P_{\mathbf{Sys}}(\langle A,Q,R \rangle) = A$ and $P_{\mathbf{Sys}}(\langle f,h \rangle) = f$.

Theorem (Transition System Dialectics). *Metrical transition systems form a dialectical base over sets; that is, the underlying set functor* $P_{\mathbf{Sys}} : \mathbf{Sys} \longrightarrow \mathbf{Set}$ *is a dialectical base.*

This theorem allows us to regard parallel interaction (of a collection of concurrent processes) as inverse dialectical flow. In particular, a discrete pair of transition systems $S = \langle S_0, S_1 \rangle$ over a pair of alphabets $A = \langle A_0, A_1 \rangle$ is a functor $S: 2 \longrightarrow \mathbf{Sys}$ satisfying the condition $S \,.\, P_{\mathbf{Sys}} = A$. A span $\sigma = \langle A_0, f_0, B, f_1, A_0 \rangle$ specifies parallel interaction, or inverse dialectical flow, taking S to the combined system $\sigma(S): 1 \longrightarrow \mathbf{Sys}$ over alphabet B. So, binary parallel interaction $\equiv$ inverse dialectical flow $2 \Longrightarrow 1$; two systems are combined into one system.

Clearly, the category of systems **Sys** and the fibration P_{Sys} are both derivable from the more basic category of closed subspaces $\mathbf{Sub}_{\overline{P}}$ and fibration $P_{\overline{P}}$. A metrical transition system is *contracting* when the associated transition function $\gamma: Q \longrightarrow \overline{P}(A \times Q)$ is a contraction: for any symbol $a \in A$ and for any two states p and q, the sets of next states $\gamma(p,a)$ and $\gamma(q,a)$ are closed and $d(p,q) \geq d_P(\gamma(p,a),\gamma(q,a))$, where d_P is the Hausdorff metric. So, we can regard a contracting transition system to be an F–coalgebra $\gamma: Q \longrightarrow F(Q)$ for functor $F(\ ) = \overline{P}(A \times (\ ))$. If we restrict all metrical transition systems in **Sys** to be contracting, then $\mathrm{CoAlg}_A \cong \mathbf{Sys}_A = P_{Sys}^{-1}(A)$, the A–th fiber of P_{Sys}.

Let $\alpha \in \mathbf{I} = [0,1]$ be a fixed attenuation constant. For notational simplicity let us define the following functors on **UQMet**: $E_\alpha() = \overline{P}(A \times \alpha()))$; $N_\alpha() = A \times \alpha\overline{P}()$; and $N_\alpha^{\ P}() = 1 + A \times \alpha\overline{P}()$. So that E_α and N_α are the compositions of $\overline{P}$ and $A \times \alpha()$, and $N_\alpha^{\ P}$ is the composition of $\overline{P}$ then $1 + A \times \alpha()$. The functors $A \times \alpha()$ and $1 + A \times \alpha()$ specify the "vertical" sequential aspect of trees, and the functor $\overline{P}$ specifies the "horizontal" choice aspect of trees (branching structure). The constant α represents the relative importance of vertical location in trees as measured by the asymmetric prefix metric $d_\alpha^{\ E}$: large (small) α means that outer edges or nodes are (not) almost as important as inner nodes. The choice of α does not change the underlying set of trees, but only affects the asymmetric prefix metric $d_\alpha^{\ E}$ on the set of trees. When $\alpha = 1$ the prefix metric $d_1^{\ E}$ is precisely the traditional prefix order $\leq_{pre}$.

We have shown elsewhere (Kent [1985]) that the functor E_α is at the heart of Milner's semantics of synchronous communicating systems (Milner [1983]). In particular, we have shown that:

- metrical transition systems (Milner's processes can be identified with states of these) are E_α–coalgebras;
- Milner's strong bisimulations are (the kernels of, and hence identifiable with) morphisms of E_α–coalgebras;
- the datatype of Milner's synchronization trees is the greatest fixpoint solution of the recursive domain equation $X \cong E_\alpha(X)$;
- Milner's strong congruence on a metrical transition system is the kernel of the unique (unfoldment tree) bisimulation from the system to the greatest fixpoint solution of synchronization trees;
- Milner's five basic operators of action, sum, product, morphism and restriction in SCCS, when suitably defined and extended, internally define the sequencing and branching in transition systems and externally define dialectical flow between the different contexts represented by the fibers of P_{Sys} such as CoAlg_A and CoAlg_B; and
- Milner's equational properties are properties of the above objects, morphisms, categories and functors.

For more details see Kent [1985,1986]. So our approach (Golson and Rounds [1983], Kent [1984,1985]) provides a strong link between two important and well-known theories of concurrency: the complete metric space approach of DeBakker and Zucker [1982] and Nivat [1979], and the algebraic approach of Milner [1980,1983].

Tree Datatype Theorem. *When* $F = E_\alpha$, N_α or $N_\alpha^{\,P}$, *the greatest fixpoint solution to the recursive domain equation* $X \cong F(X)$ *exists and consists of:* $T_A^{E}\alpha = \langle T_A^{E}, d_\alpha^{E}\rangle$ = *finite or infinite edge-labelled (synchronization) trees (Mealy trees);* $T_A^{N}\alpha = \langle T_A^{N}, d_\alpha^{N}\rangle$ = *finite or infinite full node-labelled trees (full Moore trees); and* $T_A^{N^P}\alpha = \langle T_A^{N^P}, d_\alpha^{N^P}\rangle$ = *finite or infinite partial node-labelled trees (Moore trees); respectively.*

All three classes of trees have branching structure which is unordered and without duplicates (multiplicity). We show that $T_A^{E}\alpha \cong \overline{P}(T_A^{N}\alpha)$, and that $T_A^{N}\alpha \cong A \times \alpha(T_A^{E}\alpha)$; that is, that Mealy trees are the same thing as closed forests of full Moore trees, and that full Moore trees are the same thing as <A-symbol,Mealy-tree> pairs.

Proof. We show that the recursive domain equation $X \cong E_\alpha(X)$ has a greatest fixpoint solution. The existence of solutions for the other two equations is just as easy to prove. For $\alpha \neq 0,1$, E_α preserves the limit of its approximation diagram $S(E_\alpha)$, since: for $\alpha \neq 0,1$ $S(E_\alpha)$, $S(N_\alpha)$ and $S(N_\alpha^{\,P})$ are Cauchy diagrams; for $\alpha \neq 0$ the functors $A \times \alpha(\,)$ and $1 + A \times \alpha(\,)$ are ω^{op}-continuous; and the functor $\overline{P}$ preserves the limit of any Cauchy diagram.

In summary then, we have defined a powerobject construction, closure powerspace, in the context of quasimetric spaces; a context generalizing both the order-theoretic and the metric contexts. We have developed the continuity properties of this construction in terms of generalized Cauchy sequences. Finally, we have applied this construction to fixpoint semantics for interacting processes. Although we have defined our concepts, stated our results and constructed our proofs in the closed preorder R of nonnegative real numbers, they should extend to any closed preorder, and possibly any closed category. Currently we are working out the relationship between the closure powerspace construction and the Plotkin and Smyth powerdomain constructions. We are also investigating the obvious topos-theoretic nature of **QMet** and closure powerspace, and generalizing various theorems. Some open problems for future research are: 1. relating the closure powerspace construction to Main's semiring module construction (Main [1984]); 2. developing a theory of guarded commands in the quasimetric context; and 3. integrating our quasimetric approach within the Lehmann-Smyth categorical theory of abstract data types (Lehmann and Smyth [1981]).

I would like to thank Daniel Lehmann for his suggestions and his encouragement.

References

1. J. Adamek and V. Koubek, Least Fixed Point of a Functor, J. Comput. System Sci. 19 (1979) 163.

2. M.A. Arbib and E.G. Manes, Parameterized Data Types Do Not Need Highly Constrained Parameters, Info. and Contr. 52 (1982) 139-158.

3. S. Bernow and P. Raskin, Ecology of Scientific Consciousness, Telos 28, Summer (1976).

4. J.W. DeBakker and J.I. Zucker, Processes and the Denotational Semantics of Concurrency, Info. and Contr. 54 (1982) 70-120.

5. W. Golson and W. Rounds, Connections between Two Theories of Concurrency: Metric Spaces and Synchronization Trees, Tech. Rep. TR-3-83, Computing Research Laboratory, University of Michigan, 1983.

6. J. Gray, Fibred and Cofibred Categories, Conference on Categorical Algebra (1965).

7. R.E. Kent, Observational Equivalence of Concurrent Processes is the Kernel of a Multiplicity Morphism of Abstract Tree Data Types, International Computer Symposium 1984, Taipei, Taiwan (1984).

8. R.E. Kent, The Metric Powerspace Construction, Tech. Rep. UIC-EECS-84-14, EECS Dept., University of Illinois at Chicago, 1984.

9. R.E. Kent, Synchronization Trees and Milner's Strong Congruence: a Fixpoint Approach, Tech. Rep. UIC-EECS-85-5, EECS Dept., University of Illinois at Chicago, 1985.

10. R.E. Kent, Dialectical Systems: Interactions and Combinations, manuscript (1986).

11. R.E. Kent, Dialectical Development in First Order Logic: I. Relational Database Semantics (1987), submitted for publication.

12. K. Kuratowski, Topology, Vol.1, (Academic Press, New York, N.Y., 1966).

13. F.W. Lawvere, Metric Spaces, Generalized Logic, and Closed Categories, Seminario Matematico E. Fisico. Rendiconti. Milan. 43 (1973) 135-166.

14. D.J. Lehmann and M.B. Smyth, Algebraic Specification of Data Types: A Synthetic Approach, Math. Systems Theory 14 (1981) 97-139.

15. M.G. Main, Free Constructions of Powerdomains, in: Proceedings of the Conference on Mathematical Foundations of Programming Language Semantics, Lec. Notes in Comp. Sci. 239, (Springer-Verlag, New York).

16. M.G. Main, Semiring Module Powerdomains, Tech. Rep. CU-CS-286-84, Department of Computer Science, University of Colorado at Boulder, 1984.

17. H. Marcuse, Zum Problem der Dialektik, Die Gesellschaft, Volume VII (1930-31); On the Problem of the Dialectic, Telos 27, Spring (1976).

18. R. Milner, A Calculus of Communicating Systems, Lec. Notes in Comp. Sci. 92, (Springer-Verlag, New York, 1980).

19. R. Milner, Calculi for Synchrony and Asynchrony, Theo. Comp. Sci. 25 (1983) 267-310.

20. M. Nivat, Infinite Words, Infinite Trees, Infinite Computations, Mathematical Centre Tracts 109 (1979) 1-52.

21. G.D. Plotkin, A Powerdomain Construction, SIAM J. Comput. 5 (1976) 452-487.

22. K. Popper, What is Dialectic?, Mind 49 (1940) 403-426.

23. M.B. Smyth, Power Domains, J. Comput. System Sci. 16 (1978) 23-36.

24. M.B. Smyth, Power Domains and Predicate Transformers: a Topological View, 10th ICALP, Barcelona, Spain, Lec. Notes in Comp. Sci. 154, (Springer-Verlag, New York, 1983).

25. M. Steenstrup, M.A. Arbib and E. Manes, Port Automata and the Algebra of Concurrent Processes, COINS Tech. Rep. 81-25, University of Massachusetts, 1981.

A POWERDOMAIN CONSTRUCTION

Karel Hrbacek
Department of Mathematics
City College of the City University of New York
New York, New York 10031

Abstract

We show that the forgetful functor from the category of nondeterministic algebraic lattices into the category of algebraic lattices has a left adjoint. The construction provides an analog of the Plotkin powerdomain entirely within the category of algebraic lattices. Similar results hold for the category of bounded complete algebraic posets and the category of continuous lattices.

Powerdomains are used in denotational semantics of programming languages for representation of nondeterminism. The Smyth (upper), Hoare (lower) and Plotkin (convex) powerdomains have been studied intensively; while the first two fit well into the general theoretical machinery of denotational semantics, Plotkin powerdomains do not. SCOTT (e.g. in 1981) defines _domains_ as bounded complete algebraic posets; however, the Plotkin powerdomain of a domain usually is not bounded complete. Two solutions to this problem offer themselves. One can enlarge the category of domains, as PLOTKIN does in his original paper (1976); the resulting category of SFP-objects is closed under the Plotkin powerdomain construction as well as other needed functors. The disadvantage of this approach is that SFP-objects are harder to understand and to work with than Scott's domains. The other solution is to complete the Plotkin powerdomain so as to obtain a bounded complete algebraic poset. We have investigated one such completion (via Frink ideals) extensively in HRBACEK (1985), (1987). The main reason for choosing the Frink completion is that it is _minimal_, i.e., it introduces as few extraneous elements as possible into the powerdomain; moreover, the new elements can be neatly characterized in terms of the old, computationally meaningful, ones. However, this construct does not have all of the desirable mathematical properties (e.g. it is not a functor in the category of bounded complete algebraic posets and continuous maps) and ad-hoc solutions have to be devised to circumvent the resulting problems (see HRBACEK (1987)).

Here we investigate another way of completing Plotkin powerdomain, this time using a modification of the maximal completion (also known as the Alexandrov completion or the completion by lower sets). It turns out that this construction is easier to work with and mathematically better behaved (e.g. it can be characterized as a left adjoint to a suitable forgetful functor, in full analogy with the Plotkin powerdomain). Its one disadvantage is a much greater profusion of extraneous elements; nevertheless, it is still possible to distinguish them from the old ones both set-theoretically and algebraically.

The first section of this paper reviews the Plotkin powerdomain construction. In the second section we present the details of our construction for the category of algebraic lattices. The modifications needed in order to obtain similar results for the category of Scott domains (bounded complete algebraic posets) or continuous lattices are outlined at the end of the section. The last section is concerned mainly with the relationship of our powerdomain construct to the Plotkin powerdomain.

1. The Plotkin Powerdomain

In this section we review basic concepts related to algebraic posets and the well-known order-theoretic construction of the Plotkin powerdomain. The reader is referred to GIERZ et al. (1980) for more information concerning the former and to PLOTKIN (1976), (1981) and SMYTH (1978) for the details of the latter.

A partially ordered set (poset) $(D,\leq)$ is complete if every directed $X \subseteq D$ has a supremum (least upper bound) sup X. An element d of a complete poset $(D,\leq)$ is compact if for every directed $X \subseteq D$, $d \leq \sup X$ implies $d \leq x$ for some $x \in X$. The set of all compact elements of D will be denoted K(D). A complete poset $(D,\leq)$ is algebraic if, for all $d \in D$, $\{x \in K(D) \mid x \leq d\}$ is directed and $d = \sup \{x \in K(D) \mid x \leq d\}$.

A function f from a complete poset $(D,\leq_D)$ into a complete poset $(E,\leq_E)$ is continuous if for each directed $X \subseteq D$, $f(X) = \{f(x) \mid x \in X\}$ is directed and $f(\sup X) = \sup f(X)$.

ALG is the category whose objects are algebraic posets and whose morphisms are continuous functions.

A structure $(D,\leq_D,\cup_D)$ where $(D,\leq_D)$ is an algebraic poset and $\cup_D$ is a binary operation on D which is continuous (in both variables), commutative, associative and absorptive (i.e., $d \cup_D d = d$ for all $d \in D$) will be called a nondeterministic algebraic poset. A function f from $(D,\leq_D,\cup_D)$ into $(E,\leq_E,\cup_E)$ is additive if $f(a \cup_D b) = f(a) \cup_E f(b)$ holds for all $a,b \in D$. The category whose objects are nondeterministic algebraic posets and whose morphisms are continuous additive functions will be denoted NALG.

Let $(D,\leq)$ be a poset; $A \subseteq D$ is a lower set if $y \leq x \in A$ implies $y \in A$. $\downarrow X = \{y \in D \mid y \leq x \text{ for some } x \in X\}$ denotes the smallest lower set containing X. Ideals of $(D,\leq)$ are directed lower sets; in particular, $\downarrow\{x\} = \downarrow x$ is the principal ideal generated by $x \in D$. The set of all ideals of D will be denoted $|D|$. The following result is well-known (e.g. GIERZ et al. (1980), 1, Prop. 4.12 and Ex. 4.30).

<u>Proposition</u>. Let $(D,\leq)$ be a poset. Then $(|D|,\subseteq)$ is an algebraic poset. The mapping $x \to \downarrow\{x\}$ of D into $|D|$ is an embedding, preserves finite sups, and its range is $K(|D|)$. $(|D|,\subseteq)$ is called the <u>ideal completion</u> of $(D,\leq)$. ∎

Now let $(D,\leq)$ be an algebraic poset. The <u>Egli-Milner ordering</u> is defined for $X,Y \subseteq D$ by

$$X \leq_{EM} Y \leftrightarrow (\forall x \in X)(\exists\, y \in Y)\; x \leq y \;\&\; (\forall y \in Y)(\exists\, x \in X)\; x \leq y$$

(strictly speaking, $\leq_{EM}$ is only a preordering; it becomes an ordering on the set of equivalence classes of the relation $X \equiv_{EM} Y \leftrightarrow X \leq_{EM} Y \;\&\; Y \leq_{EM} X$). We let $P(K(D)) = \{X \subseteq K(D) \mid X \text{ finite}, X \neq \emptyset\}$. The <u>Plotkin powerdomain</u> $P^{PL}(D)$ of D is the structure $(|P(K(D))|, \subseteq, \cup)$ where $I \cup J = \{X \cup Y \mid X \in I, Y \in J\}$ for $I,J \in |P(K(D))|$. The <u>singleton</u> embedding $\langle.\rangle_D$ of D into $P^{PL}(D)$ is given by $\langle x\rangle_D = \{Y \in P(K(D)) \mid Y \leq_{EM} \{x\}\}$. For every continuous $f: D \to E$ where $(E, \leq_E, \cup_E)$ is in <u>NALG</u> there is a unique continuous additive $f^+: P^{PL}(D) \to E$ such that the following diagram commutes:

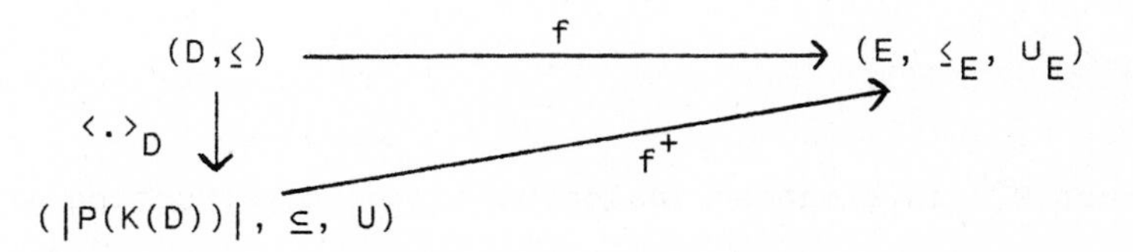

(Define $f^+(I) = \sup_E \{f(x_o) \cup_E \ldots \cup_E f(x_n) \mid \{x_o,\ldots,x_n\} \in I\}$.)
For $f: D \to E$ where $(D,\leq_D)$, $(E,\leq_E)$ are in <u>ALG</u> we let $P^{PL}f = (\langle.\rangle_E \circ f)^+$. With these definitions it is easy to verify that P^{PL} is a left adjoint to the forgetful functor from <u>NALG</u> to <u>ALG</u>. This is the category-theoretic

characterization of the Plotkin powerdomain given in HENNESSY & PLOTKIN (1979).

2. A Powerdomain Construction for Algebraic Lattices

The main result of this paper is a construction of a functor analogous to the Plotkin powerdomain functor P^{PL}, but defined in the category of algebraic lattices.

An *algebraic lattice* is an algebraic poset which is a complete lattice (i.e., every subset has a supremum). The full subcategory of ALG whose objects are algebraic lattices will be denoted AL. A *nondeterministic algebraic lattice* is a nondeterministic algebraic poset $(D, \leq_D, \cup_D)$ where $(D, \leq_D)$ is an algebraic lattice and $\cup_D$ distributes over arbitrary sups; i.e., $(\sup X) \cup_D a = \sup \{x \cup_D a \mid x \in X\}$ holds for all $X \subseteq D$, $X \neq \emptyset$, $a \in D$. An example is provided by $(D, \leq, \vee)$ where $(D, \leq)$ is an algebraic lattice and $\vee$ is the (binary) supremum. NAL is the category whose objects are nondeterministic algebraic lattices and whose morphisms are continuous, additive and sup-preserving functions.

Main Theorem. The forgetful functor from NAL to AL has a left adjoint P^{AL}.

We shall construct P^{AL} in a manner analogous to the construction of P^{PL} outlined in Section 1, but replacing the ideal completion of $(P(K(D)), \leq_{EM})$ with a modified completion by lower sets.

Let $(D, \leq)$ be an algebraic lattice with the least element 0. A lower set $A \subseteq P(K(D))$ is *union-closed* if $X, Y \in A$ implies $X \cup Y \in A$. We let $|P(K(D))|_L$ be the collection of all nonempty union-closed lower sets.

We note that $|P(K(D))| \subseteq |P(K(D))|_L$: If I is an ideal and $X,Y \in I$ then $X \lesssim_{EM} Z$, $Y \lesssim_{EM} Z$ for some $Z \in I$, hence $X \cup Y \lesssim_{EM} Z$ and $X \cup Y \in I$. The collection $|P(K(D))|_L$ is clearly closed under arbitrary intersections and $\subseteq$-directed unions. For any $F \subseteq P(K(D))$, $F \neq \emptyset$, $[F] = \cap \{A \in |P(K(D))|_L \mid F \subseteq A\} = \{Y \in P(K(D)) \mid Y \lesssim_{EM} X_o \cup \ldots \cup X_n$ for some $\{X_o, \ldots, X_n\} \subseteq F\}$ is the nonempty union-closed lower set <u>generated</u> by F. We note that $[F_1] \vee [F_2] = [F_1 \cup F_2]$ (where $\vee$ is the supremum in $(|P(K(D))|_L, \subseteq)$).

<u>Lemma 1</u>. $(|P(K(D))|_L, \subseteq)$ is an algebraic lattice; its compact elements are precisely all $[F]$ for $F \subseteq P(K(D))$, F finite and nonempty. $\langle . \rangle_D$ is a continuous embedding of $(D, \lesssim)$ into $(|P(K(D))|_L, \subseteq)$.

<u>Proof</u>. GRÄTZER (1979) or GIERZ et al. (1980). ∎

We note that $\langle . \rangle_D$ does <u>not</u> preserve finite sups: if $c = a \vee b$ for $a,b,c \in K(D)$ then $\langle a \rangle_D \vee \langle b \rangle_D = \downarrow\{\{a\}, \{b\}, \{a,b\}\} \subseteq \downarrow\{\{c\}\} = \langle c \rangle_D$, but equality does not hold if $a \neq c$, $b \neq c$.

For $A,B \in |P(K(D))|_L$ we define $A \cup B = \{X \cup Y \mid X \in A, Y \in B\}$.

<u>Lemma 2</u>. $(|P(K(D))|_L, \subseteq, \cup)$ is a nondeterministic algebraic lattice.

<u>Proof</u>. (1) $A \cup B$ is a lower set:
If $Z \lesssim_{EM} X \cup Y$ let $X_1 = \{z \in Z \mid (\exists\ x \in X)\ z \lesssim x\}$ and $Y_1 = \{z \in Z \mid (\exists\ y \in Y)\ z \lesssim y\}$. Then $Z = X_1 \cup Y_1$, $X_1 \lesssim_{EM} X$, $Y_1 \lesssim_{EM} Y$, so $Z \in A \cup B$.

(2) $A \cup B$ is union-closed: Obvious.

(3) $A \cup A = A$: We have $A \cup A \subseteq A$ because A is union-closed; the other inclusion is obvious.

(4) Commutativity and associativity of $\cup$ are trivial.

(5) $\cup$ is continuous: If $\langle A_i \mid i \in I\rangle$ is $\subseteq$-directed, the definition of $\cup$ gives immediately $(\bigcup_{i\in I} A_i) \cup B = \bigcup_{i\in I} (A_i \cup B)$.

(6) $\cup$ distributes over finite sups: On the one hand, $A \cup C \subseteq (A \vee B) \cup C$, $B \cup C \subseteq (A \vee B) \cup C$, hence $(A \cup C) \vee (B \cup C) \subseteq (A \vee B) \cup C$. On the other hand, $A \vee B = A \cup B \cup (A \cup B)$ and hence $(A \vee B) \cup C \subseteq (A \cup C) \cup (B \cup C) \cup (A \cup B \cup C) = (A \cup C) \vee (B \cup C)$.

(7) Distributivity of $\cup$ over arbitrary sups follows from (5) and (6). ∎

We denote $(|P(K(D))|_L, \subseteq, \cup)$ by $P^{AL}(D)$. If $f: D \to E$ is continuous, D,E algebraic lattices, we define $P^{AL}f : |P(K(D))|_L \to |P(K(E))|_L$ by $P^{AL}f(A) = \{Y \in P(K(E)) \mid Y \leq_{EM} f(X) \text{ for some } X \in A\}$.

<u>Lemma 3</u>. P^{AL} is a functor from <u>AL</u> into <u>NAL</u>.

<u>Proof</u>. (1) $P^{AL}f$ is continuous: Obvious.

(2) $P^{AL}f$ is additive:

$Z \in P^{AL}f(A \cup B) \leftrightarrow Z \leq_{EM} f(X \cup Y)$ for $X \in A$, $Y \in B$

$\leftrightarrow Z \leq_{EM} f(X) \cup f(Y)$ for $X \in A$, $Y \in B$

$\leftrightarrow Z = V \cup W$ for $V \in P^{AL}f(A)$, $W \in P^{AL}f(B)$

(let $V = \{z \in Z \mid z \leq f(x) \text{ for some } x \in X\}$, $W = \{z \in Z \mid z \leq f(y)$ for some $y \in Y\})$

$\leftrightarrow Z \in P^{AL}f(A) \cup P^{AL}f(B)$.

(3) $P^{AL}f$ preserves finite sups:

$Z \in P^{AL}f(A \vee B) \leftrightarrow Z \leq_{EM} f(W)$ for some $W \in A \vee B = A \cup B \cup (A \cup B)$

$\leftrightarrow Z \in P^{AL}f(A)$ or $Z \in P^{AL}f(B)$

or $Z \in P^{AL}f(A \cup B) = P^{AL}f(A) \cup P^{AL}f(B)$

$\leftrightarrow Z \in P^{AL}f(A) \vee P^{AL}f(B)$.

(4) It follows from (1), (2), (3) that $P^{AL}f$ is a morphism in the category <u>NAL</u>. Obviously $P^{AL}(1) = 1$. The only nontrivial step in

checking that $P^{AL}(g \circ f) = P^{AL}g \circ P^{AL}f$, where $f: D \to E$ and $g: E \to F$, is to show that $Z \leq_{EM} g(f(X))$ implies $Z \leq_{EM} g(Y)$ and $Y \leq_{EM} f(X)$ for some $Y \in P(K(E))$: For each pair (z,x) where $z \in Z$, $x \in X$ and $z \leq g(f(x))$ pick one $y \in K(E)$ such that $z \leq g(y)$ and $y \leq f(x)$ (using the fact that $f(x) = \sup \{y \in K(E) \mid y \leq f(x)\}$, g is continuous, and $z \in K(F)$). Then let Y be the finite set of all such y's. ∎

Proof of the Main Theorem. Let $(D, \leq_D)$ be in AL and $(E, \leq_E, \cup_E)$ in NAL. For every continuous $f: D \to E$ define $f^+: |P(K(D))|_L \to E$ by

$$f^+(A) = \sup_E \{f(x_o) \cup_E \ldots \cup_E f(x_n) \mid \{x_o, \ldots, x_n\} \in A\}.$$

(1) f^+ is continuous:

If $\langle A_i \mid i \in I\rangle$ is directed and $A = \sup_{i \in I} A_i$ then

$$f^+(A) = \sup_E \{f(x_o) \cup_E \ldots \cup_E f(x_n) \mid \{x_o, \ldots, x_n\} \in \bigcup_{i \in I} A_i\}$$

$$= \sup_{i \in I}{}_E \sup_E \{f(x_o) \cup_E \ldots \cup_E f(x_n) \mid \{x_o, \ldots, x_n\} \in A_i\} = \sup_{i \in I} f^+(A_i).$$

(2) f^+ is additive:

$$f^+(A \cup B) = \sup_E \{f(x_o) \cup_E \ldots \cup_E f(x_n) \cup_E f(y_o) \cup_E \ldots \cup_E f(y_m)$$
$$\mid \{x_o, \ldots, x_n\} \in A, \{y_o, \ldots, y_m\} \in B\}$$
$$= \sup_E \{f(x_o) \cup_E \ldots \cup_E f(x_n) \mid \{x_o, \ldots, x_n\} \in A\} \cup_E$$
$$\sup_E \{f(y_o) \cup_E \ldots \cup_E f(y_m) \mid \{y_o, \ldots, y_m\} \in B\}$$

(because $\cup_E$ distributes over $\sup_E$)

$$= f^+(A) \cup f^+(B).$$

(3) f^+ preserves finite sups:

$$f^+(A \vee B) = f^+(A \cup B \cup A \cup B)$$
$$= f^+(A) \vee_E f^+(B) \vee_E f^+(A \cup B) \text{ (definition of } f^+)$$
$$= f^+(A) \vee_E f^+(B)$$

(because $f^+(A \cup B) = f^+(A) \cup_E f^+(B) \leq f^+(A) \vee_E f^+(B)$).

(4) It follows that f^+ is a morphism in the category NAL. Clearly it is the only morphism in NAL for which the following diagram commutes:

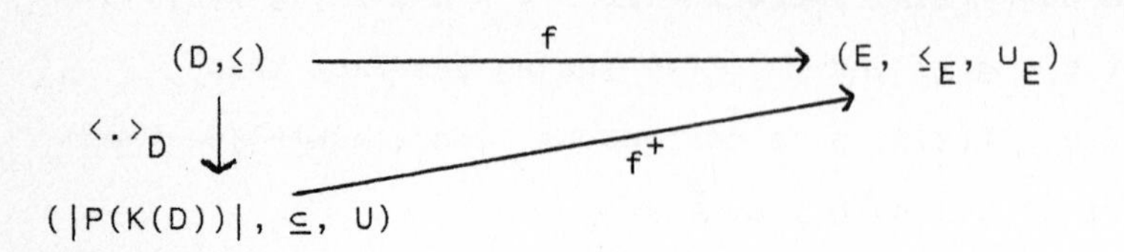

Conversely, for any morphism $F\colon |P(K(D))|_L \to E$ in NAL, $f = F \circ \langle . \rangle_D$ is the unique continuous $f\colon D \to E$ such that $F = f^+$. Hence φ defined by $\varphi(f) = f^+$ is a bijection between the set of all continuous functions from D to E and the set of all continuous, additive and sup-preserving functions from $P^{AL}(D)$ to E. It is easy to check that φ is, in fact, order-preserving and continuous as a mapping of $[D \to E]$ into $[P^{AL}(D) \to E]$ (when these are viewed as algebraic lattices with the usual pointwise ordering).

It remains to check the naturality conditions. Let D' be in AL and E' in NAL.

(5) If $h\colon D' \to D$ and $f\colon D \to E$ are continuous, then $(f \circ h)^+ = f^+ \circ P^{AL}h$:

$$
\begin{aligned}
(f^+ \circ P^{AL}h)(A') &= f^+(P^{AL}h(A')) \\
&= f^+(\{X \in P(K(D)) \mid X \leq_{EM} h(X') \text{ for some } X' \in A'\}) \\
&= \sup_E \{f(x_0) \cup_E \ldots \cup_E f(x_n) \mid \{x_0,\ldots,x_n\} \in P(K(D))\ \& \\
&\qquad \{x_0,\ldots,x_n\} \leq_{EM} h(X') \text{ for some } X' \in A'\} \\
&= \sup_E \{f(h(x_0')) \cup_E \ldots \cup_E f(h(x_m')) \mid \{x_0',\ldots,x_m'\} \in A'\} \\
&\qquad \text{(here we use continuity of } f \text{ and } \cup_E) \\
&= (f \circ h)^+(A').
\end{aligned}
$$

(6) If $f\colon D \to E$ is continuous and $K\colon E \to E'$ is continuous, additive and sup-preserving, then $(K \circ f)^+ = K \circ f^+$:

$$
\begin{aligned}
(K \circ f^+)(A) &= K(f^+(A)) \\
&= K(\sup_E \{f(x_o) \cup_E \ldots \cup_E f(x_n) \mid \{x_o,\ldots,x_n\} \in A\}) \\
&= \sup_E \{K(f(x_o)) \cup_E \ldots \cup_E K(f(x_n)) \mid \{x_o,\ldots,x_n\} \in A\} \\
&= (K \circ f)^+(A). \blacksquare
\end{aligned}
$$

We note that $\langle . \rangle$ is the unit of the adjunction established in the Main Theorem. The multiplication of the associated monad provides the "big union" operation $\uplus$, and various properties of $\langle . \rangle$ and $\uplus$ follow from this category-theoretic characterization.

Similar constructions can be given for the category of Scott's domains and the category of continuous lattices. We outline the required modifications below.

A <u>bounded complete algebraic poset</u> is an algebraic poset where every bounded (above) set has a supremum; we denote the category of such posets (and continuous maps) by <u>BCALG</u>. We say that a lower set $A \subseteq P(K(D))$ is <u>consistent</u> if $X, Y \in A$ implies $X \leq_{EM} Z$, $Y \leq_{EM} Z$ for some $Z \in P(K(D))$. Every lower set in an algebraic lattice D is consistent (let $Z = \{1\}$ where 1 is the greatest element of D). For $(D, \leq)$ in <u>BCALG</u> we let $|P(K(D))|_L$ be the collection of all nonempty consistent union-closed lower sets. The formulation and proof of the Main Theorem for <u>BCALG</u> require only cosmetic changes.

The modifications needed in the case of continuous lattices are somewhat more involved. Let $(D, \leq)$ be a continuous lattice with a way-below relation $\ll$ (see GIERZ et al. (1980)). This induces an approximating relation $\;_{EM}$ on $P(D) = \{X \subseteq D \mid X \text{ finite}, X \neq \emptyset\}$:

$$X \prec_{EM} Y \leftrightarrow (\forall x \in X)(\exists\, y \in Y)\; x \ll y \;\&\; (\forall y \in Y)(\exists\, x \in X)\; x \ll y$$

(see HRBACEK (1987)). We consider the poset $(P(D), \leq_{EM})$ and say that a lower set $A \subseteq P(D)$ is pointed if for any $X \in P(D)$, $(\forall Y \in P(D))(Y \prec_{EM} X \rightarrow Y \in A)$ implies $X \in A$. We let $\|P(D)\|_L$ be the collection of all nonempty pointed union-closed lower sets. Analogs of the results in this section can be developed for continuous lattices using $\|P(D)\|_L$ in place of $|P(K(D))|_L$.

3. Final Comments

We first list two results important for applications of P^{AL} in denotational semantics. For definitions of the concepts used in their statements see SCOTT (1981); the proofs are evident from the definitions.

Theorem 1. P^{AL} (viewed as a functor from AL into AL) is continuous on maps and monotone and continuous on domains. ∎

It follows that one can find functorial solutions to domain equations involving P^{AL} by methods expounded in SCOTT (1981).

THEOREM 2. If D is effectively given then $P^{AL}(D)$ is effectively given. ∎

The last question we touch upon here is the relationship of $P^{AL}(D)$ to the Plotkin powerdomain of D, $P^{PL}(D)$. It was pointed out earlier that $|P(K(D))| \subseteq |P(K(D))|_L$; moreover, $\lesssim$ and $\cup$ on $|P(K(D))|$ and on $|P(K(D))|_L$ coincide. Consequently $P^{PL}(D)$ is a substructure of $P^{AL}(D)$; we denote the inclusion embedding of $P^{PL}(D)$ into $P^{AL}(D)$ by Δ_D. We next show that $P^{PL}(D)$ can be defined algebraically in $P^{AL}(D)$.

Let $(D, \lesssim_D, \cup_D)$ be a nondeterministic algebraic poset; we recall that an element d of the upper semilattice $(D, \lesssim_D)$ is called coprime if $d \lesssim x \vee_D y$ implies $d \lesssim_D x$ or $d \lesssim_D y$, for all $x, y \in D$. We say that $d \in D$

is <u>indecomposable</u> if $d \lesssim_D x \cup_D y$ implies $d \lesssim_D x$ and $d \lesssim_D y$. Finally, $d \in D$ is <u>composite</u> if $d = d_o \cup_D \ldots \cup_D d_n$ where each d_i is compact, coprime and indecomposable.

<u>Lemma 3</u>. Let $A \in |P(K(D))|_L$; $A = \langle d\rangle_D$ for some $d \in K(D)$ if and only if A is compact, coprime and indecomposable in $P^{AL}(D)$.

<u>Proof</u>. If A is compact then $A = [F]$ for some nonempty finite $F \subseteq P(K(D))$ and hence $A = \sup \{[\{X\}] \mid X \in F\}$. If A is coprime, this implies that $A = [\{X\}]$ for some $X \in F$. Let $X = \{x_o,\ldots,x_n\}$; then $[\{X\}] = \langle x_o\rangle_D \cup \ldots \cup \langle x_n\rangle_D$. If A is also indecomposable, this implies that $A \subseteq \langle x_i\rangle_D$ for all $i \leq n$, i.e., $x_j \lesssim x_i$ holds for all $i,j \leq n$. Consequently there is $x \in K(D)$ such that $x = x_i$ for all $i \leq n$, and $A = \langle x\rangle_D$. The converse is straightforward. ∎

<u>Corollary 4</u>. Let $A \in P^{AL}(D)$; $A \in P^{PL}(D)$ if and only if A is a supremum of a directed set of composites in $P^{AL}(D)$. ∎

It is easy to verify that the embedding Δ is a natural transformation between P^{PL} and P^{AL} (when P^{PL}, P^{AL} are viewed as functors from <u>AL</u> into <u>ALG</u> or <u>NALG</u>): For any continuous $f: D \to E$ the following diagram commutes:

$$\begin{array}{ccc} P^{PL}(D) & \xrightarrow{\Delta_D} & P^{AL}(D) \\ {\scriptstyle P^{PL}f}\downarrow & & \downarrow{\scriptstyle P^{AL}f} \\ P^{PL}(E) & \xrightarrow[\Delta_E]{} & P^{AL}(E) \end{array}$$

Analogous natural transformations can be established for more complicated constructs involving powerdomains; for example, the diagram

$$
\begin{array}{ccccc}
P^{PL}P^{PL}(D) & \xrightarrow{P^{PL}(\Delta_D)} & P^{PL}P^{AL}(D) & \xrightarrow{\Delta_{P^{AL}(D)}} & P^{AL}P^{AL}(D) \\
\downarrow {\scriptstyle P^{PL}P^{PL}f} & & \downarrow {\scriptstyle P^{PL}P^{AL}f} & & \downarrow {\scriptstyle P^{AL}P^{AL}f} \\
P^{PL}P^{PL}(E) & \xrightarrow{P^{PL}(\Delta_E)} & P^{PL}P^{AL}(E) & \xrightarrow{\Delta_{P^{AL}(E)}} & P^{AL}P^{AL}(E)
\end{array}
$$

establishes a natural transformation Δ^2 between the functors $P^{PL} \circ P^{PL}$ and $P^{AL} \circ P^{AL}$ (where $\Delta^2_D = \Delta_{P^{AL}(D)} \circ P^{PL}(\Delta_D)$). It would be desirable to have similar results for all functors defined from P^{PL} and P^{AL} by composition with other functors useful in denotational semantics, such as $+$, $\times$, $\rightarrow$ and the least fixed point construct; however, the fact that $\rightarrow$ is not a (covariant) functor in ALG seems to complicate matters significantly, and the relationship between P^{PL} and P^{AL} in more complicated contexts remains to be fully understood.

References

GIERZ, G., HOFMANN, K.H., KIEMEL, K., LAWSON, J.D., MISLOVE, M., SCOTT, D.S., (1980), A Compendium of Continuous Lattices, Springer-Verlag, 371 pp.

GRÄTZER, G. (1979), Universal Algebra, 2nd Ed., Springer-Verlag, 581 pp.

HENNESSY, M. and PLOTKIN, G.D. (1979), Full Abstraction for a Simple Parallel Programming Language, in: MFCS 79, ed. by J. Bečvář, LNCS vol. 74, Springer-Verlag 1979, pp. 108-120.

HRBACEK, K. (1985), Powerdomains as Algebraic Lattices, in: ICALP 85, ed. by W. Brauer, LNCS vol. 194, Springer-Verlag 1985, pp. 281-289.

HRBACEK, K. (1987), Convex Powerdomains, to appear in Information and Computation.

PLOTKIN, G.D. (1976), A Powerdomain Construction, SIAM J. Comp. 5, pp. 452-486.

PLOTKIN, G.D. (1981), Unpublished 1981/82 Lecture Notes.

SCOTT, D.S. (1981), Lectures on a Mathematical Theory of Computation, Oxford Univ. Computing Lab. Prog. Research Group - 19, 148 pp.

SMYTH, M.B. (1978), Powerdomains, J. Comp. System Sci. 16, 23-36.

Closure Properties of a Probabilistic Domain Construction*

Steven K. Graham
Computer Science Program
University of Missouri - Kansas City
Kansas City, Missouri 64110

Abstract

Various closure properties of a domain theoretic construction for a probabilistic domain are considered. N. Saheb-Sjahromi's probabilistic domain construction is extended to non-algebraic domains, particularly RSFP objects. An abstract notion of a probabilistic domain is presented and the construction is shown to be free. Issues of computability and approaches to solving domain equations in probabilistic domains are addressed briefly.

0. Introduction

Investigations of probabilistic semantics are becoming more important, as current methods for formal semantics are found inadequate for dealing with probabilistic considerations. This paper focuses on an underlying foundation for probabilistic semantics. N. Saheb–Djahromi [6] introduced a complete partial order (CPO) of probability distributions on a domain, showing that from an algebraic CPO a probabilistic domain can be generated that is a CPO with a countable basis of 'finite' probability distributions (though the resulting domain is no longer algebraic). This paper examines his construction in greater depth, showing that if the original domain is finite or a sequence of finite inductive partial orders (SFP object, see Plotkin [10]) then the resulting domain is a

* This work was supported in part by NSF Grant DCR-8415919

retract of a sequence of finite inductive partial orders (RSFP object). Further, the construction is generalized to non–algebraic domains and it is shown that RSFP objects are closed under this construction.

Probabilistic algorithms form a special class of nondeterministic computations where certain of the nondeterministic branches are taken according to a known probability distribution. They arise naturally in many instances. In communications, protocol specifications often include time-outs, after which a message will be retransmitted if an acknowledgement has not been received. Such provisions are necessary to handle an imperfect transmission medium. Probability distributions can be approximated for the loss of a message or for the likelihood that an acknowledgement will not be received within the time-out period. Then the action of the sending program that retransmits if an acknowledgement has not been received can be modeled by a program that makes a stochastic branch to retransmit with the same probability that the original message was lost or the acknowledgement delayed. Communications systems are not the only systems where probabilistic concerns are manifest. Any system that uses a random number generator, real-time systems where events occur according to a probability distribution, and nondeterministic languages, where the underlying system selects nondeterministic branches based on some stochastic criteria, are also amenable to this approach. For details of some applications see the papers by Rabin [5] or Yemini [7].

Applications such as those described above can be modeled using a language that incorporates a random selection operation, denoted by:

$$\rho \rightarrow S_1, S_2$$

indicating that with probability p, S_1 is executed and otherwise S_2 is executed. This notation is also reminiscent of that occasionally used to denote a conditional operation. This is intentional, for if x has a real value drawn from a uniform distribution over the unit interval, then "if $x \leq p$ then S_1 else S_2" has the same intuitive semantics as the random selection operation described above. The notation "$p \rightarrow S_1, q \rightarrow S_2$", is sometimes used [13] for the random selection operation, but since q is determined by p ($q = 1 - p$), the "$q \rightarrow$ " is deemed unnecessary and is omitted.

A formal semantics for a language incorporating random selection must treat programs as transformations on probability distributions over the set of possible states. For example, if the value of a variable x were normally distributed around 0, and a program increments x by one, the resulting value of x would have a normal distribution about 1. Using this approach, issues such as the interaction between performance and functional criteria can be addressed allowing consideration of program correctness that is specified in terms of both attributes.

1. Background

A complete partial order (CPO) is a partially ordered set (with the order usually denoted $\leq$, subscripted where necessary for clarity) with a least element, $\perp$, and with least upper bounds for all countable increasing sequences. The upper bound of a sequence $x_0 \leq x_1 \leq x_2 \leq \ldots$, is denoted lub$\langle x_i \rangle$. CPO's are also called inductive partial orders (ipo), as in "sequences of finite inductive partial orders", see Plotkin [10]. A function between two CPO's is strict if it preserves $\perp$, and it is continuous if it preserves the least upper bounds of all ω-chains (countable increasing sequences). The collection of continuous functions on a CPO, D, is denoted $[D \to D]$. A CPO D is flat (or discrete) if $\forall x, y \in D$, $x \leq y \Rightarrow x = y$ or $x = \perp$. Such CPO's typically arise as a set with a bottom element appended. An additional relation, the way-below relation, <<, is defined on domains as,

$x << y$ iff $\forall$ H a directed subset of D, if $y \leq$ lub(H) implies $x \leq h$ for some $h \in H$. An element e is compact (also referred to as isolated or finite), if $e << e$. A subset, E of a CPO D, is called a basis if $\forall x \in D$, $\{e \in E \mid e \leq x\}$ is directed and $x =$ lub $(\{e \in E \mid e \leq x\})$. A CPO is countably based if there exists a countable subset forming a basis. A CPO is continuous when every element is the supremum of those elements which are way-below it. An algebraic CPO is one having a basis of compact elements.

The Scott topology $\sigma(D)$ is defined as follows. A subset U of D is open if and only if, whenever $x \in U$ and $x \leq y$, then $y \in U$, and whenever lub$\langle x_n \rangle \in U$ and $\langle x_n \rangle$ is increasing, then $x_m \in U$ for some m.

The Egli-Milner ordering used by Plotkin [10] to construct his powerdomain is based on the notion that a subset of states, A, is less than another subset, B, if and only if each element of A approximates some element of B and each element of B is approximated by an element of A. This is the intuitive notion and suffices for algebraic domains. For nonalgebraic domains a slightly more complex version is required, as follows:

$A \leq_m B$ iff $\forall a \in A, \forall e \leq a, \exists b \in B$, s.t. $e \leq b$, and
$\forall b \in B, \exists a \in A$, s.t. $a \leq b$.

Plotkin [10] introduced the notion of a sequence of finite inductive partial orders (SFP object). Together with continuous maps as morphisms, they form the largest cartesian closed category of algebraic domains. SFP objects are exactly the projective limits of finite posets (with $\perp$). All SFP objects are algebraic, so the category of SFP objects does not include nonalgebraic domains such as Scott's domain of real intervals. Retracts of sequences of finite inductive partial orders (RSFP objects) provide the nonalgebraic analog of SFP objects. RSFP objects are also called finitely continuous posets.

This work specializes and extends that of N. Saheb-Djahromi [6], which considers the class of probability distributions on domains and shows that when the original domain is an algebraic CPO, the resulting probabilistic domain is a countably-based CPO. Saheb–Djahromi's construction is extended to original domains that are nonalgebraic and applied to SFP, RSFP and bounded-complete domains. The notion of an abstract definition of a probabilistic domain, as well of the issues of computability, and solving domain equations will be addressed.

Let B denote the family of Borel sets of D, i.e., the σ-algebra generated by the Scott-open sets in D. A probability distribution (or measure) is a σ–additive function $P : B \to [0,1]$, with $P(D) = 1$. Given an algebraic CPO D, with compact basis Q, Pr(D) is defined to be the set of probability distributions on D. If $P \in Pr(D)$ and $x \in D$, the notation $P(x)$ rather than $P(\{x\})$ is used. The notation $|P|$ is used for $\{x \mid P(x) > 0\}$. If $|P|$ is a finite subset of Q, P is referred to as a finite distribution. F_D denotes the set of finite probability distributions; N. Saheb-Djahromi showed that F_D is a basis for Pr(D) and that $R_D = \{P \in F_D \mid \forall x, P(x) \text{ rational}\}$ is a countable basis for Pr(D), where the ordering on Pr(D) is:

$$P \leq P' \text{ iff } \forall U \in \sigma(D),\ P(U) \leq P'(U),$$

where $\sigma(D)$ is, as noted earlier, the collection of Scott-open sets on D.

A program is intuitively "better" than another program if, other things being equal, it produces a result that is "better" in terms of whatever ordering exists on the domain of answers. Considering probabilistic programs this must be modified to: *is more likely* to produce a better answer. This ordering intuitively represents a shifting of probability onto higher elements (and so into more open sets), and is shown to be consistent with the Egli-Milner ordering on non-deterministic domains. The bottom element of Pr(D) is the probability distribution with the entire mass concentrated on the bottom element of D, i.e., $\{(\perp_D,1)\}$.

Two small programs are used to motivate the selection of domain and ordering.

Program (1),

$x_1 := 0;$

$x_2 := 1;$

while x_2 do $p \to x_2 := 0, x_1 := x_1 + 1$,

operates on the domain of non-negative integers with ∞ as a top element and the usual ordering. The resulting semantics are, if $p=0$, $\{<(\infty, 1), (1, 1)>\}$ and if $p \neq 0$, $\{<(i, p * (1-p)^i), (0, 1)> \mid i \in N\}$. This example illustrates the notion of probability mass moving upward in the domain.

Program (2),

while 1 do ($\frac{1}{2}$ -> write 0, write 1)

operates on the domain of strings of binary digits with the prefix ordering. Program (2) has the following increasing sequence of probability distributions which approximate its meaning,

$\{(\perp, 1)\}$,

$\{(0, 1/2), (1, 1/2)\}$,

$\{(00, 1/4), (01, 1/4), (10, 1/4), (11, 1/4)\}, \ldots$.

Program (2) illustrates the insufficiency of discrete probability distributions, since the limit of the sequence is a probability distribution that assigns to each singleton (and all finite sets) the probability 0, yet infinite sets of interest may have non-zero probability, for example,

$P(\{w \mid \text{the initial bit of } w \text{ is } 0\}) = 0.5.$

Normally, the meaning of a program or command would be regarded as a continuous function on the domain of "states". To incorporate random selection, we must deal with functions on the domain of probability distributions on states. Letting D denote the domain of states and using $[D \to D']$ to denote the set of continuous functions from D to D', if $f \in [D \to D]$, and if π is a probabilistic computation yielding a distribution p on D, in order to compose f and π, it is necessary to define the meaning of f(p). Following N. Saheb–Djahromi [13], we define the probabilistic extension of a function.

Let $f \in [D \to D]$. The <u>probabilistic extension</u> $f' : Pr(D) \to Pr(D)$ of f, is defined as, $f'(P) = \lambda B \in B_D.\ p(f^{-1}(B))$. Since f is continuous, $f^{-1}(B) \in B_D, \forall B \in B_D$, and so f' is measurable.

When f' is so defined, f'(P) is a probability measure on B_D. The following properties hold for such extensions Theorems 7 - 12 of N. Saheb-Djahromi [13] and did not depend on the algebraic character of the domains considered:

(i) if $f \in [D \to D]$, and p is a finite distribution, then $|f'(p)| = f(|p|)$;

(ii) if $f \in [D \to D]$, $x \in D$, then $f'(\{(x,1)\}) = \{(f(x),1)\}$;

(iii) if $f \in [D \to D]$, then $f' \in [Pr(D) \to Pr(D)]$;

(iv) the mapping $\lambda f. f' : [D \to D] \to [Pr(D) \to Pr(D)]$ is continuous;

(v) if $f_1, f_2 \in [D \to D]$, then $(f_1 \circ f_2)' = f'_1 \circ f'_2$;

(vi) if $f \in [D \to D]$, $\rho \in [0,1]$, and $p, q \in Pr(D)$, then

$f'(\oplus(\rho,p,q)) = \oplus(\rho,f'(p),f'(q))$,

where $\oplus(\alpha,p,q) = \lambda B.\ \alpha * p(B) + (1 - \alpha) * q(B)$.

These results allow $[D \to D]$ to be embedded in $[Pr(D) \to Pr(D)]$ and the function $f \in [D \to D]$ to be identified with f'. This will be done whenever confusion will not result. These results will be used to specify the semantics of the following simple language.

First we will consider a deterministic language with the following abstract syntax, later adding the random selection command. This approach will highlight the extensions required for a deterministic language to accommodate random selection

Syntactic Domains and Productions:

Ide (identifiers) ::= $x_1 \mid x_2 \mid ...$

Exp (expression) ::= $\upsilon \mid \beta \mid (\tau_1+\tau_2) \mid (\tau_1-\tau_2) \mid (\tau_1/\tau_2) \mid (\tau_1*\tau_2)$

Cmd (commands)::= $(\upsilon := \tau) \mid (\pi_1; \pi_2) \mid$ (if τ then π_1 else π_2) $\mid$ (while τ do π) $\mid$ read υ $\mid$ write τ

Where β is a basic value (described below), υ ranges over Ide, τ, τ_1 and τ_2 range over Exp, π, π_1, and π_2 range over Cmd, and $\rho \in [0,1]$. We will take for our semantic domains, the following:

Semantic domains:

$U = [0,1]$	unit interval
$N = \{...,-2,-1,0,1,2,...\}$	integers
Value = U + N	basic values
State = Memory $\times$ Input $\times$ Output	states
Memory = Ide $\to$ Value	memory
Input = Value*	
Output = Value*	

We will take σ to range over states, with σ_1 designating the memory portion of the state, and β to range over basic values.

The semantic functions (of interest) for this language are:

E:Exp $\to$ State $\to$ Value , and

C:Cmd $\to$ State $\to$ State ,

Note that error handling (such as for unbound identifiers) is ignored in the interest of simplicity. Since the target of the semantic function for expressions doesn't contain State, we omit expressions with side effects. Both exclusions were made to highlight the probabilistic construct and avoid confusion due to routine but complex details.

Semantic clauses for the language are:

$E[\beta] = \beta$,

$E[\upsilon]\sigma = \sigma_1(\upsilon)$,

$E[\tau_1 + \tau_2]\sigma = E[\tau_1]\sigma + E[\tau_2]\sigma$, (other operators handled similarly)

$C[\ \pi_1;\ \pi_2\ ]\sigma = C[\pi_2](C[\pi_1]\sigma)$

$C[\text{ if } \tau \text{ then } \pi_1 \text{ else } \pi_2\ ]\sigma = \text{cond}(C[\pi_1]\sigma,\ C[\pi_2]\sigma)\ E[\tau]\sigma$,

where cond : $D \times D \to D \to D$, and

$\text{cond}(d_1, d_2)d = d_1$ if $d \neq 0$, else d_2 if $d = 0$,

$C[\ \upsilon := \tau\ ]\sigma = \text{update } \upsilon\ E[\tau]\sigma\ \sigma$,

where update : Ide $\to$ B $\to$ State $\to$ State is:

$\text{update} = \lambda i.\ \lambda b.\ \lambda s.\ (s_1[b/i],\ s_2,\ s_3)$,

and $s[b/i] = \lambda \upsilon.\ \text{cond}(b, s(\upsilon))(\upsilon = i)$

$C[\text{ while } \tau \text{ do } \pi\ ] = Y(\ \lambda f.\lambda\sigma.(\text{cond}(f(C[\pi]\sigma),\ \sigma)\ E[\tau]\sigma)\ )$,

$C[\text{ read } \upsilon\ ] = \lambda s.\ (s_1[\text{head}(s_2)/\upsilon],\ \text{tail}(s_2),\ s_3)$

$C[\text{ write } \tau] = \lambda s.\ (s_1,\ s_2,\ \text{concat}(s_3,\ E[\tau]s))$.

This is a simple direct semantics. Next, we will add random selection to the language and examine the resulting changes in semantic domains and functions. The domains and functions for the extended language will be subscripted with a 1 to distinguish them from the original language.

The abstract syntax is altered as follows:

Cmd_1 (commands)::= $(\upsilon := \tau)$ | $(\pi_1;\ \pi_2)$ | (if τ then π_1 else π_2) |
(while τ do π) | $(\rho \to \pi_1,\ \pi_2)$ | read υ | write τ

Adding the random selection feature will necessarily alter the requisite semantic domains. Suppose a program reaches a random selection "$\rho \to \pi_1,\ \pi_2$" while in state σ. Then, the meaning of the random selection command is the distribution $\{(C[\pi_1]\sigma,\ \rho),\ (C[\pi_2]\sigma,\ 1 - \rho)\}$, which results in a transformation of a state into a distribution. Since random selection commands can be nested or concatenated, the state prior to the application of such a command might already be a probabilistic distribution. Hence the state of a program must be represented as a probability distribution over what were previously the possible states. And commands must be considered as functions on the distributions, rather than as functions on the set of states. Further, expressions can no longer be treated as functions from state to value, since we no longer have a particular state available. As a result, expressions will appear as probability distributions over the domain of values. The semantic domains remain essentially the same, however the semantic functions will operate on probabilistic domains based upon the domains of states and of basic values. The modified semantic functions follow:

E_1:Exp $\to$ Pr(State) $\to$ Pr(Value) , and

C_1:Cmd $\to$ Pr(State) $\to$ Pr(State) ,

Alternatively, we can redefine the semantic domains while leaving the semantic functions (their types, at least) unchanged. The redefined semantic domains are:

$Value_1 = Pr(U + N)$		basic values
$State_1 = Pr(Memory_0 \times Input_0 \times Output_0)$		states
$Memory_1 = Pr(Ide \rightarrow Value_0)$		memory
$Input_1 = Pr(Value_0{}^*)$		
$Output_1 = Pr(Value_0{}^*)$		

The semantic functions with these changes in domain can then be expressed as in the original language. Some additional observations regarding the interaction of the probabilistic construction with other domain constructs are necessary to express the semantics of the extended language. We will identify elements of $Memory_1$ with functions from $Ide \rightarrow Pr(B)$, since given an identifier υ and $p \in Memory_1$, we readily generate the distribution $\lambda b.\ p(\{\sigma \mid \sigma(\upsilon) = b\})$, which is an element of $Pr(B)$. Keeping in mind that the semantic clauses operate on distributions over the domain of states, we will use p rather than σ as the argument. Basic values pose no difficulties, as they are mapped to the corresponding unit distribution. Variables require a more complex semantics. Remembering that $\mathbf{E}[\upsilon]$ previously yielded a function $\lambda\sigma.\ \sigma_1(\upsilon)$, an element of $State \rightarrow State$, we can substitute the corresponding probabilistic extension. Thus we will have $\mathbf{E}_1[\upsilon] = \lambda p.\ p \circ (\lambda\sigma.\ \sigma_1(\upsilon))^{-1}$, which simplifies to: $\lambda p.\ \lambda\beta.\ p(\{\sigma \mid \sigma_1(\upsilon) = \beta\})$. Semantic clauses for other features similarly revolve around the use of the probabilistic extension. The semantic clauses for the probabilistic language can, in general be expressed as:

$$\mathbf{E}_1[x] = \mathbf{E}_0[x]', \text{ or } \mathbf{C}_1[x] = \mathbf{C}_0[x]'$$

the probabilistic extension of the deterministic semantics($\mathbf{E}_0$ and $\mathbf{C}_0$ indicating the deterministic semantics). Some specific clauses are:

$\mathbf{E}_1[\beta]p = \{(\beta,1)\}$,

$\mathbf{E}_1[\upsilon]p = \lambda\beta.\ p(\{\sigma \mid \sigma_1(\upsilon) = \beta\})$, (this definition is pointwise, it is defined on sets by additivity.)

$\mathbf{C}_1[\ \upsilon := \tau\ ]p = \lambda\sigma.\ p(\{s \mid (\text{update } \upsilon\ \mathbf{E}_0[\tau]s\ s) = \sigma\})$,

$\mathbf{C}_1[\ \pi_1; \pi_2]p = \mathbf{C}_1[\pi_2](\mathbf{C}_1[\pi_1]p)$

$\mathbf{C}_1[\ \rho \rightarrow \pi_1, \pi_2\ ]p = \oplus(\rho, \mathbf{C}_1[\pi_1]p, \mathbf{C}_1[\pi_2]p)$,

where $\oplus(\rho, p, q) = \lambda B. \rho p(B) + (1-\rho)q(B)$,

2. Closure Properties of a Probabilistic Domain Construction

Prior to establishing certain closure properties for our probabilistic domain construction, it is necessary to establish a variant characterization of RSFP objects. This characterization is based upon the following observation of Plotkin, noted in Kamimura and Tang [6, Theorem 1, Corollary 1].

Theorem: A CPO D is an RSFP object iff $[D \to D]$ has a basis of functions whose ranges are finite.

Corollary: D is an RSFP object iff
$\forall n \in \omega$, $\exists f_n \in [D \to D]$ such that:
i) $f_n(D)$ is finite,
ii) $f_n << f_{n+1}$, and
iii) $id_D = lub(f_n)$

iff

$\forall n \in \omega$, $\exists f_n \in [D \to D]$ such that:
i) $f_n(D)$ is finite,
ii) $f_n \leq f_{n+1}$, and
iii) $id_D = lub(f_n)$

This characterization is used to show that RSFP objects are closed under projective limits, which allows domain equations to be solved by taking the limits of projective sequences of RSFP objects. We use a variant of their characterization to show closure under our probabilistic domain construction. Two lemmas simplify the proof. The first lemma shows that a retraction f on D induces in the natural way a mapping from $[f(D) \to f(D)]$ to $[D \to D]$ which preserves the way-below relation.

Lemma 2.1: If $f_n <<_{[f(D)\to f(D)]} f_{n+1}$ and $lub_n(f_n) = f$, with f a retraction, then $f_n \circ f <<_{[D\to D]} f_{n+1} \circ f$.

Lemma 2.2: Given a collection of sequences $<f_{n,m}>$ with $lub_m(<f_{n,m}>)=f_n$, $f_n \leq f_{n+1}$, and $f_{n,m} << f_{n,m+1}$, then subsequences $<f'_{n,m}>$ can be extracted such that $f'_{n,m} << f'_{n+1,m}$.

Theorem 2.1: If D is a CPO, and $\forall n \in \omega$, $\exists f_n \in [D \to D]$ such that:

i) $f_n(D)$ is RSFP,

ii) $f_n \leq f_{n+1}$, and

iii) $id_D = lub(f_n)$,

then D is an RSFP object.

Proof: Construct an increasing sequence of functions $<f'_n>$ in $[D \to D]$ with $f'_n(D$ finite, for all n, and with $lub(f'_n) = id_D$. Then f'_n will be of the form $f_{n,m} \circ f_n$ for some $m \geq n$, satisfying the conditions of Kamimura and Tang's characterization.

This theorem is the principal tool for proving that RSFP objects are closed under the probabilistic domain construction. It is used in the preliminary steps showing that the probabilistic domain operator applied to finite domains or to SFP objects yields an RSFP object.

The next result to be established is that when the original domain is finite, the probabilistic construction yields an RSFP object. As a preliminary, we observe that the real interval [0,1] is an RSFP object

Lemma 2.3: The real interval [0,1] with the usual ordering is RSFP.

Proof: The interval [0,1] with the usual ordering is RSFP, as can easily be seen by considering the sequence of functions:

$<trunc_n>$, where $trunc_n(x) = ceiling(10^n * x - 1)/10^n$, if x>0

0, if x=0,

and the function ceiling (x) maps a real number into the smallest integer larger than x.♦

The desired closure result for finite domains follows readily.

Theorem 2.2: If D is finite then Pr(D) is RSFP.

Proof: We can see that Pr(D) is RSFP by considering the sequence of functions $<f_n>$ defined as follows:

$$f_n(p) = \lambda x.\ trunc_n \circ p, \text{ if } x \neq \bot$$
$$1 - (\Sigma\, f_n(p)(x),\ \forall x \neq \bot\ ), \text{ if } x = \bot.$$

Use the characterization given in Theorem 2.1.

In Theorem 2.3, a sequence of functions is defined to satisfy the characterization of RSFP objects given in Theorem 2.1. Theorem 2.2 provides the necessary result that the ranges of the functions in the sequence are RSFP, allowing the application of Theorem 2.1. The following lemma is a simple result that facilitates the proof of Theorem 2.3.

Lemma 2.4: If $f \leq g$, then $\forall U \in \sigma(D),\ g^{-1}(U) \supset f^{-1}(U)$.

Theorem 2.3: If D is SFP then Pr(D) is RSFP.

Proof: Since D is SFP, there exists a sequence $<\pi_n>$ in $[D \rightarrow D_n]$ s.t. $lub(<\pi_n>)=id_D$.

Define $<f_n> : Pr(D) \rightarrow Pr(D)$ as:

$$f_n(p) = p \circ \pi_n^{-1}.$$

The sequence $<f_n>$ satisfies the conditions of Theorem 2.1, hence Pr(D) is RSFP.

The next result is the most important of this work, as it establishes the closure property of the category of RSFP objects under our probabilistic domain construction. The basic approach is to show that the probabilistic extension of a retraction is still a retraction. If D' is an RSFP object and D is an SFP object of which D' is a retract, then Pr(D) is RSFP by Theorem 2.3. The first step below shows that the probabilistic extension $r' : Pr(D) \rightarrow Pr(D')$ of the retracton $r : D \rightarrow D'$ is a retraction so that Pr(D') is a retract of the RSFP object Pr(D), and hence is also an RSFP object.

Lemma 2.5.: If r is a retraction on D then the probabilistic extension r' of r is a retraction on Pr(D).

Theorem 2.4: Let D' be an RSFP object and let D be an SFP object and $r : D' \to D$ a retraction with $D' = r(D)$. Then $Pr(r(D))$ is an RSFP object which is isomorphic to $r'(Pr(D))$, where r' is the probabilistic extension of r. Hence $Pr(D')$ is also an RSFP object.

Proof: Define $r' : Pr(D) \to Pr(D)$ as $r'(p) = p \circ r^{-1}$.

By Lemma 2.5 r' is a retraction, and $r'(Pr(D))$ is RSFP.

We claim that $r'(Pr(D)) \cong Pr(r(D))$.

To see this, define $j : r'(Pr(D)) \to Pr(r(D))$ as $j(p) = p|_{B_{r(D)}}$, and define $i : Pr(r(D)) \to r'(Pr(D))$ as $i(p) = \lambda b.\ p(b \cap r(D))$.

fig.1

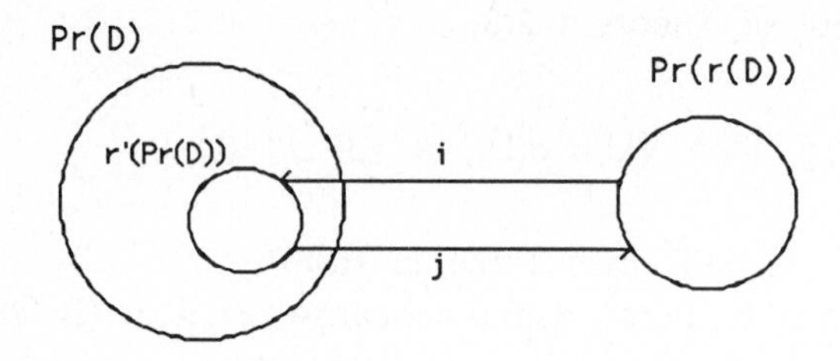

We have shown that the probabilistic extension r' of a retraction r, generates a retract $r'(Pr(D))$ that is isomorphic to $Pr(r(D))$. Since RSFP objects are closed under retractions, $r'(Pr(D))$ must be RSFP, and so must $Pr(r(D))$. Thus the collection of RSFP domains is closed under this probabilistic domain construction. To see that such closure is not generally the case, it is instructive to consider the effects of the probabilistic construction on bounded complete domains.

If the original domain, D, is bounded-complete then the probabilistic domain, $Pr(D)$, that results is not necessarily bounded-complete. A counter example is given to illustrate this fact. This is an additional substantiation of the notion that the category of bounded-complete domains is insufficient. The following domain, D, is bounded-complete:

fig. 2

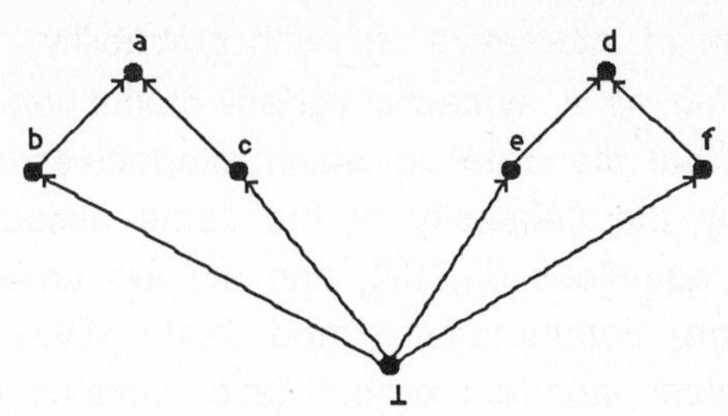

Now consider the two distributions, p and q defined on D as follows:

$$p = \{(b,1/3),\ (e,1/3),\ (\bot,1/3)\},$$

$$\text{and } q = \{(c,1/3),\ (f,1/3),\ (\bot,1/3)\}.$$

These distributions are bounded; for instance, the distribution $\{(a,1/2), (d,1/2)\}$ is an upper bound. Yet no least upper bound exists, as can be seen by considering the two following minimal upper bounds,

$$u_1 = \{(b,1/3),\ (c,1/3),\ (d,1/3)\},$$

$$\text{and } u_2 = \{(a,1/3),\ (e,1/3),\ (f,1/3)\}.$$

Next, we consider the notion of a probabilistic domain in an abstract sense, what such a collection of objects is, what axioms they satisfy, and whether the construction suggested previously is free. A good introduction for this material is Burstall and Goguen's paper [1]. The key requirement is that a probabilistic domain support some operation that introduces probabilistic nondeterminism. The simplest approach is to use the random selection operation, denoted earlier in the paper by,

$$p \to S_1, S_2.$$

The following is an axiomatization of probabilistic domains.

An <u>abstract probabilistic domain</u> is a pair $(D, \oplus)$, where D is a domain (finitely continuous cpo) . The operator, $\oplus$ is a continuous function, $\oplus : [0,1]\times D\times D \to D$, where $[0,1]$ is the unit interval of reals, $\alpha, \beta \in [0,1]$, and $\oplus$ obeys the following constraints:

i) $\oplus(1, p, q) = p$,

ii) $\oplus(\alpha, p, p) = p$,

iii) $\oplus(\alpha, p, q) = \oplus(1-\alpha, q, p)$,

iv) $\oplus(\alpha, p, \oplus(\beta, q, r)) = \oplus(\alpha+\beta - \alpha*\beta, \oplus(\alpha/(\alpha+\beta - \alpha*\beta), p, q), r)$.

These constraints are designed to capture the following intuitive ideas. First, a random selection that is determined and selects a specific alternativ is identical to the alternative. Second, random selection between identical alternatives is likewise identical to the alternatives. Third, since the random selection $\oplus(\alpha, p, q)$ designates the selection of alternative p with probability α, and the selection of alternative q with probability $1-\alpha$, we regard the right hand term of (iii) as a syntactic variant of the left hand side. And finally, in (iv), we demand that the order in which alternatives are selected be immaterial, with only the necessity of the same associated probabilities for equality. Note that equations (ii), (iii), and (iv) are analogous to the usual axioms of absorption, commutativity, and associativity.

Morphisms between abstract probabilistic domains must be continuous functions preserving $\oplus$.

<u>Lemma 2.6</u>: Pr(D) is an abstract probabilistic domain.

In the construction of Pr(D) given earlier, the operator $\oplus$ would be defined as follows:

$$\oplus(\alpha, p, q) = \lambda B.\ \alpha * p(B) + (1 - \alpha) * q(B).$$

It is readily apparent that $\oplus$ thus defined satisfies the above criteria.

The final item in this section is a proof that the construction Pr(D) is free.

<u>Theorem 2.5:</u> The construction Pr(D) is free.

<u>Proof:</u> Let D be a domain with basis E and let f be a continuous map $f : D \to A$, and let η be the embedding $\eta : D \to Pr(D)$ defined by $\eta(e) = (e,1)$. Then define $f^{\#} : Pr(D) \to A$ to be a continuous function as follows. If its argument is a unit distribution function:

$$f^{\#}(\{(e,1)\}) = f(e).$$

If the argument is a finite distribution, then it can be representedas the random selection of still simpler distributions and $f^{\#}$ is defined:

$$f^{\#}(\oplus(\alpha, p, q)) = \oplus(\alpha, f^{\#}(p), f^{\#}(q)).$$

For $p \in Pr(D)$, we have $p = lub(p_n)$, $p_n \in F_D$, and let

$$f^{\#}(p) = lub(f^{\#}(p_n)).$$

Then the following diagram commutes.

fig.3

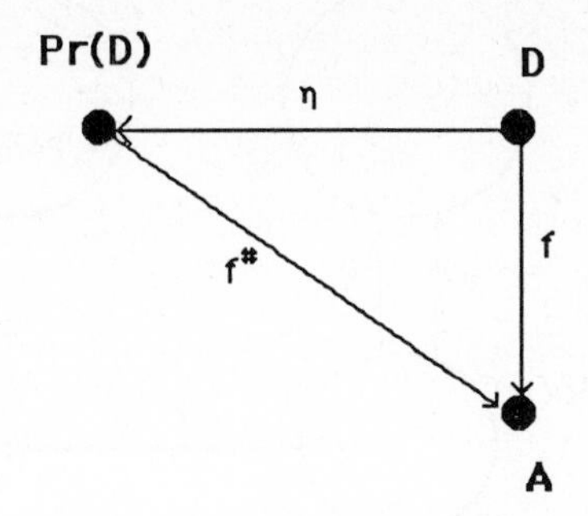

Thus we have characterized abstract probabilistic domains and found that Pr(D) is a free construction of such a domain.

3. Solving Domain Equations

In this section we address the necessity of solving recursive domain equations by showing that a universal domain approach is suitable.

In [7] Kamimura and Tang introduce a universal domain V which is the CPO of pseudo-retractions of an universal algebraic domain U. They define a transitive binary relation Δ on V, satisfying: for $r_0, r_1 \in V$,

$$r_0 \,\Delta\, r_1 \text{ iff } r_0 = r_0 \circ r_1 \circ r_0 \leq r_1.$$

Note that $r \,\Delta\, r$ if and only if r is a retraction. A function f is said to be Δ–monotonic (delta-monotonic) if it preserves the Δ relation. If f is a Δ–monotonic function then the least fixed point $r = \mathrm{lub}(f^n(\bot))$ is a retraction satisfying the equation $f(x) = x$. The usual domain constructions (e.g., +, ×, →, powerdomain) are Δ–monotonic when considered as functions on V, and this, along with closure of Δ-monotonic functions under composition, allows the solution of domain equations using the least fixed point operator.

To this end, we show that our probabilistic domain construction, considered as a function $T_{Pr} : V \to V$, is Δ–monotonic.

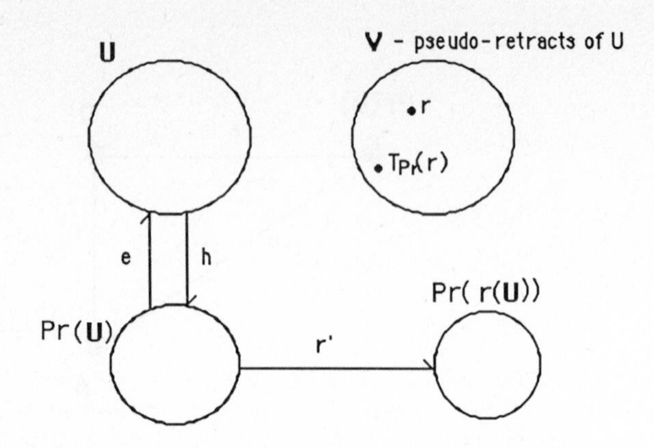

We define $T_{Pr} = \lambda r \in V.\ e \circ r' \circ h$, where r' is the probabilistic extension of r, as discussed earlier, $\lambda p.\ p \circ r^{-1}$. Remember that, if r is a retraction, then r' is as well and that $r'(Pr(D))$ is isomorphic to $Pr(r(D))$. The functions e and h are a retraction pair, hence $h \circ e = id_{Pr(U)}$. Thus $T_{Pr}(r)$ is a retraction on U and an element of V. Further, T_{Pr} is continuous andΔ–monotonic.

Lemma 3.1: T_{Pr} is continuous.

Theorem 3.1: T_{Pr} is Δ–monotonic.

Since Pr, the probabilistic domain construct, along with the standard domain constructs (+, $\times$, $\rightarrow$, P), are continuous, Δ–monotonic functions on the domain of pseudo-retractions V of the universal domain, U, we can find solutions to domain equations by taking the least fixed point of the composite function (representing the domain equation) on V.

4. Computability, Powerdomains, and Alternative Constructions

If D is an effectively given domain, the question naturally arises whether Pr(D) is also effectively given. The answer is affirmative: if D is effectively given, then so is Pr(D). This can be shown by backtracking for a moment and considering the particular recursive algebraic domain, D_0 of which D is a computable retract. Note that $Pr(D_0)$ is effectively given, and then observe that Pr(D) is a computable retract of $Pr(D_0)$, and hence is effectively given.

Plotkin [10] and others have done extensive work exploring powerdomain constructions to handle a nondeterministic union (or) operator. These powerdomains are endowed with a variety of possible orderings, the most germane being the Egli-Milner ordering. This language construct, π_1 or π_2, denotes a nondeterministic selection between alternative branches and has an intuitive relation to the random selection operator, $\rho \to \pi_1, \pi_2$. The only difference between the nondeterministic or and random selection would appear to be the parameter ρ. The function Ψ considered below maps a distribution into its set of nonzero elements. N. Saheb-Djahromi gives the following theorem in [13], that explores the relation between Plotkin's powerdomain, P_D, and the probabilistic domain Pr(D). The same result is obtained in the case of non-algebraic CPO's (this result for algebraic CPO's was Theorem 2 in [13]).

Theorem 4.1: If p, q are finite distributions, then $p \leq q$ iff

(i) For every non-empty subset A of $|p|$, there exists a subset B of $|q|$, with $A \leq_m B$ and $p(A) \leq q(B)$, iff

(ii) For every non-empty subset A of $|q|$, there exists a subset B of $|p|$, with $B \leq_m A$ and $p(B) \geq q(A)$, where $\leq_m$ is the Egli-Milner ordering.

The following corollary clearly demonstrates the relation between the ordering on finite distributions and the Egli-Milner ordering on their corresponding sets of nonzero elements.

Corollary 4.1: If p and q are finite distributions, then $p \leq q$ implies that $|p| \leq_m |q|$.

The following relation between the two domain constructs was also noted by Saheb–Djahromi as Theorem 13 of [13]. It defines a function from Pr(D) to P_D that is continuous and respects the operations of random selection and union, as well as preserving the probabilistic extensions of continuous functions on the original domain.

<u>Theorem 4.2:</u> Let D be ω-discrete (i.e. flat and countable) and let P_D be the powerdomain with the Egli-Milner ordering. Define $\Psi : Pr(D) \rightarrow P_D$ by $\Psi(p) = |p| \cup \{\perp\}$. Then:

(i) Ψ is continuous.;

(ii) if $0 < \rho < 1$, then $\Psi(\oplus(\rho, p, q)) = \Psi(p) \cup \Psi(q)$;

(iii) if $f \in [D \rightarrow D]$ and f is strict, then $\forall p \in Pr(D)$, $\Psi(f'(p)) = f(\Psi(p))$, where f' is the probabilistic extension of f.

Unfortunately, there is no canonical probability distribution that provides a suitable mapping from P_D to Pr(D). Mapping a finite element of the powerdomain to a uniform probability distribution over the elements of the set suggests itself, but is unsatisfying. Even if D is an ω-discrete domain, the mapping fails to preserve the ordering on P_D.

Dexter Kozen [8] has investigated two probabilistic semantics, one interpreting programs as partial measurable functions on a measurable space, while the other treats programs as continuous linear operators on a Banach space of measures. The focus of his work is on handling the semantics for a class of probabilistic programs whose input is allowed to vary according to some probability distribution, and that include a random assignment feature, x := **random**, that generates a number based on some distribution. No restrictions are placed upon either the input or the generated distributions. This differs from our approach, where we consider the input fixed, and generate the probabilistic character through random selection. The random selection, $p \rightarrow S_1, S_2$, can be simulated in Kozen's framework, assuming that **random** will generate a random number in [0,1], by:

x := **random**;

if $x \leq p$ then S_1 else S_2.

Kozen relates his work to the usual Scott-Strachey approach to denotational semantics by showing how a flat domain can be embedded within his approach.

Michael Main [9] introduces a semantics for probabilistic programs based on semiring modules that varies from the current treatment in a number of ways. His emphasis is on illustrating the value of the additional algebraic structure, obtained through use of semiring modules, rather than on exploring probabilistic semantics. Main's semantics addresses discrete distributions. The operations of scalar multiplication and pointwise addition replace the random selection used in our approach, and an "impossible" function that maps everything to 0 is provided as an identity for the pointwise addition. He avoids the use of a bottom element by using an auxiliary function, MASS, where a program p is total iff MASS(p) = 1. Programs have the characteristic that they will not increase mass. Suppose f, g are the meanings of two programs (or program fragments). Then there is the following restriction, $MASS(f) \geq MASS(g \circ f)$. A decrease of mass may occur within loop constructs, since there is the possibility of a nonterminating branch with zero mass.

5. Conclusions

This extension of Saheb-Djahromi's probabilistic domain construction has been satisfying, yielding the desired closure properties for SFP and RSFP objects. The lack of closure of bounded–complete domains was as expected. Closure of effectively given domains provides the basis for the a consistent notion of computability in probabilistic domains. Additionally, the delta-monotonicity of the probabilistic domain construction, considered as a function on a universal domain should allow the solution of recursive domain equations. Finally, this construction coincides with our intuitions regarding the nature and properties of a probabilistic domain considered abstractly.

The formal semantics made possible through the use of such domains will allow rigorous treatment of an area of programming that has been inadequately served. Particular applications that merit investigation would be the generation of formal semantics for languages/systems used in artificial intelligence to express the notion of uncertainty.

References

(1) Burstall, R.M., and Goguen, J.A., *Algebras, Theories and Freeness: An Introduction for Computer Scientists*, in Theoretical Foundations of Programming Methodology, ed. by Broy and Schmidt, D. Reidel, 1982, 329-348.

(2) Gordon, M. J. C., The Denotational Description of Programming Languages, Springer-Verlag, New York, 1979.

(3) Halmos, P.R., Measure Theory, Van Nostrand, New York, 1965.

(4) Kamimura, T. and Tang, A. *Finitely Continuous Posets*, University of Kansas, Dept. of Computer Science, TR-84-1, 1984.

(5) Kamimura, T. and Tang, A. *Retracts of SFP objects*, in Mathematical Foundations of Programming Semantics, Lecture Notes in Computer Science 239, Springer-Verlag, 1985.

(6) Kamimura, T. and Tang, A. *Retracts of SFP objects- projective limits, solving domain equations*, Workshop on Mathematical Foundations of Programming Semantics, Manhattan, Kansas, 1986.

(7) Kamimura, T. and Tang, A., *Domains as Finitely Continuous CPOs*, (to be presented) 3rd Workshop on Mathematical Foundations of Programming Semantics, New Orleans, 1987.

(8) Kozen, D. *Semantics of probabilistic programs*, Journal of Computer and System Sciences 22 (1981), 328–350.

(9) Main, M. G., *Free Constructions of Powerdomains*, in Mathematical Foundations of Programming Semantics, Lecture Notes in Computer Science 239, Springer-Verlag, 1985.

(10) Plotkin, Gordon, *A Powerdomain Construction*, SIAM Journal of . Computing, Vol. 5, No. 3, September 1976, 452-487.

(11) Rabin, M. *Probabilistic Algorithms*, in Algorithms and Complexity, Academic Press, 1976, 21-40.

(12) Royden, H.L., Real Analysis, MacMillan, New York, 1964.

(13) Saheb-Djahromi, N. *CPO's of measures for nondeterminism*, Theoretical Computer Science, 12 (1980), 19-37.

(14) Schmidt, D. A., Denotational Semantics, Allyn and Bacon, Inc., 1986.

(15) Smyth, M.B., *Effectively Given Domains*, Theoretical Computer Science, 5 (1977), 257-274.

(16) Tennent, R. D., *The Denotational Semantics of Programming Languages*, Communications of the ACM, Vol. 19, No. 8, Aug. 1976, 437-453.

(17) Yemini, Y. and Kurose, J.F. *Towards the unification of the functional and performance analysis of protocols or, is the alternating -bit protocol really correct?*, in Protocol, Specification, Testing, and Verification, North-Holland, 1982, 189–196.

Part III

Domain Theory

QUASI-UNIFORMITIES: RECONCILING DOMAINS WITH METRIC SPACES

M. B. Smyth

Dept. of Computing
Imperial College
London SW7

ABSTRACT

We show that quasi-metric or quasi-uniform spaces provide, inter alia, a common generalization of cpo's and metric spaces as used in denotational semantics. To accommodate the examples suggested by computer science, a reworking of basic notions involving limits and completeness is found to be necessary. Specific results include general fixed point theorem and a sequential completion construction.

I. INTRODUCTION

The metric space approach to semantics is at odds with the approach via Scott domains, simply because metric spaces are necessarily Hausdorff, whereas the natural topology of a domain, in which there are elements representing partial information, is non-Hausdorff.

If we seek a common framework for the two approaches, an obvious suggestion is to try <u>quasi</u>-metrics - that is, "metrics" that do not necessarily satisfy the symmetry axiom (see Sec. II for definitions). Any cpo $(D,\sqsubseteq)$ can be trivially represented as a quasi-metric space by putting

$$d(x,y) = \begin{cases} 0 & \text{if } x \sqsubseteq y \\ 1 & \text{otherwise} \end{cases} .$$

But what about the topology? The standard topology associated with a quasi-metric space (X,d) takes as base for the neighbourhood system at each point x the set of ε-balls $\{y \mid d(x,y) < \varepsilon\}$ $(\varepsilon > 0)$; clearly, then, our quasi-metric for a cpo will give us the Alexandroff topology (the open sets are the $\vdash$-closed subsets), whereas a computationally satisfactory topology for a cpo should at least be order-consistent ([6], [9], or Sec. VI below).

We shall meet this problem by proposing an alternative definition of the topologies associated with a quasi-metric or quasi-uniformity, which will reduce to the order-consistent topologies (and, in suitable cases, to the Scott topology) in the case of cpo's, while agreeing with the usual topology in metric spaces.

Apart from countenancing alternative topologies, we shall find reason to revise drastically the received definitions concerning convergence and completeness. The received theory suffers from a severe lack of (non-Hausdorff) examples to use for guidance. Our thesis is that domains are, or should be, a prime area for the application of quasi-uniform ideas, and can help us to get the definitions right. In fact, the received ideas give useless results when applied to domains, hence the revisions developed below.

Non-determinism and concurrency provide potential areas of application of the theory developed here. There is evidence that a convincing unification of "power domains" as studied in computer science with "hyperspaces" (topology) will need to make use of quasi-uniformities (see [14] for a partial attempt at such a unification - without the quasi-uniformities). Indeed, it has long been recognized (e.g. [1,8]) that quasi-uniformities provide a convenient setting for power space constructions. The main obstacle to integrating this material with computer science has been the lack of a decent account of completeness of quasi-uniformities; hence, again, the work reported in this paper.

Interesting computational examples of quasi-uniformities that are neither metric spaces nor cpo's arise in studies of "observational preorder" for CCS and related systems [5,11]. Hennessy-Milner logics are, from our point of view, ways of defining appropriate topologies on such quasi-uniformities. These ideas are illustrated below, in simple form (Examples 4,9).

An important aim of our work is to achieve a tie-up with sober space theory. Her we can do little more than mention some difficulties in the way of this (Sec. 5); a much fuller account is provided in [16] . Ultimately, we seek a finitary "information system" version (cf.[13,15]).

The main line of development below leads to the more modest goal of constructing the completion by Cauchy sequences of a quasi-uniformity.

Standing entirely apart from the other mathematical works cited is a very general approach to quasi-metrics proposed by Lawvere [7]. An interesting question (for future consideration) is whether Lawvere's work has any bearing on what we are doing here.

II. PRELIMINARIES

Definition 1. A quasi-uniformity on a set X is a filter $\mathcal{U}$ of relations (i.e. subsets of $X \times X$), called "entourages", such that

(a) $\forall x \in X \ \forall U \in \mathcal{U} . \ xUx$

(b) $\forall U \in \mathcal{U} \ \exists V \in \mathcal{U} . \forall x,y,z \in X . \ xVyVz \Rightarrow xUz$ (more concisely: $\forall U \ \exists V. \ V \circ V \subseteq U$).

$\mathcal{U}$ is said to be a uniformity if in addition it satisfies

(c) $\forall U \in \mathcal{U} . \ U^{-1} \in \mathcal{U}$.

A collection $\mathcal{B}$ of relations is a base of the quasi-uniformity $\mathcal{U}$ if $\mathcal{U} = \uparrow \mathcal{B}$ (that is, if $\mathcal{U} = \{S \mid \exists R \in \mathcal{B} . R \subseteq S\}$); while $\mathcal{B}$ is a subbase of $\mathcal{U}$ if the closure of $\mathcal{B}$ under finite meets is a base of $\mathcal{U}$ (equivalently, if $\mathcal{U}$ is the smallest filter containing $\mathcal{B}$).

As a useful alternative to metric intuition, (b) can be seen as a (weakened)form of transitivity: a quasi-uniformity having a singleton as base is essentially just a pre-order. Moreover we can usefully define, for any quasi-uniformity $\mathcal{U}$, the pre-order $\leq_{\mathcal{U}}$ by:

$$x \leq_{\mathcal{U}} y \quad \text{iff} \quad \forall U \in \mathcal{U} . \ xUy .$$

A quasi-metric on X is a map $d: X \times X \to [0,\infty)$ satisfying

$$d(x,x) = 0$$

$$d(x,z) \leq d(x,y) + d(y,z)$$

and - optionally -

$$d(x,y) = d(y,x) = 0 \Rightarrow x = y .$$

Each quasi-metric d determines a quasi-uniformity by taking the relations $R_\varepsilon = \{\langle x,y\rangle \mid d(x,y) < \varepsilon\}$ as a base.

For any quasi-uniformity $\mathcal{U}$ we have its conjugate $\mathcal{U}^{-1} = \{U \mid U^{-1} \in \mathcal{U}\}$; of course this corresponds to going from a quasi-metric d to its conjugate d^{-1} where $d^{-1}(x,y) = d(y,x)$. For any $\mathcal{U}$, $\mathcal{U} \cap \mathcal{U}^{-1}$ is a base of a uniformity, $\mathcal{U}^*$; and $\mathcal{U}$ is already a uniformity iff $\mathcal{U} = \mathcal{U}^{-1}$.

Examples. As already indicated, in effect, any pre-order $(D,\leq)$ may be considered as a quasi-uniformity with base $\{\leq\}$; we are particularly interested in the case in which D is a cpo. Again, metric spaces are automatically quasi-uniformities. We continue with some examples that are neither metric spaces nor (simply) pre-orders:

Example 1. The relations Q_ε , for $\varepsilon > 0$, where $Q_\varepsilon(x,y)$ iff $x < y + \varepsilon$, form a base for a quasi-uniformity Q on $\mathbb{R}$.

Example 2. For any set S, the relations P^S_a, for $a \in S$, where $A \ P^S_a \ B$

$A\ P_a^S\ B$ iff $(a \notin A \vee a \in B)$ form a subbase for a quasi-uniformity P^S on $\mathcal{P}(S)$.

Example 3. Let D be an algebraic cpo with basis B_D. The quasi-uniformity $\mathcal{U}_{B_D}$ on D will have as subbase $\{U_b | b \in B_D\}$ where

$$xU_by \quad \text{iff} \quad (b \sqsubseteq x \Rightarrow b \sqsubseteq y) .$$

As it happens, the uniformity $\mathcal{U}^*_{B_D}$ is fairly familiar from the literature; see, for example, Plotkin [9], where it is pointed out, in effect, that $\mathcal{U}^*_{B_D}$ induces the Lawson topology on D.

Example 4 : Processes. The relations $\approx_n$ used in defining observational equivalence (Milner [11]) form a base of a uniformity. As such, the definition does not allow for divergent (or partial) processes. It is well-known [5,10] that these can be accommodated by using certain pre-orders $\lessapprox_n$ in place of $\approx_n$; these will in fact give a quasi-uniformity. This may be illustrated by the following drastically simplified model of processes: a process is considered to be just a sequence of actions ($\in$ Act), which may be infinite, and may also diverge (represented by Ω):

Processes : $\quad a_1 \ldots a_n\text{NIL} \qquad (n \geq 0;\ a_i \in \text{Act})$

$\qquad\qquad a_1 \ldots a_n\ \Omega \qquad$ " $\quad$ "

$\qquad\qquad a_1 \ldots a_n \ldots \qquad (\in \text{Act}^\omega)$

Then the $\lessapprox_n$ relations, defined in the standard way, reduce in this case to the following:

$$\begin{aligned} p \lessapprox_n q \quad \text{iff} \quad & (p = a_1 \ldots a_n p',\ q = a_1 \ldots a_n q') \\ & \vee \exists k < n\ [p = a_1 \ldots a_k p',\ q = q_1 \ldots a_k q' \\ & \qquad \& \ (p' = \Omega \ \vee \ p' = q' = \text{NIL})] . \end{aligned}$$

Less formally: $p \lessapprox_n q$ iff p,q agree up to their n^{th} actions, or, for some $k < n$, they agree up to their k^{th} actions, and then either p diverges ($p' = \Omega$) or they both stop ($p' = q' = \text{NIL}$).
Taking the $\lessapprox_n$ as the base of a quasi-uniformity $\mathcal{U}$, we observe that the observational pre-order is nothing but $\leq_{\mathcal{U}}$. It is in fact the (partial) ordering of the system when regarded in the obvious way as a Scott domain.

Example 5..(Can be regarded as a metric version of Example 4, but omitting the processes which terminate with NIL.) Now let $X = \Sigma^\infty$ be the set of finite and infinite sequences in an alphabet Σ . For $x,y \in X$, let

$$d(x,y) = \begin{cases} 0 & \text{if } x \text{ is an initial segment of } y \\ 2^{-k} & \text{where } k = \text{Min}(\{n | x_n \neq y_n\}), \text{ otherwise} \end{cases}$$

The second alternative here, of course, represents a commonly used metric on the space of infinite sequences; the use of a quasi-metric allows us to accommodate also partial sequences.

III. CAUCHY SEQUENCES AND LIMITS

We want to be able to regard increasing sequences in a cpo as Cauchy sequences. A definition which agrees with this as well as with the usual notion for metric (or uniform) spaces is:

Definition 2. A sequence $(x_0, x_1, \dots)$ in $(X, \mathcal{U})$ is Cauchy if, for every $U \in \mathcal{U}$, there exists k such that $\forall m \geq k\ \forall n \geq m.\ x_m U x_n$.

Clearly, in any pre-ordered set (with the pre-order taken as base of $\mathcal{U}$), a sequence is Cauchy iff it is eventually increasing.

Definition 2 is reasonably close to what one finds in the literature. The situation is different with regard to limits, which we now consider. We begin with the idea of generalizing lubs (of chains) in a cpo. This leads us first of all to:

Definition 3. We say that x is a bound of the (Cauchy) sequence $\xi = (x_i)$ in $(X, \mathcal{U})$ if $\forall U \in \mathcal{U}\ \exists k\ \forall n \geq k.\ x_n U x$; and x is a least bound of ξ if $x \leq_{\mathcal{U}} y$ for every bound y of ξ.

It is easy to see that least bounds reduce to ordinary lubs in cpo's, and to limits in metric spaces. Unfortunately, though, least bounds are not satisfactory (as "limits" of Cauchy sequences) in all situations, and we shall need a more refined notion, having a "topological" flavour, for the general case. Indeed, from topological experience, one would perhaps expect a canonical "limit" of a sequence to arise as a greatest limit point of it rather than as a least element of some collection. Pursuing this idea for the moment, we shall try to define a limit point of a sequence as a point such that every neighbourhood of it is a neighbourhood of the sequence. But what could be meant by a "neighbourhood" of a sequence? We shall take it that "close to ξ" means "close to nearly every term of ξ". More precisely, for $U \in \mathcal{U}$, define: $U[\xi] = \{y \in X \mid \forall_\infty i.\ x_i U y\}$, where the quantifier $\forall_\infty$ means "for all but finitely many". Taking a $T_{\mathcal{U}}$-neighbourhood of ξ to be a set $S \subseteq X$ such that $U[\xi] \subseteq S$ for some $U \in \mathcal{U}$, we have:

Definition 4. A point x is a limit point of a sequence ξ in $(X, \mathcal{U})$ if every $T_{\mathcal{U}}$-neighbourhood of x is a $T_{\mathcal{U}}$-neighbourhood of ξ (notation:

$\xi \overset{\mathcal{U}}{\to} x$, or just $\xi \to x$); x is a <u>(strong) Limit</u> of ξ, written $\xi \Rightarrow x$, if x is both a bound and a limit point of ξ .

Admittedly, the resemblance with topological limits is, so far, rather superficial: a $T_{\mathcal{U}}$ -neighbourhood of a sequence (in contrast with a stronger notion of <u>uniform</u> neighbourhood, to be introduced later) need not contain any terms of the sequence, and may even be empty.

Some elementary properties of these notions are summarized in Propositions 1 and 2. The proofs are straightforward.

<u>Proposition 1</u>. Let ξ be a sequence of points in the quasi-uniformity $(X, \mathcal{U})$. Then:

(i) a point x is a bound of ξ iff every $T_{\mathcal{U}}$ -neighbourhood of ξ is a $T_{\mathcal{U}}$ -neighbourhood of x ;

(ii) if x is a limit point and y a bound of ξ, then $x \leq_{\mathcal{U}} y$;

(iii) if $\xi \Rightarrow x$, then x is both a least bound and a greatest limit point of ξ ;

(iv) if ξ is Cauchy, then x is a bound of ξ iff $\xi \overset{\mathcal{U}^{-1}}{\to} x$: thus, $\xi \Rightarrow x$ iff $\xi \overset{\mathcal{U}}{\to} x$ and $\xi \overset{\mathcal{U}^{-1}}{\to} x$. □

<u>Example</u>. In the quasi-uniformities of Examples 1,2, least bounds and greatest limit points of sequences reduce to the upper and lower limits considered in analysis.

In the remainder of the paper we shall work only with <u>Cauchy</u> sequences. For Cauchy sequences in cpo's and metric spaces, least bounds, greatest limit points, and Limits all collapse to the usual limits (or lubs). In fact, these three notions "usually" coincide; but not always so, as we see from

<u>Example 6</u>.

An algebraic cpo:

($c_k \sqsubseteq a_l$ for all $l \geq k$)

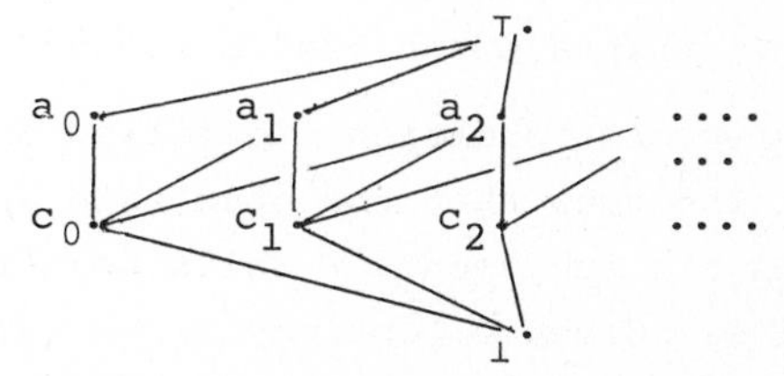

Taking this with the $\mathcal{U}_{B_D}$ quasi-uniformity (Example 3), we find that the sequence (a_i) is Cauchy and has $\top$ as least bound, but has no Limit (since $\top$ is not a limit point).

<u>Proposition 2</u>. (i) A constant sequence $(x_i = u)$ is Cauchy, and $(x_i) \Rightarrow u$.
(ii) If ξ is a Cauchy sequence with Limit x, then every subsequence of ξ (is Cauchy and) has x as Limit. The same holds with "Limit" replaced by "bound", "least bound", "limit point" , or "greatest limit point". □

IV. TOPOLOGIES

It is customary to associate with any quasi-uniformity $(X, \mathcal{U})$ the topology $T_{\mathcal{U}}$ which has $\{U[x] \mid U \in \mathcal{U}\}$ as a neighbourhood base at each point x. Here, of course, $U[x]$ means $\{y \mid xUy\}$ (which is consistent with the notation $U[\xi]$ used above, if x is identified with the constant sequence $(x,x, \ldots)$). As mentioned in the Introduction, the usual topology $T_{\mathcal{U}}$ is computationally unsatisfactory for cpo's. The approach we shall adopt here is that of developing some minimal conditions which any reasonable topology should satisfy. Eventually (Sec.VI) we reach the "strongly appropriate" topologies, among which - in suitable cases - the (generalized) Scott topology is canonical.

Appropriateness of a topology for a quasi-uniformity is best viewed in terms of the "uniform neighbourhood" or "strong inclusion" relations associated with it:

<u>Definition 5</u>. Let $(X, \mathcal{U})$ be a quasi-uniformity, T a topology on X. Given $A,B \subseteq X$, $U \in \mathcal{U}$, we say that B is a U-neighbourhood of A, written $A <_U B$, if there exist open sets O,O' such that $A \subseteq O$, $U(O) \subseteq O'$ and $O' \subseteq B$; and we say that B is a <u>uniform</u> neighbourhood of A, written $A <<_{\mathcal{U}} B$ (or just $A <<< B$), if $A <_U B$ for some $U \in \mathcal{U}$. (We are using the notation $U(A)$ for $\{y \mid \exists x \in A.\ xUy\}$, when $A \subseteq X$.)

The first condition to be imposed on a reasonable topology T for $(X, \mathcal{U})$ is :

(A) $\qquad x \in O \in T \quad \Rightarrow \quad \exists O' \in T.\ x \in O' <<< O$.

This says that the neighbourhood filter of any point is round; it fits in with the view (which we shall not be developing in this paper; but see Sec.V) that a point "is" a round Cauchy filter.

The second condition, to follow shortly, is also best understood in terms of the view that the open sets and their strong inclusion $(<_U)$ relations are fundamental, while points and their entourages are derivative. Given a topology T on $(X, \mathcal{U})$, and $U \in \mathcal{U}$, define the entourage $\hat{U}$ in terms of $<_U$ by :

$$x\ \hat{U}\ y \quad \text{iff} \quad \forall O \in T.\ x <_U O \Rightarrow y \in O .$$

(Note: we write $x <_U O$ instead of $\{x\} <_U O$)
We have, then, condition (B), to the effect that the entourages of the form $\hat{U}$ are sufficient :

(B) $\qquad \hat{\mathcal{U}} = \{\hat{U} \mid U \in \mathcal{U}\}$ is a base of $\mathcal{U}$.

Definition 6. We say that T is an appropriate topology for the quasi-uniformity $(X,\mathcal{U})$, or that $(X,\mathcal{U},T)$ is a quasi-uniform space, if conditions (A),(B) are satisfied.

Recall that the specialization (pre-)order $\leq_T$ on a topological space (X,T) is defined by :

$$x \leq_T y \quad \text{iff} \quad \forall O \in T.\ x \in O \Rightarrow y \in O .$$

As a first easy result we have :

Proposition 3. In any quasi-uniform space $(X,\mathcal{U},T)$, $\leq_T$ coincides with $\leq_{\mathcal{U}}$.

Proof. Condition (A) implies that $\leq_{\mathcal{U}} \subseteq \leq_T$, while (B) gives the reverse inclusion.

There follow some examples of appropriate topologies.

Example 7. The customary topology $T_{\mathcal{U}}$ is always appropriate.

Example 8. Let (X,T) be any topological space. Then T is appropriate for the quasi-uniformity with base $\{\leq_T\}$.

Example 9 (Extension of Example 4). A "Hennessy-Milner logic" for a system of processes ([10],[17]) may be considered as defining a base for a topology over the system. The formulas of such a logic are stratified into depths 0,1,2,... : roughly, a formula of depth n represents a property (open set) of processes which can be detected, if at all, by considering just the first n actions of the process. Moreover, we have:

$$p \lesssim_n q \quad \text{iff} \quad \text{every depth } n \text{ property of } p \text{ holds also of } q.$$

Notice in particular that, if O is a depth n open set, $\lesssim_n(O) = O$. In the light of this, conditions (A),(B) are satisfied rather trivially.

A very useful property of topologies for a quasi-uniformity, though one which may be said to be less evident than conditions (A),(B), is introduced by :

Definition 7. A quasi-uniform space $(X,\mathcal{U},T)$ is said to be interpolative if it satisfies the condition :

$$\text{(Int)} \qquad \forall U\ \exists V\ \forall A,B\ \exists C.\ A <_U B \Rightarrow A <_V C <_V B .$$

The reader may have noticed a strong resemblance between our formulations involving the $<_U$ relations and the work of Csaszar [3]. Indeed, we can state: a quasi-uniformity $(X,\mathcal{U})$ together with a topology T on X constitute an interpolative quasi-uniform space iff the family of relations $(<_U)_{U\in\mathcal{U}}$ is a syntopology in the sense of Csaszar. The syntopological point of view is very important: suitably adapted, it is just what we need for the localic version of quasi-uniform theory . Unfortunately, however, Csaszar's treatment of limits and completeness is no more satisfactory for our purposes than the others found in the literature.

V. COMPLETION BY FILTERS ?

One naturally looks for a generalization of the standard completion of uniform spaces by means of round Cauchy filters [2]. The question arises, What is a "Cauchy" filter on a quasi-uniform space? One definition which one often sees (e.g.[4],Ch.3) is: A filter F on $(X, \mathcal{U})$ is Cauchy if for each $U \in \mathcal{U}$ there exists $x \in X$ such that $U[x] \in F$. Using this definition, we find that in any quasi-uniformity having a least element w.r.t. $\leq_{\mathcal{U}}$ (the typical situation in "computing" applications), every filter is trivially Cauchy. Moreover, with the usual definitions of convergence, every (Cauchy) filter converges to the least element. This is not satisfactory as a basis for discussing completeness and completion.

An alternative approach is developed in [16]. There, we find that sobrification arises as a special case of completion (recall - Example 8 - that every space may be viewed as quasi-uniform); eventually we reach quasi-uniform locales.

In this paper, we consider only the completion by sequences. Of the filter approach we retain only the following:

Definition 8. Let $\xi = (x_i)$ be a point sequence in a quasi-uniform space $(X, \mathcal{U}, T)$. Then $S \subseteq X$ is a U-neighbourhood of ξ, where $U \in \mathcal{U}$, if S is a U-neighbourhood of $\{x_n | n \geq k\}$ for some k; S is a uniform neighbourhood of ξ if it is a U-neighbourhood of ξ for some $U \in \mathcal{U}$. The filter associated with ξ is the set of uniform neighbourhoods of ξ .

Clearly, our definitions here and elsewhere can be formulated just as easily for arbitrary nets as for sequences. We shall stick with the more computationally significant sequences, however.

Uniform neighbourhoods can be used, in place of $T_{\mathcal{U}}$ -neighbourhoods, to characterize bounds :

Proposition 4. Suppose that ξ is a Cauchy sequence in a quasi-uniform space $(X, \mathcal{U}, T)$. Then, for any $x \in X$:

x is a bound of ξ iff every uniform neighbourhood of ξ is a neighbourhood of x . □

VI. STRONGLY APPROPRIATE TOPOLOGIES

Computationally satisfactory topologies for a quasi-uniformity should - we may argue - be more than appropriate :

Definition 9. A topology T for a quasi-uniformity $(X, \mathcal{U})$ is strongly appropriate (and $(X, \mathcal{U}, T)$ is an S-quasi-uniform space) if

(i) T is appropriate for $(X, \mathcal{U})$ (Def. 5)

(ii) $(X, \mathcal{U}, T)$ is interpolative

(iii) if $x \in O \in T$, then any Cauchy sequence ξ which has x as a limit point is eventually in O .

One readily sees (using condition (A)) that the conclusion of (iii) could as well be expressed in the stronger form: O is a uniform neighbourhood of ξ . This means that an alternative form of (iii) is :

(iii') if ξ is a Cauchy sequence with x as limit point, then the filter associated with ξ refines the neighbourhood filter of x .

The impression that, in the case of a strongly appropriate topology, limit points and topological limits are closely related is well-founded, as we shall see (Proposition 6).

Recall ([6]) that a topology on a cpo D is order consistent if (i) $\forall x \in D.\ \{x\}^- = {\downarrow}x$ and (ii) lubs of increasing sequences are topological limits. The connection with strongly appropriate topologies is given by :

Proposition 5. Let $(D, \sqsubseteq)$ be a cpo considered as a quasi-uniformity with base $\{\sqsubseteq\}$. Then a topology T is strongly appropriate for D iff T is order consistent.

The computational significance of order consistent topologies has been discussed by Melton and Schmidt[9]. One observation is that continuity (of functions) with respect to order consistent topologies is sufficient for cpo-continuity (preservation of lubs of ω-chains); in the next Section we shall see how this generalizes to quasi-uniformities. Another evident remark is that, among the order consistent topologies for a cpo, the Scott topology is the finest; a (partial) generalization of this to quasi-uniformities is considered briefly in the concluding Section.

Examples. As already remarked, the Scott topology of a cpo is strongly appropriate. The usual topology of a metric space is trivially so. The Hennessy-Milner topology, Example 9, is so in suitable cases (that is, under suitable finiteness restrictions); in general, however, the interpolative condition is problematic.

When the topology is strongly appropriate, limit points (of Cauchy sequences) coincide with topological limits :

<u>Proposition 6</u>. If ξ is a Cauchy sequence in a S-quasi-uniform space $(X, \mathcal{U}, T)$ then, for any $x \in X$, x is a limit point of ξ (w.r.t. $\mathcal{U}$) iff x is a limit of ξ (w.r.t. T). Succinctly: $\xi \overset{\mathcal{U}}{\to} x$ iff $\xi \overset{T}{\to} x$.

<u>Proof</u>. ONLY IF: Trivial.

IF: Given an arbitrary $U \in \mathcal{U}$, choose V so that $\hat{V} \subseteq U$. Any V-neighbourhood of x is a V-neighbourhood of ξ (Definitions 5,8,9) . Denoting the filters of V-neighbourhoods of x, ξ by $\mathcal{N}_V(x)$, $\mathcal{N}_V(\xi)$, we thus have :

$$\begin{aligned} V[\xi] &\subseteq \bigcap \mathcal{N}_V(\xi) \qquad \text{(preamble to Def.4; Def.8)} \\ &\subseteq \bigcap \mathcal{N}_V(x) \\ &\subseteq U[x] \quad , \end{aligned}$$

showing that $\xi \overset{\mathcal{U}}{\to} x$.

VII. FUNCTIONS

<u>Definition 10</u>. A map $f: X \to X'$ of quasi-uniform spaces is <u>uniformly continuous</u> if, for every entourage U' of X' there is an entourage U of X such that, for all $x \in X$, the inverse image under f of any U'-neighbourhood of $f(x)$ is a U-neighbourhood of x. A map $f: (X, \mathcal{U}) \to (X', \mathcal{U}')$ of quasi-uniformities is <u>quasi-uniform</u> if it is quasi-uniform w.r.t. the usual topologies; that is,

$$\forall U' \in \mathcal{U}' \; \exists U \in \mathcal{U}. \; xUy \;\Rightarrow\; f(x) \; U' \; f(y)$$

or, equivalently :

$$\forall U' \; \exists U \; \forall x \in X. \; U[x] \subseteq f^{-1}(U'[f(x)]) \; .$$

<u>Proposition 7</u>. A map $f: (X, \mathcal{U}, T) \to (X', \mathcal{U}', T')$ of quasi-uniform spaces is uniformly continuous iff it is $(\mathcal{U}, \mathcal{U}')$-quasi-uniform and (T, T')-continuous.

<u>Proof</u>. We show that a uniformly continuous map is quasi-uniform (the rest of the proposition is trivial). Suppose, then, that f is uniformly continuous. Given $U' \in \hat{\mathcal{U}}'$, choose $U \in \hat{\mathcal{U}}$ such that, for all $x \in X$, the inverse image of any U'-neighbourhood of $f(x)$ is a U-neighbourhood of x. Then :

$$U[x] = \bigcap \mathcal{N}_U(x) \subseteq f^{-1}(\bigcap \mathcal{N}_{U'}(f(x))) = f^{-1}(U'[f(x)]) \; . \qquad \square$$

Quasi-uniform maps correspond to monotone maps on cpo's. They preserve bounds (but not limit points) :

Proposition 8. Let $f:(X,\mathcal{U}) \to (X',\mathcal{U}')$ be quasi-uniform, ξ a Cauchy sequence in X, x a bound of ξ. Then $f(\xi)$ is a Cauchy sequence in X' with bound $f(x)$. □

As a consequence of Proposition 6 we have :

Proposition 9. If $(X,\mathcal{U},T),(X',\mathcal{U}',T')$ are S-quasi-uniform spaces and $f:X \to X'$ is (T,T')-continuous, then f preserves limit points of Cauchy sequences in X (that is, if ξ is Cauchy and $\xi \overset{\mathcal{U}}{\to} x$ then $f(\xi) \overset{\mathcal{U}'}{\to} f(x)$). □

From Propositions 7,8,9 we have :

Proposition 10. If $(X,\mathcal{U},T),(X',\mathcal{U}',T')$ are S-quasi-uniform and $f:X \to X'$ is uniformly continuous, then f preserves Limits of Cauchy sequences; in fact, if ξ is Cauchy and $\xi \Rightarrow x$ then $f(\xi)$ is Cauchy and $f(\xi) \Rightarrow f(x)$. □

VIII. COMPLETENESS

Definition 11. A quasi-uniformity $(X,\mathcal{U})$ is (sequentially) complete if every Cauchy sequence in X has a unique Limit. A quasi-metric is complete if the quasi-uniformity determined by it is complete.

Examples. For metric spaces this evidently agrees with the usual completeness notion. For a pre-order $(D,\le)$ it amounts to: D is a "ω-dcpo" - that is, a poset with lubs of increasing sequences (equivalently, with lubs of directed sets).

IX. FIXED POINTS

We seek a decent fixed point theorem which generalizes the contraction mapping theorem for metric spaces and the least fixed point theorem for cpo's. For simplicity we consider quasi-metrics .

Definition 12. Let (X,d) be quasi-metric, and $r \in R$ with $r > 0$. We say that $f:X \to X$ is an r-contraction map if, for some $c < 1$,

$$d(x,y) \le r \quad \Rightarrow \quad d(f(x),f(y)) \le c \cdot d(x,y) ;$$

if the consequent holds unconditionally, then f is an ∞-contraction, or just a contraction map .

Moreover, for any map $f:X \to X$ and $x,y \in X$, we say that y is f-close to x provided that $\lim_{n\to\infty} d(f^n(x),f^n(y)) = 0$.

Theorem 1. Let (X,d) be complete quasi-metric, r a positive real number, and $f:X \to X'$ an r-contraction map which preserves Limits of Cauchy sequences. Then, for any $x \in X$ such that $d(x,f(x)) \le r$, there is a fixed point of f which if f-close to x and is moreover least among such fixed points (with respect to the pre-order $\le_{\mathcal{U}}$, which here takes the form: $x \le y$ iff $d(x,y) = 0$).

Remark. It should be clear that this reduces to the usual contraction mapping theorem for metric spaces on taking $r = \infty$. On the other hand, condidering a cpo $(D,\sqsubseteq)$ as a quasi-metric space (Sec.I above) and taking $r = \frac{1}{2}$, Theorem 1 reduces to: if $f:D \to D$ is continuous and $x \sqsubseteq f(x)$ then there is a fixed point of f above x, and least among such fixed points.

Outline proof of Theorem 1. Assume $x \in X$ with $d(x,f(x)) \le r$.

(1) The sequence $\sigma = (x,f(x),f^2(x),\ldots)$ is Cauchy . Given $\varepsilon > 0$ we have to find k such that, for $n \ge m \ge k$, $d(f^m(x)),f^n(x)) \le \varepsilon$; since we have $c < 1$ with $d(f^m(x),f^{m+1}(x)) \le c^{m-1}\cdot r$, it suffices to choose k so that $r\cdot c^k/(1-c) \le \varepsilon$.

(2) $\lim \sigma$ is a fixed point of f ; for, $\lim \sigma$ is also the Limit of $f(\sigma)$, and f preserves Limits of Cauchy sequences .

(3) If y is a fixed point of f, then y is a bound of σ iff y is f-close to x; for, each assertion (about y) then says :

$$\forall \varepsilon > 0\ \exists k\ \forall n \ge k.\ d(f^n(x),y) \le \varepsilon .$$

(4) Hence, $\lim \sigma$ is the least fixed point f-close to x .

X. SEQUENTIAL COMPLETION

In this Section we give a concrete construction of the completion of a quasi-uniformity $(X,\mathcal{U})$ as the T_o-ification of a suitable space of Cauchy sequences over X. We begin with a few words on T_o-ification.

One naturally tries to construct the T_o-ification of $(X,\mathcal{U})$ as the partition of X by $\equiv_{\mathcal{U}}$. This will not quite do, however, as it can happen that $x \equiv_{\mathcal{U}} y$, $z\,U\,x$ for a particular U, but $\neg zUy$. We need to use the (filter-)base $\mathcal{U}^{\equiv} = \{U^{\equiv} \mid U \in \mathcal{U}\}$, where $U^{\equiv} =_{df} \equiv_{\mathcal{U}} \circ U \circ \equiv_{\mathcal{U}}$. This is equivalent with $\mathcal{U}$ (that is, it is a base of $\mathcal{U}$), but has the advantage that each $W \in \mathcal{U}^{\equiv}$ is a <u>congruence</u> :

$$xWy, \quad x \equiv_{\mathcal{U}} x', \quad y \equiv_{\mathcal{U}} y' \quad \Rightarrow \quad x'Wy' .$$

The T_o-ification is constructed as $((X/\equiv), \mathcal{U}^{\equiv})$ (abuse of notation); we shall denote it by $(X,\mathcal{U})^{\equiv}$.

For the purposes of the following theorem, a quasi-uniform map $f:(X,\mathcal{U}) \to (Y,\mathcal{V})$ will be called <u>continuous</u> (no topologies being specified) if f preserves limit points (and hence, by Proposition 8, Limits) of Cauchy sequences .

<u>Theorem 2</u>. Let $(X,\mathcal{U})$ be a quasi-uniformity, where $\mathcal{U}$ has a countable base. For Cauchy sequences $\xi = (x_i)$, $\eta = (y_j)$, and entourage U, write $\xi\,\tilde{U}\,\eta$ iff :

$$\exists k\ \forall l \geq k\ \exists m\ \forall n \geq m.\ x_l U y_n$$

(that is, iff $\forall_\infty l\ \forall_\infty n.\ x_l U y_n$). This makes $\tilde{\mathcal{U}}$ a base of a quasi-uniformity over the Cauchy sequences; write $(\bar{X}, \bar{\mathcal{U}})$ for the T_o-ification of this quasi-uniformity. Then we have :

(i) $(\bar{X}, \bar{\mathcal{U}})$ is sequentially complete.

(ii) "Insert" X into $\bar{X}$ by $x \mapsto (x,x,\ldots)$. Then any quasi-uniform map from $(X,\mathcal{U})$ to $(Y,\mathcal{V})$, where $(Y,\mathcal{V})$ is complete, has a unique continuous quasi-uniform extension over $(\bar{X},\bar{\mathcal{U}})$.

In the course of the proof we shall understand by a <u>U-good</u> index for a sequence (x_i) in $(X,\mathcal{U})$, where $U \in \mathcal{U}$, an index i such that, for all j,k with $i \leq j \leq k$, $U(x_j,x_k)$; thus (x_i) is Cauchy iff there is a U-good index for every $U \in \mathcal{U}$.

Proof of Theorem 2. (i) Choose a (countable) base $U_0, U_1, \ldots$ of $\mathcal{U}$ such that $U_{k+1}^2 \subseteq U_k$, for all k. Let $\xi = \xi^0, \xi^1, \ldots$ be a Cauchy sequence in $(\bar{X}, \bar{\mathcal{U}})$, where each ξ^k is taken to be a Cauchy sequence

$$\xi^k = x_0^k, x_1^k, \ldots$$

in X (the choice of representative from the $\equiv_{\mathcal{U}}$-equivalence class ξ^k being immaterial). We will choose a "diagonal" sequence (y_n) from the array (x_j^k) to represent the Limit of ξ.

Specifically, we will inductively define sequences $c(.), d(.)$, in order to have $y_n = x_{d(n)}^{c(n)}$, as follows:

(1) choose $c(0) \in N$ so that $c(0)$ is $\tilde{U}_0$-good for ξ

choose $d(0)$ " $d(0)$ is U_0-good for $\xi^{c(0)}$

(2) $c(k-1), d(k-1)$ having been defined,
choose $c(k) > c(k-1)$ so that $c(k)$ is $\tilde{U}_k$-good for ξ
choose $d(k) > d(k-1)$ so that

(i) $\forall i < k\ \exists j \geq d(k-1).\ x_j^{c(i)}\ U_i\ x_{d(k)}^{c(k)}$

(ii) $d(k)$ is U_k-good for $\xi^{c(k)}$

(notice that $d(k)$ can be chosen large enough to satisfy (i), since each $c(i)$, $i \leq k$, is $\tilde{U}_i$-good for ξ).

Letting $y_n = x_{d(n)}^{c(n)}$ $(n = 0, 1, \ldots)$, we have:

(A) The sequence (y_n) is Cauchy. In fact, if $k > i > 0$, then $y_i\ U_i\ x_j^{c(i)}$ for any $j \geq d(i)$ (since $d(i)$ is U_i-good for $\xi^{c(i)}$) and $x_j^{c(i)}\ U_i\ y_k$ for some $j \geq d(i)$ (by 2(i)); hence $y_i\ U_{i-1}\ y_k$.

(B) The sequence $y = (y_n)$ (strictly, we consider $[(y_n)]_{\sim\equiv}$) is a bound of $(\xi^{c(0)}, \xi^{c(1)}, \ldots) = \xi'$. Let m be given. We claim that $\xi^{c(i)}\ \tilde{U}_m\ y$ for $i \geq m+1$. Let, then, $i \geq m+1$. Since $\xi^{c(i)}$ is Cauchy we have, for almost all n: n is U_i-good for $\xi^{c(i)}$. For any such U_i-good n, and for any $k > \max(i,n)$, we can by 2(i) find $j \geq d(k-1)$ $(\geq k-1 \geq n)$ such that $x_j^{c(i)}\ U_i\ x_{d(k)}^{c(k)}$. Since $x_n^{c(i)}\ U_i\ x_j^{c(i)}$, we have $x_n^{c(i)}\ U_m\ y_k$, as required.

(C) Finally, y is a limit point of ξ' (hence of ξ). It suffices to show that, for any $m > 0$, $\tilde{U}_{m+1}[\xi'] \subseteq \tilde{U}_m[y]$. Suppose then that

$z = (z_0, z_1, \ldots) \in \tilde{U}_{m+1}[\xi']$; specifically, that for all $j \geq k$, $\xi^{c(j)} \ \tilde{U}_{m+1} \ z$, which means that

$$\forall j \geq k \ \ \forall_\infty i \ \forall_\infty p. \ \ x_i^{c(j)} \ U_{m+1} \ z_p \ .$$

By 2(ii) :

$$\forall j \geq m+1 \ \ \forall_\infty i. \ \ x_{d(j)}^{c(j)} \ U_{m+1} \ x_i^{c(j)} \ .$$

From the two displayed assertions we infer :

$$\forall_\infty j \ \forall_\infty p. \ \ x_{d(j)}^{c(j)} \ U_m \ z_p \ \ ;$$

that is, $y \ \tilde{U}_m \ z$.

Thus, y is Limit of ξ - uniquely so, since $\tilde{\mathcal{U}}$-equivalent sequences are identified in $\bar{X}$.

(ii) Given $f:X \to Y$ quasi-uniform, unicity of its extension $\bar{f}:\bar{X} \to \bar{Y}$ is clear, since we must have :

$$\text{(3)} \qquad \bar{f}([(x_i)]) = \underset{i \to \infty}{\mathrm{Lim}} f(x_i)$$

where (x_i) is a Cauchy sequence in X .

For existence, we must check that the right hand side of (3) does not depend on the choice of the representative sequence (x_i). Suppose then that $(x_i),(u_j)$ are equivalent Cauchy sequences over X, that is

$$\forall U \in \mathcal{U} \ \ \forall_\infty i \ \ \forall_\infty j. \ \ x_i \ U \ u_j$$

and symmetrically (in $(x_i),(u_j)$). We must show that $\mathrm{Lim}\, f(x_i) = \mathrm{Lim}\, f(u_j)$ and for this it suffices to show that the (by Proposition 9, Cauchy) sequences $f(x_i), f(u_j)$ have the same bounds (noting that the Limits are the least bounds). In fact it is immediate by quasi-uniformity of f that $(f(x_i)),(f(u_j))$ are "equivalent", that is

$$\forall V \in \mathcal{V} \ \ \forall_\infty i \ \ \forall_\infty j. \ \ f(x_i) \ V \ f(u_j)$$

and symmetrically, from which the coincidence of the bounds of the two sequences is an easy consequence.

We omit the tedious verification that $\bar{f}$ is quasi-uniform and continuous. □

XI. CONCLUSION

In Theorem 2, we have a completion construction together with a reasonable universal characterization. As such, the result is an improvement on what one finds in the literature of quasi-metrics and quasi-uniformities, where the "completions" given either lack a satisfactory characterization, or (as in [4]) are actually taken with respect to the uniformity $\mathcal{U}^*$, rather than $\mathcal{U}$ itself .

Theorem 2 depends only on the theory of Cauchy convergence offered in Section III, and does not make use of the topological ideas of IV - VI. These ideas come into play in the sequel to this paper, in which we investigate the completion functor more closely. There it is shown that the notion of algebraicity generalizes from cpo,s to (complete) quasi-uniformities, and that the completion of a quasi-uniformity is best thought of as an algebraic quasi-uniform space with (generalized) Scott topology. A hint of the way in which the generalization works is as follows. The "finite" elements of the completion $(\bar{X}, \bar{\mathcal{U}})$ of a quasi-uniformity $(X, \mathcal{U})$ are, in effect, the constant sequences $\bar{a} = (a,a,\ldots)$ (where $a \in X$). The generalized Scott topology on $\bar{X}$ can be characterized in any of the following ways :

(1) the sets $\tilde{U}(\bar{a})$ $(U \in \mathcal{U}, a \in X)$ form a base of T ;

(2) a set $O \subseteq \bar{X}$ is open in T iff, for any Cauchy sequence ξ in $\bar{X}$, if $\xi \to x \in O$ then O is a uniform neighbourhood of ξ ;

(3) T is the finest strongly appropriate topology for $(\bar{X}, \bar{\mathcal{U}})$.

A final remark about the complexity of the theory outlined here : It may be felt that a theory in which we have to distinguish at least three kinds of limits of sequences (least bounds, greates limit points, Limits) is too inconvenient for everyday use. This is no doubt true; but in practice the problem is slight, since almost all the quasi-uniformities which arise in computer science are complete algebraic, for which the various limit notions coincide.

REFERENCES

[1] G.Berthiaume, On quasi-uniformities in hyperspaces, Proc. Am. Math. Soc. 66 (1977), 335-342 .

[2] N.Bourbaki, Elements of Mathematics. General Topology; Part 1 Addison-Wesley 1966 .

[3] A.Csaszar, Foundationsof General Topology, Pergamon 1963 .

[4] P.Fletcher, W.Lindgren, Quasi-uniform Spaces, Marcel Dekker 1982.

[5] M.Hennessy, G.Plotkin, A term model for CCS, Proc. 9th MFCS, LNCS 88 (1980), 261-274 .

[6] G.Gierz et al., A Compendium of Continuous Lattices, Springer 1980 .

[7] F.Lawvere, Metric spaces, generalized logic, and closed categories Rend. del Sem. Mat. e Fis. di Milano XLIII (1973), 135-166 .

[8] N.Levine, W.Stager, On the hyperspace of a quasi-uniform space, Math. J. Okayama Univ.15 (1971-2), 101-106 .

[9] A.Melton, D.Schmidt, A topological framework for cpo's lacking bottom elements, Unpublished memo (1985) .

[10] R.Milner, A modal characterization of observable machine behaviour CAAP '81, LNCS 112 (1981), 25-34 .

[11] R.Milner, A Calculus of Communicating Sequences, Springer LNCS 92 1980 .

[12] G.Plotkin, Lecture Notes on Domain Theory (Edinburgh: unpublished)

[13] D.Scott, Domains for denotational semantics, ICALP '82, LNCS 140 (1982), 577-613 .

[14] M.Smyth, Power domains and predicate transformers, ICALP '83, LNCS 154 (1983), 662-675 .

[15] M.Smyth, Finite approximation of spaces, Category Theory and Computer Programming, ed. D.Pitt et al., LNCS 240 (1986), 225-241 .

[16] M.Smyth, Quasi-uniformities as a unifying theme, Unpublished ms., 1986 .

[17] M.Hennessy, R.Milner, On observing nondeterminism and concurrency ICALP '80, LNCS 85 (1980), 299-309 .

Solving Reflexive Domain Equations in a Category of Complete Metric Spaces

Pierre America
Philips Research Laboratories
P.O. Box 80.000, 5600 JA Eindhoven, The Netherlands

Jan Rutten
Centre for Mathematics and Computer Science
P.O. Box 4079, 1009 AB Amsterdam, The Netherlands

This paper presents a technique by which solutions to reflexive domain equations can be found in a certain category of complete metric spaces. The objects in this category are the (non-empty) metric spaces and the arrows consist of two maps: an isometric embedding and a non-distance-increasing left inverse to it. The solution of the equation is constructed as a fixed point of a functor over this category associated with the equation. The fixed point obtained is the direct limit (colimit) of a convergent tower. This construction works if the functor is *contracting,* which roughly amounts to the condition that it maps every embedding to an even denser one. We also present two additional conditions, each of which is sufficient to ensure that the functor has a *unique* fixed point (up to isomorphism). Finally, for a large class of functors, including function space constructions, we show that these conditions are satisfied, so that they are guaranteed to have a unique fixed point. The techniques we use are so reminiscent of Banach's fixed-point theorem that we feel justified to speak of a category-theoretic version of it.

1980 Mathematical Subject classification: 68B10, 68C01.
1986 Computing Reviews Categories: D.1.3, D.3.1, F.3.2.
Key words and phrases: domain equations, complete metric spaces, category theory, converging towers, contracting functors, Banach's fixed-point theorem.

Note: This work was carried out in the context of ESPRIT project 415: Parallel Architectures and Languages for Advanced Information Processing — a VLSI-directed approach.

1. Introduction

The framework of complete metric spaces has proved to be very useful for giving a denotational semantics to programming languages, especially concurrent ones. For example, in the approach of De Bakker and Zucker [BZ] a process is modelled as the element of a suitable metric space, where the distance between two processes is defined in such a way that the smaller this distance is, the longer it takes before the two processes show a different behaviour.

In order to construct a suitable metric space in which processes are to reside, we must solve a reflexive domain equation. For example, a simple language, where a process is a fixed sequence of uninterpreted atomic actions, gives rise to the equation

$$P \cong \{p_0\} \,\overline{\cup}\, (A \times P).$$

(Here $\overline{\cup}$ denotes the disjoint union operation.) In [BZ] an elementary technique was developed to solve such equations. Roughly, this consisted of starting with a small metric space, enriching it iteratively, and taking the metric completion of the union of all the obtained spaces.

In many cases this technique is sufficient to solve the equation at hand, but there are equations for which it does not work: equations where the domain variable P occurs in the left-hand side of a function space construction, e.g.,

$$P \cong \{p_0\} \,\overline{\cup}\, (P \to P).$$

This kind of equation arises when the semantic description is based on *continuations* (see for example [ABKR]). In this paper we present a technique by which these cases can also be solved, at least when we restrict the function space at hand to the *non-distance-increasing* functions.

The structure of this report is as follows: In section 2 we list some mathematical preliminaries. In section 3 we introduce our category $\mathcal{C}$ of complete metric spaces, we define the concepts of converging tower and contracting functor. We show that a converging tower has a direct limit and that a contracting functor preserves such a limit. Then we see how a contracting functor gives rise to a converging tower and that the limit of this tower is a fixed point of the functor.

Section 4 presents two cases in which we can show that the fixed point we construct is the unique fixed point (up to isomorphism) of the contracting functor at hand. One case arises when we work in a base-point category: a category where every space has a specially designated base-point and where every map preserves this base-point. The other case is where the functor is not only contracting, but also hom-contracting: it is a contraction on every function space.

Finally, in section 5, we present a large class of functors (including most of the ones we are interested in), for which we can show that each of them has a unique fixed point.

Acknowledgements
We would like to thank Jaco de Bakker, Frank de Boer, Joost Kok, Frank van der Linden, John-Jules Meyer and Erik de Vink for useful discussions on the contents of this paper. We are also grateful to Marino Delusso and Eline Meijs, who have typed this report.

2. Mathematical preliminaries

In this section we collect some definitions and properties concerning metric spaces, in order to refresh the reader's memory or to introduce him to this subject.

2.1. *Metric spaces*

Definition 2.1 (Metric space)
A *metric space* is a pair (M,d) with M a non-empty set and d a mapping $d:M\times M\rightarrow[0,1]$ (a *metric* or *distance*), which satisfies the following properties:
(a) $\forall x,y\in M\,[d(x,y)=0\Leftrightarrow x=y]$
(b) $\forall x,y\in M\,[d(x,y)=d(y,x)]$
(c) $\forall x,y,z\in M\,[d(x,y)\leqslant d(x,z)+d(z,y)]$.
We call (M,d) an *ultra-metric space* if the following stronger version of property (c) is satisfied:
(c′) $\forall x,y,z\in M\,[d(x,y)\leqslant\max\{d(x,z),d(z,y)\}]$.
Note that we consider only metric spaces with bounded diameter: the distance between two points never exceeds 1.

Example
Let A be an arbitrary set. The *discrete* metric d_A on A is defined as follows. Let $x,y\in A$, then

$$d_A(x,y)=\begin{cases}0 & \text{if } x=y\\ 1 & \text{if } x\neq y.\end{cases}$$

Definition 2.2
Let (M,d) be a metric space, let $(x_i)_i$ be a sequence in M.
(a) We say that $(x_i)_i$ is a *Cauchy sequence* whenever we have:
$\forall\epsilon>0\ \exists N\in\mathbb{N}\ \forall n,m>N\ [d(x_n,x_m)<\epsilon]$.
(b) Let $x\in M$. We say that $(x_i)_i$ *converges to* x and call x the *limit* of $(x_i)_i$ whenever we have:
$\forall\epsilon>0\ \exists N\in\mathbb{N}\ \forall n>N\ [d(x,x_n)<\epsilon]$.
Such a sequence we call *convergent.* Notation: $\lim_{i\rightarrow\infty}x_i=x$.
(c) The metric space (M,d) is called *complete* whenever each Cauchy sequence converges to an element of M.

Definition 2.3
Let $(M_1,d_1),(M_2,d_2)$ be metric spaces.
(a) We say that (M_1,d_1) and (M_2,d_2) are *isometric* if there exists a bijection $f:M_1\rightarrow M_2$ such that:
$\forall x,y\in M_1\ [d_2(f(x),f(y))=d_1(x,y)]$. We then write $M_1\cong M_2$. When f is not a bijection (but only an injection), we call it an *isometric embedding.*

(b) Let $f{:}M_1 \rightarrow M_2$ be a function. We call f *continuous* whenever for each sequence $(x_i)_i$ with limit x in M_1 we have that $\lim_{i\rightarrow\infty} f(x_i)=f(x)$.

(c) Let $A \geqslant 0$. With $M_1 \rightarrow^A M_2$ we denote the set of functions f from M_1 to M_2 that satisfy the following property:
$\forall x,y \in M_1\ [d_2(f(x),f(y)) \leqslant A \cdot d_1(x,y)]$.
Functions f in $M_1 \rightarrow^1 M_2$ we call *non-distance-increasing* (NDI), functions f in $M_1 \rightarrow^\epsilon M_2$ with $0 \leqslant \epsilon < 1$ we call *contracting*.

PROPOSITION 2.4

(a) *Let* $(M_1,d_1),(M_2,d_2)$ *be metric spaces. For every* $A \geqslant 0$ *and* $f \in M_1 \rightarrow^A M_2$ *we have:* f *is continuous.*

(b) *(Banach's fixed-point theorem)*
Let (M,d) *be a complete metric space and* $f{:}M \rightarrow M$ *a contracting function. Then there exists an* $x \in M$ *such that the following holds:*
(1) $f(x)=x$ *(*x *is a fixed point of* f*),*
(2) $\forall y \in M\ [f(y)=y \Rightarrow y=x]$ *(*x *is unique),*
(3) $\forall x_0 \in M\ [\lim_{n\rightarrow\infty} f^{(n)}(x_0)=x]$, *where* $f^{(n+1)}(x_0)=f(f^{(n)}(x_0))$ and $f^{(0)}(x_0)=x_0$.

DEFINITION 2.5 (Closed subsets)

A subset X of a complete metric space (M,d) is called *closed* whenever each Cauchy sequence in X converges to an element of X.

DEFINITION 2.6

Let $(M,d),(M_1,d_1), \ldots ,(M_n,d_n)$ be metric spaces.

(a) With $M_1 \rightarrow M_2$ we denote the set of all continuous functions from M_1 to M_2. We define a metric d_F on $M_1 \rightarrow M_2$ as follows. For every $f_1,f_2 \in M_1 \rightarrow M_2$

$$d_F(f_1,f_2)=\sup_{x \in M_1}\{d_2(f_1(x),f_2(x))\}.$$

For $A \geqslant 0$ the set $M_1 \rightarrow^A M_2$ is a subset of $M_1 \rightarrow M_2$, and a metric on $M_1 \rightarrow^A M_2$ can be obtained by taking the restriction of the corresponding d_F.

(b) With $M_1 \overline{\cup} \cdots \overline{\cup} M_n$ we denote the *disjoint union* of $M_1, \ldots ,M_n$, which can be defined as $\{1\} \times M_1 \cup \cdots \cup \{n\} \times M_n$. We define a metric d_U on $M_1 \overline{\cup} \cdots \overline{\cup} M_n$ as follows. For every $x,y \in M_1 \overline{\cup} \cdots \overline{\cup} M_n$

$$d_U(x,y) = \begin{cases} d_j(x,y) & \text{if } x,y \in \{j\} \times M_j,\ 1 \leqslant j \leqslant n \\ 1 & \text{otherwise.} \end{cases}$$

(c) We define a metric d_P on $M_1 \times \cdots \times M_n$ by the following clause.
For every $(x_1, \ldots ,x_n), (y_1, \ldots ,y_n) \in M_1 \times \cdots \times M_n$

$$d_P((x_1, \ldots ,x_n),(y_1, \ldots ,y_n))=\max_i\{d_i(x_i,y_i)\}.$$

(d) Let $\mathcal{P}_{cl}(M) =^{def} \{X | X \subseteq M | X$ is closed and non-empty$\}$. We define a metric d_H on $\mathcal{P}_{cl}(M)$, called the *Hausdorff distance,* as follows. For every $X, Y \in \mathcal{P}_{cl}(M)$

$$d_H(X,Y) = \max\{\sup_{x \in X}\{d(x,Y)\}, \sup_{y \in Y}\{d(y,X)\}\},$$

where $d(x,Z) =^{def} \inf_{z \in Z}\{d(x,z)\}$ for every $Z \subseteq M,\ x \in M$.
An equivalent definition would be to set $V_r(X) = \{y \in M \mid \exists x \in X [d(x,y) < r]\}$ for $r > 0,\ X \subset M$, and then to define

$$d_H(X,Y) = \inf\{r > 0 \mid X \subset V_r(Y) \wedge Y \subset V_r(X)\}.$$

PROPOSITION 2.7
Let (M,d), $(M_1,d_1), \ldots, (M_n,d_n)$, d_F, d_U, d_P and d_H be as in definition 2.6 and suppose that (M,d), $(M_1,d_1), \ldots, (M_n,d_n)$ are complete. We have that
(a) $(M_1 \rightarrow M_2, d_F)$, $(M_1 \rightarrow^A M_2, d_F)$,
(b) $(M_1 \overline{\cup} \cdots \overline{\cup} M_n, d_U)$,
(c) $(M_1 \times \cdots \times M_n, d_P)$,
(d) $(\mathcal{P}_{cl}(M), d_H)$
are complete metric spaces. If (M,d) and (M_i,d_i) are all ultra-metric spaces these composed spaces are again ultra-metric. (Strictly spoken, for the completeness of $M_1 \rightarrow M_2$ and $M_1 \rightarrow^A M_2$ we do not need the completeness of M_1. The same holds for the ultra-metric property.)

If in the sequel we write $M_1 \rightarrow M_2$, $M_1 \rightarrow^A M_2$, $M_1 \overline{\cup} \cdots \overline{\cup} M_n$, $M_1 \times \cdots \times M_n$ or $\mathcal{P}_{cl}(M)$, we mean the metric space with the metric defined above.
The proofs of proposition 2.7 (a), (b) and (c) are straightforward. Part (d) is more involved. It can be proved with the help of the following characterization of the completeness of $(\mathcal{P}_{cl}(M), d_H)$.

PROPOSITION 2.8
Let $(\mathcal{P}_{cl}(M), d_H)$ be as in definition 2.6. Let $(X_i)_i$ be a Cauchy sequence in $\mathcal{P}_{cl}(M)$. We have:

$$\lim_{i \to \infty} X_i = \{\lim_{i \to \infty} x_i | x_i \in X_i,\ (x_i)_i \text{ a Cauchy sequence in } M\}.$$

Proofs of proposition 2.7(d) and 2.8 can be found in (for instance) [Du] and [En]. Proposition 2.8 is due to Hahn [Ha]. The proofs are also repeated in [BZ].

THEOREM 2.9 (Metric completion)
Let M be an arbitrary metric space. Then there exists a metric space $\overline{M}$ (called the completion of M) together with an isometric embedding $i: M \rightarrow \overline{M}$ such that:
(1) *$\overline{M}$ is complete*
(2) *For every complete metric space M' and isometric embedding $j: M \rightarrow M'$ there exists a unique isometric embedding $\overline{j}: \overline{M} \rightarrow M'$ such that $\overline{j} \circ i = j$.*

PROOF

The space $\overline{M}$ is constructed by taking the set of all Cauchy sequences in M and dividing it out by the equivalence relation $\equiv$ defined by

$$(x_n)_n \equiv (y_n)_n =^{\text{def}} \lim_{n\to\infty} d(x_n, y_n) = 0.$$

The metric d_c on $\overline{M}$ is defined by

$$d_c([(x_n)]_\equiv, [(y_n)]_\equiv) =^{\text{def}} \lim_{n\to\infty} d(x_n, y_n)$$

and the embedding i will map every $x \in M$ to the equivalence class of the sequence of which all elements are equal to x:

$$i(x) = [(x)_n]_\equiv.$$

It is easy to show that $\overline{M}$ and i satisfy the above properties.

3. A CATEGORY OF COMPLETE METRIC SPACES

In this section we want to generalize the technique of solving reflexive domain equations of De Bakker and Zucker ([BZ]). We shall first give an example of their approach and then explain how it can be extended.

Consider a domain equation

$$P \cong \{p_0\} \overline{\cup} (A \times P),$$

with A an arbitrary set. In [BZ] a complete metric space that satisfies this equation is constructed as follows. An increasing sequence $A^{(0)} \subseteq A^{(1)} \subseteq \cdots$ of metric spaces is defined by

$$(0)\ A^{(0)} = \{p_0\},\quad d_0 \text{ trivial},$$

$$(n+1)\ A^{(n+1)} = \{p_0\} \cup A \times A^{(n)},$$

$$d_{n+1}(p_0, q) = 1 \text{ if } q \in A^{(n+1)},\ q \neq p_0,$$

$$d_{n+1}(<a_1, p_1>, <a_2, p_2>) = \begin{cases} 1 & \text{if } a_1 \neq a_2 \\ \frac{1}{2} \cdot d_n(p_1, p_2) & \text{if } a_1 = a_2. \end{cases}$$

Note that for every $i \geqslant 0$, $A^{(i)}$ is a subspace of $A^{(i+1)}$. Their union is defined as

$$A^* = \bigcup_{n \in \mathbb{N}} A^{(n)},$$

and a domain A^∞ is defined as the *metric completion* of this union:

$$A^{\infty} = \overline{A^{\star}}.$$

It is then proved that A^{∞} satisfies the equation. (We observe that $A^{\star}$ is isometric to the set of all finite sequences of elements of A, while A^{∞} is isometric to the set of all finite and infinite sequences, in both cases with a suitable metric.)
In order to extend this approach, we shall formulate a number of category-theoretic generalizations of some of the concepts used in the construction described above.

First we shall define a *converging tower* to be the counterpart of an increasing sequence of metric spaces; then the construction of a *direct limit* of such a tower will be the generalization of the metric completion of the union of such a sequence. Finally we shall give a generalized version of Banach's fixed-point theorem.
For this purpose we define a category $\mathcal{C}$ of complete metric spaces.

DEFINITION 3.1 (Category of complete metric spaces)
Let $\mathcal{C}$ denote the category that has complete metric spaces for its objects. The arrows ι in $\mathcal{C}$ are defined as follows. Let M_1, M_2 be complete metric spaces. Then $M_1 \rightarrow^{\iota} M_2$ denotes a pair of maps $M_1 \underset{j}{\overset{i}{\rightleftarrows}} M_2$, satisfying the following properties:
(a) i is an isometric embedding,
(b) j is non-distance-increasing (NDI),
(c) $j \circ i = id_{M_1}$.
(We sometimes write $<i,j>$ for ι.) Composition of the arrows is defined in the obvious way.

REMARK
For the basic definitions from category theory we refer the reader to [ML].

We can consider M_1 as an approximation of M_2: in a sense the set M_2 contains more information than M_1, because M_1 can be isometrically embedded into M_2. Elements in M_2 are approximated by elements in M_1. For an element $m_2 \in M_2$ its (best) approximation in M_1 is given by $j(m_2)$. (The reason why j should be NDI is, at this point, difficult to motivate.)
When we informally rephrase clause (c), it states that the approximation in M_1 of the embedding of an element $m_1 \in M_1$ into M_2 is again m_1. Or, in other words, that M_2 is a consistent extension of M_1.

DEFINITION 3.2
For every arrow $M_1 \rightarrow^{\iota} M_2$ in $\mathcal{C}$ with $\iota = <i,j>$ we define

$$\delta(\iota) = d_{M_2 \rightarrow M_2}(i \circ j, id_{M_2}) \ (= \sup_{m_2 \in M_2} \{ d_{M_2}(i \circ j(m_2), m_2) \}).$$

This number plays an important role in our theory. It can be regarded as a measure of the quality with which M_2 is approximated by M_1: the smaller $\delta(\iota)$, the denser M_1 is embedded into M_2. We next try to formalize a generalization of increasing sequences of metric spaces by the following definition.

DEFINITION 3.3 (Converging tower)

(a) We call a sequence $(D_n, \iota_n)_n$ of complete metric spaces and arrows a *tower* whenever we have that
$\forall n \in \mathbb{N}\ [D_n \xrightarrow{\iota_n} D_{n+1} \in \mathcal{C}]$.

$$D_0 \xrightarrow{\iota_0} D_1 \xrightarrow{\iota_1} D_2 \rightarrow \cdots \rightarrow D_n \xrightarrow{\iota_n} D_{n+1} \rightarrow \cdots$$

(b) The sequence $(D_n, \iota_n)_n$ is called a *converging tower* when furthermore the following condition is satisfied:
$\forall \epsilon > 0\ \exists N \in \mathbb{N}\ \forall m > n \geqslant N\ [\delta(\iota_{nm}) < \epsilon]$, where $\iota_{nm} = \iota_{m-1} \circ \cdots \circ \iota_n : D_n \rightarrow D_m$.

$$D_n \xrightarrow{\iota_n} D_{n+1} \rightarrow \cdots \rightarrow D_{m-1} \xrightarrow{\iota_{m-1}} D_m \qquad (\iota_{nm} : D_n \rightarrow D_m)$$

EXAMPLE 3.4

A special case of a converging tower is a sequence $(D_n, \iota_n)_n$ that satisfies the following conditions:

(a) $\forall n \in \mathbb{N}\ [D_n \xrightarrow{\iota_n} D_{n+1} \in \mathcal{C}]$,

(b) $\exists \epsilon\ [0 \leqslant \epsilon < 1 \wedge \forall n \in \mathbb{N}\ [\delta(\iota_{n+1}) \leqslant \epsilon \cdot \delta(\iota_n)]]$.

(Note that $\delta(\iota_{nm}) \leqslant \delta(\iota_n) + \cdots + \delta(\iota_{m-1}) \leqslant \epsilon^n \cdot \delta(\iota_0) + \cdots + \epsilon^{m-1} \cdot \delta(\iota_0) \leqslant \frac{\epsilon^n}{1-\epsilon} \cdot \delta(\iota_0)$.)

EXAMPLE 3.5

Let $A^{(0)} \subseteq A^{(1)} \subseteq \cdots$ be the sequence of metric spaces defined at the beginning of this chapter. We show how it can be transformed into a converging tower, by defining a sequence of arrows $(\iota_n)_n$ (with $\iota_n = \langle i_n, j_n \rangle$) with induction on n:

$$(0)\ i_0(p_0) = p_0,\ j_0 \text{ trivial},$$

$$(n+1)\ i_{n+1} : A^{(n+1)} \rightarrow A^{(n+2)}, \text{ trivial } (i_{n+1}(p) = p),$$

$$j_{n+1} : A^{(n+2)} \rightarrow A^{(n+1)},$$

$$j_{n+1}(p_0) = p_0,$$

$$j_{n+1}(\langle a, p \rangle) = \langle a, j_n(p) \rangle \text{ for } \langle a, p \rangle \in A^{(n+2)}.$$

It is not difficult to see that we have obtained a tower

$$A^{(0)} \to^{\iota_0} A^{(1)} \to^{\iota_1} \cdots,$$

which is converging.

3.1 The direct limit construction

In this subsection we show that in our category $\mathcal{C}$ every converging tower has an *initial cone*. The construction of such an initial cone for a given tower (the *direct limit* construction) generalizes the technique of forming the metric *completion* of the union of an increasing sequence of metric spaces.

Before we treat the inverse limit construction, we first give the definition of a cone and an initial cone and then formulate a criterion for the initiality of a cone.

DEFINITION 3.6 (Cone)

Let $(D_n, \iota_n)_n$ be a tower. Let D be a complete metric space and $(\gamma_n)_n$ a sequence of arrows. We call $(D, (\gamma_n)_n)$ a *cone* for $(D_n, \iota_n)_n$ whenever the following condition holds:

$$\forall n \in \mathbb{N}\ [D_n \to^{\gamma_n} D \in \mathcal{C} \wedge \gamma_n = \gamma_{n+1} \circ \iota_n].$$

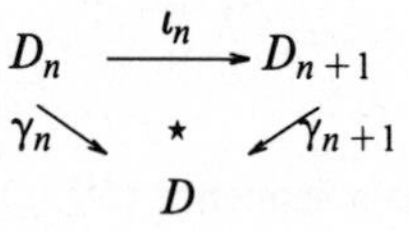

$$\begin{array}{ccc} D_n & \xrightarrow{\iota_n} & D_{n+1} \\ \gamma_n \searrow & \star & \swarrow \gamma_{n+1} \\ & D & \end{array}$$

DEFINITION 3.7 (Initial cone)

A cone $(D, (\gamma_n)_n)$ of a tower $(D_n, \iota_n)_n$ is called *initial* whenever for every other cone $(D', (\gamma'_n)_n)$ of $(D_n, \iota_n)_n$ there exists a unique arrow $\iota: D \to D'$ in $\mathcal{C}$ such that:

$$\forall n \in \mathbb{N}\ [\iota \circ \gamma_n = \gamma'_n].$$

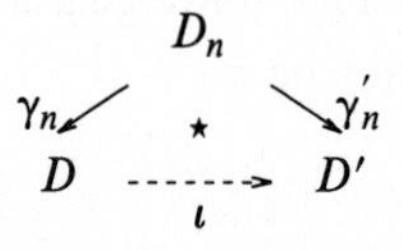

$$\begin{array}{ccc} & D_n & \\ \gamma_n \swarrow & \star & \searrow \gamma'_n \\ D & \overset{\iota}{\dashrightarrow} & D' \end{array}$$

LEMMA 3.8 (Initiality lemma)

Let $(D_n, \iota_n)_n$ be a converging tower with a cone $(D, (\gamma_n)_n)$. Let $\gamma_n = <\alpha_n, \beta_n>$. We have:

$$D \text{ is an initial cone} \Leftrightarrow \lim_{n \to \infty} \alpha_n \circ \beta_n = id_D.$$

PROOF

$\Leftarrow$

Suppose $\lim_{n\to\infty} \alpha_n \circ \beta_n = id_D$. Let $(D', (\gamma'_n)_n)$, with $\gamma'_n = <\alpha'_n, \beta'_n>$, be another cone for $(D_n, \iota_n)_n$. We have to prove the existence of a unique arrow $D \rightarrow^{\iota} D' \in \mathcal{C}$ such that

$$\forall n \in \mathbb{N} \, [\, \iota \circ \gamma_n = \gamma'_n \,] \, .$$

First we construct an embedding $i : D \rightarrow D'$, then a projection $j : D' \rightarrow D$. Next, the arrow ι will be defined as $\iota = <i, j>$.
For every $n \in \mathbb{N}$ we have

$$\alpha'_n \circ \beta_n \in D \rightarrow D'.$$

We show that $(\alpha'_n \circ \beta_n)_n$ is a Cauchy sequence in $D \rightarrow D'$ and then use the completeness of this function space to define i as the limit of that sequence.
Let $m > n \geqslant 0$. We have

$$d_{D\rightarrow D'}(\alpha'_m \circ \beta_m \, , \, \alpha'_n \circ \beta_n) =$$

$$d_{D\rightarrow D'}(\alpha'_m \circ \beta_m, \, \alpha'_m \circ i_{nm} \circ j_{nm} \circ \beta_m) =$$

$$\sup_{x \in D} \{ d_{D'}(\alpha'_m \circ \beta_m(x), \, \alpha'_m \circ i_{nm} \circ j_{nm} \circ \beta_m(x)) \} =$$

[because α'_m is isometric]

$$\sup_{x \in D} \{ d_{D_m}(\beta_m(x), \, i_{nm} \circ j_{nm} \circ \beta_m(x)) \} =$$

[because β_m is surjective]

$$\sup_{x \in D_m} \{ d_{D_m}(x, \, i_{nm} \circ j_{nm}(x)) \} =$$

$$d_{D_m \rightarrow D_m}(id_{D_m}, \, i_{nm} \circ j_{nm})) = \delta(\iota_{nm}) \, .$$

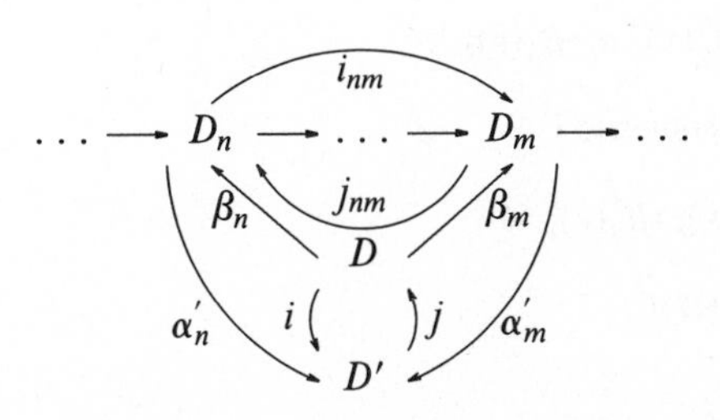

Let $\epsilon > 0$. Because $(D_n, \iota_n)_n$ is a converging tower there is an $N \in \mathbb{N}$ such that

$$\forall m > n \geqslant N \, [\, \delta(\iota_{nm}) < \epsilon \,] \, .$$

Thus $(\alpha'_n \circ \beta_n)_n$ is a Cauchy sequence. We define

$$i = \lim_{n\to\infty} \alpha'_n \circ \beta_n \, .$$

We prove that i is isometric by showing:

$$\forall x, y \in D \; [\; d_{D'} \, (i(x), \, i(y)) = d_D(x, y) \;]$$

Let $x, y \in D$, we have

$d_{D'}(i(x), i(y)) =$

$d_{D'}(\lim_{n\to\infty} \alpha'_n \circ \beta_n(x), \lim_{n\to\infty} \alpha'_n \circ \beta_n(y)) =$

$\lim_{n\to\infty} d_{D'}(\alpha'_n \circ \beta_n(x), \alpha'_n \circ \beta_n(y)) =$

[because α'_n is isometric]

$\lim_{n\to\infty} d_{D_n}(\beta_n(x), (\beta_n(y)) =$

[because α_n is isometric]

$\lim_{n\to\infty} d_D(\alpha_n \circ \beta_n(x), \alpha_n \circ \beta_n(y)) =$

$d_D(\lim_{n\to\infty} \alpha_n \circ \beta_n(x), \lim_{n\to\infty} \alpha_n \circ \beta_n(y)) =$

$d_D(x, y)$.

Thus i is isometric.

Similar to the definition of i we choose

$$j = \lim_{n\to\infty} \alpha_n \circ \beta'_n \, .$$

We have that j is NDI, because, for $x, y \in D'$:

$d_D(j(x), j(y)) =$

$d_D(\lim_{n\to\infty} \alpha_n \circ \beta'_n(x), \lim_{n\to\infty} \alpha_n \circ \beta'_n(y)) =$

$\lim_{n\to\infty} d_D(\alpha_n \circ \beta'_n(x), \alpha_n \circ \beta'_n(y)) =$

[because α_n is isometric]

$\lim_{n\to\infty} d_{D_n}(\beta'_n(x), (\beta'_n(y)) \leqslant$

[because β'_n is NDI]

$\lim_{n\to\infty} d_{D'}(x, y) =$

$d_{D'}(x, y)$.

We also show : $j \circ i = id_D$. Let $x \in D$, then

$j \circ i(x) =$

$j(\lim_{n\to\infty} \alpha'_n \circ \beta_n(x)) =$

$\lim_{n\to\infty} j \circ \alpha'_n \circ \beta_n(x) =$

$$\lim_{n\to\infty}\lim_{m\to\infty}\alpha_m \circ \beta'_m \circ \alpha'_n \circ \beta_n(x) =$$

$$\lim_{n\to\infty}\alpha_n \circ \beta'_n \circ \alpha'_n \circ \beta_n(x) =$$

$$[\text{ because } \beta'_n \circ \alpha'_n = id_{D_n}\ ]$$

$$\lim_{n\to\infty}\alpha_n \circ \beta_n(x) = x\ .$$

Now we can define

$$\iota = <i, j>\ ,$$

of which we have so far proved : $D \to^{\iota} D' \in \mathcal{C}$.

Next we have to verify that ι satisfies the condition

$$\forall m \in \mathbb{N}\ [\ \iota \circ \gamma_m = \gamma'_m\ ]\ .$$

This amounts to

$$\forall m \in \mathbb{N}\ [\ i \circ \alpha_m = \alpha'_m \wedge \beta_m \circ j = \beta'_m\ ]\ .$$

Let $m \geqslant 0$. We only prove the first part of the conjunction. We have

$$\begin{aligned} i \circ \alpha_m &= (\lim_{n\to\infty}\alpha'_n \circ \beta_n) \circ \alpha_m \\ &= (\lim_{n\to\infty}\alpha'_{n+m} \circ \beta_{n+m}) \circ \alpha_m \\ &= \lim_{n\to\infty}\alpha'_{n+m} \circ \beta_{n+m} \circ \alpha_m \\ &= \lim_{n\to\infty}\alpha'_{n+m} \circ \beta_{n+m} \circ \alpha_{n+m} \circ i_{m,\, m+n} \\ &= \lim_{n\to\infty}\alpha'_{n+m} \circ id_{D_{n+m}} \circ i_{m,\, m+n} \\ &= \lim_{n\to\infty}\alpha'_m = \alpha'_m\ . \end{aligned}$$

Finally we show that ι is *unique*. Suppose $D \to^{\iota'} D'$, with $\iota' = <i', j'>$, is another arrow in $\mathcal{C}$, that satisfies

$$\forall m \in \mathbb{N}\ [\ \iota' \circ \gamma_m = \gamma'_m\ ]\ .$$

We only show that $i' = i$, leaving the proof of $j' = j$ to the reader:

$$\begin{aligned} i' &= i' \circ id_D \\ &= i' \circ \lim_{m\to\infty}\alpha_m \circ \beta_m \\ &= \lim_{m\to\infty} i' \circ \alpha_m \circ \beta_m \\ &= \lim_{m\to\infty}\alpha'_m \circ \beta_m \\ &= i\ . \end{aligned}$$

$\Rightarrow$

Suppose now that $(D, (\gamma_n)_n)$ is an initial cone of the converging tower $(D_n, \iota_n)_n$. We have to prove that

$$\lim_{m\to\infty} \alpha_n \circ \beta_n = id_D .$$

By an argument similar to the proof for $(\alpha_n' \circ \beta_n)_n$ above, we have that $(\alpha_n \circ \beta_n)_n$ is a Cauchy sequence. We define

$$f = \lim_{n\to\infty} \alpha_n \circ \beta_n ,$$

$$D' = \{ x \mid x \in D \mid f(x)=x \}.$$

We set out to prove that $D' = D$.
The set D' is a closed subset of D, so it again constitutes a complete metric space. For each $n \in \mathbb{N}$ we have

$$\alpha_n : D_n \to D'$$

because of the following argument. Let $d \in D_n$, then:

$$f(\alpha_n(d)) =$$

$$\lim_{m\to\infty} \alpha_m \circ \beta_m(\alpha_n(d)) =$$

$$\lim_{m\to\infty} \alpha_{n+m} \circ \beta_{n+m} \circ (\alpha_n(d)) =$$

$$\lim_{m\to\infty} \alpha_{n+m} \circ \beta_{n+m} \circ \alpha_{n+m} \circ i_{n,\, n+m}(d) =$$

$$\lim_{m\to\infty} \alpha_{n+m} \circ i_{n,\, n+m}(d) =$$

$$\lim_{m\to\infty} \alpha_n(d) =$$

$$\alpha_n(d) .$$

So $f(\alpha_n(d)) = \alpha_n(d)$, and thus $\alpha_n(d) \in D'$.
Next we define, for each $n \in \mathbb{N}$:

$$\alpha_n' = \alpha_n ,$$

$$\beta_n' = \beta_n \upharpoonright D' \quad (\beta_n \text{ restricted to } D'),$$

$$\gamma_n' = <\alpha_n', \beta_n'> .$$

It is clear that $(D', (\gamma_n')_n)$ is another cone for $(D_n, \iota_n)_n$. Because $(D, (\gamma_n)_n)$ is initial, there exists a unique arrow $D \to^{\iota_1} D' \in \mathcal{C}$ with $\iota_1 = <i_1, j_1>$ such that

$$\forall n \in \mathbb{N} \, [\, \iota_1 \circ \gamma_n = \gamma_n' \,] .$$

The set D' can also be embedded into D: let $D' \to^{\iota_2} D$, with $\iota_2 = <i_2, j_2>$, be defined by

$$i_2 = id_{D'} ,$$

$$j_2 = i_1 .$$

Then $D' \rightarrow^{\iota_2} D \in \mathcal{C}$. For i_2 is isometric, j_2 is NDI and the following argument shows that $j_2 \circ i_2 = id_{D'}$. Let $d \in D'$. Then

$$j_2 \circ i_2(d) = j_2(d)$$

$$= i_1(d)$$

= [because $d \in D'$, we have $f(d) = d$;

in other words, $(\lim_{n\rightarrow\infty} \alpha_n \circ \beta_n)(d) = d$]

$$(i_1 \circ (\lim_{n\rightarrow\infty} \alpha_n \circ \beta_n))(d)$$

$$= \lim_{n\rightarrow\infty} (i_1 \circ \alpha_n \circ \beta_n)(d)$$

$$= \lim_{n\rightarrow\infty} (\alpha_n' \circ \beta_n)(d)$$

$$= \lim_{n\rightarrow\infty} (\alpha_n \circ \beta_n)(d) = d .$$

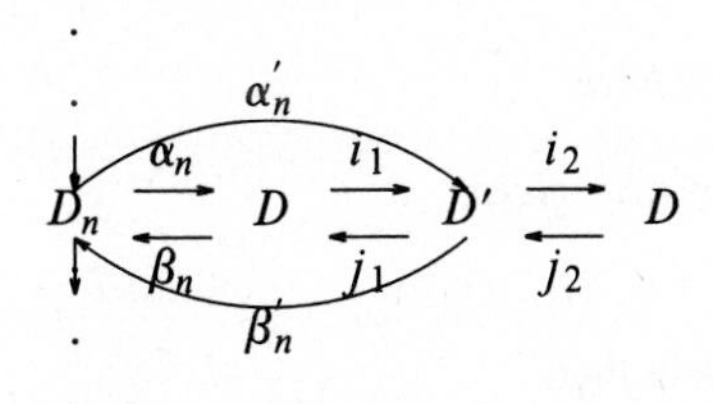

Now we are able to define $D \rightarrow^{\iota} D$ by

$$\iota = \iota_2 \circ \iota_1$$

$$= \langle i_2 \circ i_1, j_1 \circ j_2 \rangle .$$

It is easy to verify that

$$\forall n \in \mathbb{N} \, [\, \iota \circ \gamma_n = \gamma_n \,] .$$

By the initiality of D we have that

$$\iota = \langle id_D \, , \, id_D \rangle .$$

Thus $i_2 \circ i_1 = id_D$. This implies $D = D'$.
Conclusion:

$$\lim \alpha_n \circ \beta_n = id_D .$$

The initiality lemma will appear to be very useful in the sequel, where we shall construct a cone for an arbitrary converging tower and prove that it is initial.

DEFINITION 3.9 (Direct limit construction)
Let $(D_n, \iota_n)_n$, with $\iota_n = <i_n, j_n>$, be a converging tower. The *direct limit* of $(D_n, \iota_n)_n$ is a cone $(D, (\gamma_n)_n)$, with $\gamma_n = <\alpha_n, \beta_n>$, that is defined as follows:

$$D =^{def} \{(x_n)_n | \forall n \geqslant 0 [x_n \in D_n \wedge j_n(x_{n+1}) = x_n]\}$$

is equipped with a metric $d : D \times D \to [0,1]$ such that for all $(x_n)_n, (y_n)_n \in D$: $d((x_n)_n, (y_n)_n) = \sup\{d_{D_n}(x_n, y_n)\}$;
$\alpha_n : D_n \to D$ is defined by $\alpha_n(x) = (x_k)_k$, where

$$x_k = \begin{cases} j_{kn}(x) & \text{if } k < n \\ x & \text{if } k = n \\ i_{nk}(x) & \text{if } k > n; \end{cases}$$

$\beta_n : D \to D_n$ is defined by $\beta_n((x_k)_k) = x_n$.

LEMMA 3.10
Let (D,d) be as defined above. We have:

(D,d) is a complete metric space.

PROOF
Let $(x_n)_n$, $(y_n)_n \in D$. Let $m > n \geqslant 0$, then

$$\begin{aligned} d_{D_n}(x_n, y_n) &= d_{D_n}(j_{nm}(x_m), j_{nm}(y_m)) \\ &\leqslant [\text{ because } j_{nm} \text{ is NDI }] \\ &\quad d_{D_m}(x_m, y_m). \end{aligned}$$

Thus $(d_{D_n}(x_n, y_n))_n$ is an increasing sequence. It is bounded by 1, thus its supremum exists, and is equal to the limit. It is not difficult to show that d is a metric.
We shall prove the completeness of D with respect to this metric. Let $(\bar{x}^i)_i$, with $\bar{x}^i = (x_0^i, x_1^i, x_2^i, \ldots)$ be a Cauchy sequence in D. Because for all k and for all n and m:

$$\begin{aligned} d_{D_k}(x_k^n, x_k^m) &\leqslant \sup_{k \in \mathbb{N}} \{d_{D_k}(x_k^n, x_k^m)\} \\ &= d(\bar{x}^n, \bar{x}^m) \end{aligned}$$

and $(\bar{x}^i)_i$ is a Cauchy sequence, we have, for all $k \in \mathbb{N}$, that $(x_k^i)_i$ is a Cauchy sequence in D_k. For every k we set

$$x_k = \lim_{i \to \infty} x_k^i .$$

We have $j_k(x_{k+1})=x_k$, since

$$\begin{aligned} j_k(x_{k+1}) &= j_k(\lim_{i\to\infty} x_{k+1}^i) \\ &= \lim_{i\to\infty} j_k(x_{k+1}^i) \\ &= \lim_{i\to\infty} x_k^i \\ &= x_k . \end{aligned}$$

Thus $(x_k)_k$ is an element of D.
Because the convergence of the sequences $(x_k^i)_i$ for $k\in\mathbb{N}$ was uniform, we have

$$\forall\epsilon>0\ \exists N\in\mathbb{N}\ \forall k\in\mathbb{N}\ \forall n>N\ [\ d_{D_k}(x_k^n,\ x_k)<\epsilon\ ].$$

This fact implies that $(x_k)_k$ is the limit of $(\overline{x}^i)_i$, since, for $\epsilon>0$,

$$\begin{aligned} d((x_k)_k,\ \overline{x}^n) &= \sup_{k\in\mathbb{N}}\{d_{D_k}(x_k,\ x_k^n)\} \\ &\leqslant \epsilon \end{aligned}$$

for n bigger than a suitable N.

Relation between the direct limit construction and metric completion

We can look upon the construction of the direct limit for a tower $(D_n,i_n)_n$ as a generalization of taking the metric completion of the union of a sequence of metric spaces. We define

$$\begin{aligned} D_0' &= \{0\}\times D_0 \\ D_{n+1}' &= \{n+1\}\times(D_{n+1}\setminus i_n(D_n)) \cup D_n', \end{aligned}$$

and take $l_n:D_n\to D_n'$ as follows:

$$l_0(d) = <0,d> \quad \text{for } d\in D_0,$$

$$l_{n+1}(d) = \begin{cases} l_n(d') & \text{if } d=i_n(d')\in D_{n+1} \text{ with } d'\in D_n \\ <n+1,d> & \text{if } d\notin i_n(D_n). \end{cases}$$

Because each i_n is an injection, this construction works, and we see that each l_n is a bijection. Therefore, we can use $(l_n)_n$ in the obvious way to define a metric d_n' on each D_n' and suitable $i_n':D_n'\to D_{n+1}'$ and $j_n':D_{n+1}'\to D_n'$.
Now we have an isomorphic copy of our original tower, which satisfies the condition that each $i_n':D_n'\to D_{n+1}'$ is a subset embedding. From now on we leave out the primes, and just suppose that $i_n:D_n\to D_{n+1}$ satisfies this condition.
If we define U as the union of $(D_n)_n$, and $d:U\times U\to[0,1]$ by

$$d(x,y) = d_{D_k}(x,y),$$

whenever $x\in D_n, y\in D_m$ and $k\geqslant m,n$, we have that (U,d) is a metric space. Generally, it will not

be complete. The direct limit of $(D_n, i_n)_n$ can be regarded as the completion of (U,d) in the following sense.
In U we consider only such sequences $(x_n)_n$, for which:

$$\forall n \in \mathbb{N}[x_n \in D_n] \tag{1}$$

and

$$\forall n \in \mathbb{N}[x_n = j_n(x_{n+1})]. \tag{2}$$

It follows that $(x_n)_n$ is a Cauchy sequence. For $m > n$ we have

$$\begin{aligned} d(x_m, x_n) &= d_{D_m}(x_m, i_{nm}(x_n)) \\ &= d_{D_m}(x_m, i_{nm} \circ j_{nm}(x_m)) \\ &\leqslant d_{D_m \to D_m}(id_{D_m}, i_{nm} \circ j_{nm}) \\ &= \delta(\iota_{nm}). \end{aligned}$$

This number is small for large n and m, because $(D_n, i_n)_n$ is a converging tower.
For every $(x_n)_n$ and $(y_n)_n$ in U, that both satisfy (1) and (2), we have:

$$\text{if } \lim_{n \to \infty} d_{D_n}(x_n, y_n) = 0, \text{ then } (x_n)_n = (y_n)_n,$$

because of:

$$\begin{aligned} d_{D_n}(x_n, y_n) &= d_{D_n}(j_n(x_{n+1}), j_n(y_{n+1})) \\ &\leqslant d_{D_{n+1}}(x_{n+1}, y_{n+1}) \end{aligned}$$

(expressing that $(d_{D_n}(x_n, y_n))_n$ is a monotonic, non-decreasing sequence with limit 0, so all its elements are 0).
Of course it is not the case that every Cauchy sequence satisfies (1) and (2), but we can find in each class of Cauchy sequences that will have the same limit a representative sequence, which satisfies (1) and (2), and which by the above is unique. Let $(x_n)_n$ be an arbitrary Cauchy sequence in U. As a representative of the class of Cauchy sequences with the same limit as $(x_n)_n$, we take the sequence $(y_n)_n$, defined by

$$y_n = \lim_{m \to \infty} x_m^n,$$

with

$$x_m^n = \begin{cases} x_m & \text{if } x_m \in D_n \\ j_{nk}(x_m) & \text{if } x_m \notin D_n, \text{ and } k > n \text{ is the least number with } x_m \in D_k \end{cases}$$

(Remember that $k > n \Rightarrow D_k \supset D_n$). It is not very difficult to show, that we have indeed:

$$\lim_{n \to \infty} d_{D_n}(x_n, y_n) = 0,$$

and that $(y_n)_n$ satisfies (1) and (2). Finally we remark that the direct limit D of $(D_n, \iota_n)_n$ consists of exactly those sequences in U, that satisfy (1) and (2), and thus can be viewed as the metric completion of (U,d).

Remember from theorem 2.9 that the metric completion $\overline{M}$ of a metric space M is the smallest complete metric space, into which M can be isometrically embedded, in the following sense: $\overline{M}$ can be isometrically embedded into every other complete metric space with that property.
For the direct limit of a converging tower, we have a similar initiality property:

LEMMA 3.11
The direct limit of a converging tower (as defined in definition 3.9) is an initial cone for that tower.

PROOF
Let $(D_n, \iota_n)_n$ and $(D, (\gamma_n)_n)$ be as defined in definition 3.9. According to the initiality lemma (3.9), it suffices to prove

$$\lim_{n\to\infty} \alpha_n \circ \beta_n = id_D ,$$

which is equivalent to

$$\forall \epsilon>0\ \exists N\in\mathbb{N}\ \forall n>N\ [\ d(\alpha_n \circ \beta_n,\ id_D)<\epsilon\ ]$$

Let $\epsilon>0$. Because $(D_n, \iota_n)_n$ is a converging tower, we can choose $N\in\mathbb{N}$ such that

$$\forall m>n\geqslant N\ [\ d(i_{nm} \circ j_{nm},\ id_{D_m})<\epsilon\ ].$$

Let $n>N$. Let $(x_m)_m \in D$, we define

$$(y_m)_m = \alpha_n \circ \beta_n((x_m)_m).$$

For every $m>n$ we have

$$\begin{aligned} d_{D_m}(y_m, x_m) &= d_{D_m}(i_{nm}(x_n), x_m) \\ &= d_{D_m}(i_{nm} \circ j_{nm}\ (x_m), x_m) \\ &\leqslant d(i_{nm} \circ j_{nm}, id_{D_m}) \\ &< \epsilon. \end{aligned}$$

Therefore

$$d_D((y_m)_m, (x_m)_m) = \sup\{d_{D_m}(y_m, x_m)\} \leqslant \epsilon .$$

Because $(x_n)_n \in D$ was arbitrary, we have

$$d(\alpha_n \circ \beta_n, id_D)<\epsilon$$

for all $n>N$.

3.2 A fixed-point theorem

As a category-theoretic equivalent of a contracting function on a metric space, we have the following notion of a *contracting functor* on $\mathcal{C}$.

DEFINITION 3.12 (Contracting functor)

We call a functor $F:\mathcal{C}\rightarrow\mathcal{C}$ contracting whenever the following holds: there exists an ϵ, with $0\leqslant\epsilon<1$, such that for all $D\rightarrow^{\iota}E\in\mathcal{C}$ we have:

$$\delta(F\iota)\leqslant\epsilon\cdot\delta(\iota).$$

A contracting function on a complete metric space is continuous, so it preserves Cauchy sequences and their limits. Similarly, a contracting functor preserves converging towers and their initial cones:

LEMMA 3.13

Let $F:\mathcal{C}\rightarrow\mathcal{C}$ be a contracting functor, let $(D_n,\iota_n)_n$ be a converging tower with an initial cone $(D,(\gamma_n)_n)$. Then $(FD_n,F\iota_n)_n$ is again a converging tower with $(FD,(F\gamma_n)_n)$ as an initial cone.

The proof, which may use the initiality lemma, is left to the reader.

THEOREM 3.14 (Fixed-point theorem)

Let F be a contracting functor $F:\mathcal{C}\rightarrow\mathcal{C}$ and let $D_0\rightarrow^{\iota_0}FD_0\in\mathcal{C}$. Let the tower $(D_n,\iota_n)_n$ be defined by $D_{n+1}=FD_n$ and $\iota_{n+1}=F\iota_n$ for all $n\geqslant 0$. This tower is converging, so it has a direct limit $(D,(\gamma_n)_n)$. We have: $D\cong FD$.

PROOF

First we observe that $(D_n,\ \iota_n)_n$ can be proved to be a converging tower in the same way as in example 3.4. Because F preserves converging towers and their initial cones, $(FD_n,\ F\iota_n)_n$ is again a converging tower with $(FD,\ (F\gamma_n)_n)$ as an initial cone. We have that

$$(FD_n,\ F\iota_n)_n\ =\ (D_{n+1},\ \iota_{n+1})_n\ ,$$

so $(FD_n,F\iota_n)_n$ has the same direct limit (up to isometry) as $(D_n,\ \iota_n)_n$. This implies that $(D,\ (\gamma_n)_n)$ and $(FD,\ (F\gamma_n)_n)$ are both initial cones of $(D_{n+1},\ \iota_{n+1})_n$. It follows from the definition of an initial cone that D and FD are isometric.

$$\begin{array}{ccc} & D_{n+1}=FD_n & \\ \gamma_{n+1}\swarrow & & \searrow F\gamma_n \\ D & \dashrightarrow & FD \end{array}$$

REMARK
It is always possible to find an arrow $D_0 \to^{\iota_0} FD_0 \in \mathcal{C}$: Take $D_0 = \{p_0\}$; because FD_0 is non-empty we can choose an arbitrary $p_1 \in FD_0$, and put $\iota_0 = \langle i_0, j_0 \rangle$ with $i(p_0) = p_1$ and $j(x) = p_0$, for $x \in FD_0$.

4. Uniqueness of fixed points

We know that a contracting function $f: M \to M$, on a complete metric space M, has a *unique* fixed point. We would like to prove a similar property for contracting functors on $\mathcal{C}$.
Let us consider a contracting functor F on the category of complete metric spaces $\mathcal{C}$. By theorem 3.14 we know that F has a fixed point, that is there exists $D \in \mathcal{C}$ and an isometry κ such that

$$D \underset{\cong}{\overset{\kappa}{\to}} FD.$$

Suppose we have another fixed point D' with an isometry λ, such that

$$D' \underset{\cong}{\overset{\lambda}{\to}} FD'.$$

We know by the construction of D that it is the direct limit of the converging tower $(D_n, \iota_n)_n$, where $D_0 \to^{\iota_0} FD_0 \in \mathcal{C}$ is a given embedding and $D_{n+1} = FD_n$, $\iota_{n+1} = F\iota_n$.
If we have that D' is also (the endpoint of) a cone for that tower, the initiality of D implies that there exists an isometric embedding $D \to^{\iota} D' \in \mathcal{C}$. If we moreover can demonstrate that this ι is an isometry, then we can conclude that the functor F has a unique fixed point, which would be quite satisfactory.
A proof for ι being an isometry might look like:

$$\begin{aligned} \delta(\iota) &= (?)\, \delta(F\iota) \\ &\leqslant \epsilon \cdot \delta(\iota), \end{aligned}$$

implying (once the question-mark has been eliminated) that $\delta(\iota) = 0$, thus ι is an isometry.
It turns out that we can guarantee that the second fixed point D' is also a cone for the converging tower $(D_n, \iota_n)_n$ in one of *two* ways. Firstly, we can restrict our functor F to the *base-point*

category of complete metric spaces (to be defined in a moment). Secondly, we can require F to be contracting in yet another sense, to be called *hom*-contracting below.
We shall proceed in both directions, first exploring the unicity of fixed points of contracting functors on the base-point category, then focusing on functors on $\mathcal{C}$ that are contracting and hom-contracting.
In both cases it appears to be possible to prove the equality marked by (?) above. Unfortunately (for good mathematicians, who are said to be lazy), this takes some serious effort, to which the proof of the following theorem bears witness.
First we give the definition of the base-point category:

DEFINITION 4.1 (Base-point category of complete metric spaces)
Let $\mathcal{C}^*$ denote the base-point category of complete metric spaces, which has triples

$$<M,d,m>$$

for its objects. Here (M,d) is a complete metric space and m is an arbitrary element of M, called the base-point of M. The arrows in $\mathcal{C}^*$ are as in $\mathcal{C}$ (see definition 3.1), but for the constraint that they map base-points onto base-points, i.e. for $<M,d,m> \rightarrow^{<i,j>} <M',d',m'> \in \mathcal{C}^*$ we also require that $i(m)=m'$, and $j(m')=m$.

REMARK
The definitions of cone, functor etcetera can be adapted straightforwardly. Moreover, lemmas 3.8, 3.11, 3.13 and theorem 3.14 still hold.

THEOREM 4.2 (Uniqueness of fixed points)
Let F be a contracting functor $F:\mathcal{C}^* \rightarrow \mathcal{C}^*$. Then F has a unique fixed point up to isometry, that is to say: there exists a $D \in \mathcal{C}^*$ such that

(1) $FD \cong D$, and

(2) $\forall D' \in \mathcal{C}^* \; [FD' \cong D' \Rightarrow D \cong D']$.

PROOF
We define a converging tower $(D_n, \iota_n)_n$ by

$$D_0 = <\{p_0\}, d_{\{p_0\}}, p_0>,$$

$$D_{n+1} = FD_n \text{ for all } n \geqslant 0,$$

$$\iota_0 : D_0 \rightarrow D_1, \text{ trivial},$$

$$\iota_{n+1} = F\iota_n \text{ for all } n \geqslant 0.$$

Let $(D, (\gamma_n)_n)$ be the direct limit of this tower. As in theorem 3.14, we have that both $(D, (\gamma_n)_n)$

and $(FD, (F\gamma_n)_n)$ are initial cones of $(D_n, \iota_n)_n$. The initiality of $(D_n, (\gamma_n)_n)$ implies the existence of a unique arrow $D \to^{\kappa} FD$, such that for $n \geqslant 0$,

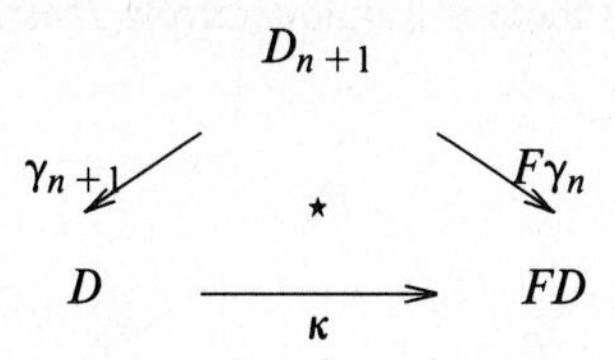

FIGURE 1

Because also $(FD, (F\gamma_n))_n$ is initial, we know that κ must be isometric.

Now let $D' \in \mathcal{C}^*$ be another fixed point of F, say $D' \underset{\cong}{\overset{\lambda}{\to}} FD'$ for an isometry λ. We define $(\tilde{\gamma}_n)_n$ such that $(D', (\tilde{\gamma}_n)_n)$ is a cone for $(D_n, \iota_n)_n$:

$\tilde{\gamma}_0 : D_0 \to D'$ is the unique arrow, which maps base-point to base-point,

$\tilde{\gamma}_{n+1} = \lambda^{-1} \circ F\tilde{\gamma}_n$.

We have that $(D', (\tilde{\gamma}_n)_n)$ is indeed a cone for $(D_n, \iota_n)_n$ because of the commutativity of the following diagram, for all $n \in \mathbb{N}$:

$$\begin{array}{ccc} D_n & \xrightarrow{\iota_n} & FD_n = D_{n+1} \\ \tilde{\gamma}_n \downarrow & \star & \downarrow F\tilde{\gamma}_n \\ D' & \xleftarrow[\lambda^{-1}]{} & FD' \end{array}$$

We prove it by induction on n :

(0) Because the arrows in $\mathcal{C}^*$ map base-points onto base-points, we have that $(\lambda^{-1} \circ F\tilde{\gamma}_0 \circ \iota_0)_1(p_0)$ and $(\tilde{\gamma}_0)_1(p_0)$ are both equal to the base-point of D', and for any $x \in D'$, that $(\lambda^{-1} \circ F\tilde{\gamma}_0 \circ \iota_0)_2(x) = (\tilde{\gamma}_0)_2(x) = p_0$.
Note that this is the *only* place, where we make use of the base-point structure of $\mathcal{C}^*$.

$(n+1)$ Suppose that we have $\lambda^{-1} \circ F\tilde{\gamma}_n \circ \iota_n = \tilde{\gamma}_n$. Then

$$\begin{aligned} \lambda^{-1} \circ F\tilde{\gamma}_{n+1} \circ \iota_{n+1} &= \lambda^{-1} \circ F(\tilde{\gamma}_{n+1} \circ \iota_n) \\ &= \lambda^{-1} \circ F(\lambda^{-1} \circ F\tilde{\gamma}_n \circ \iota_n) \end{aligned}$$

$$= \lambda^{-1} \circ F\tilde{\gamma}_n$$

$$= \tilde{\gamma}_{n+1} \, .$$

Again by the initiality of $(D, (\gamma_n)_n)$ there is a unique arrow $D \rightarrow^{\iota} D'$ such that, for all $n \in \mathbb{N}$:

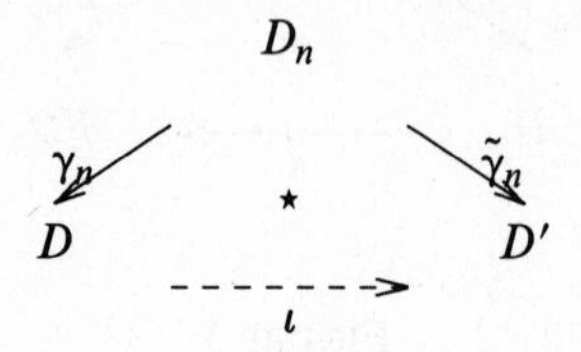

FIGURE 2

As indicated above, we now set out to prove that ι is an isometry. When we apply F to figure 2, we get

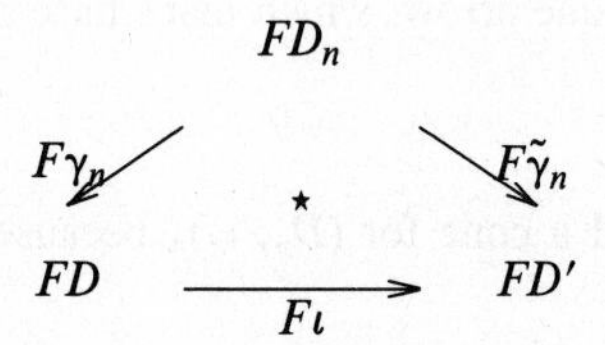

which leads to:

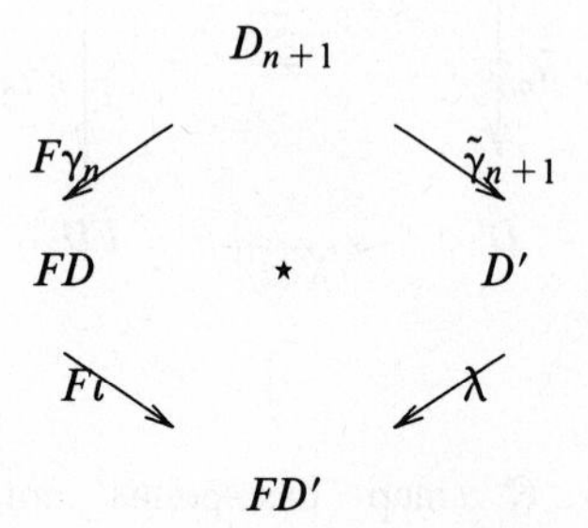

(because $\tilde{\gamma}_{n+1} = \lambda^{-1} \circ F\tilde{\gamma}_n$, so $F\tilde{\gamma}_n = \lambda \circ \gamma_{n+1}$), or, replacing λ by λ^{-1} and reversing the corresponding arrow:

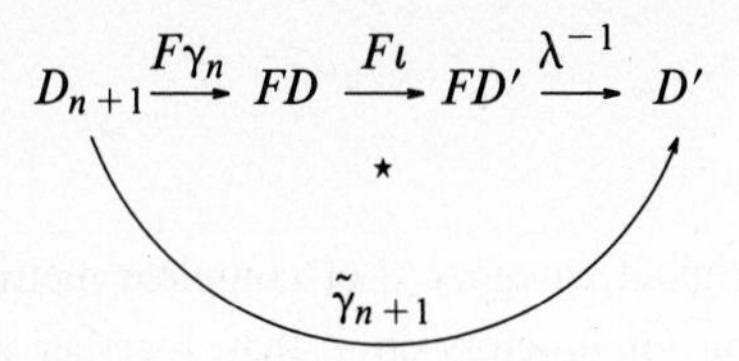

Substituting $\kappa \circ \gamma_{n+1}$ for $F\gamma_n$ (figure 1) yields:

$$D_{n+1} \xrightarrow{\gamma_{n+1}} D \xrightarrow{\kappa} FD \xrightarrow{F\iota} FD' \xrightarrow{\lambda^{-1}} D' \qquad \star \qquad \tilde{\gamma}_{n+1}$$

or: $(\lambda^{-1} \circ F\iota \circ \kappa) \circ \gamma_{n+1} = \tilde{\gamma}_{n+1}$ (this equality also holds for γ_0 and $\tilde{\gamma}_0$). But according to figure 2, ι is the only arrow with: $\forall n \in \mathbb{N}\ [\iota \circ \gamma_n = \tilde{\gamma}_n]$. Thus

$$\iota = \lambda^{-1} \circ F\iota \circ \kappa ,$$

or, in other words:

$$\begin{array}{ccc} D & \xrightarrow{\kappa} & FD \\ \iota \downarrow & \star & \downarrow F\iota \\ D' & \xrightarrow[\lambda]{} & FD' \end{array}$$

This commutativity, together with the fact that κ and λ are isometries implies:

$$\delta(\iota) = \delta(F\iota) .$$

(For the definition of δ see definition 3.2.)
Now the proof can be concluded, following the train of thought indicated above:

$$\begin{aligned} \delta(\iota) &= \delta(F\iota) \\ &\leqslant \epsilon \cdot \delta(\iota), \end{aligned}$$

for some $0 \leqslant \epsilon < 1$, since F is a contraction. This implies

$$\delta(\iota) = 0 ,$$

so (if $\iota = <i, j>$)

$$i \circ j = id_{D'} .$$

At last we can draw the desired conclusion:

$$D \underset{\cong}{\overset{\iota}{\to}} D'.$$

Now we return again to our original category $\mathcal{C}$ of complete metric spaces and provide for, as promised above, another criterion for functors on $\mathcal{C}$, that, together with contractivity, will appear to be sufficient to ensure uniqueness of their fixed points.

DEFINITION 4.3 (Hom-contractivity)
We call a functor $F:\mathcal{C}\to\mathcal{C}$ *hom-contracting,* whenever

$$\forall P \in \mathcal{C}\, \forall Q \in \mathcal{C}\, \exists \epsilon < 1\, [F_{P,Q}:(P\to^{\mathcal{C}}Q)\to^{\epsilon}(FP\to^{\mathcal{C}}FQ)]$$

where

$$P\to^{\mathcal{C}}Q = \{\iota \mid \iota:P\to Q \mid \iota \text{ is an arrow in } \mathcal{C}\},\;\; F_{P,Q}(\iota) = F\iota.$$

REMARKS
Because arrows in $\mathcal{C}$ are pairs, we have on $P\to^{\mathcal{C}}Q$ the standard metric for the Cartesian product. So let $\iota_1, \iota_2:P\to Q$, $\iota_1 = \langle i_1,j_1\rangle$ and $\iota_2 = \langle i_2,j_2\rangle$. Then their distance is defined by

$$d(\iota_1,\iota_2) = \max\{d_{P\to Q}(i_1,i_2), d_{Q\to P}(j_1,j_2)\}.$$

It is not the case that every hom-contracting functor is also contracting, which follows from the following example.
Let $A = \{0\}$ and $B = \{1,2\}$ be discrete metric spaces. We define a functor $F:\mathcal{C}\to\mathcal{C}$ as follows. For every complete metric space $P \in \mathcal{C}$ let

$$FP = \begin{cases} A & \text{if } P \text{ contains exactly 1 element} \\ B & \text{otherwise.} \end{cases}$$

For $\iota:P\to Q$ we define $F\iota$:

$$F\iota = \begin{cases} 1_A & \text{if } FP = FQ = A \\ 1_B & \text{if } FP = FQ = B \\ \iota_0 & \text{if } FP = A \text{ and } FQ = B, \end{cases}$$

where $\iota_0 = \langle i_0,j_0\rangle$, with $i_0:0\mapsto 1$, $j_0:1,2\mapsto 0$. Note that there is no $\iota:P\to Q$ if $FP = B$ and $FQ = A$. It is not difficult to verify that F is a functor, which is hom-contracting. The following argument shows that it is *not* contracting. Let $C = \{3,4\}$ with $d(3,4) = \frac{1}{2}$, and let $\kappa:A\to C$, with $\kappa = \langle k,l\rangle$ be defined by $k:0\mapsto 3$ and $l:3,4\mapsto 0$. Then we have $\delta(\kappa) = \frac{1}{2}$, but $F\kappa:FA\to FC$ is $\iota_0:A\to B$ (as defined above), for which $\delta(\iota_0) = 1$.

THEOREM 4.4
Let F be a contracting and *hom-contracting functor* $F:\mathcal{C}\rightarrow\mathcal{C}$. *Then F has a unique fixed point up to isometry, that is to say: there exists a* $D\in\mathcal{C}$ *such that*

(1) $FD\cong D$ and

(2) $\forall D'\in\mathcal{C}^{*}\ [FD'\cong D' \Rightarrow D\cong D']$.

PROOF
The proof of this theorem differs from that of theorem 4.2 only in the definition of $\tilde{\gamma}_0$. There we could take for $\tilde{\gamma}_0$ the trivial embedding of D_0 into D', mapping p_0 onto the base-point of D'. Here we have no base-points. But we can use the fact that F is hom-contracting by taking for $\tilde{\gamma}_0$ the unique fixed point of the function $G:(D_0\rightarrow^{\mathcal{C}}D')\rightarrow(D_0\rightarrow^{\mathcal{C}}D')$, that we define by: $G(\tilde{\gamma})=\lambda^{-1}\circ F\tilde{\gamma}\circ\iota_0$, for $\tilde{\gamma}\in(D_0\rightarrow^{\mathcal{C}}D')$. (Note that G is contracting because F is hom-contracting.) It follows that $\tilde{\gamma}_0$, thus defined, satisfies $\lambda^{-1}\circ F\tilde{\gamma}_0\circ\iota_0=\tilde{\gamma}_0$, which serves our purposes.

5. A CLASS OF DOMAIN EQUATIONS WITH UNIQUE SOLUTIONS

In this section we present a class of domain equations over the category $\mathcal{C}$ that have unique solutions. For this purpose we first define a set *Func* of functors on $\mathcal{C}$ and formulate a condition for its elements that implies contractivity and hom-contractivity. It then follows that every domain equation over $\mathcal{C}$ induced by a functor that satisfies this condition, has a unique solution.

DEFINITION 5.1 (Functors)
The set *Func*, with typical elements F, is defined by:

$$F ::= F_M \mid id^{\epsilon} \mid F_1\rightarrow F_2 \mid F_1\rightarrow^1 F_2 \mid F_1\overline{\cup}F_2 \mid F_1\times F_2 \mid \mathcal{P}_{cl}(F) \mid F_1\circ F_2$$

where M is an arbitrary complete metric space and $\epsilon>0$. Every $F\in Func$ is to be interpreted as a functor

$$F:\mathcal{C}\rightarrow\mathcal{C}$$

as follows. Let (P,d_P), $(Q,d_Q)\in\mathcal{C}$ be complete metric spaces. Let $P\rightarrow^{\iota}Q\in\mathcal{C}$, with $\iota=\langle i,j\rangle$. For the definition of each $F\in Func$ we have to specify:

(1) the image of P under F: FP,

(2) the image of d under F: Fd,

(3) the image of ι under F: $F\iota(=\langle Fi,Fj\rangle)$.

(a) $F = F_M$:

(1) $FP = M$,

(2) $Fd = d_M$ (the metric of M),

(3) $F\iota = <id_M, id_M>$.

We sometimes use just a set A instead of a metric space M. In this case we provide A with the discrete metric (definition 2.1).
(b) $F = id^{\epsilon}$:

(1) $FP = P$,

(2) $Fd = \epsilon \cdot d$ $(Fd(x,y) = \epsilon \cdot d(x,y)$, for $x, y \in P)$,

(3) $F\iota = \iota$.

Next we define functors that are composed. Let $F_1, F_2 \in Func$, such that

(1) $F_1P = P_1$, $F_2P = P_2$, $F_1Q = Q_1$, $F_2Q = Q_2$,

(2) $F_1d = d_1$, $F_2d = d_2$,

(3) $F_1\iota = <i_1, j_1>$, $F_2\iota = <i_2, j_2>$.

(c) $F = F_1 \rightarrow F_2$:

(1) $FP = P_1 \rightarrow P_2$,

(2) $Fd = d_F$ (see definition 2.6(a)),

(3) $F\iota = <\lambda f \cdot (i_2 \circ f \circ j_1), \lambda g \cdot (j_2 \circ g \circ i_1)>$.

($F = F_1 \rightarrow^1 F_2$ is defined similarly.)
(d) $F = F_1 \overline{\cup} F_2$:

(1) $FP = P_1 \overline{\cup} P_2$,

(2) $Fd = d_U$ (see definition 2.6(b)),

(3) $F\iota = <\lambda p \cdot$ if $p \in \{0\} \times P_1$ then $i_1((p)_2)$ else $i_2((p)_2)$ fi,

$\lambda q \cdot$ if $q \in \{0\} \times Q_1$ then $j_1((q)_2)$ else $j_2((q)_2)$ fi$>$.

(e) $F = F_1 \times F_2$:

(1) $FP = P_1 \times P_2$,

(2) $Fd = d_P$ (see definition 2.6(c)),

(3) $F\iota = <\lambda <p_1, p_2> \cdot <i_1(p_1), i_2(p_2)>, \lambda <q_1, q_2> \cdot <j_1(q_1), j_2(q_2)>>$.

(f) $F = \mathcal{P}_{cl}(F_1)$:

(1) $FP = \mathcal{P}_{cl}(P_1)$,

(2) $Fd = d_H$ (see definition 2.6(d)),

(3) $F\iota = <\lambda X \cdot \{i_1(x) \mid x \in X\}, \lambda Y \cdot closure\ \{j_1(y) \mid y \in Y\}>$.

(g) $F = F_1 \circ F_2$: the usual composition of functors on $\mathcal{C}$.

REMARK

The set *Func* contains elements of various form. We give an example. Let F_1, $F_2 \in Func$. The following functor is an element of the set *Func*, as can be deduced from its definition.

$$F_1 \rightarrow^A F_2 =^{def} id^A \circ (F_1 \rightarrow^1 (id^{\frac{1}{A}} \circ F_2)), \text{ for } A > 0.$$

LEMMA 5.2

For all $F \in Func$ we have: F is a well defined functor on $\mathcal{C}$.

PROOF

We treat only one case by way of example, being (lazy and) confident that it shows the reader how to proceed in the other cases.

Let $F = F_1 \rightarrow^1 F_2$, and suppose F_1 and F_2 are well defined. Let $(P,d_P),(Q,d_Q)$ and $P \rightarrow^{\iota} Q \in \mathcal{C}$, with $\iota = <i,j>$; furthermore, let for $k = 1,2$:

$$F_k P = P_k,\ \ F_k Q = Q_k,$$

$$F_k d_P = d_{P_k},\ \ F_k d_Q = d_{Q_k},$$

$$F_k \iota = <i_k, j_k>.$$

The functor F is defined by

(1) $FP = P_1 \rightarrow^1 P_2$,

(2) $Fd_P = d_F$,

(3) $Fi = <Fi,Fj> = <\lambda f \cdot (i_2 \circ f \circ j_1), \lambda g \cdot (j_2 \circ g \circ i_1)>$.

$$\begin{array}{ccccc} P & & & & P_1 \rightarrow^1 P_2 \\ i \downarrow \uparrow j & & \xrightarrow[F_1 \rightarrow^1 F_2]{F=} & \lambda f \cdot (i_2 \circ f \circ j_1) = Fi \downarrow & \uparrow Fj = \lambda g \cdot (j_2 \circ g \circ i_1) \\ Q & & & & Q_1 \rightarrow^1 Q_2 \end{array}$$

It follows from proposition 2.7, that $(P_1 \rightarrow^1 P_2, d_F)$ is a complete metric space, which leaves us to prove:

(a) Fi is isometric,

(b) Fj is NDI and

(c) $Fj \circ Fi = id_{FP}$.

Part (a): Let $f_1, f_2 \in P_1 \to^1 P_2$. We want to show

$$d_{FP}(f_1, f_2) = d_{FQ}(Fi(f_1), Fi(f_2)).$$

We have

$$\begin{aligned} \sup_{q \in Q_1} \{d_{Q_2}(i_2 \circ f_1 \circ j_1(q),\ i_2 \circ f_2 \circ j_1(q))\} &= \text{[because } i_2 \text{ is isometric]} \\ &\quad \sup_{q \in Q_1} \{d_{P_2}(f_1 \circ j_1(q), f_2 \circ j_1(q))\} \\ &= \text{[because } j_1 \text{ is surjective]} \\ &\quad \sup_{p \in P_1} \{d_{P_2}(f_1(p),\ f_2(p))\} \\ &= d_{P_1 \to P_2}(f_1, f_2). \end{aligned}$$

Part (b): Let $g_1, g_2 \in Q_1 \to^1 Q_2$. We want to show:

$$d_{FP}(Fj(g_1), Fj(g_2)) \leqslant d_{FQ}(g_1, g_2).$$

Let $p \in P_1$; we have:

$$\begin{aligned} d_{P_2}(Fj(g_1)(p), Fj(g_2)(p)) &= d_{P_2}(j_2 \circ g_1 \circ i_1(p), j_2 \circ g_2 \circ i_1(p)) \\ &\leqslant [j_2 \text{ is NDI}] \\ &\quad d_{Q_2}(g_1 \circ i_1(p), g_2 \circ i_1(p)) \\ &\leqslant d_{FQ}(g_1, g_2). \end{aligned}$$

Part (c): Let $f \in P_1 \to^1 P_2$. We have

$$\begin{aligned} Fj \circ Fi(f) &= j_2 \circ i_2 \circ f \circ j_1 \circ i_1 \\ &= f. \end{aligned}$$

DEFINITION 5.3 (Contraction coefficient)
For each $F \in Func$ we define its so-called *contraction* coefficient (notation: $c(F)$, with $c(F) \in [0, \infty]$), using induction on the complexity of the structure of F.

(*a*) If $F = F_M$, then $c(F) = 0$.

(*b*) If $F = id^{\epsilon}$, then $c(F) = \epsilon$.

Let F_1, $F_2 \in Func$, with coefficients $c(F_1)$ and $c(F_2)$. Then we set:

(*c*) If $F = F_1 \rightarrow F_2$, then $c(F) = \max\{\infty \cdot c(F_1),\ c(F_2)\}$.

(*d*) If $F = F_1 \rightarrow^1 F_2$, then $c(F) = c(F_1) + c(F_2)$.

(If we would restrict ourselves to ultra-metric spaces, we could write $\max\{c(F_1), c(F_2)\}$ here.)

(*e*) If $F = F_1 \overline{\cup} F_2$, then $c(F) = \max\{c(F_1), c(F_2)\}$.

(*f*) If $F = F_1 \times F_2$, then $c(F) = \max\{c(F_1), c(F_2)\}$.

(*g*) If $F = \mathcal{P}_{cl}(F_1)$, then $c(F) = c(F_1)$.

(*h*) If $F = F_1 \circ F_2$, then $c(F) = c(F_1) \cdot c(F_2)$.

(With ∞ we compute as follows: $\infty \cdot 0 = 0 \cdot \infty = 0$, $\infty \cdot c = c \cdot \infty = \infty$, if $c > 0$.)

THEOREM 5.4

For every functor $F \in Func$ we have

$$(1)\ \forall P \rightarrow^{\iota} Q \in \mathcal{C}\, [\delta(F\iota) \leqslant c(F) \cdot \delta(\iota)],$$

$$(2)\ \forall P, Q \in \mathcal{C}\, [F_{P,Q} : (P \rightarrow^{\mathcal{C}} Q) \rightarrow^{c(F)} (FP \rightarrow^{\mathcal{C}} FQ)].$$

PROOF

Let $P, Q \in \mathcal{C}$, $\iota, \iota' \in P \rightarrow^{\mathcal{C}} Q$, with $\iota = \langle i, j \rangle, \iota' = \langle i', j' \rangle$.

Case (a) $F = F_M$:

Part (a1)

$$\begin{aligned} \delta(F\iota) &= d_{FQ \rightarrow FQ}(Fi \circ Fj, id_M) \\ &= d_{FQ \rightarrow FQ}(id_M \circ id_M, id_M) \\ &= 0 = c(F) \cdot \delta(\iota). \end{aligned}$$

part (a2)

$$d_{FP \rightarrow^{\mathcal{C}} FQ}(F\iota, F\iota') = d_{M \rightarrow^{\mathcal{C}} M}(id, id) = 0 = c(F) \cdot d_{P \rightarrow^{\mathcal{C}} Q}(\iota, \iota').$$

Case (b) $F = id^{\epsilon}$:

part (b1)

$$\begin{aligned} \delta(F\iota) &= d_{FQ \rightarrow FQ}(Fi \circ Fj, id_{FQ}) \\ &= \sup_{q \in Q}\{d_{FQ}(i \circ j(q), q)\} \\ &= \sup_{q \in Q}\{\epsilon \cdot d_Q(i \circ j(q), q)\} \\ &= \epsilon \cdot \delta(\iota) \end{aligned}$$

$$= c(F)\cdot\delta(\iota).$$

Part (b2)

$$d_{FP\to^{\mathcal{C}}FQ}(F\iota,\iota') = \epsilon\cdot d_{P\to^{\mathcal{C}}Q}(\iota,\iota')$$
$$= c(F)\cdot d_{P\to^{\mathcal{C}}Q}(\iota,\iota').$$

Now let $F_1,F_2\in$ Func and suppose the theorem holds for these functors. For $k = 1,2$ we use the following notation:

$$F_k\iota = \iota_k\ ,\ F_k\iota' = \iota_k'\ ,\ F_kP = P_k\ ,\ F_kQ = Q_k\ ,$$
$$F_ki = i_k\ ,\ F_ki' = i_k'\ ,$$
$$F_kj = j_k\ ,\ F_kj' = j_k'\ ,$$

We only treat the cases that $F = F_1\to^1F_2$ and $F = F_1\times F_2$.
Case (d) $F = F_1\to^1F_2$:
Part (d1)

$$\delta(F\iota) = d_{FQ\to FQ}(Fi\circ Fj, id_{FQ})$$
$$= \sup_{g\in FQ}\{d_{FQ}(i_2\circ j_2\circ g\circ i_1\circ j_1,g)\}.$$

Let $g\in FQ = Q_1\to^1Q_2$. For $q_1\in Q_1$ we have

$$d_{Q_2}(i_2\circ j_2\circ g\circ i_1\circ j_1(q_1),g(q_1)) \leqslant d_{Q_2}(i_2\circ j_2\circ g\circ i_1\circ j_1(q_1),g\circ i_1\circ j_1(q_1))+$$
$$d_{Q_2}(g\circ i_1\circ j_1(q_1),g(q_1)).$$

(This "+" could be replaced by "max" in the case of ultra-metric spaces.)
For the first term we have

$$d_{Q_2}(i_2\circ j_2\circ g\circ i_1\circ j_1(q_1),g\circ i_1\circ j_1(q_1)) \leqslant \sup_{q\in Q_2}\{d_{Q_2}(i_2\circ j_2(q_2),q_2)\}$$
$$= \delta(F_2\iota).$$

For the second

$$d_{Q_2}(g\circ i_1\circ j_1(q_1),g(q_1)) \leqslant \text{[because } g\in Q_1\to^1Q_2\text{]}$$
$$d_{Q_1}(i_1\circ j_1(q_1),q_1)$$
$$= \delta(F_1\iota).$$

We see

$$\delta(F\iota) \leqslant \delta(F_1\iota)+\delta(F_2\iota)$$
$$\leqslant \text{[induction]}$$
$$(c(F_1)+c(F_2))\cdot\delta(\iota)$$

$$= c(F)\cdot\delta(\iota).$$

Part (d2)

$$d_{FP\to^{e}FQ}(F\iota,F\iota') = \max\{d_{FP\to FQ}(Fi,Fi'),d_{FQ\to FP}(Fj,Fj')\}.$$

For the first component, we have

$$d_{FP\to FQ}(Fi,Fi') = \sup_{f\in FP,q\in Q_1}\{d_{Q_2}(Fi(f)(q),Fi'(f)(q))\}.$$

Let $f\in FP,q\in Q_1$. Then

$$\begin{aligned} d_{Q_2}(Fi(f)(q),Fi'(f)(q)) &= d_{Q_2}(i_2\circ f\circ j_1(q),i_2'\circ f\circ j_1'(q)) \\ &\leqslant d_{Q_2}(i_2\circ f\circ j_1(q),i_2'\circ f\circ j_1(q)) + d_{Q_2}(i_2'\circ f\circ j_1(q),i_2'\circ f\circ j_1'(q)) \\ &\leqslant d_{P_2\to Q_2}(i_2,i_2') + d_{Q_2}(i_2'\circ f\circ j_1(q),i_2'\circ f\circ j_1'(q)) \\ &\leqslant [\text{because } i_2' \text{ is isometric },f\in P_1\to^1 P_2] \\ &\quad d_{P_2\to Q_2}(i_2,i_2') + d_{Q_1\to P_1}(j_1,j_1'). \end{aligned}$$

(Again, in the case of ultra-metric spaces, we would have "max" here.)
Likewise, we have for the second component

$$d_{FQ\to FQ}(Fj,Fj') \leqslant d_{P_1\to Q_1}(i_1,i_1') + d_{Q_2\to P_2}(j_2,j_2').$$

Together this implies

$$\begin{aligned} d_{FP\to^{e}FQ}(F\iota,F\iota') &\leqslant d_{P_1\to^{e}Q_1}(F_1\iota,F_1\iota') + d_{P_2\to^{e}Q_2}(F_2\iota,F_2\iota') \\ &\leqslant [\text{induction}] \\ &\quad (c(F_1)+c(F_2))\cdot d_{P\to^{e}Q}(\iota,\iota') \\ &= c(F)\cdot d_{P\to^{e}Q}(\iota,\iota'). \end{aligned}$$

Case (f) $F = F_1\times F_2$:
Part (f1)

$$\begin{aligned} \delta(F\iota) &= d_{FQ\to FQ}(Fi\circ Fj,id_{FQ}) \\ &= \sup_{\overline{q}\in FQ}\{d_{FQ}(Fi\circ Fj(\overline{q}),\overline{q})\} \\ &= \sup_{<q_1,q_2>\in FQ}\{d_{FQ}(<i_1\circ j_1(q_1),i_2\circ j_2(q_2)>,<q_1,q_2>)\} \\ &= \sup_{<q_1,q_2>\in FQ}\{\max\{d_{Q_1}(i_1\circ j_1(q_1),q_1),d_{Q_2}(i_2\circ j_2(q_2),q_2)\}\} \\ &= \max\{\sup_{q_1\in Q_1}\{d_{Q_1}(i_1\circ j_1(q_1),q_1)\},\sup_{q_2\in Q_2}\{d_{Q_2}(i_2\circ j_2(q_2),q_2)\}\} \\ &= \max\{\delta(F_1\iota),\delta(F_2\iota)\} \end{aligned}$$

$$\leqslant \text{[induction]}$$

$$(c(F_1)+c(F_2))\cdot\delta(\iota)$$

$$= c(F)\cdot\delta(\iota).$$

Part (f2)

$$\begin{aligned} d_{FP\to^{\mathcal{C}}FQ}(Fi,Fi') &= \sup_{\overline{p}\in FP}\{d_{FQ}(Fi(\overline{p}),Fi'(\overline{p}))\} \\ &= \sup_{<p_1,p_2>\in FP}\{d_{FQ}(<i_1(p_1),i_2(p_1)>,<i_1'(p_2),i_2'(p_2)>)\} \\ &= \max\{\sup_{p_1\in P_1}\{d_{Q_1}(i_1(p_1),i_1'(p_1))\},\sup_{p_2\in P_2}\{d_{Q_2}(i_2(p_2),i_2'(p_2))\}\} \\ &= \max\{d_{P_1\to Q_1}(i_1,i_1'),d_{P_2\to Q_2}(i_2,i_2')\}. \end{aligned}$$

Similarly, we have

$$d_{FQ\to FP}(Fj,Fj') = \max\{d_{Q_1\to P_1}(j_1,j_1'),d_{Q_2\to P_2}(j_2,j_2')\}.$$

Thus we obtain

$$\begin{aligned} d_{FP\to^{\mathcal{C}}FQ}(F\iota,F\iota') &= \max\{d_{P_1\to^{\mathcal{C}}Q_1}(F_1\iota,F_1\iota'),d_{Q_2\to^{\mathcal{C}}P_2}(F_2\iota,F_2\iota')\} \\ &\leqslant \text{[induction]} \\ &\quad \max\{c(F_1),c(F_2)\}\cdot d_{P\to^{\mathcal{C}}Q}(\iota,\iota') \\ &= c(F)\cdot d_{P\to^{\mathcal{C}}Q}(\iota,\iota')\cdot \end{aligned}$$

COROLLARY 5.5
For every $F\in Func$, with $0\leqslant c(F)<1$, we have

(1) *F is a contracting functor, and*

(2) *F is a hom-contracting functor.*

COROLLARY 5.6
Every reflexive domain equation over $\mathcal{C}$ of the form

$$P\cong FP,$$

for which $F\in Func$ and $c(F)<1$, has a unique solution (up to isomorphism).

6. Conclusions

We have presented a technique for constructing fixed points of certain functors over a category of complete metric spaces. This enables us to solve the reflexive domain equations associated with these functors. The technique is an adaptation of the limit construction that was first used in the context of certain partial orders (continuous lattices, complete lattices, complete partial orders). Nevertheless, we have encountered some nice metric phenomena in our metric framework. To begin with, the concept of a converging tower is an analogue to the concept of a Cauchy sequence in a complete metric space, and indeed, both have a limit. Furthermore, a contracting functor on our category of metric spaces is a concept analogous to that of a contracting function on a complete metric space, and both are guaranteed to have a fixed point. If we strengthen our requirements on the functor to include hom-contractivity (also analogous to contractivity of a function), we even know that the fixed point is unique (as is the case with a contracting function). Therefore the whole situation looks very much like Banach's theorem in a category-theoretic disguise.

A few questions remain open, however. We are still looking for a functor that is contracting but not hom-contracting, or even better for a functor that is contracting but has several non-isomorphic fixed points. Another point is what can be said about functors where the argument occurs at the left hand side of a general function space construction (*all* continuous functions, not just the NDI ones).

In any case, the class of functors (and, thus, domain equations) that we can handle is large enough, so that our technique is a useful tool in the construction of domains for the denotational semantics of concurrent programming languages.

Related Work

The subject of solving reflexive domain equations is not new. Various solutions of the kind of equations mentioned above already exist. We shall not try to give an extensive and complete bibliography on this matter and confine ourselves to the following remarks.
We mention the work of Scott ([Sc]), who uses inverse limit constructions for solving domain equations. Our method of generalizing metric notions in terms of category-theoretical notions shows a clear analogy to the work D. Lehmann ([Le]) did in the context of partial orderings. Our work is also related to the general method of solving reflexive equations of Smyth and Plotkin ([SP]). In the terminology used there, we show that our category $\mathcal{C}$ is ω-complete in the limited sense, that all converging towers have direct limits. Further we show that a certain type of ω-continuous functors (called contracting) has a fixed point. (Without having investigated the precise relationship, we also mention here the anology between their notion of an O-category, and the fact that in our category $\mathcal{C}$ the hom-sets are complete metric spaces.)

7. References

[ABKR] P. America, J. de Bakker, J. Kok, J. Rutten, *A Denotational Semantics of a Parallel Object-Oriented Language,* Technical Report (CS-R8626), Centre for Mathematics and Computer Science, Amsterdam, 1986.

[BZ] J.W. de Bakker, J.I. Zucker, *Processes and the Denotational Semantics of Concurrency,* Information and Control 54 (1982), pp. 70-120.

[Du] J. Dugundji, *Topology,* Allen and Bacon, Rockleigh, N.J., 1966.

[En] R. Engelking, *General Topology,* Polish Scientific Publishers, 1977.

[Ha] H. Hahn, *Reelle Funktionen,* Chelsea, New York, 1948.

[Le] D. Lehmann, *Categories for Mathematical Semantics,* in: Proc. 17th IEEE Symposium on Foundations of Computer Science, 1976.

[ML] S. Mac Lane, *Categories for the Working Mathematician,* Graduate Texts in Mathematics 5, Springer-Verlag, 1971.

[Sc] D.S. Scott, *Continuous Lattices,* in: Toposes, Algebraic Geometry and Logic, Lecture Notes in Mathematics 274, Springer-Verlag, 1972, pp. 97-136.

[SP] M.B. Smyth, G.D. Plotkin, *The Category-Theoretic Solution of Recursive Domain Equations,* SIAM J. Comput, Vol. 11, No. 4, 1982, pp.761-783.

Topological Completeness in an Ideal Model for Polymorphic Types†

Ernst - Erich Doberkat

Department of Computer Science
University of Hildesheim
D - 3200 Hildesheim
West Germany

ABSTRACT

We have a look at the topological structure underlying the ideal model of recursive polymorphic types proposed by MacQueen, Plotkin and Sethi. We show that their central argument in establishing a well defined semantical function, viz., completeness with respect to a metric obtained from the construction of their domain, is a special case of complete uniformities which arise in a natural way from the study of closeness of ideals on domains. These uniformities are constructed and studied, and a general fixed - point theorem is derived for maps defined on these ideals.

1. Introduction

When studying recursive polymorphic types, MacQueen, Plotkin and Sethi ([6]) had to develop an alternative to type - finding by unification, which is the usual device for identifying the type of an object in a polymorphic setting. Unification, however, leads to various complications when applied to recursive types, since the type finder may go into an infinite loop, when faced with types described directly or indirectly in terms of themselves.

In [6], the alternative was to use the Banach fixed point theorem a tool which is long established in Numerical Mathematics to make sure that numerical iterations converge: Under proper conditions, the machinery associated with this theorem works as follows: if the convergence of an iteration has to be established, this iteration is formulated in terms of a function mapping the search space into itself in such a way that the wanted value appears as a fixed point of the mapping. If the map can be shown to be a contraction, then convergence of the iterates to the fixed point follow's from Banach's theorem. The underlying mathematical structure for the Banach Theorem is a complete metric space, and the map to be considered must be a contraction, i.e. the distance between the image of two points must be by a proportionality factor smaller than the distance of these points. MacQueen, Plotkin and Sethi establish such a complete metric on the space of all σ - ideals, which are used in that paper as a type model. This metric is introduced using an arbitrary witness function, and a closer look at the situation and at the completeness proof suggests that this scenario may indeed be a special case of a far more general one, since the nature as well as properties of the

† This work has been partially supported by the ESPRIT project SED, project number 1227(1271), funded in part by the European Community

witness function are rather negligible - the only thing which is interesting here is that the witness function exists.

In this note we demonstrate that the case considered in [6] is a special case from a topological point of view. We consider the usual generalization of metric spaces - uniform spaces - and show that the method to define a metric space may be applied to a more general setting yielding complete uniform spaces. We characterize those uniformities on the set of all $\sigma-$ ideals of a domain which are generated by a rank function. Completeness was only a tool for establishing fixed points, so we discuss fixed points in greater detail, establishing from Landes' Fixed Point Theorem ([5]) for uniform spaces a characterization of those maps for which a unique fixed point exists. This is done first in the general uniform setting over the space of all ideals, and then a special case including the metric situation is derived from this.

Since fixed point constructions like those described in [6] occur in mathematical semantics rather frequently (see e.g. [1]), it is expected that the results described here may ease the application of fixed point theorems without having to struggle through arguments from metric or uniform spaces.

Acknowledgements: Benno Fuchssteiner, Thomas Landes and Marcel Ernè provided some insights into fixed points.

2. Domains

In this section, we collect some notations for later use and easier reference. The standard reference for topological and uniform spaces used here is [4], and for domains, [7] is used.

A complete partial order (abbreviated *cpo*) $(D,\leq)$ is a partial order $\leq$ on a set D which has a smallest element $\perp$ such that each increasing sequence in D has its supremum in D. An element $x \in D$ is ***compact*** iff for each increasing sequence $(y_n)_{n\in\mathbb{N}} \subset D$ with

$$x \leq \sup_{n\in\mathbb{N}} y_n$$

there exists an index k such that

$$x \leq y_k$$

holds. D is called ***algebraic*** iff there are countably many compact elements, and if for each $x \in D$ the set

$$L_x := \{a \in D; a \text{ is compact with } a \leq x\}$$

is directed with

$$x = \sup L_x$$

(a set $L \subset D$ is ***directed*** iff $\forall x,y \in L \exists z \in L: x \leq z, y \leq z$). Denote by A° all compact elements for $A \subset D$. D is called a ***domain*** iff D is an ω - algebraic cpo such that each subset bounded from above has a supremum. Fix a domain D for the rest of this paper.

A nonempty set $i \subset D$ is said to be an ***ideal*** iff $x \leq y \in i$ implies $x \in i$ for each $x \in D, y \in i$; it is called ***σ–ideal*** iff it is an ideal which contains the supremum of each of its increasing sequences. Denote by $\Upsilon(D)$, and $\Sigma(D)$ the set of all ideals, and $\sigma-$ ideals, resp., and put $\Sigma^*(D) := \Sigma(D) \cup \{\phi\}$. It is shown in [6], Proposition 1, that the map $i \to i^\circ$ maps $\Sigma(D)$ one - to - one onto $\Upsilon(D)$, with inverse map

$$\Upsilon(D) \ni i \to \{\sup_{n\in\mathbb{N}} x_n; (x_n)_{n\in\mathbb{N}} \subset i \text{ is increasing}\}$$

A subset Θ of the power set $\Pi(S)$ of a set S is a ***filterbase*** iff $\Theta \neq \phi$, $\phi \notin \Theta$, and if for every $A_1, A_2 \in \Theta$ there exists $A_3 \in \Theta$ such that $A_3 \subset A_1 \cap A_2$ holds; Θ is a ***filter*** on S iff it is a filter base with the additional property that it is a dual ideal, i. e. that $A \in \Theta, A \subset B$ implies $B \in \Theta$. A filter Γ on $S \times S$ is said to be a ***uniformity*** iff given $G \in \Gamma$, the inverse set

$$G^{-1} := \{(y,x);(x,y) \in G\}$$

is a member of Γ, if the diagonal

$$\Delta := \{(x,x);x \in S\}$$

is a member of Γ, and if for any $G \in \Gamma$ there exists $H \in \Gamma$ with $H \circ H \subset G$. Here the composition of two entourages $G, H \subset S \times S$ is defined as usual by

$$G \circ H := \{(x,z);\exists y:(x,y) \in G \text{and} (y,z) \in H\}$$

A filter base generating a uniformity is called a ***base*** for the uniformity.
If d is a metric on S, d generates a uniformity Γ_d which is defined to be the smallest filter of $S \times S$ containing the sets $\{(x,y);d(x,y) \leq r\}$ for any positive r.
Finally, if $A, B \in \Pi(S)$, the set $(A-B) \cup (B-A)$ defines the ***symmetric difference*** $A \,\Delta\, B$ of A and B. It is rather simple to see that $(\Pi(S), \Delta)$ is an Abelian group, and that

$$A \,\Delta\, B \subset (A \,\Delta\, C) \cup (C \,\Delta\, B)$$

as well as

$$(A \,\Delta\, B) \cap C = (A \cap C) \,\Delta\, (B \cap C)$$

hold (see e.g. [3], p. 6).

3. Constructing Uniformities

When defining a uniformity Γ, one has to specify which elements are considered to be close to each other. If in addition closeness can be specified in numeric terms, i.e. in terms of a distance function, a metric space is defined.
Let $r: D^\circ \to \mathbb{N}$ be a rank function on the compact elements of a domain, then the closeness $c(i,j)$ for $i,j \in \Sigma^*(D)$ is defined by

$$c(i,j) := \begin{cases} \infty & \text{if } i = j \\ \min\{r(k);k \in i^\circ \,\Delta\, j^\circ\} & \textit{otherwise} \end{cases}$$

Hence i and j are the closer the later one realizes that there is a witness in $i^\circ \,\Delta\, j^\circ$, since $i = j$ iff $i^\circ = j^\circ$ iff $i^\circ \,\Delta\, j^\circ = \phi$. The distance associated with the rank function r is then defined by

$$d(i,j) := 2^{-c(i,j)}.$$

This makes $(\Sigma^*(D), d)$ a complete (ultra -) metric space.

3.1 Definition: For $A \subset D^\circ$, define $G_A := \{(i,j) \in \Pi(D) \times \Pi(D); i^\circ \,\Delta\, j^\circ \subset A\}$

Thus we measure the closeness of two $\sigma-$ ideals in terms of the difference their trace leaves on the compact elements. This constitutes the base for a uniformity:

3.2 Lemma: Let $\Theta \subset \Pi(D^\circ)$ be a filter base.

(a) $\Gamma_\Theta := \{G_A ; A \in \Theta\}$ is the base for a uniformity $\Gamma^\bullet_\Theta$.

(b) if Ψ is another filter base with $\Theta \subset \Psi$, then $\Gamma^\bullet_\Theta \subset \Gamma^\bullet_\Psi$.

(c) if $\Phi(\Theta)$ denotes the filter generated by Θ, then $\Gamma^\bullet_{\Phi(\Theta)} = \Gamma^\bullet_\Theta$.

(d) $\Gamma^\bullet_\Theta$ is Hausdorff provided $\bigcap \Theta = \phi$.

Proof:

(a) It is enough to show that given $A \in \Theta$, there exists $B \in \Theta$ such that $G_B \circ G_B \subset G_A$. Now let $(i,j) \in G_B \circ G_B$, we know that there exists $k \in \Sigma^*(D)$ such that $i^\circ \Delta k^\circ \subset B$ and $k^\circ \Delta j^\circ \subset B$. But because $i^\circ \Delta j^\circ \subset (i^\circ \Delta k^\circ) \cup (k^\circ \Delta j^\circ)$, we see that we may take B as A. The other properties for a uniformity are straightforward.

(b) $C \in \Gamma^\bullet_\Theta$ iff there exists $G \in \Gamma_\Theta$ such that $G \subset C$. This implies the desired inclusion.

(c) $C \in \Gamma^\bullet_{\Phi(\Theta)}$ iff $G_F \subset C$ for some $F \in \Phi(\Theta)$; since $F \in \Phi(\Theta)$ iff $A \subset F$ for some $A \in \Theta$, the assertion is immediate.

(d) Let $(i,j) \in \bigcap \Gamma^\bullet_\Theta$ be a pair of ideals, then $i^\circ \Delta j^\circ \subset \bigcap \Theta$, thus $i^\circ \Delta j^\circ = \phi$, consequently $i^\circ = j^\circ$, equivalently $i = j$.■

The metric defined in [6] falls rather naturally into the realm of the uniformity defined above. Let d be as indicated derived from a rank function r, then

$$d(i,j) < \varepsilon$$

iff

$$r(k) \geq \lceil \log_2 \frac{1}{\varepsilon} \rceil$$

for each $k \in i^\circ \Delta j^\circ$, provided the $\sigma-$ ideals i and j are different. It is then easily seen that the metric uniformity Γ_r coincides with $\Gamma^\bullet_\Theta$, where

$$\Theta := \{A_n ; n \in \mathbb{N} \text{ and } A_n \neq \phi\}$$

with

$$A_n := \{r \geq n\} := \{t \in D^\circ ; r(t) \geq n\}$$

We will return to the question under which conditions $\Gamma^\bullet_\Theta$ equals Γ_r for some r in due course.

We are interested in the completeness of $\Gamma^\bullet_\Theta$; recall that a uniform space is complete iff each Cauchy net has a limit. We are going to present the representation of the limit in terms of the following set - theoretical construction. Let $(S_r)_{r \in R}$ be a net of sets, then

$$\liminf_{r \in R} S_r := \bigcup_{r \in R} \bigcap_{r_0 \geq r} S_{r_0}$$

is said to be the *limit inferior* of this net, and is the set of all elements which are eventually in all S_r, and

$$\limsup_{r \in R} S_r := \bigcap_{r \in R} \bigcup_{r_0 \geq r} S_{r_0}$$

is called the ***limit superior*** of this net, and is the set of all elements which are frequently in S_r.
It is rather easy to verify the following technical lemma:

3.3 Lemma: Let $(i_r)_{r\in R} \subset \Upsilon(D^\circ)$ be a net of ideals in D°, then we have the following:

(a) $\liminf_{r\in R} i_r$ and $\limsup_{r\in R} i_r$ are both in $\Upsilon(D^\circ)$,

(b) $\liminf_{r\in R} i_r \subset \limsup_{r\in R} i_r$,

(c) for every $r^\bullet \in R$, the following inclusions hold

$$\liminf_{r^\bullet \leq r\in R} i_r \subset \liminf_{r\in R} i_r,$$

and

$$\limsup_{r^\bullet \leq r\in R} i_r \supset \limsup_{r\in R} i_r. \blacksquare$$

This allows the representation of a limit in purely set - theoretical terms, provided one knows that it exists. Recall that a sequence in a Hausdorff space has at most one limit.

3.4 Proposition Let $(i_r)_{r\in R} \subset \Sigma(D)$ be a net converging to the ideal i with respect to $\Gamma^\bullet_\Theta$, and assume that $\Gamma^\bullet_\Theta$ is Hausdorff. Then

$$i^\circ = \liminf_{r\in R} i^\circ_r = \limsup_{r\in R} i^\circ_r.$$

Proof:

0. Convergence of the net (i_r) to i w. r. t. $\Gamma^\bullet_\Theta$ means that given $A\in\Theta$, there exists $r_0 \in R$ such that $i^\circ \,\Delta\, i^\circ_r \subset A$ for all $r \geq r_0$.
1. Fix $A\in\Theta$ and select r_0 for A according to part 0. If $r\geq r_0$, we have $i^\circ \,\Delta\, i^\circ_r \subset A$, or, equivalently, $i^\circ - A \subset i^\circ_r - A$. Consequently,

$$i^\circ - A \subset \liminf_{r\in R} i^\circ_r - A$$

Since we may infer

$$E \subset F$$

from

$$\overline{\forall A\in\Theta : E - A \subset F - A,}$$

we may conclude that

$$i^\circ \subset \liminf_{r\in R} i^\circ_r.$$

Similarly, we see that

$$\limsup_{r\in R} i^\circ_r \subset i^\circ.$$

The conclusion now follows from 3.3 b). $\blacksquare$

From 3.4, we infer completeness for the uniformity:

3.5 Proposition Let $(i_r)_{r\in R}$ be a Cauchy - net in the Hausdorff uniformity $\Gamma^\bullet_\Theta$, then (i_r) converges.

Proof:

0. We know that for every $A\in\Theta$ there exists $r^\bullet\in R$ such that $i^\circ_s \,\Delta\, i^\circ_t \subset A$ holds whenever $s,t\geq r^\bullet$.

1. Fix $A, r^\bullet$ for the moment, then

$$\forall s,t\geq r^\bullet : i^\circ_s - A \subset i^\circ_t - A,$$

hence we have for all $s\geq r^\bullet$ the inclusion

$$i^\circ_s - A \subset \liminf_{s\leq r\in R} i^\circ_r - A \subset \liminf_{r\in R} i^\circ_r - A.$$

Similarly we have

$$i^\circ_s - A \supset \limsup_{r\in R} i^\circ_r - A.$$

Now let $i\in\Sigma^\bullet(D)$ be determined by the equation

$$i^\circ = \liminf_{r\in R} i^\circ_r.$$

then we see that $i^\circ_s - A = i^\circ - A$, whenever $s\geq r^\bullet$. ∎

Let us illustrate 3.5 and our constructions with an example. It is well known that $D := \Pi(\mathbb{N})$ is a domain, using set inclusion as the order relation, and that the compact elements are just the finite subsets. Count D° using the bijection

$$\varphi : a \to \sum_{t\in a} 2^t$$

Putting

$$\Theta := \{\{\varphi\geq n\}; n\in\mathbb{N}\}$$

we obtain a complete Hausdorff uniformity $\Gamma^\bullet_\Theta$ on $\Sigma^\bullet(D)$. Now define

$$i_n := \{a\in D; a\geq n\},$$

then i_n is easily seen to be a $\sigma-$ ideal.

Claim: $(i_n)_{n\in\mathbb{N}}$ is Cauchy.

Proof:

Given $N\in\mathbb{N}$, select k with $2^k > N$. If $s,t\in\mathbb{N}$, let

$$a\in i^\circ_{k+s} \,\Delta\, i^\circ_{k+s+t} \subset i^\circ_{k+s},$$

thus

$$\varphi(a) = \sum_{m\in a} 2^m > card(a) 2^{k+s} > N,$$

so $(i_{k+s}, i_{k+s+t})\in G_{\{\varphi\geq N\}}$. Since s and t have been chosen arbitrarily, the above Claim is established. ∎

Consequence: $\Gamma_\Theta^\bullet - \lim_{n\to\infty} i_n = \phi$.■

This has been the reason for including the empty set into the set of σ– ideals we wanted to consider. Usually, one considers only the set of non - empty σ–ideals; this is justified by the fact that the bottom element in a domain constitutes an ideal.
Let us remark that φ may serve as a ranking function, so that on one hand we have the uniform space $(\Pi(\mathbb{N}),\Gamma_\Theta^\bullet)$, and on the other we have the metric space $(\Pi(\mathbb{N}),d)$ with the metric d defined as by MacQueen, Plotkin and Sethi. It is not difficult to see that the uniformity defined by the metric is just the same as $\Gamma_\Theta^\bullet$. We will denote the uniformity which is generated from the metric obtained from the rank function r by Γ_r
If Ψ is a filter, then Ψ is said to be ***countably based*** iff there exists a countable filter base Θ such that Ψ equals $\Phi(\Theta)$, the filter generated by Θ. It is the countably based filters which constitute the special case of metrizable uniformities. We return to the case of a general D.

3.6 Proposition: Let Θ be a filter base with $\bigcap\Theta = \phi$. There exists a rank function $r:D^\circ \to \mathbb{N}$ with $\Gamma_\Theta^\bullet = \Gamma_r$ iff $\Phi(\Theta)$ is countably based.

Proof:

1. If $\Gamma_\Theta^\bullet = \Gamma_r$ for some rank function r, then

$$\{A_n; n\in\mathbb{N}, A_n \neq \phi\}$$

with

$$A_n := \{r \geq n\}$$

constitute a countable base for $\Phi(\Theta)$. 2. If $\Phi(\Theta)$ has a countable base

$$\Omega = \{B_t; t\in\mathbb{N}\},$$

we may assume without loss of generality that the B_t form a descending chain

$$B_1 \supset B_2 \supset \cdots$$

(otherwise force this condition without loosing the properties of Ω). Let for $v \in D^\circ$ the rank function be defined by

$$r(v) := \mathit{card}(\{t; t\in B_t\}).$$

Then r is a finite function, because $\bigcap\Omega = \phi$, and $B_t = \{r \geq t\}$ is easily seen. This implies the desired equality $\Gamma_\Theta^\bullet = \Gamma_r$.■

4. Fixed Points

Completeness is only a necessary condition for establishing the existence of a fixed point for a map of a metric or uniform space into itself. The technical condition used in [6] was that the maps in question are contractions, so that the distance between the image of two points is bounded by a constant factor times the distance between the points themselves, the constant being strictly smaller than unity. This is shown to have the effect that successive applications of the map will eventually result in closer and closer points, thus yielding a Cauchy sequence which then must converge to the fixed point.

A similar strategy is employed in the case of a complete uniform space. Since here numerical measurements of distances are not possible, one has to use other techniques in establishing the convergence of the iterates of the map in question. A rather general fixed point theorem due to T. Landes establishes the existence of a fixed point for what is called quasi - contractive maps ([5]); here quasi - contractivity is a replacement for a uniform shrinking of distances under application of the map in question. Since this theorem is not published yet, the Appendix contains a brief discussion of the notations, techniques, and the results.

We will show here how to utilize this result in the context of $\sigma-$ ideals over a domain. It will be convenient to consider a general situation first. Let M be a complete uniform space with uniformity Ξ, and let Ξ_0 be a base for Ξ. Assume that $\varphi: M \to M$ is a quasi - contraction, and define for $V \in \Xi_0$

$$R_m(\varphi, V) := closure((\varphi^m \times \varphi^{m+1})[V]),$$

the closure being topological in the product space $M \times M$, where M is endowed with the topology canonically generated by the uniformity which is based on Ξ_0. Thus $(x,y) \in M \times M$ is in $R_m(\varphi, V)$ iff there exists a sequence $(a_n, b_n) \in V$ with the property that

$$(x,y) = \lim_{n \to \infty} (\varphi^m \times \varphi^{m+1})(a_n, b_n).$$

$R_m(\varphi, \cdot)$ maps basic neighborhoods into closed and non - empty subsets of the Cartesian product. A useful condition for a fixed point may be derived from this observation: if applying two subsequent iterations of φ to an arbitrary pair in V renders the images more and more indistinguishable, or, topologically speaking, if $R_m(\varphi, V)$ is absorbed eventually by the diagonal, then condition (A.5) establishing a fixed point is satisfied.

This condition is intuitively rather clear, and in order to formulate it we need the topological framework which is sketched in the Appendix.

4.1 Proposition: If $\lim_{m \to \infty} R_m(\varphi, V)$ exists for each $V \in \Xi_0$ and is a subset of the diagonal, then φ has a fixed point.

Proof:

0. According to the first part of Theorem A.4, is is enough to show that given $x \in M$, and $U_1 \in \Xi_0$ there exists $m \in \mathbb{N}$ such that

$$(\varphi^m(x), \varphi^{m+1}(x)) \in U_1$$

Since Ξ_0 is a base for Ξ, there exists for U_1 a symmetric entourage U such that $U \circ U \subset U_1$, and for x there exists a basic neighborhood V with the property that $(x, \varphi(x)) \in V$. Consequently,

$$\forall k \in \mathbb{N}: (\varphi^k(x), \varphi^{k+1}(x)) \in R_k(\varphi, V)$$

holds.

1. Let $\Delta^* := \lim_{n \to \infty} R_n(\varphi, V)$ be the limit with respect to the Hausdorff uniformity. Then we know that there exists an index n_0 such that

$$\forall n \geq n_0 : R_n(\varphi, V) \subset \langle U \times U \rangle (\Delta^*)$$

Thus $(z, z, \varphi^n(x), \varphi^{n+1}(x)) \in \langle U \times U \rangle$ for a suitably chosen $z \in M$ and all such n. This implies the desired property. ▪

The condition given above is a bit strong and not very practical. First, the proof shows that one need not insist of the convergence of the closures of $(\varphi^m \times \varphi^{m+1})[V]$; this has been done in order to assure a proper topological formulation of the sufficient condition (otherwise one would have to have resorted to a topology on the space of all subsets of $M \times M$, a space which is too large to be topologized properly). Secondly, the convergence is not easily checked, so the condition is not a really useful one, although it applies to the situation in question as well.

Let us return to the space $\Sigma(D)$ of all $\sigma-$ ideals of the domain D (note that we can do now without the empty set). Assume that we have a quasi contraction φ defined on $\Sigma(D)$ and that furthermore a rank function $r: D^\circ \to \mathbb{N}$ is defining the uniformity. Let Θ be the base determined by r, then $\Gamma_\Theta^\bullet$ is Hausdorff. Define for each subset $A \subset D$ its weight

$$r^\bullet(A) := \min\{r(d); d \in A^\circ\}$$

The next lemma will be an auxiliar statement establishing the existence of a fixed point. Intuitively, it describes what happens when the map φ makes the difference of the images of two ideals harder to discern than the ideals themselves.

4.2 Lemma: If

$$r^\bullet(\varphi(i)) > r^\bullet(i)$$

holds for all $i \in \Sigma(D)$, then we can find for every $C, A \in \Theta$ an index m_0 such that

$$R_m(\varphi, G_C) \subset G_A$$

holds for all $m \geq m_0$.

Proof:

0. We may and do assume that $C = \{r \geq s\}$ and $A = \{r \geq t\}$ for suitable values s and t. We show by induction on m that

$$\forall (a_1, a_2) \in R_m(\varphi, G_C): \min_{i=1,2} r^\bullet(a_i) \geq s + m$$

holds. This will establish the Lemma.

1. In fact, if $m = 0$ and $(a_1, a_2) \in (id \times \varphi)[G_C]$, the assertion is obvious, and if we take (a_1, a_2) from the topological boundary of $(id \times \varphi)[G_C]$, an application of 3.4 establishes the claim for $m = 0$.

2. The inductive step works similar. To begin with, assume that

$$(a_1, a_2) \in (\varphi^{m+1} \times \varphi^{m+2})[V]$$

hence $a_1 = \varphi(b_1)$, $a_2 = \varphi(b_2)$ with

$$(b_1, b_2) \in (\varphi^m \times \varphi^{m+1})[V].$$

Thus

$$r^\bullet(a_i) = r^\bullet(\varphi(b_i)) > r^\bullet(b_i)$$

and the conclusion follows from the induction hypothesis. If (a_1, a_2) is a member of the boundary, we utilize 3.4 to complete the inductive step. ▪

We are now in a position to formulate a fixed point theorem adequate for the situation considered.

4.3 Proposition: Let $\varphi:\Sigma(D)\to\Sigma(D)$ be a weight increasing map (hence

$$r^{\bullet}(\varphi(i)) > r^{\bullet}(i)$$

holds for all $i\in\Sigma(D)$). Then φ has a fixed point.

Proof:

We infer from 4.2 that for any $i\in\Sigma(D)$ and for any $C\in\Theta$ there exists $m\in\mathbb{N}$ with

$$(\varphi^m(i),\varphi^{m+1}(i))\in G_C,$$

thus condition (A.5) is satisfied. To establish the existence of a unique fixed point, it has to be shown that the condition (A.6) is satisfied, too. For this, let i and j be $\sigma-$ ideals. Because $\Gamma^{\bullet}_{\Theta}$ is Hausdorff, we can find $C\in\Theta$ so that G_C does not contain the pair (i,j) and we may find $B\in\Theta$ with $(\varphi(i),j)\in G_B$. From 4.2 we see that eventually $R_m(\varphi,G_B)\subset G_C$ holds, so we find m such that

$$(\varphi^{m+1}(i),\varphi^{m+1}(j)) = (\varphi^m\times\varphi^{m+1})(\varphi(i),j)\in G_C,$$

thus condition (A.6) holds indeed.
2. It remains to be shown that φ is a quasi contraction. Define for $U\in\Theta$ the map $F:\Theta\to\Theta$ by setting

$$F(U) := \{(\varphi(i),\varphi(j));(i,j)\in U\}$$

then by construction $(\varphi(i),\varphi(j))\in F(U)$ holds whenever $(i,j)\in U$. We have, however, to establish the properties (A.1) as well as (A.2). Since given $d\in\varphi(i)^{\circ}$ there exists $d_0\in i^{\circ}$ such that $r(d) > r(d_0)$, we have for $A := \{r\geq n\}$ the property

$$i^{\circ}\ \Delta\ j^{\circ}\subset A \textit{ implies } \varphi(i)^{\circ}\ \Delta\ \varphi(j)^{\circ}\subset A,$$

which in turn establishes (A.1). In order to establish (A.2), we first observe that

$$F(U\circ U') = F(U)\circ F(U'),$$

hence that F is a homomorphism with respect to composing relations. From this, (A.2) is easily inferred, since from what we have shown it is plain that

$$U^N\subset U$$

holds for each $U\in\Theta$.∎

Closing this note, we want to briefly discuss one particular aspect of our results' applicability. MacQueen, Plotkin and Sethi show that sum, product and exponentiation of ideals are contractive maps ([6], Theorem 7), when considering a domain V that satisfies the domain equation

$$V = \{\textbf{true},\textbf{false}\}+\mathbb{N}+(V\to V)+(V\times V)+(V+V)+\{\textbf{wrong}\},$$

the latter component denoting type errors at run time. Here $V\to V$ ($V\times V$, and $V+V$) is the function space (the Cartesian product, and the direct sum, resp.) associated with V.
Technically, sum, product and exponentiation are maps working on a pair and producing a single result, so a fixed point theorem is not directly applicable. It is this technical matter that we want to discuss briefly, and we restrict our

attention to the case of the product of two $\sigma-$ ideals. For this, we assume that the domain D satisfies the domain equation

$$D = Y+(D\times D);$$

to be more precise, that there is an embedding

$$\beta: Y+(D\times D) \rightarrow D$$

preserving and reflecting the order and all existing least upper bounds which in addition preserves the compact elements (that β is an ***embedding*** means that there exists a map

$$\gamma: D \rightarrow Y+(D\times D)$$

such that both

$$\gamma \circ \beta = id_{Y+(D\times D)}$$

and

$$\beta \circ \gamma \leq id_D$$

hold). γ is called the projection for β . If π_i is the ith projection on $D\times D$, and if out_R is the projection of the direct sum $Y+(D\times D)$ onto $D\times D$, we claim that

$$\tau_i := \pi_i \circ out_R \circ \gamma$$

maps $\Sigma(D)$ into itself ($i = 1, 2$). This is easily proved.
Putting

$$\vartheta := (\tau_1, \tau_2),$$

we construct a map

$$\Sigma(D) \rightarrow \Sigma(D)\times\Sigma(D),$$

for which we can find another map

$$\xi: \Sigma(D)\times\Sigma(D) \rightarrow \Sigma(D)$$

so that both

$$\vartheta \circ \xi = id_{\Sigma(D)\times\Sigma(D)},$$

and

$$\xi \circ \vartheta \supset id_{\Sigma(D)}$$

hold. The map ξ is constructed as follows: denoting by in_R the injection of $D\times D$ into $Y+(D\times D)$, then we set

$$\xi(i,j) := in_R[i\times j]$$

(The pair (ϑ,ξ) has all the desired properties of an embedding and its projection, except that the inclusion in

$$\xi \circ \vartheta \supset id_{\Sigma(D)}$$

needs to be reversed).
The desired map

$$\rho: \Sigma(D) \rightarrow \Sigma(D)$$

for which one wants to establish the fixed point is then defined by

$$\rho := in_R \circ prod \circ \vartheta,$$

where

$$prod{:}\Sigma(D)\times\Sigma(D) \to \Sigma(D\times D)$$

is the map assigning each pair of $\sigma-$ ideals their product.

It is now not too difficult to establish that ρ is weight increasing. Similar constructions apply to the basic operations of constructing the direct sum, and the exponentiation of two $\sigma-$ ideals.

Appendix

This paper proposes a generalization for the metric structure used in [6], and it extends a fixed point argument of the Banach type. Recently, Thomas Landes gave a suitable generalization of Banach's celebrated theorem which fits into the framework considered here, and which has been of considerable use in the arguments above. Since this result has not been published yet, and in order to make the present paper self contained, we give Landes' result in greater detail and sketch its proof (the arrangement is a bit different from the one given by Landes for technical reasons; the proofs are entirely due to Landes, however).

Fix for the remainder of this appendix a uniform space M with separated and complete Hausdorff uniformity Ξ which is supposed to have Ξ_0 as a base. A map $F{:}\Xi_0 \to \Xi_0$ is said to be an ***attractor*** iff

(A.1)$\forall U\in\Xi_0{:}F(U)\subset U,$

(A.2)$\forall U\in\Xi_0 \exists N\in\mathbb{N}\forall k\in\mathbb{N}{:}F^N(U)\circ \cdots \circ F^{N+k}(U) \subset U$

Here N in condition (A.2) depends of course on the neighborhood U, and on F; F^N is the Nth composition of F with itself. Now call a map $f{:}M\to M$ a *quasi contraction* iff there exists an attractor F_f such that

(A.3)$\forall U\in\Xi_0{:}(x,y)\in U$ *implies* $(f(x),f(y))\in F_f(U)$

The following Theorem asserts existence and uniqueness of a fixed point. This is done separately, so that the different assumptions may be differentiated more effective.

A.4 Theorem: 1. Let f be a quasi contraction with the following property

$$(A.5)\quad \forall x\in M\forall U\in\Xi_0\exists m\in\mathbb{N}{:}(f^m(x),f^{m+1}(x))\in U$$

then there exists a fixed point x^* for f, and

$$x^* = \lim_{n\to\infty} f^n(x) \text{ for } any\ x$$

2. If in addition

$$(A.6)\quad \forall x\neq y\exists U\in\Xi_0\exists m\in\mathbb{N}{:}(x,y)\notin U \text{ and } (f^m(x),f^m(y))\in U,$$

then x^* is the unique fixed point.

Proof:

1. If $x^* = \lim_{n\to\infty} f^n(x)$ is a fixed point, and if (A.6) holds, then x^* is plainly uniquely determined.

2. Fix $x\in M$, then it is sufficient to demonstrate that $(f^n(x))_{n\in\mathbb{N}}$ is a Cauchy sequence. Given $U\in\Xi_0$, we know that there exists $m\in\mathbb{N}$ such that

$$(f^m(x),f^{m+1}(x))\in U.$$

Now let $k \geq 1, n \geq N+m$, where N is chosen according to condition (A.2) for the attractor F_f, then

$$(f^n(x), f^{n+1}(x)) \in F_f^N(\underline{U}) \circ \cdots \circ F_f^{N+k}(U) \subset U.$$

This establishes the Theorem. ▪

We need to characterize two constructions from topology: constructing the product uniformity, and constructing the Hausdorff uniformity on the space of all non - empty closed subset of a uniform space. Let us construct the product uniformity on $M \times M$ first. Define for $W \subset M^4$ the set

$$<W> := \{(a,c,b,d);(a,b,c,d) \in W\}$$

Then

$$\{<A \times B>: A, B \in \Xi_0\}$$

defines the basis for the product uniformity $\Xi \times \Xi$ on $M \times M$ (intuitively, two points in the Cartesian product are close iff both their components are close). This makes $M \times M$ a complete uniform space again.

Now denote by $CL(M)$ the set of all non - void subsets of M. If $U \in \Xi_0$, define for $A \subset M$ the sets

$$U(A) := \{y \in M; \exists x \in A: (x,y) \in U\}$$

and

$$[U] := \{(A,B) \in CL(M); A \subset U(B) \text{ and } B \subset U(A)\}$$

Hence $U(A)$ measures the closeness of a set B to the set A in terms of the entourage U . It is not too difficult to see that

$$\{[U]; U \in \Xi_0\}$$

defines the basis for a complete and separated uniformity on $CL(M)$ ([2], Theorem II - 12, p. 45).

References

[1] de Bakker, J. W., Kok, J. N.: Towards a Uniform Topological Treatment of Streams and Functions on Streams. In W. Brauer (Ed.): Proc. 12th ICALP. Lecture Notes in Computer Science 194, Springer - Verlag, 1985, 140 - 148

[2] Castaing, C., Valadier, M.: Convex Analysis and Measurable Multifunctions. Lecture Notes in Mathematics 580, Springer - Verlag, Berlin, 1977

[3] Hewitt, E., Stromberg, K.: Real and Abstract Analysis. Springer - Verlag, Berlin, 1969

[4] Kelly, J.L.: General Topology. Van Nostrand Reinhold, New York, 1955

[5] Landes, Th.: The Banach Fixed Point Principle in Uniform Spaces. Preprint, Department of Economics, University of Paderborn, Sept. 1986

[6] MacQueen, D., Plotkin, G., Sethi, R.: An Ideal Model for Recursive Polymorphic Types. Expanded Version (presented in condensed form at the 11th Ann. ACM Symp. Princ. Progr. Lang., Salt Lake City, UT, 1984)

[7] Stoy, J. E.: Denotational Semantics: The Scott - Strachey Approach to Programming Language Theory. MIT Press, Cambridge, MA, 1977.

New Results on Hierarchies of Domains

Achim Jung
Fachbereich Mathematik
Technische Hochschule Darmstadt
D-6100 Darmstadt
West Germany

Abstract

The relation between a cpo D and its space $[D \longrightarrow D]$ of Scott-continuous functions is investigated. For a cpo D with a least element we show that if $[D \longrightarrow D]$ is algebraic then D itself is algebraic. Together with a generalization of Smyth's Theorem to strict functions this implies that $[D \longrightarrow D]$ is ω-algebraic whenever $[D \xrightarrow{s} D]$ is. It is an open question, whether $[D \longrightarrow D]$ is algebraic whenever $[D \xrightarrow{s} D]$ is algebraic.

Smyth's Theorem is extended to the class of cpo's with no least element assumed. Our main result asserts that in this context the profinites again form the largest cartesian closed category of domains. The proof also yields the following: if the functionspace of a cpo D is algebraic, then D has infima for filtered sets. The question, whether an ω-algebraic functionspace implies that D is profinite, remains open.

1 Introduction

Profinite domains were introduced by Gordon Plotkin in [4] in order to have a category of ordered structures which is closed not only under the usual domain constructions such as product, function space, etc., but also under the (convex) powerdomain construction. It turned out that this category was very handy to work with and that most domain equations one could think of had nontrivial profinite solutions. Subsequently M.B. Smyth showed in [5] that this category is also in some sense inevitable as every category of domains closed under the function space operator is contained in it. The work presented here continues this line of investigation.

We show that the algebraicity of the function space implies the algebraicity of the domain itself, so we can remove one of the hypotheses in Smyth's Theorem. As a closer examination shows, only strict functions (those, which respect the join of the empty set) are used in the proof, which implies that the profinites form the largest full subcategory of the ω-algebraic cpo's closed under the strict function space operator. Combining these results we obtain as a corollary that the full function space of a cpo is ω-algebraic, if and only if its strict function space is ω-algebraic. As simple and natural this sounds, there seems to be no direct proof. To our knowledge it is also the first application of Smyth's Theorem apart from his own maximality result.

Removing the requirement that a domain must have a least element leads again to the question, whether there is a "largest" category of domains. The proofs are getting rather intricate now: we show that an ω-algebraic function space implies that the domain itself must be ω-algebraic. But the direct analogue to Smyth's Theorem appears to be too hard to prove. We establish the following similar result: if $\big[[D \to D] \longrightarrow [D \to D]\big]$ (the second-order function space) is ω-algebraic, then D is profinite. This suffices to show that the profinites form the largest cartesian closed full subcategory of the ω-algebraic cpo's without least element.

2 Domains with a least element

Notation and Definitions. A subset A of a partially ordered set D is *directed* if it is nonempty and every pair of elements of A has an upper bound in A. A poset D is *complete* (is a *cpo*) if every directed subset A of D has a least upper bound $\bigvee A$ in D. An element x of a cpo D is *compact* if whenever $A \subseteq D$ is directed and $x \leq \bigvee A$ then $x \leq y$ for some $y \in A$. The set of compact elements in D is denoted by $B(D)$. (The *base* of D.) A cpo D is *algebraic* if every element of D is the supremum of a directed set of compact elements. If in addition the base of D is countable, then D is said to be *ω-algebraic.*

A cpo D is a *profinite domain* if it is isomorphic to the inverse limit of a directed system of finite posets. We omit details here since we give the corresponding intrinsic characterisation for profinite domains below in Theorem 7.

For A a subset of a partially ordered set D we denote by $mub(A)$ the set of minimal upper bounds of A in D. If every upper bound of A is above some minimal upper bound of A, we say that $mub(A)$ *is complete*. If there is a least element in D, we denote it by $\perp$.

Let D and E be cpo's. A function f from D to E is *continuous* if for every directed subset A of D, $f(A)$ is directed and $f(\bigvee A) = \bigvee f(A)$. A function is said to be *strict* if it is continuous and the least element of D is mapped onto the least element of E. The constant function with image $\{x\}$ is denoted by c_x. We denote the set of all continuous functions from D to E by $[D \longrightarrow E]$, the set of all strict functions by $[D \overset{s}{\longrightarrow} E]$.

Fact 1 *If $f : D \to E$ is continuous then*

$$\tilde{f}(x) = \begin{cases} f(x), & \text{if } x \neq \perp_D; \\ \perp_E, & \text{otherwise} \end{cases}$$

is the largest strict function below f.

Fact 2 *If D and E are cpo's, then so are $[D \longrightarrow E]$ and $[D \overset{s}{\longrightarrow} E]$.*

Fact 3 *If D and E are (ω-) algebraic, then $[D \longrightarrow E]$ and $[D \overset{s}{\longrightarrow} E]$ need not be (ω-) algebraic again.*

The last statement raises the question, which cpo's have an algebraic function space. This question was answered by M.B. Smyth in 1983:

Theorem 1 (M.B. Smyth) *If D and $[D \longrightarrow D]$ are ω-algebraic, then D is a profinite domain.*

Smyth used strict functions in his proof, except for "Lemma 4" where the constant functions "$\lambda x.m$" are used (and these can be replaced by the strict functions $\tilde{c}_m$ which map everything except $\perp$ to m), so we have the following variation of Theorem 1 (this observation is due to C.Gunter [2]).

Theorem 2 *If D and $[D \stackrel{s}{\longrightarrow} D]$ are ω-algebraic cpo's, then D is a profinite domain.*

We will show now that the first hypothesis is redundant:

Theorem 3 *If D is a cpo and $[D \stackrel{s}{\longrightarrow} D]$ is (ω-) algebraic, then D is (ω-) algebraic.*

Proof. If D equals the one-point domain, then there is nothing to prove. So assume there is an element $b \neq \bot$. Let a be any element of D. We will show that a is the limit of a directed set of compact elements.

Let g be a compact function below the strict constant function $\tilde{c}_a$. We will prove that $g(b)$ is compact in D. First assume that $g(b)$ is the limit of a directed set $(x_i)_{i \in I}$. Define

$$h_i(x) = \begin{cases} a, & \text{if } x \not\leq b; \\ x_i, & \text{if } \bot \neq x \leq b; \\ \bot, & \text{if } x = \bot. \end{cases}$$

Clearly, each h_i is continuous and $\tilde{c}_a$ is the limit of the h_i, so there must be an index $j \in I$, for which h_j is greater or equal to g. This implies $x_j = h_j(b) = g(b)$. So $g(b)$ cannot be approximated from below. On the other hand $\tilde{c}_{g(b)}$ is the supremum of compact functions and by what we just proved, there must be a compact function h below $\tilde{c}_{g(b)}$, which maps b onto $g(b)$. Assume now that $\bigvee_{i \in I} x_i \geq g(b)$. This implies that $\bigvee_{i \in I} \tilde{c}_{x_i} \geq \tilde{c}_{g(b)} \geq h$, so there must be an index j, for which $\tilde{c}_{x_j} \geq h$ holds and we get $x_j = \tilde{c}_{x_j}(b) \geq h(b) = g(b)$ as desired. So $g(b)$ is a compact element below a. Everything else follows easily from this. ∎

Corollary 1 *If D is a cpo and $[D \longrightarrow D]$ is (ω-) algebraic, then D is (ω-) algebraic.*

Proof. This follows from Theorem 3 because the strict function space is a Scott-closed subset of the full function space and thus is algebraic, whenever the latter is. ∎

Note, however, that the category of profinite domains with strict functions as arrows is *not* cartesian closed, although it is closed with respect to a different product, namely the "smash"-product, where all elements of the form $(\bot, y)$ or $(x, \bot)$ in the ordinary product are identified with the bottom element. This product is in accordance with one possible philosophy about the least element, namely, that a function of several variables should be undefined, whenever at least one of the arguments is undefined.

We now use the results above to show that $[D \longrightarrow D]$ and $[D \stackrel{s}{\longrightarrow} D]$ are simultaneously ω-algebraic or not, and this happens precisely when D is profinite.

Theorem 4 *Let D be a cpo. Then the following conditions are equivalent:*

(i) $[D \longrightarrow D]$ is ω-algebraic.
(ii) $[D \stackrel{s}{\longrightarrow} D]$ is ω-algebraic.
(iii) D is a countably based profinite domain.

Proof. (i) $\Longrightarrow$ (ii) See the proof of Corollary 1.
(ii) $\Longrightarrow$ (iii) Theorem 2.
(iii) $\Longrightarrow$ (i) This is well known; see [1], for example. ∎

Problem. Is $[D \longrightarrow D]$ algebraic when and only when $[D \stackrel{s}{\longrightarrow} D]$ is algebraic?

Remark. All results of this section remain true, if we replace the space $[D \xrightarrow{s} D]$ of strict functions by the space $[D \xrightarrow{r} D]$ of retractions or even $[D \xrightarrow{sr} D]$, the space of strict retractions. They become false, if we consider the space $[D \xrightarrow{p} D]$ of projections (in the sense of [1]) instead. There is no connection between the algebraicity of D and the algebraicity of $[D \xrightarrow{p} D]$.

3 Domains without a least element

We now turn our attention to the case, where a least element is no longer required. Why, one might ask, since a least element can always be added to a domain? The answer is that this less restrictive concept of a domain allows new constructions to be added such as a coproduct, which meet the needs of new programming concepts. We analyse this situation in a similar fashion as above.

Theorem 5 *Let D be a cpo with an (ω-) algebraic function space. Then D itself is (ω-) algebraic.*

With a little care, a proof of this can be carried out similar to the proof of Theorem 3. We prefer to deduce it from the following stronger result:

Theorem 6 *Let D be a cpo with an algebraic function space. Then every compact function maps all of D into the base of D.*

With other words: *For every $x \in D$ and every compact $g \in [D \longrightarrow D]$, $g(x)$ is compact in D.*

Proof. Suppose $x \in D$ and $g \in [D \longrightarrow D]$, g compact. First we alter g a little bit:

$$\tilde{g}(e) = \begin{cases} g(e), & \text{if } e \leq x; \\ g(x), & \text{otherwise.} \end{cases}$$

There is a compact function h below $\tilde{g}$, which maps x onto $g(x)$: $\tilde{g}$ is the supremum of compact functions $\tilde{f}_i$, $i \in I$, and g can be approximated similarly:

$$f_i(e) = \begin{cases} \tilde{f}_i(e), & \text{if } e \leq x; \\ g(e), & \text{otherwise.} \end{cases}$$

(Each f_i is continuous, because $\tilde{f}_i\,|_{\downarrow x} \leq \tilde{g}\,|_{\downarrow x} = g\,|_{\downarrow x}$.) Since g is compact, there is an index j, such that $g = f_j$. This implies $g(x) = f_j(x) = \tilde{f}_j(x)$. So $\tilde{f}_j$ is the required compact function h. Now assume that $\bigvee_{i \in I} x_i \geq g(x)$. Then $\bigvee_{i \in I} c_{x_i} \geq c_{g(x)} \geq h$, so there is an index j, such that $c_{x_j} \geq h$, which means that $x_j = c_{x_j}(x) \geq h(x) = g(x)$. So $g(x)$ is compact. ∎

Underlying this proof is the fact that for any two cpo's D and E and elements $x \in D$, $y \in E$, $[\downarrow x \longrightarrow \downarrow y]$ is algebraic, whenever $[D \longrightarrow E]$ is. We omit a proof of this here.

The proof of Theorem 5 is now short enough. Choose any x in D. Then the set $\{g(x) \,|\, g \in [D \longrightarrow D], g \leq c_y\,, g \text{ compact}\}$ consists of compact elements, is directed and approximates y. ∎

Recall the definition of the minimal upper bound operator U from [1] or [5]: Let A be a subset of an ordered set P. We define

$$\begin{aligned} U^0(A) &= A, \\ U^{n+1}(A) &= \{x \in P \,|\, x \text{ is a minimal upper bound for} \\ &\qquad\qquad \text{some finite subset of } U^n(A)\}, \\ U^\infty(A) &= \bigcup_{n \in N} U^n(A). \end{aligned}$$

We prove Smyth's Theorem for bottomless domains by going through roughly the same three lemmas Smyth himself used. This is possible, because the Characterization Theorem for profinite domains is still valid in our case:

Theorem 7 (G. Plotkin) *A cpo D is profinite if and only if $B(D)$ satisfies the following two conditions for any finite set $A \subset B(D)$:*

(i) $mub(A)$ is complete,

(ii) $U^\infty(A)$ is finite.

(For a proof see [1].)

But we take the opportunity to prove stronger versions of Smyth's lemmas, which give us additional insight into the structure of profinite domains. Smyth's proof of "Lemma 1" for example can be carried over to the bottomless case without difficulty (see [1] for such an approach). Instead we show the following:

Theorem 8 *If $[D \longrightarrow D]$ is an algebraic cpo, then D has infima for filtered sets.*

Proof. By an application of Iwamura's Lemma (see [3]) we know that it is sufficient to show that every chain has an infimum. We can further restrict our attention to downward well-ordered chains since every chain must contain a coinitial downward well-ordered chain. So assume there is a monotone mapping s from an ordinal α^{op} into D such that there is no infimum for the image in D. Clearly, every such s is continuous, because α^{op} consists of compact elements only. We also assume that s is injective.

We can distinguish two different areas in D:

$$\begin{aligned} A &= \{x \in D \mid x \text{ is a lower bound for all } s(\gamma), \gamma \in \alpha\} \\ B &= D \backslash A \end{aligned}$$

B can be retracted onto α^{op} by defining

$$r(x) = \inf\{\gamma \in \alpha^{op} \mid s(\gamma) \geq x\}.$$

Because α is an ordinal, there exists no strictly increasing infinite sequence of elements in α^{op}. Thus r is continuous.

For every monotone function $\sigma : \alpha^{op} \to \alpha^{op}$ we get a continuous mapping $\tilde{\sigma}$ from D into itself by setting

$$\tilde{\sigma}(x) = \begin{cases} x, & \text{if } x \in A; \\ s \circ \sigma \circ r(x), & \text{otherwise.} \end{cases}$$

We want to show now that $\widetilde{id}$ is not approximated by the functions way below it. We proceed in two steps:

1. A function below $\widetilde{id}$, which maps the image of α^{op} under s into B, is not way below $\widetilde{id}$.
2. If $\widetilde{id}$ equals the supremum of a directed set of functions $(f_i)_{i \in I}$ there must be an index j, from which on every f_i maps the image of α^{op} under s into B.

Together this says that the functions way below $\widetilde{id}$ cannot approximate $\widetilde{id}$. This contradicts the algebraicity of $[D \longrightarrow D]$ and so establishes the theorem.

ad 1. Let $f \leq \widetilde{id}$ map the image of α^{op} under s into B. Consider the successor function τ on α^{op}, defined by $\tau(\gamma) = \gamma + 1$. Then the functions

$$g_\beta(x) = \begin{cases} \widetilde{id}(x), & \text{if } x \in B \text{ and } r(x) \geq \beta; \\ \bar{\tau} \circ f \circ \widetilde{id}(x), & \text{if } x \in B \text{ and } r(x) < \beta; \\ x, & \text{if } x \in A \end{cases}$$

approximate $\widetilde{id}$, but none of them dominates f, since $\widetilde{id}(x) = x$ and $\bar{\tau}(x) \not\geq x$ holds for all x in the image of s.

ad 2. If A is empty, we are done. In the other case $A \cap B(D)$ cannot be directed, because then its supremum would be the infimum for the chain. So there are compact elements c_1 and c_2, which belong to A and for which there is no upper bound in A. Any continuous function, which maps c_1 and c_2 onto themselves must map upper bounds of $\{c_1, c_2\}$ onto upper bounds again and in particular the image of s into B. But because of the compactness of c_1 and c_2, this must happen from some index on in every approximating family of functions for $\widetilde{id}$. ∎

Note that with minor changes in the last paragraph this proof is also valid, if D and $[D \longrightarrow D]$ are continuous cpo's only.

From this theorem it follows easily that every subset of a cpo D has a complete set of minimal upper bounds provided that $[D \longrightarrow D]$ is an algebraic cpo.

Recall that the *root* $R(D)$ of a cpo D is defined as $U^\infty(\emptyset)$. (cf. [1].) In general the root of a cpo can be empty, but under the hypotheses of Theorem 8 there is a minimal element under each element of D. In case D has a least element, the root of D is the one-element set containing just this element. If D is algebraic, every element of the root is compact.

Lemma 1 *If D is a cpo and $\big[[D \to D] \longrightarrow [D \to D]\big]$ is ω-algebraic, then the root of D is finite.*

Proof. Because of Theorem 5, both D and $[D \longrightarrow D]$ are also ω-algebraic. Assume now that $R(D)$ is infinite. We define for every $r \in R(D)$ a continuous mapping f_r from D into itself:

$$f_r(x) = \begin{cases} x, & \text{if } x \leq r; \\ r, & \text{otherwise.} \end{cases}$$

(This is the canonical retraction onto $\downarrow r$.) Such a function may itself not be minimal, but it is easy to see that no two such functions have a common lower bound. Indeed, every function below f_r must keep all elements of the set $\downarrow r \cap R(D)$ fixed as is seen by induction on n in the definition of U^∞. For $r \neq r'$ and g a function which keeps $\downarrow r \cup \downarrow r'$ fixed, we have either $g(r) = r \not\leq r' = f_{r'}(r)$ or $g(r') = r' \not\leq r = f_r(r')$, so g is not a lower bound of $\{f_r, f_{r'}\}$. Under each f_r we have a minimal function in $[D \longrightarrow D]$, so there are infinitely many minimal elements in $[D \longrightarrow D]$. We construct continuum many minimal elements of $\big[[D \to D] \longrightarrow [D \to D]\big]$.

Choose g and g' minimal in $[D \longrightarrow D]$ such that they have a minimal upper bound h. (Choose them arbitrarily, if no two minimal elements have an upper bound.) For any subset S of minimal elements of $[D \longrightarrow D]$ define F_S by

$$F_S(f) = \begin{cases} g, & \text{if all minimal elements below } f \text{ belong to } S; \\ g', & \text{if no minimal element below } f \text{ belongs to } S \text{ and} \\ h, & \text{otherwise.} \end{cases}$$

Each F_S is a minimal element in $\big[[D \to D] \longrightarrow [D \to D]\big]$ and for $S \neq S'$ we have $F_S \neq F_{S'}$. This contradicts the assumption that the second-order function space be countably based. ∎

Lemma 2 *Under the assumptions of Lemma 1, D has a finite set of minimal elements for each finite set of compact elements. ("D has property M")*

Proof. (Because D may have no least element, a straightforward generalization of the proof of Smyth's "Lemma 3" is not possible in this setting.) It suffices to prove property M for pairs of compact elements (cf. [5]) so assume the set $\{x, x'\}$ had an infinite set B of minimal upper bounds. There are minimal elements $m \leq x$ and $m' \leq x'$, for which the elements of B are upper bounds also. Since the root of D is finite by Lemma 1, there is a minimal upper bound n of $\{m, m'\}$, which is below an infinite subset B' of B. We distinguish three cases:

1. $n \leq x$: Then m' is a lower bound for x and x'.
2. $n \leq x'$: Then m is a lower bound for x and x'.
3. $n \not\leq x, n \not\leq x'$: In this case B' is a set of upper bounds for $\{x, n\}$. Now either $\{x, n\}$ has an infinite set B'' of minimal upper bounds (and m is a lower bound for them) or there is a minimal upper bound p of $\{x, n\}$, which is below an infinite subset B'' of B'. The elements of B'' are then minimal upper bounds for $\{p, x'\}$ and m' is a lower bound for them.

We see that in each case we get a two-element set $\{a, b\}$, an infinite set A of minimal upper bounds for $\{a, b\}$ and a lower bound c for it. Smyth's proof of property M for domains can now be mimicked with c playing the role of the least element. ∎

It remains to show that $U^\infty(F)$ is finite, whenever F is a finite set of compact elements:

Lemma 3 *Under the assumptions of Lemma 1, U^∞ maps finite sets of compact elements onto finite sets.*

Proof. Assume $U^\infty(F)$ is infinite for F a finite subset of $B(D)$. By the same arguments as Smyth used in his "Lemma 4" we get an infinite chain $(m_n)_{n \in N}$ sitting inside $U^\infty(F)$. The limit $m = \bigvee_{n \in N} m_n$ of this chain is mapped onto itself by any function below the identity, which keeps F pointwise fixed. (This is easily shown by induction.) And there *is* a compact function below the identity which keeps the elements of F fixed, so we get a contradiction to Theorem 6. ∎

Now the proof of our main result is complete:

Theorem 9 *A cpo D for which $\big[[D \to D] \longrightarrow [D \to D]\big]$ is ω-algebraic is a profinite domain.*

The maximality result for the categories is also not hard to show:

Theorem 10 *Any cartesian closed full subcategory of ω-**ALG** (no least element assumed) is contained in the category of profinite domains.*

Proof. It must be shown that in a full subcategory of ω-**ALG** the terminal object, the product and the function space are what one would expect them to be, namely, the one-point domain, the cartesian product and the space of Scott-continuous functions, respectively. This can be done exactly like Smyth did it with two comparable compact elements playing the rôle of "$\{x, \bot\}$" there. ∎

References

[1] **C.Gunter:** *Profinite Solutions for Recursive Domain Equations.* Doctoral Dissertation, University of Wisconsin,Madison 1985.

[2] **C.Gunter:** *Comparing Categories of Domains.* Technical report, Carnegie-Mellon University, Pittsburgh 1985.

[3] **G.Markowsky:** *Chain-complete p.o. sets and directed sets with applications.* Algebra Universalis **6** (1976), p.53–68.

[4] **G.Plotkin:** *A powerdomain construction.* SIAM J. Comput. **5** (1976), p.452–488.

[5] **M.B.Smyth:** *The Largest Cartesian Closed Category of Domains.* Theoretical Computer Science **27** (1983), p.109–119.

Part IV

Domain Theory and Theoretical Computation

SEMANTICALLY BASED AXIOMATICS

Stephen D. Brookes
Carnegie-Mellon University
Department of Computer Science
Pittsburgh
Pennsylvania 15213

ABSTRACT

This paper discusses some fundamental issues related to the construction of semantically based axiomatic proof systems for reasoning about program behavior. We survey foundational work in this area, especially early work of Hoare and Cook on while-programs, and we try to pinpoint the principal ideas contained in this work and to suggest criteria for an appropriate generalization (faithful to these ideas) to a wider variety of programming languages. We argue that the adoption of a mathematically clean semantic model should lead to a natural choice of assertion language(s) for expressing properties of program terms, and to syntax-directed proof systems with clear and simple rules for program constructs. Hoare's ideas suggest that in principle syntax-directed reasoning is possible for all syntactic categories (declarations, commands, even expressions) and all semantic attributes (partial correctness of commands, aliasing properties of declarations, L- or R-values of expressions, proper use of variables, and so on). Semantic insights may also influence assertion language design by suggesting the need for certain logical connectives at the assertion level. This point is obscured by the fact that Hoare's logic for while-programs needed no assertion connectives (although of course the usual logical connectives are permitted inside pre- and post-conditions), but an application of our method to a class of parallel programming languages brings out the idea well: semantic analysis suggests the use of conjunctions at the assertion level. We argue that this method can lead to proof systems which avoid certain inelegant features of some earlier systems: specifically, we avoid the need for "extra-logical" and "non-compositional" notions such as interference checks and auxiliary variables. We also discuss the author's applications of these techniques to other programming languages, and point to some future research directions continuing this work. Although we do not have a completely satisfactory general theory of semantically based axiomatization, and consequently some of our techniques may seem rather *ad hoc* to the reader, we hope that our ideas have some merit.

INTRODUCTION

Our main point in this paper is to develop the argument that axiomatic reasoning should be semantically based, one of the important ideas behind Hoare's early work on proving partial correctness of programs. Although this may sound obvious and in fact most existing proof systems could be claimed to be based on some more or less explicit semantics, we believe that previously proposed proof systems for program properties have sometimes failed to take full advantage of the benefits of well chosen semantic bases. If the underlying semantics is unnecessarily complex, it is more likely that an attempt to design a proof system based upon it will result in errors or undesirable features such as very complicated inference rules with complex side conditions on applicability (or, worse still, unsoundness). We argue that the adoption of a mathematically simple semantic model should lead to a natural choice of an assertion language for expressing program properties, and then to a syntax-directed proof system with clean and simple rules for program constructs. The semantic structure should influence the choice of assertion language, and may even suggest the need for logical connectives such as conjunction at the assertion level, although this is not needed in Hoare's original proof system for while-programs because of their especially simple semantics.

To be more precise, we mean that firstly, when desiring to reason about a certain class of program properties (e.g. partial or total correctness, deadlock-freedom, etc.), one should begin by formulating a semantic model for the programming language which is *adequate* for that program property. By adequacy we mean that the semantics must be able to distinguish between program terms which can be used in some program context to induce different program behaviors. This is one half of the full abstraction definition [17]: full abstraction requires that the semantics should distinguish between terms if and only if they can induce different program behaviors in some program context. Thus, a fully abstract semantics would certainly be adequate; however, full abstraction may be difficult to achieve for certain languages and classes of properties (see [15,23] for instance), while adequacy is rather easier to achieve and suffices for accurate support of syntax-directed reasoning, since it permits the replacement of semantically equal terms inside a program without affecting the overall properties of the program.

Secondly, the semantic model should have a clean mathematical structure. For instance, the standard partial function semantics for while-programs is certainly clean and simple and is at the right level of abstraction to be adequate for partial correctness or, for that matter, for total correctness. Thirdly, the very structure of the semantic model should guide the choice of assertion language(s) for expressing properties of program terms. We illustrate this last idea with examples, although as yet we do not have a completely worked out general theory which would describe for an arbitrary semantic model how to construct a corresponding assertion language. We have more to say on this in the conclusion, where we also draw attention to other work in this direction.

A particularly important facet of this last idea is that the structure of a semantics may necessitate the use of logical connectives in an assertion language; while it may seem trite to a logician to argue for the inclusion of logical connectives in what is, after all, a logical formalism, Hoare's logic for while-programs contained no explicit connectives. In that case, there was actually no need: a simple semantic justification can be given that shows that one does not need to form (for instance) conjunctions of Hoare logic assertions in order to obtain a complete proof system. However, this property need not hold for more intricate programming languages, and it certainly fails with parallel programs. Again this point is brought out fully by our example. We also point to some directions for further investigation. Firstly, however, we give some historical remarks to set the scene.

HOARE'S LOGIC

The ideas of using axiomatic techniques to reason about program behavior go back to the pioneering work of Floyd [13] and Hoare [14]. In 1966, Hoare proposed an axiomatic basis for computer programming, in a paper of that title. The main idea was very simple but powerful: to use partial correctness assertions (pca's) to specify program properties, and to design a proof system in which pca's about compound programs can be deduced from pca's about their syntactic constituents. In popular terms, Hoare proposed a "syntax-directed" logic for partial correctness. A partial correctness assertion takes the form $\{P\}C\{Q\}$, where P and Q are *conditions* and C is a program. This is intended to be interpreted informally as saying that if C is executed from an initial state satisfying P, then if execution terminates, it does so in a final state satisfying Q. The conditions used by Hoare were drawn from a first order logical language for arithmetic, a natural enough choice since the expressions of his programming language denoted integers and boolean values.

In Hoare's proof system there are axioms for each atomic form of program (e.g., assignment) and inference rules for each program construct. For instance, the following *while*-rule from [14] is well known by now:

$$\frac{\{P \,\&\, B\}C\{P\}}{\{P\}\text{while } B \text{ do } C\{P \,\&\, \neg B\}}$$

This rule allows us to deduce a (special type of) pca about a while-loop from a (special kind of) pca about the loop body. In addition, and what might seem incongruous if one is looking for a purely syntax-directed proof system, Hoare's proof system included a so-called *rule of consequence*:

$$\frac{P' \Rightarrow P \quad \{P\}C\{Q\} \quad Q \Rightarrow Q'}{\{P'\}C\{Q'\}}$$

This rule, applicable to all while-programs C (not just to loops), allows us to "strengthen" a pre-condition and "weaken" a post-condition. Note also that some such rule is necessary in any case because of the special restrictions on the assertions deducible from the syntax-directed rules; without some general rule like this, one would only be able to prove assertions about loops with the fixed format $\{P \& B\}$**while** B **do** $C\,\{P \& \neg B\}$, and one would be unable to manipulate conditions inside assertions.

Thus, it is sometimes stated that Hoare's proof system has two parts: a syntax-directed part containing the axioms and rules for the programming constructs, and a "logical part" containing (at least) the rule of consequence. A "logical" rule is applicable to all programs in the programming language, whereas a syntax-directed rule is only applicable to programs built using a specific construct.

SOUNDNESS AND COMPLETENESS

Cook [11] established the soundness of the Hoare system, and also gave what is now a standard definition of completeness for program logics ("relative completeness"). It is useful for our purposes to summarize the essential ideas. Fuller details are in [2] and [11].

Firstly, one needs to make the reasonable assumption that the condition language be chosen to be sufficiently powerful to contain all intermediate conditions needed during a proof. Cook therefore defines an "expressivity" criterion. Bearing in mind that the programming language in question here (while-programs) has a partial function semantics, expressivity boils down to the following property: for every condition P in the condition language and every program C, there must be a condition Q in the condition language characterizing the set of states

$$\{\, s \mid \mathcal{M}(C)s \text{ is defined and satisfies } P\,\},$$

where $\mathcal{M}(C)$ is the partial function denoted by C. Loosely speaking, expressivity amounts to closure under (semantic) weakest pre-condition (Dijkstra [12]). An equivalent formulation can be given in terms of strongest post-condition (Clarke [9,10]). The first order language for arithmetic used by Hoare is certainly expressive for his application, and we will not focus on this issue further at this point.

Completeness in the standard sense is clearly impossible, given the well known fact that the validity problem for pca's is undecidable ([2]): there is no effective procedure for testing the validity of partial correctness assertions. This holds even if the condition language is trivially small, as long as it contains **true** and **false**, so that halting problems become expressible as pca's. It is therefore impossible to obtain a complete proof system in the usual sense of completeness, since the set of provable assertions would then be recursive, while the set of valid pca's is recursively enumerable and not recursive. Cook wanted to isolate any incompleteness due to the syntax-directed proof rules and axioms. Correspondingly, he specified that Hoare's logic should be regarded as "relatively" complete if the following condition holds: for every valid pca there is a proof of that pca in which we are allowed to use as axioms or unproved assumptions any valid *condition*. The idea is to allow the proof system to access an "oracle" able to answer validity questions about conditions (and implications between conditions). The primary place where this oracle is useful, of course, is the rule of consequence.

Cook proved the following completeness theorem: for any expressive condition language, Hoare's proof system for while-programs is relatively complete. The proof given by Cook is itself illuminating. He showed that, for every valid pca $\{P\}C\{Q\}$, we can find an assignment of pre-conditions and post-conditions to the subterm occurrences of C which can be used in a natural, syntax-directed manner, to prove $\{P\}C\{Q\}$. Since a particular subterm may occur several times inside a given program, and in general each occurrence serves a different purpose semantically, one needs to be able to say something different about each occurrence; hence the reference to subterm *occurrences* rather than simply to subterms. [This preoccupation with subterm occurrences rather than subterms will recur throughout this paper.] Let $\prec$ denote the relation "is an immediate subterm occurrence of"; the transitive closure of this relation is the subterm occurrence relation, denoted $\prec^*$; at the moment we will only refer to subterms which are commands, although of course subterms may be expressions or declarations. If we let *pre* and *post* be the functions mapping subterm occurrences of C to pre- and post-conditions respectively, the idea is that the set

$$\{\{pre(C')\}C'\{post(C')\} \mid C' \prec^* C\}$$

can be used in a syntax-directed manner to build a proof of $\{P\}C\{Q\}$. The choice of the functions *pre* and *post* obviously depends on P, C, and Q.

To be precise about the requirements on these *pre* and *post* functions, and explain exactly what is intended by "using" this set of pca's in a syntax-directed manner, one would need to provide a collection of constraints. Typical of these is the constraint imposed by a subterm occurrence built by sequential composition: if $C' = C_1; C_2$ is a subterm occurrence of C we require that

$$\begin{aligned} pre(C') &\Rightarrow pre(C_1) \\ post(C_1) &\Rightarrow pre(C_2) \\ post(C_2) &\Rightarrow post(C'). \end{aligned}$$

These constraints allow a proof of

$$\{\,pre(C')\,\}C'\{\,post(C')\,\}$$

from the pair of assertions

$$\{\,pre(C_i)\,\}C_i\{\,post(C_i)\,\}, \quad i = 1, 2$$

by using Hoare's rule for sequential composition and the rule of consequence. There are similar constraints on *pre* and *post* for the other program constructs. In each case, the constraints allow the deduction of the assertion $\{\,pre(C')\,\}C'\{\,post(C')\,\}$ from the corresponding assertions about the (immediate) syntactic subterms of C' and the rule of consequence. This formalizes the notion of a "standard" method of syntax-directed proof.

The assignment of pre- and post-conditions to the subterm occurrences of C in order to prove $\{\,P\,\}C\{\,Q\,\}$ is known as a "proof outline", for obvious reasons. It is important that a single pca suffices for each subterm occurrence; this is the crucial property that enables us to achieve (relative) completeness with a proof system in which each inference rule has a single premise for each immediate syntactic component of the program appearing in the conclusion to the rule: i.e. the rules have fixed finite numbers of premises whose structure directly corresponds to the syntactic structure of the conclusion.

Because of the close connection between the idea of weakest pre-conditions and the construction of the *pre* and *post* functions in Cook's theorem, Cook's proof of relative completeness can be paraphrased as showing that weakest pre-conditions can be used in a straightforward way to prove any valid pca, provided we assume that all necessary reasoning inside the condition language can be carried out. Dijkstra [12] showed that a syntactic definition of weakest pre-conditions can be given for while-programs. Clarke [10] established the connection between Cook's notion of expressibility and the ability of the condition language to contain weakest pre-conditions and/or strongest post-conditions.

LOGICAL RULES AND CONNECTIVES

To a logician there might appear some *ad hoc* features in Hoare's "logic" for while-programs. If pca's are the objects of proof (*cf.* formulas or theorems), where are the logical connectives? One never forms the conjunction of two pca's in Hoare's logic, nor the disjunction, and certainly not the negation of a pca! Apart from the so-called logical rules, traditional logical connectives appear to play a very minor role in Hoare's proof system. Certainly logical combinations of conditions are manipulated (e.g. in the rules for loops and conditionals, and in the rule of consequence), but the same is emphatically not true for pca's.

The role of the rule of consequence in the completeness property for Hoare's logic is crucial. Other sound logical rules, such as

$$\frac{\{P_1\}C\{Q_1\} \quad \{P_2\}C\{Q_2\}}{\{P_1 \,\&\, P_2\}C\{Q_1 \,\&\, Q_2\}}$$

(or, for that matter, the similar rule involving $\vee$) are not needed to achieve completeness, although it could be argued that they are useful pragmatically. Note also that this rule is not really related to the familiar &-introduction rule of natural deduction, despite its superficial resemblance in that it introduces conjunction in both pre- and post-conditions: the assertion $\{P_1 \& P_2\}C\{Q_1 \& Q_2\}$ does not behave logically as a conjunction of assertions, since the fact that C satisfies it does not logically imply that either of the assertions $\{P_i\}C\{Q_i\}$ is also valid. Of course, in propositional or predicate calculus each of ϕ and ψ is a logical consequence of $\phi \,\&\, \psi$. Thus, even rules such as this which superficially seem to involve connectives used with assertions about the same program are not truly treating assertions as objects of a boolean algebra.

Yet the rule of consequence is in reality a disguised form of *modus ponens*. If we define the obvious notion of implication for pca's, i.e. that

$$\{P\}C\{Q\} \Rightarrow \{P'\}C\{Q'\} \quad \Leftrightarrow \quad (P' \Rightarrow P) \,\&\, (Q \Rightarrow Q'),$$

the connection with *modus ponens* becomes clearer. Note, nevertheless, that this "implication" is not a true logical connective on pca's, because it only applies to pca's about the same program. If we reformulate Hoare's ideas without using sugar, writing

$$C \models (P, Q) \quad \text{for} \quad \{P\}C\{Q\}$$

and writing $(P, Q) \Rightarrow (P', Q')$ for $(P' \Rightarrow P)$ & $(Q \Rightarrow Q')$, we have for the rule of consequence:

$$\frac{C \models (P, Q) \qquad (P, Q) \Rightarrow (P', Q')}{C \models (P', Q')}$$

This merely syntactic reformulation of the rule involves less syntactic sugar and, I feel, emphasizes better the logical structure of the rule. We will see later that this version generalizes to other settings.

The reason that no logical connectives (other than implication) on assertions were necessary is summarized as follows. In order to prove a (single) pca about a program C, we can find an assignment of pca's to the syntactic subterm occurrences of C from which a proof can be constructed. This is the content of the "proof outline theorem" as described earlier. If, on the contrary, one might have needed several assertions for each subterm of C, it might have been necessary to allow conjunction of pca's (rather than allowing rules with arbitrarily many premises). Crucially for while-programs, we can get by with a single assertion about each piece of a program.

However, there is no reason why this should hold when we move to more complicated programming languages. Firstly, if a language has a more complicated semantic model (than partial functions from states to states) there is no particular reason to suppose that the pca format should lend itself naturally to syntax-directed reasoning. Essentially, the problem is that pca's may no longer have a structure that "fits" the semantic structure adequate for partial correctness. For example, in [8] we reported on a syntax-directed proof system for a simple block structured language with aliasing introduced by declarations, and we found indeed that the pca format needed to be modified to allow a good fit with semantic structure.

In addition, there may be quite different notions of program correctness which are of interest, and then it is not always most natural to use pca's to express program properties. By way of example, again from [8], when reasoning about the correctness of block structured programs in a setting where aliasing may occur among program variables, it was found necessary to axiomatize the (purely declarative) aliasing properties of declarations in addition to their imperative effects; although the aliasing properties were conveniently expressed in a notation which superficially resembles that of pca's, the "pre- and post-conditions" were drawn from a very simple language solely chosen to allow succinct descriptions of aliasing relationships and did not need the full power of a language for arithmetic. For another example, in which a much more radical departure from the pca format is suggested by the semantic structure, see the application to parallel programs in the next section of this paper.

Finally the need for connectives on assertions should be re-examined for an application to more complicated languages. Again, the parallel programming example brings this out well, and we now examine this in more detail.

AN EXAMPLE

To illustrate our points more concretely, we summarize the application of our ideas to a simple parallel programming language, essentially the language discussed by Owicki in her thesis [19] and in the paper [20]. We provide here a condensed development of this work; a full presentation appeared in [6]. Many of the details are suppressed to avoid excessive duplication of effort.

The programming language is a standard while-loop imperative language to which we add a parallel composition $C_1 \parallel C_2$ and a conditional critical region **await** B **then** C, in which the body C must be executable as a single atomic unit. We regard assignments and boolean expression evaluations as atomic. [In fact, it is reasonable to constrain the body of a critical region to be a finite sequence of assignments, although this is not crucial to our discussion.] The interpretation of parallel composition will be nondeterministic interleaving of atomic actions of the two parallel processes C_1 and C_2 until both have terminated. A conditional critical region **await** B **then** C can only be executed when B is true, and its effect is to perform all of C without allowing interruption from any other (parallel) process. When its test condition is false, an await command is unable to progress and must, as its name suggests, wait for the state to be changed (by another program executing in parallel) to one satisfying the condition. When a command is unable to progress but has not yet properly terminated (typically because of an await) it is said to be deadlocked. Thus, for instance, the effect of

$$x{:=}0;[(\textbf{await}\ x = 1\ \textbf{then}\ y{:=}2) \parallel x{:=}1]$$

is to (finally) set x to 1 and y to 2; on the other hand, the effect of

$$x{:=}1;[(\textbf{await}\ x = 1\ \textbf{then}\ y{:=}2) \parallel x{:=}0]$$

is either to deadlock (with x set to zero) or to terminate with $x = 0$ and $y = 2$. Full semantic details appear in [6].

It is well known that for a treatment of partial correctness and deadlock this language requires a more sophisticated semantic model than the obvious modification of state-transformations to relational semantics. Using the operational presentation of Hennessy and Plotkin [15] it is convenient to describe the semantics of this language in terms of the computations of an abstract machine whose configurations have the form $\langle C, s\rangle$ (where C is a command and s is a state), and whose transitions (one-step of a computation) involve a single atomic action and are described by a family of transition relations $\rightarrow^{\alpha}$ (where α ranges over the set of atomic actions). A transition of the form $\langle C, s\rangle \rightarrow^{\alpha} \langle C', s'\rangle$ represents the ability of C, in state s, to execute the atomic action α and that immediately afterwards the state is s' and the remaining command is C'. There are two types of configuration from which no transition is possible: deadlocked and successfully terminated configurations. A full semantic description of the language gives definitions of the transition relations and of two sets *DEAD* and *TERM* of configurations. The partial correctness and deadlock behavior of a program C can then be described as a function

$$\mathcal{M}(C) : S \rightarrow (\mathcal{P}(S) \times \mathcal{P}(S))$$

which is itself presentable in the following form, by means of two auxiliary functions

$$\begin{aligned}
&\mathcal{S}(C) : S \to \mathcal{P}(S)\\
&\mathcal{S}(C)s = \{\, s' \mid \langle C, s\rangle \to^* \langle C', s'\rangle \in \mathit{TERM}\}\\
&\mathcal{D}(C) : S \to \mathcal{P}(S)\\
&\mathcal{D}(C)s = \{\, s' \mid \langle C, s\rangle \to^* \langle C', s'\rangle \in \mathit{DEAD}\}\\
&\mathcal{M}(C)s = \langle \mathcal{S}(C)s, \mathcal{D}(C)s\rangle
\end{aligned}$$

Here $\to$ denotes the one-step transition relation, so that $\langle C, s\rangle \to \langle C', s'\rangle$ represents that for some α we have $\langle C, s\rangle \to^{\alpha} \langle C', s'\rangle$. We use $\to^*$ for the transitive closure of this relation. Thus, $\mathcal{S}(C)s$ is the set of states in which some finite sequence of transitions may terminate from the initial configuration $\langle C, s\rangle$. However, it is impossible to construct this function in a purely denotational, syntax-directed manner. The reason is that we are ignoring intermediate states and the potential for interference between parallel processes. As Hennessy and Plotkin established, it is necessary instead to use a semantics based on a more intricate structure (resumptions). Ignoring for the moment the fact that the resumptions are recursive objects from a recursively defined domain, we can design a semantics

$$\begin{aligned}
&\mathcal{R} : \mathbf{Com} \to R\\
&R = [S \to \mathcal{P}^*(R \times S)]\\
&\mathcal{R}(C)(s) = \{\, \langle \mathcal{R}(C'), s'\rangle \mid \langle C, s\rangle \to \langle C', s'\rangle\,\}
\end{aligned}$$

We use here $\mathcal{P}^*$ to denote a simple variant of the finite powerset constructor in which there are two versions of the empty set, which we denote $\bullet$ and $\circ$, representing respectively termination and deadlock. Thus, if $\mathcal{R}(C)(s) = \bullet$ we say that C has terminated in state s, and similarly for deadlock. In this structure it is important to note that the sets $\mathcal{R}(C)(s)$ are always finite, and indeed that each member of such a set is the result of a unique atomic action (occurrence) from the text of C which is "enabled" in state s. Again, we refer the reader to [6] for more details. The rigorous mathematical justification for the use of recursively defined domains here is not germane to the paper, although of course justification is necessary for the semantic definitions to make sense.

Now, in the same way that Hoare's syntax for pca's "fits" the semantic structure $[S \to S]$, once we have chosen a syntax for conditions describing S, we may design an assertion language for the structure R. Let ϕ be an assertion describing a resumption r. We need in ϕ to be able to describe, given some information about an initial state s (i.e., given a condition P), whether or not (the command whose resumption is) r deadlocks or terminates, and if not, some information about each of the possible results of the atomic actions enabled in s. Each of these results requires another assertion about a resumption (describing the resulting command) and a "post"-condition describing the resulting state. Thus we are led, with some syntactic sugaring, to the following "grammar" for ϕ:

$$\phi ::= P\circ \mid P\bullet \mid P\sum_{i=1}^{n} \alpha_i P_i \phi_i$$

We have chosen to include atomic actions in the syntax of assertions purely for ease of relation to other proof systems, and we have chosen a linear notation for the general form of assertion because it resembles very closely Milner's format for synchronization trees [16]; in fact, it is convenient to think of an assertion as a synchronization tree with conditions before and after each of its arcs. Our aim will now be to produce a proof system for establishing properties of the form $C \models \phi$. The interpretation of such a property is, of course, closely based on the semantics.

Now that we have an assertion language whose structure closely models the semantic structure of the objects it describes, it is clearly going to be possible to reason about these objects in a syntax-directed manner, just as the denotational semantics builds meanings of commands in a syntax-directed manner. In fact, just as there is a semantic operation $\|$ on resumptions such that

$$\mathcal{R}(C_1 \| C_2) = \mathcal{R}(C_1) \| \mathcal{R}(C_2)$$

we can introduce a "semantic connective" $\|$ on assertions with the intended property that whenever ϕ_1 describes C_1 and ϕ_2 describes C_2, then also $\phi_1 \| \phi_2$ describes $C_1 \| C_2$. The required definition, taken from [6], is for the "base cases":

$$\begin{aligned}(P\bullet) \| (Q\bullet) &= \{P \,\&\, Q\}\bullet \\ (P\bullet) \| (Q\circ) &= \{P \,\&\, Q\}\circ \\ (P\circ) \| (Q\circ) &= \{P \,\&\, Q\}\circ .\end{aligned}$$

For $\phi = (P\sum_{i=1}^{n} \alpha_i P_i \phi_i)$ and $\psi = (Q\sum_{j=1}^{m} \beta_j Q_j \psi_j)$,

$$\phi \| \psi = \{P \,\&\, Q\}(\sum_{i=1}^{n} \alpha_i P_i [\phi_i \| \psi] + \sum_{j=1}^{m} \beta_j Q_j [\phi \| \psi_j]).$$

Given this definition, which we regard as constituting a logical characterization of $\|$ as a connective on assertions, we can use the following proof rule for reasoning about parallel programs:

$$\frac{C_1 \models \phi_1 \quad C_2 \models \phi_2}{[C_1 \| C_2] \models [\phi_1 \| \phi_2]}$$

Although we have not given the details here, the soundness of this rule is obvious, because the definition of $\phi \| \psi$ is essentially a rephrasing of the semantic clause defining $\mathcal{R}(C_1 \| C_2)$ from $\mathcal{R}(C_1)$ and $\mathcal{R}(C_2)$. Since the assertion language and rules are semantically based, the soundness proofs are made by appeal to the semantic definitions.

Using syntax-based rules of the above kind for the programming constructs, it is straightforward to design a simple proof system for properties of the form $C \models \phi$, in which the premises in each rule involve assertions about the principle subterms of the conclusion.

At this point, one might ask if the assertion language and proof system are sufficiently powerful. The answer is no, if we want to achieve the desired completeness properties. The first reason is that, as with Hoare's logic for while-programs, we need an analogue of the rule of consequence to allow us to manipulate conditions. There is in fact a very natural generalization of the rule of consequence, which we will embody as a form of *modus ponens*. Firstly, it is possible to define an *implication* on assertions: implication $\Rightarrow$ is characterized by the properties:

$$\begin{array}{rcl} (P\bullet) \Rightarrow (Q\bullet) & \Leftrightarrow & (Q \Rightarrow P) \\ (P\circ) \Rightarrow (Q\circ) & \Leftrightarrow & (Q \Rightarrow P) \\ (P\sum_{i=1}^{n} \alpha_i P_i \phi_i) \Rightarrow (Q \sum_{i=1}^{n} \alpha_i Q_i \psi_i) & \Leftrightarrow & (Q \Rightarrow P)\ \& \\ & & \bigwedge_{i=1}^{n} [(P_i \Rightarrow Q_i)\ \&\ (\phi_i \Rightarrow \psi_i)] \end{array}$$

Note that we use $\Rightarrow$ to denote assertion implication and the usual implication on conditions; the context makes it clear which is intended. It should be clear that this definition of implication is the obvious extension to our more highly structured assertion language of the (implicit) notion of implication for ordinary pca's as described earlier. The analogue in this setting to the rule of consequence is:

$$\frac{C \models \phi \qquad (\phi \Rightarrow \psi)}{C \models \psi}$$

Even with the inclusion of this rule, there is not yet any analogue of the Cook proof outline property. There is a second reason for incompleteness. There are simple examples of commands C and assertions ϕ such that $C \models \phi$ is valid, but is not deducible by first proving a *single* assertion for each syntactic subterm occurrence of C. A rather elementary example is:

$$[x{:=}x+1 \mid\mid x{:=}x+2].$$

No pair of single assertions (one each) about $x{:=}x+1$ and $x{:=}x+2$ can be combined to prove (the assertion which states) that this program increases x by 3. Instead, in this case, we need to be able to make *two* assertions about each subterm occurrence (one for use when the term is executed before the other, and one after). The general scheme is that one needs in principal to allow arbitrary finite conjunctions of assertions about each subterm occurrence. Thus, we can recover the "one assertion for each subterm" property by throwing in conjunction at the assertion level! (This also necessitates a careful axiomatization of the interaction of conjunction with parallel composition of assertions, as discussed in [6]. It is also necessary to specify the obvious implicational properties of conjunctions.) If we do this, as shown in [6], we obtain the expected results: a sound and relatively complete proof system. The point to emphasize here is that semantic considerations have led us to include conjunction in the assertion language.

The analogue of Cook's proof outline theorem is then: for every valid assertion $C \models \phi$ there is an assignment *assert* of assertions to subterm occurrences of C such that $C \models \phi$ is deducible in a standard manner from the premises

$$\{C' \models assert(C') \mid C' \prec^* C\}.$$

Again, the assignment of assertions to subterms must satisfy certain requirements for this standard deduction to be possible. For instance, given a subterm occurrence $C_1 \parallel C_2$ of C, we require that

$$(assert(C_1) \parallel assert(C_2)) \Rightarrow assert(C_1 \parallel C_2).$$

As with the proof outline constraints, each type of syntactic construct imposes a constraint on the assert assignment. Fuller details will be given in an expanded version of [6]. The main point is that assertions about a compound command should be deducible from assertions about its immediate subterms.

It should be noted that the use of conjunction means that we do not need recourse to auxiliary variables; and that the careful definition of parallel composition of assertions was made in order to avoid the need for interference-freedom checks [19,20]. Both of these points are elaborated in more detail in [6] and [7]. It is in avoiding recourse to these rather extra-logical features that we see the principal advantages conferred by our approach.

GENERALIZING HOARE'S LOGIC

We have surveyed early developments in axiomatization and given an example which we believe generalizes the important ideas appropriately to a significantly more complicated programming language. Although we would not claim to have worked out a complete theory of semantically based axiomatization applicable to all possible programming languages and classes of program properties, we feel that several essential ideas are suggested by our experience. The following points summarize these ideas well.

- Assertion language(s) should be designed to fit semantic model(s).

- Syntax-directed reasoning is (in principle) feasible for all syntactic classes.

- Logical connectives should be included in an assertion language if the semantic properties imply their utility.

The first point was, as we have already remarked, certainly satisfied by Hoare's choice of syntax for pca's, in that a pca contains two conditions, one for the initial and one for the final state. The other two points are perhaps not so clearly visible in Hoare's while-program logic, since the principal topic of his work was the axiomatization of the partial correctness properties of commands alone; there was no need then to axiomatize expressions or declarations, and no need for conjunctions of assertions. Indeed, to some extent, the syntactic sugar used in the pca format obscures the logical structure.

As a point often taken for granted, note that even the choice of *condition* language ought to be influenced by semantic properties. In particular, we know that while-programs suffice to describe all partial recursive functions on the integers (assuming that all expressions are integer-valued). Moreover, if a condition language contains the usual logical connectives (a natural enough property!) and contains conditions of the form $I = E$, where I is a program variable or identifier, and E is an (integer-valued) expression, any finite state (i.e., any finite function from a finite set of identifiers to integers) can be described uniquely by a "characteristic condition", of the form

$$I_1 = v_1 \ \& \ I_2 = v_2 \ \& \ \ldots \ \& \ I_k = v_k.$$

Since programs describe partial recursive functions, the (sets of states described by) conditions must be closed under image and pre-image of partial recursive functions if we are to be able to express all necessary intermediate conditions. This conclusion is obviously related to the results established in [3], where an analysis is given of the suitability and expressive power of recursive and recursively enumerable condition languages.

The second point made above is that syntax-directed reasoning is possible for any syntactic category, not solely for programs. This was already implicit in Hoare's early work [14], but I do not believe that later developments took up the idea to its fullest extent. For one example, some early work on axiomatizing languages with declarations as well as commands utilized the pca format $\{P\}C\{Q\}$ again, but with the pre- and post-conditions also expressing declarative properties such as aliasing relationships. It is conventional to give semantics by means of separate (but related) environments and stores, as for example in Stoy [24] and in Strachey [25]. The meaning of a command is then a function from environments to partial store-transformations, and a declaration denotes an environment-transformation. When this is done, the pca format no longer fits as well: from the format of a pca for commands it appears that we will have to keep *proving* that the command has no effect on the environment, because the post-condition mentions properties of the environment. But only declarations change the environment, so it is actually more convenient to design two proof systems in tandem: one for declarations and one for commands. The assertions for commands should involve two separate pre-conditions and a single post-condition, with a pre-condition for the environment and one also for the store. A detailed exposition of such an approach can be found in [8].

Next we expand on the idea that semantic structure should influence the design of assertion languages and proof systems. We attempt a general definition of what a semantically based, syntax-directed proof system should be.

In a denotational semantic definition for a programming language the meaning of a term is constructed from the meanings of its parts. Given the abstract syntax of a programming language, in particular the set of syntactic *types* (e.g. **Com, Exp, Ide, BExp**) and the syntactic constructors (e.g., ";" of type $\mathbf{Com} \times \mathbf{Com} \rightarrow \mathbf{Com}$), a denotational semantics consists of a family of semantic functions, (usually) one for each type, each mapping terms of a particular type into meanings appropriate for that type. These functions are usually presented by a structural induction on the syntax. It is usual to pick out a collection of semantic domains, one for each type, so that if τ ranges over the syntactic types, we can write D_τ for the semantic domain appropriate for type τ and define a semantic function $\mathcal{M}_\tau : \tau \rightarrow D_\tau$. The denotational condition that a term's meaning is determined from the meanings of its subterms then becomes reflected as follows. For each program construct *op* of type $\tau_1 \times \cdots \times \tau_n \rightarrow \tau$ there is a semantic operation F_{op} which constructs the meaning $\mathcal{M}_\tau(op(t_1, \ldots, t_n))$ from the meanings $\mathcal{M}_{\tau_i}(t_i)$. With the usual notion of a subterm occurrence, and of immediate subterm occurrence, written $t' \prec t$ as before (or $(t' : \tau') \prec (t : \tau)$ if we want to indicate the types), each t_i is an immediate subterm occurrence of $op(t_1, \ldots, t_n)$. This means that $\mathcal{M}_\tau(t)$ is constructed from the set $\{ \mathcal{M}_{\tau'}(t') \mid (t' : \tau') \prec (t : \tau) \}$ in a standard way depending on the operator used to construct t. The definition of $\mathcal{M}_\tau(t)$ is commonly referred to as a *semantic clause.*

The relevance of this general definition when we try to formalize what is the essence of syntax-directed, semantically based axiomatics should be clear. For each syntactic construct *op* and its corresponding semantic clause there should be an inference rule. There should be assertion languages and proof systems for each type τ, so that if we write $(t \models A) : \tau$ to indicate a typical assertion about a term t of type τ, this inference rule typically would take the form:

$$\frac{\{(t' \models A') : \tau' \mid (t' : \tau') \prec (t : \tau)\}}{(t \models A) : \tau}$$

with a premise for each immediate subterm occurrence of t. Although this formulation may look somewhat awkward, it collapses to the usual Hoare rules when applied to while-programs and partial correctness semantics, with the adoption of the usual syntactic sugaring and suppression of syntactic types. And for each type τ it should be possible to define an appropriate notion of implication on assertions about objects of that type, so that the proof system for that type would include a *modus ponens* rule. Of course, the proof systems for the various types fit together in a hierarchical manner which mimics the syntactic structure of the programming language; and the various implication relations at each type may require a mutually recursive definition.

An appropriate generalization of the use of proof outlines in the Cook theorem and its proof would then be the following. If $(t \models A) : \tau$ is a valid assertion (validity being defined by appeal to the semantics of the language, of course), then there is an (type-respecting) assignment *assert* of *assertions* to the (immediate) subterm occurrences of t such that $(t \models A) : \tau$ is deducible from the application of this rule to the premises

$$\{(t' \models assert(t')) : \tau' \mid (t' : \tau') \prec (t : \tau)\}$$

This again collapses to the Cook theorem for while-programs in that simple case. And in the example of parallel programs this seems an appropriate generalization of the relevant theorem.

RELATED WORK AND FURTHER RESEARCH

We have not described a general-purpose technique for constructing semantically based, syntax-directed proof systems. Rather, our as yet limited experience gained while investigating some particular programming languages and program properties has led us to make some (we hope) reasonably coherent guidelines. Our formulation of the denotational setting and its corresponding axiomatic analogue in the previous section is intended as a first step towards a general theory. We believe that the ideas are much more widely applicable than we have been able to indicate here, and a general theory would be very worthwhile. The recent work of Abramsky [1] may turn out to be an important contribution towards such a general theory; he aims at a logical presentation of the domain theoretic constructs prevalent in denotational semantics. In a similar vein we also mention the recent developments of Robinson, also exploring the axiomatics of denotational semantics ([21]). The full implications of this work need to be worked out, and the connections with existing proof systems could be interesting. One particularly interesting example should be provided by the work of Stirling [22], who has developed a compositional (syntax-directed) formulation of the Owicki-Gries proof system involving a different type of assertion from ours.

There are several interesting issues from an axiomatic point of view which remain to be explored. A fairly simple example occurs in proof systems for dealing with arrays (see [4] for example) where it is common to find an axiom for assignment to an array position which superficially looks as simple as Hoare's axiom for "ordinary" assignment to a variable:

$$\{ [E \setminus A[E_0]]P \} A[E_0] := E \{ P \}.$$

However, the syntactic definition of what it means to substitute an expression for $A[E_0]$ in a condition is somewhat involved. I believe it would be worth investigating an alternative proof system in which a component of the proof system involves reasoning about the (R-)value ([24]) of an (index) expression. It may be useful to allow reasoning about assertions such as "the value of E_0 lies in a certain range", with the obvious intention. One might then design a proof rule of the following form.

$$\frac{\{ E_0 : X \} \quad \bigwedge_{v \in X}([E \setminus A[v]]Q \Rightarrow P)}{\{ P \} A[E_0] := E \{ Q \}}$$

where $E_0 : X$ is an assertion saying that the value of E_0 is in a range described by the set X (a finite subset of the integers). Since the substitution now only involves "simple" array variables $A[v]$ where v is known, it ought to be possible to give a more straightforward syntactic definition. We have ignored issues pertaining to range checking for indices out of bounds. And of course this idea itself brings other problems, such as the possible need for more complex assertions for E_0, and the choice of a syntax for describing finite subsets of integers. Nevertheless, our version of the rule could be argued as more accurately reflecting the semantics and our operational intuitions about what happens in an array assignment. We do not want to get involved here in the details, but we propose to investigate this issue elsewhere.

As we outlined above, syntactic classes other than commands are candidates for syntax-directed reasoning, and an approach that systematically investigates the possibilities may lead to some revision of our current ideas as to the "best" way to reason about programs. Notable work along these lines is reported in [5], where an axiomatization is given for expressions with side-effects.

Much more challenging applications of our ideas are provided by procedural languages, especially when including higher types (procedures as parameters to procedures); the semantic structures necessary to describe partial correctness of programs then become intricate. It will be interesting to see if any benefits can be gleaned from a semantically based approach.

REFERENCES

[1] Abramsky, S., Domain Theory in Logical Form, Proc. Symposium on Logic in Computer Science, Ithaca, NY, IEEE Computer Society Press (1987) 47-53.

[2] Apt, K. R., Ten Years of Hoare's Logic: A Survey, ACM TOPLAS, Vol. 3 (1981) 431-483.

[3] Apt, K. R., Bergstra, J. A., and Meertens, G. L. T., Recursive Assertions are not enough—or are they?, TCS 8 (1979) 73-87.

[4] de Bakker, J. W., Mathematical Theory of Program Correctness, Prentice-Hall (1980).

[5] Boehm, H.-J., Side-effects and Aliasing can have Simple Axiomatic Descriptions, ACM TOPLAS, vol. 7, no. 4 (1985) 637-655.

[6] Brookes, S. D., An Axiomatic Treatment of a Parallel Language, Proc. Symposium on Logics of Programs, Springer LNCS 193 (1985) 41-60.

[7] Brookes, S. D., A Semantically Based Proof System for Deadlock and Partial Correctness in CSP, Proc. Symposium on Logic in Computer Science, IEEE Computer Society Press (1986) 58-65.

[8] Brookes, S. D., A Fully Abstract Semantics and a Proof System for an ALGOL-like Language with Aliasing, Proc. Conference on Mathematical Foundations of Programming Semantics, Manhattan, Kansas, Springer LNCS 239 (1985) 59-100.

[9] Clarke, E. M., The Characterization Problem for Hoare's Logic, in: Mathematical Logic and Programming Languages, eds. C. A. R. Hoare and J. C. Shepherdson, Prentice-Hall (1986) 89-103.

[10] Clarke, E. M., Programming Language Constructs For Which It Is Impossible To Obtain Good Hoare Axiom Systems, JACM Vol. 26 No. 1 (January 1979) 129-147.

[11] Cook, S., Soundness and Completeness of an Axiom System for Program Verification, SIAM J. Comput 7 (1978) 70-90.

[12] Dijkstra, E. W., A Discipline of Programming, Prentice-Hall (1976).

[13] Floyd, R., Assigning Meanings to Programs, in: J. T. Schwartz, ed., Mathematical Aspects of Computer Science, Proc. Symp. Applied Math. (American Math. Soc. Providence) Vol. 19 (1967) 19-32.

[14] Hoare, C. A. R., An Axiomatic Basis for Computer Programming, CACM 12 (1969) 576-580).

[15] Hennessy, M. C. B., and Plotkin, G. D., Full Abstraction for a Simple Parallel Language, Proc. MFCS 1979, Springer LNCS 74 (1979) 108-120.

[16] Milner, R., A Calculus of Communicating Systems, Springer LNCS 92 (1980).

[17] Milner, R., Fully Abstract Models of Typed Lambda-Calculi, Theoretical Computer Science vol. 4 no. 1 (1977) 1-22.

[18] O' Donnell, M., A Critique of the Foundations of Hoare-style Programming Logic, CACM vol. 25 no. 12 (December 1982) 927-934

[19] Owicki, S. S., Axiomatic proof techniques for parallel programming, Ph.D. thesis, Cornell University (1975).

[20] Owicki, S. S., and Gries, D., An Axiomatic Proof Technique for Parallel Programs, Acta Informatica 6 (1976) 319-340.

[21] Robinson, E., Axiomatic Aspects of Denotational Semantics, preprint, Cambridge University (1986).

[22] Stirling, C., A Compositional Reformulation of Owicki-Gries's Partial Correctness Logic for a Concurrent While Language, Proc. ICALP 1986, Springer LNCS 226 (1986) 407-415.

[23] Stoughton, A., Fully Abstract Models of Programming Languages, Ph. D. thesis, Department of Computer Science, Edinburgh University (1986).

[24] Stoy, J., Denotational Semantics, MIT Press (1977).

[25] Strachey, C., The Varieties of Programming Language, Proceedings of International Computing Symposium, Cini Foundation, Venice (1972) 222-233.

Metric spaces as models for real-time concurrency

G.M. Reed and A.W. Roscoe[1]

Oxford University Computing Laboratory
7-11 Keble Road, Oxford OX1 3QD, U.K.

ABSTRACT. *We propose a denotational model for real time concurrent systems, based on the failures model for CSP. The fixed point theory is based on the Banach fixed point theorem for complete metric spaces, since the introduction of time as a measure makes all recursive operators naturally contractive. This frees us from many of the constraints imposed by partial orders on the treatment of nondeterminism and divergence.*

1 Introduction

Real time has generally been considered to be too much an implementation matter to include in abstract models of concurrency. Most existing theories view of time is restricted to the relative order of events and to high level concepts such as 'eventually' and 'forever'. Nevertheless there are several reasons why it is desirable to have the ability to reason about real times. Most obviously, it is likely that anyone specifying a real system will wish to impose constraints on its running speed (and perhaps more detailed timing matters concerning its external communications). But perhaps more importantly from a theoretical point of view, there are several concepts commonly used in concurrent languages, such as interrupts and priority, which do not fit easily or at all into untimed models.

We therefore believe that there is a need for models of real-time parallel computation. But since it is likely to remain easier and cleaner, where possible, to do analysis in an untimed framework, it is important that a real-time model has well-understood links with an untimed theory. We have chosen to base our work on (extensions of) the theoretical version of CSP and to try to discover models that have links with that language's untimed theory. In an earlier paper [RR] we showed how the traces model [H1] could be expanded to include time. In this paper we give a timed version of the 'failures' model (including divergence) described in [BR,H2].

One of the main purposes of this paper is to show how real time gives a particularly natural measure for comparing processes: we can think of two processes as being t-alike if they are indistinguishable up to time t. This notion is easily formalised as a metric over the space of processes which provides a natural fixed point theory, seemingly with few of the disadvantages of the traditional ways of defining fixpoints in untimed models. In particular, we are able to deal with the problems of unbounded nondeterminism.

In the next section we present the model and the semantics of CSP. The difficulties of time mean that the model is quite complex and some of the semantic operators quite subtle; unfortunately time constraints for publication mean we cannot motivate or explain our definitions as

[1]The work reported in this paper was supported by the U.S. Office of Naval Research under grant N0014-85-G-0123.

thoroughly as we would have liked. (We aim to give a fuller presentation of our work in the near future.) Section 3 shows how we have treated nondeterminism and divergence and how the introduction of time frees us from difficulties found in the construction of untimed models. Finally we present our conclusions and outline some ways in which our work can be extended.

2. The timed failures-stability model for CSP

2.1 Objectives of Timed CSP

Our objective is the construction of a timed CSP model which provides a basis for the definition, specification, and verification of real-time processes with an adequate treatment of divergence and deadlock. Furthermore, we wish the model to be a "natural" extension of existing untimed models, and in particular, it should contain the timed equivalents of those CSP constructs modelled in [BHR,BR].

2.2 Abstract syntax for TCSP (Timed Communicating Sequential Processes)

We shall essentially extend the abstract syntax for untimed CSP from [BHR,BR] (with the addition of $\perp$, the diverging process which engages in no event visible to the environment). We use P, Q, R to range over syntactic processes; a, b over the alphabet Σ; X, Y over subsets of Σ; f over the set of finite-to-one functions from Σ to Σ; and F over "appropriate" compositions of our syntactic operators.

The basic requirement for analysing real-time programming languages is the ability to model time-outs and interrupts. This can be accomplished in CSP simply by the addition of a process *WAIT* t for each real number $t \geq 0$: the process which engages in no visible event to the environment and which terminates successfully after t units of time. Intuitively, *SKIP* should coincide with *WAIT* 0.

TCSP

$$P ::= \perp \mid STOP \mid SKIP \mid WAIT\ t \mid (a \rightarrow P) \mid P \Box Q \mid P \sqcap Q \mid P \parallel Q \mid$$
$$P\ {}_X\|_Y\ Q \mid P \,|||\, Q \mid P;Q \mid P \setminus X \mid f^{-1}(P) \mid f(P) \mid \mu p.F(p)$$

2.3 Timing Postulates

The following are our basic assumptions about timing in a distributed system.

(1) **A global clock.** We assume that all events recorded by processes within the system relate to a *conceptual* global clock.

(2) **A system delay constant.** We realistically postulate that a process can engage in only finitely many events in a bounded period of time. The structure of our timed models allows several parameters by which to ensure adherence with this postulate. In the current presentation, for simplicity we assume the existence of a single delay constant δ such that:

a) For each $a \in \Sigma$ and each process P, the process $(a \rightarrow P)$ is ready to engage in P only after a delay of time δ from participation in the event a.

b) A given recursive process is only ready to engage in an observable event after a delay of δ time from making a recursive call.

Inevitably, our semantics are influenced by these and other decisions about the implementation of the language. We imagine, however, that the semantics presented below could be modified to take account of different decisions, or even of a nondeterministic choice of possible implementations.

(3) **Hiding.** We wish $(a \rightarrow P)$ to denote the process that is willing at *any* time to engage in the event a and then to behave like the process P. Clearly, if $P = a \rightarrow P$, we then wish $P \backslash a = \bot$. However, consider $P = a \rightarrow STOP$ (the process that is willing to engage in a at any time ≥ 0 and then to deadlock). What do we wish $P \setminus a$ to denote?

By hiding, we remove external control. Hence, any time a process is willing to engage in an internal action, it is permitted to do so. Thus, we assume that each hidden event has taken place as soon as such event was possible. In the above example, we would wish:

$$(a \rightarrow STOP) \setminus a = WAIT\ \delta; STOP$$

(4) **Timed stability.** In untimed CSP, it is only necessary to know that a given process can or cannot diverge after engaging in a trace s; in the timed models, it is necessary to know (if the process cannot diverge after s) *when* it will again be ready to respond to the environment. This analysis leads us to consider the untimed divergence models [Ros,Brookes,OH,BR]) as providing discrete information for a given trace s ("0" cannot diverge, "∞" can diverge), and our corresponding timed model as providing continuous information ($\alpha \in [0, \infty]$ such that the process is guaranteed to be stable within α time after engaging in s). Our topological models will be based on this notion of *stability*, which is the dual of divergence.

We will model a timed CSP process as a specified set of ordered 3-tuples $(s, \alpha, \aleph)$, where s is a *timed trace* of the process, $\aleph$ is a timed refusal of the process (a subset of Σ in which the process can fail to engage over a specifed time interval), and α is the time at which the process is guaranteed to be stable after the "observation" of s and $\aleph$. If $(s, \alpha, \aleph)$ is in the process P and $\alpha < \infty$, then the next observable event in the life of the process following s may occur at any time on or after time α at the discretion of the environment, and the set of possible next events must be the same at *all* such times after stability. Clearly no event can *become* available after α.

We think of timed stability as a red light on the outside of a process which goes off when the process can make no more internal progress.

(5) **Termination and sequential composition.** The sequential composition operator treats the termination of its first argument very much as a hidden event. That is, we postulate in $P; Q$ the process P *must* terminate as soon as it can not refuse to do so. Thus, we assume that participation in the "hidden" event $\surd$ has taken place as soon as such participation was possible.

For example,

$$((a \rightarrow STOP) \square\ WAIT\ 1); b \rightarrow STOP$$

is the process that is prepared to participate in the event a for the first unit of time and then deadlock, or to wait one unit of time and then participate in the event b and then deadlock; after time 1, it is no longer able to participate in the event a.

It is the above assumption about termination that allows us to model interrupts in Timed CSP.

2.4 Notation

The set (alphabet) of all communications (untimed events) will be denoted Σ. A *timed event* is an ordered pair (t, a), where a is a communication and $t \in [0, \infty)$ is the time at which it occurs. The set $[0, \infty) \times \Sigma$ of all timed events is denoted $T\Sigma$. The set of all *timed traces* is

$$(T\Sigma)^*_{\leq} = \{s \in T\Sigma^* \mid \text{if } (t, a) \text{ precedes } (t', a') \text{ in } s, \text{ then } t \leq t'\}.$$

If $s \in (T\Sigma)^*_{\leq}$, we define $\#s$ to be the length (i.e., number of events) of s and $\Sigma(s)$ to be the set of communications appearing in s (i.e., the second components of all its timed communications).

$begin(s)$ and $end(s)$ are respectively the earliest and latest times of any of the timed events in s. (For completeness we define $begin(\langle\rangle) = \infty$ and $end(\langle\rangle) = 0$.)

If $X \subseteq \Sigma$, $s \upharpoonright X$ is the maximal subsequence w of s such that $\Sigma(w) \subseteq X$; $s \setminus X = s \upharpoonright (\Sigma - X)$. If $t \in [0, \infty)$, $s \upharpoonright t$ is the subsequence of s consisting of all those events which occur no later than t. If $t \in [-begin(s), \infty)$ and $s = \langle (t_0, a_0), (t_1, a_1), \ldots, (t_n, a_n) \rangle$,

$$s + t = \langle (t_0 + t, a_0), (t_1 + t, a_1), \ldots, (t_n + t, a_n) \rangle \ .$$

If $s, t \in (T\Sigma)^*_{\leq}$, we define $s \cong t$ if, and only if, t is a permutation of s (i.e., events that happen at the same time can be re-ordered).

If $s, w \in (T\Sigma)^*_{\leq}$, $Tmerge(s, w)$ is defined to be the set of all traces in $(T\Sigma)^*_{\leq}$ obtained by interleaving s and w. (Note that this is a far more restricted set than in the untimed case, as the times of events must increase through the trace. In fact, $Tmerge(s, w)$ only contains more than one element when s and w record a pair of events at exactly the same time.)

Let $TSTAB = [0, \infty] = [0, \infty) \cup \{\infty\}$. This is the set of all "timed stability values".

Define:

$$\begin{array}{lcll} I \in TINT & = & \{ [l(I), r(I)) \mid 0 \leq l(I) < r(I) < \infty \} & (Time\ Intervals) \\ T \in RTOK & = & \{ I \times X \mid I \in TINT \wedge X \in P(\Sigma) \} & (Refusal\ Tokens) \\ \aleph \in RSET & = & \{ \bigcup Z \mid Z \in p(RTOK) \} & (Refusal\ Sets) \end{array}$$

1) $\forall \aleph \in RSET$,

$$\begin{array}{rcl} \Sigma(\aleph) & = & \{a \in \Sigma \mid \exists t \in [0, \infty) \ such\ that\ (t, a) \in \aleph\} \\ I(\aleph) & = & \{t \in [0, \infty) \mid \exists a \in \Sigma \ such\ that\ (t, a) \in \aleph\} \\ begin(\aleph) & = & min(I(\aleph)),\ \forall \aleph \neq \emptyset \\ end(\aleph) & = & sup(I(\aleph)),\ \forall \aleph \neq \emptyset \\ begin(\aleph) & = & \infty, \quad for\ \aleph = \emptyset \\ end(\aleph) & = & 0, \quad for\ \aleph = \emptyset \\ \forall t \geq -begin(\aleph), \quad \aleph + t & = & \{(t' + t, a) \mid (t', a) \in \aleph\}. \end{array}$$

2) $\forall S \subseteq (T\Sigma)^*_{\leq} \times TSTAB \times RSET$,

$$\begin{array}{rcl} Traces(S) & = & \{s \mid \exists \alpha \in TSTAB,\ \aleph \in RSET \ such\ that\ (s, \alpha, \aleph) \in S\} \\ Stab(S) & = & \{(s, \alpha) \mid \exists \aleph \in RSET \ such\ that\ (s, \alpha, \aleph) \in S\} \\ Fail(S) & = & \{(s, \aleph) \mid \exists \alpha \in TSTAB \ such\ that\ (s, \alpha, \aleph) \in S\} \\ SUP(S) & = & \{(s, \alpha, \aleph) \mid (s, \aleph) \in Fail(S) \\ & & \wedge\ \alpha = sup\{\beta \mid (s, \beta, \aleph) \in S\}\} \end{array}$$

If $\aleph \in RSET$ and $t \in [0, \infty)$, let $\aleph \upharpoonright t$ denote $\aleph \cap ([0, t) \times \Sigma)$.

2.5 The evaluation domain TM_{FS}

We formally define TM_{FS} to be those subsets S of $(T\Sigma)^*_{\leq} \times TSTAB \times RSET$ satisfying:

1. $\langle\rangle \in Traces(S)$
2. $(s.w, \aleph) \in Fail(S) \Rightarrow (s, \aleph \upharpoonright begin(w)) \in Fail(S)$
3. $(s, \alpha, \aleph), (s, \beta, \aleph) \in S \Rightarrow \alpha = \beta$
4. $(s, \alpha, \aleph) \in S \wedge s \cong w \Rightarrow (w, \alpha, \aleph) \in S$
5. $\forall t \in [0, \infty), \exists n(t) \in \mathbf{N}$ such that $\forall s \in Traces(S), (end(s) \leq t \Rightarrow \#s \leq n(t))$
6. $(s, \alpha, \aleph) \in S \Rightarrow end(s) \leq \alpha$
7. $(s, \alpha, \aleph) \in S \Rightarrow$ if $t > \alpha$, $t' \geq \alpha$, $a \in \Sigma$ and $w \in (T\Sigma)^*_{\leq}$ is such that $w = \langle(t, a)\rangle.w'$, then $(s.w, \alpha', \aleph') \in S \wedge \aleph \subseteq \aleph' \upharpoonright t \Rightarrow$ $\exists \gamma \geq \alpha' + (t' - t). (s.(w + (t' - t)), \gamma, \aleph_1 \cup \aleph_2 \cup (\aleph_3 + (t' - t))) \in S$, where $\aleph_1 = \aleph' \upharpoonright \alpha$, $\aleph_2 = [\alpha, t') \times \Sigma(\aleph' \cap ([\alpha, t) \times \Sigma))$, and $\aleph_3 = \aleph' \cap ([t, \infty) \times \Sigma)$.
8. $(s, \alpha, \aleph) \in S \wedge (s.\langle(t, a)\rangle, \aleph) \in Fail(S) \wedge t > t' \geq \alpha \wedge t \geq end(\aleph) \Rightarrow (t', a) \notin \aleph$
9. $(s, \alpha, \aleph) \in S \wedge \aleph' \in RSET$ such that $\aleph' \subseteq \aleph$ $\Rightarrow \exists \alpha' \geq \alpha$ such that $(s, \alpha', \aleph') \in S$
10. $(s.w, \alpha, \aleph) \in S \wedge \aleph' \in RSET$ is such that $end(s) \leq begin(\aleph') \wedge$ $end(\aleph') \leq begin(w) \wedge (\forall (t, a) \in \aleph', (s.\langle(t, a)\rangle, \aleph \upharpoonright t) \notin Fail(S))$ $\Rightarrow (s.w, \alpha, \aleph \cup \aleph') \in S$
11. $(s, \alpha, \aleph) \in S \Rightarrow$ $\forall I \subseteq [\alpha, \infty), (s, \alpha, \aleph \cup (I \times \Sigma(\aleph \cap ([\alpha, \infty) \times \Sigma)))) \in S$

Though some of these axioms appear complex, each reflects a simple healthiness property. They are explained as follows.

1. Every process has initially done nothing at all.
2. If a process has been observed to communicate $s.w$ while refusing $\aleph$ then at the time when the first event of w occurred the pair $(s, \aleph \upharpoonright begin(w))$ had been observed.
3. There is only one stability value for each trace/refusal pair: the least time by which we can guarantee stability after the given observation.
4. Traces which are equivalent (are the same except for the permutation of events happening at the same times) are interchangeable.
5. The process cannot perform an infinite number of events in a finite time.
6. The time of stability is always after the end of the trace.
7. After stability the same set of events is available at all times. Furthermore the behaviour of a process after such an event does not depend on the exact time at which it was executed. Thus the trace w and the corresponding part of the refusal may be translated so as to make the first event of w now occur at time t'. The stability value γ correponding to

the translated behaviour may, in general, be greater than the obvious value because the translated behaviour may in some circumstances be possible for other reasons.

8. A stable process cannot communicate an event which it has been seen to refuse since stability.

9. If a process has been observed to communicate s while refusing $\aleph$ then it can communicate the same trace while refusing any subset of $\aleph$. This simply reflects the fact that the environment might offer it less and so have less refused. However, because less has been observed, the stability value can, in general, be greater.

10. Any set of impossible events *must* be refused if offered. Such observation does not give any extra information to the observer, so the stability value is not affected.

11. Something that is refused at one time on or after stability is refused at all such times.

2.6 The complete metric on TM_{FS}

If $S \in TM_{FS}$ and $t \in [0, \infty)$, we define

$$\begin{aligned} S(t) &= \{(s, \alpha, \aleph) \in S \mid \alpha < t \wedge end(\aleph) < t\} \\ &\cup \{(s, \infty, \aleph) \mid end(s) < t \wedge end(\aleph) < t \wedge \exists \alpha \geq t.\, (s, \alpha, \aleph) \in S\}. \end{aligned}$$

This is just a standard representation of the behaviour of S up to time t: S and T have identical behaviours up to t if and only if $S(t) = T(t)$.

The complete (ultra-)metric on TM_{FS} is defined:

$$d(S_1, S_2) = inf\{2^{-t} \mid S_1(t) = S_2(t)\}$$

The completeness of this metric is a consequence of the fact that, if one of the above axioms is not satisfied by some set S of behaviours, this will show up in some finite time t. Specifically, if $S(t) = T(t)$ then T does not satisfy the axiom either.

2.7 The semantic function $\mathcal{E}$

We now define the semantic function $\mathcal{E} : TCSP \to TM_{FS}$. The definitions of most of the operators are closely modelled on their untimed counterparts. The only new operator is *WAIT* t, which becomes stable and able to communicate $\surd$ at time t.

Except in the case of hiding the use of the *SUP* operator simply reflects the fact that there can only be one stability value record for each trace/refusal pair, and some such pairs can get into the sets for several different reasons.

The definition of hiding is surprisingly simple, but the way stability is handled is rather subtle. The fact that X can be refused thoughout the behaviour ensures that hidden events occur as soon as they can. The βs in the set are all times which are demonstrably lower bounds for the time of stability, and with thought it can be seen that applying the *SUP* operator gives exactly the correct stability value.

Each operator has the important property that the behaviour of $F(P)$ up to time t depends only on the behaviour of P up to time t. It is this which makes all operators non-expanding. See the next section for more discussion of this.

$$\mathcal{E}[\![\bot]\!] = \{(\langle\rangle, \infty, \aleph) \mid \aleph \in RSET\}$$

$$\mathcal{E}[\![STOP]\!] = \{(\langle\rangle, 0, \aleph) \mid \aleph \in RSET\}$$

$$\begin{aligned}\mathcal{E}[\![SKIP]\!] = \; & \{(\langle\rangle, 0, \aleph) \mid \surd \notin \Sigma(\aleph)\} \\ & \cup\{(\langle(t,\surd)\rangle, t, \aleph_1 \cup \aleph_2) \mid t \geq 0 \wedge (I(\aleph_1) \subseteq [0,t) \wedge \surd \notin \Sigma(\aleph_1)) \\ & \quad \wedge I(\aleph_2) \subseteq [t,\infty)\}\end{aligned}$$

$$\begin{aligned}\mathcal{E}[\![WAIT\, t]\!] = \; & \{(\langle\rangle, t, \aleph) \mid \aleph \cap ([t,\infty) \times \{\surd\}) = \emptyset\} \\ & \cup\{(\langle(t',\surd)\rangle, t', \aleph_1 \cup \aleph_2 \cup \aleph_3) \mid t' \geq t \wedge I(\aleph_1) \subseteq [0,t) \\ & \quad \wedge(I(\aleph_2) \subseteq [t,t') \wedge \surd \notin \Sigma(\aleph_2)) \wedge I(\aleph_3) \subseteq [t',\infty)\}\end{aligned}$$

$$\begin{aligned}\mathcal{E}[\![a \rightarrow P]\!] = \; & \{(\langle\rangle, 0, \aleph) \mid a \notin \Sigma(\aleph)\} \\ & \cup\{(\langle(t,a)\rangle.(s + (t+\delta)), \alpha + t + \delta, \aleph_1 \cup \aleph_2 \cup (\aleph_3 + (t+\delta))) \mid t \geq 0 \\ & \quad \wedge(I(\aleph_1) \subseteq [0,t) \wedge a \notin \Sigma(\aleph_1)) \wedge I(\aleph_2) \subseteq [t, t+\delta) \wedge (s,\alpha,\aleph_3) \in \mathcal{E}[\![P]\!]\}\end{aligned}$$

$$\begin{aligned}\mathcal{E}[\![P \,\Box\, Q]\!] = \; & SUP(\{(\langle\rangle, max\{\alpha_P,\alpha_Q\}, \aleph) \mid (\langle\rangle,\alpha_P,\aleph) \in \mathcal{E}[\![P]\!] \;\wedge\; (\langle\rangle,\alpha_Q,\aleph) \in \mathcal{E}[\![Q]\!]\} \\ & \quad \cup\{(s,\alpha,\aleph) \mid s \neq \langle\rangle \;\wedge\; (s,\alpha,\aleph) \in \mathcal{E}[\![P]\!] \cup \mathcal{E}[\![Q]\!] \\ & \quad \wedge (\langle\rangle, \aleph \upharpoonleft begin(s)) \in Fail(\mathcal{E}[\![P]\!]) \cap Fail(\mathcal{E}[\![Q]\!])\})\end{aligned}$$

$$\mathcal{E}[\![P \sqcap Q]\!] = SUP(\mathcal{E}[\![P]\!] \cup \mathcal{E}[\![Q]\!])$$

$$\begin{aligned}\mathcal{E}[\![P \| Q]\!] = \; & SUP(\{(s, max\{\alpha_P,\alpha_Q\}, \aleph_P \cup \aleph_Q) \mid (s,\alpha_P,\aleph_P) \in \mathcal{E}[\![P]\!] \\ & \quad \wedge(s,\alpha_Q,\aleph_Q) \in \mathcal{E}[\![Q]\!]\})\end{aligned}$$

$$\begin{aligned}\mathcal{E}[\![P \,{}_X\|_Y Q]\!] = \; & \{(s, max\{\alpha_P,\alpha_Q\}, \aleph_P \cup \aleph_Q \cup \aleph_Z) \mid \exists(s_P,\alpha_P,\aleph_P) \in \mathcal{E}[\![P]\!], \\ & (s_Q,\alpha_Q,\aleph_Q) \in \mathcal{E}[\![Q]\!] \; with \; \Sigma(\aleph_P) \subseteq X \wedge \Sigma(\aleph_Q) \subseteq Y \; such\ that \\ & s \in (s_P \,{}_X\|_Y s_Q) \wedge \Sigma(\aleph_Z) \subseteq (\Sigma - (X \cup Y))\} \\ & \text{where} \\ & v \,{}_X\|_Y w = \{s \in (T\Sigma)^*_{\leq} \mid s\upharpoonleft(X \cup Y) = s \;\wedge\; s\upharpoonleft X = v \;\wedge\; s\upharpoonleft Y = w\}\end{aligned}$$

$$\begin{aligned}\mathcal{E}[\![P \,|||\, Q]\!] = \; & SUP(\{(s, max\{\alpha_P,\alpha_Q\}, \aleph) \mid \exists(u,\alpha_P,\aleph) \in \mathcal{E}[\![P]\!], \; (v,\alpha_Q,\aleph) \in \mathcal{E}[\![Q]\!] \\ & \quad such\ that \; s \in Tmerge(u,v)\})\end{aligned}$$

$$\begin{aligned}\mathcal{E}[\![P;Q]\!] = \; & SUP(\{(s,\alpha,\aleph) \mid \surd \notin \Sigma(s) \;\wedge\; \forall I \in TINT \\ & \quad (s,\alpha,\aleph \cup (I \times \{\surd\})) \in \mathcal{E}[\![P]\!]\} \\ & \quad \cup\{(s.(w+t), \alpha + t, \aleph_1 \cup (\aleph_2 + t)) \mid \; \surd \notin \Sigma(s) \;\wedge\; end(\aleph_1) \leq t \\ & \quad \wedge (s.\langle(t,\surd)\rangle, \aleph_1 \cup ([0,t) \times \{\surd\})) \in Fail(\mathcal{E}[\![P]\!]) \\ & \quad \wedge (w,\alpha,\aleph_2) \in \mathcal{E}[\![Q]\!]\})\end{aligned}$$

$$\begin{aligned}\mathcal{E}[\![P \setminus X]\!] = \; & SUP\{s \setminus X, \beta, \aleph) \mid \; \exists \alpha \geq \beta \geq end(s). \\ & \qquad (s,\alpha,\aleph \cup ([0, max\{\beta, end(\aleph)\}) \times X)) \in \mathcal{E}[\![P]\!]\}\end{aligned}$$

$$\mathcal{E}[\![f^{-1}(P)]\!] = \{(s,\alpha,\aleph) \mid (f(s),\alpha,f(\aleph)) \in \mathcal{E}[\![P]\!]\}$$

$$\mathcal{E}[\![f(P)]\!] = SUP(\{(f(s),\alpha,\aleph) \mid (s,\alpha,f^{-1}(\aleph)) \in \mathcal{E}[\![P]\!]\})$$

$\mathcal{E}[\![\mu p.F(p)]\!] =$ The unique fixed point of the contraction mapping $\hat{C}(Q) = C(WAIT\,\delta; Q)$, where C is the mapping on TM_{FS} represented by F.

3. Remarks on the model

3.1 Hiding and recursion

To illustrate the intuitive appeal of topological limits in the analysis of CSP processes, consider the following untimed example.

$$\begin{array}{rclcrcl} Q & = & b \rightarrow Q & \qquad & P_0 & = & Q \\ P & = & a \rightarrow P & \qquad & \forall n \geq 1, P_n & = & a \rightarrow P_{n-1} \end{array}$$

Recall that, by $P = a \rightarrow P$, we mean $P = \mu p.F(p)$ where $F(R) = a \rightarrow R$ is an appropriate mapping on our semantic domain.

Clearly, an observer looking at behaviours on traces of length $\leq n$ cannot distinguish between P and P_n. Hence it seems intuitive that $\lim_{n \rightarrow \infty} P_n = P$. Indeed this is the case under a complete metric structure. However, with the standard complete partial order structure for the (untimed) failures model, $\lim_{n \rightarrow \infty} P_n$ does not exist. When we move to the timed CSP models, this situation becomes critical. If an observer looking at a record of all traces completed in n units of time cannot distinguish between P_n and P, we would certainly expect $\lim_{n \rightarrow \infty} P_n = P$. In particular, we wish $\lim_{t \rightarrow \infty}(\mathit{WAIT}\ t) = \perp \neq \mathit{STOP}$.

Several authors, for example [N,Ros,BZ,GR,Rou], have considered untimed models of concurrency as metric spaces. The metrics have generally been based on equivalence up to a certain number of steps or communications in much the same way as ours has been based on indistinguishability up to a certain time. However the fact that hiding deletes communications means that a model with a metric of *visible* actions will have a discontinuous hiding operator. For example, in a topological traces model,

$$\lim_{n \rightarrow \infty}(P_n \setminus a) = Q \neq (P \setminus a) = \{\langle\rangle\} = \mathit{STOP}$$

Also, recursions are not defined unless they represent contraction maps: something which is by no means automatic, especially when hiding is involved. For example,

$$\mu p.a \rightarrow (p \backslash a) \quad \text{is undefined.}$$

Some authors have chosen to retain hidden actions in their models (often synchronisation trees). This avoids the above problems, but leads to models which are insufficiently abstract for many purposes (indeed, semantics of this type are often termed operational).

We had none of these problems in constructing the present model because time cannot be hidden, and yet we would expect an observer (with a clock) to be able to observe it. There is no operator (even hiding) whose behaviour up to time t depends on the behaviours of its operands after t: no reasonable operator can be expected to see into the future. Thus every operator represents at worst a nonexpansive function of the metric space. (Note that a nonexpansive function is always continuous.)

Consider P, Q, and P_n as defined above in the timed failures-stability model (with the appropriate change in P_n to reflect the delay induced by each recursive call in P).

$$\begin{array}{rclcrcl} Q & = & b \rightarrow Q & \qquad & P_0 & = & Q \\ P & = & a \rightarrow P & \qquad & \forall n \geq 1,\ P_n & = & a \rightarrow (\mathit{WAIT}\ \delta\ ;\ P_{n-1}) \end{array}$$

Now, $\lim_{n \rightarrow \infty} P_n = P$ and $\lim_{n \rightarrow \infty}(P_n \setminus a) = P \setminus a = \perp \neq \mathit{STOP}$.

We make every recursion into a contraction by observing that, realistically, a recursion will always take a small amount of time to unwind. For example:

$$\begin{aligned} \mu p.p &= \mathit{fix}(\hat{C}), \quad \text{where } \hat{C}(Q) = \mathit{WAIT}\,\delta; Q \\ &= \bot \end{aligned}$$

3.2 Divergence

The timed failures-stability model (like the timed stability model of [RR]) differs from previous CSP models relevant to divergence in that $(s, \infty) \in \mathit{Stab}(P)$ does not imply that $(s.w, \infty) \in \mathit{Stab}(P)$ for all traces w. That is, just because a process *may* diverge after engaging in a given trace, it does not mean that some time later after extending the trace, the process might not again become stable. For example, let $P = a \rightarrow P$ and consider the process $R = (P \setminus a) \sqcap (b \rightarrow (b \rightarrow \mathit{STOP}))$. Both $(\langle\rangle, \infty)$ and $(\langle(t, b)\rangle, t + \delta) \in \mathit{Stab}(R)$ (for any $t \geq 0$). This process can diverge on the empty trace; however, once we observe a b, we know that we are safe. Although it is possible to modify our model to conform to the untimed models in this regard, we choose to allow the finer distinction of CSP processes made possible by the topological structure of our evaluation domain.

These distinctions were not made in the failures-divergence model [BR,H2] because of the use of least fixed points to define recursions. In any model where the partial order is based on nondeterminism or definedness ($P \sqsubseteq Q$ iff Q is more deterministic than P), the least fixed point of $\mu p.p$ (operationally, a simply diverging process) is the most nondeterministic process. We are thus forced to identify the diverging process with one that can do anything (including diverge). This is closely related to the philosophy of the Smyth powerdomain (the powerdomain of dæmonic nondeterminism).

The failures-divergence model's axioms essentially state that we cannot specify anything about a process' behaviour after the possibility of divergence. This is tantamount to saying that we will never be prepared to accept, for any practical purpose, a process that can diverge. Some authors have disliked being forced to take this very strict view, and would have prefered a theory more like that of the timed model. This has lead them to use alternative fixed point theories such as *optimal* fixed points [Broy] (usually using more than one partial order). By using a metric space we have been able to achieve the same effect more easily.

3.3 Infinite hiding

The reader will note that the axiom of bounded nondeterminism from the failures model in [BR] is not included in our model:

$$(\forall Y \in p(X),\ (s, Y) \in S) \Rightarrow (s, X) \in S$$

or, in a timed context,

$$\begin{aligned} I \in \mathit{TINT} \wedge X \in P(\Sigma)\ \textit{such that}\ (\forall Y \in p(X),\ \exists \alpha\ \textit{such that}\ (s, \alpha, \aleph \cup (I \times Y)) \in S) \\ \Rightarrow \exists \alpha'\ \textit{such that}\ (s, \alpha', \aleph \cup (I \times X))) \in S \end{aligned}$$

Operationally, a process can be said to be boundedly nondeterministic if, at each point, it has only finitely many internal choices which may affect its future behaviour. It has generally proved much easier to model boundedly nondeterministic processes than to consider the more general case. In partial orders one often takes limits by intersecting the nondeterministic choices that can be made in an increasingly deterministic sequence of processes. Where processes have an infinite number of options it is possible to construct such an infinite sequence with empty intersection. (For example, suppose P_i is a process which nondeterministically communicates any integer $j \geq i$ on its first step.) It becomes necessary to introduce compactness assumptions, such as the axiom above, which are not natural in every circumstance.

Since the convergence in our metric space is independent of nondeterminism we have the choice whether to have such an axiom or not.

'Non-compact' unbounded nondeterminism can enter models of concurrency through declining to ignore what might happen after divergence and also through assumptions such as fairness. But the clearest source of unbounded nondeterminism is infinite hiding, where an infinite set of external choices are made internal. It has usually been necessary to exclude $P \backslash X$ (X infinite) from CSP because of this problem.

For example, let Q denote the process which is prepared to input any odd integer n and then to participate in the event b and then to output $n + 1$. (See [H2] for the obvious generalization to our syntax.)

$$Q = n : Odds \rightarrow (b \rightarrow (n + 1 \rightarrow STOP))$$

Now, Q is certainly definable in the complete partial order failures-divergence model as well as the timed model. However, $Q \setminus X$ is not allowed in the failures-divergence model, since $\{(b, X) \mid X \in p(Evens)\} \subseteq Q \setminus Odds$, but $(b, Evens) \notin Q \setminus Odds$.

Given that in the timed model we have chosen not to ignore what might happen after possible divergence, it might appear that it is possible to obtain unbounded nondeterminism from only finite hiding. For example, consider $P_0 \backslash a$, where

$$\begin{aligned} P_n \quad = \quad & (a \rightarrow P_{n+1} \\ & \Box \\ & b \rightarrow (n \rightarrow STOP)) \end{aligned}$$

Such a process can be defined in the topological models by use of infinite mutual recursion (easily added to the semantics). After communicating a b, this process can apparently choose to communicate any natural number, but cannot refuse them all. It would be problematic in an untimed model, but is not in our one because only finitely much nondeterminism is exhibited up to any finite time (as only finitely many hidden 'a's will have occurred).

It will be consistent to bring in an axiom of bounded nondeterminism if, and only if, we do not want to model any operator which, like infinite hiding, has the potential of introducing infinitely many choices in a finite time.

3.4 Choice between waiting and participation

Let us postulate the effect of the environment being given the choice of participating in a given process or of waiting. For example, $P = ((a \rightarrow STOP) \Box \; WAIT \; 1)$ offers the environment the initial choice of participating in the event a or of terminating successfully after 1 second. Again, what do we wish $P \setminus a$ to denote? Since we have assumed that the hidden event takes place as soon as possible, we would expect (under the assumption that $\delta < 1$):

$$((a \rightarrow STOP) \Box \; WAIT \; 1) \setminus a \; = \; WAIT \; \delta ; STOP$$

Similarly, we would wish:

$$(((a \rightarrow SKIP) \Box WAIT\,1)\ ;\ b \rightarrow STOP) \setminus a \quad = \quad WAIT\,\delta\ ;\ b \rightarrow STOP$$
$$(((a \rightarrow STOP) \Box WAIT\,1)\ ;\ b \rightarrow STOP) \setminus a \quad = \quad WAIT\,\delta\ ;\ STOP$$

Note from examination of for the above processes, it is clear that to achieve our intuitive semantics, when defining $P \setminus X$ in the context of $\Box$, we must be able to exclude some traces in P from consideration based on information about their possible refusals prior to stability. In fact, the situation is even more complicated. Consider:

$$P_1 \quad = \quad ((a \rightarrow STOP) \Box (b \rightarrow STOP)) \sqcap (a \rightarrow c \rightarrow STOP)$$
$$P_2 \quad = \quad ((a \rightarrow c \rightarrow STOP) \Box (b \rightarrow STOP)) \sqcap (a \rightarrow STOP)$$

Such processes would seem free of our current concern since they do not involve either hiding or delays. However, let

$$Q = (\,WAIT\,1 \Box (b \rightarrow STOP))\ ;\ a \rightarrow c \rightarrow STOP$$

Operationally, as indicated in our assumption (5) from section 2.3, we would expect:

$$(P_1 \parallel Q) \setminus b \neq (P_2 \parallel Q) \setminus b$$

In particular, we would expect:

$$\langle (1, a)(1+\delta, c) \rangle \in Traces((P_1 \| Q) \setminus b) \;\; but \;\; \langle (1, a)(1+\delta, c) \rangle \notin Traces((P_2 \| Q) \setminus b)$$

Hence, in our timed failures-stability model, we must distinguish between processes such as P_1 and P_2. Note that it is impossible to make such a distinction based on what a process can refuse *after* a given trace has been achieved. Hence, it is necessary not only to record refusals on a given trace prior to stability, but also to record what refusals were involved in the state changes which led to the final state witnessed by the trace. This seems to be a crucial issue in achieving a successful semantics for real-time parallel languages. The distributivity of the hiding operator over $\sqcap$ depends on the subtle resolution of this issue.

It is the fact that we record refusals throughout a process' history that has allowed us to dispense with the 'hatted' events of [RR]. (There, $\hat{a}$ represented the communication of an event a at the instant when it became available. These communications were essential for the correct definition of hiding and sequential composition.) For it is now apparent that an event has just become available if it communicated immediately after it has been refused.

4. Conclusions

We have seen that using time as the basis of a metric space allows one to be freed from the constraints of complete partial orders without losing generality or abstractness. The reader should compare the model presented here to timed models for concurrency based on complete partial orders in [J,KSRGA,Bo]. The main difficulty that remains (under both structures) is that we still cannot describe properties (such as fairness) which are only detectable over infinite time spans.

The ideas behind our model are conceptually reasonably straightforward: a process is just modelled by the records of experiments that an observer can carry out on it (communications accepted and communications refused). The fact that refusals must be recorded all the way through a trace is, as was explained in 3.4, a consequence of the way timed processes interact: in some sense they can perform more delicate experiments on each other than untimed processes. Refusals only after traces are no longer properly compositional. Other authors [J,Bo] have remarked on this and suggested or introduced similar solutions (based on partial orders rather than metric spaces). Phillips [P] has studied the corresponding untimed congruence.

The reader may note that the *traces* of $P\backslash a$ depend in a crucial way on the (timed) *failures* of P. (This is clearly illustrated by the second example of 3.4.) An immediate consequence of this is that the traces of a CSP program modelled in the timed traces model [RR] may be a strict superset of the traces that can be extracted from the failures model. Because the earlier did not contain enough information to accurately predict the possible traces of $P\backslash a$, it was necessary to give an upper bound. This is unlike the case with untimed CSP, where the traces predicted by the failures semantics are always precisely those predicted by the trace semantics. (The problems with hiding in the timed trace semantics arise from its failure to distinguish $P \sqcap Q$ and $P \Box Q$: in reality the traces of $(P \Box Q)\backslash a$ may be a strict subset of those of $(P \sqcap Q)\backslash a$.)

We believe that a version of our present model with stablity omitted is the simplest equivalence which is a full and natural congruence with respect to all the usual CSP operators. This congruence does *not* exist in the untimed case, since there consideration of divergence is necessary if one is to consider failures.

Thus our inclusion of stability in the present paper has been in some sense optional. Aside from the well-known arguments for wishing to distinguish a deadlocked process from a diverging one, there is another good reason for our inclusion of it. It is our declared aim to build a hierarchy of models, both timed and untimed, with well-understood links between them. Stability allows natural links to be formed with the failures-divergence model, since the liveness properties predicted by that model can be inferred from the time of stability on.

We expect to be able to exploit this link by using reasoning in two simpler models (the timed stablilty model of [RR] and the failures-divergence model) to infer total correctness properties in the timed failures-stablity model (where we expect detailed computations to be rather complex). Such reasoning can be expected to be sufficient when proving simple properties of processes that do not depend on the details of timed interaction to achieve 'untimed' correctness.

The facts that our model is a complete metric space and all recursions are contraction mappings makes it a natural vehicle for correctness proofs using the form of recursion induction described in [Ros,RR]. (A predicate that represents a non-empty closed subset and which is preserved by a recursion must contain the unique fixed point.) The introduction of stability seems to enhance the range of useful predicates which represent closed sets, since it (to a limited extent) allows us to look into the future.

The semantics we gave for CSP is by no means the only possible one that is reasonable, for any such semantics must make specific timing assumptions about the language and its implementation. We assumed that all events take exactly δ, while in practice each event a might take its own duration $\delta(a)$ or even a time chosen nondeterministically from some interval. In the last case our new-found ability to cope with unbounded nondeterminism would be essential. We also assumed that none of the operators except recursion consumed any time by running (i.e, there was never any setting-up or "overhead" time). Also both the parallel operators we gave were *true* parallel operators, in that the time taken by the two operands was not summed: one might well need time-sliced pseudo-parallel operators in applications. In a particular application one will have to decide on the "right" timed semantics, but there should never be any problem in accomodating it in the model TM_{FS}. This is a topic for further research.

As indicated earlier, we intend shortly to give a fuller presentation of the timed failures-stability model and its CSP semantics. Some of the above issues will be investigated further, and also others such as full abstraction and the nondeterminism partial order.

5. References

[Bo] A. Boucher, *A time-based model for occam,* Oxford University D.Phil. thesis 1986.

[Brookes] S.D. Brookes, *A model for communicating sequential processes,* Oxford University D.Phil. thesis 1983.

[Broy] M. Broy, *Fixed point theory for communication and concurency,* TC2 Working Conference on Formal Description of Programming Concepts II, Garmisch, 1982.

[BHR] S.D. Brookes, C.A.R. Hoare and A.W. Roscoe, *A theory of communicating sequential processes,* JACM 31 (1894), 560-599.

[BR] S.D. Brookes and A.W. Roscoe, *An improved failures model for communicating processes,* Proceedings of the Pittsburgh Seminar on Concurrency, Springer LNCS 197 (1985).

[BZ] J.W. de Bakker and J.I. Zucker, *Processes and the denotational semantics of concurrency,* Information and Control 54 (1982), 70-120.

[GR] W.G. Golson and W.C. Rounds, *Connections between two theories of concurrency: metric spaces and synchronisation trees,* Information and Control 57 (1983), 102-124.

[H1] C.A.R. Hoare, *A model for communicating sequential processes,* On the construction of programs CUP (1980), 229-248.

[H2] C.A.R. Hoare, *Communicating sequential processes,* Prentice-Hall International, 1985.

[J] G. Jones, *A timed model for communicating processes,* Oxford University D.Phil thesis, 1982.

[KSdRGA-K] R. Koymans, R.K. Shyamasundar, W.P. de Roever, R. Gerth and S. Arun-Kumar, *Compositional semantics for real-time distributed computing* Faculteit der Wiskunde en Natuurwetenschappen, Katholieke Universiteit, Nijmegen, Technical report 68, 1985.

[N] M. Nivat, *Infinite words, infinite trees, infinite computations,* Foundations of Computer Science III (Math. Centre Tracts 109, 1979), 3-52.

[OH] E.R. Olderog and C.A.R. Hoare, *Specification-oriented semantics for communicating processes,* Springer LNCS 154 (1983), 561-572.

[P] I. Phillips, *Refusal testing,* Proceedings of ICALP'86, Springer LNCS 226 (1986), 304-313.

[Ros] A.W. Roscoe, *A mathematical theory of communicating processes,* Oxford University D.Phil thesis 1982.

[Rou] W.C. Rounds, *Applications of topology to the semantics of communicating processes,* Proceedings of the Pittsburgh Seminar on Concurrency, Springer LNCS 197 (1985).

[RR] G.M. Reed and A.W. Roscoe, *A timed model for communicating sequential processes,* Proceedings of ICALP'86, Springer LNCS 226 (1986), 314-323.

DI-DOMAINS AS A MODEL OF POLYMORPHISM

Thierry Coquand, Carl Gunter and Glynn Winskel

Computer Laboratory, University of Cambridge, Cambridge CB2 3QG, England

In this paper we investigate a model construction recently described by Jean Yves Girard. This model differs from the models of McCracken, Scott, *etc.* in that the types are interpreted (quite pleasingly) as *domains* rather than closures or finitary projections on a universal domain. Our objective in this paper is two-fold. First, we would like to generalize Girard's construction to a larger category called *dI-domains* which was introduced by Berry [2]. The dI-domains possess many of the virtues of the domains used by Girard. Moreover, the dI-domains are closed under the *separated sum* and *lifting* operators from denotational semantics and this is *not* true of the domains of Girard. We intend to demonstrate that our generalized construction can be used to do denotational semantics in the ordinary way, but with the added feature of type polymorphism with a "types as domains" interpretation. Our second objective is to show how Girard's construction (and our generalization) can be done *abstractly*. We also give a *representational* description of our own construction using the notion of a *prime event structure*.

1 Introduction

The polymorphic λ-calculus was discovered by Girard [6] and later rediscovered by Reynolds [14]. As was the case with the simple untyped λ-calculus, the syntax of the calculus was, at first, understood better than its semantics. A model for the polymorphic calculus was first presented by McCracken [10] based on the cpo of closures over the algebraic lattice of subsets of the natural numbers. A similar technique can be used [1] to build models for the polymorphic calculus using finitary projection models such as the ones described by Scott [15] and Gunter [8]. More recently still there has been progress in saying what a model of the polymorphic calculus is in general. As with the simple untyped calculus, this can be done through the use of environment models [3] or categorically [16].

In this paper we investigate a model construction recently described by Girard [7]. This model differs from the models of McCracken, Scott, *etc.* in that the types are interpreted (quite pleasingly) as *domains* rather than closures or finitary projections on a universal domain. The construction is carried out over an interesting cartesian closed category of algebraic cpo's called *qualitative domains* which satisfy a very strong finiteness property. Our objective in this paper is two-fold. First, we would like to generalize Girard's construction to a larger category called *dI-domains* which was introduced by Berry [2]. The dI-domains possess many of the virtues of the qualitative domains. In addition, the dI-domains are closed under the *separated sum* and *lifting* operators from denotational semantics and this is *not* true of the qualitative domains. We intend to demonstrate that our generalized construction can be used to do denotational semantics in the ordinary way, but with the added feature of type polymorphism with the "types as domains" interpretation. For example, we will be able to interpret data types such as trees ($T \cong T + T$) and S-expressions ($S \cong \mathit{Atoms} + (S \times S)$) in the way they are ordinarily interpreted in the Scott-Strachey theory. Other useful types based on the lift operation (such as the solution to the domain equation $X \cong X_\perp$) will also be available with our approach. As with the qualitative domains we will also be able to obtain solutions for equations (such as $L \cong \mathit{Atoms} + L \to L$) with higher types. Our second objective is to show how Girard's construction (and our generalization) can be done *abstractly*. An ultimate result might carry out these constructions for "qualitative categories" and "dI-categories". For the purposes of this paper, however, we will (usually) restrict ourselves to posets. We also give a *representational* description of our own construction using the notion of a *prime event structure* which was introduced by Nielsen, Plotkin and Winskel [11] and Winskel [18].

The paper is divided into four sections. In the second section we present background definitions for dI-domains, event structures, *etc.* and demonstrate some basic properties. The third section gives the basic model construction in abstract and representational styles. In the fourth section we discuss the calculus we seek to model which we call the *polymorphic fixedpoint calculus.*

We would like to accord significant credit to Jean-Yves Girard and Gérard Berry for the ideas of this paper. In fact, the idea of developing a theory which includes a separated sum was suggested by Girard in Annex B of [7] (although the specific choice of dI-domains is our own). We also received valuable assistance and encouragement from Martin Hyland, Eugenio Moggi and Pino Rosolini.

2 DI-domains and event structures

A poset $\langle D, \sqsubseteq \rangle$ having a least element $\perp$ is said to be *complete* (and we say that D is a *cpo*) if every directed subset $M \subseteq D$ has a least upper bound $\bigsqcup D$. A monotone function $f : D \to E$

between cpo's D and E is *continuous* if $f(\bigsqcup M) = \bigsqcup f(M)$ for any directed $M \subseteq D$. A point x of a cpo D is said to be *isolated* if, for every directed collection $M \subseteq D$ such that $x \sqsubseteq \bigsqcup M$, there is a $y \in M$ such that $x \sqsubseteq y$. Let $\mathbf{B}_D$ denote the collection of isolated elements of D. The cpo D is *algebraic* if, for every $x \in D$, the set $M = \{x_0 \in \mathbf{B}_D \mid x_0 \sqsubseteq x\}$ is directed and $x = \bigsqcup M$. We will just call algebraic cpo's *domains*. A cpo D is *bounded complete* if every bounded subset of D has a least upper bound. In particular, if a pair $\{x, y\}$ is bounded then we will write $x \uparrow y$. If $x \uparrow y$ then $\{x, y\}$ has a least upper bound which we write as $x \sqcup y$. In a bounded complete cpo any pair $\{x, y\}$ has a greatest lower bound which we write as $x \sqcap y$. We will say that a point $x \in D$ is *very finite* if there are only finitely many points $y \sqsubseteq x$.

Definition: A *dI-domain* is a bounded complete domain D which satisfies

- *axiom d:* for every $x, y, z \in D$, if $y \uparrow z$ then $x \sqcap (y \sqcup z) = (x \sqcap y) \sqcup (x \sqcap z)$ and
- *axiom I:* every isolated point is very finite. ▮

The dI-domains were introduced in Berry's thesis [2], where he made the discovery that they could be made into a cartesian closed category by choosing appropriate continuous functions as morphisms. At first this seemed surprising because until then the only cartesian closed category of domains known were those in which exponentiation was the Scott function space, consisting of all continuous functions ordered pointwise, and this construction certainly leads to domains failing axiom I. Trying to solve the full-abstraction problem for typed λ-calculi, in order to capture certain operational features in denotational semantics, Berry was led to the definition of stable functions and the stable order between them. The stable order does not relate functions in a solely pointwise fashion but also takes into account the manner in which they are computed. For this reason stable functions on dI-domains possess a function space different from Scott's, one which obeys axiom I.

Definition: Let D, E be dI-domains. A function $f : D \to E$ is *stable* iff it is continuous and satisfies

$$x \uparrow y \Rightarrow f(x \sqcap y) = f(x) \sqcap f(y). \quad ▮$$

Definition: We define **DI** to be the category with objects the dI-domains and morphisms the stable functions under the usual function composition. ▮

Theorem 1 *The category* **DI** *is cartesian closed; products are formed as cartesian products ordered coordinatewise and the function space of dI-domains D and E consists of the set of stable* *functions $f : D \to E$ ordered by the stable ordering i.e. for stable functions $f, g : D \to E$ we put $f \sqsubseteq g$ iff*

$$\forall x \in D.\ f(x) \sqsubseteq g(x) \text{ and}$$
$$\forall x, y \in D.\ x \sqsubseteq y \Rightarrow f(x) = f(y) \sqcap g(x). \blacksquare$$

A great deal of the usual style of denotational semantics can be done in this category including the solving of recursive domain equations involving for example product, sum and function space. The solving of domain equations depends on a more restricted definition of embedding than is usual; embeddings must be ***rigid*** in the sense of Kahn and Plotkin [9] so that dI-domains are closed under direct limits.

Definition: Let D and E be domains. Let $f : D \to E$ be a continuous function. Say f is a *rigid embedding* iff there is a continuous function $g : E \to D$, called a ***rigid projection***, such that $g \circ f = id$ and and $f \circ g \sqsubseteq id$. ▮

The rigid embeddings in this definition do not quite correspond to embeddings in the sense of Smyth and Plotkin [17]. We have added the modifier "rigid" to emphasize the fact that the condition $f \circ g \sqsubseteq id$ is being taken with respect to the *stable ordering* on functions. The following lemma should help clarify the significance of this assumption.

Lemma 2 *Let D and E be domains and $f : D \to E$ a continuous function. Then f is a rigid embedding iff there is a continuous function $g : E \to D$ such that*

$$g \circ f(d) = d \text{ for all } d \in D \text{ and}$$
$$f \circ g(c) \sqsubseteq c \text{ for all } c \in E \text{ and}$$
$$c \sqsubseteq f(d) \Rightarrow f \circ g(c) = c. \blacksquare$$

Notice that any function greater or equal to the identity function with respect to the stable ordering is actually equal to the identity; indeed, if $id \leq f$, then for any x, we have $x \leq f(x)$, and hence, by definition of the stable ordering, $x = f(x) \sqcap f(x)$, *i.e.* $x = f(x)$. It follows that the rigid embeddings are exactly those stable functions $f : D \to E$ on dI-domains for which there is a stable function $g : E \to D$ such that $g \circ f \sqsupseteq id$ and and $f \circ g \sqsubseteq id$ with respect to the stable order, in contrast to the case of embeddings with respect to the Scott order. (As Andy Pitts has remarked, in the appropriate 2-category setting this means that rigid embeddings are precisely those morphisms which have right adjoints.)

Instead of showing directly that dI-domains have direct limits of rigid embeddings we shall work with a representation of dI-domains by prime event structures. The represention shows clearly the link with Girard's qualitative domains; prime event structures are like qualitative domains but with an extra partial order structure.

Definition: Define a *(prime) event structure* to be a structure $E = (E, Con, \leq)$ consisting of a set E, which are partially ordered by $\leq$, and a predicate Con on finite subsets of E, the *consistency relation*, which satisfy

$$\{e' \mid e' \leq e\} \text{ is finite,}$$
$$\{e\} \in Con,$$
$$Y \subseteq X \in Con \Rightarrow Y \in Con,$$
$$X \in Con \;\&\; \exists e' \in X.\ e \leq e' \Rightarrow X \cup \{e\} \in Con$$

for all $e \in E$, and finite subsets X, Y of E.

Define its *consistent left-closed subsets*, $\mathcal{L}(E)$, to consist of those subsets $x \subseteq E$ which are

- *consistent:* $\forall X \subseteq x.\ X \in Con$ and
- *left-closed:* $\forall e, e'.\ e' \leq e \in x \Rightarrow e' \in x$.

In particular, define $\downarrow e = \{e' \in E \mid e' \leq e\}$. ▮

Remark: Event structures often appear as a basic model of parallel processes when the set E is thought of as a set of event occurrences, the partial order $\leq$ as a relation of causal dependency, and the consistency relation as expressing what events can occur together. Then the configurations, the consistent left-closed sets of events, are thought of as states.

The configurations of an event structure form a dI-domain when ordered by inclusion. In this domain the configurations $\downarrow e$, for an event e, are characterized as special kinds of isolated elements, the complete primes.

Definition: Let D be a bounded complete cpo. A *complete prime* of D is an element p such that, for any bounded subset X of D,

$$p \sqsubseteq \bigsqcup X \Rightarrow \exists x \in X.\ p \sqsubseteq x. \quad ▮$$

Theorem 3 *Let E be a event structure. Then $(\mathcal{L}(E), \subseteq)$ is a dI-domain. The domain $(\mathcal{L}(E), \subseteq)$ has as complete primes those elements of the form $\downarrow e$ for $e \in E$.* ▮

Conversely, as we have indicated, any dI-domain is associated with an event structure in which the events are its complete primes, as was shown *eg.* in [19].

Definition: Let D be a dI-domain. Define $\mathsf{Pr}(D) = (P, Con, \leq)$, where P consists of the complete primes of D,

$$p \leq p' \iff p \sqsubseteq p',$$

for $p, p' \in P$, and

$$X \in Con \iff X \text{ is bounded}$$

for a finite subset X of P. ▮

Theorem 4 *Let D be a dI-domain. Then $\mathsf{Pr}(D)$ is a event structure, with $\phi : D \cong (\mathcal{L}\mathsf{Pr}(D), \subseteq)$ giving an isomorphism of partial orders where $\phi(d) = \{p \sqsubseteq d \mid p \text{ is a complete prime}\}$ with inverse $\theta : \mathcal{L}\mathsf{Pr}(D) \to D$ given by $\theta(x) = \bigsqcup x$.* ▮

Rigid embeddings between dI-domains are represented by embeddings between event structures which reduce, in the case where the embeddings are inclusions, to a substructure relation between event structures.

Definition: Let $E_0 = (E_0, Con_0, \leq_0)$ and $E_1 = (E_1, Con_1, \leq_1)$ be event structures. An *embedding* of E_0 in E_1 is a 1-1 total function $f : E_0 \to E_1$ on events such that

$$X \in Con_0 \iff f(X) \in Con_1$$

$$f(\downarrow e) = \downarrow f(e),$$

for any $X \subseteq E_0$ and $e \in E_0$. When $E_0 \subseteq E_1$ and the inclusion map $\iota : E_0 \hookrightarrow E_1$ is an embedding we write $E_0 \trianglelefteq E_1$, and say E_0 is a *substructure* of E_1.

We shall use $\mathbf{E}$ for the category of event structures with embeddings. ▮

Proposition 5 *Let $f : E_0 \to E_1$ be an embedding between event structures. Then the function*

$$f^L : (\mathcal{L}(E_0), \subseteq) \to (\mathcal{L}(E_1), \subseteq)$$

given by $f^L(x) = f(x)$, is a rigid embedding with projection

$$f^R : (\mathcal{L}(E_1), \subseteq) \to (\mathcal{L}(E_0), \subseteq)$$

given by $f^R(y) = f^{-1}(y)$.

A rigid embedding $h : D \to E$ between dI-domains restricts to an embedding $h' : \mathsf{Pr}(D) \to \mathsf{Pr}(E)$ between event structures, where $h'(p) = h(p)$ for complete primes p of D. ▮

Using the event structure representation it is now easy to see two properties of dI-domains which are crucial to the development that follows.

Proposition 6 *The category* $\mathbf{E}$ *of event structures with embeddings has direct limits and pullbacks. The category* $\mathbf{DI}^L$ *of dI-domains and rigid embeddings is equivalent to* $\mathbf{E}$ *and so has direct limits and pullbacks.*

Proof: It is sufficient to consider a family of event structures $\{E_i \mid i \in I\}$ indexed by some directed set $(I, \leq)$ so that $i \leq j \Rightarrow E_i \trianglelefteq E_j$. With the understanding that each E_i has the form $(E_i, Con_i, \leq_i)$, form the union $E = (\bigcup_{i\in I} E_i, \bigcup_{i\in I} Con_i, \bigcup_{i\in I} \leq_i)$. With the inclusion maps $E_i \hookrightarrow E$, this forms a direct limit.

In showing the existence of pullbacks it suffices to consider embeddings which are inclusions. Let $f : Y \trianglelefteq X$ and $g : Z \trianglelefteq X$. The intersection $W = (Y \cap Z, Con_Y \cap Con_Z, \leq_Y \cap \leq_Z)$ is an event structure for which the inclusions $f' : W \trianglelefteq Z$ and $g' : W \trianglelefteq Y$ form a pullback of f and g. ∎

Following well-known lines (see *e.g.* [17] or [7]) we can make function space, product and sum into functors on event structures with embeddings, or equivalently dI-domains with rigid embeddings. These functors can then be shown to have the property that they preserve direct limits and pullbacks, a property important in what follows. Moreover, this category satisfies properties [17] sufficient for solving recursive domain equations.

3 Modelling polymorphism

In this section we show how Girard's results in [7], on types in the polymorphic λ-calculus, generalize to dI-domains. We show this in two ways. We first give an elementary proof using the representation of dI-domains by the equivalent category of event structures. The proofs are then straightforward generalizations of Girard's in [7]; we need only take account of the extra partial order structure present in event structures but absent in qualitative domains. Following this we give a more abstract proof, less directly linked to Girard's constructions, but informative and useful, we hope, as a step in understanding polymorphism more abstractly.

Two central notions in [7] are that of *variable type* and *objects of a variable type.* Adopting the idea for event structures, a variable type is a functor $T : \mathbf{E} \to \mathbf{E}$ which preserves direct limits and pullbacks. An *object of* T is defined to be a function t on event structures such that $t_X \in T(X)$, for all event structures X, and $t_X = f^R(t_Y)$ for all embeddings $f : X \to Y$ of event structures. Ordered pointwise, *i.e.* taking

$$t \sqsubseteq t' \text{ iff } t_X \sqsubseteq t'_X \text{ for all event structures } X$$

these form a partial order. In fact this partial order can be represented as the set of configurations of an event structure, and so forms a dI-domain, and this is the key to a treatment of universal types in polymorphic λ-calculus.

As in [7] we use the notion of a trace of such a functor.

Theorem 7 *Let $T : \mathbf{E} \to \mathbf{E}$ be a functor on the category of event structures with embeddings which preserves direct limits and pullbacks. Let X be a event structure and let e be an event of $T(X)$. Then there is a finite event structure X_0, an event e_0 of $T(X_0)$ and an embedding $f : X_0 \to X$ such that $e = T(f)(e_0)$ and for any event structure X', an embedding $f' : X' \to X$ and e' an event of $T(X')$ such that $T(f')(e') = e$ there is a unique embedding $h : X_0 \to X'$ such that*

$$e' = T(h)(e_0) \text{ and } f = f' \circ h.$$

Proof: The proof of theorem 2.5 given by Girard in [7] applies almost literally here in the wider context of event structures in place of his qualitative domains. ▮

Definition: Let $T : \mathbf{E} \to \mathbf{E}$ be a functor which preserves direct limits and pullbacks. A *trace* of T is defined to be a set A of pairs (X, e) with X a finite event structure and e an event of $T(X)$ with the property that for any event structure X' and event e' of $T(X')$ there is a unique (X_0, e_0) in A and an embedding $h : X_0 \to X'$ such that $e' = T(h)(e_0)$. ▮

By Theorem 7, we know that a trace as defined above always exists. Choosing one particular trace we can define an event structure associated with a functor T on event structures preserving direct limits and pullbacks. It is slightly simpler to first define when a subset of a trace is inconsistent.

Definition: Let $T : \mathbf{E} \to \mathbf{E}$ be a functor which preserves direct limits and pullbacks. Let A be a trace of T. Say a subset of A is *inconsistent* when it includes a finite subset $\{(X_i, e_i) \mid i \in I\}$ for which there are embeddings $f_i : X_i \to X$, for $i \in I$, into some event structure X, so that

$$\{T(f_i)(e_i) \mid i \in I\} \notin Con_{T(X)},$$

where $Con_{T(X)}$ is the consistency predicate in $T(X)$. Now, define $\mathcal{E}(T) = (E, Con, \leq)$ where

- $E = \{(X, e) \in A \mid \{(X, e)\}$ is not inconsistent$\}$,
- Con consists of those finite subsets $\{(X_i, e_i) \mid i \in I\}$ of E which are not inconsistent, and

- $\leq$ is a binary relation on E given by

$$(X', e') \leq (X, e) \text{ iff } T(f)(e') \leq_{T(X)} e$$

for some embedding $f : X' \to X$ into the event structure $X = (X, Con_X, \leq_X)$. ∎

Thus the structure $\mathcal{E}(T)$ is constructed out of the the "self-consistent" elements of a trace of T, those elements (X, e) for which there are no two embeddings $f_1, f_2 : X \to Y$ for which $\{T(f_1)(e), T(f_2)(e)\}$ is not consistent in $T(Y)$. It is unique to within isomorphism by the properties of a trace. It is an event structure, to verify which we shall use the following lemma.

Lemma 8 *For embeddings $f : X \to Y$ and $g : X \to Z$ of event structures there is an event structure W and embeddings $f' : Z \to W$ and $g' : Y \to W$ such that $g' \circ f = f' \circ g$.*

Proof: Assume $X = (X, Con_X, \leq_X)$, $Y = (Y, Con_Y, \leq_Y)$ and $Z = (Z, Con_Z, \leq_Z)$. It suffices to consider the case where $f : X \trianglelefteq Y$ and $g : X \trianglelefteq Z$, that is when the embeddings are inclusions, and where we further assume that $X = Y \cap Z$. Then we define

$$W = (Y \cup Z, Con_Y \cup Con_Z, \leq_Y \cup \leq_Z),$$

the union of Y and Z. Taking f' and g' to be the inclusions $f' : Z \trianglelefteq W$ and $g' : Y \trianglelefteq W$ fulfils the requirements of the lemma. ∎

Lemma 9 *Let $T : \mathbf{E} \to \mathbf{E}$ be a functor which preserves direct limits and pullbacks. The structure $\mathcal{E}(T)$ defined above is an event structure.*

Proof: The only difficulty comes in showing that $\mathcal{E}(T)$ satisfies the property

$$\{(X_i, e_i) \mid i \in I\} \in Con \;\&\; (X'_j, e'_j) \leq (X_j, e_j), \text{ for } j \in I, \Rightarrow \{(X_i, e_i) \mid i \in I\} \cup \{(X'_j, e'_j)\} \in Con.$$

Suppose otherwise, *i.e.* that $\{(X_i, e_i) \mid i \in I\} \in Con$ & $(X'_j, e'_j) \leq (X_j, e_j)$, for $j \in I$, while $\{(X_i, e_i) \mid i \in I\} \cup \{(X'_j, e'_j)\}$ is inconsistent. Then there is a subset $K \subseteq I$ with embeddings $f_k : X_k \to X$ and $f'_j : X'_j \to X$ into some event structure X so that $\{T(f_k)(e_k) \mid k \in K\} \cup \{T(f'_j)(e'_j)\}$ is not consistent in $T(X)$. Also there is an embedding $g : X'_j \to X_j$ such that $T(g)(e'_j) \leq_{X_j} e_j$. By lemma 8, there are embeddings $g' : X \to W$ and $f_j : X_j \to W$, for some event structure W, for which $g' \circ f'_j = f_j \circ g$. However this yields embeddings $g' \circ f_k : X_k \to W$, for $k \in K$, and $f_j : X_j \to W$ for which $\{T(g' \circ f_k)(e_k) \mid k \in K\} \cup \{T(f_j)(e_j)\} \notin Con_{T(W)}$. This contradicts the consistency of $\{(X_i, e_i) \mid i \in I\}$. Hence the property is proved. ∎

We now show how the event structure associated with a variable type T has a domain of configurations isomorphic to the partial order of objects of type T.

Theorem 10 *Let $T : \mathbf{E} \to \mathbf{E}$ be a functor which preserves direct limits and pullbacks. Let $\mathcal{E}(T)$ have the form $(E, Con, \leq)$. There is an isomorphism between the domain of configurations $\mathcal{L}(\mathcal{E}(T))$, ordered by inclusion, and the objects of T, ordered pointwise; the isomorphism is determined as follows:*

1. *An object t of variable type T determines a configuration a of $\mathcal{E}(T)$ where*

$$a = \{(X, e) \in E \mid e \in t_X\}.$$

2. *A configuration a of $\mathcal{E}(T)$ determines an object t of variable type T which acts so*

$$t_X = \{T(f)(e) \mid \exists X_0.\ (X_0, e) \in a \ \&\ f : X_0 \to X \text{ is an embedding}\}$$

for all event structures X.

Proof: Again the proof more or less follows Girard's in [7]; the additional partial order structure causes no real difficulties. ▮

Now we give the more abstract proof.

Definition: Let $\mathbf{C}$ be a category—we shall call $\mathbf{C}$ a *finitary category* if, and only if, $\mathbf{C}$ has the following properties

1. $\mathbf{C}$ has pull backs
2. there exists a *set* S of objects of $\mathbf{C}$ such that every object of $\mathbf{C}$ is the direct limit of objects in S. ▮

Note the following fundamental property:

Proposition 11 *The category $\mathbf{DI}^L$ is finitary. Furthermore, the product of two finitary categories is finitary.*

Proof: In this case, the set S is any small dense subcategory of the finite dI-domains. ▮

Definition: Let $\mathbf{C}$ be a category. Let F be a functor from $\mathbf{C}$ to $\mathbf{DI}^L$. We say F is a variable type (or sometimes a *stable functor*) when it commutes with pullbacks. We say that a family (t_X), such that $t_X \in F(X)$ for all objects X in $\mathbf{C}$, is an *object of variable type* F (or sometimes a *uniform family of* F) if, and only if, for every pair of objects X and Y in $\mathbf{C}$, and every morphism $f \in \mathbf{C}(X,Y)$, we have $F(f)^R(t_Y) = t_X$. We shall write $\Pi(F)$ for the collection of all objects of variable type F. ∎

Note that the collection of uniform families of a given functor from $\mathbf{C}$ into $\mathbf{DI}^L$ is in general a class and not a set, if $\mathbf{C}$ is a large category. However, we shall see that this difficulty does not really happen in the cases we consider. Note the following presentation of objects of variable type.

Proposition 12 *Let* $\mathbf{C}$ *be a category. Let* F *be a functor from* $\mathbf{C}$ *to* $\mathbf{DI}^L$*. Then a family* (t_X) *such that* $t_X \in F(X)$ *for all objects* X *in* $\mathbf{C}$ *is a uniform family of* F *if, and only if, for every pair of objects* X *and* Y *in* $\mathbf{C}$*, and every morphism* $f \in \mathbf{C}(X,Y)$*, we have*

$$\forall p \in \mathsf{Pr}(F(X)).\ p \sqsubseteq t_X \Leftrightarrow F(f)^L(p) \sqsubseteq t_Y. \ ∎$$

We can now state Girard's discovery in our framework:

Theorem 13 *Let* $\mathbf{C}$ *be a finitary category, and* F *a functor from* $\mathbf{C}$ *to* $\mathbf{DI}^L$ *which stable. Then* $\Pi(F)$ *is a set, and it is a dI-domain in the pointwise ordering.* ∎

Proof: That $\Pi(F)$ is a set comes directly from the fact that $\mathbf{C}$ is set-generated and that F preserves directed limits. The verification of the fact that it is a bounded complete cpo uses directly our reformulation of what is a uniform family of F. As infs and sups are computed pointwise, the distributivity axiom is satisfied. Finally, the hard point is to find a basis with very finite and isolated elements. We shall give only the construction. Let (t_X) be a given object of variable type F, and A an element of S. For each very finite element $a \sqsubseteq t_A$ of $F(A)$, we shall show how to build a very finite and isolated object (u_X), such that $a \leq u_A$ and (u_X) is less than or equal to (t_X) for the pointwise ordering (which corresponds to the usual construction of step functions).

Define first the relation $(X,x) \sqsubseteq (Y,y)$ if, and only if, X, Y are objects of $\mathbf{C}$, and $x \in F(X)$, $y \in F(Y)$, and there exists a morphism $f \in \mathbf{C}(X,Y)$ such that $F(f)^L(x) \sqsubseteq y$. This is a transitive relation. We note that, for every object Y, the following subset of $F(Y)$

$$\{F(f)^L(x) \mid (X,x) \sqsubseteq (A,a) \text{ and } f \in \mathbf{C}(X,Y)\}$$

is bounded by t_Y. Since $F(Y)$ is bounded complete, we can define

$$u_Y = \bigsqcup\{F(f)^L(x) \mid (X,x) \sqsubseteq (A,a) \text{ and } f \in \mathbf{C}(X,Y)\}.$$

Then, we can check that this family has all the wanted properties. Let us prove that this family is uniform, which is the key point. We take $f \in \mathbf{C}(X,Y)$ and we want to show that $F(f)^R(u_Y) = u_X$. Actually, we know, by the proposition 12, that it is enough to show that is p is a prime element of $F(X)$, then $p \leq u_X$ if, an only if, $F(f)^L(p) \leq u_Y$. If $p \leq u_X$, then, by definition of u and since p is prime, there exists $(B,b) \leq (A,a)$ and $g \in \mathbf{C}(B,X)$ such that $p \leq F(g)^L(b)$. Then we have $F(f)^L(p) \leq F(f \circ g)^L(b)$, $(B,b) \leq (A,a)$ and $f \circ g \in \mathbf{C}(X,Y)$ so that $F(f)^L(p) \leq u_Y$. Conversely, if $F(f)^L(p) \leq u_Y$, then, since $F(f)^L$ preserves primes, $F(f)^L(p)$ is prime and there exists $(B,b) \leq (A,a)$ and $g \in \mathbf{C}(B,Y)$ such that $F(f)^L(p) \leq F(g)^L(b)$. Now, from $F(f)^L(p) \leq F(g)^L(b)$ we can deduce that there exists $b_1 \leq b$ such that $F(f)^L(p) = F(g)^L(b_1)$ (b_1 is simply $F(g)^R(F(f)^L(p))$). Hence, if $h \in \mathbf{C}(D,X)$ is the pull-back of g along f, there exists $d \in D$ such that $F(h)^L(d) = p$ and $(D,d) \leq (B,b)$. We then have $(D,d) \leq (A,a)$ and $p \leq F(h)^L(d)$, hence $p \leq u_X$. ∎

The importance of this proof is that it gives an insight as to how this model can be extended to a model of Fω. This is the remarkable closure property of dI-domains which will allow us to interpret the abstraction relatively to *types* (types as parameters). The key point is the Proposition 3. Note that in the case where the category **C** is the poset of natural numbers, then the construction is the usual inverse limit construction! However, it is important to emphasize that, in general, $\Pi(F)$ is *not* the limit of the functor F.

We will apply these constructions to provide a semantics for the calculus which is described in the following section. For now we shall only give an outline of how to give an interpretation of second-order calculus. This is already done in Girard's paper [7]. Since dI-domains form a cartesian closed category (with stable functions), it is well-known how to interpret the usual application and abstraction (see [2]) Closed types will be interpreted as dI-domains and, more generally, a type with n variables will be interpreted as a functor from $(\mathbf{DI}^L)^n$ to $\mathbf{DI}^L$ which is stable. Closed terms will be interpreted as elements of the domains and, more generally, a term which depends on type variables will be interpreted as a uniform family of the functor associated to its types. One interesting feature is that the interpretation is extensional.

4 The polymorphic fixedpoint calculus

In this section we will describe the syntax of a calculus which we wish to interpret using the constructions set out in the previous sections. It is a fragment of the language of McCracken [10] and is closely related to Fairbairn's programming language Ponder [12]. We call this language the *(pure) polymorphic fixedpoint calculus.* Its types have the following abstract syntax:

$$\sigma ::= \sigma_1 \rightarrow \sigma_2 \mid \alpha \mid \mu\alpha.\, \sigma \mid \Pi\alpha.\, \sigma,$$

where α is a type variable, and it has the following terms:

$$M ::= x \mid \lambda x : \sigma.\, M \mid M_1(M_2) \mid \Lambda\alpha.\, M \mid M\{\sigma\} \mid \mathsf{intro}^{\mu\alpha.\,\sigma}(M) \mid \mathsf{elim}^{\mu\alpha.\,\sigma}(M) \mid \mu x : \sigma.\, M$$

where x is a variable.

4.1 Typing and equational rules.

A closed term will be assigned a unique type by a system of typing rules. Typing sequents have the form $H \vdash_\Sigma M : \sigma$ where H is a (possibly empty) list of hypotheses of the form $x : \sigma$. We assume that a list H has no repetitions of variables x. Axioms are given by the scheme:

$$H_1,\ x : \sigma,\ H_2 \vdash_\Sigma x : \sigma$$

where x does not appear in H_1 or H_2 and Σ is a (possibly empty) set of type variables. There are the following rules for introducing and eliminating function types:

$$\frac{H,\ x : \sigma_1 \vdash_\Sigma M : \sigma_2}{H \vdash_\Sigma \lambda x : \sigma_1.\, M : \sigma_1 \rightarrow \sigma_2} \qquad \frac{H \vdash_\Sigma M_1 : \sigma_1 \rightarrow \sigma_2 \qquad H \vdash_\Sigma M_2 : \sigma_1}{H \vdash_\Sigma M_1(M_2) : \sigma_2}$$

There are the following rules for introducting and eliminating Π:

$$\frac{H \vdash_{\Sigma,\,\alpha} M : \sigma}{H \vdash_\Sigma \Lambda\alpha.\, M : \Pi\alpha.\, \sigma} \qquad \frac{H \vdash_\Sigma M : \Pi\alpha.\, \sigma_1}{H \vdash_\Sigma M\{\sigma_2\} : [\sigma_2/\alpha]\sigma_1}$$

where the first rule is subject to the condition that α is not free in the type of any free term variable of M and $[\sigma_2/\alpha]\sigma_1$ is the expression that results from substituting σ_2 for α in σ_1 (where bound variables in σ_1 are renamed to avoid capturing free variables of σ_2). Recursion can occur both at the level of types and at the level of terms. The rule for typing a recursive term is

$$\frac{H,\ x : \sigma \vdash_\Sigma M : \sigma}{H \vdash_\Sigma \mu x : \sigma.\, M : \sigma}$$

Introduction and elimination for recursive types use the operators intro and elim respectively.

$$\frac{H \vdash_\Sigma M : [(\mu\alpha.\, \sigma)/\alpha]\sigma}{H \vdash_\Sigma \mathsf{intro}^{\mu\alpha.\,\sigma}(M) : \mu\alpha.\, \sigma} \qquad \frac{H \vdash_\Sigma M : \mu\alpha.\, \sigma}{H \vdash_\Sigma \mathsf{elim}^{\mu\alpha.\,\sigma}(M) : [(\mu\alpha.\, \sigma)/\alpha]\sigma}$$

There is a collection of equations which an interpretation of the polymorphic fixedpoint calculus must satisfy. We assume throughout that all terms are type-sensible. First of all, there are the basic equality rules:

$$H \vdash_\Sigma M = M \qquad \frac{H \vdash_\Sigma M_1 = M_2}{H \vdash_\Sigma M_2 = M_1} \qquad \frac{H \vdash_\Sigma M_1 = M_2 \qquad H \vdash_\Sigma M_2 = M_3}{H \vdash_\Sigma M_1 = M_3}$$

and also rules for application:

$$\frac{H \vdash_\Sigma M_1 = M_2}{H \vdash_\Sigma M_1(M_3) = M_2(M_3)} \qquad \frac{H \vdash_\Sigma M_1 = M_2}{H \vdash_\Sigma M_3(M_1) = M_3(M_2)}$$

The calculus must also satisfy the β-rule:

$$H \vdash_\Sigma (\lambda x : \sigma.\ M_1)(M_2) = [M_2/x]M_1$$

and the ξ-rule

$$\frac{H,\ x : \sigma \vdash_\Sigma M_1 = M_2}{H \vdash_\Sigma \lambda x : \sigma.\ M_1 = \lambda x : \sigma.\ M_2}$$

Finally, we also require the η-rule:

$$\lambda x : \sigma.\ M(x) = M$$

(where x does not appear free in M). For type application we have, of course, the following:

$$\frac{H \vdash_\Sigma M_1 = M_2}{H \vdash_\Sigma M_1\{\sigma\} = M_2\{\sigma\}}$$

We also have the β-rule

$$H \vdash_\Sigma (\Lambda\alpha.\ M)\{\sigma\} = [\sigma/\alpha]M$$

the ξ-rule

$$\frac{H \vdash_{\Sigma,\ \alpha} M_1 = M_2}{H \vdash_\Sigma \Lambda\alpha.\ M_1 = \Lambda\alpha.\ M_2}$$

and the η-rule

$$\Lambda\alpha.\ M\{\alpha\} = M$$

(where α does not appear free in M). There are some basic rules for the recursive type operators:

$$\frac{H \vdash_\Sigma M_1 = M_2}{H \vdash_\Sigma \mathsf{intro}^{\mu\alpha.\ \sigma}(M_1) = \mathsf{intro}^{\mu\alpha.\ \sigma}(M_2)} \qquad \frac{H \vdash_\Sigma M_1 = M_2}{H \vdash_\Sigma \mathsf{elim}^{\mu\alpha.\ \sigma}(M_1) = \mathsf{elim}^{\mu\alpha.\ \sigma}(M_2)}$$

and a pair of equations asserting an isomorphism:

$$H,\ x : [(\mu\alpha.\ \sigma)/\alpha]\sigma \vdash_\Sigma \mathsf{intro}^{\mu\alpha.\ \sigma}(\mathsf{elim}^{\mu\alpha.\ \sigma}(x)) = x$$

$$H,\ x : \mu\alpha.\ \sigma \vdash_\Sigma \mathsf{elim}^{\mu\alpha.\ \sigma}(\mathsf{intro}^{\mu\alpha.\ \sigma}(x)) = x$$

4.2 An extended calculus

For the purposes of denotational semantics, it is useful to have the additional type constructors of product $\times$ and sum $+$. We interpret the $+$ operator as the usual separated sum of denotational semantics. Given dI-domains $D_1, \ldots, D_n$, the sum $+(D_1, \ldots, D_n)$ is defined to be the disjoint sum of the domains D_i together with a new element $\perp$ which is taken to be the least element of $+(D_1, \ldots, D_n)$. We emphasize the point which we mentioned in the Introduction that this construction is *not possible* over the category of qualitative domains. We therefore propose the an extension of the pure language which will include sums. The extended language has the following types:

$$\sigma ::= \sigma_1 \times \sigma_2 \mid \sigma_1 + \sigma_2 \mid \sigma_1 \to \sigma_2 \mid \alpha \mid \mu\alpha.\, \sigma \mid \Pi\alpha.\, \sigma.$$

The terms of the extended language are given as follows:

$$\begin{aligned} M ::= \; & \langle M_1, M_2 \rangle \mid \mathsf{proj}_i(M) \mid \\ & \mathsf{in}_1^{\sigma_1,\sigma_2}(M) \mid \mathsf{in}_2^{\sigma_1,\sigma_2}(M) \mid \mathsf{cases}\ M\ \mathsf{of}\ x_1 : \sigma_1.\, M_1,\ x_2 : \sigma_2.\, M_2 \mid \\ & x \mid \lambda x : \sigma.\, M \mid M_2(M_1) \mid \\ & \Lambda\alpha.\, M \mid M\{\sigma\} \mid \\ & \mathsf{intro}^{\mu\alpha.\,\sigma}(M) \mid \mathsf{elim}^{\mu\alpha.\,\sigma}(M) \mid \\ & \mu x : \sigma.\, M \end{aligned}$$

Typing rules for the extended language are the same as those for the pure language together with the rules for products:

$$\frac{H \vdash_\Sigma M_1 : \sigma_1 \qquad H \vdash_\Sigma M_2 : \sigma_2}{\langle M_1, M_2 \rangle : \times(\sigma_1, \sigma_2)} \qquad\qquad \frac{M : \sigma_1 \times \sigma_2}{\mathsf{proj}_i(M) : \sigma_i}$$

and for sums:

$$\frac{M : \sigma_i}{\mathsf{in}_i^{\sigma_1,\sigma_2}(M) : \sigma_1 + \sigma_2} \qquad \frac{M : \sigma_1 + \sigma_2 \qquad H,\ x_1 : \sigma_1 \vdash_\Sigma M_1 : \sigma \qquad H,\ x_2 : \sigma_2 \vdash_\Sigma M_2 : \sigma}{\mathsf{cases}\ M\ \mathsf{of}\ x_1 : \sigma_1.\, M_1,\ x_2 : \sigma_2.\, M_2 : \sigma}$$

$(i = 1, 2)$. There are also equations to be satisfied by the new constructs. For the product, there are some basic equations

$$\frac{H \vdash_\Sigma M_1 = M_1' \qquad H \vdash_\Sigma M_2 = M_2'}{H \vdash_\Sigma \langle M_1, M_2 \rangle = \langle M_1', M_2' \rangle} \qquad\qquad \frac{H \vdash_\Sigma M = M'}{H \vdash_\Sigma \mathsf{proj}_i(M) = \mathsf{proj}_i(M')}$$

$(i = 1, 2)$ and equations

$$H \vdash_\Sigma \langle \mathsf{proj}_1(M), \mathsf{proj}_2(M) \rangle = M$$

$$H \vdash_\Sigma \mathsf{proj}_i(\langle M_1, M_2 \rangle) = M_i$$

($i = 1, 2$) which assert that the operator $\times$ is to be interpreted as a categorical product. For the sum we also require some basic equations:

$$\frac{H \vdash_\Sigma M = M'}{H \vdash_\Sigma \mathsf{proj}_i(M) = \mathsf{proj}_i(M')}$$

$$\frac{H,\ x_1 : \sigma_1 \vdash_\Sigma M_1 = M_1' \qquad H,\ x_2 : \sigma_2 \vdash_\Sigma M_2 = M_2' \qquad H \vdash_\Sigma M = M'}{H \vdash_\Sigma \mathsf{cases}\ M\ \mathsf{of}\ x_1 : \sigma_1.\ M_1,\ x_2 : \sigma_2.\ M_2 = \mathsf{cases}\ M'\ \mathsf{of}\ x_1 : \sigma_1.\ M_1',\ x_2 : \sigma_2.\ M_2'}$$

and a pair of equations

$$H,\ x_i : \sigma_i \vdash_\Sigma \mathsf{cases}\ \mathsf{in}_i(x_i)\ \mathsf{of}\ x_1 : \sigma_1.\ M_1,\ x_2 : \sigma_2.\ M_2 = M_i$$

($i = 1, 2$).

However, we do *not* have the following additional equations

$$H,\ x_i : \sigma_i \vdash_\Sigma \mathsf{in}_i(\mathsf{cases}\ x_i\ \mathsf{of}\ x_1 : \sigma_1.\ M_1,\ x_2 : \sigma_2.\ M_2) = M_i$$

($i = 1, 2$) which would assert that $+$ is a categorical coproduct. Surprisingly, this last equation is satisfied only by a trivial interpretation of the calculus! Under our interpretation, however, these equations are *almost* true, in the sense that they *do* hold when $x_i \neq \bot$.

5 Semantics of the polymorphic lambda calculus

5.1 Some preliminary results

We now describe our model. We omit the semantics for the recursion and the extended language and concentrate on the polymorphic lambda calculus. We give, by structural induction, the denotations $[\![\Sigma \vdash \sigma]\!]$ and $[\![H \vdash_\Sigma M : \sigma]\!]$. For this, we need first some general results.

Definition: Let F be a continuous and stable functor from $\mathbf{DI}^L$ to $\mathbf{DI}^L$. A continuous stable family with respect to F is a family (t_X) indexed over dI-domains X such that

- $t_X \in F(X)$ for all X,
- if $f \in \mathbf{DI}^L(X, Y)$, then $F(f)^L(t_X) \sqsubseteq t_Y$ (monotonicity),
- if $f_i \in \mathbf{DI}^L(X_i, X)$ is such that X is the directed colimit of the system (X_i, f_i), then t_X is the sup of the directed family $F(f_i)^L(t_{X_i})$,
- if $u_1 \in \mathbf{DI}^L(X_3, X_1)$ and $u_2 \in \mathbf{DI}^L(X_3, X_2)$ define a pull-back of $f_1 \in \mathbf{DI}^L(X_1, X)$ and $f_2 \in \mathbf{DI}^L(X_2, X)$ and we write $f_3 = f_1 \circ u_1 = f_2 \circ u_2$, then $F(f_3)^L(t_{X_3}) = F(f_1)^L(t_{X_1}) \wedge F(f_2)^L(t_{X_2})$ (stability).

Note that the two last conditions express that (t_X) commutes with directed colimits and pull-backs. For checking that these properties hold for a given family, it is convenient to use the characterisation of directed colimit of [17]: if X is the directed colimit of the system (X_i, f_i) then $f^L \circ f^R$ is the sup of the directed family $(F(f_i)^L(t_{X_i}))$, and the corresponding characterisation of pull-backs: with the notation of the definition,we have $f_3^L \circ f_3^R = f_1^L \circ f_1^R \wedge f_2^L \circ f_2^R$. ∎

To prove the next proposition, the following lemma which is derived from a more general theorem due to Eugenio Moggi, is convenient.

Lemma 14 *Let F be a continuous and stable functor from $\mathbf{DI}^L$ to $\mathbf{DI}^L$. An indexed family (t_X) is continuous stable if, and only if, it is uniform.* ∎

We will also call such a family a *continuous stable section* (or just section) of the functor F. With this result, it is possible to generalize the definition of $\Pi(F)$.

Proposition 15 *Let $\mathbf{C}$ be an arbitrary category, and F a continuous and stable functor from $\mathbf{C} \times \mathbf{DI}^L$ to $\mathbf{DI}^L$. Define $\Pi(F)$, functor from $\mathbf{C}$ to $\mathbf{DI}^L$, so that, for an object A of $\mathbf{C}$, $\Pi(F)(A)$ is $\Pi(F(A, \_))$ as defined above[1], and if $f \in \mathbf{C}(A, B)$, then $\Pi(f)^L((t_X)) = (F(f, id_X)^L(t_X))$, and $\Pi(f)^R((u_X)) = (F(f, id_X)^R(u_X))$. Then $\Pi(F)$ is a continuous and stable functor.* ∎

Indeed, it is straightforward to check, for instance, that $(F(f, id_X)^L(t_X))$ is a continuous stable family in X (and, from Moggi's result, we deduce the somewhat surprising fact that this family is also uniform). We will use the continuous and stable functors $\rightarrow$ and $\times$, from $\mathbf{DI}^L \times \mathbf{DI}^L$ to $\mathbf{DI}^L$. If F and G are two functors from the same category $\mathbf{C}$ to $\mathbf{DI}^L$, we write $F \rightarrow G$ for $\rightarrow \circ \langle F, G \rangle$, and $F \times G$ for $\times \circ \langle F, G \rangle$. The same method may be applied to prove the next result.

Proposition 16 *Let F, G and H be three continuous and stable functors from the same category $\mathbf{C}$ into $\mathbf{DI}^L$, and K a continuous stable functor from $\mathbf{C} \times \mathbf{DI}^L$ to $\mathbf{DI}^L$. We can define the operators* app, App, curry *and* Curry *on sections and continuous stable functors.*

- *if $t = (t_X)$ is a section for $F \rightarrow (G \rightarrow H)$ and $u = (u_X)$ a section for $F \rightarrow G$, then $\mathsf{app}(t, u)$ is the section $(\lambda x.\ (t_X(x))(u_X(x)))$ of the functor $F \rightarrow H$,*
- *if $t = (t_X)$ is a section for $F \times G \rightarrow H$, then $\mathsf{curry}(t)$ is the section $(\lambda x.\ \lambda y.\ t_X(x, y))$ of the functor $F \rightarrow (G \rightarrow H)$,*
- *if $t = (t_X)$ a section of the functor $F \rightarrow \Pi(K)$, then $\mathsf{App}(t, G)$ is the section of the functor $F \rightarrow (K \circ \langle Id, G \rangle)$ defined as the family $(\lambda x.\ (t_X(x))_{G(X)})$,*

[1] Indeed, $F(A, \_)$ is a continuous and stable functor from $\mathbf{DI}^L$ to $\mathbf{DI}^L$.

- *if $t = (t_X)$ is a section of the functor K, then* $\mathsf{Curry}(t)$ *is the section of the functor* $\Pi(K)$*, from* $\mathbf{C}$ *to* $\mathbf{DI}^L$*, defined by* $\mathsf{Curry}(t)_X = (t_{(X,Y)})$. ∎

5.2 The model

As was said before, the meaning $[\![\Sigma \vdash \sigma]\!]$ and $[\![H \vdash_\Sigma M : \sigma]\!]$ are given by structural induction.

First, we define $[\![\Sigma \vdash \sigma]\!]$ of a type (we'll write $[\![\sigma]\!]$ whenever Σ is clear enough). It is a continuous and stable functor from $(\mathbf{DI}^L)^\Sigma$ to $\mathbf{DI}^L$ (in particular, if Σ is empty, we see that the semantics of a type is a dI-domain). The definition is by cases on σ

- $\sigma \equiv \alpha$, then $[\![\Sigma \vdash \sigma]\!]$ is the α'th projection functor,
- $\sigma \equiv \sigma_1 \to \sigma_2$, then $[\![\sigma]\!]$ is $[\![\sigma_1]\!] \to [\![\sigma_2]\!]$ as defined above,
- $\sigma \equiv \Pi\alpha.\ \sigma_1$, then we have $\Sigma, \alpha \vdash \sigma_1$, and $[\![\sigma_1]\!]$ is a functor from $(\mathbf{DI}^L)^\Sigma \times \mathbf{DI}^L$ to $\mathbf{DI}^L$. We take $[\![\sigma]\!] = \Pi([\![\sigma_1]\!])$ as defined above.

Next, we define the meaning of $H \vdash_\Sigma M : \sigma$ (we'll write $[\![M]\!]$ if Σ and H are clear enough). For this, we remark that if $H = x_1 : \sigma_1, \ldots, x_n : \sigma_n$, then we have $\Sigma \vdash \sigma_1, \ldots, \Sigma \vdash \sigma_n$. Furthermore, we know that $\Sigma \vdash \sigma$. We thus have $\Sigma \vdash \sigma_1 \times \cdots \times \sigma_n \to \sigma$. Hence, $[\![\sigma_1 \times \cdots \times \sigma_n \to \sigma]\!]$ is a continuous and stable functor from $(\mathbf{DI}^L)^\Sigma$ to $\mathbf{DI}^L$. The denotation $[\![M]\!]$ is then a continuous and stable section (or, equivalently, a uniform section) of the functor $[\![\sigma_1 \times \cdots \times \sigma_n \to \sigma]\!]$. The definition of $[\![M]\!]$ is by cases on M.

- $M \equiv x_i$, then σ is σ_i and $[\![M]\!]$ is the uniform family of i'th projections,
- $[\![M_1(M_2)]\!] = \mathsf{app}([\![M_1]\!], [\![M_2]\!])$,
- $[\![\lambda x : \tau.\ M]\!] = \mathsf{curry}([\![M]\!])$,
- $[\![M\{\tau\}]\!] = \mathsf{App}([\![M]\!], [\![\tau]\!])$,
- $[\![\Lambda(M)]\!] = \mathsf{Curry}([\![M]\!])$.

This presentation of the semantics is perhaps a little sketchy. The full treatment given in [5] may be helpful. An alternative description of this kind of model, using the categorical presentation of Seely [16], is given in [4].

References

[1] Amadio, R., Bruce, K. B., Longo, G., *The finitary projection model for second order lambda calculus and solutions to higher order domain equations.* In: **Logic in Computer Science,** edited by A. Meyer, IEEE Computer Society Press, 1986, pp. 122–130.

[2] Berry, G., *Stable models of typed λ-calculi.* In: **Fifth International Colloquium on Automata, Languages and Programs,** Springer-Verlag, **Lecture Notes in Computer Science,** vol. 62, 1978, pp. 72–89.

[3] Bruce, K. and Meyer, A., *The semantics of polymorphic lambda-calculus.* In: **Semantics of Data Types,** edited by G. Kahn, D.B. MacQueen and G. Plotkin, **Lecture Notes in Computer Science,** vol. 173, Springer-Verlag, 1984, pp. 131–144.

[4] Coquand, T., and Ehrhard, T., *An equational presentation of higher-order logic.* To appear in: **Category theory and computer science, Lecture Notes in Computer Science,** Springer-Verlag, 1987.

[5] Coquand, T., Gunter, C., and Winskel, G., Domain theoretic models of polymorphism. In preparation. To appear as: University of Cambridge Computer Laboratory Technical Report, 1987.

[6] Girard, J. Y., **Interprétation fonctionelle et élimination des coupures de l'arithmétique d'ordre supérieur.** Thèse d'Etat, Université Paris VII, 1972.

[7] Girard, J. Y., *The system F of variable types, fifteen years later.* **Theoretical Computer Science,** vol. 45, 1986.

[8] Gunter, C. A., *Universal profinite domains.* **Information and Computing,** vol. 72 (1987), pp. 1–30.

[9] Kahn, G., and Plotkin, G., *Domaines concrets.* Rapport IRIA Laboria, no. 336, 1978.

[10] McCracken, N., **An Investigation of a Programming Language with a Polymorphic Type Structure,** Doctoral Dissertation, Syracuse University, 1979.

[11] Nielsen, M., Plotkin, G., Winskel, G., *Petri nets, event structures and domains.* **Theoretical Computer Science,** vol. 13, 1981.

[12] Fairbairn, J., *Design and implementation of a simple typed language based on the lambda-calculus.* University of Cambridge Computer Laboratory Technical Report, no. 75, 1985, 107pp.

[13] Reynolds, J. C., *Polymorphism is not set-theoretic.* In: **Semantics of Data Types,** edited by G. Kahn, D.B. MacQueen and G. Plotkin, **Lecture Notes in Computer Science,** vol. 173, Springer-Verlag, 1984, pp. 145–156.

[14] Reynolds, J. C., *Towards a theory of type structures.* In: **Colloque sur la Programmation,** Springer-Verlag, **Lecture Notes in Computer Science** 19, 1974, pp. 408–425.

[15] Scott, D. S., *Some ordered sets in computer science.* In: **Ordered Sets,** edited by I. Rival., D. Reidel Publishing Company, 1981, pp. 677–718.

[16] Seely, R., *Categorical semantics for higher order polymorphic lambda calculus.* Manuscript, 1986, 33pp.

[17] Smyth, M. B. and Plotkin, G. D, *The category-theoretic solution of recursive domain equations.* **SIAM Journal of Computing,** vol. 11 (1982), pp. 761–783.

[18] Winskel, G., **Events in Computation.** Doctoral Dissertation, University of Edinburgh, 1980.

[19] Winskel, G., *A representation of completely distributive algebraic lattices.* Report CS-83-154 of the Computer Science Department, Carnegie-Mellon University, 1983.

[20] Winskel, G., *Event structures.* University of Cambridge Computer Laboratory Technical Report, no. 95, 1986, 69pp.

Continuous Auxiliary Relations*

Tsutomu Kamimura
Tokyo Research Laboratory
IBM Japan, Chiyoda-Ku
Tokyo 102, Japan

Adrian Tang
Department of Computer Science
University of Missouri, Kansas City
Kansas City, Missouri 64110

Abstract

In this article, we introduce finite auxiliary relations and show that in a domain given by a retract of Plotkin's SFP object, Scott's way-below relation can be obtained as the union of an ascending chain of finite auxiliary relations.

0. Introduction

Scott's way-below relation $\ll$ [6] plays an important role in the formulation of continuous cpos. The Compendium [1] introduced auxiliary relations and showed that $\ll$ is the smallest approximating auxiliary relation with the interpolation property. Thus approximating interpolative auxiliary relations provide a top-down characterization of $\ll$. In this article, we provide a bottom-up characterization of $\ll$ using finite auxiliary relations. We show that in a domain given by a retract of Plotkin's SFP object [5] the relation $\ll$ can be obtained as the union of an ascending chain of finite auxiliary relations. For this reason, retracts of SFP objects are called finitely continuous cpos. Detailed study of finitely continuous cpos can be found in [3] and [4].

* This work was supported by National Science Foundation Grant DCR-8415919.

1. Continuous Auxiliary Relations

Definition

An auxiliary relation on a cpo D is a binary relation $\prec$ satisfying:

$$\bot \prec x$$

$$x \prec y \Rightarrow x \sqsubseteq y$$

$$x \sqsubseteq y \prec z \sqsubseteq w \Rightarrow x \prec y$$

$$\{y \in D \mid y \prec x\} \text{ is } \sqsubseteq\text{-directed}$$

for all x,y,z,w in D. It is a continuous auxiliary relation (abbreviated as CAR) if it also satisfies the following continuity property:

$$x \prec \bigsqcup H \Rightarrow x \prec h \text{ for some } h \in H$$

for every directed set H.

There is a duality between CAR's on D and continuous functions $f: D \to Id(D)$ satisfying $f(x) \subseteq \downarrow x$ for all $x \in D$ (where Id(D) is the poset of ideals in D ordered by subset inclusion). For a given CAR $\prec$, we can define a function $f: D \to Id(D)$ as follows:

$$f(x) = \{y \in D \mid y \prec x\}$$

Continuity of f follows immediately from the continuity property of $\prec$. On the other hand, given a continuous function $f: D \to Id(D)$ which satisfies $f(x) \subseteq \{y \in D \mid y \sqsubseteq x\}$ for all x in D, we can define CAR $\prec$ on D as follows:

$$x \prec y \text{ iff } x \in f(y)$$

The CAR's that are defined by the finitely valued continuous functions from D to Id(D) are called finite auxiliary relations (abbreviated as FAR's).

A CAR $\prec$ on D is approximating if $x = \sqcup\{y \in D \mid y \prec x\}$ for every x in D. D is a continuous cpo (abbreviated as ccpo) if there is an approximating CAR on D. If D is a ccpo, we show that Scott's way-below relation $\ll$ [6] is the only approximating CAR. It is well-known that $\ll$ is an approximating CAR. Let us show that every approximating CAR $\prec$ must be equal to $\ll$. The continuity property of $\prec$ implies that $\prec$ is contained in $\ll$. Now assume $x \ll y$. Since $y = \sqcup\{z \in D \mid z \prec y\}$ and the set $\{z \in D \mid z \prec y\}$ is directed, there exists some $z \in D$ such that $x \sqsubseteq z \prec y$. Hence $x \prec y$, concluding $\prec = \ll$. From now on, we assume that all the cpos are continuous.

Let CAR_D be the poset of all the CAR's on a ccpo D ordered by subset inclusion. CAR_D is a cpo where the top element is Scott's way-below relation. A Galois connection [1] exists between CAR_D and $[D \xrightarrow{sub} D]$, the subposet of $[D \to D]$ consisting of all the subidentity maps, i.e. maps that are less than the identity id_D. We define maps $g: CAR_D \to [D \xrightarrow{sub} D]$ and $d: [D \xrightarrow{sub} D] \to CAR_D$ as follows:

$$g(\prec)(x) = \sqcup\{y \in D \mid y \prec x\}$$

$$\text{and } d(f)(x) = \{y \in D \mid y \ll f(x)\}$$

(Note that $d(f)$, as a member of CAR_D, is defined as a function from D to Id(D)). Let us observe the following properties:

(i) $d \circ g(\prec)$ is contained in $\prec$; and

(ii) $g \circ d(f)(x) = \sqcup\{y \in D \mid y \ll f(x)\}$

$= f(x)$

hence $g \bullet d(f) = f$

Thus (g,d) forms a (upper,lower) adjoint pair, showing that $[D \xrightarrow{sub} D]$ is a projection of

CAR_D. Here are a few more properties:

(iii) $g(\ll) = id_D$ and $d(id_D) = \ll$;

(iv) If $\prec$ is a FAR, then $g(\prec)$ is a finite subidentity map; and if f is a finite subidentity map, then d(f) is a FAR. Consequently, Scott's way-below relation $\ll$ is an ascending union of FAR's iff the identity map id_D is the lub of some ascending chain of finite maps.

Definition

A finitely continuous cpo D is a ccpo where $\ll$ is the union of some countable ascending chain of FAR's (or equivalently, id_D is the lub of some countable ascending chain of finite continuous maps).

Propostion 1 Finitely continuous cpos are closed under retracts.

Proof: Suppose that r is a retraction on a finitely continuous cpo D where $id_D = \sqcup g_n$ for some countable ascending chain of finite maps. Then $id_{r(D)} = \sqcup r \circ g_n \circ r$ where $\{r \circ g_n \circ r\}_n$ is a countable ascending chain of finite maps on the retract r(D).

□

Proposition 2 D is finitely continuous iff $[D \to D]$ is finitely continuous.

Proof: (⇒)

Suppose that $id_D = \sqcup g_n$ for some countable ascending chain $\{g_n\}_n$ of finite maps. Then $id_{[D \to D]} = \sqcup f_n$ where for each n, f_n: $[D \to D] \to [D \to D]$ is defined by: $f_n(h) = g_n \circ h \circ g_n$. Note that every f_n is a finite map.

(⇐)

D is a projection of $[D \rightarrow D]$, hence is finitely continuous by Proposition 1. □

Corollary 3

The category of finitely continuous cpos and continuous maps is the largest cartesian closed category of ccpos where the function space objects have a basis of finite functions.

Proposition 4 Let D be an ω-algebraic ccpo. Then D is finitely continuous iff D is SFP.

Proof: $(\Rightarrow)$

Let $id_D = \sqcup f_n$ for some countable ascending chain of finite functions f_n's on D. For each n, define E_n to be $\{x \in D \mid f_n(x) = x\}$. Note that E_n is contained in E_{n+1}.

Claim: $E = \cup E_n$ where E is the compact basis

Proof: Assume $x \in E$. Since $x = \sqcup f_n(x)$ and x is compact, $x = f_n(x)$ for some n, hence $x \in E_n$. On the other hand, given $x \in E_n$ and $x \sqsubseteq \sqcup H$ for some directed set H. Then $x = f_n(x) \sqsubseteq f_n(\sqcup H) = \sqcup f_n(h)$. Since f_n is a finite function, $x \sqsubseteq f_n(h) \sqsubseteq h$ for some h, showing that x is compact. □ of claim.

Claim: E_n is a projection of E_{n+1}

Proof: It suffices to show that for every x in E_{n+1}, the set $\{y \in E_n \mid y \sqsubseteq x\}$ has a maximal element. For every $k \in \omega$, let $(f_n)^k$ be $f_n \circ f_n \circ \ldots \circ f_n$ (composed k times). Since f_n is a finite map less than the identity, there must exist some k such that $(f_n)^k(x) = (f_n)^{k+1}(x)$. Then $(f_n)^k(x)$ is the maximal element in the set $\{y \in E_n \mid y \sqsubseteq x\}$. □ of claim.

The last two claims establish that the E_n's form a projective sequence and hence D is SFP.

$(\Leftarrow)$

Suppose that D is SFP obtained by a sequence of finite projections π_n's on D. Then $id_D = \sqcup \pi_n$. □

<u>Proposition 5 (Plotkin)</u> D is finitely continuous iff D is a retract of some SFP object.

<u>Proof</u>: (⇐)

This follows from Propositions 1 and 4.

(⇒)

Let $id_D = \sqcup f_n$ for some countable ascending chain of finite continuous maps on D. Define the cpo D' as follows:

$$D' = \{\langle x_n \rangle_n \mid x_n \in \cup\{f_i(D) \mid i \leq n\} \text{ and } x_n \sqsubseteq x_{n+1}\}$$

Under the pointwise ordering, D' is SFP. Next define r: D' → D and e: D → D' as follows:

$$r(\langle x_n \rangle) = \sqcup x_n$$

$$e(x) = \langle f_n(x) \rangle_n$$

Note that $r \circ e = \sqcup f_n = id_D$, hence D is a retract of D'. □

<u>Propostion 6 (Gunter [2])</u> D is finitely continuous iff D is a projection of some SFP object.

<u>Proof</u>: In the proof of Proposition 5, replace D' by another cpo D" where

$$D" = \{\langle x_n \rangle_n \mid x_n \in \cup\{f_i(D) \mid i \leq n\}, x_n \sqsubseteq x_{n+1}, \text{ and } f_n(\sqcup x_n) \sqsubseteq x_{n+1}\}.$$

Then D" is SFP and furthermore, $e \circ r \sqsubseteq id_{D"}$. □

Given $\prec_1, \prec_2$ in CAR_D, we say that $\prec_1$ is <u>interpolative</u> in $\prec_2$ if for any x,y in D, $x \prec_1 y$ implies $x \prec_2 z \prec_2 y$ for some z in D. Then:

<u>Propostion 7</u> In a finitely continuous D, ≪ is the union of a countable ascending chain of FAR's $\{\prec_n\}_n$ where $\prec_n$ is interpolative in $\prec_{n+1}$ for every n.

<u>Proof</u>: It was shown in [3] that $id_D = \sqcup f_n$ where $\{f_n\}$ is an ascending chain of finite

continuous maps satisfying $f_n \sqsubseteq f_{n+1}^2$ for every $n \in \omega$. For each n, define the FAR $\prec_n$ to be: $x \prec_n y$ iff $x \sqsubseteq f_n(y)$. Clearly $\ll = \bigcup \prec_n$.

Claim: $\prec_n$ is interpolative in $\prec_{n+1}$.

Proof: Assume $x \prec_n y$, hence $x \sqsubseteq f_n(y)$. Let z be $f_{n+1}(y)$. Since $f_n(y) \sqsubseteq f_{n+1}(y)$, we have $z \prec_{n+1} y$. Also, $x \sqsubseteq f_n(y)$ implies $x \sqsubseteq f_{n+1}{}^2(y) = f_{n+1}(z)$, i.e. $x \prec_{n+1} z$.

□ of claim and proposition

$\prec$ is an <u>interpolative CAR</u> if $\prec$ is interpolative in itself. Then:

Proposition 8 (g,d) is a Galois connection between interpolative CAR's and projections, i.e. idempotent subidentity maps.

Proof: It suffices to show the following:

(i) if $\prec$ is interpolative, then $g(\prec)$ is a projection; and

(ii) if f is a projection, then d(f) is interpolative.

For (i):

$$\begin{aligned} g(\prec) \circ g(\prec)(x) &= \bigsqcup\{y \mid y \prec g(\prec)(x)\} \\ &= \bigsqcup\{y \mid y \prec \bigsqcup\{z \mid z \prec x\}\} \\ &= \bigsqcup\{y \mid y \prec z \prec x \text{ for some } z\} \\ &= \bigsqcup\{y \mid y \prec x\} \\ &= g(\prec)(x) \end{aligned}$$

For (ii):

$$\begin{aligned} y \in d(f)(x) &\text{ iff } y \ll f(x) \\ &\text{ iff } y \ll f \circ f(x) \\ &\text{ iff } y \ll \bigsqcup\{f(z) \mid z \ll f(x)\} \\ &\text{ iff } y \ll f(z) \sqsubseteq z \ll f(x) \text{ for some } z \\ &\text{ iff } y \in d(f)(z) \text{ and } z \in d(f)(x) \text{ for some } z \end{aligned}$$

□

Ershov [7] showed that there is a finitely continuous D where $\ll$ is not the union of any countable ascending chain of interpolative FAR's. For this reason, interpolative FAR's are not the right CAR's to study $\ll$ in a bottom-up fashion.

References

[1] G. Gierz, K. Hofman, K. Kiemel, J. Lawson, M. Mislove and D. Scott, A Compendium of Continuous Lattices, Springer Verlag, 1981.

[2] C. Gunter, Profinite solutions for Recursive Domain Equations, Ph.D. dissertation, University of Wisconsin, 1985.

[3] T. Kamimura and A. Tang, Retracts of SFP objects, Lecture Notes in Computer Science 239, 1986, 135-148.

[4] T. Kamimura and A. Tang, Domains as Finitely Continuous CPOs, Technical Report, 1987.

[5] G. Plotkin, A Powerdomain Construction, SIAM Journal of Computing, vol. 5, no. 3, 1976, 452-487.

[6] D. Scott, Continuous Lattices, Lecture Notes in Mathematics 274, 1970, 97-136.

[7] D. Scott, Data Types as Lattices, SIAM Journal of Computing, vol. 5, no. 3, 1976, 522-587.

Computable One-to-one Enumerations of Effective Domains

Dieter Spreen[1]

Dipartimento di Informatica - Universita' di Pisa - Corso Italia, 40 - I-56100 Pisa - Italy

Abstract

In this paper we consider ω-algebraic complete partial orders where the compact elements are not maximal in the partial order. Under the assumption that the compact elements admit a one-to-one enumeration such that the restriction of the order to them is completely enumerable, it is shown that the computable domain elements also can be effectively enumerated without repetition. Such computable one-to-one enumerations of the computable domain elements are minimal among all enumerations of these elements with respect to the reducibility of one enumeration to another. The admissible indexings which are usually used in computability studies of continuous complete partial orders are maximal among the computable enumerations. As it is moreover shown, admissible numberings are recursively isomorphic to the directed sum of a computable family of computable one-to-one enumerations. Both results generalize well known theorems by Friedberg and Schinzel, respectively, for the partial recursive functions. The proof uses a priority argument.

1. Introduction

In programming one has to deal with several different data structures, not just the natural numbers or strings of binary digits. As has been discovered by Scott (1970), the underlying structure of these domains is that of an algebraic complete partial order. His basic insight was that all the domains we deal with in programming contain some (finite) basic elements by which all other elements can be approximated. If this approximation process is effective, then, of course, the approximated element is computable. The set of such elements with the inherited partial order is an effective domain.

There are many ways to enumerate the elements of an effective domain. Each such numbering can be thought of as a programming system for the effective generation of approximations. The indices are obtained by coding the programs. If this generation can be done in a uniform way, then the corresponding numbering or programming system is called computable. In the case of the domain of partial recursive

---

[1]On leave from Siemens Corporate Laboratories for Research and Technology, Munich, West Germany. Supported by a grant of the Italian C.N.R. to work at the Computer Science Department of the University of Pisa and by the Siemens Corporate Laboratories for Research and Technology.

functions the computable numberings are those that have a computable universal function. In the literature, admissible numberings are most often used (cf. e.g. Weihrauch, Deil 1980) which generalize the Gödel numberings of the partial recursive functions. They are computable and any computable indexing can be effectively translated into (reduced to) each of them. However, they are extremely non one-to-one: each domain element has infinitely many indices (names). So two questions arise: first, does a given effective domain also have a computable numbering to which no other numbering can be reduced, except equivalent ones, and, second, can the domain elements also be enumerated in a computable and one-to-one manner. It is easy to show that each computable one-to-one enumeration is minimal among the indexings of a given effective domain with respect to reducibility. This solves the first problem. It remains to solve the second one.

For some special cases such as recursively enumerable sets and partial recursive functions, the second problem has been solved by Friedberg and others (Friedberg 1958; Khutoretskiĭ 1970; Mal'cev 1980). In this paper we generalize these results to a large class of effective domains. This class is characterized by the requirements that the basic elements are not maximal in the approximation ordering of the domain and are effectively enumerable without repetition. The proof of our result uses a priority argument.

As has already been mentioned, computable one-to-one enumerations are minimal among the enumerations of an effective domain, with respect to reducibility, and admissible enumerations are maximal among the computable ones. We prove a connection between these two types of numberings for the class of effective domains as mentioned above: namely, each admissible numbering of the domain elements is recursively isomorphic to the directed sum of a computable family of computable one-to-one numberings of these elements. This theorem generalizes a result of Schinzel (1977) for the partial recursive functions. The proof uses a refinement of the above mentioned priority argument.

2. Basic Definitions and Properties

Let $P = (P, \sqsubseteq)$ be a partial order. It is *complete*, if every directed subset S of P has a least upper bound $\bigsqcup S$ in P. Let $\bot = \bigsqcup \emptyset$. Moreover, let « denote the *way-below relation* on P, i.e., let x«y iff for any directed subset S of P, from $y \sqsubseteq \bigsqcup S$ it always follows that $x \sqsubseteq u$, for some $u \in S$.

A subset B of P is a *basis* of *P*, if for any $x \in P$ the set $B_x =_{df} \{u \in B \mid u \ll x\}$ is directed and $x = \bigsqcup B_x$. If *P* has a basis, then it is called *continuous*, and if $\{x \in P \mid x \ll x\}$ is a basis of *P*, then *P* is said to be *algebraic*. Observe that $B \supseteq \{x \in P \mid x \ll x\}$, for any basis B of *P*, and that a complete partial order is algebraic precisely when x«y means there is some $z \in B$ with $x \sqsubseteq z \sqsubseteq y$.

There have been many suggestions in the literature as to which effectivity requirements a complete partial order (cpo) should satisfy in order to develop a sufficiently rich computability theory for these structures (cf. Scott 1970; Egli, Constable 1976; Smyth 1977; Sciore, Tang 1978a,b; Kanda, Park 1979; Kanda 1980; Kreitz 1982). Here we use a very general approach which is due to Weihrauch (cf. Weihrauch, Deil 1980).

Let $\langle\, ,\, \rangle: \omega^2 \rightarrow \omega$ be a pairing function, i.e. a total recursive bijection, and let (π_1, π_2) with $\pi_i(\langle x_1, x_2 \rangle) = x_i$ be its inverse. Then a continuous cpo $P = (P, \sqsubseteq, B)$ is *effective*, if it has a surjective coding function $\beta:\omega\to B$ of the basic elements such that $\{<i,j> \mid \beta_i \ll \beta_j\}$ is recursively enumerable (r.e.). Note that this definition is independent of the chosen pairing function. An element x of an effective cpo $P=(P,\angle,B,\beta)$ is *computable*, if $I_\beta(x)=_{df}\{i \mid \beta_i \ll x\}$ is r.e. Let P_c denote the set of all computable elements of P, P_c can be characterized as that subset K of P for which, for every directed subset S of B, $\bigsqcup S\in K$ iff $\{i \mid \beta_i \in S\}$ is r.e.

We shall now restrict ourselves to algebraic cpo's. Moreover, we are only interested in the constructive part of such cpo's. Let $(P,\angle,B,\beta)$ be an algebraic effective cpo; then $(P_c,\angle,B,\beta)$ is called an *effective domain.*

There are many ways to enumerate the elements of an effective domain $D=(D,\angle,B,\beta)$. Let $Nm_D=\{\eta:\omega\to D \mid \eta \text{ onto}\}$ be the set of all such enumerations. In general there is no uniform effective way to generate the basic elements approximating a domain element from an index of this element. But if we think of a numbering η as being obtained from a programming system by some coding, then there should exist such a uniform generation procedure or, equivalently, $\{<i,j> \mid \beta_i \ll \eta_j\}$ should be r.e. Numberings with this property are called *computable.*

In the literature, a special type of computable numberings is used, namely, admissible numberings. These are a generalization of the Gödel numberings of the partial recursive functions. Let W be a canonical indexing of the r.e. sets (cf. Rogers 1967). A computable numbering $\eta\in Nm_D$ is called *admissible,* if there is a total recursive function d with $\eta_{d(i)}=\bigsqcup\beta(W_i)$, for any $i\in\omega$ such that $\beta(W_i)$ is directed.

In (Riccardi 1980) other types of indexings have been studied in order to show the independence of certain control structures. All indexings in Nm_D are interrelated by the following reducibility (or translatability) relation: δ is *reducible* to η $(\delta\leq\eta)$ iff there is a total recursive function f such that $\delta_i=\eta_{f(i)}$, for all $i\in\omega$. If f in addition is one-to-one, then δ is called *1-reducible* to η $(\delta\leq_1\eta)$, and if f is both one-to-one and onto, then δ and η are said to be*recursively isomorphic* $(\delta\equiv\eta)$. It is easy to check that $\leq$ and $\leq_1$ are preorders on Nm_D and $\equiv$ is an equivalence relation. By the Theorem of Myhill-Eršov $\delta\equiv\eta$ iff $\delta\leq_1\eta$ and $\eta\leq_1\delta$ (cf. Eršov 1973). Moreover, we have

Lemma 2.1. Let $\delta,\eta\in Nm_D$ such that η is admissible. Then

(a) δ is computable iff $\delta\leq\eta$ iff $\delta\leq_1\eta$,

(b) $|\eta^{-1}(x)|=\aleph_0$, for any $x\in D$.

Proof. The first part of (a) follows from (Weihrauch, Deil 1980). There it is moreover shown that η is precomplete, which means that for any partial recursive function p there is a total recursive function g with $\eta_{p(i)}=\eta_{g(i)}$, for all $i\in\omega$ such that p(i) is defined. It is proved in (Eršov 1973) that such numberings have property (b). Furthermore, there is a total recursive one-to-one function h with $\eta_{h(i,j)}=\eta_i$, for all $i,j\in\omega$. Now, let $\delta\leq\eta$ via f, then g with g(i)=h(f(i),i) witnesses that $\delta\leq_1\eta$. From this the second part of (a) follows.

Thus, the admissible numberings of D are maximal among the computable ones with respect to reducibility, but they are extremely non one-to-one. This leads to the question whether there exist minimal computable indexings of D and whether the elements of D can also be enumerated in a computable and one-to-one manner?

Lemma 2.2. Let $\delta \in Nm_D$ be computable and one-to-one. Then for all $\eta \in Nm_D$ with $\eta \leq \delta$ we also have that $\delta \leq \eta$.

Proof. Let $\eta \leq \delta$ via f and define the recursive function g by $g(y)=\mu x$: $f(x)=y$. Since δ is one-to-one, f is onto. Hence, g is total. Moreover, $\delta \leq \eta$ via g.

This shows that computable one-to-one numberings of D are minimal with respect to reducibility, not only among the computable numberings but among all numberings of D. Thus, in order to answer the two questions posed above, it suffices to give a positive answer to the second one. For some special cases such as the r.e. sets and the partial recursive functions this has been done by Friedberg and others (Friedberg 1958; Khutoretskiĭ 1970; Mal'cev 1970). As we shall see now, computable one-to-one numberings do exist for a much greater class of effective domains.

3. Computable One-to-one Numberings

Let $MAX=\{x \in D \mid x \text{ is maximal with respect to } \sqsubseteq\}$.

Theorem 3.1. Let $(D,\sqsubseteq,B,\beta)$ be an effective domain such that β is one-to-one and $B \cap MAX=\emptyset$. Then one can construct a computable one-to-one numbering δ of D.

Proof. Let η be an admissible indexing of D. The basic idea of the proof is to represent each $x \in D$ by the r.e. set $I_\beta(x)$ $(=\{i \mid \beta_i \ll x\})$ and to construct a computable one-to-one numbering δ' of the recursively enumerable family $\{I_\beta(\eta_i) \mid i \in \omega\}$ of r.e. sets such that δ defined by

$$\delta_i = \bigsqcup \beta(\delta'_i)$$

is a computable one-to-one indexing of D. A computable one-to-one numbering of $\{I_\beta(\eta_i) \mid i \in \omega\}$ can be constructed with the help of a priority argument. This makes use of the fact that each r.e. set is the union of an effective nondecreasing sequence of finite subsets which are enumerated at subsequent time steps, one at each step. In our construction we use finite subsets A for which in addition $\beta(A)$ is directed. As usual, the priority construction consists of a simultaneous enumeration of all sets $I_\beta(\eta_i)$ $(i \in \omega)$ and a strategy which prevents sets in this family from being enumerated more than once. This strategy is such that in cases where the finite parts of the sets $I_\beta(x)$ $(x \in D)$ which have been generated in t steps indicate that some set G will be listed at least twice, one of the enumerations of G is stopped at step t. Thus, in the enumeration δ' which we are going to construct there will appear not only the sets $I_\beta(x)$ $(x \in D)$, but also finite subsets A of them. Since the images under β of these subsets A are directed, $\bigsqcup \beta(A)=\max\beta(A)$ exists and is a basic element of the domain *D*. Hence, in the enumeration δ' some basic elements u are represented twice, namely by $I_\beta(u)$ and by some finite set A with $u=\max\beta(A)$, which means that the induced enumeration δ of D is no longer one-to-one. Thus the priority construction must be such that not only the enumerated sets are different but that for all finite sets X,Y appearing during the construction we also have that $\max\beta(X) \neq$

$\max\beta(Y)$. Since always $u\in B_u$, for $u\in B$, in the nondecreasing sequence of finite sets by the help of which $I_\beta(u)$ is enumerated there is some set H with $u=\max\beta(H)$. Hence the extra condition prevents u from being represented twice.

Now, let $T=\{<i,j> \mid \beta_i \angle \beta_j\}$. Then T is r.e. Since η is admissible, $\{<i,j> \mid \beta_i \angle \eta_j\}$ is r.e. too. Thus, there is some total recursive function r with $W_{r(j)}=\{i \mid \beta_i \angle \eta_j\}$. Observe that $W_{r(j)}$ is nonempty, for all $j\in\omega$, since $\perp\in B$. In the first construction step we fix a class of finite subsets of ω by which we shall approximate the sets $W_{r(j)}$. To this end, let E be a canonical enumeration of the finite subsets of ω (cf. Rogers 1967), and for each r.e. set C, let C_t be the finite subset of C which is enumerated in t steps with respect to a fixed enumeration.Then define

$$Z = \{i \mid \exists\, a,j,t\, [E_i \subseteq W_{r(j),a} \wedge E_i \neq \emptyset \wedge \forall\, m,m'\in E_i\ \exists\, m''\in E_i\ <m,m''>,<m',m''>\in T_t\}.$$

It follows that Z is infinite and r.e. Moreover,

(i) $\beta(E_i)$ is directed, for each $i\in Z$,

(ii) for each finite subset A of $W_{r(j)}$ there is some $i\in Z$ with $A\subseteq E_i \subseteq W_{r(j)}$, and

(iii) for each $i\in Z$ there is some $j\in Z$ such that $E_i \subset E_j$ and $\max\beta(E_i)\angle\max\beta(E_j)$.

The last property follows from the supposition that no basic element is maximal with respect to $\angle$. Note that since $\beta(E_i)$ is finite and directed, it has a maximal element.

In what follows we construct sets $A_j^0 \subseteq A_j^1 \subseteq \ldots$ and $C_i^0 \subseteq C_i^1 \subseteq \ldots$ and a partial recursive function f such that $A_j^t, C_i^t \in E(Z)\cup\{\emptyset\}$ and $A_j =_{df} \bigcup_t A_j^t = W_{r(j)}$. Moreover, the sets $C_i =_{df} \bigcup_t C_i^t$ will be such that $\bigsqcup\beta(C_0), \bigsqcup\beta(C_1), \ldots$ is an enumeration of D without repetition.

If f(n,t) is defined and f(n,t)=a, we say that a is a *follower of* n *at time* t. In the case that a=f(n,t-1) but f(n,t) is not defined, a is said to be *freed* (*from* n) *at time* t. The equation a=f(n,t) means that at time t and at the following steps we try to construct the sets C_a^m in such a way that finally $C_a=A_n$. Only under certain circumstances can we later on be forced not to follow this enumeration and to free a from n. Numbers which are no follower at time t are called *free at time* t. In the following construction we shall consider the index $n_t =_{df} \pi_1(t)$ at time t, i.e., we shall follow the enumeration $W_{r(n_t)}$. In this way each index is considered infinitely often. The sets A_n^t and C_a^t and the function f which we are going to construct will satisfy the following requirements:

(1) If $A_n^t \neq \emptyset$, then $A_n^t \in E(Z)$. Moreover, $A_n^{t-1} \subseteq A_n^t$.

(2) At any time t, a number a cannot be the follower of more than one number n. If a is freed at time t, then a remains free at all times $t'\geq t$.

(3) $C_a^{t-1} \subseteq C_a^t$, for all $a,t\in\omega$. If a has been free for all times $t'<t$, then $C_a^{t-1}=\emptyset$.

(4) If $C_a^t \neq \emptyset$, then $C_a^t \in E(Z)$. Furthermore, $C_a^{t-1} \subseteq A_{n_t}^t$, for $a = f(n_t,t)$.

(5) At any time t, $\max\beta(C_a^t) \neq \max\beta(C_b^t)$, if $a \neq b$, $C_a^t \neq \emptyset$ and $C_b^t \neq \emptyset$.

Now, for $a,n\in\omega$ set $A_n^{-1} = C_a^{-1} = \emptyset$ and let f(n,-1) be undefined. In order to define A_n^t, C_a^t and f(n,t) for $t\geq 0$, assume that A_n^m, C_a^m and f(n,m) are already known and satisfy conditions (1) - (5) for all $m<t$ and all $n,a\in\omega$. We first define A_n^t. For $n>t$ we set $A_n^t = \emptyset$. For $n\leq t$ we define A_n^t in the following way: Let g be a fixed total recursive function that enumerates Z. Now, successively enumerate the sets $E_{g(0)}, E_{g(1)}, \ldots$ and enumerate the set $W_{r(n)}$. Moreover, for each c = 0, 1, ..., compare the sets $E_{g(0)}$,

$E_{g(1)}, \ldots, E_{g(c)}$ with the given sets A_n^{t-1}, $W_{r(n),t}$ and the growing set $W_{r(n),c}$. Because of property (ii) we can find an index b such that

$$W_{r(n),t} \cup A_n^{t-1} \subseteq E_{g(b)} \subseteq W_{r(n),c}.$$

Let b' be the smallest such b, and we define $A_n^t = E_{g(b')}$. Obviously, the sets A_n^t constructed in this way fulfil condition (1).

Next we define the sets C_a^t and f(n,t). This will be done by considering the following cases.

CASE I. $f(n_t,t-1)=b$ and for some $n<n_t$

$$\{0,1,\ldots,b\} \cap A_n^t = \{0,1,\ldots,b\} \cap A_{n_t}^t.$$

In this case we free b and let $f(n_t,t)$ be undefined. Moreover, we set $C_a^t = C_a^{t-1}$ and $f(n',t) = f(n',t-1)$, for $a \in \omega$ and $n' \neq n_t$.

CASE II. Case I does not hold and there is some index i such that $\max\beta(A_{n_t}^t) = \max\beta(C_i^{t-1})$ and in addition

(a) $i=f(n,t-1)$, for some $n \leq n_t$

or

(b) i is free at time t-1 and $i \leq n_t$

or

(c) i is free at time t-1 and has been displaced by n_t at some time t'<t. (The displacement operation is defined in case III(C).)

In this case nothing is changed, i.e., for $a,n \in \omega$ we set $C_a^t = C_a^{t-1}$ and $f(n,t) = f(n,t-1)$.

CASE III. Neither case I nor case II holds. Then we successively perform the following operations:

(A) If $f(n_t,t-1)$ is defined, set $f(n_t,t) = b = f(n_t,t-1)$. Otherwise, let b be the smallest number that has not been a follower at any time t'<t and define $f(n_t,t) = b$.

(B) Set $C_b^t = C_b^{t-1} \cup A_{n_t}^t$. (Because $C_b^{t-1} \subseteq A_{n_t}^t$, by (3),(4), it follows that $C_b^t = A_{n_t}^t$ after the execution of (B).)

(C) If there is some i with $i \neq b$ and $\max\beta(C_i^{t-1}) = \max\beta(A_{n_t}^t)$ (by (5) there is at most one such index), then set $C_i^t = E_{g(j)}$ where j is the smallest number for which $C_i^{t-1} \subset E_{g(j)}$ and $\max\beta(E_{g(j)}) \notin \{\max\beta(C_b^t), \max\beta(C_c^{t-1}) \mid c \in \omega \wedge C_c^{t-1} \neq \varnothing\}$. (Because of (iii) such a j always exists.)

If (C) is performed, we say that i is *displaced by* n_t *at time* t.

(D) If the index i which is displaced by n_t at time t is not free at time t-1, free it now. Moreover, set $C_a^t = C_a^{t-1}$ and $f(n,t) = f(n,t-1)$, for those a,n that are not listed in (A) - (C).

It follows from the definition that the sets A_n^t and C_a^t and the function f satisfy requirements (1) - (4). We now show that (5) also holds.

To this end, let $a,b\in\omega$ with $a\neq b$, $C_a^t \neq \emptyset$ and $C_b^t \neq \emptyset$, and let us assume that $\max\beta(C_a^{t-1}) \neq \max\beta(C_b^{t-1})$ provided that $C_a^{t-1} \neq \emptyset$ and $C_b^{t-1} \neq \emptyset$. Obviously, we only have to consider the cases that $f(n_t,t)\in\{a,b\}$ and that a or b is displaced by n_t at time t. Let us start with the first case. Without restriction let $b=f(n_t,t)$. Then $C_b^t = A_{n_t}^t$. If a is displaced by n_t at time t, it follows from step III(C) that $\max\beta(C_a^t) \neq \max\beta(C_b^t)$. Otherwise we have that $\max\beta(C_a^{t-1}) \neq \max\beta(C_b^t)$ and $C_a^t = C_a^{t-1}$. Let us next consider the case that $f(n_t,t)\notin\{a,b\}$ and, without restriction, b is displaced by n_t at time t. Our inductive assumption implies that a cannot also be displaced by n_t at time t in this case. Hence $C_a^t = C_a^{t-1}$ and $\max\beta(C_b^t) \neq \max\beta(C_a^{t-1})$, which shows that (5) holds in this case, too.

Let us see which further properties the above constructed sets have. First note that $C_a \neq \emptyset$, for each $a\in\omega$. The reason is that $A_n^t \neq \emptyset$, for all $t\geq 0$, and every a will become a follower in step III(A), since the family $(W_{r(j)})_{j\in\omega}$ is infinite. Moreover, we have

(6) Each number can be displaced only finitely often.

Assume that a is displaced at time t and is not free at time t-1. Then a is free at time t, because of (D), and it remains free at all later times, by (2). Now suppose that a is again displaced, by n_m, at some later time $m>t$. As we have already seen, a is free at time m-1. If in addition $a\leq n_m$ or if a has been displaced before, at some earlier time than m, then case II holds. This is impossible, since case III already holds. Thus, at all later times m a can only be displaced by numbers $n_m<a$ and by each such number at most once. This shows that a can be displaced only finitely often.

(7) $\bigsqcup\beta(C_a)$ exists in D, for each $a\in\omega$. If C_a is infinite, then from some time t, a is the constant follower of some n and $C_a = A_n$. Hence $\bigsqcup\beta(C_a) = \bigsqcup\beta(A_n) = \eta_n$.

If C_a is finite, then $C_a = C_a^t$, for some t. Thus $C_a\in E(Z)$, which implies that $\beta(C_a)$ is directed. Therefore $\bigsqcup\beta(C_a)$ exists in D.

If, on the other hand, C_a is infinite, then for infinitely many t, $C_a^{t-1} \neq C_a^t$. This can only happen, if C_a^t is constructed according to case III(B) (b=a) or if a is displaced by n_t at time t. By (6) a can be displaced only finitely often. It follows that in the construction of C_a case III(B) with $a=f(n_t,t)$ appears for infinitely many t. Since a cannot be the follower of different numbers, there is some n such that, for infinitely many t, $n=n_t$ and hence $a=f(n,t)$ and $C_a^t = A_n^t$. Thus, from some time t, a is the follower of n and $C_a = A_n = W_{r(n)}$. Because $\beta(A_n)$ is directed, the same holds for $\beta(C_a)$. Hence, $\bigsqcup\beta(C_a)$ exists in D.

(8) If $a\neq b$, then $\bigsqcup\beta(C_a) \neq \bigsqcup\beta(C_b)$.

Let us assume that $a\neq b$, but $\bigsqcup\beta(C_a) = \bigsqcup\beta(C_b)$. If C_a and C_b are both infinite, by (7) there exist numbers i and j with $C_a = A_i = W_{r(i)}$, $C_b = A_j = W_{r(j)}$, $\bigsqcup\beta(C_a) = \eta_i$ and $\bigsqcup\beta(C_b) = \eta_j$. Then it follows from $\bigsqcup\beta(C_a) = \bigsqcup\beta(C_b)$ that $C_a = W_{r(i)} = W_{r(j)} = C_b$. Since from some time t_0, a is the constant follower of i and b is the constant follower of j and, since $a\neq b$, we have that $i\neq j$ too. Let $i<j$. Then it follows from $A_i = A_j$ that from some time t_1 for all $t\geq t_1$

$$\{0,1,\ldots,b\} \cap A_i^t = \{0,1,\ldots,b\} \cap A_j^t.$$

Now, choose t such that $n_t=j$, $t>t_0$ and $t>t_1$. Then case I holds at time t, which implies that b has to be freed at this time, contradicting the fact that b is a constant follower of j from time t_0. In the case that C_a and C_b are infinite we therefore have that $\bigsqcup\beta(C_a) \neq \bigsqcup\beta(C_b)$.

If C_a and C_b are both finite, there is some t such that $C_a = C_a^t$ and $C_b = C_b^t$. Hence our assumption contradicts property (5).

If, finally, only one of the two sets is finite, say C_a, then there is some t with $C_a = C_a^{t''}$, for all $t''\geq t$, and some j with $C_b = A_j = W_{r(j)}$. Since $\bigsqcup\beta(C_a) = \max\beta(C_a^t) \in B$ in this case, it follows from our asssumption that $\bigsqcup\beta(C_a) \in \beta(A_j)$. Thus, there is some $t'\geq t$ such that $\bigsqcup\beta(C_a) \in \beta(A_j^{t'}) = \beta(C_b^{t'})$, from which we obtain that $\max\beta(C_a^{t'}) = \bigsqcup\beta(C_a) = \max\beta(A_j^{t'}) = \max\beta(C_b^{t'})$, contradicting property (5).

(9) For each $x \in D$ there is some a with $x = \bigsqcup\beta(C_a)$.

Let m be the smallest index of x with respect to η. Then $x = \eta_m = \bigsqcup\beta(A_m)$ and for all $m'<m$, $\bigsqcup\beta(A_{m'}) = \eta_{m'} \neq \eta_m = \bigsqcup\beta(A_m)$. Hence, there is some $\bar{t}$ such that for all $t',t''\geq\bar{t}$ and $m'<m$

$$\max\beta(A_{m'}^{t'}) \neq \max\beta(A_m^{t''}). \tag{1}$$

From $\eta_{m'} \neq \eta_m$ we obtain that $A_{m'} \neq A_m$. Thus, there is some $c_{m'}$ which belongs to one of the two sets but not to the other. Let $\hat{t}$ be such that each of the numbers $c_0,\ldots, c_{m-1}$ is enumerated in one of the sets $A_0^{\hat{t}}, \ldots, A_m^{\hat{t}}$ and let $c = \max\{c_0,\ldots,c_{m-1}\}$. Then it follows that for all $\hat{t}',\hat{t}'' \geq \hat{t}$, $b \geq c$ and $m' < m$

$$\{0,1,\ldots,b\} \cap A_{m'}^{\hat{t}'} \neq \{0,1,\ldots,b\} \cap A_m^{\hat{t}''}. \tag{2}$$

Let $\tilde{t}$ be the maximum of $\bar{t}$ and $\hat{t}$.

Now, assume that at some time t_0 number j obtains the follower a which has not been a follower before. This can only happen in step III(A). Therefore

$$n_{t_0} = j,\ a = f(j,t_0) \text{ and } C_a^{t_0} = A_j^{t_0}.$$

Let us study how C_a^t changes as t grows from t_0 to some time ν in which a is freed. Since a is not displaced at times $t<\nu$, C_a^t can only be changed by operation III(B), at times t at which

$$n_t = j,\ a = f(j,t) \text{ and } C_a^t = A_j^t$$

Thus, as t grows from t_0 to ν, then C_a^t is successively equal to $A_j^{t_0}, A_j^{t_1}, \ldots, A_j^{t_\sigma}$ which means that at time ν-1 we have that $C_a^{\nu-1} = A_j^{t_\sigma}$ with $t_\sigma \geq t_0$. Hence

$$\max\beta(C_a^{\nu-1}) = \max\beta(A_j^{t_\sigma}). \tag{3}$$

The problem we shall treat now is whether m may obtain a follower and lose it afterwards infinitely often. A number that was a follower and was then freed, never again can become a follower. Thus, if the question has a positive answer, then at some time $t_0\geq\tilde{t}$, m will obtain a follower $a\geq c$ and will loose it at some later time $\nu\geq\tilde{t}$. We show that this is impossible.

At time ν a can be freed only in two cases, after it has been displaced by n_ν, or if $n_\nu = m$ and case I holds. Since for $n_\nu = m$

$$a \geq c,\ a = f(m,\nu\text{-}1) \text{ and } \nu \geq \bar{t},$$

it follows from inequality (2) that case I cannot hold at this time. Let us therefore assume that a is displaced by n_ν at time ν. Then

$$\max\beta(C_a{}^{\nu-1}) = \max\beta(A_{n_\nu}{}^{\nu}),\ a = f(m,\nu\text{-}1) \text{ and } n_\nu < m.$$

The last inequality holds since case II does not apply at time ν. With equation (3) (j=m) it follows that

$$\max\beta(A_{n_\nu}{}^{\nu}) = \max\beta(A_m{}^{t_\sigma})$$

with $t_\sigma \geq t_0 \geq \bar{t}$, $\nu \geq \bar{t}$ and $n_\nu < m$, in contradiction to inequality (1).

This shows that there is some time $w \geq \bar{t}$ from which m cannot lose its follower. There are infinitely many $t \geq w$ with $n_t = m$. Since at any such moment m cannot loose its follower, case I does not hold at these times. Let us assume that case III applies at infinitely many of them. Then for infinitely many $t \geq w$ we have that

$$m = n_t,\ a = f(m,t\text{-}1) \text{ and } C_a{}^t = A_m{}^t.$$

Since, beginning with time w, m does not loose its follower, a does not depend on t. Thus $C_a{}^t = A_m{}^t$, for infinitely many $t \geq w$, which implies that $C_a = A_m$. It follows that $x = \eta_m = \bigsqcup\beta(A_m) = \bigsqcup\beta(C_a)$.

Let us next assume that case III holds at only finitely many of the times considered above. Then at least one of the cases II(a), II(b) and II(c) must apply infinitely often.

If II(b) holds infinitely often, then for each such time t we have that

$$\max\beta(C_a{}^{t-1}) = \max\beta(A_m{}^t),$$

for some $a \leq m$. In this case a depends on t. But since $a \leq m$, there are only finitely many such a. Hence there must be some a such that for infinitely many t, $\max\beta(C_a{}^{t-1}) = \max\beta(A_m{}^t)$. For this a it follows that $\bigsqcup\beta(C_a) = \bigsqcup\beta(A_m) = x$.

If case II(c) applies infinitely often, then for each such time t we have that

$$\max\beta(C_a{}^{t-1}) = \max\beta(A_m{}^t), \tag{4}$$

for some a which has been displaced by m at some earlier time than t. Since a number can only be displaced by m, if case III holds, and since this will happen only finitely often according to our assumption, there are only finitely many a but infinitely many t for which equation (4) holds. Hence, there is some a that satisfies (4) for infinitely many t. For this a it follows that $\bigsqcup\beta(C_a) = \bigsqcup\beta(A_m) = x$.

Let us finally suppose that II(a) holds infinitely often. Then for each such time $t \geq w$ there is some $n \leq m$ with

$$\max\beta(C_a{}^{t-1}) = \max\beta(A_m{}^t) \tag{5}$$

where a=f(n,t-1). Since n is bounded by m, equation (5) must hold for some fixed $n \leq m$ at infinitely many times $t \geq w$. Let $\hat{n}$ be such an index n.

We first consider the case that $\hat{n}=m$. Since from time w, m does not lose its follower, the index a in equation (5) does not depend on t. Therefore $\max\beta(C_a^{t-1}) = \max\beta(A_m^t)$, for infinitely many t, which implies that $\bigsqcup\beta(C_a) = \bigsqcup\beta(A_m) = x$.

If $\hat{n}<m$ and a can take only finitely many values as t varies, there must be some a that satisfies equation (5) for infinitely many t. Thus, $\bigsqcup\beta(C_a) = \bigsqcup\beta(A_m) = x$.

If, on the other hand, $\hat{n}<m$ but the set of values $f(\hat{n},t-1)$ can take, if t varies, is not bounded, $\hat{n}$ must obtain a new follower infinitely often. Let us assume that a becomes a follower of $\hat{n}$ at time $t_0 \geq w$ and is still a follower of $\hat{n}$ at time $t>t_0$ at which it satisfies equation (5). By (3) with $j=\hat{n}$ we then obtain that

$$\max\beta(C_a^{t-1}) = \max\beta(A_{\hat{n}}^{t_\sigma}),$$

for some $t_\sigma \geq t_0$, from which it follows with (5) that

$$\max\beta(A_{\hat{n}}^{t_\sigma}) = \max\beta(A_m^t).$$

Since $t, t_\sigma \geq \bar{t}$ and $\hat{n} < m$, this is impossible by inequality (1). Thus, the last case will not appear. In every other case we have seen that there is some a such that $x = \bigsqcup\beta(C_a)$.

From properties (7) - (9) it now follows that δ defined by

$$\delta_a = \bigsqcup\beta(C_a)$$

is a one-to-one enumeration of D. It remains to show that δ is computable, i.e., that $\{\langle i,a\rangle \mid \beta_i \angle \delta_a\}$ is r.e.

Since $\{\langle i,j\rangle \mid \beta_i \angle \beta_j\}$ is r.e., there is partial recursive function p with $\beta_{p(i)} = \max\beta(E_i)$, for all $i \in \omega$ such that $\beta(E_i)$ is directed. From the above construction, there is a total recursive function h with $E_{h(a,t)} = C_a^t$. Because $\bigsqcup\beta(C_a) = \bigsqcup_t \max\beta(C_a^t)$, it follows that

$$\beta_i \angle \delta_a \;\Leftrightarrow\; \exists t\; \beta_i \angle \max\beta(C_a^t) \;\Leftrightarrow\; \exists t\; \beta_i \angle \beta_{ph(a,t)},$$

which shows that $\{\langle i,a\rangle \mid \beta_i \angle \delta_a\}$ is r.e. Thus, δ is also computable. This completes the proof of Theorem 3.1.

For some special effective domains of the kind considered in the theorem above it has been shown that there exist infinitely many computable one-to-one numberings which are pairwise incomparable with respect to the reducibility preorder (cf. Pour-El 1964; Khutoretskiĭ 1969). It would be interesting to know whether this is true for all effective domains of the above kind.

As we have seen above, for any effective domain *D* the computable one-to-one indexings of D are minimal among all numberings of D with respect to reducibility, whereas the admissible numberings of D are maximal among the computable ones. We now establish a relationship between both types of indexings which says that any admissible numbering of D can be decomposed into a computable family of computable one-to-one numberings of D.

Let $(\delta^i)_{i\in\omega}$ be a family of indexings of D. Then $(\delta^i)_{i\in\omega}$ is called *computable*, if $\{<n,i,j> \mid \beta_n \ll \delta^i_j\}$ is r.e. Moreover, let $\oplus_{i\in\omega}\delta^i$ be defined by

$$(\oplus_{i\in\omega}\delta^i)(<m,n>) = \delta^m_n.$$

Then $\oplus_{i\in\omega}\delta^i$ is an indexing of D, which is the directed sum of $(\delta^i)_{i\in\omega}$. If $(\delta^i)_{i\in\omega}$ is computable, then each δ^i as well as $\oplus_{i\in\omega}\delta^i$ is also computable.

Theorem 3.2. Let $(D, \sqsubseteq, B, \beta)$ be an effective domain such that β is one-to-one and $B\cap MAX=\emptyset$. Then for each admissible indexing η of D, one can construct a computable family $(\delta^i)_{i\in\omega}$ of computable one-to-one indexings of D such that

$$\eta \equiv \oplus_{i\in\omega}\delta^i.$$

Proof. The proof of this theorem is a modification of the proof of Theorem 3.1. Again we represent each $x\in D$ by the r.e. set $I_\beta(x)$, but now we only use the sets $I_\beta(\eta_n)$ with $n\geq i$ for the construction of δ^i. Let τ be a computable enumeration of $\{<i,n> \mid i\leq n\}$ such that each pair is enumerated infinitely often. The δ^i are constructed simultaneously. If $\tau(t)=<i,n>$, then at step t the construction of δ^i is considered. In what follows we use the notation of the above proof. Just as in that proof, we construct sets $A_n^0 \subseteq A_n^1 \subseteq \ldots$ and $C_{a^i}^0 \subseteq C_{a^i}^1 \subseteq \ldots$ and a partial recursive function f, which now is a function of the three parameters n, i, t, such that $A_n^t, C_{a^i}^t \in E(Z)\cup\{\emptyset\}$ and $A_n =_{df} \bigcup_t A_n^t = W_{r(n)}$. The aim is to construct the sets $C_{a^i} =_{df} \bigcup_t C_{a^i}^t$ in such a way that $(\delta^i)_{i\in\omega}$ with $\delta^i_{a^i} = \bigsqcup\beta(C_{a^i})$ is a computable family of computable one-to-one numberings of D which is recursively isomorphic to η. If f(n,i,t) is defined and $a^i = f(n,i,t)$, then a^i is called an *i-follower of* n *at time* t.

The construction of the sets A_n^t is the same as in the proof of Theorem 3.1.Except for minor changes the same is true for the sets $C_{a^i}^t$. If $\tau(t)=<i,n_t>$, then instead of the condition $n<n_t$, in cases I and II(a) one has to use the restriction $i\leq n<n_t$. In case II(b) the number c^i with $\max\beta(A_n^t) = \max\beta(C_{c^i}^{t-1})$ has to be free at time t-1 and $c^i \leq n_t$-i-1. In case III(A), finally, if $f(n_t,i,t-1)$ is undefined, then in order to define $f(n_t,i,t)$ one has to take the smallest number b^i which up to time t has not been an i-follower. All other cases remain unchanged. It follows that properties (1) - (8) of the above proof hold for these sets and the function f. In addition the following modification of property (9) holds.

(9') If $x\in D$, then for any i there is some a^i with $x = \bigsqcup\beta(C_{a^i})$.

The proof is the same as that of property (9). One only has to let m be the smallest index of x under η with $m\geq i$ in this case and to observe that the inequalities (1) and (2) in that proof now only hold for numbers m' with $i\leq m'<m$.

Because of these properties we obtain that δ^i defined by

$$\delta^i_{a^i} = \bigsqcup\beta(C_{a^i})$$

is a one-to-one enumeration of D. Since there is total recursive function k with $E_{k(a^i,i,t)} = C_{a^i}{}^t$, it follows that each indexing δ^i as well as the family $(\delta^i)_{i\in\omega}$ of all these indexings is computable.Hence, $\oplus_{i\in\omega}\,\delta^i$ is also computable. By Lemma 2.1(a) we therefore have that

(10) $\oplus_{i\in\omega}\,\delta^i \leq_1 \eta$.

Because of the Theorem of Myhill-Eršov it remains to show that the converse reduction also holds. To this end we once more consider the above construction.

(11) For any m there is some i such that, from some time on, m has a constant i-follower.

Let t be the smallest number with $\tau(t)=\langle m,m\rangle$. We shall now show that, at this time, m obtains an m-follower, independently of whether it already has a j-follower, for some $j<m$. At time t case I does not hold, since there is no n with $m\leq n<m$. Moreover, case II(a) does not hold. Its condition may only be satisfied by $c^m = f(n,m,t-1)$ with n=m, but f(m,m,t-1) is not defined by the choice of t. For the same reason, there cannot be a c^m that is free at time t-1, but which has been displaced by m at some earlier time than t. Hence, case II(c), too, does not hold. Finally, there is no $c^m < m-m = 0$, which means that also case II(b) cannot hold. Thus, case III holds and m obtains an m-follower b^m, which will not be freed later on. The reason is that case I also cannot hold at any time $t'>t$ with $\tau(t')=\langle m,m\rangle$. Moreover, b^m cannot be displaced at any time $t'>t$. This would be the case if $\max\beta(C_{b^m}{}^{t'-1}) = \max\beta(A_{n_{t'}}{}^{t'})$. Since case II(a) and case III cannot both hold at the same time, it would follow that $n_{t'}<m$, which is impossible, since only pairs $\langle m,n_{t'}\rangle$ with $m\leq n_{t'}$ are considered. Thus, from time t, b^m is a constant m-follower of m.

(12) $\eta \leq_1 \oplus_{i\in\omega}\,\delta^i$.

Obviously, the time t above can be computed from m. As we have seen in this proof, f(m,m,t) is defined and $b^m = f(m,m,t)$ is a constant m-follower of m. Since f is partial recursive, there is a total recursive function h with $b^m = h(m)$. Then $\eta_m = \bigsqcup\beta(A_m) = \bigsqcup\beta(C_{b^m}) = \delta^m{}_{b^m} = \delta^m{}_{h(m)}$, which shows that $\eta \leq_1 \oplus_{i\in\omega}\,\delta^i$ via $\lambda m.\langle m,h(m)\rangle$. This proves Theorem 3.2.

Acknowledgement

I would like to thank the Computer Science Department of the University of Pisa for its hospitality. It was the pleasant atmosphere at Pisa which made it possible to finish this paper. Moreover, I am indebted to the Consiglio Nazionale delle Ricerche and the Siemens Corporate Laboratories for Research and Technology for supporting my stay at Pisa. Last but not least thanks are due to the unknown referee for his help in improving the English of my manuscript.

References

Egli, H. and R. L. Constable (1976), Computability concepts for programming language semantics, *Theoret. Comp. Sci.* **2**, p. 133-145.

Eršov, Yu. .L (1972), Theorie der Numerierungen I, *Zeitschr. f. math. Logik u. Grundl. d. Math.* **19**, p. 289-388.

Friedberg, R. M. (1958), Three theorems on recursive enumeration. I. Decomposition. II. Maximal sets. III. Enumeration without duplication, *J. Symb. Logic* **23**, p. 309-316.

Kanda, A. (1980), *Gödel Numbering of Domain Theoretic Computable Functions* , Report Nr. 138, Dept. of Computer Science, University of Leeds.

Kanda, A. and D. Park (1979), When are two effectively given domains identical? *Theoretical Computer Science, 4th GI Conference, Aachen 1979* (Weihrauch, K, ed.), p. 170-181, Lec. Notes Comp. Sci. 67, Springer, Berlin.

Khutoretskii, A. B. (1970), On the reducibility of computable enumerations, *Algebra and Logic* **8**, p. 145-151.

Kreitz, Ch. (1982), *Zulässige cpo's, ein Entwurf für ein allgemeines Berechenbarkeitskonzept*, Schriften zur Angew. Math. u. Informatik Nr. 76, RWTH Aachen.

Mal'cev, A. I. (1970), *Algorithms and Recursive Functions*, Wolters-Noordhoff, Groningen.

Pour-El, M. B. (1964), Gödel numberings versus Friedberg numberings, *Proc. Amer. Math. Soc.* **15**, p. 252-256.

Riccardi, G. A. (1980), *The Independence of Control Structures in Abstract Programming Systems*, PhD Thesis, State Univ. of Buffalo.

Rogers, H. Jr. (1967), *Theory of Recursive Functions and Effective Computability*, McGraw-Hill, New York.

Schinzel, B. (1977), Decomposition of Gödelnumberings into Friedbergnumberings, *Zeitschr. f. math. Logik u. Grundl. d. Math.* **23**, p. 393-399.

Sciore, E. and A. Tang (1978a), Admissible coherent c.p.o.'s, *Automata, Languages and Programming* (Ausiello, G and Böhm, C, eds.), p. 440-456, Lec. Notes Comp. Sci. 6,. Springer, Berlin.

Sciore, E. and A. Tang (1978b), Computability theory in admissible domains, *10th Annual Symp. on Theory of Computing, San Diego 1978*, p. 95-104, Ass. Com. Mach., New York.

Scott, D. S. (1970), *Outline of a Mathematical Theory of Computation,* Techn. Monograph PRG-2, Oxford Univ. Comp. Lab.

Smyth, M. B. (1977), Effectively given domains, *Theoret. Comp. Sci.* **5**, p. 257-274.

Weihrauch, K. and Th. Deil (1980), *Berechenbarkeit auf cpo-s*, Schriften zur Angew. Math. u. Informatik Nr. 63, RWTH Aachen.

Part V

Implementation Issues

Extended Abstract of

MIX: A SELF-APPLICABLE PARTIAL EVALUATOR FOR EXPERIMENTS IN COMPILER GENERATION[†]

Neil D. Jones
Peter Sestoft
Harald Søndergaard[‡]

DIKU, University of Copenhagen
Universitetsparken 1, DK-2100 Copenhagen Ø, Denmark

INTRODUCTION: Since the early seventies it has been known that in theory, the program transformation principle called *partial evaluation* can be used for compiling and compiler generation, and even for the automatic generation of a compiler generator. A successful implementation of an experimental partial evaluator able to generate stand-alone compilers and compiler generators, however, had not been carried out prior to 1984 when the first mix system was brought to work at the University of Copenhagen.

In this paper we discuss partial evaluation and its applications to compiler generation. We also sketch the current mix system (mid 1987). The results we report are sufficiently remarkable to justify further research into using partial evaluation for compiler generation purposes.

A partial evaluator may be thought of as a "smart interpreter". If an interpreter is given a program and only *part* of this program's input data, it will leave the program unevaluated and report an error. A partial evaluator will attempt to evaluate the given program as far as possible.

In our terminology, partial evaluation of a *subject program* with respect to known values of some of its input parameters results in a *residual program*. By definition, running the residual program on any remaining input yields the same result as running the original subject program on all of its input. Thus a residual program is a *specialization* of the subject program to known, fixed values of some of its parameters. A *partial evaluator* is a program that performs partial evaluation given a subject program and fixed values for some of its parameters.

The relevance of partial evaluators for compilation, compiler generation, and compiler generator generation stems from the following fact. Consider an interpreter for a given programming language S. The specialization of this interpreter to a known source program s (written in S) *already is* a target program for s, written in the same language as the interpreter. Thus, partial evaluation of an interpreter with respect to a fixed source program amounts to compiling the source program. From this viewpoint then, partial evaluation and compilation are nothing but special cases of program transformation for the purpose of optimization.

Furthermore, partially evaluating a partial evaluator with respect to a fixed interpreter yields a compiler for the language implemented by the interpreter. And partially evaluating the partial evaluator with respect to itself yields a compiler generator, namely, a program that transforms interpreters into compilers.

In the rest of the paper we will make this a little more formal, give an example of partial evaluation, and sketch the structure of the partial evaluator mix.

[†]The full version of this paper is to appear in the journal *LISP and Symbolic Computation*
[‡]Current address: Computer Science Dept., Univ. of Melbourne, Parkville Vic. 3052, Australia

1. PRELIMINARIES

In this section a framework is set up for discussing partial evaluation and its applications. Our definition of a programming language may seem a bit pedantic at first sight. A precise notation is necessary, however, since more than one language may be discussed at the same time, and programs can play multiple roles: sometimes as active agents, sometimes as passive data, and even sometimes as both at once.

We assume there is given a fixed set D whose elements may represent programs in various languages, as well as their input and their output. The set D should be closed under formation of sequences $<d_1,...,d_n>$ of elements of D. The set D may be the set of all character sequences, the set of all Lisp lists, *etc.*

Parentheses will usually be put to use only when necessary to disambiguate expressions. We write $X \to Y$ to denote the set of all total functions from X to Y, and $X - \to Y$ for the partial functions. A function type expression $X \to Y \to Z$ is parenthesized as $X \to (Y \to Z)$, and a double function application f x y is read (f x) y (where f, x, and y have types $f: X \to Y \to Z$, x: X, y: Y for some X, Y, and Z).

We identify a *programming language* L with its semantic function on whole programs (assumed to be computable):

$$L: D - \to D - \to D.$$

The *well-formed* L-programs are those to which L assigns a meaning:

$$\text{L-programs} = \text{domain } L.$$

The *input-output function* computed by $\ell \in$ L-programs is $(L\ \ell): D - \to D$ (which is partial since ℓ may loop). Thus $L\ \ell <d_1,...,d_n>$ denotes the output (if any) obtained by running the L-program ℓ on input data $<d_1,...,d_n>$. For an example, consider the following program power to compute x to the n'th power:

```
power =   f(n, x)  =  if n = 0 then 1
                      else if even(n) then f(n/2, x)^2
                      else x*f(n-1, x)
```

The result of running the program power is the result of applying its first function f (the goal function) to the program's input values. For example, L power <3, 2> = 8.

2. PARTIAL EVALUATION

We proceed to give formal definitions of residual programs and partial evaluation. Below, the equality sign shall always mean strong equality: either both sides are undefined, or else they are defined and equal.

Definition 2.1. Let ℓ be an L-program and let $d_1, d_2 \in D$. Then an L-program r is a *residual program for* ℓ *with respect to* d_1 iff

$$L\ \ell\ \langle d_1, d_2 \rangle = L\ r\ d_2$$

for all $d_2 \in D$. □

Definition 2.2. A P-program p is an L-*partial evaluator* iff

$$P\ p \langle \ell, d_1 \rangle \text{ is a residual L-program for } \ell \text{ with respect to } d_1$$

for all L-programs ℓ and values $d_1 \in D$. □

We refer to the program ℓ as the *subject program.* So a partial evaluator takes a subject program and part of its input and produces a residual program; the residual program applied to any remaining input produces the same result as the subject program applied to all of its input. The characteristic equation for the partial evaluator p therefore is:

$$L\ \ell\ \langle d_1, d_2 \rangle = L\,(P\ p \langle \ell, d_1 \rangle)\ d_2. \tag{2.1}$$

For a simple example, let the L-program ℓ be power from Section 1, and suppose we are given that n equals 5. A trivial residual program resid_1 may easily be constructed by adding a single equation to power:

$$\text{resid}_1 = \boxed{\begin{array}{lcl} g(x) & = & f(5, x) \\ f(n, x) & = & \textit{if}\ n = 0\ \textit{then}\ 1 \\ & & \textit{else if}\ \text{even}(n)\ \textit{then}\ f(n/2, x)^2 \\ & & \textit{else}\ x * f(n-1, x) \end{array}}$$

The general possibility of partial evaluation in recursive function theory is known as the S-m-n Theorem. In fact that theorem is often *proved* by the very possibility of adding equations as above [Rogers 1967].

A less trivial residual program may be obtained by symbolic evaluation of the program $resid_1$. This is possible since the program's control flow is completely determined by n, and it yields an equivalent program with only one equation:

$$resid_2 = \boxed{g(x) \quad = x*(x^2)^2}$$

Partial evaluation thus can be viewed as substitution of known values for some parameters, possibly followed by equivalence preserving program transformations, and can result in residual programs which are faster (though sometimes larger) than the original. Examples may also be found in [Beckman *et al.* 1976] and in [Emanuelson and Haraldsson 1980].

3. COMPILATION AND COMPILER GENERATION

We now turn to the applications of partial evaluation to compiler generation. First we give simple formal definitions of interpreters and compilers.

3.1 Interpreters and Compilers

Let L and S be programming languages (S is intended to be a source language).

Definition 3.1. An L-program int is an S-*interpreter* iff

$$\text{L int} <\text{s, d}> = \text{S s d} \tag{3.1}$$

for all S-programs s and data d ∈ D. □

By this definition, an interpreter takes as input both the program to be interpreted and its input data. We shall call int an L-*self-interpreter* iff S = L (sometimes the term "metacircular interpreter" is used). The set of interpreters for the language S (written in L) is denoted by

$$\boxed{\begin{array}{c} S \\ L \end{array}}$$

Now let T be a programming language (intended to be a target language).

Definition 3.2. An L-program c is an S-to-T *compiler* iff

1) L c s ∈ T-programs for all S-programs s, and
2) T (L c s) d = S s d for all S-programs s and data d ∈ D. □

The result t = L c s of running a compiler thus is a (target) T-program t with the same input-output behavior as the (source) program s. The set of S-to-T compilers written in L is denoted by

S → T
L

3.2 Compilation by Partial Evaluation of an Interpreter

Let the P-program p be an L-partial evaluator. If an S-interpreter int is partially evaluated with respect to a given S-program s, the result will be an L-program with the same input-output behavior as s, since

S s d = L int <s, d> - by (3.1)
= L (P p <int, s>) d - by (2.1)

Note that the last line describes the application of a certain L-program (namely the program P p <int, s>) to the input d. The result of this is the same as the result of applying the S-program s to d, and therefore we may reasonably call the resulting program target:

target = P p <int, s>

since it is an L-program with the same input-output behavior as S-program s. In other words we have compiled the source S-program s into an L-program target by partially evaluating the S-interpreter with respect to the source program s. For a concrete example, see Figure 5.1 (source program s), 5.2 (interpreter int), and 5.3 (target program target) below.

3.3 Compiler Generation

Definition 3.3. An L-program mix is an L-*autoprojector* iff it is an L-partial evaluator. We shall refer to the language L as the *subject language.* □

An autoprojector is thus a partial evaluator for the language in which it is itself written. The term is due to [Ershov 1982] and as to etymology, "auto" comes from the program's self-applicability, and "projector" from the fact that partially evaluating a program for f(x, y) with respect to x yields a program computing f's projection along the x-axis, in an analytical geometry sense.

In the following we will assume that an autoprojector mix is given. Letting mix play the role of the partial evaluator p from Section 2, it holds that

target = L mix <int, s>. (3.2)

This application does not depend on mix's self-applicability, but for the following it is essential that mix is an autoprojector. A compiler from S to L may be generated by computing:

comp = L mix <mix, int> (3.3)

that is, by partially evaluating the autoprojector itself with respect to the S-interpreter. To see this, observe that

L comp s	=	L (L mix <mix, int>) s	- by (3.3)
	=	L mix <int, s>	- by (2.1)
	=	target	- by (3.2)

so comp is a stand-alone compiler that given s will produce a target program for s, that is:

comp ∈ [S → L / L]

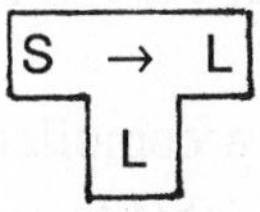

Note that we now have two possibilities of compiling s by means of partial evaluation: either by running mix on <int, s>, or by generating comp (using mix) and applying that to s.

Applying the compiler comp to s can be expected to be more efficient than computing L mix <int, s>. The reason is that mix is a general-purpose partial evaluator, while comp is a rather specialized version of mix, predisposed to partially evaluate a fixed interpreter with respect to varying S-programs given as known input. The presumption that comp is faster is well borne out by the experimental results reported in Section 8.

3.4 Compiler Generator* Generation

By a similar reasoning a compiler generator may be obtained:

cocom = L mix <mix, mix>. (3.4)

To see this, observe that

L cocom int	=	L (L mix <mix, mix>) int	- by (3.4)
	=	L mix <mix, int>	- by (2.1)
	=	comp	- by (3.3)

The function computed by the L-program cocom thus transforms an interpreter int into a compiler comp that defines the same language:

L cocom : [S / L] → [S → L / L]

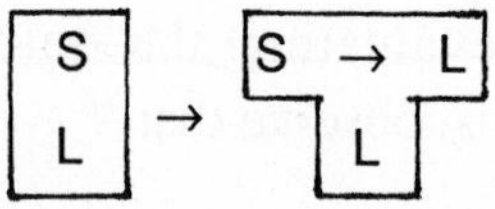

Letting rep(A → B) denote the set of program representations of functions from A to B, it is interesting to compare the types of the functions computed by mix and by cocom. As can be seen, the function L cocom is a curried version of L mix:

L mix : rep(X × Y → Z) × X → rep(Y → Z)
L cocom : rep(X × Y → Z) → rep(X → rep(Y → Z)).

In fact, cocom is more than a compiler generator. It is a realization of a general intensional currying function, able to transform a program for a two-place function f(x, y) into a program which, when given data $x{=}x_0$, will yield as output a program for the function $\lambda y.f(x_0, y)$. In particular, cocom transforms an interpreter into its curried form, a compiler. Also note that cocom = L cocom mix. In this sense cocom can be seen as a compiler generator generator generator ....

4. HISTORICAL NOTES

Theory: The concept of partial evaluation is certainly very old and has seeds from the lambda calculus and recursive function theory. To our knowledge the first explicit statement of its possibility was given when Kleene formulated and proved the S-m-n Theorem [Kleene 1952]. An early use of partial evaluation as a programming aid was suggested in Lombardi's paper on incremental computation [Lombardi 1967].

Futamura saw that compiling may in principle be done by partial evaluation, and also that compilers may be generated by self-application of the partial evaluator [Futamura 1971]. Turchin was probably the first to realize that even a compiler generator could be built automatically by applying a partial evaluator to itself [TsNIPIASS 1977]. In any case these applications seem to have been independently discovered in the USSR, Japan, and Sweden in the mid-seventies and subsequently communicated in [Beckman *et al.* 1976], [Ershov 1978], and [Turchin 1979]. It was however not until Ershov's expository paper that the ideas became widely accessible in the West [Ershov 1982]. Ershov coined the term "mixed computation" for what we call partial evaluation.

Practice: Projects aimed at putting partial evaluation to practical use are described in [Beckman *et al.* 1976], [Haraldsson 1978], [Emanuelson and Haraldsson 1980], and [Turchin, Nirenberg and Turchin 1982]. Partial evaluation also has been used as a tool for building compiler generators [Mosses 1979] [Paulson 1982].

Partial evaluation related to Prolog is described in [Komorowski 1981] and [Kahn 1982]. Partial evaluation of an imperative language was addressed in [Ershov 1978] and [Bulyonkov 1984].

Autoprojectors: Autoprojectors have been reported in [Venken 1984], [Takeuchi and Furukawa 1986], [Safra and Shapiro 1986].

To our knowledge all of these systems require considerable human assistance. None of them appear to have been successfully self-applied, or at least no such self-application has yet been reported.

A non-trivial self-applicable partial evaluator was developed in 1984 by the authors and communicated in [Jones, Sestoft and Søndergaard 1985]. The system was called mix (following Ershov's terminology) and was a preliminary version of the system described in the present paper. It generated good compilers by self-application, with the proviso that the user had to annotate function calls to indicate whether they were intended to be residual or not. A detailed description was given in [Sestoft 1986].

5. THE LANGUAGE MIXWELL

5.1 Description of Mixwell

The subject language of mix is called Mixwell and may be thought of as essentially a subset of (pure) Scheme [Rees and Clinger 1986]. A Mixwell program takes the form of a system of recursive equations as shown here in abstract form:

$$
\begin{array}{lcl}
f_1(x_1,\ldots,x_m) & = & e_1 \\
\cdots & & \cdots \\
f_h(x_1,\ldots,x_p) & = & e_h
\end{array}
$$

Expression values range over D = {d | d is an S-expression} and the expressions are constructed from variables (atoms) and constants of form (*quote* d) by operators: *car*, *cdr*, *cons*, *equal*, and *atom* (as known from Lisp) in addition to *if* and *call*. The operator *if* is used in a conditional (*if* e_0 e_1 e_2), whereas *call* is used in a function call (*call* f_j e_1 e_2 ... e_n). The variant *callx* is used to call external functions (*e.g.* gensym).

Programs are statically scoped. All operators are strict, except *if*, which is strict only in the first operand. In particular, *call* is strict, which implies a call-by-value semantics. Mixwell is first order: functions may *not* be manipulated as data objects. The program's input is through the first function's variables.

5.2 Mixwell+

For the sake of partial evaluation it is important that Mixwell be simple, but such simplicity may impair readability. We resolve this dilemma by allowing certain forms of simple syntactic extensions, translatable by machine into Mixwell. We call the extended language Mixwell+.

5.3 An Example

In order to give a non-trivial example program in Mixwell+, we present an interpreter for another simple language M. This will also give us the opportunity to show an input/output example for a compilation done by partial evaluation of an interpreter. The language M has a syntax defined by the following grammar:

```
<program>    ::=  (read <variable> and evaluate <expression>)
<expression> ::=  <variable>
              |   (con <constant>)
              |   (<operator> <expression> <expression>)
              |   ( if <expression> <expression> <expression>)
              |   (min <variable> such that <expression> = 0)
<operator>   ::=  + | - | *
<variable>   ::=  <Lisp atom>
```

We hope the intended semantics of M is clear from the syntax as well (**min** abbreviates "the minimum"). In the **if** expression, a positive first operand is regarded as denoting "true". For simplicity the syntactic category <constant> denotes lists of 1's: the value m is indicated in unary as a list of m 1's. Note that "–" denotes *cut-off subtraction*: x–y is zero if and only if $y \geq x$.

Figure 5.1 shows an example program written in M. Given a value $x \in \mathbb{N}$, the program evaluates *min* $\{y \in \mathbb{N} \mid x^2-(y^2+5)=0\}$, provided the value exists.

```
(read x and evaluate (min y such that (- (* x x) (+ (* y y) (con (1 1 1 1 1)))) = 0))
```

Figure 5.1: A program written in M

Figure 5.2 shows an interpreter for M written in Mixwell$^+$. The kernel is the function eval which uses a traditional interpretation loop. Again, values are represented by lists. The mutually recursive functions f and g implement the "iterating" expression (**min** <variable> **such that** <expression> = **0**). To keep the example simple, very little checking is done by the interpreter: it gives meaningful results only on M programs that are (syntactically) well-formed, and in which every variable v used is declared by an enclosing "**min** v **such that** ..." or "**read** v **and evaluate** ..." expression.

```
(
  (run (P  X)  =  (let  (read N and evaluate E) = P in  (call eval E (list N) (list X))))

  (eval (E Ns Vs)
  = (case E of
       (atom? N)                         : (call lookup N Ns Vs)
       ('con C)                          : C
       ('+ E1 E2)                        : (call  add (call eval E1 Ns Vs)  (call eval E2 Ns Vs))
       ('- E1 E2)                        : (call  sub (call eval E1 Ns Vs)  (call eval E2 Ns Vs))
       ('* E1 E2)                        : (call  mul (call eval E1 Ns Vs)  (call eval E2 Ns Vs))
       ('if E0 E1 E2)                    : (if  (call eval E0 Ns Vs)  then  (call eval E1 Ns Vs)
                                                                      else  (call eval E2 Ns Vs))
       ('min N such that E = 0)          : (call  f E (N :: Ns) ('nil :: Vs))        ; Bind N to 0
     otherwise                           : 'error))

  (lookup (N Ns Vs)
  = (let   (N1 . Nr) = Ns
           (V1 . Vr) = Vs  in
     (if    (null  Ns)  then    'error
      elsf  (N  =  N1)  then    V1
      else  (call  lookup N Nr Vr))))

  (f (E Ns Vs)  =  (call g (call eval E Ns Vs) E Ns Vs))                     ; Evaluate E

  (g (W E Ns Vs)
  = (let   (V1 . Vr) = Vs  in
     (if    (null W)    then    V1                                            ; Exit if E = 0, else
      else  (call  f E Ns (('1 :: V1) :: Vr)))))                              ; increment N by 1

  (add (X Y)  =  ... )

  (sub (X Y)  =  ... )

  (mul (X Y)  =  ... )
)
```

Figure 5.2: Interpreter for M written in Mixwell⁺

Compilation from M to Mixwell can now be done by applying mix to the above interpreter, or by using cocom to generate an M-to-Mixwell compiler. As an example of a target program, Figure 5.3 shows the (machine-produced) result of compiling the example program from Figure 5.1 by partially evaluating the interpreter in Figure 5.2. It is given in Mixwell+ for readability.

```
(
   (run (X) = (call f (list 'nil X)))

   (f (Vs)
   = (if  ('nil = (call sub  (call mul  (cadr Vs) (cadr Vs))
                             (call add  (call mul (car Vs) (car Vs))
                                        '(1 1 1 1 1)))

          then   (car Vs)
          else   (call  f (('1 :: (car Vs)) :: (cdr Vs)))))

   (add (X Y)  = ... )

   (mul (X Y)  = ... )

   (sub (X Y)  = ... )
)
```

Figure 5.3: Target program in Mixwell+

To see how much the running time may be reduced by using the target program (Figure 5.3) instead of the interpreter and the source program (Figures 5.2 and 5.1), we may count 1 time step for each *car*, *cdr*, *cons*, *quote*, =, and *call*, and the functions add, sub, and mul are counted as taking one time step for each call to any of them. By this scheme the target program executes 20 steps per iteration of its main loop, while the interpreter executes 119 steps, the ratio being 6.0.

This ratio tends to grow as the source program or the source language get larger, since relatively more interpretation time is needed for syntax analysis and environment references. Speedup factors between 30 and 200 are reported in [Emanuelson and Haraldsson 1980] for a pattern matching language.

6. METHODS AND PROBLEMS

We now turn to the basic principles and problems involved in partially evaluating sets of function equations.

6.1 Specializing Functions by Symbolic Evaluation

The residual program for a system of equations, each of form

$$f\,(x_1,...,x_n) \quad = e$$

is naturally another system of equations, each of form

$$f^*\,(y_1,...,y_m) \quad = \quad e^*$$

where f* represents a *specialized* version of f, and the variables of f* form a subset of the variables of f. For example, if it is discovered that in one call to f, the first argument always has value 5, the partial evaluator can exploit this fact by constructing an f-variant f* with the first variable removed, and in which e* is a simplified version of e. A function f may have several specialized versions, each corresponding to a possible tuple of values of its known arguments, or none.

The body e* of a specialized version of a function $f\,(x_1,...,x_n) = e$ is obtained by *symbolic evaluation* of its body expression e. Symbolic evaluation deals with expressions (*i.e.*, pieces of Mixwell programs) as values, and is always done in a *symbolic environment* which is a set of bindings of variables to expressions:

$$env = \{\, x_1 \mapsto e_1,...,x_k \mapsto e_k \,\}.$$

In this way, the symbolic environment in mix contains information known at function specialization time about the arguments $x_1,...,x_n$ of f (namely, their symbolic values). In a more sophisticated partial evaluator, symbolic environments containing other kinds of information may be used. For an example, see [Beckman *et al.* 1976].

For a simple example of symbolic evaluation, let [e] represent the result of symbolically evaluating expression e in a given symbolic environment. A natural way to symbolically evaluate (*cdr* e) is:

$$[(\mathit{cdr}\ \mathsf{e})] = \textbf{case}\ [\mathsf{e}]\ \textbf{of} \quad \begin{array}{ll} [\![(\mathit{quote}\ (\mathsf{hd}\ .\ \mathsf{tl}))]\!] & : [\![(\mathit{quote}\ \mathsf{tl})]\!] \\ [\![(\mathit{cons}\ \mathsf{e}_1\ \mathsf{e}_2)]\!] & : [\![\mathsf{e}_2]\!] \\ \textbf{otherwise} & : [\![(\mathit{cdr}\ [\mathsf{e}])]\!] \end{array} \tag{6.1}$$

Note that the result is an *expression, i.e.*, a piece of text. So if the result [e] of symbolically evaluating expression e is the expression (*cons* e_1 e_2) for some expressions e_1 and e_2, then [(*cdr* e)] is the expression e_2. The semantic brackets "$[\![$" and "$]\!]$" are used in the usual way: metalanguage variables (hd, tl, e, e_1, e_2) may appear inside the brackets, and then stand for the expressions they are bound to, whereas object language operators stand for themselves.

If an expression e to be symbolically evaluated contains a function call, it must be decided whether the call should be *unfolded* or *suspended*. Unfolding the call means replacing it by the called function's body, with argument expressions substituted for variables. If suspended, the call will appear in the residual program.

Those variables of f that are present in f* are called *dynamic variables,* and the others are called *static*. By the techniques used by mix, all the specialized versions of a function f have the same sequence of dynamic variables.

6.2 The Treatment of Function Calls

Some partial evaluators determine for each defined function whether all calls to it should be unfolded or suspended during partial evaluation. Other partial evaluators make this decision for each call encountered during function specialization. A key feature of the mix approach is to make a decision on this for each function call appearing in the text of the subject program, so that the decision may be made in advance of function specialization.

Consider a function call (*call* f e_1 ... e_n). Two obvious possibilities are: either to produce a residual call (to a specialized version of f), or to unfold the call. To do the unfolding, the equation $f(x_1,...,x_n) = e$ defining f is found, and (*call* f e_1 ... e_n) is replaced by the result of symbolically evaluating e in the local symbolic environment $\{ x_1 \mapsto [e_1],...,x_n \mapsto [e_n] \}$.

The problem: The problem of finding a good call unfolding strategy is very subtle. In this context there are at least three pitfalls to avoid. First, a too conservative strategy leads to enormous or trivial residual programs as shown in Section 2. Second, a too liberal strategy leads to unfolding loops in which the same function is unfolded infinitely. Finally, the residual programs can easily turn out to be *less* efficient than the original subject programs, owing to the call-by-name nature of symbolic evaluation.

6.3 Some Principles for Call Unfolding

Consider a call appearing in a recursively defined function:

$$f(x_1,...,x_n) \;=\; ...\; f(e_1,...,e_n)\; ...$$

If there exists an argument x_i which always decreases (according to some well-founded partial ordering), then the call may safely be unfolded, provided x_i is evaluable to a constant at partial evaluation time. For example, the recursive calls to f in program power in Section 1 may be unfolded when n is known, but not when n is unknown.

This applies also to partial evaluation of interpreters. Consider the computation of

target = L mix <int, s>.

For the great majority of programming languages, an interpreter can perform/evaluate some commands/expressions on the basis of their subcomponents, without reference to other parts of the enclosing program s (except that it uses the information carried by the environment). Recursive calls realizing such "descents" by the interpreter to smaller parts of s may always safely be unfolded, but whenever the interpreter shifts its attention to a different or a larger part of s, its corresponding call may not be unfolded, owing to the risk of infinite expansion. This justifies marking individual calls rather than entire functions.

For example, in Figure 5.2 some of the calls implementing **min...** should *not* be unfolded because of the risk of infinite expansion, but most of the others may. The problem arises because of the mixture of static actions (dependent only on the source program) and program execution actions usually found in interpreters.

One possible unfolding strategy is rather simple, but works surprisingly well in practice. A rather conservative strategy is used in a preprocessing step, and further unfoldings are based on an analysis of the *call graph* of the first version of the residual program. We return to this analysis in Section 6.5.

The conservative strategy for call annotation is to mark a call as "to be suspended" unless 1) it can be seen that all its arguments are static (totally known), or 2) a static (totally known) argument is bound to a proper substructure of itself in a directly recursive call. If by this, infinite unfolding results during function specialization, then the subject program already contained a function that would be infinitely evaluated for any value of the program's dynamic parameters (though this does not imply that the subject program would run forever: the function might never be called).

6.4 Preprocessing: Binding Time Analysis

Clearly some kind of static analysis is required to gather the information about which arguments will have known values at partial evaluation time. This turns out to have other uses as well, and seems to be absolutely necessary for successful application of mix to itself. (The rather subtle and intricate arguments for this appear in the full paper). The preprocessing, which we call *binding time analysis*, can be done by *abstract interpretation* of the subject program [Cousot and Cousot 1977]. The program is evaluated on the data domain {Static, Dynamic} to yield information about which arguments to functions will be definitely known at partial evaluation time, and which are possibly unknown. Function variables corresponding to argument positions can thereby be classified as static or dynamic. These variable descriptions are obtained for the interpreter given in Figure 5.2: run(S,D), eval(S,S,D), lookup(S,S,D), f(S,S,D), g(D,S,S,D), add(D,D), sub(D,D), mul(D,D), where S = Static and D = Dynamic.

This information is used to guide specialization of functions, and for the preprocessor's (conservative) call annotation: calls having only static arguments, and calls one of whose static arguments is broken down recursively, are marked for subsequent unfolding.

Further, every operator (*e.g.*, *cdr*) is annotated during this preprocessing, either as *static* (*e.g.*, *cdrs*, meaning "static *cdr* operator") or *dynamic* (*e.g.*, *cdrd*), resulting in a heavily annotated version of the subject program.

6.5 Postprocessing: Call Graph Analysis

The strategy for marking calls during binding time analysis is rather conservative, so it is profitable to do more unfolding after the specialization phase. Hence partial evaluation has three phases: preprocessing, specialization, and final unfolding.

Due to the conservative strategy for call marking, many of the generated function body expressions will be fairly simple, often just a call to another specialized function. Such a call may be replaced by the called function's body (with appropriate substitution of argument expressions for variables), since this reduces the number of functions and calls. The call-by-name nature of such unfolding may make the residual program terminate more often than the subject program. This is regarded to be of minor concern.

For the final unfolding and reduction step, an analysis of the intermediate residual program must be done. This analysis works by finding a cutpoint in each elementary cycle in the program's call graph (one that does not properly contain another cycle). A cutpoint is a function name, and the intention is that all calls to such a function should be suspended (*i.e.* should not be unfolded).

Call unfolding can now be done as another symbolic evaluation: a call is suspended only if it was selected for suspension by the call graph analysis, or if unfolding would result in call duplication. By selecting a cutpoint from each elementary cycle, infinite unfolding is prevented, and hence the method is safe.

7. THE ALGORITHMS USED IN MIX

Partial evaluation using mix is most easily understood as a sequence of phases, each performing a translation, an analysis, or a transformation of the subject program (*i.e.*, the program to be partially evaluated). In this section we will first give a brief overview of the structure of mix, and then describe some of its phases. Our partial evaluation algorithm can be thought of as proceeding in 6 phases:

1) Translation from Mixwell$^+$ to Mixwell : *desugar*

---

2) Binding time analysis : *bta*
3) Program annotation : *ann*

---

4) Function specialization : *fsp*

---

5) Call graph analysis : *cga*
6) Call unfolding and reduction : *unf*

The first phase, *desugar*, is a simple translation from Mixwell$^+$ to Mixwell. The remaining phases constitute three program transformation steps, the first one and the last one consisting of an analysis phase and a synthesis phase. The purpose and gross input/output behavior of each is briefly described here.

Suppose we partially evaluate an L-program ℓ (recall that L = Mixwell) with respect to known argument d_1. This yields a residual program $r = L\ mix <\ell, d_1>$, satisfying $L\ r\ d_2 = L\ \ell <d_1, d_2>$ for all d_2.

The first transformation consists of binding time analysis and annotation. Its output is an *annotated* version of ℓ, namely, a copy ℓ_a of the subject program, marked with additional information to make function specialization faster and simpler.

The second transformation is the function specialization phase, which is the heart of partial evaluation. It produces an intermediate residual program r' for ℓ, given the annotated program ℓ_a and the input available at PE-time.

The third transformation comprises call graph analysis together with call unfolding and reduction. To produce a (better) final residual program r, more function calls are unfolded and redundant code is reduced in the intermediate residual program. Call graph analysis of r' yields as output a set *cga* r' of function names. The idea is that avoiding unfolding of calls to all these will prevent infinite expansion. The final phase applies this information to r', yielding the final residual program r=*unf* <r',*cga* r'>.

Below we describe the binding time analysis phase, the annotation phase, and the function specialization phase of mix in more detail.

7.1 Binding Time Analysis

Input is the subject program ℓ (the program to be partially evaluated) and a description of which of its parameters will be available (known) during partial evaluation and which will not. The net result of the binding time analysis is used for annotating the subject program.

Binding time analysis is based on an abstract interpretation using the two-value domain {Static, Dynamic}. The result of this analysis describes every variable x_{ij} of every function f_i as either Static or Dynamic. Here Static means that the possible values of the variable (definitely) depend only on the input available during partial evaluation. Conversely, Dynamic means that the possible values (may) depend also on input unavailable during partial evaluation.

The binding time analysis keeps a partial description of the variables, initially describing all variables as Static, except those variables of the goal function whose values are unavailable during partial evaluation. The program is abstractly interpreted starting with the goal function, and if it is discovered that a variable v currently described as Static may take on a value dependent on a Dynamic variable, then v's description is changed to Dynamic, and the descriptions of all variables that depend on v are recomputed. Every recomputation is preceded by the change of at least one variable from Static to Dynamic, and since there are finitely many variables and they never change back, the analysis will terminate.

The abstract interpretation classifies an expression e as static if and only if:

- e is a constant (*quote* d), d an S-expression, or
- e is a static variable, or
- e has form (*call* f e_1 ... e_n) and e_1,...,e_n are all static, or
- e has form (*op* e_1 ... e_n), where *op* ≠ *call*, and e_1,...,e_n are all static.

7.2 The Annotation Phase

Input is the subject program ℓ and the variable description just computed. Output is an ℓ-version $ℓ_a$ in which all expressions and function calls are *annotated* for use by the function specialization phase.

Operator annotation: An operator in an expression other than a call is marked as static if it can be applied at partial evaluation time, namely, if its performance depends only on the available input. Otherwise the operator is marked as dynamic.

Function call annotations: A function call (*call* f e_1 ... e_n) is marked as unfoldable if there is no risk of infinite expansion during function specialization, and residual otherwise. The simple (and rather conservative) scheme for recognizing this was mentioned in Section 6.3.

7.3 The Function Specialization Phase

Input to this phase consists of the annotated subject program $ℓ_a$ together with actual values for some of the subject program's parameters. Output is an intermediate residual program r' which is a system of specialized versions of ℓ's functions.

A function that is specialized from the function $f(sv_1,...,sv_m, dv_1,...,dv_n) = e$ has form

$$f_{svv}(dv_1,...,dv_n) = [e]$$

where f_{svv} is a new function name composed from the name f of the function in $\mathfrak{L}$ and a sequence svv of values of f's static variables $sv_1,...,sv_m$. The variables of f_{svv} are f's dynamic variables $dv_1,...,dv_n$, and [e] is the result of symbolically evaluating the right-hand side e of the original function definition f(...) = e using the known values svv for f's static variables $sv_1,...,sv_m$.

The classification of all variables as static or dynamic makes it easier to build residual programs, and the basic transformation rules become very simple. For example, the argument a static cdr operator *cdrs* will always be a known value, and so *cdrs* can be evaluated during partial evaluation, whereas a *cdrd* expression will be left as residual. Therefore the **case** expression (6.1) reduces to two simpler versions:

$$[(cdrs\ e)] = [\![(quote\ tl)]\!] \text{ where } (hd\ .\ tl) = [e], \text{ and}$$
$$[(cdrd\ e)] = [\![(cdr\ [e])]\!].$$

8. MIX ASSESSMENT

In this section we evaluate the structure and performance of mix, and we mention some of the tasks to which mix has been applied.

8.1 Ideas behind Mix

The aim has been to construct an *autoprojector*, well suited for the special purpose of compiler generation, rather than a general-purpose partial evaluator. This makes the task easier in some respects, but in general the development of a good autoprojector is harder than that of a partial evaluator. The reason is that - owing to the self-application - an attempt to increase the *quality* (by including more powerful transformations) very often implies overwhelming penalties as regards *efficiency*.

The use of binding time analysis appears to be novel in comparison to other approaches to program transformation. It serves three purposes: to classify function variables, thereby determining the list of variables for each residual function; to annotate all operators as "static" or "dynamic"; and to gather information used to attach unfold/suspend annotations to function calls. As a result of binding time analysis, we have been able to reduce the transformation rules used to an extremely simple subset. If binding time analysis is not applied, the generated compilers in our experiments have turned out to be typically two orders of magnitude larger, and much less efficient.

The call unfolding strategy seems appropriate, and when given suitably written subject programs, mix gives good results. The target programs and compilers produced are reasonably small and efficient. While they sometimes contain inelegant code, they contain little unnecessary code, as is witnessed by the fact that compiling speed is of the right order of magnitude, about 100 lines/second on a VAX 785 for a toy language. On the whole they look like traditional recursive descent compilers, except that more optimization is done while generating code than usual in compilers. Since the compiler is a specialized version of mix, it inherits the transformational capabilities built into the partial evaluator.

8.2 Performance of Mix

To illustrate the performance of mix in compilation and compiler generation we give some tables of program size and run times. In particular, we give run times for examples of compilation by partial evaluation, compiler generation, and compiler generator generation corresponding to the runs (3.2), (3.3), and (3.4) discussed in Section 3. In addition, it is shown how the run time is composed of preprocessing time (binding time analysis and annotation), function specialization time, and postprocessing time (call graph analysis and unfolding).

The interpreter int used in the runs below interprets a tiny imperative language (called MP) with assignment, a conditional, a while-loop, and with S-expressions as the only data type. The MP source program s used computes x to the y'th power by enumerating all different tuples of length y with elements chosen from a set of cardinality x. The runs involving s and target below compute 5^5=3125.

The *size* of a Mixwell program is given by two measures below: the number of functions in the program, and the length in lines when translated into Lisp and "prettyprinted". The programs generated by mix are seen to have a very manageable size, considering that target results from "combining" int and the source program s, comp results from combining fsp and int, and cocom from combining two copies of fsp.

Program	No. of functions	No. of lines	Ratio (lines)
s target	- 6	approx. 30 36	1.2
int comp	9 24	176 303	1.7
fsp cocom	27 49	533 1062	2.0

Figure 8.1: Size of programs

The *run time* results given below were obtained with the Franz Lisp system running under Unix on a Vax11/785. The Mixwell programs were translated into applicative Lisp programs and compiled to have fast (direct) function calls. Run times are given in the form: processing time + garbage collection time = total run time (in cpu seconds). The left hand side of the table shows the bare run time of the function specialization phase fsp and of the residual programs (comp and cocom) derived from it, whereas the right hand side gives the additional time spent on pre- and postprocessing and the total run times. The corresponding speed-up ratios are also given, and are seen to be all greater than 1.

Run	Run time (cpu secs.) processing+g.c.=total	Speed -up	Plus run time for pre- and post-processing	Total	Speed -up
output = L int<s,data>	19.62+ 2.20=21.82	8.1		**21.82**	**8.1**
= L target data	0.56+ 2.14= 2.70			**2.70**	
target = L fsp<int_a,s>	0.66+ 0.00= 0.66	1.9	preproc.(int): 0.50 postproc.(target): 0.26	**1.42**	**2.4**
= L comp s	0.34+ 0.00= 0.34		postproc.(target): 0.26	**0.60**	
comp = L fsp<fsp_a,int_a>	7.56+ 3.00=10.56	2.3	preproc.(int): 0.50 preproc.(fsp): 2.60 postproc.(comp): 2.02	**15.68**	**2.2**
= L cocom int_a	3.18+ 1.42= 4.60		preproc.(int): 0.50 postproc.(comp): 2.02	**7.12**	
cocom = L fsp<fsp_a,fsp_a>	37.32+21.72=59.04	1.6	preproc.(fsp): 2.60 postproc.(cocom):10.70	**72.34**	**1.5**
= L cocom fsp_a	19.84+16.46=36.30		preproc.(fsp): 2.60 postproc.(cocom):10.70	**49.60**	

Figure 8.2: Run times

The run time results in Figure 8.2 show that

- The overhead of interpretation is removed by compiling the source program s into a target program target. The speed-up is more than 8 times, which is quite satisfactory.
- Compilation by a mix-generated stand-alone compiler is more than twice as fast as compilation by partial evaluation.
- Generating a compiler using the mix-generated compiler generator is faster than generating a compiler by partially evaluating the partial evaluator with respect to the interpreter.
- Similarly, regenerating the compiler generator cocom by using the compiler generator is faster than generating it using mix alone.

The results, and in particular the run time results, justify our approach: compiling by means of a mix-generated compiler *is* faster than compiling using a general partial evaluator, as it is done in [Kahn and Carlsson 1984] or [Takeuchi and Furukawa 1986].

8.3 Applications of Mix

Mix has been used on a variety of problems, all of an experimental nature but some more applied than others. Mostly it has been used to generate compilers and target programs for various languages (imperative, functional and pattern matching).

One larger application has been context-free parsing [Dybkjær 1985]. A general-purpose context-free parser resembling Earley's was partially evaluated with respect to a fixed grammar G, automatically yielding a much more efficient parser, specialized to the syntax defined by G. Further, application of cocom to the general parser yielded a parser generator:

```
specificparser   =  L mix <generalparser, G>
parsergenerator  =  L mix <mix, generalparser>
                 =  L cocom generalparser
```

Another application has been the improvement of the important but computation-intensive ray-tracing technique of computer graphics [Mogensen 1986]. Here the ray-tracer was partially evaluated with respect to a given scene. For this purpose, Mogensen has written a rather larger version of mix than the one described here, using C as implementation language instead of Lisp. The input language is still functional and allows computation with real numbers. Mogensen's version of mix is also self-applicable, but does not automatically determine call unfolding. Significant improvements in computation time have been reported.

9. DISCUSSION

9.1 Subject Language

We think that the following characteristics of the subject language Mixwell of mix have contributed much to the practicability of the project:

- Programs can accept programs as input data and produce them as output.
- Mixwell's simple semantics makes it easy to perform symbolic evaluation and to design a good binding time analysis. In particular, good unfolding properties seem essential.
- The recursion natural to the partial evaluation process is easy to program.
- The language being applicative facilitates specialization of an arbitrary program part without disturbing other parts.

It would be very desirable to have a self-applicable partial evaluator for an imperative language, because target programs would then come out in a language that we know how to implement efficiently. It seems, however, more difficult to build an auto-projector for an imperative language, and the problem is still open as far as we know. One difficulty is recognizing "descents" by an imperative program into a smaller part of a structured static argument. Also, the ubiquitous state component in such a language necessitates more involved partial evaluation time environments and more sophisticated partial evaluation techniques.

A self-applicable partial evaluator for a higher-order functional language, or one for a language that includes function invocation by pattern matching would also be very desirable, owing to the power and conciseness of such languages. It would be harder to write than one based on Mixwell, because of more complex control flow and data descriptions needed for these.

Logic programming languages also seem to have all of the above mentioned useful characteristics, so a non-trivial self-applicable partial evaluator for Prolog should be possible. Performing constant propagation in a Prolog program is not hard, but unfolding problems become more difficult than in Mixwell, owing to Prolog's more complicated parameter concept and control flow.

9.2 Applications of Partial Evaluation

In the paper we have focussed on the applications of partial evaluation to compilation and compiler generation. Below we will exemplify three more general ways to use a partial evaluator:

- program generation
- program specialization
- avoiding inefficiency in metaprogramming

Program generation includes compilation, compiler generation, and compiler generator generation. In addition, certain source-to-source program transformations may be done by partial evaluation of a self-interpreter sint, *i.e.*, an L-interpreter written in L. Given an L-program p, we may obtain a transformed program p' = L mix <sint, p> that has the same input-output behavior as p. All other properties (size, speed, etc) of p' are however rather implicitly determined by the way mix and sint work. By going a step further, one may obtain an automatic program transformer trans = L mix <mix, sint> such that p' = L trans p.

Program specialization may be applied for maintaining a family of related programs, such as control programs for set of similar machines with different parameters. Rather than running a slow and overly general program on every machine, or writing efficient special versions for each of them, one may write a common general version of the program and obtain efficient specialized versions automatically by partial evaluation. This has obvious advantages for program maintenance and for keeping the specialized versions consistent with each other.

Metaprogramming has become a popular tool for implementation of problem-oriented languages and for extension of existing language processors to include new features such as tracing of program execution. This is however often achieved by an interpretive implementation, thus incurring a loss of efficiency. If multiple layers of interpretation are used, the efficiency problems may become severe. Again, partial evaluation may provide a solution, since it will effectively eliminate a layer of interpretation, so that metaprogramming can be used without serious loss of efficiency.

SUMMARY

We have discussed partial evaluation of programs in statically scoped Lisp-like languages and described a self-applicable partial evaluator, mix, that has been successfully applied to generate compilers for toy languages, and even to generate a compiler generator. We assessed mix and gave tables of running times and space usage to illustrate its behavior.

As a basis for this, we introduced a formal framework for partial evaluation, compiling, and compiler generation which enabled the presentation of mix's applications. We also described and discussed the language Mixwell that was designed as the subject language for mix.

ACKNOWLEDGEMENTS

Many people have contributed to the mix project in many different ways. Special thanks go to: Nils Andersen, Anders Bondorf, Olivier Danvy, Andrei Ershov, Niels Carsten Kehler Holst, Kim Høglund, Torben Mogensen, Marek Rycko, Rodney Topor, and Valentin Turchin.

REFERENCES

Beckman, L. *et al.*,
A partial evaluator, and its use as a programming tool,
Artificial Intelligence **7**, 4 (1976) 319-357

Bulyonkov, M. A.,
Polyvariant mixed computation for analyzer programs,
Acta Informatica **21** (1984) 473-484

Cousot, P. and R. Cousot,
Abstract interpretation: a unified lattice model for static analysis of programs by construction or approximation of fixpoints,
Proc. Fourth ACM POPL Symp., Los Angeles, California 1977, 238-252

Dybkjær, H.,
Parsers and partial evaluation: An experiment.
DIKU student report No. 85-7-15, University of Copenhagen, Denmark, 1985

Emanuelson, P. and A. Haraldsson,
On compiling embedded languages in Lisp,
Proc. 1980 Lisp Conf., Stanford, California (1980) 208-215

Ershov, A. P.,
On the essence of compilation,
Formal Description of Programming Concepts (ed. E. J. Neuhold),
North-Holland 1978, 391-418

Ershov, A. P.,
Mixed computation: Potential applications and problems for study,
Theoretical Computer Science **18** (1982) 41-67

Futamura, Y.,
Partial evaluation of computation process - an approach to a compiler-compiler,
Systems, Computers, Controls **2**, 5 (1971) 45-50

Ganzinger, H. and N. D. Jones (eds.),
Programs as Data Objects, Lecture Notes in Computer Science **217**,
Springer-Verlag 1986

Haraldsson, A.,
A partial evaluator, and its use for compiling iterative statements in Lisp,
Proc. Fifth ACM POPL Symp., Tucson, Arizona 1978, 195-202

Jones, N. D., P. Sestoft, and H. Søndergaard,
An experiment in partial evaluation: The generation of a compiler generator,
Rewriting Techniques and Applications (ed. J.-P. Jouannaud),
Lecture Notes in Computer Science **202**, Springer-Verlag 1985, 124-140

Kahn, K. M.,
A partial evaluator of Lisp programs written in Prolog,
Proc. First Int. Logic Programming Conf. (ed. M. Van Caneghem),
Marseille, France 1982, 19-25

Kahn, K. M. and M. Carlsson,
The compilation of Prolog programs without the use of a Prolog compiler,
Proc. Int. Conf. Fifth Generation Computer Systems, Tokyo, Japan 1984, 348-355

Kleene, S. C.,
Introduction to Metamathematics, Van Nostrand 1952

Komorowski, H. J.,
A Specification of an Abstract Prolog Machine and Its Application to Partial Evaluation,
Linköping Studies in Science and Technology Dissertations **69**,
University of Linköping, Sweden 1981

Kugler, H.-J. (ed.),
Information Processing 86, Proc. IFIP 86 Conf., North-Holland 1986

Lombardi, L. A.,
Incremental computation, *Advances in Computers* **8**
(ed. F. L. Alt and M. Rubinoff), Academic Press 1967, 247-333

Mogensen, T. Æ,
The Application of Partial Evaluation to Ray-Tracing,
Master's thesis, University of Copenhagen, Denmark, 1986

Mosses, P. D.,
SIS - Semantics Implementation System, Reference Manual and User Guide,
DAIMI Report MD-30, University of Aarhus, Denmark 1979

Paulson, L.,
A semantics-directed compiler generator,
Proc. Ninth ACM POPL Symp., Albuquerque, New Mexico 1982, 224-233

Rees, J. and W. Clinger (eds.),
Revised[3] report on the algorithmic language Scheme,
SIGPLAN Notices **21**, 12 (1986) 37-79

Rogers, H.,
Theory of Recursive Functions and Effective Computability, McGraw-Hill 1967

Safra, S. and E. Shapiro,
Meta interpreters for real, in [Kugler 1986], 271-278

Sestoft, P.,
The structure of a self-applicable partial evaluator,
in [Ganzinger and Jones 1986], 236-256

Takeuchi, A. and K. Furukawa,
Partial evaluation of Prolog programs and its application to meta programming,
in [Kugler 1986], 415-420

TsNIPIASS,
Bazisnyi Refal i yego Realizatsiya na Vychislitelnykh Mashinakh,
TsNIPIASS, Gosstroi SSSR, Moscow 1977

Turchin, V. F.,
A supercompiler system based on the language Refal,
SIGPLAN Notices **14**, 2 (1979) 46-54

Turchin, V. F., R. M. Nirenberg, and D. V. Turchin,
Experiments with a supercompiler,
Proc. 1982 ACM Symp. Lisp and Functional Programming,
Pittsburgh, Pennsylvania 1982, 47-55

Venken, R.,
A Prolog meta-interpreter for partial evaluation and its application
to source to source transformation and query-optimisation,
Proc. ECAI-84, Pisa, Italy (ed. T. O'Shea), North-Holland 1984, 91-100

Semantics-Based Tools
for a Specification-Support Environment

Joylyn Reed
Programming Research Group, Oxford University
8-11 Keble Road, Oxford, OX1-3QD, U.K.

Abstract

We describe recently developed semantics-based support tools for Z, a mathematical specification language based on typed set theory. Z has proven very useful and popular with a number of industrial as well as academic software developers. These tools are components of Forsite, a support environment currently under development to integrate languages and operations with formally defined semantics and implementable operations. We anticipate that these tools will impose a *de facto* standard for the language.

Z has undergone a noteworthy chronology of development and use. It was developed in stages by mathematicians who extended and adapted it according to the needs and experiences of industrial as well as academic users. Care was taken to maintain mathematical soundness, and when the language had stabilized, a denotational semantics for it was defined. Subsequently, support tools entirely based on this semantics were and are being developed. We describe the implemented type checker, which gives the sense of directly transliterating the formal type semantics, and preview the proof checker. We discuss the benefits of such a chronology for language and methodology development.

I. Introduction

The Forsite project, aimed at developing a specification support environment, began in 1985. This Alvey funded project is a three-year collaborative effort involving two industrial groups (Racal, ITD and Systems Designers) and two university groups (Oxford PRG and Surrey). The immediate objective is to provide mechanical support for creating and reasoning about specifications written in Z, but the broader goal is to develop an environment into which can be integrated languages and operations with formally defined semantics and implementable operations, e.g., CSP specifications, refining specifications into programs, verifying that a design satisfies its requirements or that a program implements its design, etc. By August 1986 a "preprototype" had been implemented which provided wysiwyg ("what you see is what you get") editing, parsing, and type-checking for Z. The next version, planned for 1987, will include a mechanical proof checker designed to aid program development from initial requirements to code generation.

Z (pronounced "zed") is a language for specifying and reasoning about programs. A Z specification contains notation from ordinary logic and set theory, together with specialized notation convenient for abstracting the essence of programs and minimizing formal clutter. The "Z method" encourages requirements formulation followed by transformations of specifications into programs. The beginnings of Z can be found in [ASM80,Abriel81], but the syntax and well-understood (but informally expressed) meaning of the language evolved to its present form as a result of contributions of various researchers at Oxford [Morgan84, KiSoWo86, MoSu82, SSSW86, Spivey85, Spivey86, Sufrin83, Sufrin86]. A commonly held criticism of formal, mathematically based specification methods is that, while they are theoretically interesting, they have proven infeasible to use in actual software development. Z has been applied to a variety of life-sized problems [Hayes87], and a number of U.K. industry and government groups (e.g., IBM UK, RSRE) now routinely use Z to express specifications during the requirements and design phases of their software development. These practitioners were keen to use Z, even though computerized support for such work has until now been limited to editors adapted to provide Z symbols.

In 1985 J.M. Spivey [Spivey85] defined a denotational semantics for Z based on typed set theory, which formalized the semantics generally understood by Z researchers and practitioners. The Forsite project undertook to provide a Z support environment based on this semantics.

The first such semantics-based tool to be implemented is a type checker. The implementation language, SML, proved very convenient for expressing the type semantics of Z. SML is an extension of ML, a language developed to implement a proof checker for denotational semantics [Harper85, GMW78, Milner]. Since many of the formalisms in Spivey's semantics were found to be directly expressible in SML, the code for the type checker is in some sense a direct transliteration of the formally defined semantics.

Z has undergone, and continues to undergo, a noteworthy chronology of development and use. Mathematicians developed the language and a "Z style of program development" in stages. They extended and adapted according to the needs of an initially small, but *real* user community, while taking care to retain mathematical soundness. A consequence of the concern for mathematical soundness is that some of the language becomes arcane to the ordinary Z user, but the demonstrated strength of Z is that it is sufficiently rich to offer a usable and complete set of features to software developers. It was strongly felt that standards for the language should be avoided until Z stabilized and struck an optimal balance between mathematical simplicity and practical use. Since mechanical tools have the effect of producing *de facto* standards, development of such tools have awaited a sound formal basis. We will now monitor with great interest what impact these tools will have on Z users who previously applied Z by hand, with no automated checks for consistency or correctness but with no notification of incorrect use.

Section II provides an overview of Z and a formal semantics for it. Section III describes the type checker. Section IV previews the theorem checker. Section V presents conclusions.

II. Overview of Z and a Note on its Formal Semantics

The following brief overview is taken from [SSSW86]. Z is often described as a combination of a set theoretic language with typing of a sort familiar to programmers together with a schema language. The set theoretic language is similar to a conventional set theoretic notation, but extended with a small number of concepts which allow the specifier to deal easily with concepts and rules commonly encountered during the development of software systems. The schema language is a simple language used to factor out common parts of a specification so that designs can share components, proofs share arguments, and theories share abstractions in such a way that existing parts can be modified easily.

Experience shows that the use of mathematics such as set theory often leads to prohibitively complex descriptions when applied to practical systems. The most important part of the Z approach is the conventions and style which have been adopted for using the schema language when presenting descriptions in the set theoretical language. These conventions encourage separation of concerns and allow for concepts of the system which is being documented to be formally explained in the simplest possible context. The presentation in a Z document is a mixture of informal text (e.g. English, pictures) and mathematical text. This style has proven useful to users, technical writers and managers as well as to designers and specifiers.

For example, we can define a useful relation operator *domain corestriction* and a functional operator *override* with straightforward mathematics:

```
═[ X,Y ]══════════════════════════════════════
│ _ ⩤ _ : ℙ X × (X × Y) → (X ↔ Y)
│ _ ⊕ _ : (X ⇸ Y) × (X ⇸ Y) → (X ⇸ Y)
├────────────────────
│ ∀ S : ℙ X; R : X ↔ Y •
│     S ⩤ R = {x:X; y:Y | x ∉ S ∧ x R y • (x,y)
│ ∀ f, g : X ⇸ Y •
│     f ⊕ g = ( (dom g) ⩤ f) ∪ g
└──────────────────────────────────────────────
```

Description: The *domain corestriction* S ⩤ R of a relation R to a set S relates x to y iff R relates x to y and x is not in S. The function f ⊕ g is defined on the union of the domains of f and g. On the domain of g it agrees with g, and elsewhere on its domain it agrees with f.

This definition contains some declarative information above the horizontal dividing line, and some further information conveyed by the axioms below it. The axiom forming the *predicate* below the dividing line is explained with English commentary. The declarative part of the definition forming its *signature* above the line introduces the following information:

- The generic parameters X and Y.

- The variable name ⊕.

- Some type information: ⊕ is an operator on two functions from X to Y, which returns a function from X to Y.

- Some syntactic information: ⊕ is an infix operator.

Frequently in specifications a certain pattern occurs many times - a mathematical structure which describes some variables whose values are constrained in some way. We call this introduction of variables under some constraint a *schema*.

For example, we might model a symbol table as a *schema* containing a mapping between symbols (SYM) and values (VAL), together with a set of reserved symbols not to be assigned values:

```
┌─ Symtab ──────────────────────────
│ st : SYM ⇸ VAL
│ keywds : ℙ SYM
├───────────────
│ keywds ∩ dom st = {}
└───────────────────────────────────
```

All symbol table operations deal with a before state (Symtab) and an after state (Symtab′). We can factor the information to be duplicated with a schema, which can be incorporated into all operations:

```
┌─ ΔSymtab ──────
│ Symtab
│ Symtab′
└────────────────
```

which is equivalent to:

```
┌─ ΔSymtab ──────────────────────────
│ st, st' : SYM ⇸ VAL
│ keywds, keywds' : ℙ SYM
├──────────────────
│ keywds ∩ dom st = {}
│ keywds' ∩ dom st' = {}
└────────────────────────────────────
```

Operations which do not change key words share the following, which is ΔSymtab with an additional axiom:

```
┌─ Stabletab ─────
│ ΔSymtab
├──────────────────
│ keywds' = keywds
└──────────────────
```

One such operation is updating the table so that symbol s? is associated with the value v?. Any previous value (if any) assigned to s? is lost. The partial operation can proceed only if s? is not a keyword :

```
┌─ UpdateOK ─────────────────
│ Stabletab
│ s? : SYM
│ v? : VAL
├──────────────────
│ s? ∉ keywds
│ st' = st ⊕ { (s?, v?) }
└────────────────────────────
```

The error case is:

```
┌─ UpdateError ──────────────
│ Stabletab
│ s? : SYM
│ v? : VAL
├──────────────────
│ s? ∈ keywds   ∧   st' = st
└────────────────────────────
```

Combining to obtain the full operation:

```
Update ≙ UpdateOK ∨ UpdateError
```

The disjunction indicates that the signatures of the component schemas are merged, common variables are identified, and the predicate parts are disjoined.

The *schema calculus* offers a number of useful operations for defining and manipulating schemas in order to build large specifications in stages. There are logical operations, relational compositional operations, hiding operations (for simplifying the signature interface), and a *precondition operator* which when applied to an operation yields the set of states for which the operation is specificed to guarantee termination. Theorems can be concisely stated. For each operation that we introduce on a system state, we should prove that it is applicable. For example when expanded, the following notation becomes a mathematical statement asserting that Update is a total operation defined on symbol tables:

Symtab ⊢ *pre* Update

A Formal Semantics for Z

In Spivey's semantics, the meaning of a Z specification is the collection of all its models. The key concept involved in his construction of semantic domains is the "variety", which is a signature together with a class of structures for the signature. Signatures contain an alphabet of generic parameters or "given sets", and an alphabet of typed variable names, and structures interpret these as sets. The denotational semantics consists of definitions for semantic functions which link the constructs of an abstract syntax with operations on the semantic domains.

A *structure* takes certain given-set names (global to a specification or local to a generic definition) and variables and gives them values in a universe of sets. A requirement on the structures for a schema is that they be consistent with the signature of the schema: the value of each variable must be in the "carrier" for the corresponding type (a set of elements formed by interpreting Z type constructors as operations in the world of sets). A *variety* - the meaning of a schema - is defined as a signature together with a set of structures for the signature which satisfy the axioms of the schema.

These concepts are used to build semantic domains for a denotational semantics for Z consistent with its "common and everyday" use. Treatment of such Z constructs as generics and references to instances of schemas might be described as disappointingly complex, but the ordinary Z user need not be aware of these subtleties in order to apply Z correctly.

III. A Z Type Checker

Variables in signatures (or quantifiers) are introduced with type information. A Z declaration for a variable is of the form `id : term`, where `term` can be any term referring to a set. There are several advantages in separating the declarative information found in signatures from the axioms of the associated predicates :

- The signature of a schema may be regarded as an interface by means of which it may be assembled with other schemas to form larger specifications.
- The theory of signatures is decidable, and the well-formedness of specifications in terms of signatures is particularly suited to mechanical checking.
- The type information contained in signatures has a pragmatic value in preventing errors in specification.

For a given set of identifiers (ID) we constructively define an associated set of "types". For a **Z** specification, the ID's would be the given-set names, and at certain localities, the generic parameters. Below on the left is a **Z** recursive definition of such a set, and on the right an SML implementation:

```
TYPE ≙ ident << ID >>           datatype TYPE = ident of ID
     | powerset << TYPE >>                    | powerset of TYPE
     | product <<seq TYPE>>                   | product of TYPE
                                                          list
     | schema <<ID ⇸ TYPE>>                   | schema of
                                                (ID * TYPE)list
```

Intuitively, given sets (or generics) are types, and powersets, products, and schema bindings of id's to types are types.

The key to type checking a **Z** specification is a collection of functions which return the type of a term with respect to an environment of *normalized* declarations, i.e. declarations of the form `id :: type`. We derived these functions directly from a set of inference rules specifying the same in Spivey's semantics. For example, the inference rule for ordinary set comprehension (with certain technical niceties omitted) states that the type of a set comprehension term is the powerset of the type of the defining term in the original environment "enriched" with the enclosed declarations:

```
newenv = enrich (env,decs)
newenv ⊢  term :: itstype
_____________________________________

env  ⊢  {decs | pred • term}  :: ℙ itstype
```

The corresponding SML function is:

```
fun typeofset  (env:ENV) (setcomp(decs,pred,term)) =
   let val newenv = enrich env decs;
       val itstype = typeof newenv term
    in (powerset itstype)
    end
```

An important function is norm, which takes an environment and a set of declarations and returns the *normalized* version of the declarations, i.e., a set of signatures associating each id with its proper type. A Z specification of this function (using curried functions and functional composition) is:

```
norm : ENV → DEC ⇸ NORMDEC
─────────────────────────────
∀ env : ENV ; decs : DEC •
    norm env decs = decs ⨾ type env ⨾ powerset⁻¹
```

where ENV and DEC are the sets of partial functions from ids to types and ids to terms, resp., and type is a function taking an environment and a term returning the corresponding type of the term.

Thus to normalize a declaration, find the type of the term on the right of the colon and then strip off the powerset. For example, for the declaration

$\mathrm{poseven} : \{y : \mathbb{N} \mid y \geq 0 \bullet 2y\}$

the right side has type $\mathbb{P}\ \mathbb{N}$, and the normalized version is:

$\mathrm{poseven} :: \mathbb{N}$

The key SML function to implement the above is:

```
fun normadec (env:ENV) ((id,term) : DECPAIR) : SIG =
    (id, invpowerset (typeof env term))
```

where,

```
fun invpowerset (powerset ty) = ty
```

An interesting consequence of a strongly typed language such as Z is that the predicate $x \in x$ does not type check. It follows from the type rules for Z that the type of the right side must be a powerset of the type of its left side -clearly violated by this predicate. The strong typing of

Z does not however protect against such meaningless (but type correct declarations) as

$$x : \{ y : \mathbb{N} \mid y > 0 \land y < 0 \}.$$

SML proved very succesful for directly translating a formal specification of the denotational semantics into executable code. The type checker component of Forsite takes as input an abstract syntax tree (constructed by the parser and translated to SML structures) and produces error diagnostics or an "ok" sign. The SML program consists of approximately 700 lines of code.

IV. Preview of a Theorem Checker

Work is underway to derive a logical calculus for reasoning about specifications. Such a calculus should be provably sound from the denotational semantics in the way that the Hoare style proof rules for a programming language are provably sound from the denotational semantics for the language.

Proof rules have recently been developed which allow a Z specification to be translated into Dijkstra's "do-od" language. Also recently developed are proof obligations for performing data refinement from an abstract specification into a more concrete one. Simple semantics have been given for the use of schemas as predicates, and this notion is useful for stating theorems about individual specifications as well as for stating general proof obligations. For example, one of the data refinement rules demands that if the precondition for the abstract operation AOP holds, and the abstract and concrete states are related by R, then the precondition for the concrete operation COP holds. This proof obligation is met whenever the following concisely stated theorem is proved:

$$R \land \mathit{pre}\ AOP \ \vdash \ \mathit{pre}\ COP$$

The logical calculus for Z should provide the ability to reason at the level of the schema calculus rather than at the level of the traditional mathematics sublanguage. We will develop the rules of reasoning in conjunction with a mechanical proof checker designed to understand these rules. The proof checker is to be in the style of a proof checker built by J.R. Abrial. Abrial's tool is a simple natural deduction system in which axioms and inference rules can be expressed in the traditional way, rather than in a style dictated by a programming language. While not offering sophisticated proof strategies, it applys rules in a goal-directed fashion which can be easily interrogated and understood by its user.

V. Conclusions

Several benefits have resulted from the success of Z's developers in maintaining both mathematical simplicity and practicality. The carefully controlled evolution of Z has led to its increasing popularity and use among software developers. The formal semantics for Z has played a major role in its evolution.

The formal semantics revealed that certain use of the language was not as had been expected, while justifying the intuitive intent for others. For example, Spivey imposed constraints for generic definitions, which were not previously recognized but which do not practically limit their use. He also demonstrated that our intuitive notion that composite models of specifications should contain certain inherent aspects of their components beyond simple textual ones is sound.

The formal semantics provided the basis for a rigorous development of tools essential to the task of specification and program development. We are encouraged by our experience so far that current implementation languages enable "direct transliterations" of Z's formal semantics.

We expect the currently implemented Forsite tools to impose on Z users the mathematical standard which has so carefully been developed. We hope the immediately forthcoming tools such as a theorem checker will encourage a software development discipline based on formal principles of specification and refinement.

ACKNOWLEDGEMENTS -We gratefully acknowledge Bernard Sufrin, whose ideas provided initial inspiration for the above work, and Jane Sinclair, who constructed the interface. We also acknowledge the valuable work of our Forsite colleagues.

VII. References

[ASM80] J-R. Abrial, S.A.Schuman, and B. Meyer, "Specification Language", in *Construction of Programs*, ed. R.M. McKeag & A.M. MacNaughten, Cambridge Univ. Press, 1980.

[Abriel81] J-R. Abrial, "A course on System Specification", Lecture notes, Programming Research Group, Univ. of Oxford, 1981.

[GMW79] M. Gordon, R. Milner, and C. Wadsworth, *Edinburgh LCF: A Mechanical Logic of Computation*, Lecture Notes Com. Sc.,78, Springer-Verlag, 1979.

[Harper86] R.W. Harper, D. MacQueen, and R. Milner, *Standard ML*, Edinburgh University Internal Report ECS-LFCS-86-2, 1986.

[Hayes87] *Specification Case Studies*, ed. I. Hayes, Prentice-Hall Inter., London, 1987.

[KiSoWo86] S. King, I.H. Sorenson and J. Woodcock, "A Syntax for the Z Notation, Programming Research Group, Univ. of Oxford, 1986.

[Morgan84] C.C. Morgan, "Schemas in Z: a Preliminary Reference Manual", Programming Research Group, Univ. of Oxford,1984.

[Milner] R. Milner, "A proposal for Standard ML", *Proc. ACM Symp. LISP and Functional Programming*, Austin, Texas, 1984, pp. 184-197.

[MoSu82] C.C. Morgan and B.A. Sufrin, "Specification of the UNIX Filing System", *IEEE Trans. Soft.Eng.*, v.10,2 1983, pp 128-42.

[SSSW86] J. Sanders, I.H. Sorenson, B.A. Sufrin, and J. Woodcock, *Notes for Software Engineering*, Z Course given at Wolfson College, Oxford, 1986.

[Spivey84] J.M. Spivey, "Towards a Formal Semantics for the Z Notation", Tech. Mon. PRG-41, Programming Research Group, Univ. of Oxford, 1984.

[Spivey85] J.M. Spivey, *Understanding Z: A Specification Language and its Formal Semantics*, D.Phil. Thesis, Univ. of Oxford, 1985 (to be published by Cambridge Univ. Press).

[Spivey86] J.M. Spivey, *The Z Library, A Reference Manual*, 1986.

[Sufrin83] B.A. Sufrin, "Formal Specification of a Display-oriented Text Editor", *Science Com. Prog.*, v.1, 1983, pp 157-202.

[Sufrin86] B.A. *Z Handbook*, Draft 1.1, March 1986.

[Harper85] R.W. Harper, D. MacQueen, and R. Milner, *Standard ML*, Edinburgh University Internal Report ECS-LFCS-86-2, 1985.

A Treatment of Languages with Stages of Evaluation

Laurette Bradley
Dept. of Computer Science and Engineering, C-014
University of California, San Diego
La Jolla, California 92093

Abstract

The notion of languages that inherently have multiple stages of evaluation is introduced. Typically, evaluation is done in stages so that evaluation in some one stage is able to be done very efficiently, even at the expense of prior stages of evaluation. A key novel feature in such languages is that constructs may have appropriate times of meaning as well as appropriate meanings. While it is possible to give semantics to such languages without regard to times of meaning, it is shown that certain semantic related concepts, such as translation, cannot be adequately described without reference to times of evaluation. The major contribution provided here to the study of complex languages with multiple stages of evaluation is the development of a method of describing such languages so that semantics reflecting various times of evaluation can be derived from the descriptions.

1. Times of Evaluation

1.1. Introduction

The notation of programming languages is inherently a specification of a computation that is done in stages. For instance, the declarative phrases in a language expression are *intended* to be evaluated at one stage - compile time - whereas the assignment statements are *intended* to be evaluated at another stage - run time. This facet of the structure of programming languages is very much at odds with the structure of classically studied languages such as those from mathematical logic or linguistics. This stratified evaluation scheme is founded in notions of efficiency, since the only reason to evaluate declarations, say, at different time than assignment statements is to make the evaluation of the assignment statements more efficient. In this paper we will introduce the notion of languages that specify computations to be done in stages, and develop an appropriate method of definition for such notations. We will show that while it is possible to give semantics for complex notations such as this denotationally, that is, without reference to the various intended stages of evaluation, there are some semantics related concepts which cannot be described denotationally for such languages.

The paper has four sections. In the remainder of the first we discuss the notion of languages whose constructs have *times* of meanings as well as meanings. We will begin with a brief examination of the classical concepts of semantics that programming languages inherited from mathematics and linguistics. Then we will give examples to show why programming languages are crucially different. In the second section we develop a method of describing languages such that we can derive various semantics which embody different notions of time for the meanings. In the third section we give some examples of language design. In the fourth section we give our conclusions.

1.2. Past treatments of semantics

Ideally, semantics should provide us with a complete understanding of language meaning. It should be possible to examine the semantic model for a language and deduce answers to all meaningful questions about the language. In the past it has been sufficient to assume that meaning could be attached to linguistic expressions "all at once", or denotationally, since this was the nature of the languages studied. In mathematical logic a language is specified by giving the *symbols* of the language and rules for combining the symbols into expressions of the language, which are terms and formulas. Semantics is given for these languages by specifying a *structure* which consists of a universe of individuals and a meaning for each function and predicate symbol of the language. Meanings of expressions are then defined denotationally, that is, inductively. (See, for example, [Schoenfield 67]). In the classical development of the semantics of natural language ([Frege 92], [Carnap 47], [Montague 74]) the principle that language should be defined denotationally is referred to as the Principle of Compositionality, or Frege's Principle.

1.3. Staged computations

Why is it important to introduce a notion of time into semantics? We give two examples here. One involves the difficulties arising from treating translation as a denotational process, the other considers the difficulty of introducing even a simple notion of arrays into a language without some conception of times of meaning.

1.3.1. Compilers as homomorphisms

Past studies ([Morris 73], [ADJ 80]) of the structure of compilers have developed the idea that a compiler is, or should be, describable as a homomorphism between two Σ-algebras, one of which is the initial Σ-algebra whose elements correspond to parse trees of expressions in the language to be compiled, and the other of which is a derived Σ-algebra over the original target-code algebra. The studies concentrate on the development of a compiler for an exceedingly simple language which, importantly, is almost completely bereft of any typing, block structure, or overloading. It is precisely because of these aspects of programming languages that their translation to machine code is not well described as homomorphic. It is a trivial matter to describe the translation of an if-then-else statement, or a loop statement into derived operations over some basic machine code, but it is not trivial, and more importantly, not *desirable* to describe the translation of declarations into derived operations over some basic machine code. Typing information is not intended to be executed at run-time. To arrange for overloading resolution, type checking and storage allocation of simple variables to be done at run-time is grossly at odds with the usual notion of compilation. Furthermore, the notion that translation is homomorphic precludes, or makes very difficult, development of optimization within the framework, since it is unnatural to embody optimization actions within each construct of the language.

The mistake made in the approach to compilers as homomorphisms is to assume that translation can be attached uniformly, at one time, to all phrases in a programming language. Inherent in programming languages is a structure that assumes that various phrases of language expressions will be evaluated at compile time, while other phrases will be evaluated at run time. A correct model of translation is one that reflects the compiler as an active evaluator of part of the language expression; the translator is not a "passive" homomorphism that "pastes together" target code fragments to obtain a translated version of the program. Rather, phrases of the language expression are evaluated at translation time, and the meanings of these expressions determine, in part, the translation of the other phrases in the expression.

1.3.2. The introduction of simple arrays into a language

In our next example we will look at times of evaluation from a slightly different perspective, that of the language designer. If phrases such as declarations are intended to be evaluated at a different stage of computation than phrases such as assignment statements, how does the designer of a language view the meaning of language expressions, with all these various times of evaluation in mind? What is an appropriate notion of times of evaluation?

Consider the language designer who wishes to introduce notation for array manipulation into a language. A crucial aspect of the design of the notation is the decision on how much information about the array objects needs to be present in the language, and at what other stages of computation information may be added. The classical decisions that have been made in the past are that

- All information on the array object, must be determinable from examination of the language expression.
- Bounds information need not be expressed in the language, but must be available at the time of entrance to the block in which the array is declared.
- Bounds information can be determined and altered, repeatedly, dynamically.

Each of these decisions results in vastly different compile-time and run-time structures for arrays. In order to reason effectively about the languages which result from these decisions, that is, in order to understand the effect on the efficiency of the computation which is to be performed at various stages (compile time and run time) it is essential that a language designer work with a semantic model in which such effects can be expressed. A language designer might well need an answer to the question "What are array meanings at compile time, pre-block-entrance time, and post-block-entrance time?" even though such a question cannot be phrased in a denotational context.

1.4. Summary

We have discussed the need for a description of languages that includes some notion of *time* of evaluation. Such descriptions would be pointless for languages in the classical sense of mathematical logic, but could be very useful in understanding aspects of programming languages. As we will argue in the following sections, this is because classical languages are assumed to represent their underlying structures in a very simple way, whereas programming languages have a much broader, more complex, range of representation schemes.

When we say that a compiler is not well described as a homomorphism from the source language considered as a many-sorted algebra to a derived algebra over the target code, we are actually claiming that the translation of programming languages has a very special property; namely, that portions of a language expression are evaluated during translation, and the resulting values partially determine the translation of the remainder of the expression.

Similarly, when we imagine allowing arrays with incomplete bounds information as objects in a language, we are assuming that, at some later time in evaluation, bounds information will be "added". This is a remarkable situation, since it implies that the language expression itself only partially represents the "underlying object", and that the rest of the information is represented in other expressions, outside of the language.

In the next section we will attempt to precisely capture the uniqueness of these situations by defining languages in terms of how they represent underlying semantic domains. Essentially, a language that represents a semantic domain in a "simple" fashion will have a "simple" definition, and a simple semantics, while languages that represent their semantic domains in more complex fashions will have unusual translations (when their underlying domains are other languages) and unusual semantics.

2. Semantics with Times of Evaluation

Having described the significance of times of evaluation to programming language semantics, we now would like a method of describing languages such that the meaning of language phrases at various times of evaluation could be expressed. It has long been informally acknowledged that in order to use a semantics to *implement* a programming language some notion of times of evaluation of the constructs in the language needs to be introduced. From the beginning ([LLS 68]) many operational semantics have involved times of execution in the form of compiling and run-time machines, but these are very concrete. There are, of course, many ways to induce a notion of times of evaluation into a denotational semantics. For instance, it is common to induce a notion of compile-time into a semantics by evaluating meanings with respect to certain arguments ([Mosses 79], [Paulson 82]); the result of this evaluation is the compiled version of the program. The results of evaluation with respect to the rest of the arguments corresponds to execution of the program. [Jorring 86] gives an account of staging transformations that induce pass separation, similar to that in a compiler, in a semantics. [Nielson 86] describes a two-level metalanguage for separating compile time and run time, and [Lee 87] separates semantics into macrosemantics and microsemantics which correspond to compile time and run time (and which embody a "well engineered" semantics in other respects as well). Of all of these, the results of our approach are most similar to those of Jorring. However, our method is very different. We will discuss the relationship of our work with that of others again in the last section of this paper, but for now we point out that, except for Jorring, all these approaches *fix* various times of evaluation, namely compile time and run time, then express semantics with respect to them. None of these approaches is well suited to handling a multiplicity of times of evaluation, whereas ours is. Rather than fixing some times of evaluation (compile time, run time, pre-block-entrance time, post-block-entrance time, etc) and then demanding that semantics always be expressed within a framework including these times, we will instead develop a very general technique for describing any language, such that multiple semantics can be derived from a description, each of which gives the meanings for language phrases at various times of evaluation.

As an introduction to our approach, consider a typical language X. In the compilation of X we know that certain phrases, such as declarations, will be evaluated by the compiler and no code will be emitted for them. We also know that type checking for some phrases will be done by the compiler, whereas for others this will not be possible. Intuitively then, we break up evaluation of language expressions into two times: compile time and run time. A key question then is, what evaluation can be done at compile time and what must be done at run time. Also, given that we have an answer to that question, another question of interest is, how can we change the language so that more evaluation can be done at compile time and less at run time. More generally, suppose we have a language that represents a computation to be done in stages. How can we alter the language so that as much computation as possible is done in the early stages?

Our goal here is to develop a method of defining languages that is useful in addressing questions such as this. This method of definition must describe times of evaluation, and allow easy experimentation with the language so that alternative languages, with other stages of evaluation, can be examined.

We will *define* languages by giving *formal designs* for them. This description of a language in terms of how it describes its underlying "semantic domain" can be used to derive various kinds of semantics for the language, each of which reflects a different notion of time of evaluation for the constructs in the language.

2.1. Appropriate languages

We use the expression "semantic domain" to mean what is classically called the universe of discourse of a language; it is what the language refers to -- the objects and functions. Our approach is to view languages and semantic domains as algebras, and to view language design (and so, also, language definition) as a process of "reshaping" a semantic domain algebra into an appropriate language algebra. This has the following intuition: We can use the language of simple arithmetic expressions such as "1+1" or "(2+3)*4" for expressing arithmetic on a stack machine. These terms *represent* the machine code language terms "push 1; push 1;add" and "push 2; push 3;add; push 4;mult". Considered as algebras, the set of integers with operations "+", "-", "*", and "/" and the machine algebra for a stack machine are very different. We can *define* the use of simple arithmetic as a "stack machine language" by showing how to apply transformations to the stack algebra to reshape it into an algebra of the same form as the algebra for simple addition. This reshaped algebra must, of course, give the "right" semantics for the constants "1", "2". etc. as well as the operations, "+", "-", "*", "/". Also, importantly, while we have a simpler language for the stack machine, we now have to do some work to translate the terms of this language into the terms of the original stack machine, so we incur a translation cost.

Since we will be "reshaping" semantic domains to get languages in an algebraic framework, we must start by settling on definitions for *semantic domain, language*, and *language for a semantic domain.* A *semantic domain* is a many-sorted algebra (see [ADJ 77]). We denote a signature, Σ, for an algebra by a triple, $<S,O,F>$, where we use the convention that S is the set of sort names, O is the set of operation names, and F gives the functionality of the operation names as a tuple whose first element is in S^*, and whose second element is in S. For signature $\Sigma=<S,O,F>$, Σ-algebra A, $s \in S$, $\sigma \varepsilon O$, A_s denotes the carrier of sort s in A, A_σ denotes the operation named σ in A. A *language* is a term algebra; the freely generated algebra for some signature. We use the notation T_Σ for the term algebra for the signature Σ, and $T_\Sigma(X)$ for the algebra freely generated over some set of generators X. Finally, we identify signatures with grammars as in ([ADJ 77]).

It remains to define the notion of *language for a semantic domain.* This is the heart of our approach and the essense of our departure from earlier work. As described in ([ADJ 77]), the only appropriate language for a Σ-algebra is T_Σ. This is the algebraic expression of the classical notion of compositionality of meaning. The underlying assumption is that the only reasonable language for a semantic domain is a language that has an identical algebraic structure. Our thesis is that languages may have very different structures from underlying semantic domains, since the language may be specifying computations which occur at different stages of evaluation, and at none of these stages of evaluation need the semantic domain have an identical structure to that of the language.

Intuitively we want to consider a language, T_Σ, as appropriate for a semantic domain which is a Σ'-algebra if we can find some way of *representing* all the "terms of interest" in the natural language of the semantic domain (that is, $T_{\Sigma'}$) by terms in T_Σ, *and* we can find some way of *interpreting* T_Σ so that the values of $T_{\Sigma'}$ terms and their representatives in T_Σ are the same. For instance, in the above example, the sort of terms of interest can be characterized informally as the set of all normal arithmetic expressions; we are not interested in representing "push 1; push 33; push 17; push 9" in our language, for example, since it does not use the machine to do arithmetic. In the definition below, we call this set of terms of interest "I".

Definition 1: Let Σ_1=<S_1,O_1,F_1> and Σ_2=<S_2,O_2,F_2> be signatures, let A be a Σ_1-algebra, and let I $\subseteq T_{\Sigma_1}$. A *language view of A w.r.t I in* T_{Σ_2} is given by <sm,tm,B> where

- sm: $S_1 \rightarrow S_2$ ("$\rightarrow$" means partial function)
- tm: $\{tm_s : T_{\Sigma_{1_s}} \rightarrow P(T_{\Sigma_{2_{sm(s)}}})\}$, $s \in S_1$, $sm(s)$ defined (P means power set)
- B is a Σ_2-algebra such that

  $A_\gamma = B_\delta$ for each $\gamma \in T_{\Sigma_{1_s}}$, $\delta \in tm_s(\gamma)$

  $tm_s(\gamma) \neq \emptyset$, for all $\gamma \in I \cap T_{\Sigma_{1_s}}$, for all $s \in S_1$.

A language view of A w.r.t I in T_{Σ_2} *represents* terms from the natural language of A with sets of terms in T_{Σ_2} so that every term in I is represented by a non-empty set of terms; it also gives an interpretation of T_{Σ_2}, B, so that the interpretation of any I term in A equals the interpretation of any of its T_{Σ_2} representatives in B. From this we naturally have:

Definition 2: Let A be a Σ_1-algebra, $I \subseteq T_{\Sigma_1}$. If there is a Σ_2-algebra B such that for some sm and tm, <sm,tm,B> is a language view of A w.r.t. I in T_{Σ_2}, then we say that T_{Σ_2} is a *language for A w.r.t. I* and B is a *language view interpretation* for A.

We note that, when there is no confusion, we will frequently use "tm" alone when we should more properly be using "tm_s".

2.2. Language definition by design

The above definitions allow a wide range of languages to be used for a semantic domain. How do we find them? We now define a kind of tool box for the language designer. It consists of (an infinite number of) "constructors" which map algebras into language views, as in figure 2-1.

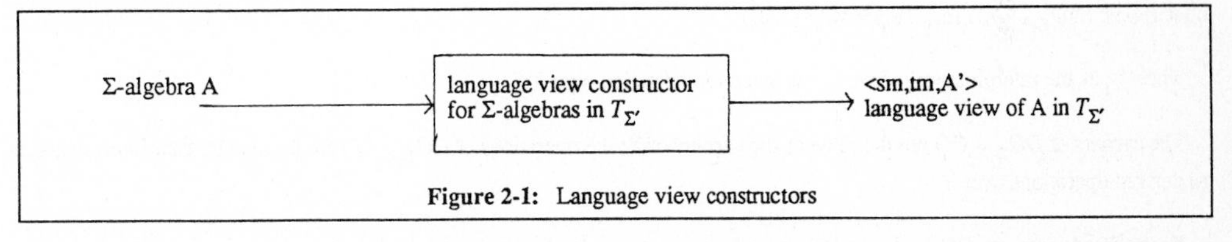

Figure 2-1: Language view constructors

Speaking loosely, a constructor "reshapes" the structure of A into that of a Σ'-algebra that correctly interprets $T_{\Sigma'}$. In this paper we will define classes of constructors reflecting generally useful "reshapings". [Bradley 85] contains many more.

If we had a Σ-algebra A and a constructor of language views for Σ-algebras in $T_{\Sigma'}$, then we could apply the constructor to get a language view of A in $T_{\Sigma'}$. Of course, it might turn out that the tm associated with the language view was not defined on all of the set of terms, I, in which we were interested. In that case we would have to try another constructor of language views for Σ-algebras in $T_{\Sigma'}$. Supposing we were able to find a constructor which produced a language view <sm,tm,A'>, where tm was defined on all of I, we could then apply another constructor, of language views of Σ'-algebras in $T_{\Sigma''}$, to A', producing a language view <sm',tm',A''> in $T_{\Sigma''}$. If tm' were defined on all $\gamma \in$ tm(δ), for all $\delta \in I$, then we could use the "composition" of these two language view constructors to define a language view from A to $T_{\Sigma''}$, namely <sm°sm',tm°tm',A''>. In this way we can define a language by giving a semantic domain and a sequence of constructors to be applied to the domain to produce a language view from the domain to T_{Σ_n} for some signature Σ_n. This property of compositionality of constructors is proved in section 2.5.

A *constructor* will be given by a triple $<sm,tm,\phi>$, where ϕ is a mapping from Σ-algebras to Σ'-algebras. A constructor always constructs a view with respect to a set of terms of interest, I, which is just the union of the ranges of all functions in tm. We will not explicitly state the I unless it is of interest. We call ϕ a *transformation*; it produces the correct interpretation of the new language -- it does the "reshaping". In the following we first describe some classes of constructors, then further expand our ability to design languages with the notion of *splitting*.

2.3. Classes of constructors: derived operations, renamings, merged operations and additions

2.3.1. Derived operations

The following description of the class of derived operation constructors assures us that we can always construct a language for some part of a semantic domain by using derived operations over the original domain as basic operations in the new language; there will be constructors which we can apply to our original domain to form a correct interpretation for the language which includes the derived operations. In terms of the example we gave earlier, it assures us that we can use simple addition expressions such as "1+1" to "program" a stack machine whose own code for this would be "push 1;push 1;add", roughly because we can interpret the operation "+" in our language of choice as meaning the derived operation "push x;push y;add" while the constant operation "1" continues to mean the integer 1. Put another way, if we want the luxury of using the expression "1+1" to program a stack machine then we must accept the need to "add" the new (derived) operation to our stack machine algebra so that we can interpret the simpler language. The previous work which we earlier contrasted with ours ([Morris 73], [ADJ 80]) develops a notion of language that can be phrased entirely in terms of this constructor.

Let $\Sigma=<S,O,F>$ be a signature. The *derived operation signature* of Σ, Σ_{DO} is $<S,O_{DO},F_{DO}>$ where $O_{DO}=T_\Sigma(X)$, and $F_{DO}(\gamma)$ gives the functionality of γ considered as a derived operation in $T_\Sigma(X)$. That is, Σ_{DO} is a new signature in which all the derived operation have been added with their correct functionality.

Let $\Sigma_1=<S_1,O_1,F_1>$ and $\Sigma_2=<S_2,O_2,F_2>$ be signatures such that Σ_2 is a subsignature of $\Sigma_{1_{DO}}$. We define the transformation $DO_{\Sigma_1\Sigma_2}(Y)$ which produces a Σ_2-algebra for any Σ_1-algebra Y by describing the Σ_2-algebra produced:

carriers: $(DO_{\Sigma_1\Sigma_2}(Y))_s = Y_s$ for all $s \in S_2$

operations: $(DO_{\Sigma_1\Sigma_2}(Y))_\gamma(e_1,...,e_n) = \gamma_Y(e_1,...,e_n)$

where γ_Y is the interpretation of $\gamma \in T_{\Sigma_1}$ as a derived operation over Y.

The carriers of $DO_{\Sigma_1\Sigma_2}(Y)$ are the same as the carriers of Y; the operations of $DO_{\Sigma_1\Sigma_2}(Y)$ are defined by their interpretations as derived operations over Y.

Proposition: $<sm,tm,DO_{\Sigma_1\Sigma_2}>$ is a language view constructor for Σ_1-algebras in T_{Σ_2}, with sm = identity on S_2, $tm_s(\sigma(e_1,...,e_n)) = \{\alpha(d_1,...,d_m) \in T_{\Sigma_2} \mid \alpha_{T_{\Sigma_1}}(d_{1_{T_{\Sigma_1}}},...,d_{m_{T_{\Sigma_1}}}) = \sigma(e_1,...,e_n)\}$

Proof: We need to show that $X_\gamma = X'_\delta$, for $\delta \in tm(\gamma)$ for all Σ_1-algebras X, and for all $\gamma \in I$, letting $X' = DO_{\Sigma_1\Sigma_2}(X)$.

Let $\gamma = \sigma(e_1,...,e_n) \in I$, and let $\delta =$ "$\alpha(d_1,...,d_m)$" $\in tm(\gamma)$. Then $X'_{\alpha(d_1,...,d_m)} = \alpha_X(d_{1_X},...,d_{m_X}) = X_{\alpha_{T_{\Sigma_1}}(d_{1_{T_{\Sigma_1}}},...,d_{m_{T_{\Sigma_1}}})} =$ $X_{\sigma(e_1,...,e_n)}$

Other strategies for forming a new language are just as natural.

2.3.2. Renaming

Any time we have a signature Σ and another signature Σ' which renames any of the operations or sorts in Σ, then for any Σ-algebra A, we can construct an interpretation for $T_{\Sigma'}$ as a language for A; we just give all the operations with new names the values of the operations with the old names in A.

Definition 3: Let $\Sigma_1 = \langle S_1, O_1, F_1 \rangle$, and $\Sigma_2 = \langle S_2, O_2, F_2 \rangle$, be signatures. Σ_2 is a *renaming* of Σ_1 iff there are

$$\tau_S : S_1 \to S_2$$
$$\tau_O : O_1 \to O_2$$

such that τ_S and τ_O are both 1-1 and onto, and

$$F_2(\tau_O(\sigma)) = \langle \tau_S(s_1)\ldots\tau_S(s_n), \tau_S(s) \rangle, \text{ for all } \sigma \in O_1, \text{ where } F_1(\sigma) = \langle s_1,\ldots,s_n, s \rangle$$

Let $\Sigma_1 = \langle S_1, O_1, F_1 \rangle$ and $\Sigma_2 = \langle S_2, O_2, F_2 \rangle$ be signatures such that Σ_2 is a renaming of Σ_1. For any Σ_1-algebra X, define the Σ_2-algebra $REN_{\Sigma_1\Sigma_2}(X) = X'$ by

carriers:
$X'_{s'} = X_s$, where $s' = \tau_S(s)$

operations:
$X'_{\sigma'} = X_\sigma$, where $\sigma' = \tau_O(\sigma)$

Proposition: $\langle sm, tm, REN_{\Sigma_1\Sigma_2} \rangle$ is a language view constructor for Σ_1-algebras in T_{Σ_2}, with sm = τ_S, $tm(\sigma) = \tau_O(\sigma)$, for all 0-ary operations σ, and $tm(\sigma(e_1,\ldots e_n)) = \tau_O(\sigma)(\delta_1,\ldots,\delta_n)$, where $\delta_i = tm(e_i)$, for all other σ.

Proof: Obvious.

2.3.3. Merging

It is easy to justify the general idea of merged operations. A machine may have operations "$+_{Int}$", "$+_{Real}$", "$+_{Longreal}$", yet it may be more convenient to have just one "merged" operations "+" in the language which is disambiguated in the context of its use in a program. There are many ways in which merging of operations could be done (i.e. there are many classes of constructors for merging). We describe a particular kind of merge below which is useful to us in the examples of the next section.

This constructor assures us that, under suitable conditions, it will be possible to use a language for a semantic domain that has an operation that "merges" together operations and sorts. We begin with a definition of merged signature.

Definition 4: Let $\Sigma_1 = \langle S_1, O_1, F_1 \rangle$ be a signature. Let O be an equivalence relation on operators in O_1, and let S be an equivalence relation on sorts in S_1, such that if $F_1(\sigma_1) = F_2(\sigma_1)$ then $(\sigma_1,\sigma_2) \notin O$, and if $\sigma_1, \sigma_2 \in O_1$ with $F_1(\sigma_1) = \langle s_1 \ldots s_n, s \rangle$, $F_1(\sigma_2) = \langle t_1 \ldots t_n, t \rangle$, and $\sigma_1 \, O \, \sigma_2$, then $s_1 \, S \, t_1, \ldots, s_n \, S \, t_n$, sSt. Now define $F : O \to S^* \times S$, as follows. If $F_1(\sigma) = \langle s_1 \ldots s_m, s_0 \rangle$, t_i is the equivalence class of s_i for all i, and $\sigma \in \{\sigma_1,\ldots,\sigma_n\} \in O$, then $F(\{\sigma_1,\ldots,\sigma_n\}) = \langle t_1 \ldots, t_m, t_0 \rangle$. Then the signature $\Sigma = \langle S, O, F \rangle$ is called a *merge* of the signature Σ_1.

Let $\Sigma_1 = \langle S_1, O_1, F_1 \rangle$ and $\Sigma_2 = \langle S_2, O_2, F_2 \rangle$ be signatures such that Σ_2 is a merge of Σ_1. For any Σ_1-algebra X, define the the Σ_2-algebra $M_{\Sigma_1\Sigma_2}(X)$ by

carriers

$(M_{\Sigma_1\Sigma_2})_s = \prod^{l}_{s' \in s} X_{s'_\lambda}$
where $X_{s'_\lambda} = X \cup \{\lambda\}$

Before we give the operators we must introduce some notation. To demonstrate it we will use this example: let $\Sigma = \langle \{Int, Color\}, \{\{+, mix\}, F \rangle$, where $F(+) = \langle IntInt, Int \rangle$ and $F(mix) = \langle ColorColor, Color \rangle$. Also, let the equivalence relations O and S be given by {Int,Color} and {+,mix}. Let A be a Σ-algebra where $3,4 \in A_{Int}$, and $red, blue \in A_{Color}$. We will write elements of a product positionally when there is no confusion in this, but update a product element, and extract components of it by name association. For example, the value of the expression "Int.<3,red>" is "3", and the value of "Color.<3,red>" is "red". Also, the value of "<3,red> [Int$\leftarrow$4]" is "<4,red>". Finally, we can update more than one component at a once; The value of "<3,red>[Int←4,Color←blue]" is "<4,blue>".

operators

Take $e_i \in M_{\Sigma_1\Sigma_2}(X)_{s_i}$, and $\sigma \in O_2, F_2(\sigma)=\langle s_1 ... s_k, s\rangle$, where $\sigma=\{\sigma_1,...,\sigma_m\}, F_1(\sigma_i)=\langle r^i{}_1 ... r^i{}_k, r^i\rangle$, $s_i=\{s^i{}_1,...,s^i{}_{n_i}\}$.

$(M_{\Sigma_1\Sigma_2}(X))_\sigma(e_1,...e_k)= \langle\lambda,...,\lambda\rangle[\langle r^i \leftarrow X_{\sigma_i}(r^i{}_1.e_1,...r^i{}_k.e_k)\rangle | r^i{}_j \neq \lambda$ for all j], if $r^i \neq r^j$ for any i,j, $\langle\lambda,...,\lambda\rangle$ otherwise

Proposition: $M_{\Sigma_1\Sigma_2}(X)$ is a language view of X w.r.t. I where $I=\{\gamma \in T_{\Sigma_1} \mid (M_{\Sigma_1\Sigma_2}(X))_\delta = X_\gamma, \delta \in tm(g'g)\}$, with the view sm(s)= equivalence class of s in S, tm(σ) = the equivalence class of σ in O, for all 0-ary operations σ, and tm("$\sigma(e_1,...e_n)$")=τ(tm(e_1),...,tm(e_n)), where τ is the equivalence class of σ in O.

Proof: Obvious from definition of I.

The carriers of the language view algebra in this case consist of tuples of elements from the merged carriers of the original algebra. The element "λ" is added as a tuple element so that some tuples will contain, for instance, just one "good" component, as in "<3,λ>. The sets of operators which constitute the operations in this new algebra are then defined so that when applied to tuple elements of carriers each operator acts on appropriate elements of the tuples. The resulting tuple reflects the joint behavior of all the operations.

2.3.4. Additions

As well as simplifying the language for a semantic domain, it is also quite natural to make a language more complex than the domain it represents. Examples of this are the use of constructs such as declarations and equivalence statements in programming languages. Typically, there are no declaration or aliasing operations in the machine code to which these are translated, rather declarations and aliasing operations are an additional language operations whose meaning is fully determined during compilation. In [Bradley 85] we describe several families of constructors of this sort.

2.3.5. Splittings

All of the above constructors reshape one language for a domain into another. We will now expand our abilities to define languages to include the case that two languages can be used *jointly* to represent a semantic domain. The reader should think about this as follows. In many situations in computer science, unlike classical linguistics or mathematics, either because the semantic domains are so complex that they can not be feasibly represented with one language, or, because time of usage is important, multiple languages are used to represent a domain. In any one of these languages, the meaning of an expression is not an object of ultimate concern; instead, the meanings of language expressions are "associated" with the meanings of *other* language expressions. For example: A program P, (*source language*) is compiled into P_{code} (*machine language*). An expression "run P_{code}, input=P_{input}, output=P_{output}" (*operating system language*) is then interpreted. This interpretation can be viewed as the application of P_{code} to an expression in the *language of input files*, P_{input}. This produces an expression *in the language of output files*, P_{output}. The operating system may then supply P_{output} as the input to a printer process which eventually produces paper with ink on it, which *is* the object of our ultimate concern.

Splitting then, is the technique of design that we will use to produce and coordinate the various languages we need to express objects of ultimate concern. Our immediate concern here is not with all of computing (linkers, loaders, printers, etc.) but more modestly with the use of splitting to produce interesting language constructs, as in the examples of the next section.

We introduce here only a simple notion of splitting, though [Bradley 85] and [Bradley 87] contain a more complete version. The idea is to remove information from one language and represent it in an associated language. We will use splitting here to induce a notion of time of evaluation. For example we will design a language which has "dynamic" arrays for a domain which has only "static" arrays by splitting bounds information from language expressions into a separate language. In general, neither language resulting from a split will be a complete representation since the values of expressions in neither language will be in the original domain.

The examples of the next section should make this clearer, but we can give a simple example here. Suppose one began with the usual language for representing the addition of two numbers, namely the language with expressions such as "1+1". Now, suppose that instead of using this language we decided to represent the addition of two numbers by two languages jointly. One would contain expressions such as "1+x" while the other contained expressions such as "1". That is, one language contains expressions that denote unary functions which add some constant to their arguments, while the other language contains the arguments. Two points are important here. First, perhaps obviously, this is a perfectly adequate scheme of representation. Provided that we have expressions in each of the two languages, we have a clear representation of the addition of two numbers. Nothing about it inherently requires a single language of representation. Secondly, while the values of expressions in our original, simple language for addition were integers, the values of expressions in one of our new languages are not; they are functions. These expressions can not be translated into equivalent expressions in the original language. This miniature example demonstrates what we mean by "removing information from one language and representing it in an associated language". We now treat splitting more formally.

Suppose Σ=<S,O,F> is a signature, X some set of variables, $\{x_1,...,x_n\}$, and A is a Σ-algebra. A *splitting* of T_Σ as a language for A is given by $s \in \Sigma$. The languages resulting from the splitting are

$$L_1 = T_{\Sigma_x}$$
$$L_2 = \Pi(T_{\Sigma_s})$$

where Σ_x=<S_x,F_x,O_x> is the signature given by

$S_x = S$
$O_x = O \cup X$
$F_x(\sigma)$ = if $\sigma \neq x_i$ for any i, then $F(\sigma)$ else $<\lambda,s>$

and $\Pi(T_{\Sigma_s})$ is the set of all tuples of elements from the carriers T_{Σ_s}.

In Σ_x we have added a set of 0-ary operations of sort s. (We assume that none of the original operations has the name "x_i" for any i.) Terms involving the x's are to be interpreted as derived operations over A. A term involving $x_{i_1},..., x_{i_k}$, where $x_{i_a} \leq x_{i_b}$ for $a \leq b$, denotes a function which takes as argument an i_k-tuple of values from A_s and produces the value gotten by substituting the jth tuple component for x_{i_j} wherever it appears in the term. The meanings of expressions in L_2 are tuples of elements in A_s. The *evaluated meaning* of any (γ_1,γ_2) for $\gamma_1 \in L_1$, $\gamma_2 \in L_2$ is given by the meaning of γ_1 applied to the meaning of γ_2, if applicable, undefined otherwise.

2.4. Summary of language design

We have described various ways of constructing a language for a semantic domain. A ***language design*** is given by

$$(A,s,C_1,...,C_n)$$

where A is a semantic domain, s gives a splitting and $C_1,...,C_n$ are constructors that will be applied to the language L_1 of the splitting. Either the splitting or the sequence of constructors may be omitted.

The reader should note that this description of language design could easily be made more general. We could, for example allow more than one opportunity for splitting, and a more general notion of each instance of splitting; and, of course, we could present more constructors than we have given here. (See [Bradley 85] and [Bradley 87].) Our goal is to introduce this method of language definition by design and to show its usefulness. We are limiting ourselves to a restricted setting only because the basic ideas are easier to present in this setting, not because they are limited to this setting.

2.5. Derivation of semantics with times of evaluation

In the above we showed how to define a language by giving a ***design*** for it that consists of a sequence of transformations on a starting semantic algebra. These designs can now be used to derive various semantics for the language each of which reflects a different arrangement of times of evaluation. The following results capture this observation.

We begin with a theorem that proves that sequences of constructors actually do define a language, because transformations are composable. After the theorem we describe how various sorts of semantics can be derived from a definition by design.

Definition 5: Composition of constructors. Suppose $\Sigma_1, \Sigma_2, \Sigma_3$ are signatures, $C_1 = \langle sm_1, tm_1, \tau_1 \rangle$, $C_2 = \langle sm_2, tm_2, \tau_2 \rangle$, are constructors such that $\tau_i : \Sigma_i$-algebras $\rightarrow \Sigma_{i+1}$-algebras, for i=1,2. We define the *composition* of C_1 and C_2, $C_2 \circ C_1$, by $\langle sm, tm, \tau \rangle$ where

$sm(s) = sm_2(sm_1(s))$, if defined, undefined otherwise,

$tm_s(\gamma) = \{\delta \in T_{\Sigma_3} \mid \delta \in tm_{2_{sm_1(s)}}(\alpha)$, for $\alpha \in tm_{1_s}(\gamma)\}$

$\tau(A) = \tau_2(\tau_1(A))$, for Σ_1 algebra A

Theorem 1: Suppose $\Sigma_1, \Sigma_2, \Sigma_3$ are signatures, $C_1 = \langle sm_1, tm_1, \tau_1 \rangle$, $C_2 = \langle sm_2, tm_2, \tau_2 \rangle$, are constructors such that $\tau_i : \Sigma_i$-algebras $\rightarrow \Sigma_{i+1}$-algebras, for i=1,2, as in the above definition. Then $C_2 \circ C_1$ is a constructor for Σ_1-algebras in T_{Σ_3}, w.r.t. $I = \{\delta \in T_{\Sigma_3} \mid \delta \in tm_{2_{sm_1(s)}}(\alpha)$, for $\alpha \in tm_{1_s}(\gamma)$, $\gamma \in T_{\Sigma_{1_s}}$, s a sort of $\Sigma_1\}$

Proof:

Let $C_2 \circ C_1 = \langle sm, tm, \tau \rangle$ as in the above definition. We will show that, for any Σ_1-algebra A, $\langle sm, tm, \tau(A) \rangle$ is a language view of A w.r.t. I as given.

We need to show that $A_\gamma = \tau(A)_\delta$, for each $\gamma \in T_{\Sigma_{1_s}}$, $\delta \in tm_s(\gamma)$.

Since C_1 is a constructor for Σ_1-algebras in T_{Σ_2}, we have that $A_\gamma = [\tau_1(A)]_\alpha$, for $\alpha \in tm_{1_s}(\gamma)$. Since C_2 is a constructor for Σ_2-algebras in T_{Σ_3}, we have that $[\tau_1(A)]_\alpha = [\tau_2(\tau_1(A))]_\delta$, for each $\delta \in tm_{2_{sm_1(s)}}(\alpha)$. And so, $A_\gamma = [\tau(A)]_\delta$, for each $\gamma \in$ I.

Done.

We now describe the usefulness of definitions by design for deriving various semantics for a language. In a design, $\langle A, s, C_1 \ldots C_n \rangle$, the sequence of constructors could be applied to all Σ_x-algebras. In particular, they can be applied to T_{Σ_x}. The result of this application is an algebra of interpretation for the defined language in which the meanings of operations are *compiling actions* for the translation of language expressions to T_{Σ_x} expressions. Further, we can apply subsequences of these constructors to T_{Σ_x} to define an algebra of compiling actions for the translation of the defined language to languages representing intermediate stages of evaluation. We sum this up in the following observation.

Let $\Sigma_0, \ldots, \Sigma_n$ be signatures, and let A be a Σ_0-algebra, and $C_1, \ldots, C_n$ be constructors, where, C_i is a language view constructor, $\langle sm, tm, \tau_i \rangle$, $\tau_i : \Sigma_{i-1}$-algebras $\rightarrow \Sigma_i$-algebras, for $0 \leq i \leq n$, and s is a sort name in Σ_0. Then, $\tau_n(\tau_{n-1}(\ldots(\tau_1(A))\ldots))$ gives a denotational semantics for T_{Σ_n}, and $\tau_n(\tau_{n-1}(\ldots(\tau_1(T_{\Sigma_1}))\ldots))$ gives a translation of T_{Σ_n} to T_{Σ_1}. The natural interpretation of the meanings assigned to operations in this interpretation of T_{Σ_n} is that they are the compiling actions necessary for the translation of T_{Σ_n} to T_{Σ_1}. Further, each of $\tau_n(T_{\Sigma_{n-1}})$, $\tau_n(T_{\Sigma_{n-2}})$, ..., gives a translation for T_{Σ_n} into the intermediate languages $T_{\Sigma_{n-1}}$, $T_{\Sigma_{n-2}}$ and so on, each of which have the following interpretations as languages for A: $\tau_{n-1}(\ldots(\tau_1(A)\ldots))$, $\tau_{n-2}(\ldots(\tau_1(A)\ldots))$, ...

The implication of this result is that once we have given a definition of a language by design, we can use this definition to derive various semantics for a language. One of the semantics will be a traditional denotational semantics, wherein all language operations are assigned meanings in one evaluation step. But we will also be able to derive semantics in which the original language expression is translated into expressions in other languages. In each of these translation steps the structures of the source and target languages are different. We will discuss this observation further in the context of specific examples in the next section.

3. Examples of Language Design

In this section we present language design as described above as a method of language definition and compiler specification. In section 1 above, we outlined reasons why times of evaluation are important when describing programming languages. Recall that a compiler is inappropriately described as a homomorphism because this ignores the important aspect of language structure that some phrases of language expressions are intended to be evaluated *during* compilation, rather than be mapped homomorphically to target code. Also recall that some notion of time of evaluation would be useful to the language designer when introducing constructs into a language. In section 2 we described a method of defining languages by giving designs for them, and observed that one can derive a variety of semantics and translations from a design. In this section we give two examples of language design that connect formal designs with times of evaluation more vividly, although these examples are too brief to bring out the full flavor of the definitional technique described above. Rather, they are the simplest possible introduction. Once again, [Bradley 85] upon which this is based, contains more details.

3.1. Array example

We will describe a very simple algebra for arrays. We will then design three languages for this algebra. In our language designs we will freely use the fact that constructors are composable as shown in the previous section.

The Σ-algebra A is given in figure 3-1. We do not show the 0-ary operations on the sort Int, but we assume that we have available some range of integers.

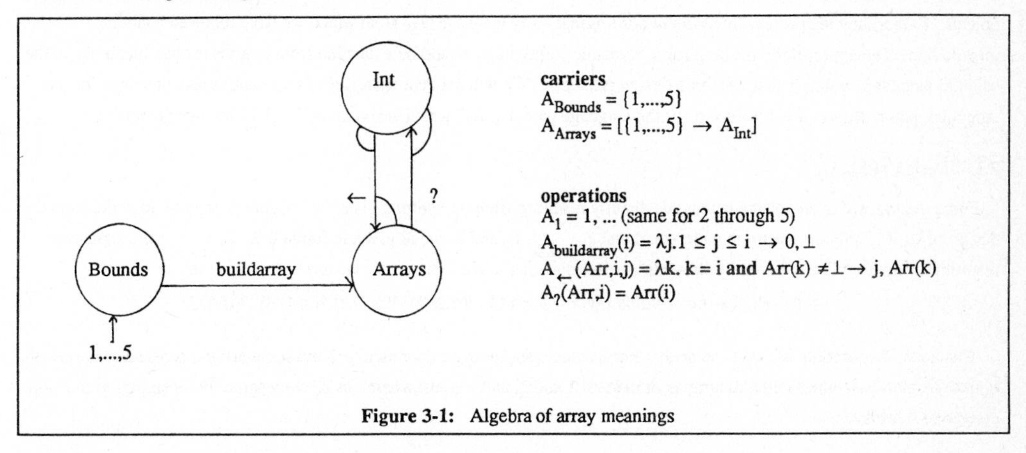

Figure 3-1: Algebra of array meanings

In the algebra we can create array objects with lower bound 0 and upper bound from 1 to 5 (buildarray), update them (←), and enquire for the value at some index value (?).

To understand the differences between the languages we will design for this algebra it is most instructive to see sample expressions from each language (complete definitions follow). For languages which come in pairs, always think of the unprimed language as the "programming" language, and the primed language as its associated language of inputs. In L_1 we have an expression for enquiring about the value of an array of size 5 at position 4 after updating it at four positions with various values. In L_2 we have an expression for the same computation, except that now the value of the size of the array has been removed from the language expression and placed in another language L_2'. In L_3 we have an expression for the same computation except that in this language we allow arrays to be resized with the operation "shrink". L_3' thus contains not only the value of the original bound for the array, but a second value as well.

L_1:
```
?(4,
←(3,91,
←(5,43,
←(2,19,
←(1,82,
buildarray(5))))))
```

L_2:
```
?(4,
←(3,91,
←(5,43,
←(2,19,
←(1,82,
buildarray(x1))))))
```

L_2':
```
5
```

L_3:
```
?(4,
shrink(
←(3,91,
←(5,43,
←(2,19,
←(1,82,
buildarray(x1)))))))
```

L_3':
```
5 4
```

L_1 is simply the natural language for the algebra. (L_2, L_2') is a pair of languages where L_2 allows arrays to be "dynamic" (the actual bound for buildarray is in the input language). (L_3, L_3') is a pair of languages where L_3 allows arrays to be dynamic *and* to get new bounds during execution with "shrink".

We will now give the designs for each language. We will then describe each of the designs in detail.

L_1 : (A)
L_2, L_2' : (A, bounds)
L_3, L_3' : (A, bounds, $DO_{\Sigma_1\Sigma_2}$, $REN_{\Sigma_2\Sigma_3}$)

3.1.1. Design of L_1

L_1 is defined by the trivial design which contains no design steps. So L_1 is simply T_Σ.

3.1.2. Design of L_2, L_2'

The language pair L_2, L_2', is defined by a design consisting of a single step which is a split. This captures our desire to have a language in which arrays can be "dynamic". That is, the expressions in L_2 need not contain full information about array objects. Rather they may represent terms in our original array language (T_Σ) from which we have removed bounds information. The appropriate meaning for a "dynamic" array in L_2 should be a function from bounds to array meanings in the original language, which is just what this design produces. We will not give examples of expressions and meanings for this language, but in section 3.1.4 we do so for the language pair (L_3, L_3') which is similar to (L_2, L_2') but more complex.

3.1.3. Design of L_3, L_3'

First note that L_3' is one of the languages directly resulting from the splitting given by "bounds". In order to understand the design of L_3 it is only necessary to understand that $\Sigma_{1_{bounds}}$, Σ_2 and Σ_3 are as given in figure 3-2. $\Sigma_{1_{bounds}}$ is the signature resulting from the splitting given by "bounds". Also, note that in the signature Σ_2 we use "σ" to stand for

$$\leftarrow(\leftarrow(\leftarrow(\leftarrow(\leftarrow(\text{buildarray}(x_1),?(\alpha,1),1),?(\alpha,2),2),?(\alpha,3),3),?(\alpha,4),4),?(\alpha,5),5)$$

This derived operation "shrinks" an array. For instance, applying the operation to 2 and some array α produces an array of bound 2 which has values equal to array α at indexes 1 and 2, and is $\perp$ elsewhere. In Σ_3 we rename this operation to the more congenial "shrink".

This coupled with our discussion on the classes of constructors given above, fully explicates the design of L_3 and L_3'. To see this recall, that, for example, since Σ_2 differs from $\Sigma_{1_{bounds}}$ only in that a derived operation has been added to the signature, we know from the results of the previous section that there is a constructor, which we will call $DO_{\Sigma_1\Sigma_2}$, which we can apply to A to produce a language view of it in Σ_2. (Note that, by our notational conventions, we should name this constructor "$DO_{\Sigma_{1_{bounds}}\Sigma_2}$". We have given it the shorter name here since it is shorter and there can be no confusion as to which class of algebras it applies to.) Similarly for Σ_3: it is a renaming of Σ_2.

3.1.4. Analysis of languages

In this section we apply the results of section 2 to the designs of array languages above and show that we can derive from a single design a variety of semantics and translations for the language with "dynamic arrays" (L_3 and L_3').

How do our definitions by design for languages for the array algebra capture the idea of stages of evaluation, and what new information about the languages do they provide us with? Consider figures 3-3 and 3-4. These figures show how we can derive both "interpretation" and "translation" semantics for the array language L_3. In each diagram begin at the upper left-hand corner. Note that A′ is the algebra that gives the interpretation for language L_1 resulting from the split. In figure 3-3 the notation "$A'/T_{\Sigma_{1_{bounds}}}$" indicates that as we move down the left-hand side of the diagram we will be applying constructors to the $\Sigma_{1_{bounds}}$-algebra, A′, and as we move across the top of the diagram we will be applying the same constructors to the $\Sigma_{1_{bounds}}$-algebra $T_{\Sigma_{1_{bounds}}}$. Recall that this is a signal aspect of constructors as we have defined them; they produce language views of *all* $\Sigma_{1_{bounds}}$-algebras. The effect of applying $DO_{\Sigma_1\Sigma_2}$ to A′ and $T_{\Sigma_{1_{bounds}}}$ is to produce the Σ_2-algebras $DO_{\Sigma_1\Sigma_2}(A')$

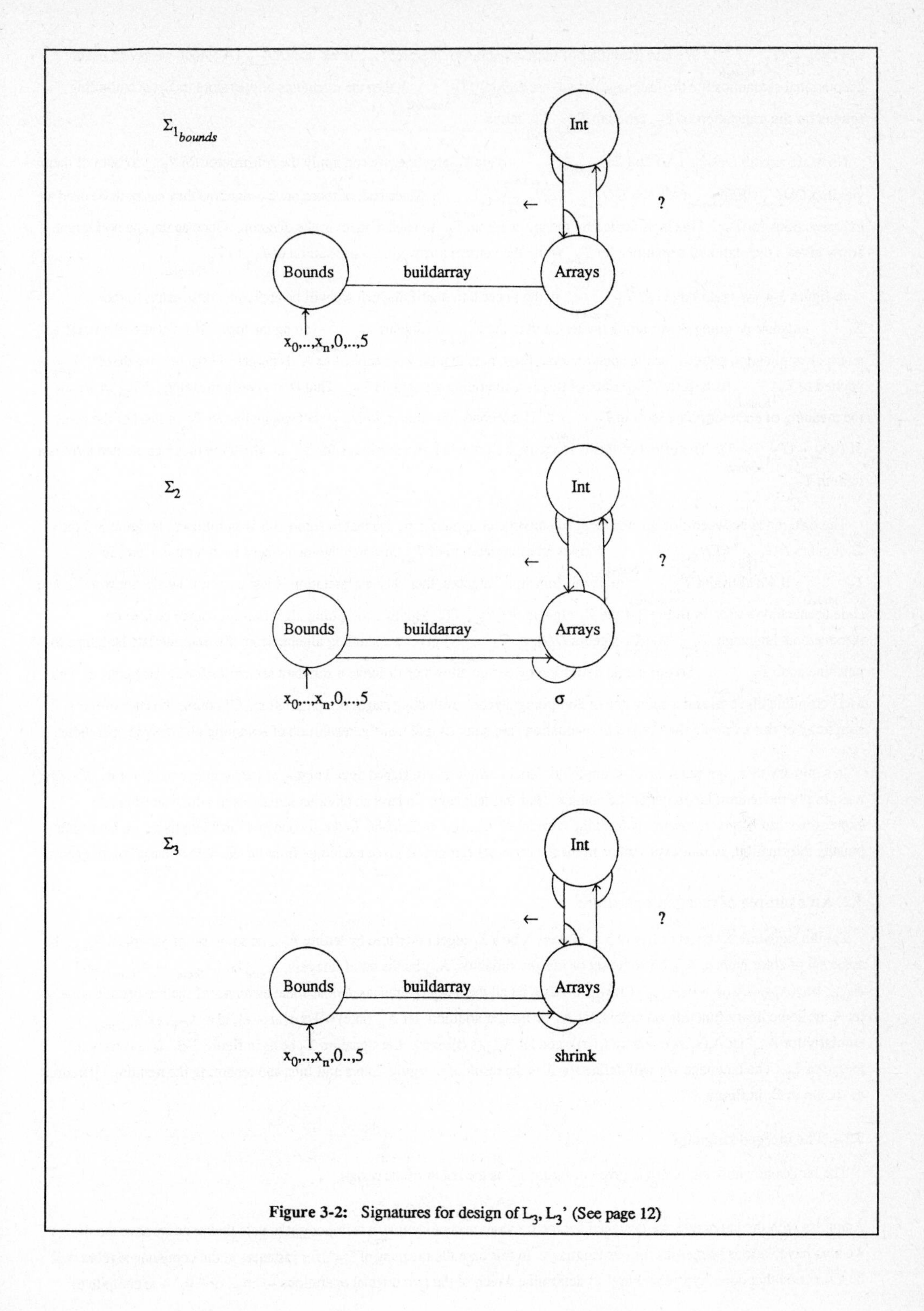

Figure 3-2: Signatures for design of L_3, L_3' (See page 12)

and $DO_{\Sigma_1\Sigma_2}(T_{\Sigma_{1_{bounds}}})$. We can use either of these algebras to interpret T_{Σ_2}. If we use $DO_{\Sigma_1\Sigma_2}(A')$, then we have a usual denotational semantics for the language. If we use $DO_{\Sigma_1\Sigma_2}(T_{\Sigma_{1_{bounds}}})$, then the meanings of operators in Σ_2 are compiling actions for the translation of T_{Σ_2} terms to $T_{\Sigma_{1_{bounds}}}$ terms.

Next, since both $DO_{\Sigma_1\Sigma_2}(A')$ and $DO_{\Sigma_1\Sigma_2}(T_{\Sigma_{1_{bounds}}})$ are Σ_2-algebras, we can apply the constructor $REN_{\Sigma_2\Sigma_3}$ to both of them yielding $DO_{\Sigma_1\Sigma_2}{}^{\circ}REN_{\Sigma_2\Sigma_3}(A')$ and $DO_{\Sigma_1\Sigma_2}{}^{\circ}REN_{\Sigma_2\Sigma_3}(T_{\Sigma_{1_{bounds}}})$. Since both of these are Σ_3-algebras they can both be used as interpretations for T_{Σ_3}. This is indicated by the arrows from T_{Σ_3} to each of them in the diagram. Once again, one horizontal arrow gives a denotational semantics for T_{Σ_3} while the vertical arrow gives a translation of T_{Σ_3} to $T_{\Sigma_{1_{bounds}}}$.

In figure 3-4 we again have "$A'/T_{\Sigma_{1_{bounds}}}$" in the upper left, signifying that we will be applying constructors to the $\Sigma_{1_{bounds}}$-algebra A′ going down along the left, and to the $\Sigma_{1_{bounds}}$-algebra $T_{\Sigma_{1_{bounds}}}$ along the top. This diagram illustrates an alternate application pattern for the constructors. First, note that they are applied to A′ as before. Next, observe that $DO_{\Sigma_1\Sigma_2}$ is applied to $T_{\Sigma_{1_{bounds}}}$ as before. The result of this is a compiling algebra for T_{Σ_2}. That is, it gives a meaning for T_{Σ_2} in which the meaning of each term is a term in $T_{\Sigma_{1_{bounds}}}$. The second constructor, $REN_{\Sigma_2\Sigma_3}$ is then applied to T_{Σ_2} instead of the result of $DO_{\Sigma_1\Sigma_2}(T_{\Sigma_{1_{bounds}}})$. The effect of this is to create a Σ_3-algebra as a semantics for T_{Σ_3} in which the meaning of each term is a term in T_{Σ_2}.

The difference between this arrangement of constructor applications and that in figure 3-3 is as follows. In figure 3-3 the Σ_3-algebra $DO_{\Sigma_1\Sigma_2}{}^{\circ}REN_{\Sigma_2\Sigma_3}(T_{\Sigma_{1_{bounds}}})$ gives an interpretation of T_{Σ_3} in which the meaning of each term is a term in $T_{\Sigma_{1_{bounds}}}$. If we consider $T_{\Sigma_{1_{bounds}}}$ to be the "machine" algebra, then this interpretation is just a compiling algebra from source code to machine code. In figure 3-4 the Σ_3-algebra, $REN_{\Sigma_2\Sigma_3}(T_{\Sigma_2})$ gives a compiling algebra from source code to the *intermediate* language, T_{Σ_2}. The Σ_2-algebra $DO_{\Sigma_1\Sigma_2}(T_{\Sigma_{1_{bounds}}})$ gives a compiling algebra from this intermediate language to machine code $T_{\Sigma_{1_{bounds}}}$. So our single definition by design allows us to derive a standard semantics for the language, as well as a compiling algebra, and a sequence of compiling algebras reflecting stages of compilation. Of course, because of the simplicity of this example the "stages of evaluation" are quite simple namely, resolution of renaming and derived operations.

In summary then, we began with a simple domain of arrays and designed three languages (or language pairs) for it. The first was simply the natural language of the domain. For this language we have an obvious semantics in which all language expressions can be given meanings from the carriers of the original domain. In the second and third languages we have split bounds information, so that expressions in these languages can not be given meanings from the carriers of the original algebra.

3.2. An example of merged operators

Let the signature Σ_1 be as in figure 3-5, and let A be a Σ_1-algebra defined by letting A_{CV} be some set of variables, A_{Color} be some set of color names, A_{IV} be some set of integer variables, A_{Int} be the set of integers, A_{Stmt} be $[A_{Store} \rightarrow A_{Store}]$, and A_{Store} be $[(A_{CV} \cup A_{IV}) \rightarrow (A_{Color} \cup A_{Int})]$. Also, let all the 0-ary operations denote some elements of their respective sorts, let A_f be some unary function on colors, let A_+ be integer addition, let $A_{:=_C}(id,e) = \lambda\sigma.\sigma[id \leftarrow e]$, $id \in A_{CV}$, $e \in A_{Color}$, similarly for $A_{:=_I}$, let $A_;(s_1,s_2) = \lambda\sigma.s_2(s_1(\sigma))$, and let $A_{sval}(s,\sigma) = s(\sigma)$. Let signature Σ_2 be as in figure 3-6. Σ_2 is clearly a merge of Σ_1. The language we will define for A is the result of merging Colors and Ints, and renaming the resulting operations as shown in Σ_3 in figure 3-7.

3.2.1. The merged language

The language whose signature is given in figure 3-7 is the result of the design

$$<A, M_{\Sigma_1\Sigma_2}, REN_{\Sigma_2\Sigma_3}>.$$

Again, as with the languages we designed for the previous array example, it is important to note that once we have the design we also have various semantics for our language. In this case the meaning of ":=", for instance, in the compiling algebra will be a function that does "type checking" to determine which of the two original operations -- "$:=_C$" or "$:=_I$" -- to compile to.

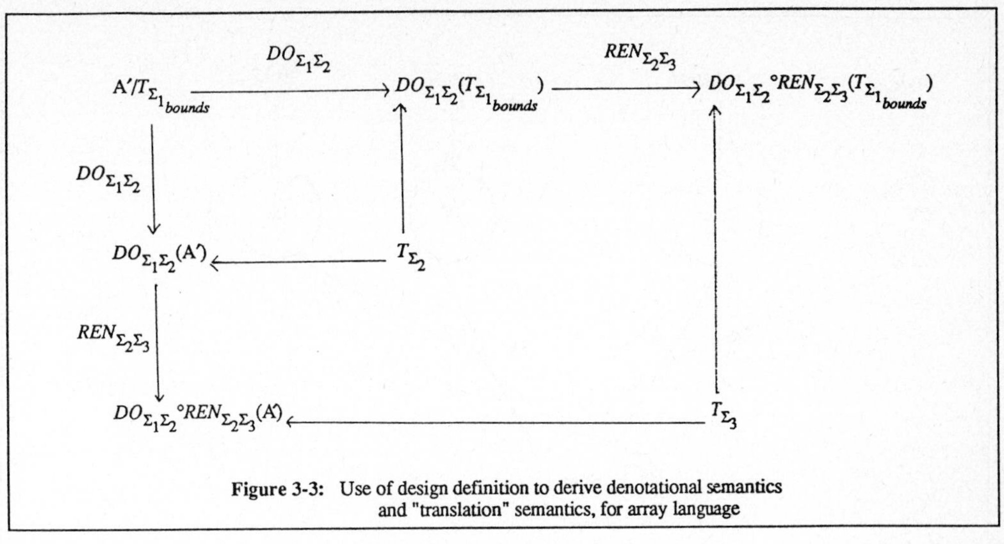

Figure 3-3: Use of design definition to derive denotational semantics and "translation" semantics, for array language

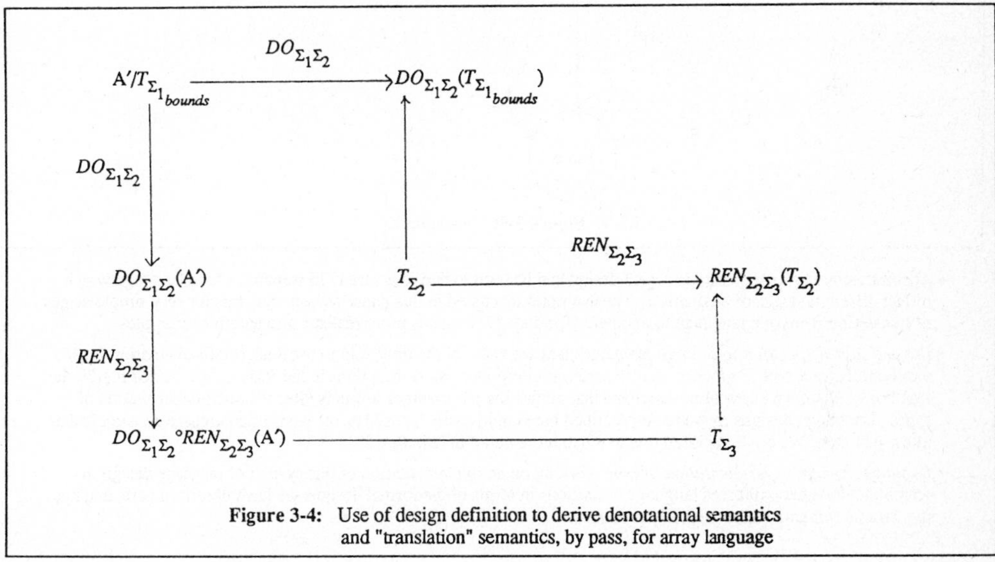

Figure 3-4: Use of design definition to derive denotational semantics and "translation" semantics, by pass, for array language

4. Conclusions; Future and Related Work

We have described the importance of the notion of times of meaning for languages which specify computations to be done in stages. Also, we have given the mathematical foundation for a method of definition of such languages from which various sorts of semantics can be derived which embody different notions of the stages of evaluation of the language. We have given several simple examples of this style of definition.

4.1. Uses for the designs

Formal designs of languages have a variety of uses:

- *Definitions.* Designs are definitions. They precisely constrain the form and meaning of language expressions.

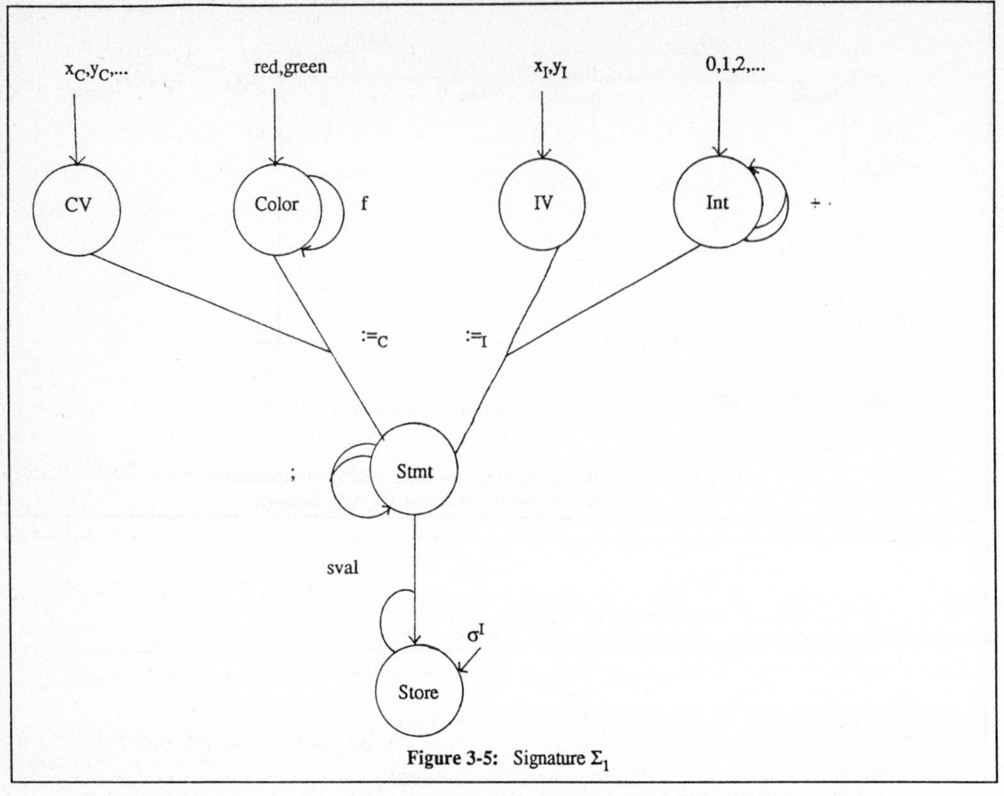

Figure 3-5: Signature Σ_1

- *Abstract compiler specifications.* Each design can be used to derive a variety of semantics for a language which reflect different stages of evaluation. The examples discussed in this paper demonstrate this for very simple stages of translation from one language to another. [Bradley 87] contains more realistic and lengthier examples.
- *I/O and literal specification.* In all languages that we know of the only way to use a user-defined notation for I/O expressions for a type is to write an interpreter/uninterpreter pair of functions in the form of specialized read/write functions. Also, we know of *no* language that allows the programmer to freely alter the notation for literals of types. Language designs as we have described here could easily be used as the basis of a language constructs that allow just this. We have explored this in much more detail in [Bradley 87].
- *Language design experimentation* We are working on an implementation of this system of language design in which the designer expresses language definitions in terms of the formal designs we have described here, and has the various semantics automatically produced.

4.2. Related work

Citing only representative papers, much work has been done on the topics of language definition ([Schoenfield 67], [ADJ 77]), use of formal definitions to implement a language([Mosses 79], Paulson 82]), the mathematical structure of a compiler([Morris 73], [ADJ 80]), and the representation in a semantics of stages of compilation as well as compile time and run time distinctions. ([Jorring 86], [Nielson 86], [Lee 87]).

Concerning language definition, we follow a markedly approach than previous work. We believe that programming languages are very different from the languages of traditional mathematics and philosophy, and that the approach to definition used in those fields is simply not informative enough for programming languages. Programming languages are routinely translated into languages of radically different algebraic structure. As an artifact of this, programming language expressions incorporate phrases that are intended to be *evaluated* by the translator, rather than translated. Also, programming languages exist in a rich environment of other languages each of which specifies a different aspect of the total computation. Finally, programming languages are *designed* and must represent algorithms *efficiently*. All of these aspects are strikingly unique.

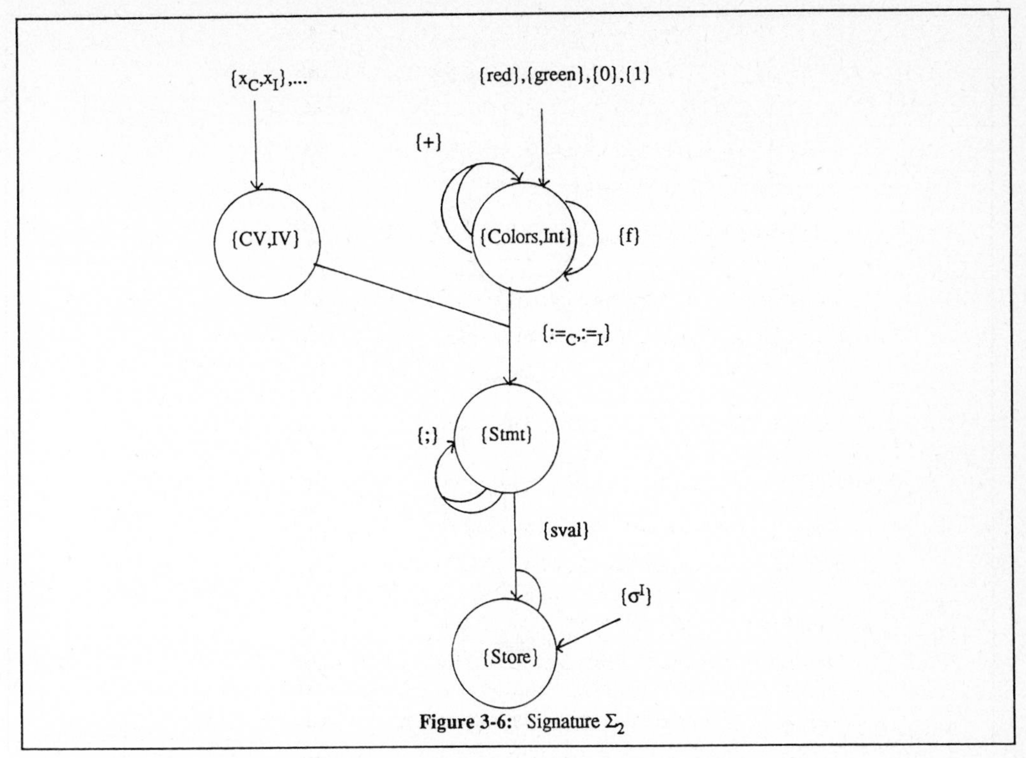

Figure 3-6: Signature Σ_2

They call for a style of definition in which properties of a design are manifest and redesign, and comparisons of designs, are simple. These have been the aims of the style of definition we have presented here.

Also, we believe that much previous work on the mathematical nature of compilation has actually been descriptive of a class of languages without the complexities of programming languages; namely the traditional languages of mathematics and philosophy. Compilation is not well described as a homomorphic mapping between source and target languages. We feel that any accurate model of it must account for the stages of evaluation which are so crucial in programming language translation and interpretation.

The use of formal design, as described here, to implement a language shows promise chiefly because it provides such a flexible mechanism for representation of stages of compilation and compile time and run time distinctions. Loosely, it contrasts with [Jorring 86] in that they apply program transformations to a denotational definition to obtain passes, whereas we begin with a semantic domain and then apply constructors, yielding a language definition with both the denotational and by-pass definitions derivable.

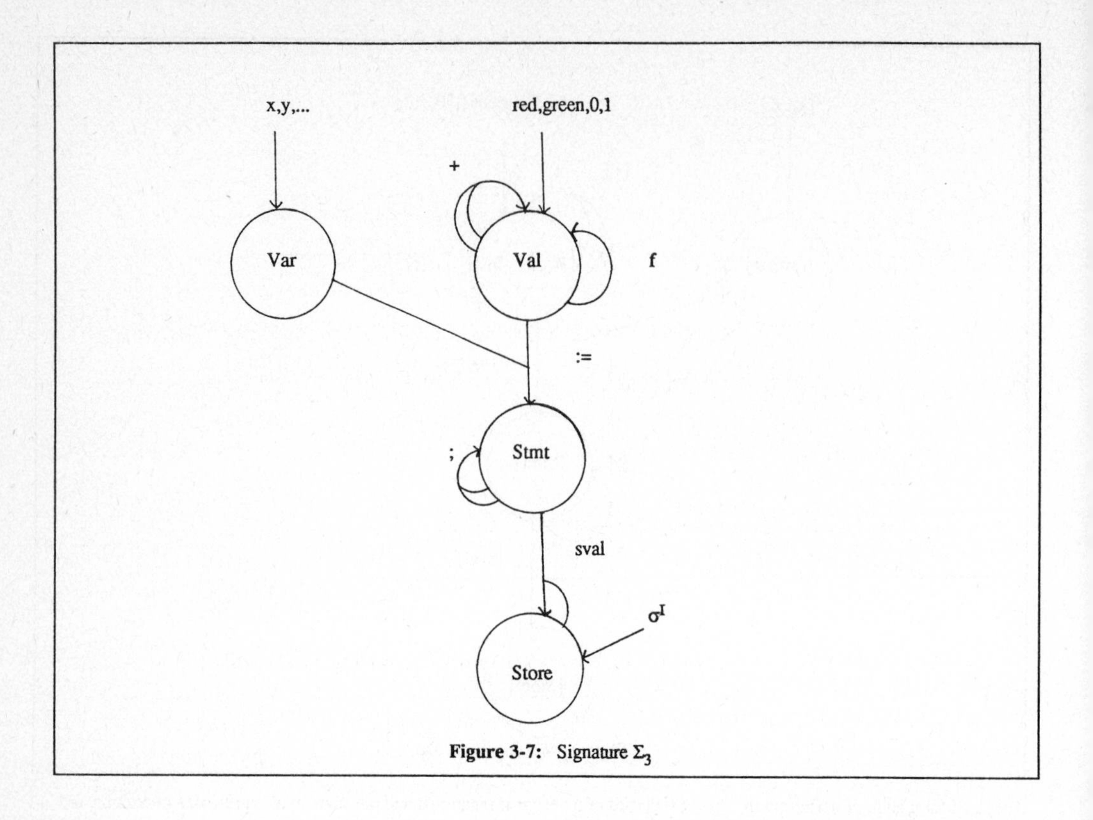

Figure 3-7: Signature Σ_3

Bibliography

[ADJ 77] Thatcher, J.W., Wagner, E.G., Wright, J.B., "Initial algebra semantics and continuous algebras", JACM Vol 24, No 1, January 1977, p 68-95.

[ADJ 80] Thatcher, J.W., Wagner, E.G., Wright, J.B., "More Advice on Structuring Compilers and Proving Them Correct", In: *Semantics Directed Compiler Generation*, N. Jones editor, Springer-Verlag, 1980, p. 165-189.

[Bradley 85] Bradley, Laurette, *A Study of Language Representation of Semantic Domains with Applications to Language Design and Definition*, Dissertation, University of Michigan, 1985.

[Bradley 87] Bradley, Laurette, "Timely Semantics: Definitions for Languages in Complex Settings", in preparation.

[Carnap 47] Carnap, Rudolf. *Meaning and Necessity*. University of Chicago, 1947.

[Frege 92] Frege, Gottlob. Ueber Sinn und Bedeutung in *Zeitshr. fur Philos. und philos Kritik.* 100 (new ser. 1892) 25-50. English translation in *Translations from the Philosophical Writings*, translated by P. Geach and M. Black, Oxford, 1952.

[Jorring 86] Jorring, Ulrik and Scherlis, William. "Compilers and Staging Transformations" Thirteenth Symposium on Principles of Programming Languages, pp 86-96, 1986.

[Lee 87] Lee, Peter and Pleban, Uwe. "A Realistic Compiler Generator Based on High-Level Semantics", Fourteenth Symposium on Principles of Programming Languages, pp 284-295, 1987.

[LLS 68] Lucas, P., Lauer, P, Stigleitner, H., "Method and Notation for the Formal Definition of Programming Languages", Technical Report 25.087, IBM Laboratories, Vienna Austria.

[Montague 74] Montague, Richard. *Formal Philosophy*, Yale University, 1974.

[Morris 73] Morris, F.L., "Advice on Structuring Compilers and Proving Them Correct", Proceedings ACM Symposium on Principles of Programming Languages, 1973, p. 144-152.

[Mosses 79] Mosses, P., "SIS-reference manual and user's guide", DAIMI MD-30, Aarhus University, Aarhus, Denmark.

[Nielson 86] Nielson, H. R. and Nielson F. "Semantics directed compiling for functional languages", Proc. 1986 ACM Conf. LISP and Functional Programming, p. 249-257. DAIMI MD-30, Aarhus University, Aarhus, Denmark.

[Paulson 82] Paulson, L., *A semantics-directed compiler generator*, Ninth Symposium on Principles of Programming Languages, pp 224-233, 1982.

[Schoenfield 67] Schoenfield, Joseph R. *Mathematical Logic*, Addison-Wesley, 1967.

Operational semantics and a distributed implementation of CSP

Gerlinde Schreiber (*)

Abstract

In this paper an operational semantics for Hoare's CSP is presented. This semantics definition is used to develop an interpreter for CSP in a distributed environment. The correctness of this interpreter is proved.

Introduction

To increase the efficiency of computers, developments in computer architecture suggest the use of multiprocessor systems instead of monoprocessor machines. There are two fundamentally different ways to work with such a multiprocessor system:

- hide the architecture of the machine from the user (e.g. by a multiprogrammed operating system)
- leave the use of the multiprocessor to the programer by providing special parallel programming languages.

This work was developed at the University of Kiel, West Germany, and was partially supported by the DFG.

(*) The author is employed at Siemens AG, E STE 366, 8520 Erlangen, West Germany.

In his paper "Communicating Sequential Processes" [HO87] Hoare has made a widely regarded proposal for the main features of such a parallel language. CSP is designed for a system of similar prosessors, each of them working on its own private memory. The processors are connected by named channels. CSP offers a parallel command for the creation of processes: ||<process>$_1$..<process>$_n$ denotes the generation of n processes that are to be executed in parallel on different processors. As the processes can't communicate by shared variables, Hoare introduces special input and output commands. In these commands a channel connecting the two communicating processes is referenced: <channelid>?<varid> means the assignment of the value presented on the channel <channelid> to the variable <varid>. Besides the exchange of data between two processes, the communication commands support their synchronisation: an input command in one process and an output command in another process concerning the same channel are executed at the same time. The execution of one communication command is delayed until a corresponding communication command is ready. The attempt to communicate with a terminated process ends in a deadlock.
Another main feature of CSP are nondeterministic commands:
□<guard>$_1$; <commandlist>$_1$..<guard>$_n$; <commandlist>$_n$ denotes the nondeterministic choice between n alternatives that is controlled by guards. These guards are constructed of a boolean expression and a communication command. Only such an alternative can be chosen whose boolean guard evaluates to true and whose communication guard can be executed. If all guards fail (i.e. the boolean guards evaluate to false or the communication partners are terminated) the nondeterministic command fails.

Up to now a lot of work has been done in specifying the semantics of CSP [HO80], [HO81], [HBR81], [CH81], [PL83], [OH84]. A few CSP-like parallel languages have been developed and implemented [FS81], [MS84], [RE83], the most famous of them being OCCAM®. Of these implementations, just one [RE83] is based on a formal semantics description, namely on the model of the interleaving semantics by Hoare and Olderog [OH84]. This model uses as one component of the semantics definition the *observation of program-execution*. In a parallel program many observable actions are executed at the same time, but at one time just one observation can be made. In the model, the parallel actions are represented by a sequence of observations made in arbitrary order. Thus parallelism is represented in a sequential way. The implementation of the model makes use of this property and realizes CSP on a monoprocessor machine.

In this paper the semantics of CSP is defined by labelled transitions. As in [OH84], one component of the semantics definition is observation - but instead of the observation of program-execution the observation of one single process is considered. This approach leads to a semantics definition that can be transformed into an interpreter for CSP in a multiprocessor environment. Some examples illustrate the problems of such a distributed implementation. Up to now these problems have led to constraints in the implemented CSP-dialect [MS84] or they could only be managed by additional requirements on the architecture of the multiprocessor system [FS81] yielding complex implementation structures.

In this proposal the implementation problems have been solved by some amendments to the transformation system used for the semantics definition. Thus the implementation can be proved to be correct according to the given semantics. The basic ideas of the correctness proof of the CSP-interpreter are presented. (For a detailed solution see [GS85].)

The CSP-interpreter has been implemented at the Department of Computer Science, University of Kiel. The underlying multiprocessor architecture had to be simulated on a monoprocessor machine. The resulting system is currently in use as a basis for practical exercises in parallel programming.

Operational semantics for CSP

Before defining the operational semantics of the implemented CSP-dialect, some remarks on its syntactic properties and the static semantics should be made.
The syntax of CSP is simply extended by declarations of variables and channels. The static semantics make some requirements on the structure of CSP-programs by excluding "unreasonable" constructions.
The most important restrictions are:

- one channel connects exactly two processes in one direction
- processes must not be generated inside a recursion.

These requirements can be checked by a static analysis of a program. The following operational semantics definition is given for programs that are correct according to the static semantics. This semantics definition uses as one basic component the observation of the execution of one process. Actions that are considered to be observable are those that have nonlocal effects on their process, that is

- a parallel command
- a communication
- the choice of one alternative in a guarded nondetermnistic command
- the termination of a process
- the deadlock of a process.

In the first step of the following semantics definition we will consider CSP-commands and their effect on one process. In the second step the results wil be combined to specify the semantics of a whole CSP-program. Thus we will first split parallel problems into more familiar sequential reasoning. Then a parallel solution will be constructed out of the sequential results.

Definition 1: The state of the execution of a process p is the 3-tuple
p = (pid, obs-hist, cl) with the components pid = processid, obs-hist = observation of the previous computation, cl = remaining commandlist.

A process is executed in an environment consisting of variable-ids, channel-ids and commandlist-ids (these are necessary for constructing recursion). The environent env_p of a process p is built straightforwardly out of his declarations.

Definiton 2: A configuration of a process p is a 2-tuple
$K_p = (p, env_p)$ with the components p = state of the execution of the process and env_p = environment of the process.

Definition 3: The start configuration of a process p is the 2-tuple
$K_p^{start} = (p, env_p)$ with p = (pid, ε, commandlist of the process).

The actual values of the environment are given by the state δ from the following domains:

$\mathcal{D}_{var}$ = val, where val is a set of values including *true*

$\mathcal{D}_{cl}$ = {CSP-commandlist} and $\mathcal{D}_{chan}$ = {opened, closed}.

In the starting state δ^{start} all channel-ids are initialized with "opened".

The execution of a process in a CSP-program is formalized as the tranformation of the process configuration and the state, beginning with the start configuration and the starting state. A parallel command results in generating the start configurations of the new processes.

Definition 4: A configuration of a program Π is an n-tuple
$K_\Pi = (K_{p1}, .., K_{pn})$ of configurations of the previously generated processes.

Now we have got all the terms necessary for the definition of the transformations caused by CSP-commands. The transformations are formally represented as a relation $\longrightarrow$ between process configurations and states. As we are just considering the commands' effect on single processes, the synchronizing effect of i/o-commands can't yet be represented. This will be done in a second step by extending our concept to a relation $\longrightarrow_\Pi$ between program configurations and states.

First step: definition of ⟶

The definition of ⟶ is presented for the most important CSP-commands. The rest can easily be defined.

Let K_p be a process configuration, K_Π a program configuration.

- assignment *<varid> : = <exp>*

  Let's start with one of the most simple commands. As can be expected, the assignment results in a substitution of the state:

  (((pid, obs-hist, <varid>: = <exp>), env>, δ)

  ⟶ (((pid, obs-hist, ε), env), δ[δ(<exp>)/<varid>])

- output command *<channel-id>!<exp>*

  An output command results in a deadlock, iff the partner process is terminated, i.e. the addressed channel is closed. Otherwise the output command leads to a waitstate for the synchronisation.

  (((pid, obs-hist, <channel-id>!<exp>), env), δ)

  ⟶ {
  (((pid, obs-hist•deadlock, ε), env), δ)
      - iff δ(<channel-id>) = closed
  (((pid, obs-hist, wait<channel-id>!<exp>), env), δ)
      - otherwise

- waiting for synchronization *wait <channel-id>!<exp>*

  This waitstate can be left on two ways:

  - either by communicating with the partner process
  - or by reaching a deadlock

  A deadlock is reached, iff the partner process is terminated, i.e., the channel is closed. The communication is executed, iff an input command concerning the same channel is ready in the partner process.

  Let pr_{bool}(guard) denote the boolean part, pr_{comm}(guard) denote the communication part of a guard.

  (((pid, obs-hist, wait<channel-id>!<exp>), δ)

  ⟶ {
  (((pid, obs-hist•deadlock, ε), env), δ)
      - iff δ(<channel-id>) = closed
  (((pid, obs-hist•<channel-id>!δ(<exp>), ε), env), δ)
      - iff [∃(pid$_2$, obs-hist$_2$, wait<channel-id>?<varid>•<cl$_2$>), env$_2$) ∈ K_Π
      or ∃((pid$_2$, obs-hist$_2$, □<guard>$_1$;<cl>$_1$...<guard>$_n$;<cl>$_n$),env$_2$) ∈ K_Π
      with pr_{comm}(<guard>$_i$) = <channel-id>?<var-id>
      and δ (pr_{bool}(<guard>$_i$)) = true for an i< = n]
      and δ (<channel-id>) = opened

- parallel command $\| <process>_1 .. <process>_n$

  A parallel command results in the generation of the start configurations of the listed processes. The generating process enters a waitstate for the termination of the new processes.

  The parallel command can be observed, as well in the generating process as in the generated processes.

  $$(((pid, obs\text{-}hist, \| <process>_1 ... <process>_n), env), \delta)$$

  $$\longrightarrow \begin{Bmatrix} (((pid, obs\text{-}hist \bullet obpar <process>_1 ... <process>_n, \\ \quad \underline{waittermination} <process>_1 ... <process>_n), env), \delta) \\ (((<process>_1 id, i_1, <cl>_1), env_1), \delta) \\ ... \\ (((<process<_n id, i_n, <cl>_n), env_n), \delta) \end{Bmatrix}$$

- *termination*

  A process that has executed its commandlist without entering a deadlock closes its channels and terminates. The termination can be observed.

  Let conchan(pid) be the set of channels connected to the process pid and let lastobs(obs-hist) be the last observation in the observation-history.

  $$(((pid, obs\text{-}hist, \varepsilon), env), \delta)$$

  $$\longrightarrow (((pid, obs\text{-}hist \bullet termination, \varepsilon), env), \delta[closed/<chid>, \forall <chid> \in conchan (pid)]$$

  $$\text{iff } lastobs (obs\text{-}hist) \notin \{deadlock, termination\}$$

Second step: *definition of* $\longrightarrow_\Pi$

$\longrightarrow_\Pi$ is the relation between program configurations and states that is induced by the relation $\longrightarrow$. With the help of $\longrightarrow_\Pi$ the notion of time can be modelled.

A transformation that is caused by a statement cmd will be called a *cmd-transformation*. If the execution of cmd includes a synchronisation, $\overline{cmd}$ will denote a possible partner statement. Nondeterministic commands may have several possible cmd-transformations.

Definition 5: Let K_Π be a progam configuration, let $K_p = (p, obs\text{-}hist_p, cl_p)$ be one component of K_Π. The set posact (K_p, K_Π, δ) is the set of all possibilities to proceed in the execution of the process p,

$$posact (K_p, K_\Pi, \delta) := \begin{cases} \{nil\} & \text{if } lastobs(obs\text{-}hist_p) = terminated \\ & \text{or } lastobs (obs\text{-}hist_p) = deadlock \\ & \text{or } nextcommand(cl_p) \text{ can't be executed according to} \longrightarrow \\ \{act_1, ..., act_n\} & \text{if these are the possible tranformations caused by the execution of } nextcommand(cl_p) \text{ according to} \longrightarrow \end{cases}$$

Definition 6: $(a_1,...,a_n) \in$ posact $(K_{p1}, K_\Pi, \delta) \times (K_{p2}, K_\Pi, \delta) \times .. \times (K_{pn} K_\Pi, \delta)$ is called a tuple of executable transformations to proceed in the execution of the program $\Pi \Leftrightarrow_{df}$ for all transformations a_i including a synchronisation ,$1 <= i <= n$, $a_i \neq nil$, there is a partner-transformation $a_j = \overline{a_i}$.

Definition 7: $(K_\Pi, \delta) = (K_{p1},..,K_{pn}), \delta) \rightarrow_\Pi (K_\Pi', \delta') = (K'_{p1},..,K'_{pn},..,K_{pm}), \delta')$ by executing the tuple of executable transformations $(a_1,..,a_n)$

$\Leftrightarrow_{df}$
- $K'_{pi} = K_{pi}$ iff $a_i = nil$ and $i <= n$
- K'_{pi} is the result of the a_i-transformation on K_{pi} iff $a_i \neq nil$ and $i <= n$
- K_{pi} is the start configuration of a process that is generated in a parallel command $a_j, j <= n$ iff $i > n$

and δ' is the state resulting from δ by the necessary substitutions.

According to this definition a $\rightarrow_\Pi$-transformation of a program configuration represents the following:

- all processes that can execute their next command do so,
- statements including synchronization are executed at the same time as the corresponding command, and newly generated processes are included in the program configuration.

This definition doesn't take into account the different speed of processes: while one process executes several statements, another process won't perhaps proceed at all though he could do so. This behaviour can easily be modelled by introducing additional "nil"-transformations of processes.

The execution of a program can thus be represented by a sequence of program configurations, starting with a program configuration consisting of start configurations of processes and ending with the first program configuration whose components can't proceed any more (i.e. for all process configurations only "nil"-transformations are possible).

Definition 8: The semantics of a CSP-program Π is defined to be the set of all sequences of program configurations of Π according to $\rightarrow_\Pi$.

The implementation

In a perfect environment CSP-programs are run on a multiprocessor system, where each process is executed by a separate processor.

The actions of such a processor are determined by the CSP-interpreter. The interpreter realizes the presented $\rightarrow$relation and additionally the synchronisation of communicating processes that are formalized in the relation $\rightarrow_{\Pi}$.
$\rightarrow$-transformations that are independent of the state of other processes can easily be implemented. The other $\rightarrow$-transformations (i.e. communication, termination and external nondeterminism) are hard to perform in a distributed system.
Communication and termination can be handled by the exchange of special handshake- and termination-signals. The external nondeterminism is more complicated, because it is possible that two parallel processes are working on such a command and that they have to synchronize with each other. The following example will illustrate the problem.

Example 1

Let Π be a CSP-program of the following form

Π: ... (|| (process p1 ... (□ (test true;c!e ∘$<cl>_{11}$) (test true;d?x∘$<cl_{12}$))...)
(process p2 ... (□ (test true;f!e ∘$<cl>_{21}$) (test true;c ?y∘$<cl_{22}$))...)
(process p3 ... (□ (test true;d!e ∘$<cl>_{31}$) (test true; f ?z∘$<cl_{32}$))...)

that is

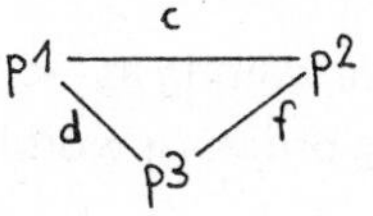

c
p1 p2
d f
p3

According to the semantics definition each process will choose an alternative with a guard whose boolean part evaluates to true and whose communication part can be executed. Thus exactly one of the three prossible communications will happen.
An implementation has to cope with the following problems:

- Each process chooses an alternative whose guard doesn't correspond to the guards of the alternatives selected by the other processes. The execution of Π ends in a deadlock that is caused by a wrong implementation.
- Each process chooses an alternative just for a short time that it spends waiting for a partner, without success. The execution of Π ends in a livelock.

The problems can be solved by defining a reflexive and total ordering on the channel-ids (e.g. the lexicographic ordering on the channel-ids). According to this ordering the communication guards of a nondeterministic command are tested to be executable and the corresponding channels are labelled. The following algorithm specifies the procedure.

Algorithm 1:

- The test of the guards starts with that guard that addresses the "smallest" channel according to the ordering. The smallest channel is labelled with "<", the channels of all other guards are labelled with ">".

- If the channel of the tested guard is already labelled with "<" by the partner process, the tested guard is chosen and is treated like a simple communication command. The channel is labelled with "waiting".
- If the channel of the tested guard is labelled with ">" by the partner process or isn't labelled at all, the interpreter will either
  - test the "next" guard and label the actual guard with ">", if there is a "next" guard according to the order or
  - start again to test the "smallest" guard and label the channels of all other guards with "<", otherwise.
- If the channel of the tested guard is labelled with "waiting" by the partner process, the communciation can be executed.

Lemma 1:

The execution of a guarded nondeterministic command according to the algorithm will not lead to a deadlock.

Proof 1:

The execution is deadlocked, iff all processes are waiting at "<"-labels.
A channel is labelled with "<", iff the labelling process is working on the corresponding or on a "smaller" guard.
The set of processes in a program is finite (we don't allow recursion on parallel commands). Therefore a process is waiting at the "smallest" channel among the channels where processes are waiting.
In the case of a deadlock this process must be waiting for a process that is waiting at an even "smaller" channel.
But this is a contradiction!

The freedom of livelock can be proved in an analogous way. There are still more problems left. The proof just guarantees that an implementation working according to this algorithm will find an executable guard if there is one. Besides this existence criterion the completeness criterion must be fulfilled. The following example illustrates the problem left.

Example 2:

Let Π be a CSP-program of the following form

Π:

The semantics definition assures that two communications take place. The presented algorithm just guarantees one communication and has to be modified slightly [GS85].

The correctness of the implementation

The presented algorithm can formally be represented as a refinement of the $\longrightarrow$-relation, the so called $\longrightarrow^{int}$-relation. $\longrightarrow^{int}$ models the activity of one processor executing one process. The activity of several processors working in parallel according to $\longrightarrow^{int}$ yields a relation $\longrightarrow_{\Pi}{}^{int}$ between program configurations and states. In contrary to the semantics relation $\longrightarrow_{\Pi}$, the definition of $\longrightarrow_{\Pi}{}^{int}$ doesn't have to cope with the problems of interaction between processors (e.g. communications, terminations), because they are already considered in the definition of $\longrightarrow^{int}$. Thus we end up with two sets of sequences of program configurations. The first (set $\longrightarrow_{\Pi}$) is given by the relations $\longrightarrow$ and $\longrightarrow_{\Pi}$ and defines the semantics of a CSP-program, the second (set $\longrightarrow_{\Pi}{}^{int}$) is given by the relations $\longrightarrow^{int}$ and $\longrightarrow_{\Pi}{}^{int}$ and specifies the transformations made by an implementation.
The aim is to prove that $\longrightarrow^{int}$ induces a correct distributed implementation of CSP. In terms of the derived formalism it has to be shown that the two sets (set $\longrightarrow_{\Pi}$ and set $\longrightarrow_{\Pi}{}^{int}$) *correspond* to each other.
At first sight *correpondance* should simply be equality. Because of the additional transformations made by $\longrightarrow^{int}$ for the exchange of synchronisation signals, equality of the two sets is impossible. Therefore *correspondence* is defined to be equality modulo an equivalence relation that allows to neglect these additional transformations.
This correspondence criterion is met:

- all computations made according to $\longrightarrow^{int}$ are allowed by the semantics definition (this "soundness"of the implementation can easily be proved: set $\longrightarrow_{\Pi}{}^{int} \subset$ set $\longrightarrow_{\Pi}$).
- all computations required by the semantics definition are executed by the implementation (this "completeness" of the implementation has been shown with the proof of deadlock- and livelock-freedom: set $\longrightarrow_{\Pi}{}^{int} \supset$ set $\longrightarrow_{\Pi}$)

Finally we have found a formal specification of a CSP-interpreter that is proved to generate exactly those computations that are equivalent to the one's defined by the CSP-semantics. Thus the full scope of CSP can be implemented correctly in a distributed system.

References

[CH81] Zhou Chaochen, C.A.R. Hoare, "Partial correctness of CSP", in: Proc. of 2nd Intern. Conference on Distributed Computing, Paris, Computer Society Press, pp. 1-12, 1981

[FS81] L.Shrira, N. Francez, "An experimental implementation of CSP", in: Proc. 2nd intern. Conference on Distributed Computing, Paris, Computer Society Press, 1981

[GS85] G.B. Schreiber, "Operational semantics and a distributed implementation of CSP", Diplomarbeit, Department of Comp.Sc., Univ. Kiel, 1985 [in German]

[H078] C.A.R. Hoare, "Communicating sequential processes", Comm. ACM vo. 21, no. 8, pp. 666-677, 1978

[H080] C.A.R. Hoare, "A model for CSP", in: R.M. McKeag, A.M McNeghton (Eds.) On the Construction of Programs, Cambridge University Press, pp. 229-243, 1980

[H081] C.A.R. Hoare, "A calculus of total correctness for CSP", Science of Computer Programming 1, pp. 49-72, 1981

[HBR81] C.A.R. Hoare, S.D. Brookes, A.W. Roscoe, "A theory of CSP", Techn. Monograph PRG-16, Oxford Univ., Program Research Group, Oxford, 1981

[MS84] D. May, R. Shepherd, "The Transputer implementation of OCCAM", in: Proc. intern. Conference on Fifth Generation Computer Systems, 1984

[OH84] E.-R. Olderog, C.A.R. Hoare, "Specification-oriented semantics for CSP", Techn. Monograph PRG-37, Oxford Univ., Program Research Group (1984)

[PL83] G.D. Plotkin, "An operational semantics for CSP", in: D. Bjørner (Ed.), Formal Description of Programming Concepts II, Amsterdam, North Holland, pp. 199-223, 1983

[RE83] R. Reinecke, "Networks of communicating processes: a functional implementation", Manuscript, Department of Comp.Sc., Univ. Kaiserslautern, 1983

The Semantics of Miranda's Algebraic Types*

Kim B. Bruce and Jon G. Riecke†
Department of Computer Science
Williams College
Williamstown, MA 01267

Abstract

Miranda has two interesting features in its typing system: implicit polymorphism (also known as ML-style polymorphism) and algebraic types. Algebraic types create new types from old and can operate on arbitrary types. This paper argues that functions on types, or *type constructors*, best represent the meaning of algebraic types. Building upon this idea, we develop a denotational semantics for algebraic types. We first define a typed lambda calculus that specifies type constructors. A semantic model of type constructors is then built, using the ideal model as a basis. (The ideal model gives the most natural semantics for Miranda's implicit polymorphism.) The model is shown to be sound with respect to this lambda calculus. Finally, we demonstrate how to use the model to interpret algebraic types, and prove that the translation produces elements in the model.

1 Introduction

Recently, much of the theoretical and practical research in programming languages has focused on extensions to *strong-typing*, where programs may be checked before run-time for a consistent use of types. One of these extensions, *polymorphism*, has been studied extensively. In a polymorphic language, parameters to procedures or functions may accept arguments of more than one type. (This notion of polymorphism can be traced to [Str67].) For example, one may write a "sort" routine that sorts lists of integers, characters, or strings. Compared with a strongly-typed but non-polymorphic language like Pascal, the ability to write one "sort" routine can be quite powerful. In Pascal, one would have to write three different sort routines for integers, characters, and strings – even though the algorithms are the same – since parameters can match arguments of only one type. Polymorphism adds convenience, expressibility, and security to a strongly-typed language, and for those reasons, it has been incorporated into ML [Mil78], HOPE [BMS80], Russell [DD85], and Ada [DoD83] (in the form of *generics* that accept types as parameters).

Because of its growing importance in the design of new languages, researchers in denotational semantics have examined various interpretations for polymorphism. The work has mainly dealt with versions of the

*Partially supported by NSF Grants MCS-8402700, DCR-8603890, and DCR-8511190. The second author is supported by an NSF Graduate Fellowship.

†Current address: MIT Laboratory for Computer Science, 545 Technology Square, Cambridge, MA 02139

lambda calculus, in order to study polymorphism on an elementary and "pure" level. Our goal was to study a "real" polymorphic language and see what complexities arose. We chose Miranda[1] [Tur85a,Tur85b] as our language. The most interesting aspect of the study was not, however, polymorphism, but *algebraic types*, another extension to types found in Miranda. In this paper, we describe how to interpret algebraic types in a model that can also interpret Miranda's polymorphism.

Miranda is descended from SASL, another language invented by Turner (see [Tur85a,Tur85b] for a history of Miranda). While there are minor syntactic differences between the two languages, the most important new features (at least in the view of the authors) are polymorphism and algebraic types, which create new types from old. [KS85] provides a denotational semantics for SASL, which can serve as a basis for much of the semantics of Miranda [Rie86]. ([Jon87] has also given a denotational semantics for Miranda, albeit an *untyped* semantics. Our work focuses on the meaning of types, and so differs from [Jon87] in its goal.) The semantics of Miranda will be given with respect to the *ideal* model of [MS82,MPS84,MPS86], which extends the untyped semantics of SASL to interpret Miranda's polymorphism.

The ideal model cannot, however, adequately interpret Miranda's algebraic types, and this paper shows how to extend the ideal model to interpret algebraic types. The approach here makes use of *kinds*, a concept introduced in [McC79] for models of the second-order lambda calculus. Kinds describe higher-order type constructors which, we argue, best represent the meaning of algebraic types. A careful development of the semantics of higher-order type constructors provides a way to interpret algebraic types, while using the ideal model as a basis for the semantics of Miranda. A soundness theorem shows that this extended semantics matches the type inference mechanism of the language.

This paper may be divided into three parts. We begin by introducing Miranda: Section 2 provides a brief description of Miranda's polymorphism and algebraic types, and Section 3 explores various possibilities for interpreting algebraic types and justifies the approach taken in this paper. Then, in Sections 4 through 6, we sketch the details of the semantics of algebraic types. Finally, Section 7 shows how these semantics fit into the rest of Miranda's semantics, while Section 8 discusses remaining problems.

2 Miranda's Type System

2.1 Implicit Polymorphism

In most strongly-typed languages, a program must contain declarations of variables and their types. Miranda does not follow this convention, for types may be omitted from a Miranda program. Miranda uses an *implicit* typing system, similar to ML's [Mil78], in which the type-checker for the language deduces the types of variables from the context. As long as there are no conflicts in the types deduced for variables, a program is well-typed.

As an example of how the type-checker works, consider the Miranda function

```
add x y = x + y
```

[1] "Miranda" is a trademark of Research Software Ltd.

The parameters to add are x and y, and the body is x + y. In order to assign a type to these parameters, the type-checker examines the body of the function and finds that x and y must both have type Num for the body to be well-typed. Since we can assign consistent types to x and y, the function is well-typed. Functions are first-class objects in Miranda, and so we must also assign a type to add; its type is $Num \rightarrow Num \rightarrow Num$, because it takes two numbers and returns another number. (Note that "$\rightarrow$" associates to the right, so $Num \rightarrow Num \rightarrow Num$ is read as $Num \rightarrow (Num \rightarrow Num)$.) Types for functions are written in this curried notation, since applying one function of type $Num \rightarrow Num \rightarrow Num$ to a number, as in

```
add3 = add 3
```

returns a function of type $Num \rightarrow Num$.

An implicit type-checker cannot always deduce a unique type for a variable. For example, consider the Miranda function

```
id x = x
```

which is the identity function. Nothing in the body of this function constrains the type of x, so the type-checker would assign a type variable (such as α) as the type of x. To deduce a type for id, notice that it takes a value of type α and returns a value of type α. The type-checker then infers the type $\forall\alpha.\alpha \rightarrow \alpha$ for id, where the "$\forall$" symbol means that any type may replace α. id is an example of a *polymorphic* function, a function that accepts arguments of more than one type.

Miranda is considered an *implicit polymorphic* language, because polymorphic types are introduced by the type-checker, not by the programmer. The other major form of polymorphism is *explicit polymorphism*, in which types can be passed to functions. This form of polymorphism may be found in the second-order lambda calculus [Gir71,Rey74] and in Russell [DD85].

2.2 Algebraic Types

Another novel feature of Miranda's typing system is algebraic types. Algebraic types allow a programmer to define the data domains of the program [Tur85a]. For example, suppose we want to describe binary trees. One could represent binary trees in lists, as is done typically in Lisp, with the first element being the value at a node, and the next two elements being the left and right branches. This representation can be confused with other lists, though. Algebraic types allow the programmer to distinguish lists and trees through the creation of new data types, such as a type called "tree" which contains all binary trees.

Algebraic types can define completely new values. For instance, we can create a new type called "fruit":

```
fruit ::= Apple | Cherry | Peach
```

Simple algebraic types like this one resemble Pascal's enumerated types: the type "fruit" includes the values Apple, Cherry, and Peach. The name of the new type appears on the left of the "::=", and the names on the right are the values in this type, separated by a "|" which is read as "or". Any expressions in the program can use the values in fruit once it has been defined.

Algebraic types can also build new types from types already present in Miranda. Consider, for instance, the following algebraic type:

```
numchar ::= Partnum num  |  Partchar char
```

This algebraic type may be read as "numchar is comprised of the values Partnum followed by a number, or Partchar followed by a character." A program containing this algebraic type could use values like

```
Partnum 5
Partchar 'a'
```

both of which would be assigned the type *numchar* by the type-checker. Values like Partnum 'a' are not allowed, since the algebraic type specifies that a value of type Num must follow any occurence of the identifier Partnum.

The expressive power of algebraic types comes from the ability to use *recursion* inside definitions. Trees are one example of a recursive type, since trees are defined in terms of themselves, storing two subtrees at each internal node. In Miranda, one could create the algebraic type

```
treenum ::= Leaf num  |  Node num treenum treenum
```

to describe numeric binary trees. In this case, the algebraic type appears in its own definition – the identifier Node must be followed by a value of type Num, and two values of type *treenum*.

In Miranda, we can generalize this algebraic type. Suppose we wanted binary trees with numbers or characters at the nodes. One could define two different algebraic types, or alternatively, one could write

```
tree * ::= Leaf *  |  Node * (tree *) (tree *)
```

The * stands for "any type", and so this algebraic type describes a whole class of types, with instances being *tree* Num, *tree* $Char$, or even *tree* (*tree* Num). Values in these new types include:

```
Node 5 (Node 1 (Leaf 0) (Leaf 3)) (Leaf 11)
Node 'a' (Leaf 'b') (Leaf 'c')
```

with types *tree* Num and *tree* $Char$. A value not described by this algebraic type is

```
Node 12 (Leaf 'a') (Leaf 6)
```

because Leaf 'a' has type *tree* $Char$, which is inconsistent with Node 12. Miranda's type-checker would reject such a value.

Miranda allows more than one type variable (such as ** or ***) in algebraic types, as in

```
tagtype * ** ::= Left *  |  Right **
```

In addition, mutual recursion between algebraic types is also permitted, as in

```
tree1 * ::= Leaf1 *  |  Node1 * (tree2 *) (tree2 *)
tree2 * ::= Leaf2 *  |  Node2 * (tree1 *) (tree1 *)
```

Algebraic types are thus quite general, and this generality will complicate their semantics.

In the next section, we shall examine a few ways to interpret algebraic types. For simplicity, we assume that all algebraic types appearing on the right side of a "::=" have the appropriate number of types following them. For instance, an algebraic type like

```
tree * ::= Leaf *  |  Node * (tree) (tree)
```

should be considered illegal, because `tree` is used incorrectly on the right side. ([Rie86] describes a way to check algebraic types for these errors.) We also will not consider a further extension to algebraic types, in which "laws" are associated with an algebraic type [Tur85a]. In Section 8, we will briefly address how these *unfree* algebraic types may be incorporated into the semantics.

3 Interpretation of Algebraic Types

For any strongly-typed language like Miranda, a well-typed program always produces answers of the appropriate type. The denotational semantics of a typed language must mirror this operational behavior, or in other words, the meaning of any well-typed program must be in the appropriate type in the model. This result is commonly called a "soundness theorem", and in order to verify it, we must have an intepretation for types.

Given the extensive type system of Miranda, finding intepretations for every type is difficult. Without algebraic types, the task is easier – one can use the ideal model of [MS82,MPS84,MPS86], for it can interpret all the non-algebraic types, including polymorphic types. One would then like to find meanings for algebraic types in this model. At first glance, it does not seem possible to add algebraic types to *any* model, since they appear to create new types during the execution of a program. We must either extend the ideal model, or examine the model more closely, to make sure that types produced by algebraic types have a meaning.

At least three methods can be used to interpret algebraic types within the ideal model. One method involves describing a *collection of models* for each set of algebraic types. The model for a set of algebraic types would contain all types described by the algebraic types. Furthermore, each model would be an ideal model, so that the implicit polymorphism of Miranda could be interpreted. Finding the meaning of a particular program under these semantics would involve choosing a model that contained the appropriate algebraic types.

This method, though, requires an infinite collection of models, one for each possible set of algebraic types; it would be better if we could incorporate all of the algebraic types into *one* model. An alternative method involves finding a model in which all types described by algebraic types may be found. For example, consider the algebraic type

```
tree * ::= Leaf *  |  Node * (tree *) (tree *)
```

One can think of this algebraic type as defining two new functions, `Leaf` and `Node`, where `Leaf` takes a value of some type t and returns a value of type $tree\,t$, and `Node` takes arguments of types t, $tree\,t$, and $tree\,t$, and returns a value of type *tree t*. One can model `Leaf` as merely returning its value, and `Node` as returning a tuple of three items. To find a meaning for `tree`, we simply combine the values returned by `Leaf` and `Node`:

$$tree\ t = t + t \times (tree\ t) \times (tree\ t)$$

(where "+" constructs a disjoint union, and "×" constructs a cross product of the types). In [MPS84,MPS86], a method is given for interpreting such recursive type equations, so for any type t, there is a meaning for *tree t* in the model. This method avoids the infinite collection of models of the first method.

The second method has another flaw – no meaning is found for the algebraic type itself, only for its instantiations. A third method, and the one we chose, treats algebraic types as *type constructors*, that is, functions which take types as arguments and return types. For example, one can think of the algebraic type `tree` as a function which takes one type and returns a new type, with the value of `tree` being

$$tree = \lambda t.t + t \times (tree\ t) \times (tree\ t)$$

This method seems to capture the meaning of an algebraic type better than the first two methods – an algebraic type has a meaning, not simply as a collection of related types. While this method requires adding another layer of values to the semantic objects to the model (the "kind" hierarchy of functions on types), the method is more semantically natural. Given the structure of Miranda, a semantics of Miranda should assign a meaning to every piece of the program, so algebraic types should have a denotational meaning. The following sections outline one method for achieving this goal.

4 Overview of Semantics of Algebraic Types

The decision to model algebraic types as type constructors requires that, in addition to ordinary, first-order functions, we include the meaning of higher-order functions (the functions on types) in our model. Incorporating higher-order functions is not a new idea: they were used in the first model of the second-order lambda calculus [McC79], and have subsequently been employed in other models of the second-order lambda calculus [BMM85,ABL86]. Interestingly, the ideal model originally given in [MS82] included higher-order functions to solve certain recursive type equations (This approach was abandoned in a later paper [MPS84]). Our semantics extends the use of higher-order functions to model a particular language construct, algebraic types.

Higher-order functions, or type constructors, may be classified like first-order functions – instead of "types," the classifiers are called *kinds* [McC79]. *Kinds* are syntactic objects that classify type constructors according to the number of types they accept as arguments, and are specified by the grammar

$$\kappa ::= T \mid \kappa \Rightarrow \kappa \mid (\kappa)$$

The T in the grammar is shorthand for "type", while "$\Rightarrow$" denotes a function space. Accordingly, any simple type expression will have kind T, and type constructors will have more complicated kinds. The binary tree type constructor, for example, has kind $T \Rightarrow T$, since it returns a new type when given a type. The standard type constructors $\times$ and $+$ also have a kind – kind $T \Rightarrow T \Rightarrow T$ – since they construct a new type from two types. (Note that kinds, like types, are specified in curried form.)

In addition to classifying type constructors, kinds will also classify the sets that contain the meaning of type constructors. These sets in the model, denoted $Kind^{\kappa}$, will contain the meaning of all type constructors with kind κ. Our model will include an infinite number of these sets, one for each possible kind.

Along with kinds, we will use another concept, *constructor expressions*, adapted from [McC79] and [BMM85]. Constructor expressions are lambda terms specified by the grammar

$$\begin{aligned} cexpr \ ::= \ & Int \mid Real \mid Num \mid Bool \mid Char \mid Trivial \mid id \mid + \mid \times \mid List \mid \forall id^{T}.cexpr \mid (cexpr) \mid \\ & cexpr \rightarrow cexpr \mid \lambda id^{\kappa}.cexpr \mid cexpr\ cexpr \end{aligned}$$

where *id* represents any identifier. Constructor expressions will denote the meanings of type constructors in the model, serving the same purpose as the lambda calculus in denotational semantics. Notice that constructor expressions are *typed* lambda terms: any time a variable is bound by a lambda, a kind must be given for the variable.

The syntax of constructor expressions expresses more functionals than just the type constructors, since the syntax allows functions that accept other type constructors as arguments. One example of such a constructor expression is $\lambda t^{T \Rightarrow T}.t$, a function that takes a function of kind $T \Rightarrow T$ and returns it. The kind of this constructor expression is $(T \Rightarrow T) \Rightarrow (T \Rightarrow T)$. We will take advantage of higher-kinded bound variables in the definition of the *List* type constructor.

Having these ideas in mind, we now give the semantics of algebraic types. First, in Section 5, we will consider the semantics of constructor expressions. Using these semantics, we will show how to convert algebraic types into constructor expressions in Section 6.

5 Constructor Expressions

Both constructor expressions and algebraic types represent type constructors. Constructor expressions, however, have a more uniform syntax and provide insight into type constructors. Before considering the semantics of algebraic types, we carefully consider the semantics of constructor expressions. We devise a system that type-checks, or rather *kind-checks*, each constructor expression. (Algebraic types will be converted into well-kinded constructor expressions in the next section.) We then define a model and give the semantics of constructor expressions. In this section we also sketch a key result, namely that any well-kinded constructor expression has a meaning in the appropriate kind set.

5.1 Kind-Checking

Due to the presence of constants in the syntax for constructor expressions, not all constructor expressions have a meaningful interpretation. The syntax can generate constructor expressions like

$$\lambda t^{T \Rightarrow T}.t + t$$

that misuse these constants. (This expression does not respect the meaning of +, in that + should be applied to two types, not two objects of kind $T \Rightarrow T$.) In order to detect such illegal expressions, we devise a deduction system that infers the kinds of expressions. Any expression for which a type can be inferred will be called *well-kinded.* If the system cannot infer a kind for an expression, the expression will be illegal.

The kind-checking system resembles many deduction systems. We first need a notion of *syntactic kind assignments.* During the inference of kinds, the system must be able to remember the kinds of free variables in subexpressions; syntactic kind assignments "remember" the kinds of these variables. Formally,

Definition 1 *A* syntactic kind assignment *is a partial function that assigns kind expressions to identifiers.*

To change an arbitrary syntactic kind assignment A, we use the notation

$$A[v_1 : \kappa_1, \ldots, v_n : \kappa_n]$$

which denotes the same syntactic kind assignment as A except at the variables $v_1, \ldots, v_n$ which are assigned kinds $\kappa_1, \ldots, \kappa_n$.

Like other formal systems, the kind-checking system consists of axioms and rules. We begin with the axioms of the system that determine the kinds of "simple" constructor expressions. We will use the standard logic notation

$$A \vdash \mu : \kappa$$

to mean that "assuming the free variables of μ have kinds determined by A, μ has kind κ."

1. $A \vdash \tau : T$, where $\tau \in \{Int, Real, Num, Bool, Char, Trivial, NIL\}$

   This axiom states that the base types have kind T.

2. $A \vdash v : (A\ v)$

   Any free variable has a kind determined by the syntactic kind assignment.

3. $A \vdash List : T \Rightarrow T$

   "List" is a special type constructor that builds lists of a given type. Its kind is $T \Rightarrow T$.

4. $A \vdash +, \times : T \Rightarrow T \Rightarrow T$

   The standard type constructors + (disjoint union) and $\times$ have kind $T \Rightarrow T \Rightarrow T$ as noted before. As is customary, + and $\times$ will be written infix.

In addition to these rules, we will need four rules of inference. Typically, a typed lambda calculus only requires two rules of inference, one for lambda abstractions and one for applications. Two additional rules are necessary for constructor expressions, in order to guarantee that the set of well-kinded constructor expressions can be interpreted by the semantics.

1. $\frac{A[t:\kappa_1] \vdash \mu : \kappa_2}{A \vdash \lambda t^{\kappa_1}.\mu : \kappa_1 \Rightarrow \kappa_2}$, ($t$ not free in the left argument of a "$\rightarrow$" or in any subexpression $(\forall s^T.\alpha)$ of μ.)

   This rule assigns a kind for lambda abstractions, and the restriction is necessary to accommodate the ideal model. Our model only considers continuous functions on types; unfortunately, "$\rightarrow$" is not a continuous function in the ideal model (for reasons which are given below) so we must not treat "$\rightarrow$" as a function that can be applied like "+" and "$\times$". (In the final section, we will consider models in which "$\rightarrow$" is continuous.) Also, notice that "$\forall$" is not considered a function on types. We do not know if this constructor is continuous in the ideal model.

2. $\frac{A \vdash \mu_1 : \kappa_1 \Rightarrow \kappa_2, A \vdash \mu_2 : \kappa_1}{A \vdash \mu_1\mu_2 : \kappa_2}$

   An application of a constructor expression with kind $\kappa_1 \Rightarrow \kappa_2$ to another with kind κ_1 is a constructor expression with kind κ_2.

3. $\frac{A[s:T] \vdash \mu : T}{A \vdash \forall s^T.\mu : T}$

   This rule says that if a constructor expression has kind T if a variable t is assumed to have kind T, then using "$\forall$" produces a constructor expression that is a type. This rule is necessary since the constant "$\forall$" is not treated as a function on types.

4. $\frac{A \vdash \mu_1 : T, A \vdash \mu_2 : T}{A \vdash \mu_1 \rightarrow \mu_2 : T}$

   As with "$\forall$", this rule is needed since "$\rightarrow$" is treated as a special constant.

By way of example of how these rules can be used, consider the derivation of the kind for $\lambda t^T.t \times t$:

Deduction	Reason
$\emptyset[t:T] \vdash t : T$	Axiom 2
$\emptyset[t:T] \vdash \times : T \Rightarrow T \Rightarrow T$	Axiom 4
$\emptyset[t:T] \vdash t \times t : T$	Rule 2
$\emptyset \vdash \lambda t^T.t \times t : T \Rightarrow T$	Rule 1

These axioms and rules completely define the kind-checking of constructor expressions.

5.2 Semantic Sets and Semantics of Constructor Expressions

Now we present a semantic model to interpret constructor expressions. In order to build the model, we will begin with the semantic model of Miranda without algebraic types. The ideal model makes it convenient to interpret other Miranda statements, as was stated earlier.

The model for the base language is constructed via a sequence of mutually-recursive domain equations. *Domains* are complete partial orders (cpo's), i.e. partial orders in which every directed set has a least upper bound, or *sup*, and that are additionally:

1. *consistently-complete* (every bounded set has a sup)
2. *ω-algebraic* (every element is a sup of basic elements)

(see [ABL86,Sco82]). The domain of values for the base language is constructed using the standard domain constructors +, ×, and → (as opposed to type constructors), using the series of equations:

$$\begin{aligned}
Value &\cong Bool + Num + Char + NIL + Trivial + Newtype + Function\\
Bool &\cong flat\ cpo\ of\ truth\ values\\
Int &\cong flat\ cpo\ of\ integers\\
Real &\cong flat\ cpo\ of\ representable\ reals\\
Num &\cong Int + Real\\
Char &\cong flat\ cpo\ of\ characters\\
NIL &\cong flat\ cpo\ of\ the\ set\ \{nil\}\\
Trivial &\cong flat\ cpo\ of\ the\ set\ \{()\}\\
Newtype &\cong Value + (Value + Value) + (Value \times Value)\\
Function &\cong Value \to Value
\end{aligned}$$

(A *flat cpo* is a cpo in which $x \leq y$ if and only if $x = \perp$ or $x = y$.) These equations may be solved using the general methods given in [ABL86] and [Sco82]. The domain $Value$ is essentially the same domain used in the ideal model of [MS82,MPS84], which allows us to use their results.

Using this domain as a basis, we build a series of domains, the kind sets, that will contain the meaning of all constructor expressions. We first recall that in the ideal model, types are *ideals* which are nonempty, consistently-closed, downward closed subsets of a cpo [MS82]. We then let

$$Kind^T = \{A \subseteq Value \mid A\ is\ an\ ideal\}$$

This set includes the meanings of all type expressions. Furthermore, under the partial order of inclusion, $Kind^T$ is a domain [MS82]. Using this fact we build the other kind sets using the equation

$$Kind^{\kappa_1 \Rightarrow \kappa_2} = Kind^{\kappa_1} \to Kind^{\kappa_2}$$

which is the domain of all continuous functions (under the Scott topology) from the domain $Kind^{\kappa_1}$ to the domain $Kind^{\kappa_2}$. For example, the kind set $Kind^{T \Rightarrow T}$ is specified by the domain equation

$$Kind^{T \Rightarrow T} = Kind^T \to Kind^T$$

These domains, it will be shown, contain the meanings of all well-kinded constructor expressions.

Using these kind sets, we may now specify the semantics of constructor expressions. We define two semantic sets upon which the semantics will be defined:

$$\begin{aligned}
Cvalue &= \bigcup_{\kappa \in KindExpr} Kind^{\kappa}\\
Cenv &= Ide \to Cvalue
\end{aligned}$$

$Cvalue$ contains the meaning of all constructor expressions, and $Cenv$ is the set of all *constructor environments.* A constructor environment allows the semantics to remember the values of free variables when interpreting constructor expressions. We will use the symbol η to represent an arbitrary constructor environment. In order to update a constructor environment, we use the notation

$\eta[\mu_1/v_1, \ldots, \mu_n/v_n]$

which means that the new environment has the same values as the old environment η except at the variables $v_1, \ldots, v_n$ which are assigned values $\mu_1, \ldots, \mu_n$.

Now we may define *Cval*, the semantic function that interprets constructor expressions. The interpretation of a constructor expression is determined from the following semantic clauses that specify *Cval*:

$Cval : ConsExpr \rightarrow Cenv \rightarrow Cvalue$

1. $Cval[\![Int]\!]\eta = \mathbf{Int}$, where **Int** denotes the ideal corresponding to the integers in the domain *Value*.

   Thus the base type *Int* has a meaning as an ideal. There are similar clauses for the other base types (*Real*, *Bool*, *Char*, *NIL*, *Trivial*) as well.

2. $Cval[\![Num]\!]\eta = Cval[\![Int + Real]\!]\eta$

   The type *Num* is the disjoint union of **Int** and **Real** in the model.

3. $Cval[\![v]\!]\eta = \eta\, v$, where v is an identifier.

   The meaning of a free variable is determined by looking up the variable in the constructor environment.

4. $Cval[\![+]\!]\eta = \oplus$, where $\oplus$ represents the disjoint union of ideals and is written in infix notation.

5. $Cval[\![\times]\!]\eta = \otimes$, where $\otimes$ represents the direct product of ideals and is written in infix notation.

6. $Cval[\![\mu_1 \rightarrow \mu_2]\!]\eta = \{f \in Function \mid f(Cval[\![\mu_1]\!]\eta) \subseteq Cval[\![\mu_2]\!]\eta\}$

   This definition of "→" comes from [MS82], and is a fundamental definition in the ideal model. It states that any function in the domain *Value* that takes objects in type μ_1 to objects in the type μ_2 is in the type $\mu_1 \rightarrow \mu_2$. This definition is quite intuitive, but there are problems with it that we discuss below.

7. $Cval[\![\forall t^T.\mu]\!]\eta = \bigcap_{a \in Kind^T} Cval[\![\mu]\!](\eta[a/t])$

   This definition is another fundamental definition in the ideal model. To see how it works, consider the polymorphic identity function. It lies in

   $$\begin{array}{lcl} Char & \rightarrow & Char \\ Int & \rightarrow & Int \end{array}$$

   and so on, according to the definition of "→" given above. The identity function lies in all of these ideals, and so it lies in the intersection of the ideals. Polymorphic types are therefore modelled as intersections of ideals.

8. $Cval[\![List]\!]\eta = \mathbf{List} = \mathbf{fix}(\lambda l^{T \Rightarrow T}.\lambda t^T.\mathbf{NIL} \oplus (t \otimes l\ t))$, where **fix** is the least fixpoint operator in $Kind^{((T \Rightarrow T) \Rightarrow (T \Rightarrow T)) \Rightarrow (T \Rightarrow T)}$. The operator **fix** is defined by the equation

$$\mathbf{fix}(f) = \bigsqcup_{i<\omega} f^i(\bot)$$

where $f^i = f \circ f \ldots f$ with f appearing i times.

This clause defines the *List* type constructor that builds homogeneous lists (i.e. those with elements drawn from a single type). This definition was given in [MS82].

9. $Cval[\![\lambda v^\kappa.\mu]\!]\eta = \lambda u \in Kind^\kappa.Cval[\![\mu]\!](\eta[u/v])$

   The meaning of a constructor expression representing a function is a function accepting arguments from the appropriate kind set.

10. $Cval[\![\mu_1\mu_2]\!]\eta = Cval[\![\mu_1]\!]\eta\, (Cval[\![\mu_2]\!]\eta)$

    A constructor expression representing an application is intepreted by applying the meaning of the function to the meaning of the argument.

Note that "$\rightarrow$" is not a function in these semantics; this is because it is not a continuous function, and so does not lie in any kind set. To see that "$\rightarrow$" is not continuous, consider the two ideals

$Value \rightarrow Int$

$Int \rightarrow Int$

Any function in the first ideal is also in the second, so

$Value \rightarrow Int \subseteq Int \rightarrow Int$

However, $Int \subseteq Value$, so "$\rightarrow$" is *antimonotonic* in its first argument. Since any antimonotonic function is not continuous, "$\rightarrow$" is not continuous.

5.3 Soundness of Constructor Semantics

Having defined the semantics of constructor expressions and kind-checking rules for them, we must now show that they match via a soundness theorem. The theorem basically states that the meaning of any well-kinded constructor expression is in the appropriate kind.

The theorem relies on a connection between syntactic kind assignments and constructor environments. Intuitively, the value of a free variable, as defined by an environment, should be in the kind set determined by the syntactic kind assignment. We call this concept *compatibility*:

Definition 2 *Let A be a syntactic kind assignment, and η be a constructor environment. Then η is* compatible *with A, written $\eta \models A$, if whenever $v \in Ide$ and v is in the domain of A, $\eta\, v \in Kind^{(A\, v)}$.*

If $\eta \models A$, we know that the semantics of constructor expressions of the form v, where $v \in Ide$, are sound since their meaning is given by $\eta\, v$ which lies in the appropriate kind set $(A\, v)$.

We now state the theorem:

Theorem 1 *If $A \vdash \mu : \kappa$ and $\eta \models A$, then $Cval[\![\mu]\!]\eta \in Kind^{\kappa}$.*

Proof: (Sketch) We use induction on the length of the deduction of $A \vdash \mu : \kappa$. Beginning with the base case, when the length is 1, we notice that only axioms can give a valid proof of length 1. We can check all of the axioms to ensure that they satisfy the conclusion.

For the induction case, where $n \geq 2$, we assume that all proofs of length less than n satisfy the conclusion. We can then show, by case analysis, that if any rule is used as the last step of a deduction, then the conclusion holds. This proves the theorem.

6 Semantics of Algebraic Types

Having built up a theory of type constructors, we apply this theory to study algebraic types, the goal of this paper. Using the function *Cval*, we convert constructor expressions representing algebraic types to elements in the model. We show that the semantic functions produce elements in the model, ensuring that the semantics is well-defined.

6.1 Conversion to Constructor Expressions

We need a general technique to convert algebraic types to constructor expressions. The main difficulty lies in interpreting mutually-recursive algebraic types. Our method is based on a method used in [KS85] (which uses the method to solve sets of mutually-recursive, first-order functions in SASL). The idea is to construct a chain of environments, with the meaning of each algebraic type determined from its definition and the last environment. Taking the sup of the chain results in finding the sup of the approximations to each algebraic type.

To do this construction, we need two semantic functions *Usertype* and *Settype*. *Settype* builds an environment in the chain that contains an approximation to the meaning of a set of algebraic types. *Usertype* actually builds the chain of environments and takes the sup of the chain, using *Settype* to build each approximation in the chain. The definitions for these functions are:

$$Usertype : TypeDefs \rightarrow Cenv \rightarrow Cenv$$
$$Usertype[\![AlgTypes]\!]\eta = fix(\lambda\eta'.Settype[\![AlgTypes]\!]\eta'\,\eta)$$

$$Settype : TypeDefs \rightarrow Cenv \rightarrow Cenv \rightarrow Cenv$$

1. $Settype[\![AlgType; AlgTypes]\!]\eta'\,\eta = Settype[\![AlgTypes]\!]\eta'\,(Settype[\![AlgType]\!]\eta'\,\eta)$
2. $Settype[\![\tau\; t_1 \ldots t_n ::= id_1\; arg_{1,1} \ldots arg_{1,j_1} \mid \ldots \mid id_k\; arg_{k,1} \ldots arg_{k,j_k}]\!]\eta'\,\eta = \eta[e/\tau]$

$$\begin{aligned} where\; e &= Cval[\![\lambda t_1^T. \ldots \lambda t_n^T.subtype_1 + \cdots + subtype_k]\!]\eta' \\ for\; subtype_i &= Trivial\; if\; j_i = 0 \\ &\quad arg_{i,1} \times \cdots \times arg_{i,j_i}\; otherwise \end{aligned}$$

Suppose, for example, that we wanted to interpret the following two mutually-recursive algebraic types using this method:

```
f * ::= Nil  | Push * (g *)
g * ::= Cons * (f *)
```

To build a chain of environments, we start with

$$\eta_0 = \eta[Bot_{T \Rightarrow T}/f, Bot_{T \Rightarrow T}/g]$$

and define, for all $i \geq 0$,

$$\eta_{i+1} = Settype[\![AlgTypes]\!]\eta_i\,\eta$$

where Bot_κ is the bottom element in the domain $Kind^\kappa$. (Note that $Bot_{T \Rightarrow T}$ applied to anything of kind T returns Bot_T.) Constructing the chain of environments yields

$$\eta_1 = \eta[(\lambda t_1^T.Trivial + (t_1 \times Bot_T))/f, (\lambda t_1^T.t_1 \times Bot_T)/g]$$

$$\eta_2 = \eta[(\lambda t_1^T.Trivial + t_1 \times t_1 \times Bot_T)/f, (\lambda t_1^T.t_1 \times (Trivial + t_1 \times Bot_T))/g]$$

and so on. Taking the sup of these environments is equivalent to calculating

$$fix(\lambda\eta\prime.Settype[\![AlgTypes]\!]\eta\prime\,\eta)$$

which is the definition of $Usertype$.

6.2 Well-Definedness of Semantics

We now show that our semantic functions for interpreting algebraic types produce valid constructor environments:

Theorem 2 *Let AlgTypes be a set of algebraic types defined*

$$AlgTypes = \tau_1\, t_{1,1} \ldots t_{1,j_1} ::= id_{1,1} arg_{1,o_1} \ldots arg_{1,p_1} \mid \ldots \mid id_{1,m_1} arg_{1,q_1} \ldots arg_{1,r_1}$$

$$\vdots$$

$$\tau_k\, t_{k,1} \ldots t_{k,j_k} ::= id_{k,1} arg_{k,o_k} \ldots arg_{k,p_k} \mid \ldots \mid id_{k,m_k} arg_{1,q_k} \ldots arg_{1,r_k}$$

If η is a constructor environment, then $(Usertype[\![AlgTypes]\!]\eta)\,\tau_i \in Kind^{\kappa_i}$, for $\kappa_i = T \Rightarrow \cdots \Rightarrow T$ with $j_i + 1$ T's.

Proof: (Sketch) To show that

$$(Usertype[\![AlgTypes]\!]\eta)\,\tau_i \in Kind^{\kappa_i}$$

we verify that for each η_j in the chain, that

$\eta_j\ \tau_i \in Kind^{\kappa_i}$

Taking the sup of this chain of environments is the same as taking the sup of each of the elements in the environments, so

$$(Usertype[\![AlgTypes]\!]\eta)\ \tau_i = \bigsqcup_{j<\omega}(\eta_j\ \tau_i)$$

Since $Kind^{\kappa_i}$ is a domain,

$$\bigsqcup_{j<\omega}(\eta_j\ \tau_i) \in Kind^{\kappa_i}$$

hence

$$(Usertype[\![AlgTypes]\!]\eta)\ \tau_i \in Kind^{\kappa_i}$$

which proves the theorem.

7 Connection to Miranda's Semantics

Algebraic types represent a significant part of the full semantics of Miranda, and the rest of the semantics must be defined so that the algebraic types fit into the model. In this section, we give an overview of the Miranda's semantics, tying it in to our work on algebraic types.

Specifying the semantics of Miranda, a typed language, consists of defining a set of type-checking rules, giving the semantics, and showing that the semantics is sound with respect to the type-checking system. Our type-checking system for Miranda was patterned after the systems found in [Mil78] and [MS82], since these type-checkers work with implicit polymorphic languages. The semantics for Miranda is based on the semantics given for SASL in [KS85].

Like the kind-checking system given before, the type-checking system for Miranda is composed of axioms and inference rules. An example of such an inference rule is

$$\frac{B \vdash g : \tau_2 \rightarrow \tau_3, B \vdash f : \tau_1 \rightarrow \tau_2}{B \vdash g.f : \tau_1 \rightarrow \tau_3}$$

where B is a *syntactic type assignment* that assigns types to variables, and "." is the operator in Miranda that composes two functions. This inference rule infers a type for a composition of two functions. With implicit polymorphic functions, the rules can get quite complicated; this rule, however, gives a flavor of the style.

To specify the semantics of Miranda, we use the semantics of SASL as a basis. A typical semantic clause looks like

$$Eval[\![g.f]\!]\rho = \lambda x.Eval[\![g]\!]\rho\ ((Eval[\![f]\!]\rho)x)$$

This clause interprets function composition, and the "ρ" is an environment that assigns values to identifiers.

The most important semantic clause, for the purposes of this paper, is the one that interprets algebraic types. Algebraic types define a new type *and* functions on that new type. These new functions must be added to the environment of a program, which the following semantic function *Tdecl* does:

$Tdecl : TypeDefs \rightarrow Env \rightarrow Env$

1. $Tdecl[\![\tau\, t_1 \ldots t_n ::= id_1\, arg_{1,1} \ldots arg_{1,j_1} \mid \ldots \mid id_k\, arg_{k,1} \ldots arg_{k,j_k}]\!]\rho = \rho[f_1/id_1, \ldots f_k/id_k]$

   where $f_i = \lambda e_1 \ldots . e_{j_i} .(e_1, \ldots e_{j_i})$

2. $Tdecl[\![AlgType; AlgTypes]\!]\rho = Tdecl[\![AlgTypes]\!]\,(Tdecl[\![AlgType]\!]\rho)$

The functions thus defined by algebraic types return tuples when given the appropriate number of arguments. This is the way we thought of these functions in Section 3.

The model for this semantics uses the underlying ideal model "Value". Given the semantics and the type-checking rules, one can prove the key theorem

Theorem 3 *If M is a Miranda expression with type τ, then*

$Eval[\![M]\!]\rho \in Cval[\![\tau]\!]\eta$

(For a more precise rendering of this theorem and its proof, see [Rie86].) Note that in order to prove this theorem, we need an intepretation for type expressions and, consequently, the algebraic types that appear in type expressions. It was the necessity to prove this theorem that generated the interest in modelling algebraic types.

8 Remaining Problems and Relevance to Previous Work

Our semantics has raised a number of issues in attempting to find a meaning for algebraic types. In this section, we clarify some of these issues, and propose solutions to them.

One issue that has been raised by Albert Meyer is the issue of structural equivalence versus name equivalence of types. Suppose, for example, we have the following two algebraic types in a program:

```
f ::= Nil1  |  Cons1 int f
g ::= Nil2  |  Cons2 int g
```

The type-checker will not allow us to write a value like

```
Cons2  1  Nil1
```

since `Nil1` is not of type `g`, so the expression would be rejected. Our semantics, however, does not distinguish between the two types – they are *structurally equivalent.* Operationally, Miranda distinguishes the values in the two types by tags – that is, Miranda uses *name equivalence* of types. One could, perhaps, tag every type in the model with an appropriate tag, but there may be complications in solving domain equations with tags.

An alternative approach is to consider the difference between structural and name equivalence to be simply a syntactic issue. That is, the type inference mechanism screens out all terms which are illegal under a name equivalence approach to typing. After that it does not matter if the semantics of the two types are different, since type-checking has eliminated the terms which would have been problematic. We leave open the inclusion of name equivalence and the issues it raises in semantics.

Another open problem is the incorporation of *unfree algebraic types* [Tur85a] into the semantics. Unfree algebraic types are algebraic types with laws for reducing elements to a normal form. For example, one could build a type consisting only of ordered lists:

```
olist  ::=  Onil  |  Ocons num olist
Ocons a (Ocons b x) => Ocons b (Ocons a x), a>b
```

This example is taken from [Tur85b]. The second line is the law for reducing an object of type *olist* to a normal form, and states that if a>b, reverse the order of b and a in the expression. [Tho87] gives one possible interpretation for an unfree algebraic type, by defining a set of functions that correspond to the reduction rules. For this example, we need only one function for the one rewrite rule:

```
Ocons' a Onil = Ocons a Onil
Ocons' a (Ocons b x) = Ocons b (Ocons' a x), a>b
                     = Ocons a (Ocons' b x)
```

One can then transform any program written with Ocons into one written with Ocons' by replacing any outer Ocons with an Ocons' and defining Ocons' in the program. One would then want to show that the values produced by Ocons' is a type in the model. For this example, the values produced by Ocons' are indeed a type in the model. (Proof hint: Without the law, the meaning of olist is a flat cpo, and so any subset of it including the bottom element is an ideal.) What is not known, however, is if this function construction can be generalized to all normalizing rewrite rules, and if the functions always produce values that form a type.

There is another alternative for adding unfree algebraic types. Since all terms of unfree types are required to have a normal form, the semantics could take advantage of this fact. We would find the interpretation of the corresponding free type and interpret the terms as the interpretations of their normal forms in that type. Like name equivalence, however, we leave the incorporation of unfree types as an open problem.

Our main problems, though, arise from problems in the ideal model, especially by the fact that "$\rightarrow$" is not a continuous function on types. If we allow algebraic types like

```
fun * ::= * -> num
```

in the language, we cannot interpret them. Using this example, the corresponding constructor expression would be

$$\lambda t^T.t \rightarrow Num$$

which is not continuous since it is antimonotonic in its first argument. There are polymorphic models, though, that can interpret constructor expressions like these: the closure model [McC79] and the finitary projection model [ABL86]. An interesting exercise would involve writing the semantics of Miranda using one of these two models as a basis.

We have, however, shown how to interpret many of Miranda's type constructors within the ideal model. The most important result in this paper is the practical use made of higher-order functions; these functions describe algebraic types very naturally, and are included in the semantic model. We hope that the use of higher-order functions on types in semantics will now be justified.

9 Acknowledgements

We would like to thank Albert Meyer for his suggestions on structural equivalence and name equivalence, and Matt Kaufmann for helpful comments on a preliminary version of this work.

References

[ABL86] Roberto Amadio, Kim B. Bruce, and Giuseppe Longo. The finitary projection model for second order lambda calculus and solutions to higher order domain equations. In *First Annual Symposium on Logic in Computer Science*, Cambridge, MA, 1986.

[BMM85] Kim B. Bruce, Albert R. Meyer, and John C. Mitchell. The semantics of second-order lambda calculus. 1985. To appear, *Information and Computation.*

[BMS80] R. M. Burstall, D. B. MacQueen, and D. T. Sannella. Hope: an experimental applicative language. In *Proceedings of the First International LISP Conference*, Stanford, CA, 1980.

[DD85] Alan Demers and James Donahue. Data types are values. *ACM Transactions on Programming Languages and Systems*, 7:426–445, July 1985.

[DoD83] U.S. Department of Defense. *Reference Manual for the Ada Programming Language*, Springer-Verlag, New York, 1983.

[Gir71] J.-Y. Girard. Une extension de l'interpretation de Gödel à l'analyse, et son application à l'élimination des coupures dans l'analyse et al théorie des types. In J.E. Fenstad, editor, *Second Scandanavian Logic Symposium*, pages 63–92, North Holland, Amsterdam, 1971.

[Jon87] Simon L. Peyton Jones. *The Implementation of Functional Programming Languages*, Prentice-Hall, Englewood Cliffs, NJ, 1987.

[KS85] Matthew Kaufmann and Douglas Surber. *Syntax, Semantics, and a Formal Logic for SASL.* Technical Report ARC 85-03, Burroughs Austin Research Center, January 1985.

[McC79] Nancy McCracken. *An Investigation of a Programming Language with a Polymorphic Type Structure.* PhD thesis, Syracuse University, 1979.

[Mil78] Robin Milner. A theory of type polymorphism. *Journal of Computer and System Sciences*, 17:348–375, 1978.

[MPS84] D. B. MacQueen, Gordon Plotkin, and Ravi Sethi. An ideal model for recursive polymorphic types. In *Proceedings of the Eleventh ACM Symposium on the Principles of Programming Languages*, Salt Lake City, UT, pages 165–174, 1984.

[MPS86] D. B. MacQueen, Gordon Plotkin, and Ravi Sethi. An ideal model for recursive polymorphic types. *Information and Control*, 71, 1986.

[MS82] D. B. MacQueen and Ravi Sethi. A semantic model of types for applicative languages. In *1982 ACM Symposium on Lisp and Functional Programming*, Pittsburgh, PA, pages 243–252, 1982.

[Rey74] John C. Reynolds. Towards a theory of type structure. In *Proceedings Colloque sur la Programmation, Lecture Notes in Computer Science 19*, pages 408–425, Springer-Verlag, New York, 1974.

[Rie86] Jon G. Riecke. A denotational approach to the semantics of polymorphic languages. B.A. Honors Thesis, Department of Computer Science, Williams College. 1986.

[Sco82] Dana S. Scott. Domains for denotational semantics. In M. Nielsen and E. M. Schmidt, editors, *Automata, Languages, and Programming, Lecture Notes in Computer Science 140*, pages 577–613, Springer-Verlag, New York, 1982.

[Str67] Christopher Strachey. Fundamental concepts in programming languages. Lecture notes for International Summer School in Computer Programming, Copenhagen. August 1967.

[Tho87] Simon Thompson. Lawful types in Miranda. Unpublished manuscript. 1987.

[Tur85a] David A. Turner. Functional programs as executable specifications. In Hoare and Shepherdson, editors, *Mathematical Logic and Programming Languages*, pages 29–54, Prentice-Hall, Englewood Cliffs, NJ, 1985.

[Tur85b] David A. Turner. Miranda: a non-strict functional language with polymorphic types. In *Proceedings IFIP International Conference on Functional Programming Languages and Computer Architecture, Nancy, Lecture Notes in Computer Science 201*, Springer-Verlag, New York, 1985.

Part VI

New Directions

Path Semantics

Adrienne Bloss
Paul Hudak*
Yale University
New Haven, Connecticut 06520

Abstract

Knowledge of *order of evaluation of expressions* is useful for compile-time optimizations for lazy sequential functional programs. We present *path semantics*, a non-standard semantics that describes order of evaluation for a first-order functional language with lazy evaluation. We also provide an effective abstraction of path semantics that provides compile-time information. We show how path semantics may be used in strictness analysis, process scheduling in parallel systems, and optimized self-modifying thunks.

1 Introduction

Functional programming languages are becoming increasingly popular in the research community because they offer clean semantics, lazy evaluation, and no side effects. However, their acceptance into the "real world" has been hindered by their typically slow execution relative to that of imperative languages. *Lazy evaluation* has been recognized as one source of inefficiency; the compiler doesn't know which arguments to a function will be evaluated, and so must create a *delayed representation* or "thunk" for each argument so that it may be evaluated at an arbitrary time in the future. In applicative-order evaluation, typical of imperative languages, this is not an issue since all arguments to a function are evaluated at the time of the function call. Lazy evaluation has such advantages as terminating in some cases when applicative-order evaluation does not and allowing the use of "infinite" data structures, but due to the overhead of building and invoking thunks it is frequently less efficient than applicative-order evaluation.

One attempt to optimize lazy evaluation has been through *strictness analysis*. A function f is strict in its ith argument if and only if $f(...x_{i-1}, \perp, x_{i+1}...) = \perp$ for all values of x_j, $j \neq i$. If this can be inferred at compile-time, then the compiler may always evaluate f's i^{th} argument at the time of the call to f without fear of changing the termination properties of the program. Strictness analysis has been widely studied [2,3,6,7,9,10], and for the case where the x_i take values from a flat domain the problem appears to be well understood.

*This research was supported in part by the National Science Foundation under grant DCR-8451415

There is another inefficiency that arises from lazy evaluation that has not been so widely explored. Applicative-order evaluation typically tells the compiler not only that a function's arguments are evaluated at the time of the function call, but also the *order* in which those arguments are evaluated. While the semantics of the language may not specify any particular order of evaluation for function arguments, an ordering will usually be fixed by the compiler; this increases the compiler's knowledge about the state of the machine when any given expression is evaluated and allows various optimizations. In lazy evaluation, however, there is no such ordering; even after strictness analysis, the compiler knows only *what* will be evaluated, not *when* it will be evaluated.

In this paper we present *path semantics*, a simple but powerful analysis that addresses the issue of *order of evaluation* for a first-order functional language with sequential lazy evaluation. We show that path semantics not only subsumes first-order strictness analysis, but can also be applied to problems such as scheduling process creation in parallel systems and optimizing self-modifying thunks.

The next section gives an intuitive overview of path semantics and discusses how it relates to previous work. Section 3 gives formal definitions for an exact path semantics (which describes precisely the path taken through a function when it is executed on known arguments) and an abstract path semantics (which describes the set of all possible paths that might be taken through a function without knowledge of its arguments). In Section 4 we suggest some applications of this work, and in Section 5 we present our conclusions and discuss implementation issues and future work.

2 Preliminaries

2.1 Related Work

As previously noted, strictness analysis has been studied extensively for first-order and higher-order functional languages on flat domains. Issues involving order of evaluation have also been studied in several places [1,5,11]. However, to our knowledge, no *general* analysis such as the one presented here has been done in the past. The most closely related work is that of [1], in which abstract interpretation is used to determine which expressions are (or might be) evaluated before (or after) a given expression. While the spirit of that work is similar to this, it is limited in that it requires a separate analysis for each type of information desired, and the analyses themselves are somewhat complex. Path semantics subsumes that work in scope and is at the same time simpler.

The work of Raoult and Sethi [11] is related to ours in that they attempt to derive an optimal order of evaluation by "pebbling" a program graph. Their work assumes applicative-order evaluation and is targeted at *specifying* an order of evaluation so as to minimize storage requirements of a program. Our work, however, assumes normal-order evaluation and attempts to *infer* the order of evaluation information that is already implicit in a program. However, we may eventually layer an "optimizer" *on top* of our analysis to select an "optimal" path by some criterion (e.g., minimizing storage requirements).

2.2 Intuitive Description

A *path* through a function f is defined as an ordering on the evaluation of f's bound variables. Thus at run-time there is exactly one path through each invocation of f, but at compile-time we can only infer that there is a *set* of possible paths through f, one of which will actually be chosen for each invocation during program execution. For example, the "wrapped-up conditional" function "$f(x,y,z) =$ if x then y else z" has two possible paths, $\langle x,y\rangle$ and $\langle x,z\rangle$.

Since we are using lazy evaluation, a bound variable is evaluated at most once. Thus a path through f need not include all of f's bound variables, and it cannot include any bound variable more than once. This second property guarantees that there is a finite number of paths through any function, which leads to our termination property.

2.3 Notation and Abstract Syntax

Our analysis is for a first-order functional language with lazy evaluation. The abstract syntax for our language is given below:

$$\begin{array}{rcll} c & \in & Con & \text{constants} \\ x & \in & Bv & \text{bound variables} \\ p & \in & Pf & \text{primitive functions} \\ f & \in & Fv & \text{function variables} \\ e & \in & Exp & \text{expressions, where } e = c \mid x \mid p(e_1,...,e_n) \mid f(e_1,...,e_n) \\ pr & \in & Prog & \text{programs, where } pr = \{f_i(x_1,...,x_n) = e_i\} \end{array}$$

We use double brackets to surround syntactic objects, as in $\mathcal{P}[\![x_i]\!]$, and single brackets to indicate environment update, as in $env[y/x]$; $[y_i/x_i]$ is shorthand for $\perp[y_1/x_1,...,y_n/x_n]$, where the subscript bounds are inferred from context. Angle brackets surround paths, as in $\langle x,y\rangle$. If a function f has an argument x_i that is used in k places in f we say that x_i has k *occurrences*, denoted $x_{i1},...,x_{ik}$.

3 First-Order Path Semantics

We begin by defining an exact path semantics. Note that this is a non-standard semantics, *not* an abstraction of the standard semantics, since it contains information not found in the standard semantics.

3.1 Exact Path Semantics

We introduce D, the set of all variables, and $Path$, the flat domain of paths with bottom element $\perp_p$. $Path$ is defined as follows:

$$Path = \{\perp_p\} \cup \{\langle d_1, ..., d_n\rangle \mid n \geq 0, \forall i, 1 \leq i < n, d_i \in D\}$$

Note that the empty path $\langle\rangle$ is simply the path in which no bound variables are evaluated, and is not to be confused with the bottom path $\perp_p$, which represents non-termination.

We define "::" as an infix path append operator as follows: $\forall p \in Path,\ x_i \in D,\ 1 \leq i \leq n,\ n > 0$

$$\begin{aligned}
p :: \perp_p &= \perp_p \\
\perp_p :: p &= \perp_p \\
p :: \langle\rangle &= p \\
\langle\rangle :: p &= p \\
\langle x_1, ..., x_m\rangle :: \langle x_{m+1}, ..., x_n\rangle &= \text{if } \quad x_{m+1} \in \{x_1, ..., x_m\} \\
& \text{ then } \langle x_1, ..., x_m\rangle :: \langle x_{m+2}, ..., x_n\rangle \\
& \text{ else } \langle x_1, ..., x_m, x_{m+1}\rangle :: \langle x_{m+2}, ..., x_n\rangle
\end{aligned}$$

Note that all but the first occurrence of a given bound variable are removed from a path; as discussed earlier, this is because a path reflects the order of *evaluation* of bound variables, and in lazy evaluation a bound variable is evaluated at most once.

Semantic Domains

$Path$,		the flat domain of paths with bottom element $\perp_p$
$Pfun$	$= \cup_{n=1}^{\infty}(Path^n \to Path)$,	the function space mapping tuples of paths to paths
$Penv$	$= Fv \to Pfun$,	the function variable environment
Bve	$= Bv \to Path$,	the bound variable environment

Semantic Functions

$$\begin{aligned}
\mathcal{P} &: Exp \to Bve \to Penv \to Path \\
\mathcal{K}_p &: Pf \to Pfun \\
\mathcal{P}_{pr} &: Prog \to Penv
\end{aligned}$$

$$\begin{aligned}
\mathcal{K}_p[\![+]\!] &= \lambda(x, y).\ x :: y \\
\mathcal{K}_p[\![\text{IF}]\!] &= \lambda(p, c, a).\ p :: (oracle \to c, a)
\end{aligned}$$

$$\begin{aligned}
\mathcal{P}[\![c]\!]bve\ penv &= \langle\rangle \\
\mathcal{P}[\![x]\!]bve\ penv &= bve[\![x]\!] \\
\mathcal{P}[\![p(e_1, ..., e_n)]\!]bve\ penv &= \mathcal{K}_p[\![p]\!](\mathcal{P}[\![e_1]\!]bve\ penv, ..., \mathcal{P}[\![e_n]\!]bve\ penv) \\
\mathcal{P}[\![f(e_1, ..., e_n)]\!]bve\ penv &= penv[\![f]\!](\mathcal{P}[\![e_1]\!]bve\ penv, ..., \mathcal{P}[\![e_n]\!]bve\ penv) \\
\mathcal{P}_{pr}[\![\{f_i(x_1, ..., x_n) = e_i\}]\!] &= penv \text{ whererec}
\end{aligned}$$

$$penv = [(\lambda(y_1, ..., y_n).\mathcal{P}[\![e_i]\!][y_j/x_j]\ penv)/f_i]$$

The meaning of these equations should be clear. A path is a tuple of bound variables. Thus there is no path associated with a constant; the path associated with a bound variable is its value in the bound variable environment; the path of a primitive operator depends on its definition in $\mathcal{K}_p$ applied to the paths of the arguments; the path of a function call is the meaning of the function in the function environment applied to the paths of the arguments; and the meaning of a program is a function environment which, when given a function name and a path for each argument, returns the path through that function.

The definitions of $\mathcal{K}_p$ reflect the run-time behavior of our language. $\mathcal{K}_p[\![+]\!]$ takes two paths and concatenates them, removing duplicates (thus we are assuming left-to-right evaluation for the strict operator "+"). $\mathcal{K}_p[\![IF]\!]$ takes three paths, corresponding to the predicate, consequent, and alternate, and appeals to an oracle to decide if it should return the concatenation of the predicate and the consequent or the predicate and the alternate. The oracle could be removed by embedding the standard semantics in with the path semantics, but that unnecessarily complicated the presentation.

3.2 Abstract Semantics

The abstraction of the exact semantics is straightforward. We lift the domain of paths to the *powerdomain* of paths (using the Egli-Milner powerdomain construction), and lift our functions to map sets of paths to sets of paths, instead of paths to paths. Note that our method corresponds to the *relational attribute* method described in [8] and [9]. Also note that we have lifted the restriction of left-to-right evaluation in the primitive operators; this yields a more general analysis, with the potential for the compiler to *choose* an optimal ordering. However, the analysis still reflects sequential evaluation, as modeling parallel evaluation would require *interleaving* the paths returned by the operator's arguments.

Semantic Domains

$Path$,		the domain of paths
$P(Path)$,		the powerdomain of $Path$
$Pfun$	$= \bigcup_{n=1}^{\infty}(P(Path^n) \to P(Path))$,	the function space mapping paths to paths
$Aenv$	$= Fv \to Pfun$,	the function environment
Bve	$= Bv \to Path$,	the bound variable environment

Semantic Functions

$$\begin{aligned} \mathcal{A} &: Exp \to Bve \to Aenv \to P(Path) \\ \mathcal{K}_a &: Pf \to Pfun \\ \mathcal{A}_{pr} &: Prog \to Aenv \end{aligned}$$

$$\begin{aligned} \mathcal{K}_a[\![+]\!] &= \lambda s.\{\langle x :: y\rangle, \langle y :: x\rangle \mid (x,y) \in s\} \\ \mathcal{K}_a[\![IF]\!] &= \lambda s.\{\langle p :: c\rangle, \langle p :: a\rangle \mid (p,c,a) \in s\} \end{aligned}$$

$$\begin{aligned} \mathcal{A}[\![c]\!]\; bve\; aenv &= \{\langle\rangle\} \\ \mathcal{A}[\![x]\!] bve\; aenv &= \{bve[\![x]\!]\} \\ \mathcal{A}[\![p(e_1,...,e_n)]\!] bve\; aenv &= \mathcal{K}_a[\![p]\!](\mathcal{A}[\![e_1]\!] bve\; aenv \times ... \times \mathcal{A}[\![e_n]\!] bve\; aenv) \\ \mathcal{A}[\![f(e_1,...,e_n)]\!] bve\; aenv &= aenv[\![f]\!](\mathcal{A}[\![e_1]\!] bve\; aenv \times ... \times \mathcal{A}[\![e_n]\!] bve\; aenv) \\ \mathcal{A}_{pr}[\![\{f_i(x_1,...,x_n) = e_i\}]\!] &= aenv \text{ whererec} \end{aligned}$$

$$aenv = [(\lambda s. \bigcup\{\mathcal{A}[\![e_i]\!]\; [y_j/x_j]\; aenv \mid (y_1,...,y_n) \in s\})/f_i]$$

Theorem 1 *$\mathcal{A}_{pr}[\![pr]\!]$ is computable for any finite program pr.*

Proof: Using the standard iterative method for computing Kleene's ascending chain, our first approximation is $aenv^0 = \mathcal{A}^0_{pr}[\![pr]\!] = [(\lambda s.\{\perp_p\})/f_i]$. For each subsequent approximation $aenv^n$, $aenv^n = \mathcal{A}^n_{pr}[\![pr]\!] = (\lambda(y_1,...,y_n).\mathcal{A}[\![e_i]\!]\; [y_j/x_j]\; aenv^{n-1})/f_i]$. Since all operators are monotonic and the domains are finite, we are guaranteed to arrive at a least upper bound in a finite number of steps. □

The paths defined by the semantics above are paths of *bound variables.* Thus they yield information about the order in which distinct bound variables are evaluated, but not about the order in which the *occurrences* of a *particular* bound variable are demanded. Yet for some optimizations (such as efficient self-modifying thunks), this is precisely the information required.

Fortunately, the problem is easily solved. For each function f_i in program *pr*, define a new version f_i' which is identical to f_i except that each *occurrence* of a bound variable in f_i becomes a unique bound variable in f_i'. Thus if f_i has j bound variables, each of which has k occurrences, then f_i' will have $j * k$ bound variables, each of which occurs once. Note that the body of f_i' is otherwise identical to the body of f_i — for example, if f_i is recursive, then f_i' will call f_i internally, *not* f_i'. Letting pr' be pr extended in this way, if $aenv = \mathcal{A}_{pr}[\![pr]\!]$, we simply need to evaluate $aenv[\![f_i']\!]\{(\langle \mathrm{x}_{11}\rangle, ..., \langle \mathrm{x}_{1k}\rangle, ..., \langle \mathrm{x}_{j1}\rangle, ..., \langle \mathrm{x}_{jk}\rangle)\}$ to get the path through the *occurrences.* Note that the first occurrence of each bound variable in a path corresponds to that variable's *evaluation,* while subsequent occurrences correspond to *uses.* Intuitively, we can say that $aenv[\![f_i]\!]$ characterizes f_i's *external* behavior, while $aenv[\![f_i']\!]$ characterizes f_i's *internal* behavior.

4 Applications

4.1 Strictness Analysis

Paths are useful for a variety of optimizations. For example, once we have computed the set of all possible paths through a function f, computing the strictness properties of f is easy.

Consider the definition of strictness:

$$f \text{ is strict in its } i^{th} \text{ argument} \iff \forall x_j, j \neq i, f(...x_{i-1}, \bot, x_{i+1}...) = \bot$$

In terms of paths, we can say that f is strict in its ith argument x_i if and only if x_i appears in every terminating path through f. That is,

$$\forall p \in aenv[\![f]\!]\{(\langle x_1\rangle, ..., \langle x_n\rangle)\}, (x_i \in p) \vee (p = \bot_p)$$

For example, consider the following "wrapped-up conditional" function:

$$f\ x\ y\ z == x \rightarrow y, z$$

The paths through f are $\{\langle x, y\rangle, \langle x, z\rangle, \bot_p\}$. f is strict in x, as it appears in both terminating paths, but not in y (since it is not in $\langle x, z\rangle$) or z (not in $\langle x, y\rangle$).

It is clear that we can do *some* strictness analysis with paths, but we would like to show that we can do a *good* analysis. Specifically, we show here that we do precisely the same analysis that Hudak and Young do in their first-order work [6]. More formally:

Let *senv* be the strictness environment found by Hudak and Young's first-order strictness analysis; *senv* takes a function variable f_i and a list of sets $(s_1, ..., s_n)$, where s_j represents the set of variables in which f_i's j^{th} argument is strict, and returns the strictness properties of f_i (that is, those variables in which f_i is strict) in terms of $(s_1, ..., s_n)$. Then

$$x \in senv[\![f_i]\!](s_1, ..., s_n) \iff x \in F(penv[\![f_i]\!]\{(\langle y_1\rangle, ..., \langle y_n\rangle)\})(y_1, ..., y_n)(s_1, ..., s_n)$$

where F is defined as follows:

$$F\{p_1, ..., p_m\}(y_1, ..., y_n)(s_1, ..., s_n) = p'_1 \cap p'_2 \cap ... \cap p'_m$$

$$\text{where } p'_i = \bigcup\{s_j \mid y_j \in p_i\}$$

Thus F takes a set of paths, a list of the elements from which those paths are composed, and a list of the sets passed to the strictness environment and "translates" from the path model to the strictness model.

As an example, again let $f(x, y, z) = x \rightarrow y, z$. Then using Hudak and Young strictness, we get:

$$senv[\![f]\!](s_1, s_2, s_3) = s_1 \cup (s_2 \cap s_3)$$

Using paths, we get:

let

$$P = penv[\![f]\!]\{(\langle y_1\rangle, \langle y_2\rangle, \langle y_3\rangle)\} = \{\langle y_1, y_2\rangle, \langle y_1, y_3\rangle\}$$

then

$$F(P)(y_1, ..., y_n)(s_1, ..., s_n) = (s_1 \cup s_2) \cap (s_1 \cup s_3)$$
$$= s_1 \cup (s_2 \cap s_3)$$

In fact, this equality holds in general.

Theorem 2 *Let "senv" be the first-order strictness environment found by Hudak and Young's analysis of [6], "penv" the path environment defined in our Section 3.2, and the function "F" defined as above. Then*

$$x \in senv[\![f_i]\!](s_1, ..., s_n) \iff x \in F(penv[\![f_i]\!]\{(\langle y_1\rangle, ..., \langle y_n\rangle)\})(y_1, ..., y_n)(s_1, ..., s_n)$$

Proof: See Appendix.

4.2 Optimizing Self-Modifying Thunks

Another interesting application is the optimization of "self-modifying thunks". In lazy evaluation, the value of an actual parameter is computed at most once; that value is then stored and returned when subsequently demanded. This requires a check at each reference to the variable to see if it has already been evaluated — if not, the value is computed; if so, the stored value is returned. However, a variable may be referenced many times, and it may be possible to infer at compile-time which occurrence will be demanded first, forcing computation, and which occurrences will be demanded later, requiring no computation. With this knowledge we can avoid the check in the self-modifying thunk. Path semantics can be used to infer this information as follows:

If function f has an argument x_i that has j occurrences, then:

1. If x_{ik} is the first occurrence of x_i in *every* path through f, then the "thunk" for x_i may be invoked when x_{ik} is referenced without checking to see if the value was already computed.
2. If x_{ik} is the first occurrence of x_i in *no* path through f, then its value must have already been computed, and thus may be returned immediately, again without doing a status check.
3. If neither (1) nor (2) holds, a runtime check must be made.

4.3 Scheduling Processes in a Parallel System

In many parallel systems a scheduler attempts to "optimally" allocate resources, that is, to assign processors to processes in such a way as to minimize wait time. However, the scheduler frequently has no information about how long the processes will take to run; thus a reasonable heuristic is to assume that they all require the same computation time and to give highest priority to those processes whose results are demanded first [4]. In functional languages, at each call to a function f it is natural to create a process for each strict argument to f, and so the order in which f will demand its arguments should determine the order in which the argument processes are created. Of course, this ordering is precisely the information contained in the paths through f.

5 Discussion and Future Work

While path semantics is an elegant and powerful way of capturing the notion of order of evaluation of expressions, an efficient implementation is critical to its use in a compiler. Our current implementation is designed for research purposes only, yet we are encouraged by its speed, and we are optimistic that an optimized implementation will be fast enough for practical use. An obvious concern about computation of paths is the exponential nature of the number of paths. Fortunately, functions tend not to permute their arguments, and we have found that if we fix the order of evaluation for the strict primitive operators such as "+", we get modest growth in execution time with the number of arguments.

We believe that path semantics is a useful tool for a variety of compile-time optimizations. Its simplicity makes it easy to reason about while its generality makes it applicable to many problems. We have implemented first-order path semantics and have applied it to strictness analysis and the "before" and "after" analyses described in [1]. We are currently exploring higher-order path semantics; while the higher-order extension appears to be straightforward, we suspect that, due to the usual problem with higher-order analyses, path semantics will have to be combined with a collecting interpretation to give useful results. This is particularly true because, in general, safety considerations require that we find *all* possible paths through expressions, which means that in the absence of information (*e.g.* in the case of a functional argument) we must conservatively assume that all paths are possible. It is unfortunate that the "no information" result is so cumbersome (as opposed to strictness, *e.g.*, where an unknown function must be assumed to be strict in nothing), but it makes collecting that much more appealing.

We are also exploring ways of extending path semantics so that it may be useful for a wider range of optimizations. For example, we are considering paths through *labels* instead of bound variables; this would allow us to reason directly about the use of any expression, not just bound variables.

Finally, eventually we hope to investigate the problem of choosing an *optimal* path; that is, given a set of possible paths through each function in a program, choose those paths that yield the "best" code in some way (for example, we might choose the paths that require the fewest copies of arrays). This is related to the pebbling problem discussed in [11], but we think it is worth investigating in the framework of paths.

Appendix: Proof of Relationship Between Path Semantics and First-Order Strictness Analysis

Here we show that we can use paths to do exactly the same strictness analysis done in the first-order work by Hudak and Young. We use a small trick in that we show that the paths derived using the *independent* attribute method yield the same analysis, instead of those derived from the *relational* attribute method that we present in this paper. We do this because the form of the independent attribute method more closely matches that of Hudak and Young's strictness analysis, making comparison of the two analyses easier. While the relational attribute method yields a more precise (smaller) set of paths (and thus might seem to give a *better* strictness analysis), it is easy to show that the two methods behave identically when the strictness function F (defined below) is applied to the resulting sets. (To see why, consider the extra paths generated by the independent method. Every such path must contain a path generated by the relational method, and thus performing the intersection in F with the sets derived from these extra paths can have no effect on the result.)

The independent method is very much like the relational method, except that bound variables are bound to *sets of paths*, instead to a single paths as in the relational method. The differences are apparent in the domains and equations shown below (only those equations that differ from the relational method are shown):

Domains

$D,$		the domain of path elements
$Path,$		the domain of paths
$P(Path),$		the powerdomain of $Path$
$Pfun$	$= \bigcup_{n=1}^{\infty}((P(Path))^n \rightarrow P(Path)),$	the function space mapping paths to paths
$Penv$	$= Fn \rightarrow Pfun,$	the function environment
Bve	$= Bv \rightarrow P(Path),$	the bound variable environment

Functions

$$\begin{aligned} \mathcal{A} &: Exp \rightarrow Bve \rightarrow Penv \rightarrow P(Path) \\ \mathcal{K}_p &: Pf \rightarrow Pfun \\ \mathcal{P}_{pr} &: Prog \rightarrow Penv \end{aligned}$$

$$\begin{aligned} \mathcal{P}[\![f(e_1,...,e_n)]\!]\ bve\ penv &= penv[\![f]\!](\mathcal{P}[\![e_1]\!]\ bve\ penv,...,.\mathcal{P}[\![e_n]\!]\ bve\ aenv) \\ \mathcal{P}[\![p(e_1,...,e_n)]\!]\ bve\ penv &= \mathcal{K}_p[\![f]\!](\mathcal{P}[\![e_1]\!]\ bve\ penv,...,.\mathcal{P}[\![e_n]\!]\ bve\ penv) \\ \mathcal{K}_p[\![IF]\!] &= \lambda(p,c,a).p * (c \cup a) \\ \mathcal{K}_p[\![+]\!] &= \lambda(x,y).x * y \\ \mathcal{P}_{pr}[\![\{f_i(x_1,...,x_n) = body_i\}]\!] &= penv \text{ whererec} \end{aligned}$$

$$penv = [(\lambda(y_1,...,y_n).\mathcal{P}[\![body_i]\!]\ [y_j/x_j]\ penv)/f_i]$$

The "cross-product" operator $*$ is defined in terms of the path-append operator "::" (defined in the relational semantics) as follows:

$$\{p_1,...,p_m\} * \{p_{m+1},...,p_n\} =$$

$$\{p_1 :: p_{m+1}, p_1 :: p_{m+2},...,p_1 :: p_n,...,p_m :: p_{m+1},...,p_m :: p_n\}$$

We now proceed with our proof.

Let $\perp_{Sv}$ be the bottom element in the strictness domain (following [6]), $\perp_p$ the bottom element in the path domain, and $\{\perp_p\}$ the bottom element in the powerdomain of paths (as in Section 3.2). Note that $\perp_{Sv}$ represents the set of all variables V, and $\{\perp_p\}$ represents the set containing only the non-terminating path. Define

$$senv_0 = [(\lambda(z_1, ..., z_n).\perp_{Sv})/f_i]$$

$$penv_0 = [(\lambda(z_1, ..., z_n).\{\perp_p\})/f_i]$$
$$senv_k = [(\lambda(z_1, ..., z_n).\mathcal{S}[\![e_i]\!][z_j/x_j]senv_{k-1})/f_i]$$
$$penv_k = [(\lambda(z_1, ..., z_n).\mathcal{P}[\![e_i]\!][z_j/x_j]penv_{k-1})/f_i]$$

Let

$$\begin{array}{lcl} p_i & \in & Path \\ y_i & \in & D \text{ (path elements)} \\ s_i & \in & P(V) \text{ (strictness sets)} \end{array}$$

To "translate" from paths to strictness sets, we define the function F as follows:

$$F\{p_1, ..., p_m\}(y_1, ..., y_n)(s_1, ..., s_n) = p'_1 \cap p'_2 \cap ... \cap p'_m$$
$$\text{where } p'_i = \bigcup\{s_j \mid y_j \in p_i\}$$

Let $senv$ be the strictness environment as found by Hudak and Young, and $\mathcal{S}$ be the strictness function; also let $penv$ be the path environment, and $\mathcal{P}$ be the path function. Then we claim that if the u_i are strictness expressions composed of s_i, and the v_i are sets of paths that are composed of y_i,

$$x \in senv[\![f_i]\!](u_1, ..., u_n) \iff x \in F(penv[\![f_i]\!](v_1, ..., v_n))(y_1, ..., y_n)(s_1, ..., s_n)$$
$$\text{where } u_j = F\ v_j\ (y_1, ..., y_n)(s_1, ..., s_n)$$

Let $\Psi(senv, penv)$ be the above predicate. (Since $senv_*$ and $penv_*$ are constructed from finite domains and monotonic operators, every chain $senv_0, senv_1, ..., penv_0, penv_1, ...$ is guaranteed to be of finite height, and thus the predicate is admissible.) Also let $senv_*$ be the least fixed point of the strictness equations, and $penv_*$ the least fixed point of the path equations, then we will use structural and fixpoint induction to show that $\Psi(senv_*, penv_*)$ holds.

First, consider $\Psi(senv_0, penv_0)$:

$$\Psi(senv_0, penv_0) = x \in \perp_{Sv} \iff x \in F\{\perp_p\}(y_1, ..., y_n)(s_1, ..., s_n)$$

$\perp_{Sv}$ is defined to be V, the set of all strictness elements, and by definition $\forall i, y_i \in \perp_p$. Thus we have

$$\Psi(senv_0, penv_0) = x \in V \iff x \in \bigcup_i s_i$$

which is true, since here V is defined to be precisely $\bigcup_i s_i$.

Now consider the case of $\Psi(senv_k, penv_k)$:

$\Psi(senv_k, penv_k) =$

$$x \in \mathcal{S}[\![e_i]\!]senv_{k-1}[u_j/x_j] \iff x \in F(\mathcal{P}[\![e_i]\!]penv_{k-1}[v_j/x_j])(y_1, ..., y_n)(s_1, ..., s_n)$$

This requires structural induction on e_i. Let $a = senv_{k-1}[u_j/x_j]$, $b = penv_{k-1}[v_j/x_j]$, $c = (y_1, ..., y_n)(s_1, ..., s_n)$.

1. e_i is a constant. Then $\Psi(senv_k, penv_k)$ becomes

$$x \in \{\} \iff x \in F\{\langle\rangle\}c$$

which, applying F, becomes

$$x \in \{\} \iff x \in \{\}$$

2. e_i is a bound variable x_j. Then $\Psi(senv_k, penv_k)$ becomes

$$x \in u_j \iff x \in F\ v_j\ c$$

By the structural induction hypothesis,

$$u_j = F\ v_j\ c$$

so trivially the implication holds.

3. $e_i = f(e_1, ..., e_n)$. Then $\Psi(senv_k, penv_k)$ becomes

$$x \in senv_{k-1}[\![f]\!](\mathcal{S}[\![e_1]\!]a, ..., \mathcal{S}[\![e_n]\!]a) \iff x \in F(penv_{k-1}[\![f]\!](\mathcal{P}[\![e_1]\!]b, ..., \mathcal{P}[\![e_n]\!]b))c$$

By the structural induction hypothesis,

$$\mathcal{S}[\![e_j]\!]a = F(\mathcal{P}[\![e_j]\!]b)c$$

but then the fixpoint induction hypothesis immediately applies.

4. $e_i = IF(e_1, e_2, e_3)$. Then $\Psi(senv_k, penv_k)$ becomes

$$x \in \mathcal{S}[\![IF(e_1, e_2, e_3)]\!]a \iff x \in F(\mathcal{P}[\![IF(e_1, e_2, e_3)]\!]b)c$$

Applying the definitions for $\mathcal{S}$ and $\mathcal{P}$, we get

$$x \in (\mathcal{S}[\![e_1]\!]a \cup (\mathcal{S}[\![e_2]\!]a \cap \mathcal{S}[\![e_3]\!]a)) \iff x \in (F(\mathcal{P}[\![e_1]\!]b \cup (\mathcal{P}[\![e_2]\!]b * \mathcal{P}[\![e_3]\!]b))c)$$

Let $Q_1 = \mathcal{S}[\![e_1]\!]a \cup (\mathcal{S}[\![e_2]\!]a \cap \mathcal{S}[\![e_3]\!]a)$, $Q_2 = \mathcal{P}[\![e_1]\!]b \cup (\mathcal{P}[\![e_2]\!]b * \mathcal{P}[\![e_3]\!]b)$. Thus we are trying to show that

$$(x \in Q_1 \iff x \in F\ Q_2\ c)$$

Note that by the definition of F,

$$x \in (F\ paths\ c) \iff (\forall p \in paths)(\exists y_j \in p)\ x \in s_j \tag{1}$$

Clearly,

$$x \in Q_1 \iff (x \in \mathcal{S}[\![e_1]\!]a) \vee (x \in \mathcal{S}[\![e_2]\!]a \wedge x \in \mathcal{S}[\![e_3]\!]a) \tag{2}$$

By (1) and the definition of $*$, we can see that

$$\begin{aligned} x \in F\ Q_2\ c \iff & ((\forall p \in \mathcal{P}[\![e_1]\!]b)(\exists y_j \in p)\ x \in s_j) \vee \\ & (((\forall p \in \mathcal{P}[\![e_2]\!]b)(\exists y_j \in p)\ x \in s_j) \wedge ((\forall p \in \mathcal{P}[\![e_3]\!]b)(\exists y_j \in p)\ x \in s_j)) \end{aligned}$$

which, by (1), implies that

$$x \in Q_2 \iff (x \in F(\mathcal{P}[\![e_1]\!]b)c) \vee ((x \in F(\mathcal{P}[\![e_2]\!]b)c) \wedge (x \in F(\mathcal{P}[\![e_3]\!]b)c)) \tag{3}$$

But by the structural induction hypothesis, we know that

$$\mathcal{S}[\![e_i]\!]a = F(\mathcal{P}[\![e_i]\!]b)(y_1, ..., y_n)(s_1, ..., s_n)$$

and so (2) and (4) are precisely equivalent and the implication holds.

The proofs of the other primitive functions take the same form, and their details are left to the reader.

References

[1] A. Bloss and P. Hudak. Variations on strictness analysis. In *Proc. 1986 ACM Conf. on LISP and Functional Prog.*, pages 132–142, ACM, August 1986.

[2] G.L. Burn, C.L. Hankin, and S. Abramsky. *The theory and practice of strictness analysis for higher order functions.* Technical Report DoC 85/6, Imperial College of Science and Technology, Department of Computing, April 1985.

[3] C. Clack and S.L. Peyton Jones. Strictness analysis – a practical approach. In *Functional Programming Languages and Computer Architecture*, pages 35–49, Springer-Verlag LNCS 201, September 1985.

[4] B. Goldberg. *Multiprocessor Execution of Functional Programs.* PhD thesis, Yale University, Department of Computer Science, September 1987.

[5] P. Hudak and A. Bloss. The aggregate update problem in functional programming systems. In *12th ACM Sym. on Prin. of Prog. Lang.*, pages 300–314, ACM, 1985.

[6] P. Hudak and J. Young. Higher-order strictness analysis for untyped lambda calculus. In *12th ACM Sym. on Prin. of Prog. Lang.*, pages 97–109, January 1986.

[7] T. Johnsson. *Detecting when call-by-value can be used instead of call-by-need.* Laboratory for Programming Methodology Memo 14, Chalmers University of Technology, Dept. of Computer Science, October 1981.

[8] N.D. Jones and S.S. Muchnick. *Complexity of flow analysis, inductive assertion synthesis and a language due to Dijkstra*, pages 380–393. Prentice-Hall, 1981.

[9] A. Mycroft. *Abstract Interpretation and Optimizing Transformations for Applicative Programs.* PhD thesis, Univ. of Edinburgh, 1981.

[10] A. Mycroft. The theory and practice of transforming call-by-need into call-by-value. In *Proc. of Int. Sym. on Programming*, pages 269–281, Springer-Verlag LNCS Vol. 83, 1980.

[11] J-C. Raoult and R. Sethi. The global storage needs of a subcomputation. In *11th ACM Sym. on Prin. of Prog. Lang.*, pages 148–157, ACM, January 1984.

The formal description of data types using sketches

Charles Wells
Department of Mathematics and Statistics
Case Western Reserve University
Cleveland, OH 44106, USA

and

Michael Barr
Department of Mathematics and Statistics
McGill University
805 Sherbrooke St. West
Montréal, Québec
Canada H3A 2K6

Abstract: This paper is an exposition of the basic ideas of the mathematical theory of sketches and a detailed description of some of the ways in which this theory can be used in theoretical computer science to specify datatypes. In particular, this theory provides a convenient way of introducing datatypes which have variants, for example in case of errors or nil pointers. The semantics is a generalization of initial algebra semantics which in some cases allows initial algebras depending on a parameter such as a bound for overflow.

1. Introduction

In this article, we describe a new technique for specifying abstract data types using sketches as developed in [Barr & Wells, 1985] and [Barr, 1986]. We define the concept of sketch precisely later in the paper. To understand the definition, you need to be familiar with categories and functors; all other concepts are defined.

The sketch (for trees or for natural numbers, for example) is the syntax for the specification of the data type. Each sketch has a category of models, and we propose the initial model or family of models (a generalization of the concept studied by [Meseguer and Goguen, 1985]) as the semantics.

The elements of the initial model for the sketch of a given data type are the possible values (configurations) of a variable of that type. As an example, the sketch we give for trees is a recursive description of a binary rooted tree with data of unspecified type D stored in its nodes, and the elements of the initial model are all the trees of that type.

The sketch for the natural numbers which we give has a family of initial models, instead of just one. One of the initial models, has short (two byte) natural numbers as elements, another has four byte natural numbers, and a third has all the natural numbers as models. All the possibilities for overflow arise naturally in the semantics in this way, strictly as a result of including a single error sort in the description (i.e., in the sketch).

When describing data types using sketches, the syntax is given by a graph depicting the sorts and operations together with certain data (diagrams, cones, cocones) which assert equations and specify that some sorts are constructed in specific ways from others (cartesian products, equationally defined subsets, disjoint unions). By contrast, in classical logic the formal object describing the syntax consists of variables, formulas and equations.

We discuss three types of sketches: finite product (FP) sketches; finite limit (FL) sketches, and finite discrete (FD) sketches. FL sketches and FD sketches each have FP sketches as special cases. We also mention more briefly the finite-limit-and-sum (FLS) sketches, which combine the expressive power of FL and FD sketches, primarily because that is the natural level of generality of our treatment of initial algebras (Theorem 10.4).

FP sketches have essentially the same expressive power as multisorted universal equational theories in logic, and the initial-model semantics in that case is the same as that of [Meseguer and Goguen, 1985]. They allow specification in terms of n-ary operations and universal equations involving those operations.

FL sketches can express anything expressible by universal Horn theories, and in addition allow one to consider partial operations whose domains can be described equationally in terms of other operations.

FD sketches allow one to express strict disjunctions in addition to the equational conditions of FP sketches. These sketches (and FLS sketches) do not correspond to any widely-studied type of theory in the classical approach to models.

Although we have compared the expressive power of various types of sketches to certain types of theories in the sense of mathematical logic, the difference between using sketches for syntax and the classical methods of logic is more (we assert) than a difference in notation. We list several advantages:

In the first place, the sketch approach is variable-free and applicative in nature: formulas with variables are replaced by composites of operations pictured as arrows, and equations are replaced by commutative diagrams. Thus the sortedness of an operation is apparent from the syntax so the reader does not have to remember it from the context. Variables play no role in the syntax. They do not correspond to anything in the models and introduce extraneous technicalities concerning freeness and boundedness. Nevertheless, we introduce an informal way of describing the data in a sketch in terms of equations (not using variables) involving composites and tuples which is useful when it is inconvenient to draw diagrams. The difference between sketches and the classical approach is that for sketches this is just an informal description; the formal data consist of graphs, diagrams, and other graph-like objects.

The interesting article by Mateti and Hunt [1985] in which an primitive editor is described using a software design language makes evident the belief of the authors of that paper that the software theoretician must make a choice between rigor and ready comprehensibility. If indeed the only rigorous model were that of first order logic, we would share this view. One of the main purposes of this paper is to introduce a formalism which is equally as rigorous as first order logic while being much more comprehensible.

Another advantage to using sketches is that the classification of theories according to the kind of sketch leads to a more natural division of types of theories than does the classification using formal languages. In particular, we believe that FL sketches are a more natural class than universal Horn theories, which as presented in the literature do not allow the introduction of equationally defined subsorts. McLarty [1986] presents an extension of Horn logic which does this. Without such an extension, universal Horn logic is strictly weaker than FL sketches, since for example the category of small categories is the category of models of an FL sketch but not of a universal Horn theory (see 10.2).

The advantages we have claimed so far have to do with the superiority of sketches over the linguistic approach of classical logic as a method for description. We would like computer scientists to know about them for that reason alone. However, our treatment of initial-model semantics for FD sketches shows another advantage: FD sketches are the right way to handle many types of error.

By generalizing somewhat the notion of initial algebra (to initial families), one can readily incorporate error semantics into the theory. As an example, an FD sketch allows you to state in a natural way that division (say, of integers) is defined on the set of all ordered pairs of integers for which the second coordinate is nonzero — note that this not an equation but the negation of one — or alternatively that it is defined on all the ordered pairs but when the second coordinate is zero it takes its value in an error sort. Although an FD sketch may have an initial family of models instead of a single initial one, these initial algebras correspond in many cases to a natural classification. In many other cases, as in the tree example in Section 11, there is still a single initial algebra.

The possible lack of a single initial algebra in the situation where disjunctions are allowed has apparently dissuaded computer scientists from considering those more general theories in the past. For example, we find in [Goguen, 1978]: "Any abstract data type can be specified purely equationally." This claim is directly contradicted in the body of that very interesting paper in which the Horn theories are modified in order to bring the errors into the analysis. Of course, in that case, the usual initiality results fail and Goguen was forced to reprove them in that very precise context. What he did not do was modify the mathematical context to bring the error analysis directly into the underlying theory. Doing just that is one aim of this paper.

There are other advantages to sketches which this paper will not bring out. For one thing, subsuming both equationally defined subobjects and the construction of tuples under the same concept of limit cone makes proofs easier to state and to understand. This may be seen by comparing the two proofs by [Makowsky, 1985] that models of horn theories have initial algebras and by [Barr, 1986] that models of FL sketches have initial algebras. The proof by Barr is much shorter and (we believe) more conceptual, although of course the latter depends to some extent on one's background.

Another aspect not apparent in this paper is that a sketch can have models in a fairly general kind of category; here we discuss only models in sets. Classical logic treats only set-valued models. Modeling a theory in many different categories has been one of the major uses of category theory in pure mathematics.

Sketches were originally invented by Charles Ehresmann for the purpose of describing algebraic structure. A good account of the early developments can be found in [Bastiani & Ehresmann, 1968]. More recently the theory has been highly developed in the work of René Guitart and Christian Lair both joint and separate. Much of their work appeared in the publication **Diagrammes** which is virtually unobtainable in North America. A survey and some references can be found in [Guitart, 1986]. It came to our attention after this paper was written that most of the actual mathematical results therein can be found, albeit from a different point of view, in [Guitart & Lair, 1980].

In addition to this Yves Diers has carried out some very closely related work, espcially for the FD sketches, again from a rather different viewpoint. See especially [Diers, 1977].

The first author of this paper was supported by the Ministère de l'Education du Québec through FCAR grants to the Groupe Interuniversitaire en Etudes Catégoriques and by an individual operating grant from the National Science and Engineering Research Council.

2. Various approaches

Before we define the concept of sketch, we will illustrate it and contrast with more traditional methods of defining an abstract structure. Specifically, we will contrast the model theorist's definition of a monoid with that of a category theorist.

A classical textbook definition would say that a monoid is a set M with a specific element $e \in M$ and a binary operation $\mathrm{mult}: M \times M \longrightarrow M$ subject to the axioms

$$\forall x(\mathrm{mult}(e,x) = \mathrm{mult}(x,e) = x) \tag{2.1}$$

and

$$\forall x, \forall y, \forall z(\mathrm{mult}(x,\mathrm{mult}(y,z)) = \mathrm{mult}(\mathrm{mult}(x,y),z)). \tag{2.2}$$

Of course, these are usually expressed with infix notation.

In classical model theory, this kind of definition is *turned into a mathematical object*. You take "M", "e", "mult" and the variables as symbols in a formal language, along with other standard symbols such as " = ", and you regard the equations 2.1 and 2.2 as formal objects to be interpreted as requirements on the models. A model, which in this case is a monoid, is given as a set corresponding to M with an element corresponding to e, and a function corresponding to mult, for which standard interpretations of the equations becomes true statements. Note that nothing in the model corresponds to the variables. The only purpose of the variables x, y and z in 2.1 and 2.2 is to hold places in the functions.

By contrast, the category theorist describes a monoid as an object M of a category together with a morphism mult:$M \times M \longrightarrow M$ and a morphism $e{:}1 \longrightarrow M$ such that the diagrams

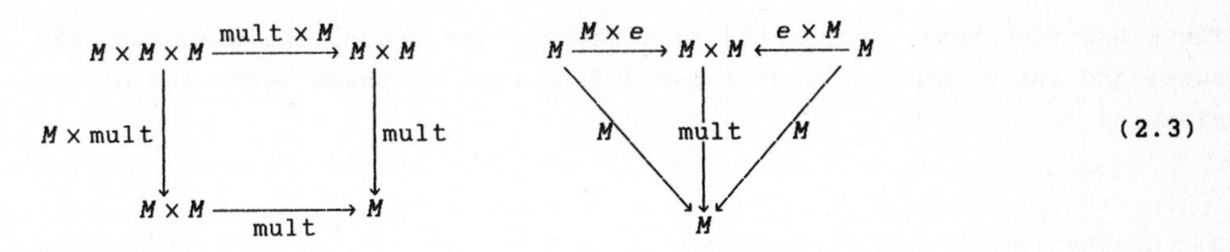

commute.

As is common in category theory, the identity function on a set M is denoted by the same name M. If A, B, C and D are sets and $f{:}A \longrightarrow C$ and $g{:}B \longrightarrow D$ are functions, then $f \times g{:}A \times B \longrightarrow C \times D$ is the function whose value at an ordered pair $\langle a,b \rangle \in A \times B$ is $\langle f(a),g(b) \rangle$. Thus $M \times \mathrm{mult}{:}M \times M \times M \longrightarrow M \times M$ is the function defined by $(M \times \mathrm{mult})(r,s,t) = \langle r,\mathrm{mult}(s,t) \rangle$. Note that this requires that the set $M \times M \times M$ to be identified with the set $M \times (M \times M)$, which is reasonable since they represent the same data. This identification is an example of the canonical isomorphism between two categorical products of the same discrete diagram ([Barr & Wells, 1985], Section 1.7, Proposition 1.) We will return to this when we discuss cones.

Similarly, $\mathrm{mult} \times M{:}M \times M \times M \longrightarrow M \times M$ takes $\langle r,s,t \rangle$ to $\langle \mathrm{mult}\langle r,s \rangle,t \rangle$, which this time identifies $M \times M \times M$ with $(M \times M) \times M$.

The statement that the diagram commutes means that any two paths from one node to another give the same result, where following an arrow means evaluating the function it represents. Thus, that the left diagram in 2.3 commutes is simply the statement that equation 2.2 holds. We give a formal definition of the concept of commutative diagram in the next section.

Just as a model theorist takes the symbols representing the data and the string of symbols representing the equation as the formal mathematical objects to be studied, the categorist takes the nodes and arrows representing the functions and operations, and the diagrams representing equations (and some other things — "cones" -- which in the present example allow the specification of the cartesian products) to be the formal objects to be studied. These data are formalized in the concept of sketch. The sketch abstracts the syntactical part of the categorist's description. Thus a sketch can best be thought of as a formal graphical language invented to state categorical properties. This will become clearer after we give the formal definition of sketch.

There are four basic components to a sketch: The graph, the diagrams, the cones and the cocones. We will now define each of these terms and give examples.

3. Graphs

What we call a graph can be best thought of as a category without composition. To a graph theorist, it is a directed multigraph with loops. Formally, a **graph** consists of a set of **nodes** (or **objects**) and a set of **arrows**. Each arrow has a **source** (or **domain**) node and **target** (or **codomain**) node. The source and target of a given arrow need not be distinct (hence the loops). The notation "$f\colon a \longrightarrow b$" means that f is an arrow and a and b are its source and target, respectively.

There may be one or more arrows or none at all with given nodes as source and target. A graph is called **discrete** if it has no arrows. Such a graph is essentially a set; discrete graphs and sets are usefully regarded as the same thing for most purposes. A graph is **finite** if the number of nodes *and* arrows is finite.

The data for a graph can be represented in the following noncommutative diagram, in which N is the set of nodes and A is the set of arrows.

$$A \underset{\text{target}}{\overset{\text{source}}{\rightrightarrows}} N \qquad (3.1)$$

A **graph morphism** from a graph G to a graph H is a function which takes nodes to nodes, arrows to arrows and preserves source and target. This can be summed up by saying that a morphism m consists of two arrows m_N and m_A which make the following diagram **commute serially**, meaning that the diagram obtained by deleting the top (respectively bottom) of each pair of horizontal arrows commutes.

(3.2)

There is a graph underlying any category, obtained by ignoring its composition law. A morphism from a graph to a category is understood to be a morphism to the graph underlying the category.

Another example of a graph is the graph with just one object, which we will call S, and one arrow we will call end: $S \longrightarrow S$. (An arrow from an object to itself is called an **endomorphism.**) A diagram M in the category of sets based on this sketch is simply a set $M(S)$ together with a function $M(\text{end}): M(S) \longrightarrow M(S)$. Without more structure, we cannot be more specific about the nature of the endomorphism. The remaining items in a sketch (diagrams and cones) allow us to be more precise.

4. Diagrams

We now give a formal definition of the concept of diagram we have already used informally: A morphism $D: \mathcal{I} \longrightarrow \mathcal{G}$ from a graph $\mathcal{I}$ to a graph $\mathcal{G}$ is called a **diagram,** more precisely a diagram of shape $\mathcal{I}$ in $\mathcal{G}$. $\mathcal{I}$ is called the **shape graph** of the diagram. Diagram 3.2 is an example of a diagram in the category of sets, based on a shape graph with four nodes and six arrows as shown.

The word "diagram" is often qualified to reflect properties of $\mathcal{I}$. For example, a diagram $D: \mathcal{I} \longrightarrow \mathcal{G}$ is called a **discrete diagram** in $\mathcal{G}$ if $\mathcal{I}$ is discrete (i.e. consists of a set of nodes with no arrows), a **finite diagram** if $\mathcal{I}$ is finite, and so on.

The formal definition of a diagram as a morphism $D: \mathcal{I} \longrightarrow \mathcal{G}$ of a graph to another graph specifically includes the possibility that D is noninjective on nodes or arrows or both. This allows diagrams which, when drawn, have more than one node or more than one arrow with the same label.

The second ingredient in a sketch is a set $\mathcal{D}$ of diagrams, all taking values in the graph $\mathcal{G}$ of the sketch. There is in general no restriction on the number or shape of diagrams specified to be in $\mathcal{D}$.

4.1. Commutative diagrams. The notion of commutative diagram makes sense only for a diagram in a category, in other words for a diagram $D: \mathcal{I} \longrightarrow \mathbf{C}$ where C is a category ($\mathcal{I}$ most emphatically need *not* be a category). To make this precise, we need a preliminary idea. Let $\mathcal{I}$ be a graph. A **path** in $\mathcal{I}$ between two nodes i and j is a sequence of arrows $\langle a_1, a_2, \ldots, a_n\rangle$ such that

(i) the source of a_1 is i,

(ii) the target of a_n is j, and

(iii) for each $0 < k < n-1$ the target of a_k is the source of a_{k+1}.

The concept of path of arrows is what the categorist uses instead of the concept of word or string of symbols. In particular, we must allow a path to be empty just as mathematical linguists must take into account the empty string. An empty path has no arrows in it. Such a path is necessarily from a node i to itself. Thus for the categorist, there is a separate empty path for each type of data. We denote the empty path from i to i by $\langle\rangle$.

Let $D: \mathcal{I} \longrightarrow \mathbf{C}$ be a diagram. D is a **commutative diagram** if whenever i and j are nodes in $\mathcal{I}$ and $\langle a_1, a_2, \ldots, a_n\rangle$ and $\langle b_1, b_2, \ldots, b_m\rangle$ two paths from i to j, then

$$D(a_n)\circ D(a_{n-1})\circ\cdots\circ D(a_1) = D(b_m)\circ D(b_{m-1})\circ\cdots\circ D(b_1).$$

A **graph with diagrams** is a pair $(\mathcal{G}, \mathcal{D})$ where $\mathcal{G}$ is a graph and $\mathcal{D}$ is a set of diagrams in $\mathcal{G}$. A **morphism of graphs with diagrams** from $(\mathcal{G}_1, \mathcal{D}_1)$ to $(\mathcal{G}_2, \mathcal{D}_2)$, is a morphism $M: \mathcal{G}_1 \longrightarrow \mathcal{G}_2$ such that whenever $D: \mathcal{I} \longrightarrow \mathcal{G}_1$ is in $\mathcal{D}_1$, then $M\circ D: \mathcal{I} \longrightarrow \mathcal{G}_2$ is in $\mathcal{D}_2$.

Any category C will be regarded as a graph with diagrams in a standard way. The graph is the underlying graph of the category gotten by forgetting the composition of the category and the diagrams will always be understood to be all the commutative diagrams. In particular, a morphism from a graph with diagrams $(\mathcal{G}, \mathcal{D})$ to a category regarded as a graph with diagrams must take every diagram in $\mathcal{D}$ to a *commutative* diagram. A morphism from a graph with diagrams to the category of *sets* is a **model** of the graph with diagrams. The intention of including a diagram in a graph with diagrams is to make the corresponding equation true in the models.

For example, consider the graph with a single node n and two arrows e, $f: X \longrightarrow X$. A morphism from this graph to the category of sets is a set together with two arbitrary endomorphisms on that set. Now add to the graph the diagram based on the graph with four objects and four morphisms as shown on the left below

$$\begin{array}{ccc} R & \xrightarrow{a} & S \\ b\downarrow & & \downarrow c \\ T & \xrightarrow[d]{} & U \end{array} \qquad\qquad \begin{array}{ccc} X & \xrightarrow{e} & X \\ f\downarrow & & \downarrow f \\ X & \xrightarrow[e]{} & X \end{array} \tag{4.2}$$

with $D(R) = D(S) = D(T) = D(U) = X$, $D(a) = D(d) = e$, $D(b) = D(c) = f$, typically drawn as shown on the right. A morphism of this graph with diagrams into the category of sets is now a set with a pair of *commuting* endomorphisms.

4.3. Notation. We find it convenient to introduce the following notation for graphs with diagrams. We should emphasize that this notation is used only when convenient. Suppose $\mathfrak{A}$ is the graph:

$$i \begin{array}{c} \xrightarrow{a_1} i_1 \xrightarrow{a_2} i_2 \xrightarrow{a_3} \cdots \longrightarrow i_n \xrightarrow{a_n} \\ \xrightarrow[b_1]{} j_1 \xrightarrow[b_2]{} j_2 \xrightarrow[b_3]{} \cdots \longrightarrow j_m \xrightarrow[b_m]{} \end{array} k \tag{4.4}$$

then we indicate that $D: \mathfrak{A} \longrightarrow \mathcal{G}$ is a diagram in the set $\mathfrak{D}$ of diagrams by saying that

$$D(a_n) \circ D(a_{n-1}) \circ \cdots \circ D(a_1) = D(b_m) \circ D(b_{m-1}) \circ \cdots \circ D(b_1).$$

is a diagram in $\mathfrak{D}$. This equation must be satisfied in any model of the graph with diagrams, but of course in the graph itself there is no composition. We will introduce more semantically inspired informal notation for the syntactic objects in a sketch later. We might also say that

$$D(i) \begin{array}{c} \xrightarrow{D(a_1)} D(i_1) \xrightarrow{D(a_2)} D(i_2) \xrightarrow{D(a_3)} \cdots \longrightarrow D(i_n) \xrightarrow{D(a_n)} \\ \xrightarrow[D(b_1)]{} D(j_1) \xrightarrow[D(b_2)]{} D(j_2) \xrightarrow[D(b_3)]{} \cdots \longrightarrow D(j_m) \xrightarrow[D(b_m)]{} \end{array} D(k)$$

is a diagram in $\mathfrak{D}$.

A special case of this occurs when $m = 0$ in which case $i = k$. The interpretation of the statement

$$D(a_n) \circ D(a_{n-1}) \circ \cdots \circ D(a_1) = \langle\rangle$$

in a category is that the indicated composite is the identity of $D(i) = D(k)$. In particular, if $\mathcal{I}$ consists of the single object i with one endomorphism $a: i \longrightarrow i$, then the diagram $D: \mathcal{I} \longrightarrow \mathrm{C}$ in the category C commutes if and only if $D(a)$ is the identity arrow of $D(i)$. A plain endomorphism can be described by a diagram D' based on the shape graph with *two* objects i and j and an arrow between them, subject to the proviso that $D(i) = D(j)$. We are indebted to Robert Paré for pointing out this way of dealing with identities.

5. Cones and cocones

The other concepts required to define a sketch are cones and cocones. In a graph $\mathcal{G}$, a **cone** consists of

(i) a diagram $D: \mathcal{I} \longrightarrow \mathcal{G}$, called the **base** of the cone;

(ii) an object g of $\mathcal{G}$, called the **vertex** of the cone; and

(iii) a family $p = \{p_i: g \longrightarrow D(i)\}$ of arrows of $\mathcal{G}$, indexed by the nodes of $\mathcal{I}$.

Such a cone will be called a **cone from** g **to** D and will be referred to as a cone $p: g \longrightarrow D$; if i is a node of $\mathcal{I}$, then $p_i: g \longrightarrow D(i)$ is the arrow given by (iii). The arrows given by (iii) are the **coordinates** of the cone (we will see why shortly). In the categorical literature they are more commonly called the projections.

A cone $p: g \longrightarrow D$ is indicated by a diagram of the form

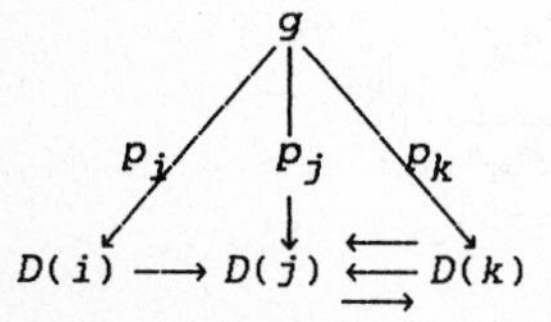

$$\begin{array}{ccccc} & & g & & \\ & \swarrow^{p_i} & \downarrow^{p_j} & \searrow^{p_k} & \\ D(i) & \longrightarrow & D(j) & \overset{\longleftarrow}{\underset{\longrightarrow}{\longleftarrow}} & D(k) \end{array} \tag{5.1}$$

in which we have indicated some typical arrows of the diagram along the bottom. When the base diagram is drawn in perspective the "cone" looks rather more like a pyramid, but "cone" is the standard name.

A cone $p: G \longrightarrow D$ *in a category* is a **limit cone** if it has the following two properties:

(i) ("The sides of the cone commute.") For every arrow $a: i \longrightarrow j$ in $\mathbb{A}$, $D(a) \circ p_i = p_j$. A cone which satisfies this requirement is a **commutative cone**.

(ii) ("Every commutative cone over a diagram maps uniquely to a limit cone over the diagram.") If $q: H \longrightarrow D$ is another commutative cone, there is a unique arrow $u: H \longrightarrow G$ such that $p_i \circ u = q_i$ for all nodes i of $\mathbb{A}$. (We sometimes say that a limit cone is a **final cone**.)

As you can see, this definition requires composition and so is not applicable for cones in a graph. Note carefully that the *base* of a limit cone, such as the diagram D above, need *not* commute.

The vertex of a limit cone over a diagram D is called a **limit** of the diagram. This terminology must be used with care, since it is in fact the vertex *with its coordinate maps* which is the limit.

The cones over a given diagram form a category, with maps being maps between the vertices which commute with the coordinates in the sense of (ii); in this sense, (ii) says a limit cone over a diagram is a terminal object in the category of cones over the diagram.

The concept of limit cone allows a uniform formulation of constructions which occur in mathematical practice in many different categories.

In the category of sets, cones in which the base diagram is discrete have as limits the cartesian product of the objects in the base. In any category, a limit cone with discrete base is called a **product cone** and the vertex is called the **product** of the objects in the base.

Other types of limits are equalizers and pullbacks, which allow the specification of objects in a category as subobjects of products satisfying equational conditions. The equalizer of two arrows u and v in a category (if the equalizer exists) is given as the limit cone

(5.2)

which is usually presented as

$$E \xrightarrow{e} B \overset{u}{\underset{v}{\rightrightarrows}} C \qquad (5.3)$$

since the arrow from E to C is determined by definition of commutative cone to be the composite $u \circ e = v \circ e$.

A pullback of $f: A \longrightarrow C$ along $g: B \longrightarrow C$ is a limit cone over the diagram

$$A \xrightarrow{f} C \xleftarrow{g} B \qquad (5.4)$$

It is symmetric and so is also a pullback of g along f.

Products, equalizers and pullbacks are discussed in detail in [Barr and Wells, 1987], or (in less detail) in [Mac Lane, 1971], or [Barr and Wells, 1985].

5.5. Cocones. In a graph $\mathcal{G}$, a cocone consists of a diagram $D: \mathbb{A} \longrightarrow \mathcal{G}$, an object g of $\mathcal{G}$, and a family $j_i: D(i) \longrightarrow g$ of arrows of $\mathcal{G}$, indexed by the nodes of $\mathbb{A}$. The object g is the vertex of the cocone, and the arrows j_i are the elements of the cocone. We will denote such a cocone as $j: D \longrightarrow g$. Thus a cocone is a cone in the opposite category. Warning: Burstall [1980] uses "cocone" for "cone".

A colimit cocone in a category C is a limit cone in the opposite category C^{op}. If $j: D \longrightarrow g$ is a colimit cocone on a diagram $D: \mathbb{A} \longrightarrow C$ and $k: D \longrightarrow h$ is another commutative cocone defined on the same diagram, then there is a unique morphism $\beta: g \longrightarrow h$ for which $k_i = \beta \circ j_i$ for each node i of the graph $\mathbb{A}$.

In this paper we will consider only colimit cocones on discrete graphs $\mathbb{A}$. We denote the colimit of a discrete diagram $\{S_1, S_2\}$ by $S_1 + S_2$. In any category it is the sum or coproduct of S_1 and S_2; in the category of sets it is the disjoint union, a phrase which is not descriptive of the sum in many other categories used by mathematicians. (For example, in the category of monoids, the sum is the free product.)

This notation is extended to more than two objects. The sum of A, B and C is $A+B+C$, for example. We will not consider infinite sums. Colimit cocones allow a categorist to express disjunctions, as we will see, and also to express existence, although we will not be dealing with the latter here.

6. Sketches

A sketch is a 4-tuple $(\mathcal{G},\mathcal{D},\mathcal{L},\mathcal{C})$ where $\mathcal{G}$ is a graph, $\mathcal{D}$ is a set of diagrams on $\mathcal{G}$, $\mathcal{L}$ is a set of cones on $\mathcal{G}$ and $\mathcal{C}$ is a set of cocones on $\mathcal{G}$. If $\mathcal{S}_1 = (\mathcal{G}_1,\mathcal{D}_1,\mathcal{L}_1,\mathcal{C}_1)$ and $\mathcal{S}_2 = (\mathcal{G}_2,\mathcal{D}_2,\mathcal{L}_2,\mathcal{C}_2)$ are sketches, a **morphism of sketches** $\mathcal{S}_1 \longrightarrow \mathcal{S}_2$ is a morphism of the graphs that takes diagrams in $\mathcal{D}_1$ to diagrams in $\mathcal{D}_2$ in the sense described in 4.1, cones in $\mathcal{L}_1$ to cones in $\mathcal{L}_2$ and cocones in $\mathcal{C}_1$ to cocones in $\mathcal{C}_2$. Thus a sketch is a graph with diagrams (as described previously) with extra structure, namely the cones and cocones.

The sketch underlying a category (which will not usually be explicitly mentioned), has as graph the graph underlying the category and has for diagrams all commutative diagrams, for cones all limit cones and for cocones all colimit cocones.

A **model** of a sketch is a sketch morphism to the category of sets. If M is such a model, then for a node s, $M(s)$ is a set. If $f:s \longrightarrow t$ in the graph of the sketch, then $M(f):M(s) \longrightarrow M(t)$ is a function. It follows from the definition of the underlying sketch of a category that a diagram of the sketch must be taken to a commutative diagram, and each (co)cone of the sketch must be taken to a (co)limit (co)cone. Models in this sense correspond to the logicians' models of the corresponding theory.

A **morphism of models** is a natural transformation, whose definition we now spell out. Let M and N be models of the same sketch $\mathcal{S}$. A morphism $\alpha:M \longrightarrow N$ consists of a collection of functions

$$\alpha(s):M(s) \longrightarrow N(s)$$

indexed by the nodes of the sketch, with the property that for every arrow $f:s \longrightarrow t$ of the sketch, this diagram commutes:

The commutativity of this diagram is what makes morphisms of models preserve the operations. An informal way to think of a morphism of models is that it must preserve all the relationships in the models which are defined in terms of the data given by the sketch.

Every sketch of a given type generates a **theory** of that type; the theory is a category generated by the sketch, in which the diagrams of the sketch commute and the (co)cones are limit (co)cones. The theory consists in a sense of all the possible expressions allowed by the sketch; the sketch is a presentation of the theory. This concept is discussed in detail in [Barr & Wells, 1985]. All such theories are embedded in toposes. The connection between toposes and logic is discussed in [Lambek & Scott, 1986] and in [Fourman & Vickers, 1986].

7. An informal notation for sketches

In Section 4, we introduced some informal notation for describing the diagrams in a graph with diagrams which was based on the fact that in a model the diagrams have to commute. Of course, this notation can be used for sketches, too. Now we introduce some analogous notation for cones and cocones which is based on the fact that they must become limiting cones or cocones in a model. All this informal notation allows one to describe a sketch in language rather similar to that used in traditional presentations of theories. One could go even further and reintroduce variables, although we think that would be a backward step. What is important to understand is that the formal structure is still in the form of graphs, diagrams and (co)cones; the expressions and equations we introduce are merely informal descriptions motivated by the semantics.

7.1. We first introduce a notation for discrete cones. When we write a node named $a \times b$ in a sketch, this implies the existence of nodes a and b, and of a cone

(7.2)

Similar remarks apply to an object named $a_1 \times a_2 \times \cdots \times a_n$. Even infinite discrete cones of this sort can be so designated using $\Pi(a_i)$, but we will not be using any such infinite cones here. Note that $a \times b$ is not actually the product of a and b in the sketch; the concept of product in a sketch does not even make sense, since it requires the concept of composition of arrows.

The requirement that a cone of the sketch become a limit cone implies that a model M of the sketch must take $a \times b$ to an isomorphic copy of the cartesian product; specifically, $M(a \times b)$ must be isomorphic to $M(a) \times M(b)$ by a bijection $\beta : M(a \times b) \longrightarrow M(a) \times M(b)$ with the property that $c_i \circ \beta = M(p_i)$, where c_i is the coordinate map from the cartesian product, for $i = 1,2$. Similar remarks apply to larger products.

7.3. A node called 1 in a sketch implies the existence of a cone with vertex 1 and an empty base. A model of a sketch must take such an object to a terminal object in the category of sets, which is a singleton set. Note that an arrow $1 \longrightarrow s$ becomes, in a model M of the sketch, an arrow $1 \longrightarrow M(s)$, that is a fixed element of $M(s)$. Such an element is normally called a constant. As in much of the literature, we thus regard a constant as a particular kind of operation, dependent on no variables.

7.4. If we name a node $a \times_c b$, this implies that we already understand nodes a, b and c and arrows $f : a \longrightarrow c$ and $g : b \longrightarrow c$ and a cone

(7.5)

This terminology $a \times_c b$ is standard notation among categorists for the pullback, which is also called the “fiber product”.

7.6. If $h:d \longrightarrow a$ and $k:d \longrightarrow b$ are arrows in the sketch, the notation $(h,k):d \longrightarrow a \times b$ is used to name a map and imply the existence of a diagram (*not* a cone)

(7.7)

where the bottom line is the cone included by definition of $a \times b$. The same notation is often used to denote an arrow $d \longrightarrow a \times_c b$, but in that case, it also implies the existence of an arrow $d \longrightarrow c$. This last arrow is rarely named explicitly.

7.8. The symbol $():a \longrightarrow 1$ stands for an arrow (necessarily the only one with domain a) with target 1.

7.9. Writing $f \times g:a \times b \longrightarrow c \times d$ implies the existence of a diagram

(7.10)

A similar convention applies to products involving more than two factors. It should be emphasized that these are not products in the sketches; these conventions insure that they become products when modeled in any category.

7.11. If we write an arrow $f:a \rightarrowtail b$, this implies the existence of a cone

It is a nice exercise to prove that such a cone in a category is a limit cone if and only if f is a monomorphism, which in the category of sets means injective.

7.13. If we write $a \simeq b$, we mean there is a cone

with vertex a. A model must take this arrow to a bijection. The inverse to the arrow is forced by the cone $\mathrm{id}_b : b \longrightarrow b$.

7.15. In an analogous way to the notation used to denote cones, when we refer to a node $a+b$, we imply the existence of a cocone

A similar convention applies to sums of more than two. All the conventions used to denote cones and arrows to them have their duals for cocones. For example, if $f : a \longrightarrow c$ and $g : b \longrightarrow c$ are arrows then the notation $\langle f,g \rangle : a+b \longrightarrow c$ indicates that there is a cocone as above as well as a diagram

Finally, an object denoted 0 will always be the vertex of the empty cocone. This forces its interpretation in any model to be the empty set.

7.18. In a number of FD sketches we have to say that a sort s is the sum of a subsort t and a constant c. Our previous notation would require us to introduce a new sort, say s_0 and use two terms to describe this, namely

$$c:1 \longrightarrow s_0,\ s = s_0 + t.$$

We therefore adopt the notation $s = \{c\} + t$ to mean the same thing. We could similarly adapt this to a situation as $s = \{c_0, c_1\} + t$. In the latter case, we would understand that the constants are distinct, so that it is an abbreviation for

$$c:1 \longrightarrow s_0,\ 1 \longrightarrow s_1,\ s = s_0 + s_1 + t.$$

8. Types of sketches

The concept of sketches admits of many special cases, depending on what restrictions one puts on the classes of cones and cocones.

We can now define the four major types of sketches which will interest us in this paper. A model of any of these types of sketches provides a set for each node, corresponding to the sorts of traditional universal algebra, and a function or operation for each arrow. Each diagram of the sketch gives an equation among the operations. As we have seen, cones allow one to specify that certain sorts in a model must be products of equationally defined subsorts, and discrete cocones allow the specification of a sort as a disjoint union of subsorts, which means allowing the negations of equations.

8.1. FP (finite product) sketches. These are sketches whose cones all have finite discrete diagrams as bases and no cocones at all. The models of an FP sketch have a sort for each node, operations for the arrows (whence arrows of a sketch are often called operations), and equations corresponding to the diagrams. The fact that they have only discrete cones means that all the operations are defined on (isomorphic copies of) cartesian products; there are no equationally-defined subsorts.

Thus the structures which constitute traditional varieties (structures with one or more sorts, with operations defined on cartesian products of these sorts, subject to equations which must hold universally) are models of FP sketches. In other words, FP sketches are equivalent to multi-sorted equational theories. There are some subtle differences between the traditional approach and the approach via sketches which will be discussed in the examples.

8.2. FL (finite limit) sketches. These are sketches whose cones all have finite bases and which have no cocones. Since more general cones are allowed, an operation can be defined on, for example, an equalizer, that is, on an equationally-defined subset of a sort. For example, you might want to define a data type "rational" whose elements are those pairs of integers $\langle n,m\rangle$ for which $m > 0$ and for which n and m have no common divisor larger than 1, i.e., satisfying the equation $\mathrm{GCD}(m,n) = 1$. Then addition, subtraction and multiplication would be defined to be the usual operations, followed by clearing out common divisors. The operation of division cannot be defined in an FL sketch, because of the impossibility of division by zero; the appropriate tool to deal with that situation is that of FLS sketches.

FL theories are called "left exact" theories in [Barr & Wells 1985].

8.3. FD (finite-discrete) sketches. These are sketches whose cones and cocones all have finite discrete bases. An operation such as "take the multiplicative inverse" on some set of numbers applies only to nonzero numbers. The set of nonzero numbers (nonzero real numbers, for example) cannot be defined by an equation; you have to specify your set of numbers as the disjoint union of the set {0} and the set of nonzero numbers. An FD sketch allows you to do this. This does not correspond to a class of theories that has been studied by traditional model theorists.

We give examples of using FD sketches for defining data types below. This type of sketch seems very natural for this purpose. The discrete cocones allow alternatives, both in the sense of describing error conditions and in the sense of defining special cases of a data type (such as an empty tree).

8.4. FLS (finite-limit-and-sum) sketches. These are sketches whose cones have finite bases and whose cocones have finite discrete bases. Thus they have the expressive power of both FD and FL sketches. They can be used to describe the rationals with division, as mentioned previously, but we give no examples of data types requiring their full expressive power. FLS sketches are the natural class whose models have initial families (Theorem

10.4). They do not correspond to any well known type of theory in classical logic, although they form a very natural class from the point of view of the properties of models ([Barr, 1986], where they are called finite-sum sketches.)

9. Natural Numbers

We now describe an FD sketch which has the natural numbers as one of its models. We will look at this example again in Section 10, after we discuss initial algebras in Section 11.

The graph has nodes 1, n_0, n_+, n_{over}, and $n = n_0 + n_+ + n_{over}$. The last equation implies the existence of a cocone from a discrete diagram with three nodes, with vertex n. There are additional nodes, but they are implied by the specification of the arrows of the graph, which we list now.

(i) An arrow $0{:}1 \doteq n_0$.

(ii) An arrow $\text{succ}{:}n_0 + n_+ \longrightarrow n_+ + n_{over}$. Note that the names "$n_0 + n_+$" and "$n_+ + n_{over}$" imply the existence of two more cocones whose vertices are named by these names.

(iii) An arrow $\text{pred}{:}n_+ \longrightarrow n_0 + n_+$.

We add an equation: $\text{succ}\circ\text{pred} = \text{id}_n$. According to our conventions, the resulting sketch has the diagram corresponding to this equation, and it has another diagram

$$n \xrightarrow{\text{id}_n} n$$

based on the graph with one node and one arrow. These two are the only diagrams of the sketch.

The natural numbers (nonnegative integers) form a model N of this sketch. $N(n_0)$ is the set $\{0\}$. Requirement (i) forces any model to make n_0 a singleton set, so our choice of $N(n_0)$ is consistent. $N(n_+)$ is the set of positive integers. $N(n_{over})$ is the empty set. Since models take cocones of the sketch to limit cocones in the category of sets, $N(n)$ is exactly the set of natural numbers. $N(\text{succ})$ and $N(\text{pred})$ can be taken to be the successor and predecessor operations, and they clearly satisfy the equation.

Two things of note about this model: First, a model can take a sort to the empty set. (Then, you ask, why have the sort n_{over}? Read on!) And second, having the ability to form disjoint unions makes it easy to define operations like predecessor which are undefined on part of the datatype. We don't have to give it some artificial value such as "error" — we just don't define it on the embarrassing part of the datatype, and in any model is it then not defined there and thus gives no trouble.

There are many other models of this sketch. One type, which is of interest in computer science, is models of bounded "natural numbers", for example the "short integers" and the "long integers" in C and other languages. They are like natural numbers up to a point of overflow. To get such a model M which overflows at 2^{16}, define $M(n_0) = \{0\}$, $M(n) = \{1,\ldots,2^{16}-1\}$, and $M(n_{over})$ $= \{\text{ERROR}\}$. $M(\text{succ})$ will be the successor function on $M(n_0+n)$, except that $M(\text{succ})(2^{16}-1)$ will be ERROR. Observe that the required equation is satisfied.

10. Initial algebras

The sketch for a data type provides its *syntax*; the *semantics* which we give are the initial algebras for the sketch, which we will discuss in this section. FL sketches always have initial algebras; FLS sketches have *families* of initial algebras, which we will argue also provide suitable semantics and that furthermore the existence of more than one semantics correctly reflects the actual situation.

An **initial object** in a graph is an object normally denoted 0 with the property that for any object c of the graph there is exactly one arrow $0 \longrightarrow c$. It follows easily from the definition that all initial objects in a *category* are isomorphic by a unique isomorphism.

In the category of sets, the unique initial object is the empty set; there is a unique empty function from the empty set to any other set. In general categories, there may be more than one initial object or no initial object. For example, in the category of monoids, every one-element monoid is an initial object. On the other hand, in an ordered set regarded as a category, an initial object is the minimum, and an ordered set needn't have a minimum. Another example is the category of fields: there is no field which maps to every field.

The relevant theorem for FL sketches is:

10.1. Theorem. *The category of models of an FL sketch has an initial algebra.*

This theorem has been well-known under various guises for years, being an immediate consequence of the main theorem of [Gabriel & Ulmer, 1973]. See [Barr, 1986] for a direct proof, or [Makowsky, 1984] for a proof of the theorem for Horn logic. Makowsky's proof is in the language of logic.

The purely categorical proof found in [Barr, 1986] may be described roughly this way: If C is the category of models of the FL sketch $\mathcal{S}$, then every diagram in C has a limit which is a model constructed by taking the limit at every node (and there is a unique such model). One uses this fact to construct a least submodel in every model as the intersection of all submodels. This will not necessarily be initial, but using it, one can construct a transfinite chain of irreducible models and epimorphisms (not necessarily surjections!) between them. A cardinality argument shows that this construction must repeat and the first time it does so, the algebra is necessarily initial.

An important subtlety complicating this proof is that the image of a submodel need not be a submodel. In this connection, we must apologize for having claimed in an earlier version of this paper that Goguen and Meseguer had erred in making the contrary claim for their order-sorted algebras. The fact is that in the special case of FL theories that they considered, the image of an algebra is an algebra. See [Goguen & Meseguer, 1985b].

10.2. Here is an example to show that the image of model is not always a submodel of an FL theory. The category of categories is itself the category of models of an FL sketch. The sketch has sorts o, a, $a \times_o a$ and $a \times_o a \times_o a$. There are operations $\text{dom}: a \longrightarrow o$ $\text{cod}: a \longrightarrow o$, $\text{comp}: a \times_o a \longrightarrow a$ and some operations defined implicitly. The cones are

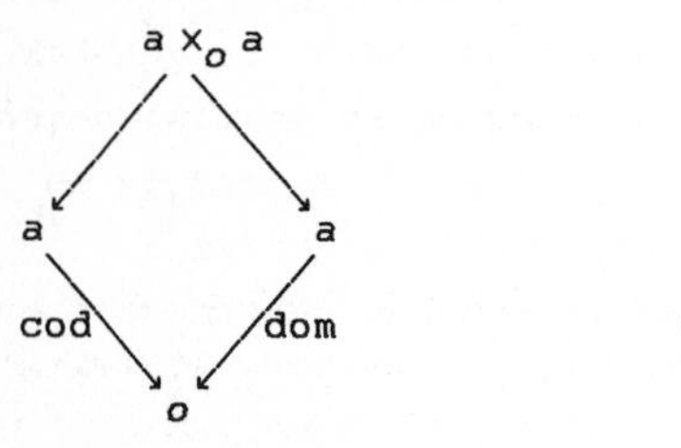

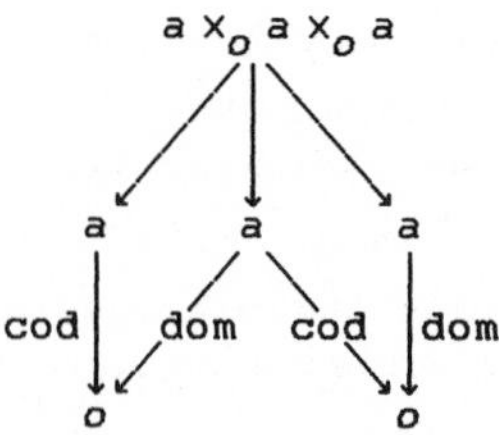

There is one diagram:

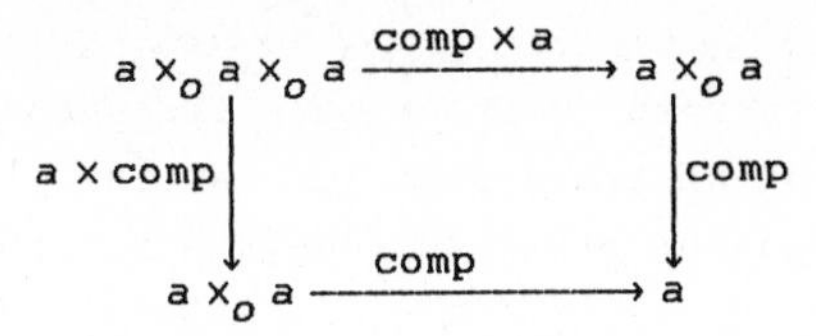

The models of this sketch are categories and the natural transformations between models are functors. If $\mathfrak{A}$ is the category with two objects, their identities and one arrow f between them and $\mathfrak{N}$ is the category with one object, its identity, an endomorphism g and all the powers of g, then the functor $F:\mathfrak{A} \longrightarrow \mathfrak{N}$ that takes both objects of $\mathfrak{A}$ to the single object of $\mathfrak{N}$ and for which $F(f) = g$ has image which is not a subcategory since g is in the image while $g \circ g$ isn't. What happens is that $f \circ f$ is not defined in $\mathfrak{A}$, but the image is defined in $\mathfrak{N}$.

10.3. Initial families. The category of models of an FLS sketch need not have a single initial algebra. What happens is that each component of the category of models has an initial algebra. A component of a category is defined just as for graphs: two objects are in the same component if there is a path from one to the other following arrows *forward* and *backward*. Another way of saying this is that a component of a category is just a component of the underlying undirected graph. In [Barr, 1986], the following is proved (cf. [Diers, 1977] and [Guitart & Lair, 1980]).

10.4. Theorem. *In the category of models for an FLS sketch each component has an initial algebra.*

For example, the integers (mod 2) are a field, and so are the rational numbers. Each is in a different component of the category of fields; in fact, each is an initial algebra for its component. A homomorphism of fields is a ring homomorphism which keeps 0 and 1 separate. It follows easily from this that all field morphisms are injective, and that there is no field morphism from the integers (mod 2) to the rational numbers or vice versa.

A set of algebras for a sketch which contains exactly one initial algebra for each component is called an **initial family** for the sketch.

Our thesis is that if $\mathcal{S}$ is an FLS-sketch regarded as the description of a data type, then an initial algebra for a component of the category of models of $\mathcal{S}$ is a semantics for the data type in the sense that a variable of that type will have exactly the elements of the initial algebra as its possible values. In the situation where the sketch is an FL sketch, there will thus be one semantics, but in general there will be a different semantics for each component.

10.5. For example, it is fairly straightforward to prove that the natural numbers are an initial model for a component of the sketch for natural numbers given in Section 9, namely the component of those models M for which the value of $M(\mathrm{succ})$ always lies in $M(n_+)$.

The model of bounded natural numbers, for a given overflow greater than 0, is initial for those models with the same overflow (the number of times $M(\mathrm{succ})$ has to be iterated starting at 0 before its value lies in $M(n_{\mathrm{over}})$.) The required morphism from the bounded natural numbers to a model with the same overflow can be defined by induction starting at 0; it will be clear that the morphism has to be unique. Note that such models do indeed give the possible values of a variable of type "natural number" with a specific overflow.

10.6. Many categories of algebraic structures, for example the category of monoids and the category of groups, have trivial one-element structures as initial algebras. Such categories can typically be sketched by FP sketches.

For example, the sketch for monoids may be constructed following the outline in Section 2. The sketch must have a node m which will become the monoid in a model. An arrow defining the binary operation requires a node labeled $m \times m$ to become its domain; the binary operation will then be given by an arrow $\mathrm{mult}: m \times m \longrightarrow m$. In accordance with the notation in Section 7, this requires a cone with vertex $m \times m$. To state the existence of an identity element requires a node 1 which becomes the terminal object, and to state the associativity law using the left diagram in Diagram 2.3 requires a node $m \times m \times m$ which becomes the triple cartesian product. That diagram and the diagram saying the identity element acts as an identity also require auxiliary diagrams for the arrows in the associativity diagram.

The category of monoids has the trivial one-element monoid as initial algebra. Thus the sketch for monoids describes a trivial data type. If the sketch is regarded as describing a mathematical structure, on the other hand, the semantics must be taken to be the whole category of models, which is not at all trivial. Similar remarks can be made about groups and rings. The category of fields provide an example of a sketch of mathematical structures which produces a series of nontrivial data types as well.

In Section 13, we describe an internal way to construct initial algebras for sketches for which everything is finite.

11. Binary trees

We now describe a sketch for ordered rooted binary trees (called trees in this discussion). "Binary" means that each node has either no children or two children, and "ordered" means that the children are designated left and right. This is an example which uses cocones to treat exceptional cases, in this case the empty tree.

Trees are parametrized by the type of data that are stored in them. We will say nothing about this type of data, supposing only that it is a type for which there is an initial model. The way in which the parametrized data type is filled in with a real one is described, for example, in [Ehrig et al., 1984]. We would like to thank Adam D. Barr for helpful discussions on how binary trees operate (especially their error states) on real machines.

The sketch will have sorts 1, t, s, d. Informally, t stands for tree, s for non-empty tree and d for datum. We have the following operations:

$$\begin{aligned} \text{empty}:1 &\longrightarrow t \\ \text{incl}:s &\longrightarrow t \\ \text{val}:s &\longrightarrow d \\ \text{left}:s &\longrightarrow t \\ \text{right}:s &\longrightarrow t \end{aligned}$$

The intended meaning of these operations is as follows:

Empty() is the empty tree; incl is the inclusion of the set of non-empty trees in the set of trees; val(S) is the datum stored at the root of S; left(S) and right(S) are the right and left branches (possibly empty) of the non-empty tree S, respectively.

We require that

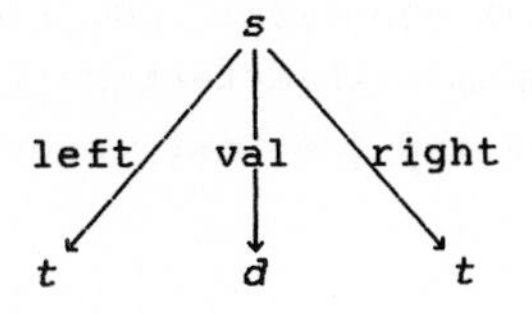

be a cone and that

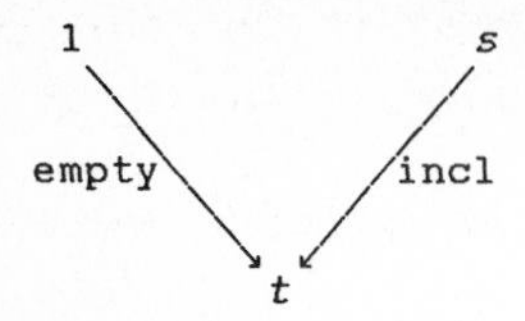

be a cocone.

There are no diagrams.

The cocone says that every tree is either empty or non-empty. This cocone could be alternately expressed $t = s + \{\text{empty}\}$. The cone says that every non-empty tree can be represented uniquely as a triplet ⟨left(S),val(S),right(S)⟩ and that every such triplet corresponds to a tree. Note that this implies that left, val and right become coordinate projections in a model.

Using this, we can define subsidiary operations on trees. For example, we can define an operation of left attachment, lat:$t \times s \longrightarrow s$ by letting lat(T,⟨left(S),val(S),right(S)⟩) = ⟨T,val(S),right(S)⟩. This can be done without elements: lat is defined in any model as the unique arrow making the following diagram commute (note that the horizontal arrows are isomorphisms):

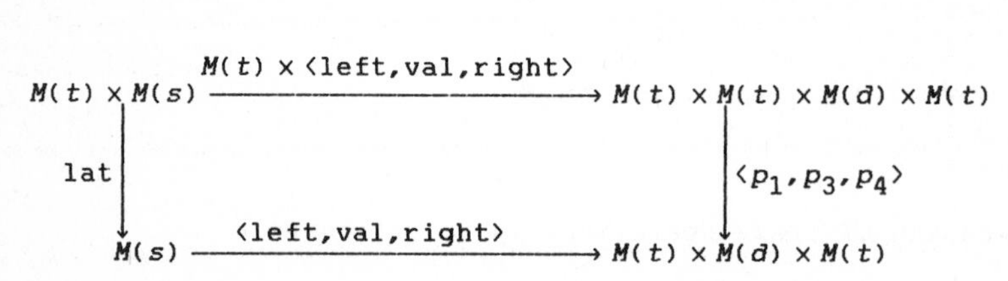

In a similar way, we can define right attachment as well as the insertion of a datum at the root node as operations definable in any tree. These operations are implicit in the sketch in the sense that they occur as arrows in the theory generated by the sketch (see 6.1), and therefore are present in every model.

It might be felt that this model of computation with trees is unrealistic. In fact, we argue that it is highly useful both from a conceptual point of view. Typically when trees are used there is a recursion that terminates at a leaf (node whose right and left branches are empty) and the test is made at every iteration. On the other hand,

11.1. Theorem. *Supposing there is an initial algebra for the data type, then the category of binary trees of that type has an initial algebra. If the data type has a unique initial algebra, then so does the corresponding category of binary trees.*

Proof. We construct the initial algebra recursively according to the rules:

(i) The empty set is a tree;

(ii) If T_l and T_r are trees and D is element of the initial term algebra for the data type, then $\langle T_l, D, T_r\rangle$ is a non-empty tree.

(iii) Nothing else is a tree.

This is a model M_0 defined by letting $M_0(s)$ be the set of non-empty trees, $M_0(t) = M_0(s) + \{\emptyset\}$ and $M_0(d)$ be the initial model of the data type. Here "+" denotes disjoint union. It is clear how to define the operations of the sketch in such a way that this becomes a model of the sketch.

Now let M be any model with the property that $M_0(d)$ is an initial model for the data type. There is a unique morphism $f(d): M_0(d) \longrightarrow M(d)$ that preserves all the operations in the data type. We also define $f(t)(\emptyset)$ to be the value of $M(\text{empty}): 1 \longrightarrow M(t)$. Finally, we define $f(s)(\langle T_l, D, T_r\rangle)$ $= \langle f(t)(T_l), f(d)(D), f(t)(T_r)\rangle$ where $f(t)$ is defined recursively to agree with $f(s)$ on non-empty trees. It is immediate that this is a morphism of models and is unique. In particular, if the data type has only one initial model then M_0 is unique. □

In Pascal textbooks a definition for a tree type typically looks like this:

```
type TreePtr = ^Tree;
     Tree = record LeftTree, RightTree : TreePtr;
                   Datum : integer
            end;
```

Note that from the point of view of the preceding sketch, this actually defines nonempty trees. The empty tree is referred to by a null pointer. This takes advantage of the fact that in such languages defining a pointer to a type D actually defines a pointer to what is in effect a variant record (union structure) which is either of type D or of "type" null. In some modern languages an exception mechanism allows one to take care of the case of the pointer being null in a more flexible way.

12. Context free grammars

In this section, we will describe a class of sketches with exactly the expressive power of context free grammars. Specifically, an initial algebra for a sketch in the class is the set of derivation trees for the language of the corresponding grammar.

A sketch is **context free** if it has no diagrams or cocones and if its nodes are divided into three mutually exclusive kinds: **vertex nodes, nonterminal nodes,** and **terminal nodes** with the following properties. Recall that a **constant** is an arrow whose domain is a node which is the vertex of an empty cone. In a model, a constant is identified with the image of the single element of its domain.

(i) A vertex node is the vertex of exactly one cone, is not in the base of any cone, and is not the target of a constant.

(ii) A nonterminal node is not the vertex of a cone and not the target of a constant.

(iii) A terminal node is not the vertex of a cone and is the target of exactly one arrow, which is a constant.

There must be at least one nonterminal node but there need not be any others. It follows from the definition that if there is a terminal node then there must be a vertex node 1 which is the vertex of an empty cone. There is clearly only one way at most that the nodes of a sketch can be divided up into three such kinds.

To avoid the reader's falling into terminal confusion, we will clue you in: the idea is that the arrows are production rules *reversed,* and vertex nodes are there to be the right sides of those production rules which have right sides of length greater than one.

By a **context free grammar** (CFG), we mean a 3-tuple $\mathcal{G} = (\mathbf{V},\mathbf{T},\mathbb{P})$ where $\mathbf{V}$ is a set called the **vocabulary,** $\mathbf{T} \subseteq \mathbf{V}$ is a subset called the **terminal vocabulary** and $\mathbb{P}$ is a set of **production rules** of the form $a: V \Rightarrow \alpha$, where $V \in \mathbf{V} - \mathbf{T}$ and $\alpha \in \mathbf{V}^{+}$, the set of nonempty strings of elements of $\mathbf{V}$. We name the production rules in order to refer to them later and also because we allow the possibility of two different rules with the same source and target.

A **derivation tree** for a CFG $\mathcal{G} = (V,T,P)$ is an ordered rooted tree (meaning the set of children of each node is totally ordered) labeled by the elements of V, defined as follows.

(i) A single node labeled by an element v of V is a derivation tree.

(ii) Let T be a derivation tree, x a leaf of T labeled by a nonterminal v, and $a{:}v \Rightarrow {'s_1 s_2 \cdots s_n'}$ a production. Construct T' from T by attaching n nodes labeled $s_1,\ldots,s_n$ *in that order* to be children of x. Then T' is a derivation tree.

If T is a derivation tree defined by (i), it is said to **derive** the string $'v'$. If T is a derivation tree deriving the string $\alpha = {'r_1 r_2 \cdots r_n'}$, $r_k = v$ is the symbol corresponding to the leaf x of T as in (ii), and T' is obtained from T by (ii), then T' derives the string obtained from α by replacing r_k by $s_1 s_2 \cdots s_n$. The function which takes a derivation tree to the string it derives is called "debracketing" in [Chomsky & Schützenberger, 1963].

A **terminal derivation tree** is a derivation tree all of whose leaves are labeled by terminals.

We define the **language** $\mathcal{L}$ generated by a CFG as follows. For each symbol $v \in V$, let $\mathcal{L}(v)$ denote the set of all strings $\alpha \in T^+$ such that there is a terminal derivation tree with root labeled v which derives α. $\mathcal{L}$ is the indexed family of the sets $\mathcal{L}(v)$ indexed by the vocabulary.

This differs somewhat from the usual treatment of CFGs in which there is an initial element of V, often denoted S, and the language generated by the CFG is the set of all strings of terminals derived from S.

We now construct, for each context free grammar $\mathcal{G}$, a context free sketch $\mathcal{S}$ with the property that the initial model of $\mathcal{S}$ is the set of derivation trees of $\mathcal{G}$ and the language generated by $\mathcal{G}$ extends to a model of $\mathcal{S}$ in such a way that the debracketing function is a quotient map of models.

Let $\mathcal{G} = (V,T,P)$ be a grammar. We will define the sketch $\mathcal{S}$ as follows. The set of sorts of the sketch is just the vocabulary V, together with such formal products of objects as are needed, including a sort 1 (unless the terminal vocabulary should be empty). Whenever there is a production $a{:}v \Rightarrow {'v_1\ v_2\ \cdots\ v_k'}$ in P, there is an operation $a^*{:}v_1 \times v_2 \times \cdots \times v_n \longrightarrow v$ in the sketch. This requires a cone to force the node $v_1 \times v_2 \times \cdots \times v_k$ to be what its name indicates. Furthermore, for each terminal element $t \in T$, let

$\langle t\rangle:1 \longrightarrow t$ be a constant. $\mathcal{S}$ is clearly a context free sketch.

We will describe a model M_0 of this sketch and show it is the initial model. For a terminal t, let $M_0(t) = \langle t\rangle$. Since no cone of the sketch has t as a vertex, this is consistent with there being no other element in $M_0(t)$ than that single constant. For a nonterminal v, let $M_0(v)$ consist of all terminal derivation trees with root labeled by v. If $v = v_1 \times \cdots \times v_n$ is a vertex node, then $M(v)$ consists of n-tuples with ith coordinate an element of v_i.

What M_0 must do to arrows corresponding to constants is clear. If $a:v \Longrightarrow {}'v_1 v_2 \cdots v_k{}'$ is a production, then $M_0(a^*)$ takes a k-tuple of trees $T_1,\ldots,T_k$ (T_i must have root v_i) to the k-branched derivation tree with root labeled by v with the trees T_i $(i = 1,\ldots,n)$ attached to the root in that order.

We must prove that M_0 is an initial model. Given a model M of $\mathcal{S}$, we construct a morphism ϕ of models by induction on the depth of a derivation tree. It extends to tuples in vertex nodes coordinatewise.

Suppose t is a terminal, hence $\langle t\rangle$ is a terminal derivation tree of depth 0. It is the only element of $M_0(t)$. There is an element of $M(t)$ corresponding to this constant; we set $\phi(t)\langle t\rangle$ to be that element.

Now suppose T is a tree with root labeled by the nonterminal v. Remove the root and you get a forest $T_1,T_2,\ldots,T_n$ of trees of smaller depth. Suppose the label of the root of T_i is s_i for $i = 1,\ldots,n$. We may suppose, by induction, that there is an element $\phi(T_i) \in M(s_i)$ for $i = 1\ldots n$. By construction there must be an arrow $a^*:s_1 \times s_2 \times \cdots \times s_n \longrightarrow v$. Let $\phi(T) = M(a^*)(T_1,T_2,\ldots,T_n)$.

The fact that this defines a morphism of models and its uniqueness are left to the reader.

The language generated by the context free grammar $\mathcal{G}$ can be extended to a model L of the sketch as follows. If v is a nonterminal, let $L(v) = \mathcal{L}(v)$. If t is a terminal, $L(t)$ is the singleton $\{t\}$. $L(s_1 \times s_2 \times \cdots \times s_k) = L(s_1) \times L(s_2) \times \cdots \times L(s_k)$. If $a:s_1 \times s_2 \times \cdots \times s_n \longrightarrow v$, then $L(a^*)$ is concatenation. It is straightforward to show that the function which takes a derivation tree to the string it derives is a morphism of models which is surjective on every sort, hence is a quotient morphism.

By reversing the process just described, given a context free sketch one can construct a context free grammar whose derivation trees are essentially the elements of the initial algebra of the sketch. The details are left to the reader. A similar construction was mentioned without much detail in [Goguen et al., 1977], Section 3.1.

13. Initial algebra as term algebra

The proof of Theorem 10.1 in [Barr, 1986] is top-down, as is that of Theorem 10.4. The interest from the point of view of data type semantics is in explicitly constructing the initial algebra from the data given in the sketch, at least for **finitary sketches** -- those with only a finite number of nodes and arrows in which all diagrams, cones and cocones are finite and there are only finitely many of each. It turns out that this can be done constructively for finitary FL sketches and constructively up to a "dæmon" for finitary FLS sketches. We give here a recursive description of the initial algebra for finitary FLS sketches which makes the constructive nature of the initial algebras clear. The construction is in terms of congruence classes of terms and for FP sketches is a generalization of that described in Meseguer and Goguen [1985]. First we need some notation and a few assumptions on finitary FLS sketches.

Let $\mathcal{S} = (\mathcal{G}, \mathcal{D}, \mathcal{L}, \mathcal{C})$ be a finitary FLS-sketch. Let λ denote the largest number of nodes in the index graph of any cone, and κ the largest number of nodes in the index graph of any cocone. If C is a cone in $\mathcal{L}$, we denote the vertex of C by v_C and its index graph by $\mathcal{I}_C$. Moreover, we assume that the set of nodes of $\mathcal{I}_C$ is an initial segment of the integers. This latter is for convenience; it avoids having double subscripts in the exposition.

We also assume the sides of a cone C (the diagrams of the form

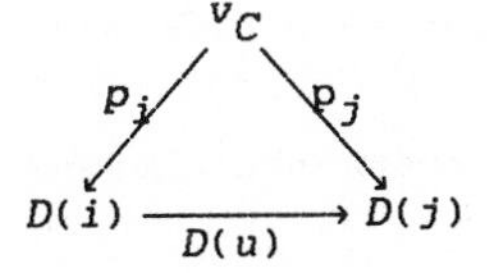

where p_i are the coordinate projections of the cone and $u: i \longrightarrow j$ is an arrow of $\mathcal{I}_C$) are in the set $\mathcal{D}$ of diagrams of $\mathcal{S}$; and similarly we assume the sides of the cocones in $\mathcal{C}$ are in $\mathcal{D}$. These assumptions are also for convenience; any model will have to take each side of a cone or cocone to a commutative diagram anyway.

Finally we make an assumption which is in fact a restriction: that no node is the vertex of more than one cocone. One can always change a sketch to a sketch with an equivalent (but not necessarily isomorphic) category of models by adding a new vertex of a cocone whose vertex is also the vertex of another cocone, together with arrows forced to be isomorphisms between them. One can also give a direct, but more complicated, description of the initial term algebra without this restriction; see [Barr & Wells, to appear].

A **compatible family** for a cone C in a model M is a function F from the set of nodes of $\mathcal{I}_C$ to the union of the value sets $M(D_C(i))$ with the property that $F(i) \in M(D_C(i))$ and if $u:i \longrightarrow j$ in $\mathcal{I}_C$ then $M(u)(F(i)) = F(j)$. It follows from the definition of limit cone that compatible families must correspond bijectively with elements of $M(v_C)$ for any cone C. In particular, for a discrete cone a compatible family is just a tuple of elements of the base, since there are no arrows $u:i \longrightarrow j$ in that case. Another fact we will use later is that if $\alpha:M \longrightarrow N$ is a morphism of models and F is a compatible family for a cone C in M, then $\alpha \circ F$ is a compatible family for C in N. This is straightforward to verify.

Note that there is no notion corresponding to compatible family for cocones.

The initial algebra for a component of the category of models of $\mathcal{S}$ will be a model I whose elements (in its various sorts) are congruence classes of strings in an alphabet $A_{\mathcal{S}}$ with respect to a congruence defined later. The alphabet consists of

(i) arrows of $\mathcal{G}$;

(ii) k-tuples of arrows of $\mathcal{G}$, denoted $\langle f_1,\ldots,f_k \rangle$, for all k between 1 and λ.

(iii) a symbol p_C for each cone C of $\mathcal{C}$.

Among strings in the alphabet $A_{\mathcal{S}}$ is an empty string $\langle\rangle$ for each node g of $\mathcal{G}$.

The congruence relation which we will define will have congruence classes denoted [-].

13.1. Dæmons. A **dæmon** for $\mathcal{S}$ is a function D from the set of all strings in the alphabet $A_{\mathcal{S}}$ to the initial segment $\{1..\kappa\}$ of the positive integers. We will use a dæmon this way: in constructing an initial algebra, if a string w must be in a sort which is the vertex of a cocone (hence in the model it must be the disjoint union of no more than κ sorts), we will choose to put it in the $D(w)$th summand. (Recall that the nodes of the index graph of the cocone are positive integers). If $D(w)$ is larger than the number of summands, the construction aborts.

Since a dæmon for an FL sketch is the constant function with value 1, it is superfluous and can be ignored in the description of the initial algebra.

13.2. Every model determines a dæmon. As we shall see, every string w which determines an element of a sort in the initial algebra corresponds in an obvious way to an element of a model. That element must be in to a unique summand; if it is the ith summand, then set $D(w) = i$. On strings not used in the construction, define $D(w) = 1$, not that it matters. This dæmon is guaranteed not to abort and will actually construct the initial model in the component of the given model.

Our definition of dæmon shows that one can attempt a construction of an initial model without already knowing models. In concrete cases, of course, it will often be possible to characterize which choices give initial models and which do not.

13.3. Initial algebra as term algebra: formal description. The following statements N-1, N-2, N-3, A-1, A-2 and C-1 through C-4 will be referred to as the RDTA (Recursive Description of the Term Algebra). For certain dæmons, the RDTA constitutes a precise description of an initial algebra I for a component of the category of models of $\mathcal{S}$. Moreover, every such initial algebra is isomorphic to one described by the RDTA for some dæmon.

In detail, the elements of the initial algebra are congruence classes of strings described by N-1 through N-3 (and no other strings) subject to the least congruence relation generated by C-1 through C-4, *provided* that no element appears in two different summands of I(s) when s is the vertex of a cocone. This provision will be satisfied if the dæmon is chosen to be that determined by a model M; then I is the initial algebra for the component of the category of models containing M. If $\mathcal{S}$ is an FL sketch, then the model I is the unique initial algebra (up to isomorphism) for the whole category of models.

The RDTA is a description of I in the sense that every statement is a true statement about I. It describes I recursively in the sense that elements or equivalences mentioned in some parts may be required by earlier or later parts.

In the RDTA, each statement is named according to whether it states a fact about the elements of the value of the model I at a Node, or about the value in the model of an Arrow, or about the Congruence relation.

N-1. If the sketch contains a node 1 (which must become a singleton in any model), then $I(1) = \{\langle\rangle_1\}$.

N-2. If $f{:}g \longrightarrow h$ in $\mathcal{G}$ and $[x] \in I(g)$, then $[fx]$ is an element of the jth summand of $I(h)$, where $j = D(fx)$.

N-3. Let C be a cone of $\mathcal{C}$ whose index graph has k nodes and let F be a compatible family for C in I. Let t denote the string $p_C\langle F(1),\ldots,F(k)\rangle$. Then $[t]$ is in the jth summand of $I(v_C)$, where $j = D(t)$.

A-1. In the notation of N-2, $I(f)[x] = [fx]$.

A-2. In the notation of N-3, $I(p_i)[t] = [F(i)]$, for $i = 1,\ldots,k$. (Note that A-1 and A-2 are forced to be consistent with each other by C-2 below.)

C-1. In the notation of N-2, if $[x] = [y] \in I(g)$, then $[fx] = [fy]$.

C-2. In the notation of N-3, $[p_i t] = [F(i)]$ for $i = 1,\ldots,k$.

C-3. In the notation of N-3, if $[x] \in I(v_C)$, then $[x] = [p_C\langle p_1 x,\ldots,p_k x\rangle]$. (It follows from C-2, C-4 and the inclusion of the sides of cones in $\mathfrak{D}$ that $F(i) := [p_i x]$ defines a compatible family.)

C-4. If $\langle f_1,\ldots,f_m\rangle$ and $\langle g_1,\ldots,g_n\rangle$ are composable paths in a diagram D of $\mathfrak{D}$ from the same node h_1 to the same node h_2 and $[x]$ is an element of $I(h_1)$, then $[f_1\circ\cdots\circ f_m x]$ and $[g_1\circ\cdots\circ g_n x]$ denote the same element of $I(h_2)$. (They are elements of $I(h_2)$ by N-2).

It is easy to see that C-2 follows from A-1 and A-2; the other requirements appear to be independent.

13.4. I is a model. The RDTA is recursive in the sense that some of the statements refer to elements generated by others (or merged by one of the requirements labeled C).

It is necessary to verify that the RDTA actually describes a model, and that the model is initial. The elements of each sort are given by N-1 through N-3. We must verify the following statements.

(i) The operations are defined on every element and have the correct sources and targets. This is easy to check.

(ii) The operations are well-defined: If the value of an operation is defined by A-1, then it is necessarily well-defined because of C-2. If the value is defined by A-2, and $[x] = [t]$ (using the notation of A-2), then successive applications of A-1, C-1 and A-1 yield: $I(p_i)[x] = [p_i x] = [p_i t] = I(p_i)[t]$.

(iii) For each vertex v_C of a cone C, $I(v_C)$ is the limit of the diagram $I \circ D$, where $D: I \longrightarrow \mathcal{G}$ is the base diagram of the cone. This is true because the map β which takes a compatible family F for C to the element $[p_C\langle F(1),\ldots,F(k)\rangle]$ is a bijection. It is surjective by N-3 and injective by C-3.

(iv) Every diagram in $\mathfrak{D}$ is taken to a commutative diagram by I. This is immediate from C-4.

(v) I is an initial algebra. To verify this, suppose $\mathcal{S}$ is a sketch as above and I is the initial term algebra determined by the RDTA and the dæmon given by a model M. We can then define the unique morphism $\alpha: I \longrightarrow M$ recursively in terms of the RDTA. Let $[y]$ be an element of a sort g. The following cases exhaust the possibilities for $[y]$.

(a) $g = 1$ and consequently $[y] = \lambda_1$. Then $\alpha[y]$ must be the unique element of $M(1)$.

(b) $[y] \in I(h)$ is of the form $[fx]$ for some $f: g \longrightarrow h$ in $\mathcal{G}$, and $\alpha[x]$ is already defined. Since $I(f)[x] = [y]$ by A-2, $\alpha[y]$ is forced to be $M(f)(\alpha[x])$ if α is going to be a morphism (the diagram corresponding to (7.1) must commute).

(c) $[y]$ is of the form $[p_C\langle F(1),\ldots,F(k)\rangle]$ for some cone C and compatible family F for I where $\alpha(F(i))$ is already defined for each i. Since $\alpha \circ F$ is a compatible family for M, there is a unique element t of $M(v_C)$ corresponding to it, and moreover $M(p_i)(t) = \alpha(F(i))$, which is necessary for α to be a morphism.

The remarks necessary for showing that α is a morphism of models are included in (a) through (c). That it is well defined follows from the fact that any string in an equivalence class in I corresponds in an obvious way to an element of *M*, and if two of them are equivalent, they are equivalent by one of C-1 through C-4, all of which must hold in any model.

14. BIBLIOGRAPHY

M. Barr, *Models of sketches.* Cahiers de Topologie et Géometrie Différentielle, 27 (1986), 93-107.

M. Barr and C. Wells. Toposes, Triples and Theories. Springer-Verlag, 1985.

M. Barr and C. Wells. Category Theory for Computer Scientists, in preparation.

A. Bastiani C. Ehresmann, *Categories of sketched structures.* Cahiers de Topologie et Géometrie Différentielle 10 (1968), 104-213.

R. M. Burstall, *Electronic category theory,* in Mathematical Foundations of Computer Science. Springer Lecture Notes in Computer Science 88 (1980).

N. Chomsky, M. P. Schützenberger, *The algebraic theory of context-free languages,* in P. Braffort, D. Hirschberg, eds. Computer Programming and Formal Systems, North-Holland, 1963.

Y. Diers, Catégories Localizables. Thèse de doctorat, Université de Paris, 1977.

J. Donahue and A. Demers, *Data types are values.* To appear in ACM Transactions on Programming Languages and Systems.

H. Ehrig, J. W. Thatcher, P. Lucas and S. N. Zilles, *Denotational and initial algebra semantics of the algebraic specification language Look.* Preprint: Technische Universität Berlin, 1982.

H. Ehrig, H.-J. Kreowski, J. Thatcher, E. Wagner and J. Wright, *Parameter passing in algebraic specification languages.* Theoretical Computer Science 28 (1984), 45-81.

H. Ehrig and B. Mahr, Fundamentals of algebraic specifications I. Springer-Verlag, 1985.

M. Fourman and S. Vickers, *Theories as categories,* in Category Theory and Computer Programming, Springer Lecture Notes in Computer Science 240. Springer-Verlag, 1986.

J. A. Goguen, *Abstract errors for abstract data types,* In E. J. Neuhold, ed. Formal Description of Programming Concepts, North-Holland, 1978.

J. A. Goguen and J. Meseguer, *Eqlog: equality, types and generic modules for logic programming.* To appear in DeGroot and Lindstrom, eds., Functional and Logic Programming, Prentice-Hall, 1985a.

J. Goguen and J. Meseguer, *Order sorted algebra I: partial and overloaded operators, errors and inheritance.* Preprint, SRI International, Menlo Park, CA 94025, 1985b.

J. Goguen, J. W. Thatcher, E. G. Wagner and J. B. Wright, *Initial algebra semantics and continuous algebras*. J. **ACM**, 24 (1977), 68-95.

J. Gray, *Categorical aspects of parametric data types.* Preprint: University of Illinois, 1985.

R. Guitart, *On the geometry of computations.* Cahiers de Topologie et Géométrie Différentielle Catégoriques, 27 (1986), 107-136.

R. Guitart and C. Lair, *Calcul syntaxique des modèles et calcul des formules internes.* Diagrammes, 4 (1980), 1-106.

J. Lambek and P. Scott, **Cartesian Closed Categories and λ-Calculus.** Cambridge Studies in Advanced Mathematics 7. Cambridge University Press, 1986.

S. Mac Lane, **Categories for the Working Mathematician.** Graduate Texts in Mathematics 5, Springer-Verlag, 1971.

J. Makowsky, *Why Horn formulas matter in computer science: initial structures and generic examples.* Technical Report #329, Department of Computer Science, Technion, Haifa, Israel, 1984.

P. Mateti and F. Hunt, *Precision descriptions of software designs: an example.* **IEEE Compsac**, 1985, 130-136.

C. McLarty, *Left exact logic.* **Journal of Pure and Applied Algebra**, 41 (1986), 63-66.

J. Meseguer and J. Goguen, *Initiality, induction and computability,* in M. Nivat and J. C. Reynolds, eds., **Algebraic Methods in Semantics**, Cambridge University Press, 1985.

D. E. Rydehead and R. M. Burstall, *The unification of terms: a category-theoretic algorithm,* in **Category Theory and Computer Programming**, Springer Lecture Notes in Computer Science 240. Springer-Verlag, 1986.

J. W. Thatcher, E. G. Wagner and J. B. Wright, *Specification of abstract data types using conditional axioms.* (Extended abstract). IBM T. J. Watson Research Center Research Report RC 6214 (#26679), 1976.

J. W. Thatcher, E. G. Wagner and J. B. Wright, *Data type specification: parametrization and the power of specification techniques.* **ACM Transactions on Programming Languages and Systems**, Vol. 4 no. 4, October, 1982.

H. Volger, *On theories which admit initial structures.* Preprint: Universität Tübingen, 1985.

E. Wagner, S. Bloom and J. W. Thatcher, *Why algebraic theories?* in M. Nivat and J. C. Reynolds, eds., **Algebraic Methods in Semantics**, Cambridge University Press, 1985.

S. N. Zilles, P. Lucas and J. W. Thatcher, *A look at algebraic specifications.* IBM T. J. Watson Research Center Research Report RJ 3568 (#41985), 1982.

Initial Algebra Semantics and Concurrency

Maria Zamfir
Computer Science Department
KSU
Manhattan, KS 66506

Abstract

The purpose of this paper is to show that *initial algebra semantics* has an immediate and useful application in the area of communicating computing systems. The major technical feature is a category of continuous many-sorted algebras ca led *parallel-nondeterministic* algebras. In this setting parallel and nondeterministic behaviors of communicating computing systems can be rigorously formulated as sequences of rewritings on abstract objects called *parallel-nondeterministic terms* or *diamonds*. It is shown that diamonds are free in the category of continuous *parallel-nondeterministic* algebras. (To demonstrate this fact, some results concerning categories of continuous algebras, which can be found in the work of the ADJ group, are presented in a self-contained form.)
Nondeterminism and parallelism are modeled explicitly by introducing a *choice* operator and a *parallel* operator, respectively.
In a companion paper [10] *flow nets* are introduced to describe parallel and nondeterministic behaviors of computing systems that communicate with each other, just as conventional flowcharts are used to describe sequential computations. In a continuous parallel-nondeterministic algebra a flow net is represented by its unfoldment - the solution of a finite system of recursive equations.

1. Introduction

Much recent research is devoted to the theory of concurrent computation. Various directions within this theory result from the differences in authors' views on concurrency and also from the problems they address. In this paper we present a minimal syntax in which parallel and nondeterministic behaviors of concurrent computations can be rigorously formulated. An important point of our formalism is that we consider sequentiality, nondeterminism, and parallelism as distinct notions.

Our approach can be outlined by analogy with the work of ADJ [2], [7] on the construction of initial continuous algebras, in which behaviors of conventional flowcharts are formulated. First, we define a category of continuous many-sorted algebras called continuous *parallel-nondeterministic* algebras. Then we show that in this category there exists a free object; we call its elements *parallel-nondeterministic* terms (expressions) or *diamonds*.

In a companion paper [10], we present *flow-of-control* algebras in which computing systems that communicate with each other are represented as generalized flowcharts, called *flow nets*. Flow-of-control algebras contain a minimal set of operations to compose flow nets, together with an equational system they must satisfy. We assume that *Input/Output capabilities (ports)* are associated with computing systems. Computing systems communicate by message passing using channels that connect their input and output ports.

In this paper we use continuous parallel-nondeterministic algebras and formulate behaviors of flow nets as diamond unfoldments. Thus, we present an algebraic framework in which behaviors of communicating computing systems are sequences of rewritings on diamonds.

In Section 2 we introduce the notation used throughout this paper and briefly recall the necessary theoretical framework on initial algebra semantics. Section 3 contains the definition of many-sorted parallel-nondeterministic algebras, a construction of a free object in the category of parallel-nondeterministic algebras, the definition of continuous many-sorted parallel-nondeterministic algebras and of many-sorted parallel-nondeterministic algebras with communication (Input/Output) capabilities. In Section 4 we introduce through an example the concept of flow net; its unfoldment in a many sorted parallel-nondeterministic algebra is given in Appendix A.

The present paper is a continuation of the work presented in dissertation [9].

1.1 An Introductory Example

As far as we are concerned, the essential features of a concurrent computing system are captured in the following example.

A multiprocessor machine composed of at least three component processors that communicate with each other is used to evaluate an expression. Let the expression be f(x) + g(y) and assume that the functions f(x) and g(y) must be evaluated in parallel for some values for x and y and that a value for y may be obtained nondeterministically from one of the three component processors. We consider that nondeterminism arises in concurrency as soon as the relative speed of concurrent processors is left unspecified.

In the work of a program for the evaluation of the above expression we distinguish two phases. In the first phase a *flow net*, which is a kind of AND/OR graph, is defined. Described briefly, a flow net is a type of graph where a node describes an independent activity of a component processor on the task and arcs represent communication links among the activities of the component processors co-operating on the task. In the second phase, the defined flow net is executed. (Flow nets are the subject of the companion paper [10].)

We introduce a language for describing the above concurrent computation by considering explicit operators for parallelism and nondeterminism. Nondeterminism is introduced by using a "choice" operator @: for example, a term x @ y represents a computation that can be pursued either as the computation of x or as the computation of y. Parallelism is introduced by using a "pairing" operator ||: thus, a term x || y represents a computation that may be thought of as the computation of x and the computation of y running in parallel (no communication between x and y is assumed). We also use the operators @ and || in the syntactic definition of functions in order to capture the nondeterminism and synchronization of their arguments. (Given a function f: A → B we write (a)f to denote the value of f in B for an argument a in A.)

In our example, we write $(y_1 @ y_2 @ y_3)g$ to express the fact that g is a function of one argument and that argument can be y_1 (corresponding to the first source), y_2 (corresponding to the second source), or y_3 (corresponding to the third source). The function g can be evaluated when a value for y_1, for y_2, or for y_3 is given. When all three arguments get values, only one argument is chosen nondeterministically and used in the evaluation of g. Clearly, one out of three evaluations may take place, yielding three possible results; thus, we have three possible computations, which are emphasized by introducing the identity

(i) $(y_1 @ y_2 @ y_3)g = (y_1)g @ (y_2)g @ (y_3)g$

We call *linear* any function g that satisfies the equality (i). Note that in the evaluation of g the order in which y_1, y_2, and y_3 are written is not important; their number is also not important. These two facts are embedded in the following identities:

(ii) $(y_1 @ y_2) @ y_3 = y_1 @ (y_2 @ y_3)$ (@ is associative), and

(iii) $y_1 @ y_2 = y_2 @ y_1$ (@ is commutative)

An extreme case arises when we consider $y_1 @ y_1$. Then, there actually exists only one computation; as a result, we state the following identity:

(iv) $y_1 @ y_1 = y_1$ (@ is idempotent)

In our example, by $(x)f \parallel ((y_1)g @ (y_2)g @ (y_3)g)$ we express the fact that f and g must be evaluated in parallel. The identity

(v) $(x)f \parallel ((y_1)g @ (y_2)g @ (y_3)g) = ((x)f \parallel (y_1)g) @ ((x)f \parallel y_2)g) @ ((x)f \parallel y_3)g)$

states that the evaluation in parallel of f and g is nothing but the evaluation in parallel of f and g for values for g coming from the first source, or the evaluation in parallel of f and g for values for g coming from the second source, and so on. In other words, we require the operation $\parallel$ to be distributive with respect to the operation @.

Then, after the evaluation of both f and g is finished, the results must be joined and used to execute the operation +. A convenient way to formulate this situation is to consider + as a function of one argument, the argument being a tuple of three elements. This requires the pairing of the results coming from different sources; the following definitions establish just that:

(vi) $(x)f \parallel (y_1)g = (x\|y_1)f\|g$, $(x)f \parallel (y_2)g = (x\|y_2)f\|g$, and $(x)f \parallel (y_3)g = (x\|y_3)f\|g$

where by definition $x\|y_1$, $x\|y_2$, and $x\|y_3$ are tuples and $f\|g$ is the pairing of functions.

Using the identities (i) - (vi) and assumming that + is linear we get a complete formal description of our problem:

$((x)f \parallel ((y_1)g @ (y_2)g @ (y_3)g))+ = ((x\|y_1)f\|g)+ @ ((x\|y_2)f\|g)+ @ ((x\|y_3)f\|g)+$

Considering the function composition $\circ$ as the sequencing of operations, we can write the above expression as $(x\|y_1)\circ(f\|g)\circ+ @ (x\|y_2)\circ(f\|g)\circ+ @ (x\|y_3)\circ(f\|g)\circ+$

It is important to mention that the above expresssion is a term in a parallel-nondeterministic initial algebra. Such an expression can be considered as a program whose execution represents the evaluation of the expression $f(x) + g(y)$ in a concurrent computing system.

We would also like to point out that the equalities (i)-(vi) considered in the example above motivate the algebraic axioms used in the definiton of parallel-nondeterministic algebras.

2. Mathematical Preliminaries

In this section we outline the notation and some definitions and results required in our work. For a detailed exposition of the algebraic background required here, see [1], [2], or [7].

Throughout this paper we use ω for the set of nonnegative integers {0,1,2,...} and [n] for {1,2,...,n}. For ordered pairs we write $<a,b>$ with angle brackets. Given the functions f: A → B and g: B → C, we write their composite f∘g: A → C, so (a)f∘g = ((a)f)g. An expression such as f∘g, which denotes a function, is called a *functional form*.

Let S be any set. Then S^n is the set of all strings over S of length n; $S^* = \bigcup_{n \geq 0} S^n$ is the set of all finite strings over S. We denote the empty string by λ and the concatenation of the strings u and v by uv. We say that u is a *prefix* of v iff $\exists$ w: v = uw and write $u \leq_{pref} v$ (for it is obviously a partial order); u is a *part* of v iff $\exists w_1, w_2$: $v = w_1 u w_2$ and write $u \leq_{in} v$. We write |w| for the length of the string w. For $L \subseteq S^*$ and $w \in S^*$, we denote by L/w the residual of L by w: $L/w = \{u \mid u \in S^* \text{ and } wu \in L\}$.

Let S be a set of symbols called *sorts*. An *S-sorted operator domain* (or *signature*) Σ is an $<S^* \times S>$-indexed family of sets $\Sigma = \{\Sigma_{w,s} \mid w \in S^*, s \in S\}$. An element $\sigma \in \Sigma_{w,s}$ is an operator symbol of type $<w,s>$, arity w, sort (or co-arity) s, and rank |w|. An *S-sorted Σ-algebra A* consists of an S-indexed family of sets $\{A_s \mid s \in S\}$, where A_s is called the *carrier* of sort s, and for each function symbol $\sigma \in \Sigma_{w,s}$ an actual function (interpretation) $\sigma_A: A^w \to A_s$ where $A^w = A_{s_1} \times ... \times A_{s_n}$ when $w = s_1...s_n$. The operation σ_A is the operation of A named by σ. For $\sigma \in \Sigma_{\lambda,s}$, $\sigma_A \in A_s$ is a *constant* symbol. It is standard practice in universal algebra to let A ambiguously denote the S-sorted carrier of the Σ-algebra as well as the algebra itself.

Let A and B be S-sorted Σ-algebras. Then a Σ-homomorphism h: A → B is an S-indexed family of functions $h_s: A_s \to B_s$ such that for $\sigma \in \Sigma_{w,s}$ the following diagram commutes:

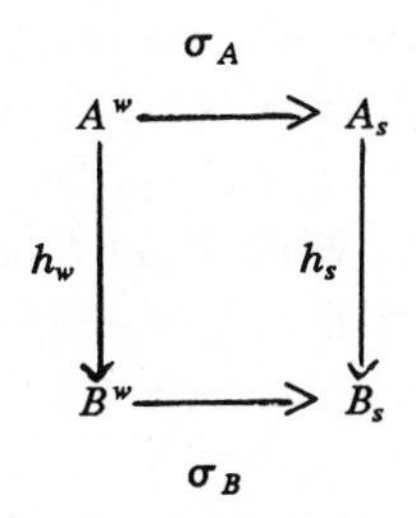

where $(a_1,...,a_n)h_w = <(a_1)h_{s_1},...,(a_n)h_{s_n}>$, $w = s_1...s_n$, and $<a_1,...,a_n> \in A^w$. Briefly, a Σ-homomorphism preserves all operations in Σ. A Σ-algebra T is said to be *initial* in the category of Σ-algebras iff for any other Σ-algebra A, there is a unique Σ-homomorphism h: T → A. The key property for the application of Σ-algebras to programming language semantics is given by the following

Proposition 2.1. For each operator domain Σ, there exists an initial Σ-algebra T_Σ. □
T_Σ is often called the Σ-*word algebra* and the carriers can be thought of as the sets of trees, or well-formed expressions, built up in the usual way from the operator symbols of Σ.

Example 2.1
If a signature Σ has the sort S = {s} with $\Sigma_{\lambda,s}$ = {c, x} and $\Sigma_{ss,s}$ = {f}, we may picture an example of a finite Σ-tree by

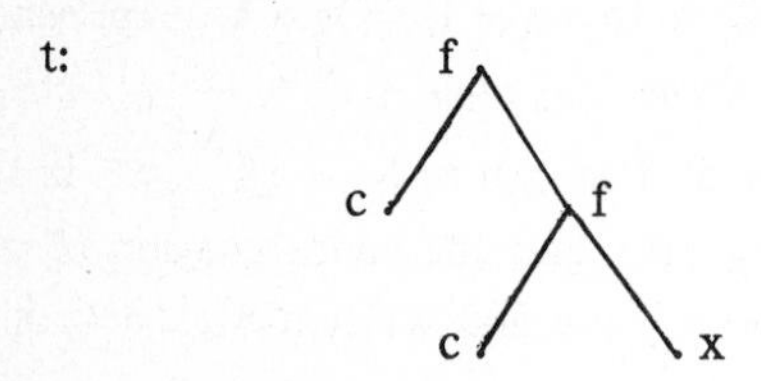

and an example of an infinite tree by

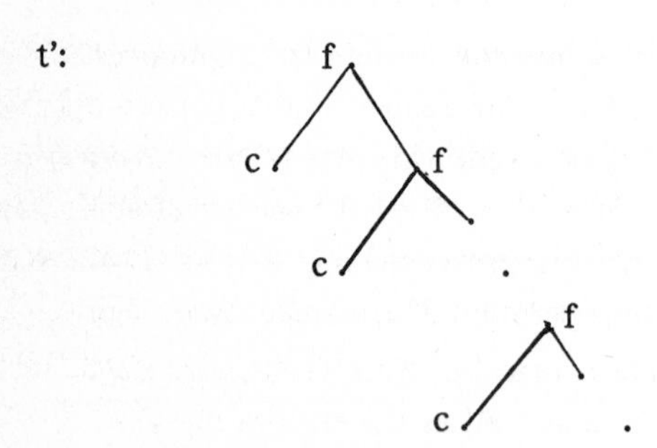

Let X be an S-indexed family of disjoint sets of *variables*, X = $\{X_s \mid s \in S\}$, where each X_s is also disjoint from each set of operator symbols in Σ. If Σ is an S-sorted operator domain, T_Σ(X) denotes the Σ-algebra *freely generated by* X. T_Σ(X) is used for presenting classes of algebras satisfying certain equational properties. A Σ-*equation of sort s* is a pair $(\forall x)t_1 = t_2$ from $T_\Sigma(X)_s$, where $(\forall x)$ is the finite set of all variables occuring in t_1 and t_2. (See [1] for details.) An *equational system (over* T_Σ(X)) is a set E of Σ-equations. What makes these pairs "equations" is how they are interpreted. For any Σ-algebra A and S-indexed family of sets of variables X, an *interpretation* or *assignment* θ: X $\rightarrow$ A of values in A_s to variables X_s is an S-indexed family of functions $\theta = \{\theta_s : X_s \rightarrow A_s\}$ for all s $\in$ S.

Proposition 2.2 . Let I_X: X $\rightarrow$ T_Σ(X) be the (S-indexed family of) injection(s) of the generators X into the carriers of T_Σ(X). Then, for any Σ-algebra A and assignment θ: X $\rightarrow$ A, there exists a unique Σ-homomorphism $\bar{\theta}$: T_Σ(X) $\rightarrow$ A that extends θ in the sense that $I_X\bar{\theta} = \theta$. □

Intuitively, θ gives values in A for the variables in a term t in $T_\Sigma(X)_s$, and A, being a Σ-algebra, gives values for the operation symbols in t; $\bar{\theta}$ is the familiar process of *evaluating* an expression and nothing more. Given a term t $\in T_\Sigma(X)_s$ and varying the assignment θ: X $\rightarrow$ A, $t_A: A^w \rightarrow$ A where $(a_1,...,a_n)t_A = (t)\bar{\theta}^w$, $<a_1,...,a_n> \in A^w$, $\bar{\theta}^w = \bar{\theta}_{s_1} \times ... \times \bar{\theta}_{s_n}$, w = $s_1...s_n$, $(x_i)\theta_{s_i} = a_i$, i $\in$ [n], is a function called the *derived operation of* t in A.

A Σ-algebra A *satisfies* $(\forall x)t_1 = t_2$ (or is a *model* for it) iff for every assignment $\theta : var_s(t_i) \rightarrow$ A, i = 1, 2, $(t_1)\bar{\theta} = (t_2)\bar{\theta}$. For an equational system *E*, a Σ-algebra A *satisfies E* iff it satisfies every equation of *E*. A Σ-algebra A satisfying *E* is called a $<\Sigma,E>$-algebra. In the category of $<\Sigma,E>$-algebras there exists an initial $<\Sigma,E>$-algebra, which is the quotient of T_Σ by the congruence relation generated by substitution instances of *E*, and which is denoted $T_{\Sigma E}$. If *E* has the Church-Rosser property and if a term has a canonical (normal) form, then the canonical form is unique.

The discrete algebras introduced above can be generalized to ω-continuous algebras. ω-continuous algebras are rational, so we can talk about solving systems of rational equations, i.e., their least fixed-point solutions. It is assumed that the reader is familiar with the elementary theory of ω-complete partially ordered sets (ω-CPOs) and ω-continuous functions. We only consider strict CPOs, i.e., CPOs with a minimum element denoted $\perp$ ("bottom").
An *ω-continuous algebra* is an algebra whose carriers are strict ω-CPOs and whose operations are ω-continuous functions. The algebra of finite and infinite partial trees plays the role of the algebra of expressions for the continuous case. Intuitively, a partial tree is a tree whose leaves can be labeled with $\perp$. Usually finite and infinite partial trees are introduced as partial functions t: ω^* -o$\rightarrow$ $\Sigma \cup \{\perp\}$ for which the domain of definition has the tree domain property, and with $\perp$ called "totally undefined". Here we adopt a convention in which nodes are represented by strings in the following way: the root is represented by λ; if a node in a tree is represented by a string u and it is labeled with the operator symbol $\sigma \in \Sigma_{w,s}$, its ith son is represented by the string uσ_i, $1 \leqslant i \leqslant |w|$ (σ indexed by which argument of σ it is). In this way we associate with Σ an alphabet W_Σ, which is called the "split alphabet": $W_\Sigma = \{\sigma_i \mid \sigma \in \Sigma_{w,s}$, and $1 \leqslant i \leqslant |w|\}$.

Example 2.1 (continued)
For Σ given by S = {s} with $\Sigma_{\lambda,s}$ = {c, x} and $\Sigma_{ss,s}$ = {f} we have $W_\Sigma = \{f_1, f_2\}$.

Finite and infinite partial trees are partial functions t: W_Σ^* -o$\rightarrow$ $\Sigma \cup \{\perp\}$ for which the domain of definition has the tree domain property:
(i) if u $\in$ (t)dom and v $\leqslant_{pref}$ u then v $\in$ (t)dom,
(ii) if uσ_i $\in$ (t)dom for $\sigma \in \Sigma_{w,s}$, |w| = k, then (u)t = σ and ((t)dom/w) $\cap W_\Sigma \subseteq \{\sigma_1,...,\sigma_k\}$.
In this definition, u $\in$ (t)dom is a node of the tree t, while (u)t is its label.

Example 2.1 (continued)
The finite tree of this example is formally defined by

$$t: \quad \begin{array}{l} (t)\mathrm{dom} = \{\lambda, f_1, f_2, f_2 f_1, f_2 f_2\} \\ \qquad\qquad\quad \downarrow \;\; \downarrow \;\; \downarrow \;\;\; \downarrow \;\;\;\; \downarrow \\ \qquad\qquad\quad f \;\; c \;\; f \;\;\; c \;\;\;\; x \end{array}$$

and the infinite tree by
$(t')\mathrm{dom} = \{u \mid \exists n \in N : u \leqslant_{pref} f_2^n f_1\}$, with $(f_2^p)t' = f$ (including $p = 0$), and $(f_2^p f_1)t' = c$.

Trees are naturally ordered as partial mappings by inclusion (of their corresponding graphs): a partial tree properly approximates another iff their corresponding nodes are identical except that the first contains $\perp$ in at least one position where the second contains a nontrivial subtree. We adopt the notation of M. Nivat [1]: $t \sqsubseteq t'$ iff $u \in (t)\mathrm{dom}$ then $u \in \mathrm{dom}(t')$ and $(u)t = (u)t'$ for $\forall u \in W_\Sigma^*$, where $\sqsubseteq$ is read "less defined than".

Example 2.1 (continued)
The infinite tree of this example is the least upper bound of the ω-chain

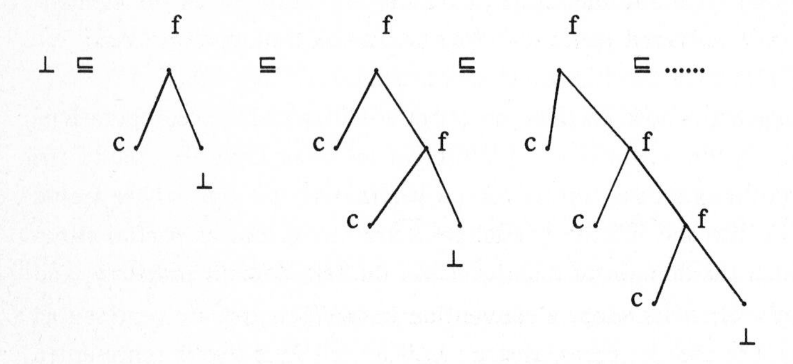

3. Parallel-Nondeterministic Algebras

The many-sorted algebras introduced in Section 2 are not adequate for modeling concurrency. The sets defining many-sorted algebras need much more complex structures for representing explicitely the "parallel" and "nondeterministic" activities of concurrent computing systems. We present in the following subsections the basic concepts of our work.

3.1 Parallel-Nondeterministic Constructions

Let $A = \{a,b,c,....\}$ be a set of atomic elements. We construct "parallel" terms (expressions) out of these elements by considering a family of n-ary "pairing" operations $||_A$ for each $n \in \omega$; $||_A a$, $a||_A b$, and $a||_A b||_A c$ are exemples of parallel terms. The name "pairing" given to the operations $||_A$ comes from their interpretation; by definition, $||_A a = \langle a \rangle$, $a||_A b = \langle a,b \rangle$, $a||_A b||_A c = \langle a,b,c \rangle$, etc... The name "parallel" given to the above expressions comes from their use to represent parallel activities of concurrent systems. We denote by $\langle A,||_A \rangle$ the set of all parallel terms over A.

We increase the complexity of the parallel terms by defining on $<A,||_A>$ a binary "choice" operation $@_A$; $(t||_A t')@_A t''$, $t||_A(t'@_A t'')$, for t, t',t'' arbitrary terms in $<A,||_A>$, are examples of "parallel-nondeterministic" terms. The name "choice" given to the operation $@_A$ comes from both its interpretation and its use to represent nondeterministic activities of concurrent systems. Operationally, the interpretation of $@_A$ is given by $t@_A t' = t$, $t@_A t' = t'$. The role of the parallel-nondeterministic terms in modeling of parallel and nondeterministic activities of concurrent systems is presented in Example 3.1.1. We now make the above precise with an inductive definition of the parallel-nondeterministic set $P\text{-}N(A) = <A,||_A,@_A>$, for a given set A, an n-ary *pairing* operation $||_A$, for each $n \in \omega$, and a binary *choice* operation $@_A$.

Definition 3.1.1 Let A be a set. The *parallel-nondeterministic (par-nd) set* $P\text{-}N(A) = <A,||_A,@_A>$ generated by A consists of:

(1) $a_1||_A \ldots ||_A a_n = <a_1,\ldots,a_n>$ for any n elements a_i in A, $n \in \omega$;

(2) $t@_A t'$ for each t, t' arbitrary terms in *P-N*(A); and

(3) $t_1||_A \ldots ||_A t_n$ for any n arbitrary terms t_i in *P-N*(A), $n \in \omega$.

The operations $||_A$ and $@_A$ are required to satisfy the following "parallel-nondeterministic" properties:

(4) $t\ @_A\ t = t$ ($@_A$ is idempotent),

(5) $t\ @_A\ t' = t'\ @_A\ t$ ($@_A$ is commutative),

(6) $t\ @_A\ (t'\ @_A\ t'') = (t\ @_A\ t')\ @_A\ t''$ ($@_A$ is associative), and

(7) $t\ ||_A\ (t'\ @_A\ t'') = (t\ ||_A\ t')\ @_A\ (t\ ||_A\ t'')$ ($||_A$ is distributive with respect to $@_A$).

where t, t', t''' are arbitrary *parallel-nondeterministic (par-nd) terms* in *P-N*(A).
The isomorphism between the sets A = {a, b, c,....} and <A> = {<a>, <b>, <c>,....} ⊂ *P-N*(A) allows us to consider the atomic elements of A as unary tuples and vice versa.
The above properties are imposed by the operational behavior of $@_A$ and its role in the representation of the nondeterministic activities within concurrent systems (see Subsection 1.1 and also Definition 3.1.3). Let $(Prop-E)_A$ denote the equational system over *P-N*(A) representing the properties (4)-(7). For convenience, we drop the subscripts of ||, @, and *(Prop-E)* when they are obvious from the context.
A set <A,||,@> is a *parallel set* by "forgetting" the @ operation; <A,||, @> is a *nondeterministic set* by forgetting the || operations.

This is an appropriate place to observe that *P-N*(A) is isomorphic to the disjoint union $\Sigma_{n\in\omega}A^n$, which follows from the property (7) and the operational behavior of @.

In *P-N*(A) we also provide a means of abstraction by which certain (compound) par-nd terms can be named and manipulated as units. The naming is done by introducing in *P-N*(A) *naming equations:* t||t' = <a> gives the name <a> to the term t||t', where t and t' are terms in <A,||,@>, and a is an element of A; t and t' are called *components* of <a>. (See Definition 3.1.2 for an explanation of a term t||t'.) A term t@t' cannot be manipulated as a unit because it would contradict the interpretation given to @; therefore, it cannot be named. Let *Name*(A) denote the set of names used in $P\text{-}N(A)$. Clearly, $Name(A) = \langle A \rangle$. Let $(Name-E)_A$ denote the set of naming equations in $P\text{-}N(A)$. Section 4 contains an example in which a naming equation is introduced.

Definition 3.1.2 Let A and B be par-nd sets. Given the functions $f_i: A_i \to B_i$, $i \in [n]$, we define their *pairing* $f_1 \| ... \| f_n : A_1 \times ... \times A_n \to B_1 \times ... \times B_n$ by $(a_1 \| ... \| a_n)f_1 \| ... \| f_n =$ $(a_1)f_1 \| ... \| (a_n)f_n$, for $a_i \in A_i$, $i \in [n]$. $f_1 \| ... \| f_n$ expresses the *concurrency* in our model: the functions f_i are evaluated in parallel. $f_1 \| ... \| f_n$ is a function of one argument. As mentioned in Subsection 1.1, this is an important point in our development: we consider that all functions have at most one argument, which may be a tuple of elements. The evaluation of $f_1 \| ... \| f_n$ (i.e., of $f_1,...,f_n$ in parallel) is possible only when all the components of the argument $a_1 \| ... \| a_n$ are available (see Subsection 1.1).

Definition 3.1.3 A function f: A → B is *linear (with respect to @)* iff $(a_1 @ a_2)f = (a_1)f @ (a_2)f$, for $a_1, a_2 \in A$ (it preserves the operation @). As stated in Subsection 1.1, this condition expresses the *nondeterminism* in our model: the evaluation of $(a_1 @ a_2)f$ is possible when one argument a_1 or a_2 is available; an arbitrary choice between a_1 and a_2 is made when both arguments are available.

Let S be a set (of sorts); S* denotes the set of strings $s_1...s_n$ over S, and S** denotes the set of tuples $\langle u_1,...,u_n \rangle$ of strings over S. Let $P\text{-}N(S^*) = \langle S^*, \|, @ \rangle$ denote the par-nd set over S*; we call its elements par-nd strings. $u_1 \| ... \| u_n = \langle u_1,...,u_n \rangle$, where $n \in \omega$, $u_i \in S^{**}$, $i \in [n]$ is an example of a par-nd string; $v_1 @ v_2$, where $v_j \in S^{**}$, $j \in [2]$ is another example of a par-nd string. We use $P\text{-}N(S^*)$ for indexing of the carriers of many-sorted parallel-nondeterministic algebras; $S^{**} \subset P\text{-}N(S^*)$.

The ADJ algebraic approach to data types, such as integers, real numbers, and stacks of integers, considers several carriers indexed by a set of sorts like S = {integer, real, stack-of-integer}. The multitude of computing facilities existing in a "true" concurrent system is reflected in our approach by the operations ‖ and also by including in S many sorts for the same data type, namely, one sort for each computing facility in the system. For example, the data type integer in a system with three computing facilities is represented by integer1, integer2, and integer3 in S.

Definition 3.1.4 An S-sorted *parallel* operator domain Σ consists of an operator domain Σ (defined in Section 2) with sorts in $P\text{-}N(S^*)$, and a pairing operation ‖ for each $n \in \omega$, such that:

(1) if $\sigma_i \in \Sigma_{u_i,v_i}$, for $u_i, v_i \in P\text{-}N(S^*)$, $i \in [n]$, then $\sigma_1 \| ... \| \sigma_n \in \Sigma_{u_1 \| ... \| u_n, v_1 \| ... \| v_n}$; we say that $\sigma_1 \| ... \| \sigma_n$ is of *arity* $u_1 \| ... \| u_n$, *co-arity* $v_1 \| ... \| v_n$, and *rank* 1;

(2) if $\sigma \in \Sigma_{u_1 \| ... \| u_n, v}$ and $u_1 \| ... \| u_n = u$ is a *(Name-E)* equation in $P\text{-}N(S^*)$, for u_i, $v \in P\text{-}N(S^*)$, $i \in [n]$, then $\sigma \in \Sigma_{u,v}$;

(3) if $\sigma \in \Sigma_{u,v}$ and $u = u_1 @ u_2$, u_i, $v \in S^*$, $i \in [2]$, then $\sigma \in \Sigma_{u_1,v}$ and $\sigma \in \Sigma_{u_2,v}$; we say that σ is of *arity* u_1 or u_2, *co-arity* v, and *rank* 1.

An operator $\sigma \in \Sigma_{u,v}$ for $u = u_1 @ u_2$, is an *overloaded* operator symbol; σ may have more than one arity. Here the operations $\|$ are *higher order* operations: they take other operations as arguments (see condition (1)).

Definition 3.1.5 An S-sorted *parallel-nondeterministic* Σ-algebra A (*par-nd* Σ-algebra) consists of a family $\{A^u \mid u \in P\text{-}N(S^*)\}$ of *nondeterministic* carrier sets A^u and an *actual function (interpretation) of at most one argument* $\sigma_A: A^u \rightarrow A^v$ for each function symbol $\sigma \in \Sigma_{u,v}$, where $u, v \in P\text{-}N(S^*)$, such that:

(1) $A^{u_1 \| \ldots \| u_n} = A^{u_1} \times \ldots \times A^{u_n}$, for $u_i \in P\text{-}N(S^*)$, $i \in [n]$;

(2) $A^{u_1 @ u_2} = A^{u_1} + A^{u_2}$, and
$(A^{u_1} + A^{u_2}) \times A^v \approx (A^{u_1} \times A^v) + (A^{u_2} \times A^v)$, and
$A^u \times (A^{v_1} + A^{v_2}) \approx (A^u \times A^{v_1}) + (A^u \times A^{v_2})$, for $u, v, u_i, v_i \in P\text{-}N(S^*)$, $i \in [2]$;
($A^{u_1} + A^{u_2}$ is the disjoint union of the sets A^{u_1} and A^{u_2});

(3) $A^{u_1} \times \ldots \times A^{u_n} \approx A^u$, for $u_1 \| \ldots \| u_n = u$ a *(Name-E)* equation in $P\text{-}N(S^*)$;
($\approx$ is the isomorphism relation on sets);

(4) $\sigma_A^1 \| \ldots \| \sigma_A^n: A^{u_1 \| \ldots \| u_n} \rightarrow A^{v_1 \| \ldots \| v_n}$, for $\sigma_A^i : A^{u_i} \rightarrow A^{v_i}$, $u_i, v_i \in P\text{-}N(S^*)$, $i \in [n]$;

(5) $(a_1 @ a_2)\sigma_A = (a_1)\sigma_A @ (a_2)\sigma_A$ for all $\sigma \in \Sigma_{u,v}$, $u = u_1 @ u_2$, $u_i, v \in P\text{-}N(S^*)$, $a_i \in A^u$, $i \in [2]$

(6) $A^{u @ u} \approx A^u$, for $u \in P\text{-}N(S^*)$.

In words, condition (1) defines the structure of carriers of sorts $u_1 \| \ldots \| u_n$; condition (2), which is a consequence of property (7) in Definition 3.1.1, defines the structure of carriers of sorts $u_1 @ u_2$, which are disjoint sets; thus, a carrier A^u is isomorphic to the disjoint union $\Sigma_{i \in [n]} A^{u_i}$, where $u = u_1 @ \ldots @ u_n$, for some $u_i \in S^{**}$. Condition (3) identifies the carriers of abstractions (naming equations). Condition (4) states that the pairing of functions is defined in a par-nd Σ-algebra A, preserving the pairing of operator symbols in Σ; condition (5) states the linearity of functions. Condition (6), which is a consequence of property (4) in Definition 3.1.1, restricts the overloading of functions to distinctive sorts.

Definition 3.1.6 Let A and B be S-sorted par-nd Σ-algebras. Then a *par-nd Σ-homomorphism* h: A $\rightarrow$ B is a Σ-homomorphism which is linear (with respect to @) and preserves the pairing operations and the *(Name-E)* equations.

We now turn to the definition of "diamonds". Diamonds are graphical representations of well-formed par-nd expressions, as trees are graphical representations of well-formed expressions of initial algebras.

Definition 3.1.7 A *directed graph* G = <E,V,s,t> consists of a set E of edges, a set V of vertices, and two functions s,t from the set of edges to the set of vertices; s(e) is the source vertex of e and t(e) is the target vertex of e. An *AND/OR directed graph* is a directed graph G together with a labeling function from V to {AND,OR}×{AND,OR}, such that:

(i) if a node n has any outgoing edges, they are either all OR edges, in which case we say that n has the *Out-type* OR, or all AND edges, in which case we say that n has the *Out-type* AND;

(ii) if a node n has any incoming edges, they are either all OR edges, in which case we say that n has the *In-type* OR, or all AND edges, in which case we say that n has the *In-type* AND.

We represent these situations pictorially by

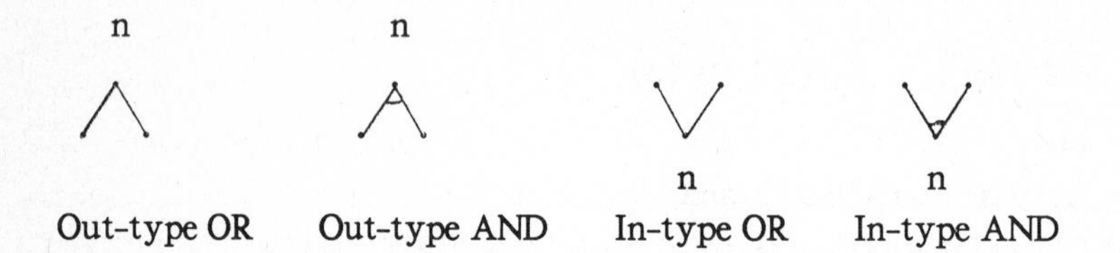

A *diamond* is a labeled AND/OR ordered directed acyclic graph, as described below.

Note: The restriction of nodes to be either of type OR or of type AND is not essential. Any directed graph containing nodes of combined types, AND and OR, can be converted to an AND/OR directed graph as that formulated in Definition 3.1.7. This restriction however helps us develop a more tractable model.

Example 3.1.1

If the parallel signature Σ has the sort set *P-N*(S*) = {s, s||s, s||s||s, s@s} with
$x, b \in \Sigma_{\lambda,s}$, $f \in \Sigma_{s||s,s}$, $g \in \Sigma_{s,s}$, and $h \in \Sigma_{s@s,s||s||s}$,
we may picture an example of a diamond with the *root* f by

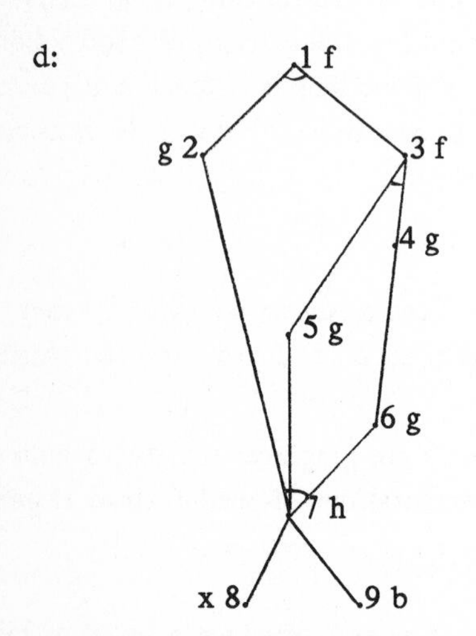

The diamond d corresponds to the well-formed par-nd term (x@b)∘h∘(g ‖ (g ‖ (g∘g))∘f)∘f: we consider the *continuations* of C. Strachey and C.P. Wadswo:th [6], and interpret an operation associated with a node in a diamond as function of the operations associated with its *successor* nodes. Note that the structural properties of diamonds point out the structures of functional forms: there is no immediate access to the arity and co-arity of the functions associated to the nodes. This kind of presentation is different from the way the conventional trees are used: the descendents of a node give a precise information on the arity and the rank of the function associated to that node.

The nodes of Out-type AND and In-type AND of a diamond provide means by which a computation can "fork" by splitting into several parallel computations which can later "join" together again; the nodes of Out-type OR provide means by which a computation can split into several computations which never join together again (we do not allow nodes of In-type OR).

The type of the operation at a node is consistent with the type of that node in the following way: f $\in \Sigma_{s\|s,s}$ is of arity s ‖ s, which is consistent with the Out-type AND of node 1 with two descendants; g $\in \Sigma_{s,s}$ is of arity s, which is consistent with the Out-type AND of nodes 2, 4, 5, and 6 with one descendant; h $\in \Sigma_{s@s,s\|s\|s}$ is of co-arity s‖s‖s, which is consistent with the In-type AND of node 7 with three predecessors, and is of arity s@s, which is consistent with the Out-type OR of node 7 with two descendants.

3.2 A Free Parallel-Nondeterministic Algebra Construction

We can now prove the main result of this paper, namely that the category of parallel-nondeterministic algebras has free objects. We begin with an inductive definition of the sets $D_{\Sigma,u}(X^u)$ of all par-nd Σ-terms (or Σ-diamonds) of co-arity u, for a given parallel signature Σ with sorts in *P-N*(S*) and a given par-nd set of variables *P-N*(X) = <X,‖,@>.

(1) $\Sigma_{\lambda,w} \subseteq D_{\Sigma,w}(X^w)$, and $X^w \subseteq D_{\Sigma,w}(X^w)$ for each w in *P-N*(S*);

(2) $(t_1 @ ... @ t_n)\sigma \in D_{\Sigma,w}(X^w)$ for each $\sigma \in \Sigma_{v,w}$ and each $t_i \in D_{\Sigma,v_i}(X^{v_i})$, i $\in$ [n], where v = $v_1 @ ... @ v_n$;

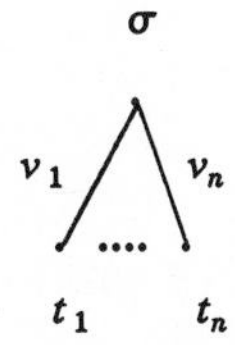

(3) $(t_1)\sigma @ (t_2)\sigma \in D_{\Sigma,w}(X^w)$ for each $\sigma \in \Sigma_{v,w}$ and each $t_i \in D_{\Sigma,v_i}(X^{v_i})$, i $\in$ [2], where v = $v_1 @ v_2$;

(4) $((t_1)\sigma_1 \| ... \| (t_n)\sigma_n)\sigma$, $((t^1 \| ... \| t^n)\sigma_1 \| ... \| \sigma_n)\sigma$, and $((x)\sigma_1 \| ... \| \sigma_n)\sigma \in D_{\Sigma,w}(X^w)$, for each $\sigma \in \Sigma_{v,w}$, $\sigma_i \in \Sigma_{u_i,v_i}$, each $t_i \in D_{\Sigma,u_i}(X^{u_i})$, and each x $\in X^u$, for i $\in$ [n], u = $u_1 \| ... \| u_n$, and v = $v_1 \| ... \| v_n$;

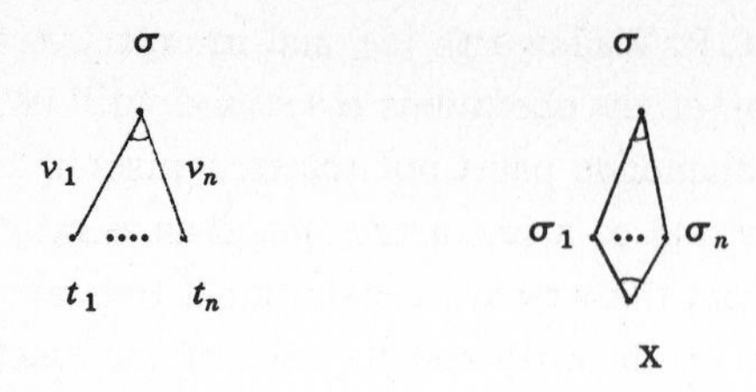

We next make the family $D_\Sigma(X) = \{D_{\Sigma,u}(X^u)\}$ into a par-nd Σ-algebra.
First, we define an interpretation, $(\sigma)\theta$, for all operation symbols σ in Σ:

(1) $(\sigma)\theta$ of σ in $\Sigma_{\lambda,u}$ is the symbol σ in $D_{\Sigma,u}(X^u)$ (it is in $D_{\Sigma,u}(X^u)$ by (1));

(2) $(\sigma)\theta$ of σ in $\Sigma_{v,w}$ for $v = v_1 @...@ v_n$ sends t_i in $D_{\Sigma,v_i}(X^{v_i})$ to the term $(t_1 @...@ t_n)\sigma$ in $D_{\Sigma,w}(X^w)$ (it is in $D_{\Sigma,w}(X^w)$ by (2));

(3) $(\sigma)\theta$ of σ in $\Sigma_{v,w}$ for $v = v_1\|...\|v_n$ sends t_i in $D_{\Sigma,v_i}(X^{v_i})$, i ∈ [n], to the term $(t_1\|...\|t_n)\sigma$ in $D_{\Sigma,w}(X^w)$ (it is in $D_{\Sigma,w}(X^w)$ by (4));

(4) $(\sigma_1\|...\|\sigma_n)\theta$ of $\sigma_1\|...\|\sigma_n$ in $\Sigma_{v,w}$, for $v = v_1\|...\|v_n$ sends t_i in $D_{\Sigma,v_i}(X^{v_i})$, i ∈ [n], to the term $(t_1\|...\|t_n)\sigma_1\|...\|\sigma_n$ in $D_{\Sigma,w}(X^w)$ (it is in $D_{\Sigma,w}(X^w)$ by (4)).

Then, in the family $D_\Sigma(X)$ we introduce the following identities, in order to satisfy the "parallel-nondeterministic"properties required by Definition 3.1.1:

(5) $(t_1 @...@ t_n)\sigma = (t_1)\sigma @...@ (t_n)\sigma$;

(6) $(t_1)\sigma @ (t_2)\sigma = (t_2)\sigma @ (t_1)\sigma$;

(7) $((t_1)\sigma_1 \|...\| (t_n)\sigma_n)\sigma = ((t^1 \|...\| t^n)\sigma_1 \| ... \| \sigma_n)\sigma$ (all terms are in $D_{\Sigma,w}(X^w)$ by (4) and (5));

(8) $((x^1\|...\| x^n)\sigma_1\| ... \| \sigma_n)\sigma = ((x)\sigma_1 \| ... \| \sigma_n)\sigma$ if there exists a naming equation (*Name-E*) such that: $x^1\|...\| x^n = x$ (all terms are in $D_{\Sigma,w}(X^w)$ by (4),(5), and (6)).

Proposition 3.2.1 $D_\Sigma(X)$ is by construction a countable union $D_\Sigma(X) = \bigcup D_\Sigma(X)^{[n]}$ of non-deterministic subsets (n represents the depth of diamonds).

(i) $D_\Sigma(X)^{[0]} = \{\Sigma_{\lambda,w} \mid w \in P\text{-}N(S^*)\} \cup \{X^w \mid w \in P\text{-}N(S^*)\}$;

(ii) $D_\Sigma(X)^{[n+1]} = D_\Sigma(X)^{[n]}$

$\cup \{\{(t_1 @...@ t_m)\sigma \mid \sigma \in \Sigma_{v,w},\ v = v_1 @ v_m,\ t_i \in D_{\Sigma,v_i}(X^{v_i})^{[n]},\ i \in [m]\}\}$

$\cup \{(t_1\|...\|t_m)\sigma \mid \sigma \in \Sigma_{v,w},\ v = v_1\|...\|v_m,\ t_i \in D_{\Sigma,v_i}(X^{v_i})^{[n]},\ i \in [m]\}\}$

$\cup \{\{((t_1\|...\|t_m)\sigma_1\|...\|\sigma_m)\sigma \mid \sigma \in \Sigma_{v,w},\ \sigma_i \in \Sigma_{u_i,v_i},$
$(t_i)\sigma_i \in D_{\Sigma,v_i}(X^{v_i})^{[n]},\ i \in [m],\ v = v_1\|...\|v_m\}\}$

$\cup \{\{(x)\sigma_1\|...\|\sigma_m)\sigma \mid \sigma \in \Sigma_{v,w},\ (x_1\|...\|x_m)\sigma_1\|...\|\sigma_m \in D_{\Sigma,v_i}(X^{v_i})^{[n]},$
$i \in [m],\ v = v_1\|...\|v_m\}\}$

Theorem 3.2.2 Let $P\text{-}N(X) = \langle X,\|,@\rangle$ be a par-nd set of variables with naming equations $(Name-E)_X$. Let I_X: $P\text{-}N(X) \rightarrow D_\Sigma(X)$ be the family of injections of $P\text{-}N(X)$ into the carriers of

$D_\Sigma(X)$, which are linear and preserve the pairing operations and the naming equations $(Name-E)_X$. Then $<D_\Sigma(X), I_X>$ is the par-nd algebra freely generated by $P\text{-}N(X)$. That is, for any par-nd algebra A and any linear map h_A: $P\text{-}N(X) \rightarrow$ A (called assignment) that preserves the pairing and the naming equations $(Name-E)_X$, there exists a unique par-nd homomorphism $\bar{h}_A$: $D_\Sigma(X) \rightarrow$ A, such that $I_X \bar{h}_A = h_A$. Pictorially, the diagram

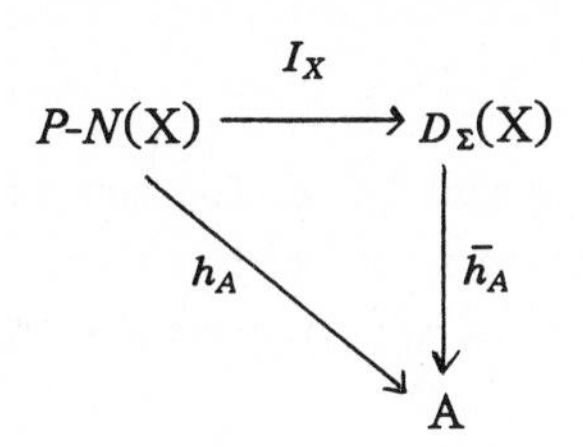

commutes.

Proof. The proof is by induction on n (introduced in Proposition 3.2.1).

Existence:

(i) We define $\bar{h}_A$ on $D_\Sigma(X)^{[0]}$ by
$(\sigma)\bar{h}_A = \sigma_A$ for each $\sigma \in \Sigma_{\lambda,w}$ and $(x)\bar{h}_A = (x)h_A$ for each $x \in X^w$;

(ii) Assuming that $\bar{h}_A$ is already defined on $D_\Sigma(X)^{[n]}$, $\bar{h}_A$ is defined on $D_\Sigma(X)^{[n+1]}$ by
$((t_1 @ t_2)\sigma)\bar{h}_A = ((t_1)\bar{h}_A @ (t_2)\bar{h}_A)\sigma_A$ ($\bar{h}_A$ is linear),
$((t_1\|...\|t_m)\sigma)\bar{h}_A = ((t_1)\bar{h}_A\|...\|(t_m)\bar{h}_A)\sigma_A$, and
$((t_1\|...\|t_m)\sigma_1\|...\|\sigma_m)\bar{h}_A = ((t_1)\bar{h}_A\|...\|(t_m)\bar{h}_A)\sigma_{1A}\|...\|\sigma_{mA}$.

Thus, $\bar{h}_A$ is defined on all of $D_\Sigma(X)$.

Uniqueness:

Suppose that $\bar{h}_A$, $\bar{h}_A$': $D_\Sigma(X) \rightarrow$ A are two par-nd homomorphisms. Then,

(i) $\bar{h}_A$, $\bar{h}_A$' coincide on $D_\Sigma(X)^{[0]}$:
$(\sigma)\bar{h}_A = \sigma_A = (\sigma)\bar{h}_A'$, and $(x)\bar{h}_A = (x)h = (x)\bar{h}_A'$

(ii) Assuming that $\bar{h}_A$, $\bar{h}_A$' coincide on $D_\Sigma(X)^{[n]}$, they coincide on $D_\Sigma(X)^{[n+1]}$:
$((t_1 @ t_2)\sigma)\bar{h}_A = ((t_1)\bar{h}_A @ (t_2)\bar{h}_A)\sigma_A = ((t_1)\bar{h}_A' @ (t_2)\bar{h}_A')\sigma_A$,
$((t_1\|...\|t_m)\sigma)\bar{h}_A = ((t_1)\bar{h}_A\|...\|(t_m)\bar{h}_A)\sigma_A = ((t_1)\bar{h}_A'\|...\|(t_m)\bar{h}_A')\sigma_A$, and
$((t_1\|...\|t_m)\sigma_1\|...\|\sigma_m)\bar{h}_A = ((t_1)\bar{h}_A\|...\|(t_m)\bar{h}_A)\sigma_{1A}\|...\|\sigma_{mA} =$
$((t_1)\bar{h}_A'\|...\|(t_m)\bar{h}_A')\sigma_{1A}\|...\|\sigma_{mA}$.
Thus, $\bar{h}_A$, $\bar{h}_A$' coincide on all of $D_\Sigma(X)$.

Proposition 3.2.3 $D_\Sigma = D_\Sigma(\Phi)$ is initial in the category of all S-sorted par-nd Σ-algebras.

Equations can be introduced over $D_\Sigma(X)$ for presenting classes of par-nd algebras satisfying certain (equational) properties. A *par-nd Σ-equation (over $D_\Sigma(X)$)* is a pair of par-nd terms from $D_\Sigma(X)$, $(\forall x)t_1 = t_2$, where $(\forall x)$ is the finite set of variables occuring in t_1 and t_2. (See [1] for details.) A par-nd algebra A *satisfies* an equation $(\forall x)t_1 = t_2$ iff for every assignment

h_A: X → A, we have $(t_1)\bar{h}_A = (t_2)\bar{h}_A$. A *par-nd equational system* (in D_Σ(X)) is a set of par-nd Σ-equations. Given a set E of par-nd Σ-equations, a par-nd Σ-algebra A *satisfies* E iff A satisfies each equation in E; let us call A a par-nd $<\Sigma,E>$-algebra. (A complete par-nd equational deduction system in D_Σ(X) consists of an equational system E, the equational systems *(P-N)* and *(Name-E)*, which are intrinsic to par-nd constructs, and the ordinary rules of equational deduction [1].) The following generalization of Theorem 3.1 says that there are free par-nd $<\Sigma,E>$-algebras.

Theorem 3.2.4 $<D_{\Sigma E}(X), I_X>$ is the par-nd algebra freely generated by X: for any par-nd algebra A that satisfies E and any linear map h_A: X → A that preserves the pairing, there exists a unique par-nd homomorphism $\bar{h}_A$: $D_{\Sigma E}$(X) → A, such that $I_X \bar{h}_A = h_A$. $D_{\Sigma E}(\Phi)$ is initial in the category of all sorted par-nd Σ-algebras satisfying E. □

G. Boudol states in [1] that problems arise in modeling explicit nondeterminism, such as that introduced by adding a new operator of choice @. Namely, if semantically @ is nothing more than set-theoretic union, whereas operationally the behavior of @ is given by x@y = x, x@y = y, then the operational semantics of nondeterministic recursive definitions is intrinsically non-continuous. We consider that semantically @ is disjoint union; this assumption leads to non-ambiguous term rewriting systems (using the terminology of G. Huet [1]). Further investigation is still needed in this direction.

3.3 Continuous Parallel-Nondeterministic Algebras.

To be able to solve systems of par-nd equations we need to extend par-nd algebras to ω-continuous par-nd algebras. An ω-continuous par-nd algebra is a par-nd algebra whose carriers are ω-CPOs, and operations are ω-continuous linear functions. The algebra of finite and infinite "partial diamonds" is initial in the category of ω-continuous par-nd algebras. Finite and infinite partial diamonds are introduced as partial functions d: $<W_\Sigma^*, \|, @>$ -o→ $\Sigma \cup \{\perp\}$ for which the domain of definition has the tree domain property:

(i) if u ∈ (d)dom and $v \leqslant_{pref} u$ then v ∈ (d)dom,

(ii) if $u\sigma_i$ ∈ (d)dom for $\sigma \in \Sigma_{w,s}$, |w| = k, then (u)d = σ and ((d)dom/w) $\cap W_\Sigma \subsetneq \{\sigma_1,\ldots,\sigma_k\}$.

In this definition, u ∈ (d)dom is a node of the diamond d, while (u)d is its label. Intuitively, a finite partial diamond is a diamond some of whose leaves are labeled with ⊥.

Example 3.1.1 (continued)
With Σ = {f, g, h, x, b} and $W_\Sigma^* = \{\lambda, f_1, f_2, f_2 f_1, f_2 f_2, f_2 f_2 g_1, f_1 g_1 \| f_2 f_1 g_1 \| f_2 f_2 g_1 g_1, (u)h_1\}$,
where $u = f_1 g_1 \| f_2 f_1 g_1 \| f_2 f_2 g_1 g_1$,
we may picture an example of a diamond by

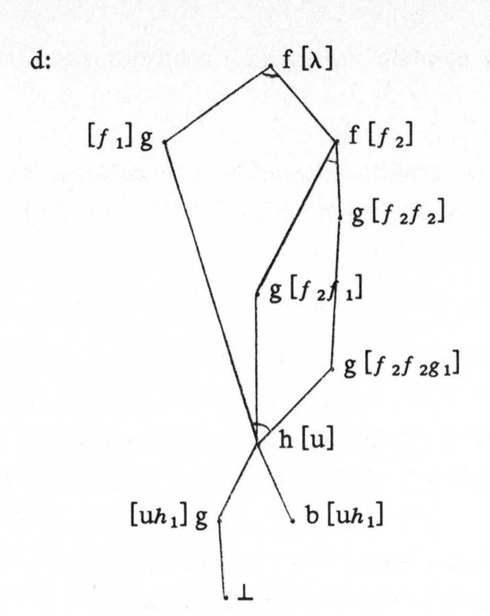

d is represented below as the partial function d: $<W_\Sigma{}^*, \|, @>$ $-o\rightarrow$ $\{f,g,h,x\}$ with $\lambda, f_2 \mapsto f$, $f_1, f_2f_1, f_2f_2, f_2f_2g_1 \mapsto g$, $u \mapsto h$, $(u)h_1 \mapsto \{g, b\}$, where $u = f_1g_1 \| f_2f_1g_1 \| f_2f_2g_1g_1$, and undefined otherwise.

The natural order relation $\sqsubseteq$, called "less defined than", which exists among trees, exists also on partial diamonds: a partial diamond properly approximates another iff their corresponding nodes are identical except that the first contains $\perp$ in at least one position where the second contains a nontrivial diamond.

3.4 Parallel-Nondeterministic Algebras with Communication Capabilities

In the companion paper [10], we distinguished the input and output operator symbols of an operator domain Σ. (The operations of an algebra A named by input and output operator symbols are identity functions; their role is to transmit values without modifying them.)

Definition 3.4.1 Let $\Gamma = \{\alpha, \beta, \gamma, ...\}$ be a finite set of names and $\overline{\Gamma} = \{\overline{\alpha}, \overline{\beta}, \overline{\gamma}, ...\}$ be a finite set of complementary names bijective with Γ such that $\gamma \in \Gamma$ implies $\overline{\gamma} \in \overline{\Gamma}$ and $\overline{\overline{\gamma}} = \gamma$. We call the names of Γ and $\overline{\Gamma}$ *input* and *output communication capabilities*, respectively; $\lambda = \overline{\lambda}$ is the *no communication capability* symbol. We have $\Gamma \cap \overline{\Gamma} = \{\lambda\}$; let $\Gamma^+ = \Gamma - \{\lambda\}$ and $\overline{\Gamma}^+ = \overline{\Gamma} - \{\lambda\}$.

Definition 3.4.2 An S-sorted operator domain Σ with *communication (I/O) capabilities* Γ is an $<S^* \times S^*>$-indexed operator domain $\Sigma = \{\Sigma_{u,v} \mid u, v \in S^*\}$, together with a labeling function from Σ onto $\Gamma \cup \overline{\Gamma}$. We call $\sigma^\alpha \in \Sigma$ an *input* operator symbol, $\sigma^{\overline{\beta}} \in \Sigma^{\overline{\Gamma}^+}$ an *output* operator symbol, and $\sigma \in \Sigma^\lambda_{u,v}$ just an *operator symbol.* We have $\Sigma^{\Gamma^+} = \bigcup_{\alpha \in \Gamma^+} \Sigma^\alpha_{u,u}$ and $\Sigma^{\overline{\Gamma}^+} = \bigcup_{\overline{\beta} \in \overline{\Gamma}^+} \Sigma^{\overline{\beta}}_{u,u}$; we may drop the subscripts u,u when we consider input and output operator symbols, and the superscript λ when we consider internal operator symbols. (Notice that the arity and co-arity of input and output operator symbols are equal: a condition resulting from their interpretation as identity functions.)

Definition 3.4.3 An S-sorted *parallel* operator domain Σ with *communication capabilities* Γ is an S-sorted parallel operator domain Σ together with a labeling function from Σ onto $\Gamma \cup \overline{\Gamma}$, such that:

(1) iff $\sigma_i \in \Sigma^{\lambda}_{u_i,v_i}$ then $\sigma_1 \| \ldots \| \sigma_n \in \Sigma^{\lambda}_{u,v}$, $u = u_1 \| \ldots \| u_n$, $v = v_1 \| \ldots \| v_n$, $u_i, v_i \in S^*$, $i \in [n]$, and

(2) iff $\sigma_i \in \Sigma^{\gamma}_{u_i,u_i}$ then $\sigma_1 \| \ldots \| \sigma_n \in \Sigma^{\gamma}_{u,u}$, $u = u_1 \| \ldots \| u_2$, $u_i \in S^*$, $i \in [n]$, $\gamma \in \Gamma^+ \cup \overline{\Gamma}^+$.

Definition 3.4.4 An S-sorted *parallel-nondeterministic* Σ-algebra *with connection capabilities* Γ, denoted $<\Sigma,\Gamma>$-algebra A, is a parallel-nondeterministic Σ-algebra A together with a family of input and output operations.

Let $\overline{A}$ and $\overline{B}$ be S-sorted par-nd $<\Sigma,\Gamma>$-algebras. Then a *par-nd* $<\Sigma,\Gamma>$*-homomorphism* $\overline{h}: \overline{A} \rightarrow \overline{B}$ is a par-nd Σ-homomorphism that preserves the input and output operations.

Proposition 3.4.1 In the category of par-nd $<\Sigma,\Gamma>$-algebras there exists an initial par-nd $<\Sigma,\Gamma>$-algebra, $D_{\Sigma,\Gamma}$. □

4. Using Continuous Parallel-Nondeterministic Algebras

In [10] a computing system is represented by a kind of graph, called *flow net*, which may be specified in terms of other flow nets and a set of operations. Like Milne and Milner in [4], Wirth [8], and Hoare [3], we consider rather separate the phase in which a composite flow net of activities is established and the phase in which the flow net is activated.

A flow net is an AND/OR graph whose nodes represent the activities that a computing system is able to perform including the communications with its environment, and whose arcs represent communication channels among its activities. We label every node of a flow net by a pair consisting of a communication name from $\Gamma \cup \overline{\Gamma}$ and an operator symbol from Σ. The communication name represents the communication capability of the computing system at that node, and the operator symbol represents the operation the computing agent is able to perform at that node. The absence of a communication capability at a node is represented by λ (the no communication capability symbol). *Input nodes* are nodes labeled by names from Γ, *output nodes* are nodes labeled by names from $\overline{\Gamma}$, and *internal nodes* are nodes labeled by λ. We will be concerned only with flow nets which have input and output nodes with distinct names; we call them *proper* flow nets. We allow nodes of complimentary names in the same flow net.

Intuitively, a node of Out-type AND and its descendants represent the splitting of a computation into several parallel computations; a node of Out-type OR and its descendants represent the splitting of a computation into several computations, such that a single computation, out of many possible ones, is chosen nondeterministically to be performed; a node of In-type AND and its predecessors represent parallel computations that join together again on a single task; a node of In-type OR and its predecessors represent several computations competing for exclusive access to a task. An example of a flow net with an input α and an output $\bar{\beta}$ is shown below.

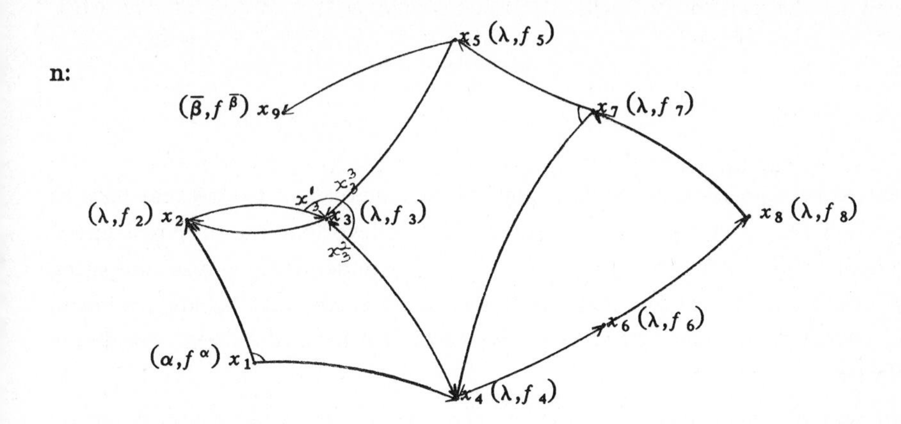

We identify the nodes of the flow net illustrated above by variables x_i. In the flow net n, the node x_4 is a node of In-type OR and Out-type OR, which has the label (λ, f_4); λ represents the absence of a communication capability of the flow net at that node with its exterior environment; the operator symbol f_4 represents the operation that can take place at that node; the node x_1 is a node of In-type AND and Out-type AND, which has the label (α, f^{α}), α representing an input communication capability of the flow net at that node, and f^{α} representing an input operator symbol; the node x_9 is a node of In-type and Out-type AND, which has the label $(\bar{\beta}, f^{\bar{\beta}})$, $\bar{\beta}$ representing an output communication capability of the flow net at that node, and $f^{\bar{\beta}}$ representing an output operator symbol.

We follow the work of ADJ on the representation of recursive flowcharts by their unfoldments: every flowchart is represented by a finite system of equations in an initial ω-continuous algebra of infinite trees. Similarly, we represent a flow net by an equational system in an initial ω-continuous par-nd algebra of par-nd expressions (or diamonds). We appeal to *continuations* of C. Strachey and C.P. Wadsworth [6], and interpret as continuations the variables x_i that we used to identify the nodes of a flow net. Thus, an operation f_i associated with a node is a function of the continuations associated with its successor nodes. For example, we associate an equation to each node of **n** in the following way:

$$x_1 = (x_2 \parallel x_4)f^{\alpha} \qquad x_4 = (x_3^2 @ x_6)f_4 \qquad x_7 = (x_4 \parallel x_5)f_7$$
$$x_2 = (x_3^1)f_2 \qquad x_5 = (x_3^3 @ x_9)f_5 \qquad x_8 = (x_7)f_8$$
$$x_3 = (x_2)f_3 \qquad x_6 = (x_8)f_6 \qquad x_9 = (x_9)f^{\beta},$$

and $x_3^1 \parallel x_3^2 \parallel x_3^3 = x_3$ (a *(Name-E)* equation)

Let *(Petri-E)* denote the system of equations given above. We "unfold" the flow net **n** by reducing $(x_2 \parallel x_4)f^{\alpha}$ using the system of equations *(Petri-E)* together with the system of equations *(P-N)* as a reduction system. See Appendix A for details. In an ω-continuous par-nd algebra a system of equations like *(Petri-E)* has a solution, namely a least fixed-point solution.

All this can be formulated clearer in the morphism of a par-nd algebraic theory of diamonds notation than in the par-nd expression notation. This is the subject of the work reported in [11].

6. Conclusions

We have introduced parallel-nondeterministic algebras of diamonds, a theory that may be useful in modeling the behavior of concurrent systems. We believe that the proposed initial algebraic semantics applied to concurrency gives a firm basis for understanding and comparing a wide variety of existing features for communication, parallelism, and nondeterminism, including deadlock, termination, interrupts, and equivalence. We have also been considering the following problems:

(I) Least fixed-point semantics is successfully used to characterize sequential computation by considering finite representations of infinite computations. It is based on "finite approximation" techniques, which in sequential programming helps the study of correctness of programs and compilers with recursive definitions. Our model creates the right framework for the least fixed-point semantics of parallel and nondeterministic computations. Future research may be done by applying it to the correctness of programs that exhibit the full power of concurrency.

(II) A difficult problem with the existing models of concurrency is their inability to handle error messages with precision. The initial algebra approach to the semantics of sequential computation made it possible to treat errors in a systematic manner [7]. We believe that our model is a proper framework for a reasonable treatment of errors in concurrent systems in an implementation independent manner.

Aknowledgments

Aside from the obvious technical debt to the entire ADJ group, the author would like to thank Dr. Joseph Goguen for his valuable suggestions and encouragement to her present research. The author would also like to thank her colleagues at KSU for their stimulation and understanding, in particular Professors David A. Schmidt and Austin Melton.

References

[1] Boudol, G., *Computational Semantics of Term Rewriting Systems*. In M. Nivat and J.C. Reynolds (editor) Algebraic Methods in Semantics. Cambridge University Press, 1985.

[2] Goguen, J.A., Thatcher, J.W., Wagner, E.G., and Wright, J.B., *Initial Algebra Semantics and Continuous Algebras*. Jour. ACM 24(1), 68-95, 1977.

[3] Hoare, C.A.R., *Communicating Sequential Processes*. Comm. ACM 21(8), 666-677, Aug. 1978.

[4] Milne, G., and Milner, R., *Concurrent Processes and their Syntax*. Jour. ACM 26(2), 302-321, April, 1979.

[5] Peterson, J.L., *Petri Net Theory and the Modelling of Systems*. Prentice-Hall, 1981.

[6] Strachey, C., and Wadworth, C.P., *Continuations - A Mathematical Semantic Model for Handling Full Jumps*. Technical Monograph PRG-11, Programming Research Group, Oxford University, Computing Laboratory, 1974.

[7] Thatcher, J.W., Wagner, E.G., and Wright, J.B., *Notes on Algebraic Fundamentals for Theoretical Computer Science*. In J.B. De Bakker (editor), Foundations of Computer Science, Part 2: Languages, Logic, Semantics. Addison-Wesley, Amsterdam, 1979.

[8] Wirth, N., *MODULA: A language for modular multiprogramming*. Res. Rep. 18, Inst. fur Informatik, Zurich, Switzerland, 1975.

[9] Zamfir, M., *On the Syntax and Semantics of Concurrent Computing*. PhD Thesis, UCLA-CSD-820819, 1982.

[10] Zamfir, M. and Martin, D., *On the Syntax and Semantics of Concurrent Computing*. Proceedings, Conference on the Mathematical Foundation of Programming Languages, KSU, April 1985, LNCS 239.

[11] Zamfir, M., *Algebraic Theories and Concurrency*. KSU Technical Report TR-CS-87-6.

Appendix A: An Example of Unfoldment

Consider the system of equations (*Petri-E*) given in Section 4, which represents the flow net n:

$$x_1 = (x_2 \parallel x_4)f^{\alpha} \qquad x_4 = (x_3^2 @ x_6)f_4 \qquad x_7 = (x_4 \parallel x_5)f_7$$
$$x_2 = (x_3^1)f_2 \qquad x_5 = (x_3^3 @ x_9)f_5 \qquad x_8 = (x_7)f_8$$
$$x_3 = (x_2)f_3 \qquad x_6 = (x_8)f_6 \qquad x_9 = (x_9)f^{\beta},$$

where $x_3^1 \parallel x_3^2 \parallel x_3^3 = <x_3^1, x_3^2, x_3^3> = x_3$

We "unfold" the flow net n by reducing $(x_2 \parallel x_4)f^{\alpha}$ using the system of equations *Petri-E* as a reduction system and having in mind that all operations are linear (with respect to @). We have:

$$(x_2 \parallel x_4)f^{\alpha}$$
$$= ((x_3^1)f_2 \parallel (x_3^2 @ x_6)f_4)f^{\alpha}$$
$$= ((x_3^1)f_2 \parallel (x_3^2)f_4)f^{\alpha} @ ((x_3^1)f_2 \parallel (x_6)f_4)f^{\alpha}$$

further,

$$((x_3^1)f_2 \parallel (x_6)f_4)f^{\alpha}$$
$$= ((x_3^1)f_2 \parallel (x_8)f_6 \circ f_4)f^{\alpha}$$
$$= ((x_3^1)f_2 \parallel (x_7)f_8 \circ f_6 \circ f_4)f^{\alpha}$$
$$= ((x_3^1)f_2 \parallel (x_4 \parallel x_5)f_7 \circ f_8 \circ f_6 \circ f_4)f^{\alpha}$$
$$= ((x_3^1)f_2 \parallel ((x_3^2 @ x_6)f_4 \parallel (x_3^3 @ x_9)f_5)f_7 \circ f_8 \circ f_6 \circ f_4)f^{\alpha}$$
$$= ((x_3^1)f_2 \parallel ((x_3^2)f_4 \parallel (x_3^3)f_5)f_7 \circ f_8 \circ f_6 \circ f_4)f^{\alpha}$$

$$@\ ((x_3^1)f_2 \parallel ((x_6)f_4 \parallel (x_3^3)f_5)f_7 \circ f_8 \circ f_6 \circ f_4)f^\alpha$$
$$@\ ((x_3^1)f_2 \parallel ((x_3^2)f_4 \parallel (x_9)f_5)f_7 \circ f_8 \circ f_6 \circ f_4)f^\alpha$$
$$@\ ((x_3^1)f_2 \parallel ((x_6)f_4 \parallel (x_9)f_5)f_7 \circ f_8 \circ f_6 \circ f_4)f^\alpha$$

now,

$$((x_3^1)f_2 \parallel ((x_3^2)f_4 \parallel (x_3^3)f_5)f_7 \circ f_8 \circ f_6 \circ f_4)f^\alpha$$
$$= ((x_3^1)f_2 \parallel (x_3^2 \parallel x_3^3)(f_4 \parallel f_5) \circ f_7 \circ f_8 \circ f_6 \circ f_4)f^\alpha$$
$$= ((x_3^1 \parallel x_3^2 \parallel x_3^3)(f_2 \parallel ((f_4 \parallel f_5) \circ f_7 \circ f_8 \circ f_6 \circ f_4)))f^\alpha$$
$$= (x_3)(f_2 \parallel ((f_4 \parallel f_5) \circ f_7 \circ f_8 \circ f_6 \circ f_4))f^\alpha$$
$$= (x_2)f_3(f_2 \parallel ((f_4 \parallel f_5) \circ f_7 \circ f_8 \circ f_6 \circ f_4))f^\alpha$$

....

We can also unfold n by substituting diamonds for variables in the following way:

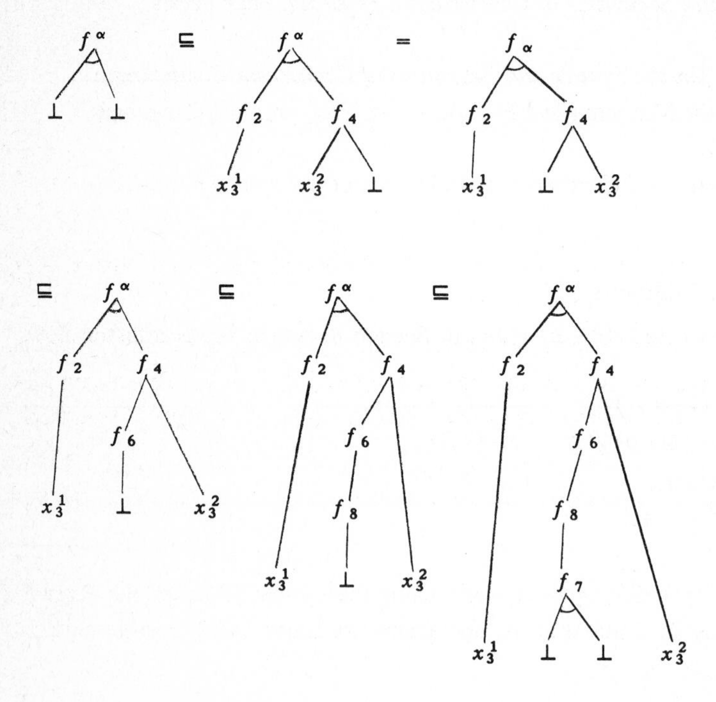

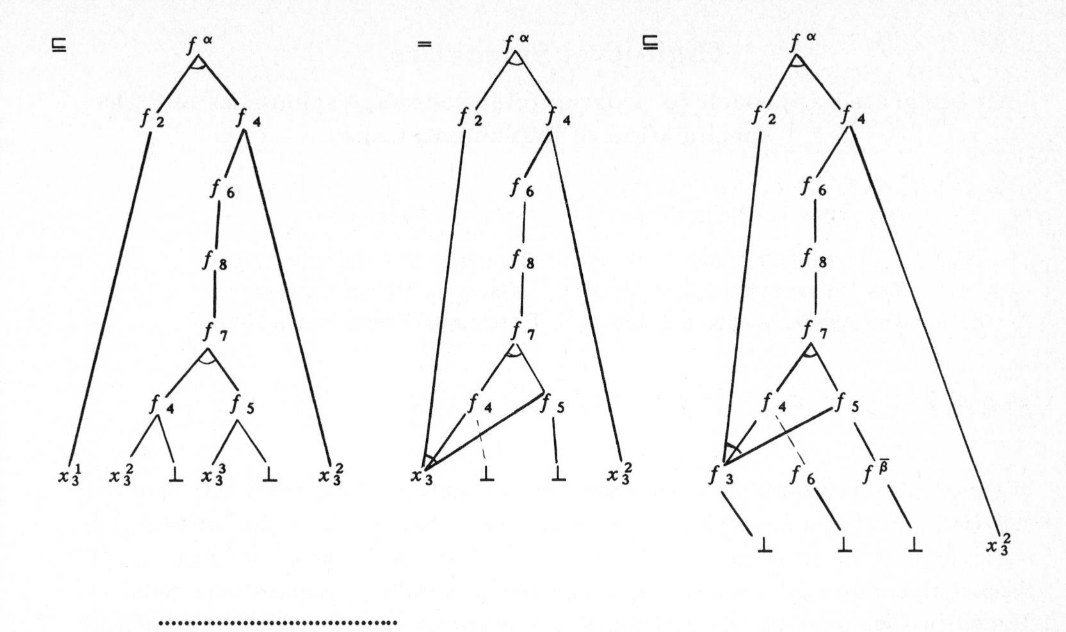

....

The least upper bound of the above chain is the unique least fixed-point solution representing the unfoldment of the flow net n given in Section 5.

An equivalence relation = on diamonds is generated by the equational system *EP* given in Section 3, the linearity of operations in par-nd algebras, and $x_3^1 \parallel x_3^2 \parallel x_3^3 = x_3$. As an example, consider the diamonds below. They are equivalent, because @ is a commutative operation: $(x_3^2 @ x_6)f_4 = (x_6 @ x_3^2)f_4$.

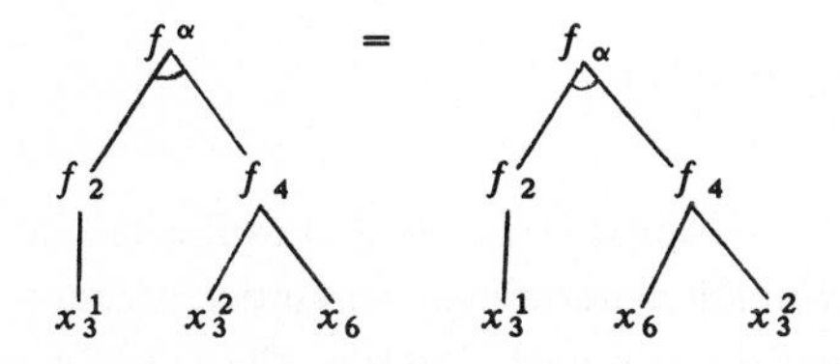

High-level Semantics

An integrated approach to programming language semantics and the specification of implementations

Uwe F. Pleban
PhiloSoft and
The University of Michigan
Ann Arbor, Michigan 48109

Peter Lee
Computer Science Department
Carnegie Mellon University
Pittsburgh, Pennsylvania 15213

Abstract

In the course of implementing a semantics directed compiler generator we have developed *high-level semantics*, a new style of semantic definition which overcomes fundamental problems concerning the specification techniques used in traditional denotational semantics. In the past, these problems have precluded the generation of realistic compilers from traditional denotational specifications. By contrast, high-level semantic specifications are suitable for both *defining* the functional *semantics* of programming languages, and *describing* realistic compiler *implementations* which are automatically generated from the semantics.

In an earlier paper we described the MESS system, a prototype implementation of a compiler generator incorporating the principles of high-level semantics. Here, we summarize the salient characteristics of our methodology. A comprehensive overview of high-level semantics and the MESS compiler generator can be found in the dissertation of the second author.

1. A Critique of Traditional Denotational Semantics

We assume the reader is familiar with traditional denotational semantics, as explicated in [MiS76], [Sto77], [Gor79], and [Sch86]. We say *traditional* denotational semantics because we are referring not only to the mathematical basis of the approach, but also the notation, writing style, and "tricks of the trade" used in denotational specifications.

Our criticism of traditional denotational semantics can be summarized as follows:

1. Rather than depending on the *properties* of the underlying semantic model, denotational definitions inextricably intertwine the architectural details of the model (such as the structure of environments and stores, the use of direct or continuation style composition of functions, etc.) with the actual semantics of the programming language. This usually necessitates a substantial reformulation of the semantic specification for a language when language features are added whose description requires a more pow-

erful semantic model.[1] In addition, certain aspects of the semantics are needlessly overspecified, such as the handling of storage allocation. By not hiding model dependent details from the actual semantics, the extensibility of a given semantic definition is severely compromised. We call this problem *lack of separability.*

2. Traditional specifications are not formulated in terms of fundamental concepts expressing the design, analysis, use, and implementation of programming languages. Instead, these concepts are always encoded via function abstraction and application. This results in convoluted functionalities, excessive use of anonymous λ-abstractions, and various "currying tricks" which make it extremely difficult to write, read, and debug semantic definitions. Moreover, the distinction between static (compile time) and dynamic (runtime) language components is blurred, as both are expressed within the same framework. Clearly, this constitutes *poor semantics engineering.*

3. One of the most striking characteristics of standard denotational semantics is that it "... reduces to a minimum the amount of substantial information that must be supplied to a state transformation, ... before an 'answer' can be obtained." [MiS76, p.11] As a consequence, the semantic aspects of certain language constructs, such as environment manipulation at routine entry and exit, the distinction between variables and parameters, and others, are either completely ignored or expressed in highly cryptic terms. We call this *minimalistic semantic explication.*

4. The style of definition in traditional denotational semantics introduces dependencies between certain parts of the semantic specification which preclude proper modularization. Principles of information hiding and data abstraction are largely ignored. Consequently, existing denotational definitions are monolithic. This is *lack of modularity.*

There are a number of important consequences of these deficiencies for the automatic generation of prototype implementations from traditional denotational specifications.

1. Expressing standard denotational descriptions in Scott's language Lambda essentially amounts to writing in a very primitive assembly language (with more or less syntactic sweetener[2]) for a λ-calculus machine, with all the accompanying drawbacks of assembly language coding. The *lack of separability* forces one to implement a traditional semantics by *emulating* a λ-calculus machine, both at compile time and at runtime. This emulation is done by means of a partial evaluator performing β-reductions. Not surprisingly, this is the approach taken by all implementation generators based on denotational semantics, including SIS [Mos79], PSP [Pau81, Pau82], SPS [Wan84], and Watt's use of ML as a semantic metalanguage [Wat84].

---

[1] A striking example of this is given in Chapters 9 through 11 in Stoy's book, where language extension forces such rewriting five times.

[2] Fortunately, SyntacSweet has zero calories.

2. The *poor semantics engineering* has two major negative effects. First, the emulation of the details of the semantic models requires an excessive number of function closures, resulting in *extremely slow execution* of the "target code" derived from the semantics of a program. Observations by Paulson [Pau81], Wand [Wan84], and Watt [Wat84], and our own extensive experiments with SIS [BoB82], PSP [Ple84a], and MESS [Lee87] indicate that such emulation of Lambda assembly code always runs between several hundred and one thousand times slower than equivalent machine code which is directly executable. Second, the blurring of static and dynamic language concepts requires the presence of a partial evaluator in a compiler generated from a traditional semantics. Consequently, the generated compilers run considerably slower than handwritten non-optimizing compilers. [LeP86a]

3. The lack of separability, the poor semantics engineering, and the minimalistic explication of standard semantic definitions make them *unsuitable for exposing the structure of realistic implementations*. Indeed, the traditional approach requires that the standard semantics be augmented by a progression of more implementation-oriented semantic descriptions (such as *store* and *stack* semantics [MiS76]) in order to obtain abstract, but reasonably realistic implementation models. These definitions must be related to one another through tortuous congruence proofs.

4. The *minimalistic semantic explication* makes it *impossible to mechanically derive efficient implementations* which exploit the richness of existing machine architectures. Curiously enough, the same problem is caused by some of the overspecification of storage models in the traditional denotational approach.

5. The lack of modularity makes it very difficult to engineer semantic specifications in the sense of writing, testing, and debugging them. For realistic languages, semantic specifications may well be several hundred pages long, and their development benefits greatly from applying to them standard software engineering principles such as modularization, information hiding, and data abstraction.

Several of our criticisms of traditional denotational semantics have already been voiced by other researchers. For example, Mosses' abstract semantic algebra approach to semantics [Mos82, Mos84] and Mosses' and Watt's *action semantics* [MoW86a] are a direct outgrowth of the lack of separability and modularity in denotational specifications. Watt [Wat82] has studied modularity issues in semantic specifications. Mosses [Mos87] has compared the modularity properties of action semantics with the lack of modularity in denotational semantics. However, to our knowledge, no comprehensive analysis of the negative impact of traditional denotational definitions on the generation of prototype implementations has ever appeared in print.

2. Characteristics of High-Level Semantics

In this section, we summarize the salient characteristics of our approach to denotational semantics, which remedies the deficiencies of the traditional approach. We call our style of definition *high-level denotational semantics.*

1. *Separability.* The semantic definition of a programming language is cleanly separated from the details of underlying semantic runtime models. This separation is enforced by requiring two distinct specifications, called *macrosemantics* and *microsemantics.*

2. *Description via action-based operators.* The macrosemantics (semantics, for short) of a language is defined by a collection of semantic functions which map syntactic phrases, compositionally, to terms of semantic algebras. Similar to action semantics, the elements of the target algebras include actions and action combinators. Conceptually, actions are abstract entities which can be performed in order to process information. Actions are classified as declarative, imperative, or value producing actions, akin to those in [Mos82] [Mos84] [MoW86a]. The actions and their combinators are intended to capture fundamental programming language as well as implementation concepts. Our operators are usually more expressive than those proposed as a basis for semantic algebras [Mos82] [Mos84] [ClF82]. In a sense, they resemble both the "basic operators" and the "special operators" of action semantics.

3. *Information hiding.* Suitable microsemantic definitions provide the interpretation for the operators. However, only the names and signatures of the operators are made available to the semantics. Thus, the principle of separability guarantees the *invariance* of the macrosemantics under different microsemantic models. We have written and successfully implemented microsemantic definitions using traditional denotational models (including both a direct and a continuation style semantics), an abstract machine model, and a code generator specification for a nontrivial language [LeP87]. We thus obtain a wide variety of prototype implementations for a language at relatively little expense.

4. *Distinction between compile time and runtime.* The separation of the semantics into macrosemantics and microsemantics automatically distinguishes compile time objects from runtime objects. In essence, all domains defined in the macrosemantics are compile time domains, while those defined in the microsemantics are runtime domains. Thus we obtain the same effect as in the two-level metalanguage TML [NiN86].

5. *Modularity.* In addition to the kind of modularity provided by the separation of the microsemantic details from the semantics, high-level semantic descriptions exhibit two other kinds of modularity. First, the static semantics may be separated from the dynamic semantics. Second, any (macro or micro) semantic specification may be written as a collection of semantic modules, whose interfaces are subject to consistency checking. For example, there may be a module for the semantics of imperative

constructs, one for expressions, and one for declarations. This allows for the incremental development of semantic definitions. As an additional benefit, microsemantic modules can be reused as "semantic libraries."

6. *Exposition of implementation structures.* The semantic operators are chosen to reflect both fundamental language as well as fundamental implementation concepts. Consequently, the operators can be efficiently implemented by interpreting them as templates of intermediate code for a code generator. We have written several specifications for code generators which produce target code for the iAPX8086 microprocessor in the IBM PC. The quality of the generated machine code rivals that of non-optimizing hand crafted compilers. In addition, the distinction between the static and dynamic components of a language obviates the need for partial evaluation by means of β-reduction during compilation. This characteristic makes our specifications suitable for generating compilers which exhibit realistic compilation performance. [LeP87, Lee87]

7. *Extensibility.* Language features requiring new operators can be readily accommodated by extending the microsemantic modules. This may require a rewriting of (parts of) the microsemantics, but always leaves the existing macrosemantics intact. The portions requiring rewriting are easily identified. [Lee87]

8. *Simplified congruence proofs.* Demonstrating the congruence of two different semantic specifications is less complex than in traditional denotational semantics, as only the microsemantic specifications need to be related. The techniques described in [Roy86] should prove particularly appropriate in this context.

9. *Readability.* Finally, our specifications are written in a readable notation based on ML [Mil85], which allows them to be processed by the MESS system.

In the next section, the characteristics of high-level semantics are contrasted with those of the traditional approach by sketching a nontrivial example.

3. Example: Semantics of Procedure Declaration and Call

3.1. High-Level Semantics

3.1.1. Overview

In order to give the reader a feeling for our techniques, we now discuss the *macrosemantics* of procedure declaration and procedure call in a hypothetical language fragment, HypoPL. This is a nontrivial example which illustrates many of the points raised in the preceding two sections. The example is complete in that it specifies both the static and dynamic aspects of procedure declaration and call.

What follows in this subsection sets the stage for the example. We would like to draw attention to two important characteristics which are exhibited by all macrosemantic definitions: first, λ-applications only involve elements of macrosemantic domains, and there are no anonymous λ-abstractions; and second, the semantics is free from details concerning storage allocation. Thus it is fair to say that a macrosemantics only depends on the structure of the semantic compile time model. In contrast, the structure of the semantic runtime model is completely hidden.

The semantic domains below describe the structure of the static environment, which may be viewed as the central compile time data structure. Note that the various sorts of *action domains* are defined in the microsemantics.

```
MODE        = union
                noneM |

                (* procedure name, parameter mode *)
                procM of (NAME * MODE) |

                (* variable and block level *)
                varM of (VAR * LEVEL) |

                valParamM of PARAM.

(* variable, parameter name and type *)
VAR         = NAME * TYPE.
PARAM       = VAR.

(* compile time types *)
TYPE        = union int_type | ....

(* block levels *)
LEVEL       = union localL | globalL.

(* identifier associations *)
ASSOC       = ID -> (MODE * BLOCK_NUMBER).

(* block counters *)
BLOCKNUMBER = INT.

(* auxiliary information *)
BLOCK_INFO  = BLOCK_NUMBER * LEVEL * ....

(* static environments *)
ENV         = ASSOC * BLOCK_INFO.
```

In the following semantic equations, note that the capitalized names, such as `DeclSeq`, `NullDecl`, `StmtSeq`, etc., are action-yielding operators imported from the microsemantics. We have ignored the definition of the auxiliary macrosemantic functions (such as `notDeclared`) because of space constraints.

The valuation D determines the meaning of declarations. Below, we define how to handle declaration lists, and the declaration of integer variables. The equation for procedure declarations is the topic of the next subsection.

```
D: Decl -> ENV -> (ENV * DECL_ACTION).

D [[ decl ";" decls ]] env =
   let (env1, declAct1) = D [[ decl ]] env in
      let (env2, declAct2) = D [[ decls ]] env1 in
         (*
          * The operator 'DeclSeq' elaborates two declaration
          * actions in sequence.
          *)
         (env2, DeclSeq (declAct1, declAct2))
      end
   end.

D [[ ]] env = (env, NullDecl).

D [[ "int" id ]] env =
   if notDeclared (id, env) then
      let b = currentBlockNumber (env).
          l = currentLevel (env).
          name = mkAlphaName (id, b).
          mode = varM ((name, int_type), l).
          newEnv = addAssoc (id, mode, b, env)
      in
         (newEnv, DeclSimpleVar (name, IntType))
      end
   else
      declError env [[ id ]] "Identifier already declared.".
```

Intuitively, the application D [[decl]] env yields an extension of env which records the "compile time" effect of the declaration, and a declaration action to be elaborated at "runtime."

The valuation S determines the meaning of statements.

```
S: Stmt -> ENV -> IMP_ACTION.

S [[ stmt ";" stmts ]] env =
   (*
    * The operator 'StmtSeq' executes two imperative actions
    * in sequence.
    *)
   StmtSeq (S [[ stmt ]] env, S [[ stmts ]] env).

S [[ ]] env = NullStmt.
```

S [[stmt]] env yields an imperative action, which is executed for effect. The equation for procedure calls is discussed in Subsection 3.1.3.

The valuation B determines the meaning of blocks. The body of programs and procedures is taken to be a block.

```
B: Block -> ENV -> IMP_ACTION.

B [[ decls "begin" stmts "end" ]] env =
   let (env1, declActs) = D [[ decls ]] env
   in
      (* The operator 'Block' first elaborates the declarative
       * action for 'decls', then executes the action for 'stmts'.
       *)
      Block (declActs, S [[ stmts ]] env1)
   end.
```

The valuation E determines the meaning of expressions. We discuss only the meaning of integer variables.

```
E: Expr -> ENV -> (TYPE * VAL_ACTION).

E [[ id ]] env =
   case lookup (id, env) of

      noneM =>
         exprError [[ id ]] "Identifier not declared." |

      varM ((name, int_type), _) =>
         ( int_type, Rvalue (Var (name, IntType)) ) |

      valParamM (name, int_type) =>
         ( int_type, Rvalue (ValParam (name, IntType)) ) |

      ...
```

3.1.2. Macrosemantics of Procedure Declaration

All identifiers in HypoPL are lexically bound. Procedures have one formal parameter of type integer, and may be recursive. The parameter is bound by value to the actual argument.

The semantic definition below captures the intuitive idea behind the procedure declaration in a straightforward way. The many details of the equation are necessitated by the static semantics of the procedure declaration.

```
D [[ "proc" procId "(" "val" "int" paramId ")" "is" body ]] env =
    if notDeclared (procId, env) then
        let oldBn = currentBlockNumber (env).
            (procBn, env') = pushNextBlockNumber (env).
            procName = mkAlphaName (procId, oldBn).
            paramName = mkAlphaName (paramId, procBn).

            (* Process the parameter specification.
             *)
            paramMode = valParamM (paramName, int_type).
            paramAction = BindValParam (paramName, IntType).

            (* Build the environment in which to process the body.
             * The existing environment is extended with
             * - the binding for the procedure identifier, and
             * - the binding for the parameter identifier.
             *)
            procMode = procM (procName, paramMode).
            procEnv = addAssoc (procId, procMode, oldBn, env').
            bodyEnv = addAssoc (paramId, paramMode, procBn, procEnv).

            (* Process the routine body.
             * Since the binding for 'procId' is visible inside the
             * body, the procedure may be called recursively.
             *)
            bodyAction = B [[ body ]] bodyEnv.

            (* Build the environment visible to the remainder of
             * the program.
             *)
            resultEnv = popBlockNumber procEnv.
        in
            ( resultEnv,
              BindProc (procName,
                  Proc (paramAction,
                      StmtSeq (bodyAction, Return)))
            )
        end
    else
         declError env [[ procId ]] "Procedure already declared.".
```

The meaning of a procedure declaration is the declarative action BindProc. When elaborating this action, the semantic name, procName, is bound in the dynamic environment to the imperative action, Proc. Executing the Proc action first elaborates the embedded binding action for the formal parameter, then executes the imperative action which is the meaning of the procedure body, and then returns.

We do not consider identifiers to be basic semantic values like integers. Thus, they are not permitted as arguments to microsemantic operators. Instead, an identifier must first be mapped to a *semantic name* from the domain NAME. This mapping is accomplished by application of one of the predefined functions, mkName or mkAlphaName.

In Algol 68 parlance, identifiers correspond to *external objects*, whereas semantic names correspond to *internal objects*. In the macrosemantics, semantic names are used as abstract names for the denotable, storable, and expressible values. Specifically, semantic names may be used to denote abstract addresses in an idealized store, although their exact structure is hidden from the macrosemantics in order to preserve separability. In a sense, the mapping from identifiers to semantic names is a "cross-over" between static and dynamic aspects of variables and other named entities in a program.

Lexical scoping is achieved by binding identifiers to α-converted names in the static environment. By merely changing the use of `mkAlphaName (paramId, int)` above to `mkName (paramId)`, we obtain *dynamic* scoping! Note that the α-conversion can in principle be accomplished by applying a function similar to the LISP `gensym` function. Instead of `gensym`, however, we use `mkAlphaName` which creates a semantic name by combining an identifier with a unique block number. This makes the compile time semantics somewhat more complicated, since the current block number must be maintained. However, the semantic names then keep a close resemblance to the original syntactic identifiers, which is useful for both experimentation and instructional purposes.

In the above equation, we have chosen to hide the unbinding of the parameter name after execution of the procedure body. This is motivated by two observations: First, the unbinding can be achieved in the microsemantics in a variety of ways, including one which leaves it implicit. Second, it does not unnecessarily restrict the implementation choices of a code generator.

3.1.3. Macrosemantics of Procedure Call

The meaning of a procedure call is obtained as follows. First, the denotation of the procedure identifier is looked up in the static environment, and it is checked that it is indeed associated with a procedure value. Second, the value action for the argument expression is determined. Finally, the `Call` operator combines the procedure name and the value action for the argument expression. The execution of this imperative action passes the expression value into the procedure, where it is bound to the formal parameter.

```
S [[ "call" procId "(" expr ")" ]] env =
    case lookup (procId, env) of

        procM (procName, _) =>
            let (_, valAction) = E [[ expr ]] env
            in
                Call (procName, valAction)
            end |

        _ => stmtError [[ procId ]] "Unknown procedure identifier."
```

3.1.4. Microsemantic Aspects

We now give a short overview of a microsemantics which properly supports the macrosemantics of procedure declaration and call. Note that we say *a* microsemantics because a variety of specifications are possible. For example, we could give an algebraic (i.e., equational) specification as in Mosses' action semantics. Since we currently do not find such specifications suitable as a basis for implementation, we prefer the functional approach and give a λ-calculus style definition. The definition below resembles the type of semantics termed continuation-style *store* semantics. [MiS76]

```
microsemantics continuation_based

semantic domains
    (* procedure invocation 'stack' *)
    DASSOC      = REC_LEVEL -> NAME -> DENOTATION.

    (* recursion level *)
    REC_LEVEL   = INT.

    (* for passing arguments *)
    ARGS        = EV list.

    (* dynamic environments *)
    DENV        = DASSOC * ARGS * REC_LEVEL *
                      INPUTFILE * ANSWER.

    DENOTATION = union unboundD | intD of INT |
                           procD of IMP_ACTION.

    (* expression values *)
    EV          = union noneE | intE of INT.

    (* continuation domains *)
    ANSWER      = OUTPUTFILE.
    CONT        = DENV -> ANSWER.     (* commands *)
    DCONT       = CONT.               (* declarations *)
    ECONT       = EV -> CONT.         (* expressions *)
```

Actions are the fundamental entities from which the macrosemantic equations are built. Loosely speaking, an action transforms a given "semantic context" into a new context. In our case, the semantic context is defined by continuations, which encode how an "answer" is obtained from a given dynamic environment.[3] The use of the recursion level counter is necessary in order to restore temporarily hidden variable bindings when exiting from recursive invocations.

---

[3]The dynamic environment is usually called "state" or "store" in conventional denotational semantics

```
action domains

    IMP_ACTION  = CONT -> CONT.           (* imperative actions *)
    DECL_ACTION = CONT -> CONT.           (* declarative actions *)
    VAL_ACTION  = ECONT -> CONT.          (* value producing actions *)
    TYPE_ACTION = union IntType | ....    (* type producing actions *)
```

The auxiliary function bind produces a declarative action by extending the dynamic environment. The operators BindProc and BindValParam are phrased in terms of bind.[4]

```
auxiliary functions

    bind: NAME * DENOTATION -> DECL_ACTION is
    bind (name, den) = fn cont (dassoc, args, rl, i, o).
       cont ([rl => [name => den] (dassoc rl)] dassoc, args).

operators

    BindProc: NAME * IMP_ACTION -> DECL_ACTION is
    BindProc (procName, procAction) =
       bind (procName, procD (procAction)).

    BindValParam: NAME * TYPE_ACTION -> DECL_ACTION is
    BindValParam (paramName, IntType) =
       fn cont (dassoc, [intE (int)], rl, i, o).
          (*
           * Consume the argument value passed by the caller, and
           * bind the 'paramName' to that value.
           *)
          bind (paramName, intD (int)) ; cont (dassoc, [], rl, i, o).
```

The Call operator obtains the procedure value associated with the procedure name, evaluates the argument expression, and triggers the execution of the procedure body.

```
    Call: NAME * VAL_ACTION -> IMP_ACTION is
    Call (procName, valAction) = fn cont (dassoc, _, rl, i, o).
       let procD (procAction) = dassoc procName
       in
          valAction ; fn v.
             (*
              * Pass the argument value to the procedure,
              * and increment the recursion level.
              *)
             procAction ; cont (dassoc, [v], rl + 1, i, o)
       end.
```

[4]Following ML, λ is spelled fn. Right associative application is expressed by the ; primitive.

The Proc operator first elaborates the parameter binding action, and then executes the body action for effect.

```
Proc: DECL_ACTION * IMP_ACTION -> IMP_ACTION is
Proc (paramAction, bodyAction) = fn cont.
   paramAction ; bodyAction cont.
```

The Return operator merely decrements the recursion level.

```
Return: IMP_ACTION is
Return = fn cont (dassoc, args, rl, i, o).
   cont (dassoc, args, rl - 1, i, o).
```

The operator StmtSeq executes two imperative actions in sequence.

```
StmtSeq: IMP_ACTION * IMP_ACTION -> IMP_ACTION is
StmtSeq (a1, a2) = fn cont. a1 ; a2 ; cont.

NullStmt: IMP_ACTION is
NullStmt = fn cont. cont.
```

3.2. Traditional Denotational Semantics

3.2.1. Semantics of Procedure Declaration and Call

An equivalent standard semantics for the HypoPL fragment is given below. It is adapted from Chapter 8 of [Gor79], but written in a more readable notation. For the sake of conciseness, the static semantics has been ignored.

```
LOC   = unspecified.                   (* locations *)
SV    = union uninit | intVal of INT.  (* storable values *)
STORE = LOC -> SV.                     (* stores *)
ANS   = unspecified.                   (* answers *)
ENV   = Ide -> DEN.                    (* environments *)
DEN   = union unbound |                (* denotations *)
              location of LOC |
              procedure of PROC.
PROC  = CONT -> LOC -> CONT.           (* procedure values *)
CONT  = STORE -> ANS.                  (* command continuations *)
DCONT = ENV -> CONT.                   (* declaration continuations *)
ECONT = EV -> CONT.                    (* expression continuations *)
EV    = unspecified.                   (* expressible values *)
```

```
D: Decl -> ENV -> DCONT -> CONT.

D [[ "procedure" procId "(" "val" "int" paramId ")" "is"
        body ]] env = fn dcont.
      (*
       * Note how the environment env is frozen inside the
       * procedure value.
       *)
      let proc = fn cont loc. B [[ body ]]
                                 ([paramId => location (loc)] env)
                                 cont
      in
         (*
          * The declaration continuation is given an environment
          * with only one entry.
          *)
         dcont ([procId => procedure (proc)] emptyEnv)
      end.

S: Stmt -> ENV -> CONT -> CONT.

S [[ "call" procId "(" expr ")" ]] env = fn cont.
   let procedure (proc) = lookup procId env in
      (*
       * The function 'ref' allocates a new location,
       * initializes it with the argument value, and passes
       * it to the procedure;
       * note the use of the currying trick.  (Which one?)
       *)
      E [[ expr ]] env ; ref ; proc cont
   end.

S [[ stmt ";" stmts ]] env = fn cont.
   S [[ stmt ]] env ; S [[ stmts ]] env ; cont.

S [[ ]] env = fn cont. cont.

B [[ decls "begin" stmts "end" ]] env = fn cont.
   D [[ decls ]] env ; fn localEnv.
      let newEnv = combine env localEnv in
         S [[ stmts ]] newEnv cont
      end.
```

3.2.2. Fundamental Problems with Environment Handling

We now discuss how the interaction of procedure declaration and call is modelled in this traditional denotational specification. We shall answer the following questions: How is the argument transmitted? How is the formal parameter bound to this value? How is the formal parameter unbound? How is lexical binding achieved? The following discussion illustrates many of the deficiencies of the traditional denotational approach mentioned earlier.

First, here is the definition of the **ref** function from Section 5.7 in [Gor79]:

```
ref: ECONT -> ECONT              (or: ECONT -> EV -> STORE -> ANS)
ref = fn econt ev store.
        if new store = error then error
        else update (new store) (econt (new store)) ev store.
```

where new allocates a new location in the store, or returns an error indication, and update is defined as follows:

```
update loc cont ev store = cont ([loc => ev] store)
```

The **ref** function initializes a newly allocated location loc to the value ev, and then executes the remainder of the program (encoded by econt) with the modified store. The newly allocated location is passed along to the expression continuation econt.

Clearly, ref "coerces" a given expression value into a location which is initialized to that value, but the equation for procedure call is still somewhat mysterious. Only after decoding the currying trick will things become clearer. Using η-conversion, the third line of the equation for procedure call can be written as follows:

```
E [[ expr ]] env1 { fn ev store. (* env1 is the caller's environment *)
    ref { proc cont1 } ev store } (* cont1 is the return continuation *)
```

Noting that the application {proc cont1} denotes an expression continuation, the application of ref can be reduced to

```
{proc cont1} loc1 store1
```

where loc1 = new store, and store1 = [loc1 => ev] store.

The effect of this is to start up the procedure proc with return continuation cont1, and the initialized location loc1 as actual argument value. This is how the actual argument is passed.

Continuing with the procedure invocation, let us assume that proc is of the form

```
proc = fn cont' loc'.
   B [[ body ]] ([paramId => location (loc')] env) cont'
```

where env is the environment at the time of procedure declaration.

Then `{proc cont1} loc1 store1` expands into

```
B [[ body ]] ([paramId => location (loc1)] env) cont1 store1
```

This shows that the environment `env` is extended by binding the `paramId` to the initialized location `loc1` and then executing the procedure body. This is how the formal parameter is bound to the actual argument. Note that at procedure entry, the environment `env` is substituted wholesale for the environment `env1` of the caller.

The unbinding of the `paramId` after execution of the body finishes is left completely implicit. Indeed, by looking at the equations for procedure declaration and call, it is not apparent at all how the caller's environment is reestablished upon procedure exit. To solve the puzzle, let us assume that the call statement is followed by another statement, say `stmt1`. Then the return continuation `cont1` for the procedure is of the form

```
cont1 = { fn store. S [[ stmt1 ]] env1 cont'' store }
```

where `env1` is the same environment as above, and `cont''` is some continuation. In the absence of jumps out of procedures the application

```
B [[ body ]] ([paramId => location (loc1)] env) cont1 store1
```

ultimately reduces to `(cont1 store2)`, where `store2` reflects the changes to `store1` made by the execution of the procedure body. Thus, execution of the statement after the call is seen to resume as follows:

```
cont1 store2 = S [[ stmt1 ]] env1 cont'' store2
```

and voilà, the caller's environment `env1` has been successfully pulled out of a continuation hat. In essence, the formal parameter is unbound by dropping the callee's environment in favor of the caller's environment.

The wholesale substitution of environments at procedure entry and exit achieves lexical binding. This implicit manipulation of environments is neither intuitive nor does it resemble any reasonable implementation of lexical binding.

In order to change the binding discipline to dynamic binding, the equations for procedure declaration and call must be rewritten, and the domain of procedure values must be redefined to become

```
PROC = CONT -> LOC -> ENV -> CONT.
```

which reflects the necessity of passing the caller's environment to the callee.

3.3. Compilation of Procedure Calls

Given the program fragment

```
...
procedure P (val int X) is
decls
   begin
      stmts
   end; /* P */
int Y;
...
call P (Y);
...
```

our compiler generated by MESS from a high-level semantics produces the following intermediate code:

```
...
DeclSeq (BindProc (P1, Proc (BindValParam (X2, IntType), <code for body>)),
     DeclSeq (
          DeclSimpleVar (Y1, IntType), NullDecl))
...
     StmtSeq (
          Call (P1, Rvalue (Var (Y1, IntType))),
          ... )
...
```

where P_1, X_2 and Y_1 are the α-converted semantic names of the identifiers P, X and Y.

Our code generator for the iAPX8086 transforms this intermediate code into the following assembler instructions:

```
        ...
        dseg
X2      equ     4              ; value of X is in [bp + X2]
        cseg
P1_:
        push    bp
        mov     bp, sp
        "body of procedure"
        pop     bp
        ret
        dseg
Y1      rb      2              ; storage for global variable Y
        cseg
        ...
```

```
push   Y1       ; push value of Y on stack (binding for X)
call   P1_      ; jump to procedure
add    sp, 2    ; unbind X
...
```

In contrast to this, a compiler generated by Paulson's system PSP produces for the call statement more than 100 instructions for the SECD machine.

4. A Note On the Choice of Microsemantic Operators

We are still in the process of designing suitable sets of operators for standard imperative languages such as C and Pascal. At the point of this writing, we have developed a complete high-level semantic specification for SOL/C, a language "sort of like C." SOL/C features, among other things, multidimensional arrays, recursive procedures with value, reference and open array parameters, standard control structures, and a rich set of integer, character, and boolean expressions. A SOL/C compiler has been successfully generated from this specification. The performance characteristics of the compiler compare favorably with those of commercially available C and Pascal compilers for the IBM PC. Additional details are contained in [Lee87].

So far, our design choices for semantic operators have been almost exclusively motivated by semantic, rather than implementation, considerations. Nevertheless, we have found that a code generator has no problems generating efficient code from the prefix operator terms describing the dynamic semantics of a program. This seems to indicate that our high-level semantic operators largely coincide with implementation concepts embodied in the intermediate languages used in contemporary compilers.

5. Related Work

Our work started after experimenting with the direct implementation of denotational specifications using SIS [BoB82] and PSP [MiP84] [Ple84a]. A first step towards high-level semantics was the development of normal form semantics [Ple84b]. This was directly inspired by Wand's work on deriving postfix code from continuation semantics [Wan82], as well as research in the area of code generator specification languages [GlG78] [Gan80] [Bir82]. The connection with Mosses' concept of action-based semantic operators [Mos82] [Mos84], although known for quite some time, was made only recently. Indeed, the continuation transformers constructed in [Ple84b] are directly analogous to microsemantic operators yielding actions.

There are two semantics-based compiler generators which are similar in spirit to our approach. The CERES system of Jones and Christiansen [JoC82] accepts semantic specifications expressed in a small number of action-oriented operators inspired by those of

Mosses. Sethi's system [Set82] generates efficient compilers by treating fundamental "run-time" operators in the semantic specification as uninterpreted symbols. His work is also motivated by that of Mosses, but still refers to microsemantic concepts such as continuations and stores. Both systems have only been used for generating compilers for languages with control structures for sequencing, looping and decision making, and simple expressions. Also, the intermediate code produced by the generated compilers must be translated by a code generator in an ad hoc manner.

The possibility of providing alternative implementations for the operators of a semantic algebra was mentioned by Watt during his experimentation with ML as a semantic metalanguage [Wat84]. However, our MESS system is the first implementation generator which enforces the separation of the microsemantics from the semantics.

Nielson and Nielson have described an approach to semantics directed compiler generation using the two level metalanguage TML which enforces the distinction between compile time and runtime domains [NiN86]. The composition of the two portions of a TML specification corresponds directly to our composition of macrosemantic and microsemantic definitions.

Finally, the work of Mosses and Watt on *action semantics* bears strong similarities with our approach. In particular, the choice of "special operators" for their Pascal action semantics [MosW86b] is surprisingly similar to the choice of operators for our SOL/C semantics. However, there are a number of differences between the two approaches: (1) actions in action semantics are defined by means of "standard" operators, which themselves have algebraic definitions; (2) compile time and runtime aspects are not distinguished in action semantics; (3) actions are used by Mosses and Watt for specifying both the static and dynamic semantics of Pascal, whereas our macrosemantic definitions still depend on the λ-machinery; and (4) data flow within an action semantics is expressed by explicit naming whereas in a high-level semantics the nesting of prefix terms determines the data flow. At the present time, action semantics has a more solid theoretical foundation than high-level semantics. We are currently investigating ways of reconciling the human engineering approach taken by Mosses and Watt with our implementation engineering approach.

References

[Bir82] Bird, P. An implementation of a code generator specification language for table driven code generators. Proc. SIGPLAN '82 Symp. Compiler Construction, SIGPLAN Notices 17, 6 (June 1982), 44-55.

[BoB82] Bodwin, J., Bradley, L., Kanda, K., Litle, D., and Pleban, U. Experience with a compiler generator based on denotational semantics. Proc. SIGPLAN '82 Symp. Compiler Construction, SIGPLAN Notices 17, 6 (June 1982), 216-229.

[ClF82] Clinger, W., Friedman, D. P., and Wand, M. A scheme for a higher-level semantic algebra. US-French Seminar on the Application of Algebra to Language

Definition and Compilation, Fontainebleau, France, June 1982.

[Gan80] Ganapathi, M. Retargetable code generation and optimization using attribute grammars. Ph. D. Thesis, Tech. Rep. 406, Computer Science Department, University of Wisconsin-Madison, 1980.

[GlG78] Glanville, R. S., and Graham, S. A new method for compiler code generation. Conf. Rec. 5th Ann. ACM Symp. Principles of Programming Languages, Tucson, AZ, Jan. 1978, 231-240.

[Gor79] Gordon, M. J. C. *The denotational description of programming languages: An introduction.* Springer-Verlag, New York, 1979.

[JoC82] Jones, N. D., and Christiansen, H. Control flow treatment in a simple semantics-directed compiler generator. Formal Description of Programming Concepts II, IFIP IC-2 Working Conference, North Holland, Amsterdam, 1982.

[Lee87] Lee, P. The automatic generation of realistic compilers from high-level semantic descriptions. Ph. D. Dissertation, The University of Michigan, April 1987.

[LeP86a] Lee, P., and Pleban, U. F. The automatic generation of realistic compilers from high-level semantic descriptions: A progress report. Technical Report CRL-TR-13-86, The University of Michigan Computing Research Laboratory, June 1986.

[LeP86b] Lee, P., and Pleban, U. F. On the use of LISP in implementing denotational semantics. Proc. 1986 ACM Conf. LISP and Functional Programming, Cambridge, MA, August 1986, 233-248.

[LeP87] Lee, P., and Pleban, U. F. A realistic compiler generator based on high-level semantics. Proc. 14th Annual ACM SIGACT/SIGPLAN Symposium on Principles of Programming Languages, Munich, W.-Germany, January 1987, 284-295.

[MiS76] Milne, R. E., and Strachey, C. *A theory of programming language semantics.* Chapman and Hall, London, 1976.

[Mil85] Milner, R. The standard ML core language. Polymorphism II, 2 (Oct. 1985).

[MiP84] Milos, D., Pleban, U., and Loegel, G. Direct implementation of compiler specifications, or: The Pascal P-code compiler revisited. Conf. Rec. 11th Ann. Symp. Principles of Programming Languages, Salt Lake City, UT, January 1984, 196-207.

[Mos79] Mosses, P. SIS—Semantics implementation system. Tech. Rep. DAIMI MD-30, Computer Science Dept., Aarhus Univ., Aug. 1979.

[Mos82] Mosses, P. Abstract semantic algebras! In: D. Bjoerner (Ed.), Formal description of programming concepts II. North Holland, Amsterdam, 1982, 63-88.

[Mos84] Mosses, P. A basic abstract semantic algebra. In: Semantics of data type, Lecture Notes in Computer Science, Vol. 173. Springer-Verlag, Berlin, 1984, 87-107.

[Mos87] Mosses, P. Modularity in action semantics (Preliminary Version). Presented at a Workshop on Semantic Issues in Human and Computer Languages, Half Moon Bay, CA, March 1987.

[MoW86a] Mosses, P., and Watt, D. A. The use of action semantics. Technical Report DAIMI PB-217, Aarhus University, Aarhus, Denmark, August 1986.

[MoW86b] Mosses, P., and Watt, D. A. Pascal: Action semantics. Draft — Version 0.3). Computer Science Dept., Aarhus University, Aarhus, Denmark, Sept. 1986.

[NiN86] Nielson, H. R., and Nielson, F. Semantics directed compiling for functional languages. Proc. 1986 ACM Conf. LISP and Functional Programming, 249-257.

[Pau81] Paulson, L. A compiler generator for semantic grammars. Ph. D. Dissertation, Stanford University, December 1981.

[Pau82] Paulson, L. A semantics-directed compiler generator. Conf. Rec. 9th Ann. ACM Symp. Principles of Programming Languages, Albuquerque, NM, Jan. 1982, 224-239.

[Ple84a] Pleban, U. Formal semantics and compiler generation. In: Morgenbrod, H., and Sammer, W. (Eds.) Programmierumgebungen und Compiler. Teubner-Verlag, Stuttgart, 1984, 145-161.

[Ple84b] Pleban, U. Compiler prototyping using formal semantics. Proc. SIGPLAN '84 Symp. Compiler Construction, SIGPLAN Notices 19, 6 (June 1984), 94-105.

[Roy86] Royer, V. Transformations of denotational semantics in semantics directed compiler generation. Proc. SIGPLAN '86 Symp. Compiler Construction, SIGPLAN Notices 21, 6 (June 1986), 68-73.

[Sch86] Schmidt, D. A. *Denotational Semantics.* Allyn & Bacon, 1986.

[Set81] Sethi, R. Control flow aspects of semantics directed compiling. Tech. Rep. 98, Bell Labs., 1981; also in Proc. SIGPLAN '82 Symp. Compiler Construction, SIGPLAN Notices 17, 6 (June 1982), 245-260.

[Sto77] Stoy, J. E. *Denotational semantics: The Scott-Strachey approach to programming language theory.* MIT Press, Cambridge, 1977.

[Wan82] Wand, M. Deriving target code as a representation of continuation semantics. ACM TOPLAS 4, 3 (July 1982), 496-517.

[Wan84] Wand, M. A semantic prototyping system. Proc. SIGPLAN '84 Symp. Compiler Construction, SIGPLAN Notices 19, 6 (June 1984), 213-221.

[Wat82] Watt, D. A. Modular language definitions. Rep. CSC/82/R3, Computing Science Department, University of Glasgow, Oct. 1982.

[Wat84] Watt, D. A. Executable semantic descriptions. Rep. CSC/84/R2, Computing Science Department, University of Glasgow, Oct. 1984; also Software - Practice and Experience 16, 1 (Jan. 1986), 13-43.

An Action Semantics of Standard ML

David A. Watt
Computing Science Department, University of Glasgow
Glasgow G12 8QQ, U.K.

Abstract

Action semantics is a form of denotational semantics that is based on abstract semantic algebras rather than Scott domains and λ-notation. It allows formal descriptions of programming languages to be written that are unusually readable and modular. This paper presents an action-semantic description of Standard ML, as evidence for the claimed merits of action semantics. Milner's structural operational semantics of the same language is used as a basis for comparison.

1. Introduction

Action semantics [Mosses83, Mosses84, Mosses88, MossesW86a, MossesW86b, MossesW87] is a form of denotational semantics that uses abstract semantic algebras (instead of Scott domains and λ-notation) to describe the values and denotations of the programming language. The motivation behind action semantics is to improve the readability and modularity of formal semantic descriptions of programming languages.

The basic notion of action semantics is that of an *action*. An action is an entity that may be *performed*, receiving information and (perhaps) producing new information. The semantics of a programming language is defined by mapping phrases (such as expessions, commands, and declarations) to appropriate actions. A standard algebra of actions has been developed [MossesW86a]. It includes *primitive actions* for computing a new value, storing a value in a storage cell, binding a token to a value, and so on; and *action combinators* for sequencing, selection, recursion, block structuring, and so on. The actions have been chosen carefully both for their close correspondence to familiar semantic concepts and for their nice algebraic properties.

The modularity of an action-semantic description stems from defining the values and denotations of the programming language, and operations on these, to be elements of abstract semantic algebras. The actions themselves form an abstract semantic algebra. This imposes a modular structure on the semantic description that allows the semantic equations to be decoupled from the way in which the values and actions are defined. Consequently the semantic description is relatively easy to modify, if the described language is changed, and relatively easy to reuse for a related language.

The readability of an action-semantic description stems primarily from the correspondence between the actions and familiar semantic concepts, and secondarily from the use of an English-like notation for the actions.

This paper attempts to justify the claimed merits of action semantics by presenting an action-semantic description of Standard ML [HarperMM86], and comparing this description with Milner's structural operational semantics of the same language [Milner85].

Section 2 is a brief overview of ML. Section 3 is a commentary on the action-semantic description of ML, which may itself be found in Appendix C. Section 4 compares the action-semantic description with Milner's operational description of ML. Section 5 summarizes other work on action semantics.

2. The ML Programming Language

ML is a polymorphic functional programming language. Here we take a brief look at some of its characteristic features. Readers familiar with ML may skip to the subsection ***Abstract Syntax***.

Type System

ML is strongly typed. It has base types (BOOL, INT, REAL, and STRING), record types, and function types. New types may be created using *value-constructors*. One example is:

```
type SHAPE  =   Point
              | Circle of INT
              | Rectangle of (INT * INT)
```

where Point, Circle, and Rectangle are value-constructors. Typical values of the type SHAPE are Point, Circle(2), and Rectangle(3,4). The latter two are examples of *constructions*, i.e., structured values. Point is an example of a *constant*, i.e., an atom. Literals like 13 and "aeiou" are just special constants, of the types INT and STRING (respectively).

Types may be polymorphic. For example:

```
type 'a LIST  =   Nil
                | Cons of {head: 'a, tail: 'a LIST}
```

where 'a is a *type-variable*, and where Nil and Cons are value-constructors. This declares 'a

LIST to be a polymorphic type; its values represent lists, whose elements are values of the type for which 'a stands. Some instances of this polymorphic type are INT LIST, SHAPE LIST, and (BOOL LIST) LIST. Note, incidentally, that the type 'a LIST is recursively defined.

ML features polymorphic type inference, using the Hindley-Milner type system [Milner78]. For example, in the following declaration:

```
val  Even  =  fn N => (N mod 2 = 0)
```

the type of Even is inferred to be INT -> BOOL. Assume now that polymorphic functions Null (of type 'a LIST -> BOOL) and Tail (of type 'a LIST -> 'a LIST) have previously been declared; then in the following declaration:

```
val rec Length  =  fn L => if Null L
                          then 0
                          else 1 + Length (Tail L)
```

the type of Length is inferred to be 'a LIST -> INT.

Functions

Functions are first-class values in ML. Examples of functions are Even, Null, Tail, and Length, illustrated above. Strict evaluation is used for function arguments.

Operators like "mod", "+", and ":=" are just (infix) functions. For example, the expression "N mod 2" is essentially 'syntactic sugar' for "mod{#1=N, #2=2}", the argument to mod being a record with fields labelled #1 and #2.

Recursion

A **val** declaration may be made recursive by inserting the keyword **rec**. An example is the function Length illustrated above. Only function values may be recursive.

Pattern Matching

Pattern matching may be used in functions and some other constructs. The Length function illustrated above may alternatively be written thus:

```
val rec Length  =  fn  Nil                        => 0
                     | Cons{head=H,tail=T}  => 1 + Length T
```

In this context the phrase "Cons{head=H,tail=T}" is an example of a *pattern*; if it matches the argument value, it binds the variables `H` and `T` to the components of the argument value. The phrase "`Nil`" in the previous line is another pattern; it matches the argument value if and only if the latter equals the constant `Nil`.

The phrase "`Cons{head=H,tail=T} => 1 + Length T`" is an example of a *match.* If the pattern on the left of "=>" matches the argument value, the expression on the right of "=>" is evaluated using the bindings established in matching the pattern (here, bindings for `H` and `T`). The phrase "`Nil => 0`" is a simpler example of a match. Matches may be combined using the symbol "`|`" — then they are applied in sequence until one is found whose pattern matches the argument value; if none of them succeeds, the standard exception `match` is raised.

Exceptions

ML features parameterized exception handling. The expression:

```
raise SyntaxError with 13
```

yields, not an ordinary result value, but a *packet* consisting of an *exception* (here the one denoted by `SyntaxError`) and an *excepted-value* (here the integer 13). Each enclosing expression also yields this packet, by default. However, an enclosing expression may *handle* the exception, for example:

```
Parse (InputString)
handle SyntaxError with N => ErrorTree (N)
```

Here the expression "`Parse(InputString)`" is evaluated. If and only if that results in a packet containing the exception `SyntaxError`, the match "`N => ErrorTree(N)`" is applied to the excepted-value in the packet; i.e., `N` is bound to 13 and the expression "`ErrorTree(N)`" evaluated.

Imperative Aspects

Imperative features have been added to ML by including *references* (to storage cells) in its value space, by including "`:=`" (assignment) and "`!`" (dereferencing) as standard functions, and by including `ref` as an (anomalous) standard value-constructor. For example:

```
let
   val  Count  =  ref 0
in
   ...   Count := !Count + 1   ...
end
```

The expression "`ref 0`" yields a reference to a newly-created cell, which is initialized to 0. The expression "`Count := !Count + 1`" is evaluated principally for its side effect, which is to increment the content of `Count`; its own result is the empty record.

Abstract Syntax

ML exists at three levels:

- The *bare language* is minimal.
- The *core language* adds a fair amount of 'syntactic sugar'. All the extensions are completely specified by syntactic transformations down to the bare language.
- The *standard language* adds input-output and a powerful system of parameterized modules.

Following [Milner85], we shall formally describe only the bare language here. We shall use the abstract syntax given as Appendix B. Note that there are several sorts of identifiers in ML:

- *Value-variables* are used to denote ordinary values (including functions). Examples are `N`, `Even`, `Length`, and `Count` above.
- *Value-constructors* are used as constants and as constructors for structured values. Examples are `Point`, `Circle`, `Rectangle`, `Nil`, and `Cons` above.
- *Exception-identifiers* are used to denote exceptions. An example is `SyntaxError` above.
- *Record-labels* are used to identify fields of records. Examples are `#1`, `#2`, `head`, and `tail` above.

These sorts of identifiers are not actually distinguished in the concrete syntax, but Milner's description [Milner85] assumes that they are distinguished in the abstract syntax. Here we shall follow Milner, for simplicity; in fact, our abstract syntax is exactly equivalent to Milner's.

Note that parts of the abstract syntax have been elided in Appendix B. Type-bindings and datatype-bindings are relevant in the static semantics, but not in the dynamic semantics which we consider here.

3. The Action-Semantic Description of ML

The dynamic semantics of the ML bare language is given as Appendix C. The semantics given here defines the behaviour of well-typed programs only. (Types play no part in the dynamic semantics of such programs.)

The action-semantic description consists of:

- *Sorts* of values and actions, and *operations* over these sorts. Some of these sorts and operations are standard, in the sense that they are used in describing many languages. Others are specific to ML.

- *Semantic functions* mapping syntactic phrases to their denotations. These are defined by semantic equations, which are expressed in terms of the various sorts and operations.

3.1. Standard Sorts and Operations

Section C.1 lists some of the standard sorts and operations of action semantics. Only those actually needed to describe ML are listed.

The standard sorts include Boolean (truth values), Action (actions, i.e., entities that may be performed), Abstraction (values that encapsulate actions), and a number of auxiliary sorts explained below. For any sort *S*, the sort *S* ? includes all the values of *S* plus the value undefined. (The notation "*S* ? ⊃ *S*" expresses the fact that *S* is a subsort of *S* ?.)

When performed, an action receives and produces information of various kinds:

- An action receives and/or produces *named values*, i.e., associations of names to values. The names are of sort Name, and the values that may be named in the actions are of sort Nameable.
- An action receives and/or produces *bindings* of tokens to values. The tokens are of sort Token, and the values to which they may be bound are of sort Bindable.
- An action receives *storage* associating cells to values, and may produce changes in storage. The cells are of sort Cell, and the values that may be stored are of sort Storable.

Note that the sorts Nameable, Bindable, and Storable correspond to some of the *characteristic domains* of denotational semantics. They are used in describing many languages, although defined differently for each language.

Terms, of sort Term, are used to compute new values from existing values. They are composed from standard and language-specific operations. The standard operation "the _" allows a term to use the value with a stated name. The standard operation "binding of _" allows a term to use the value bound to a specified token. The standard operation "contents of _" allows a term to use the value contained in a specified cell.

Actions are of sort Action. Actions are not themselves values; however, an action may be encapsulated in an *abstraction*, which is a value of sort Abstraction. An abstraction may be *enacted*, causing its encapsulated action to be performed. The encapsulated action receives no named values nor bindings, unless modified by the operation "_ with the _" or the operation "_ with all bindings" (respectively).

Formal (algebraic) definitions of actions and abstractions have been worked out by Mosses. (An outline may be found in [Mosses84], for an earlier version of the action notation.) Unfortunately, there is not enough space in this paper to discuss the definitions. Informal descriptions of the standard terms and actions are given, for easy reference, in Appendix A.

3.2. ML-specific Sorts and Operations

Section C.2 lists the sorts and operations that are specific to ML, followed by their definitions.

The sorts Constant, Construction, Record correspond to atoms, constructions, and records (respectively). The sort Value is defined to include all of these sorts, and also Abstraction (functions) and Value-Cell (references).

Values of sort Exception are identities of exceptions. Since ML exceptions are created dynamically, almost like storage cells, we use the trick of identifying each exception with a unique cell, which will contain the corresponding exception-identifier. (This might seem artificial, but the operational description of ML [Milner85] uses exactly the same trick.).

The standard sort Bindable is defined to include both Value and Exception, and correspondingly the standard sort Token is defined to include Value-Variable and Exception-Identifier, respectively. (The other sorts of identifiers — value-constructors and record-labels — are not bound to anything in this semantics.)

Operations on values are defined algebraically, in the usual way. For example, the equations under Construction state, in effect, that a value of this sort is composed using "construction (_, _)", and decomposed using "constructor of _" and "component of _". The equations under Record state, in effect, that a value of this sort is a map from values of sort Record-Label to values of sort Value.

Some of the operations are actions. These are defined by expansion into standard actions. For example, "propagate ..." abbreviates "obtain a packet from ... then escape". The action "report ..." isolates the implementation-defined action that is performed when an exception packet is propagated to the top level of an ML program, and thus improves the modularity of the semantic description. The action "establish-standard-bindings" produces bindings for standard value-variables and exception-identifiers.

3.3. Semantic Functions

Section C.3 defines the semantic functions for ML expressions, handlers, matches, patterns, declarations, and programs.

Each phrase's denotation is an action, whose outcome describes what happens when the phrase is 'evaluated'. The action might *complete*, indicating normal evaluation; it might *escape*, indicating that an exception has been raised; it might *diverge*, indicating that evaluation is non-terminating; or it might *fail*, indicating that a pattern cannot be matched or that a match cannot be applied.

The action "evaluate *E*" evaluates the expression *E*. If this action completes, it produces a resulting value (of sort Value), which is the result of evaluating the expression. (We should say, more precisely, that it produces *a value named* "resulting value".)

The action "handle *H*" receives a packet (of sort Packet), and makes *H* attempt to handle the exception contained therein; if it succeeds, it produces a resulting value.

The action "apply *M*" receives an argument value (of sort Value) and applies the match *M* to it. If this action completes, it produces a resulting value. This action will fail if *M* cannot be applied to the argument value.

The action "match *P*" receives an argument value and attempts to match that value with the pattern *P*. If this action completes, it produces bindings for any value-variables in *P*. This action will fail if *P* cannot be matched to the argument value.

The action "elaborate *D*" elaborates the declaration *D*, producing bindings. The action "elaborate *VB*" works similarly for the value-binding *VB*, and "elaborate *EB*" for the exception-binding *EB*.

The semantic functions are defined by semantic equations. The right-hand side of each equation is a particular action, of the kind characterized above. Here we shall discuss how some important features of ML are described in action semantics.

Functions

The effect of "evaluate [[**fn** *M*]]" is to produce a resulting value that is an abstraction, determined as follows. The action "apply *M* else propagate ..." applies the match *M* to an argument value, and produces a resulting value (or escapes). By means of "abstraction (_)", an abstraction encapsulating that action is formed. By means of "_ with all bindings", the encapsulated action is made to receive all current bindings, i.e., those received by "evaluate [[**fn** *M*]]". Note that this is static binding.

In "evaluate [[E_1 E_2]]", we see how a function is called. First the subexpressions E_1 and E_2 are evaluated. The result of evaluating E_1 might be either an abstraction (function) or a value-constructor. In the former case, the results of E_1 and E_2 are renamed abstraction and argument value, respectively; then, by means of "_ with the argument value", the abstraction is modified so that the argument value is received by the encapsulated action; and finally the action "enact _" is used to perform the encapsulated action.

Recursion

In the ML bare language, recursion occurs only in value-bindings of the form "**rec** *VB*". The action "elaborate *VB*" produces just the bindings programmed in *VB*. The action "recursively elaborate *VB*" is similar, except that it ensures that the bindings produced by "elaborate *VB*" are also received by "elaborate *VB*"! (See the explanation of "recursively" in Appendix A.)

Pattern Matching

In "evaluate [[*Con*]]", we see that the result of an expression consisting of a value-constructor (e.g., `Nil` or `Cons`) is just the value-constructor itself. In "evaluate [[E_1 E_2]]", such a value-constructor might turn up as the result of evaluating E_1; in that case, the result of evaluating "E_1 E_2" is a construction, which is formed from the value-constructor and the result of evaluating E_2. (The value-constructor `ref` is anomalous, and is treated specially — see ***Imperative Aspects***.)

Constructions are decomposed using patterns of the form "*Con P*". In "match [[*Con P*]]", the received argument value will normally be a construction. If the latter was constructed using the same value-constructor *Con*, pattern matching continues by matching *P* to the component of the construction. Otherwise, "match [[*Con P*]]" fails. (The received argument value might also be a reference, constructed using `ref`; that case is treated specially — see ***Imperative Aspects***.)

In general, the action "match *P*" produces bindings. The simplest case is "match [[*Var*]]", which simply binds the value-variable *Var* to the received argument value.

In "apply [[*P* => *E*]]", the action "match *P*" is performed first. If this completes, "evaluate *E*" is performed next. The "before" combinator is used to make any bindings produced by "match *P*" available to "evaluate *E*" (overriding any bindings received by the whole action). The "locally" combinator is used to localize these bindings, so that "apply [[*P* => *E*]]" as a whole produces no bindings.

In "apply [[M_1 | M_2]]", "apply M_1" is performed first. Only if it fails is "apply M_2" performed. This is ensured by the "else" combinator.

Exceptions

Evaluating an expression, elaborating a declaration, etc., might result in an exception being raised. The corresponding actions ("evaluate *E*", "elaborate *D*", etc.) would then escape, producing a packet. This is seen in "evaluate [[**raise** *Exn* **with** *E*]]", where a packet is obtained by combining an exception (the one to which *Exn* is bound) and the result of evaluating *E*.

Action combinators such as locally, or, and, then, etc., have the property that if a component action escapes, so too does the combined action. The only combinator not having this property is "then exceptionally", whose second component action is performed if the first one escapes. We use these properties of the action combinators to describe the behaviour of ML exceptions. In most cases, if evaluation of a subexpression escapes (producing a packet), then evaluation of the whole expression also escapes (producing the same packet). In "evaluate [[*E* **handle** *H*]]", however, the "then exceptionally" combinator is used to intercept any escape. The action "evaluate *E*" is performed first. Only if that action should escape is the action "handle *H*" performed next, and the packet produced by "evaluate *E*" is made available to "handle *H*".

In "handle [[*Exn* **with** *M*]]", a check is made to determine whether the exception contained in the received packet is the same as the one bound to *Exn*. If so, the match *M* is applied to the excepted-value contained in the packet; if not, the action fails. The action "handle [[? => *E*]]" never fails — this form handles any exception. In "handle [[H_1 || H_2]]", the action "handle H_1" is performed first; should that fail (because H_1 does not handle the exception), the action "handle H_2" is performed instead.

Imperative Aspects

In "evaluate [[E_1 E_2]]", the result of evaluating E_1 might turn out to be the value-constructor `ref`. In that case a cell is created, the result of evaluating E_2 is stored in that cell, and the result of evaluating the whole expression is a reference to that cell.

In "establish-standard-bindings" (see C.2), ":=" is bound to an abstraction. The encapsulated action expects to receive an argument value that is a record, the first field of that record being a reference. The action stores the second field in the cell designated by that reference.

ML specifies left-to-right evaluation to ensure that side effects (and raising of exceptions) are predictable. For that reason, the semantic equation for "evaluate [[E_1 E_2]]", uses the "before" combinator between the evaluations of E_1 and E_2. (Instead, the "and" combinator would imply implementation-dependent order of evaluation.) The alert reader might notice one or two other places in the semantic equations where left-to-right evaluation is enforced.

4. The Operational Description of ML

It is interesting to compare our action-semantic description of ML with Milner's operational description [Milner85]. The operational description covers only the bare language directly; the remainder of the core language is described by syntactic transformation down to the bare language. Thus the two descriptions are readily comparable.

The formalism used in Milner's description is *structural operational semantics* [Plotkin81]. It has been applied to describe CCS as well as ML, and is currently being used (in a modified form) to describe the concurrent aspects of Ada.

A structural operational semantics defines transitions in the 'evaluation' of a program, by means of *axioms* and *inference rules*. The simplest sentences are of the form:

$$env \vdash evaluand \Rightarrow result$$

where *env* is an environment, *evaluand* is the entity to be 'evaluated', and *result* is what is produced by 'evaluating' *evaluand* in *env*. The nature of result depends on the kind of *evaluand*. In the ML semantics, for example:

- If *evaluand* is an expression, then *result* will be a value (or packet).
- If *evaluand* is a match paired with a value, then *result* will be a value (or packet).
- If *evaluand* is a declaration, then *result* will be an environment (or packet).

More complicated sentences are needed to take account of side effects:

$$env \vdash \frac{evaluand}{store} \Rightarrow \frac{result}{store'}$$

where *store* and *store′* are the stores before and after 'evaluation' of *evaluand*. This notation can be very tedious, since most rules are not directly concerned with the store. Therefore Milner uses an abbreviation whereby the *store* component may be omitted in an inference rule where the subphrases are 'evaluated' in a left-to-right order. (He uses a similar abbreviation for exception propagation.)

A few of the inference rules for evaluating ML expressions (*E*) are shown in Appendix D. (The notation used in [Milner85] has been changed in minor respects, to facilitate comparison.)

The rules are hard to read, in our opinion. The main reason for this seems to be the need to name explicitly the environments and stores occurring in each rule. The abbreviation (mentioned above) that allows the store component to be elided from most of the rules does not entirely solve this problem; besides, the abbreviation is clearly *ad hoc*, and not generally applicable. In action semantics, by contrast, environments and stores never have to be named explicitly; the action combinators take care of the various ways in which bindings and storage changes are sequenced or merged. These combinators are suitable for describing a wide variety of programming languages, not just ML with its left-to-right evaluation order.

Noteworthy in the operational description of ML is the somewhat clumsy treatment of functions, and especially of recursive functions. A function is described by a closure of the form closure (*M*, *env*, *valenv*), where *M* is the function body (a match — note that this is a *syntactic* object); *env* is the environment in which the function was defined; and *valenv* is a second environment, which is empty in the case of a non-recursive function, but otherwise contains bindings for the function itself and any other functions with which it is mutually recursive. *valenv* is 'unrolled' (using the auxiliary function rec) and merged with *env*, at the point of *call* of the function. By contrast, the treatment of abstractions in action semantics is more general, in that they can have bindings attached to them at any time (using "_ with all bindings"); moreover, the treatment of abstractions is orthogonal to the treatment of recursive bindings.

Finally, the English-like notation of action semantics is somewhat easier on the reader, in our opinion, than the inference-rule notation of structural operational semantics. Readability is a subjective issue, of course, being necessarily influenced by the reader's familiarity with the particular notation. However, we are convinced that a reader unfamiliar with action semantics can quickly reach the stage of understanding an action-semantic description, at least at a superficial level; and understanding of the semantic description will increase smoothly as familiarity with the notation is gained. On the other hand, a reader unfamiliar with structural operational semantics has to work very hard to extract any information at all from the semantic description. (The same is true for other semantic notations like conventional denotational notation and VDM.)

5. Other Related Work

The semantic description of ML reported here is one of a series of experiments to explore how suitable action semantics is for describing a wide variety of 'real' programming languages. A description of Pascal (both static and dynamic semantics) has largely been completed by Mosses and Watt [MossesW86a]. A description of the dynamic semantics of ML has independently been written by Mark, and large parts of it reused in a description of Amber [Mark86]. Descriptions of a sequential subset of Ada, of the logic programming language Prolog, and of the concurrent programming language Joyce are also being undertaken at Aarhus and Glasgow Universities. (Although not mentioned in this paper, there are standard actions for describing communication between processes.)

The problem of describing the static semantics of ML has not been considered here, but is left to a future paper. This will be an interesting trial for action semantics, because of the need to describe polymorphic type inference.

6. Conclusions

We have presented an action-semantic description of the dynamic semantics of ML. In our own opinion, the description is highly readable, because of the English-like notation and the close links between action notation and well-known semantic concepts. It is easily modifiable, in the sense that language changes (such as removal or addition of imperative features, or changes in evaluation order) would require only commensurate changes to the description — rather than forcing it to be completely rewritten. It is reusable, in the sense that large parts of it could be used in the descriptions of other functional languages. The reader is invited to judge these claims for himself or herself.

Acknowledgments

I am very happy to acknowledge the work of Peter Mosses, who deserves the lion's share of the credit for the development of action semantics, and who provided useful comments on an earlier version of this paper. I would also like to acknowledge Jan Mark's independently-written description of ML, which despite cross-fertilization contains both similarities and differences to the present one. Finally, I wish to thank Neil Jones and Uwe Pleban for their encouraging comments, and Richard Kieburtz whose remark "Action semantics is the COBOL of semantic notations!" at the Workshop was taken in the spirit in which it was intended.

References

[HarperMM86] R. Harper, D. MacQueen, and R. Milner: Standard ML. Report ECS-LFCS-86-2, Computer Science Department, University of Edinburgh (March 1986).

[Mark86] J. Mark: Action semantics of ML and Amber. Report DAIMI IR-66, Computer Science Department, Aarhus University (November 1986).

[Milner78] R. Milner: A theory of type polymorphism in programming. *J. Computer and System Sciences* 17, 3, 348–375 (1978).

[Milner85] R. Milner: The dynamic operational semantics of Standard ML. Computer Science Department, University of Edinburgh (April 1985).

[Mosses83] P. D. Mosses: Abstract semantic algebras! In *Formal Description of Programming Concepts II* (ed. D. Bjørner), North-Holland, Amsterdam (1983).

[Mosses84] P. D. Mosses: A basic abstract semantic algebra. In *Semantics of Data Types* (ed. G. Kahn, D. B. MacQueen, and G. Plotkin), Lecture Notes in Computer Science 173, Springer, Berlin (1984).

[Mosses88] P. D. Mosses: Modularity in action semantics. In *Workshop on Semantic Issues in Human and Computer Languages*, MIT Press, Cambridge, Massachussets (forthcoming, 1988).

[MossesW86a] P. D. Mosses and D. A. Watt: Pascal action semantics — towards a denotational description of ISO Standard Pascal using abstract semantic algebras. Draft 0.30, Computer Science Department, Aarhus University (1986)

[MossesW86b] P. D. Mosses and D. A. Watt: The potential use of action semantics in standards. Report CSC/86/R1, Computing Science Department, University of Glasgow (March 1986).

[MossesW87] P. D. Mosses and D. A. Watt: The use of action semantics. In *Formal Description of Programming Concepts III* (ed. M. Wirsing), North-Holland, Amsterdam (1987).

[Plotkin81] G. D. Plotkin: Structural operational semantics. Report DAIMI-FN19, Computer Science Department, Aarhus University (1981).

Appendix A. Summary of Action Semantics

A.1. General

An *action* is an entity that can be *performed*, receiving and (possibly) producing information. The outcome of performing an action is one of the following:

- *Completion* — the action terminates normally, producing information.
- *Escape* — the action terminates exceptionally, producing information.
- *Failure* — the action is aborted, producing nothing.
- *Divergence* — the action does not terminate.

An action may be analyzed into several independent *facets*, as follows. An action's behaviour in one facet in general does not affect its behaviour in another facet. Primitive actions in general have an effect in only one facet.

- *Functional facet* — the action receives and produces associations of names to values. (These associations are passed directly from action to action, and disappear unless explicitly propagated.) The names given to values reflect their sorts in an obvious way, e.g., "thing" and "1st thing" are possible names for values of sort Thing.
- *Binding facet* — the action receives and produces bindings of tokens to values. (These bindings propagate over a certain scope.)
- *Imperative facet* — the action receives storage associating cells to values, and may produce changes in storage. (The contents of storage remain undisturbed unless an action explicitly stores to a cell, or creates or destroys a cell.)

An *abstraction* is a value that encapsulates an action. (Actions are not themselves values.)

A *term* specifies the computation of a value. A term may invoke standard and language-specific operations. In general, the value of a term depends on information received by the action containing that term.

The following five subsections summarize the standard action-semantic notation used in this paper. (Square brackets enclose optional notation.) Each primitive action completes, unless otherwise stated. Each constituent of a combined action receives the same named values, bindings, and storage as the combined action itself, unless otherwise stated. When named values or bindings are 'merged', the combined action fails if there is any conflict. If a constituent action escapes, diverges or fails, the combined action does likewise, unless otherwise stated below.

A.2. Functional Facet

the *N*	This term — where *N* is a name — yields the value named *N*.
obtain a[n] *N* from *T*	This action produces the value of the term *T*, naming that value *N*; however, it fails if the value is not of the sort implied by *N*. *Usage:* computing and naming a value; renaming a value; checking that a value is of a particular sort.
copy the *N*	This action produces just the received value named *N*. *Usage:* explicitly propagating a specific named value.
check *T*	This action — where *T* is a term yielding a truth value — produces nothing if that value is true, but fails if that value is false. *Usage:* testing an assertion; guarding alternative actions.

A.3. Binding Facet

binding of *T*	This term — where *T* is a term yielding a token — yields the value bound to that token.
bind T_1 to T_2	This action — where T_1 is a term yielding a token, and where T_2 is a term yielding a bindable value — produces a binding of the token to the bindable value. *Usage:* declarations.
copy all bindings	This action produces the same bindings as it receives. *Usage:* explicitly propagating bindings.
locally *A*	This action performs *A*, but discards any bindings produced by *A*. *Usage:* block structuring.
recursively *A*	This action performs *A* (which must complete, and produce only bindings), producing the bindings produced by *A*. *A* receives the bindings received by the combined action, overridden by those produced by *A* itself. *Usage:* recursive declarations.

A.4. Imperative Facet

contents of *T*	This term — where *T* is a term yielding (the identity of) a storage cell — yields the storable value contained in that cell.
create a[n] *N*	This action creates a new storage cell, and produces a value, named *N*, that is the identity of that cell. *Usage:* allocating storage.
store T_1 in T_2	This action — where T_1 is a term yielding (the identity of) a storage cell, and

where T_2 is a term yielding a storable value — changes the contents of the cell to be the storable value. *Usage:* assignments.

A.5. General Actions

skip — This action produces nothing at all. *Usage:* null action; preventing propagation of values.

escape — This action simply escapes, producing all received named values. *Usage:* raising exceptions; exit jumps; terminal errors.

[either] A_1 or A_2 — This action chooses between performing A_1 or performing A_2; but if the chosen action fails, it performs the other action instead. *Usage:* deterministic choice between mutually exclusive alternative actions; implementation dependent choice between non-exclusive actions.

[preferably] A_1 else A_2 — This action performs A_1; but if that fails, it performs A_2 instead. *Usage:* deterministic choice between non-exclusive alternative actions.

[both] A_1 and A_2 — This action performs A_1 and A_2 together, with arbitrary interleaving of their primitive actions, and merges any named values and bindings produced by A_1 and A_2. *Usage:* independent actions; implementation dependent ordering of actions.

[first] A_1 then A_2 — This action performs A_1 followed by A_2 (if A_1 completes). It produces only the named values produced by A_2, which receives only the named values produced by A_1. It merges any bindings produced by A_1 and A_2. *Usage:* sequencing actions; composing functions.

[first] A_1 then exceptionally A_2 — This action performs A_1 first; if A_1 escapes, A_2 is performed, receiving the named values produced by A_1, and the escape is not propagated. *Usage:* handling exceptions and exit jumps.

[first] A_1 before A_2 — This action performs A_1 followed by A_2 (if A_1 completes). It produces the bindings produced by A_1 overridden by those produced by A_2, which receives the bindings received by the combined action overridden by those produced by A_1. It merges any named values produced by A_1 and A_2. *Usage:* sequencing declarations; sequential evaluation of expressions.

[first] A_1 within A_2 — This action performs A_1 followed by A_2 (if A_1 completes). It produces only the bindings produced by A_2, which receives only the bindings produced by A_1. It merges any named values produced by A_1 and A_2. *Usage:* private declarations.

A.6. Abstractions

abstraction (A) — This term yields an abstraction encapsulating the action A, which will receive no named values or bindings. *Usage:* treating procedures as values.

T with the N — This term — where T is a term yielding an abstraction a — yields a new abstraction in which the action encapsulated by a will receive the value currently named N. *Usage:* passing arguments to procedures.

T with all bindings — This term — where T is a term yielding an abstraction a — yields a new abstraction in which the action encapsulated by a will receive all current bindings. *Usage:* dynamic binding ("enact ... with all bindings"); static binding ("abstraction (...) with all bindings").

enact T — This action — where T is a term yielding an abstraction a — performs the action encapsulated by a. The encapsulated action will receive the storage received by the enact action, but no named values other than those supplied by "_ with the _", and no bindings other than those supplied by "_ with all bindings". The "enact" action produces whatever the encapsulated action produces. *Usage:* calling procedures.

A.7. Layout

The grouping of complicated actions is indicated by indentation, for example:

```
first  A1
then   both  A2
       and   A3
```

To avoid excessive indentation in very complicated actions, punctuation may be used instead: an infix combinator with no punctuation has higher precedence than one preceded by a comma, which in turn has higher precedence than one preceded by a semicolon. For example, "A_1, then A_2 and A_3" means the same as the above action. All the above infix action combinators are associative, so explicit grouping in actions like "A_1 and A_2 and A_3" is unnecessary.

Appendix B. ML Abstract Syntax

Value-Variable
Var ::= ...

Value-Constructor
Con ::= ...

Exception-Identifier
Exn ::= ...

Record-Label
Lab ::= ...

Expression
E ::= *Var* | *Con* | **fn** *M* | $E_1 E_2$ | **let** *D* **in** *E* **end** |
{ } | { *RE* } | *E* **handle** *H* | **raise** *Exn* **with** *E*

Record-Expression
RE ::= *Lab* = *E* | RE_1 , RE_2

Handler
H ::= *Exn* **with** *M* | ? => *E* | H_1 || H_2

Match
M ::= *P* => *E* | M_1 | M_2

Pattern
P ::= _ | *Var* | *Con* | *Con P* | *Var* **as** *P* |
{ } | { ... } | { *RP* } | { *RP* , ... }

Record-Pattern
RP ::= *Lab* = *P* | RP_1 , RP_2

Declaration
D ::= **val** *VB* | **exception** *EB* |
type *TB* | **datatype** *DB* | **abstype** *DB* **with** *D* **end** |
local D_1 **in** D_2 **end** | *empty* | D_1 D_2

Value-Binding
VB ::= *P* = *E* | VB_1 **and** VB_2 | **rec** *VB*

Exception-Binding
EB ::= *Exn* | Exn_1 = Exn_2 | EB_1 **and** EB_2

Type-Binding
TB ::= ...

Datatype-Binding
DB ::= ...

Program
Prog ::= *empty* | *Prog D* ;

Appendix C. ML Action Semantics

C.1. Standard Sorts and Operations

Signature

Boolean

false, true	:	→ Boolean
not _	: Boolean	→ Boolean
etc.		

S ?
(for any sort S)

S ?	⊃ *S*	
undefined	:	→ *S ?*

Nameable

*(S-)*Name
(for any sort S such that Nameable ⊃ *S)*

Name	⊃ *S*-Name	

*(S-)*Term
(for any sort S such that Nameable ⊃ *S)*

Term	⊃ *S*-Term	
the _	: *S*-Name	→ *S*-Term

(for any sorts $S, S_1, \ldots, S_n$, *and for any operation* $o : S_1, \ldots, S_n \rightarrow S$ *)*

o	: S_1-Term, ..., S_n-Term	→ *S*-Term
obtain a[n] _ from _	: Name, Term	→ Action
copy the _	: Name	→ Action
check _	: Term	→ Action

Bindable

*(S-)*Token
(for any sort S such that Bindable ⊃ *S)*

Token	⊃ *S*-Token	
binding of _	: *S*-Token-Term	→ *S*-Term
bind _ to _	: *S*-Token-Term, *S*-Term	→ Action
copy all bindings	:	→ Action
locally _	: Action	→ Action
recursively _	: Action	→ Action

Storable

*(S-)*Cell
(for any sort S such that Storable ⊃ *S)*

Cell	⊃ *S*-Cell	
contents of _	: *S*-Cell-Term	→ *S*-Term
create a[n] _	: *S*-Cell-Name	→ Action
store _ in _	: *S*-Term, *S*-Cell-Term	→ Action

Action

skip	:	→ Action
escape	:	→ Action
[either] _ or _	: Action, Action	→ Action
[preferably] _ else _	: Action, Action	→ Action
[both] _ and _	: Action, Action	→ Action
[first] _ then _	: Action, Action	→ Action
[first] _ then exceptionally _	: Action, Action	→ Action
[first] _ before _	: Action, Action	→ Action
[first] _ within _	: Action, Action	→ Action

Abstraction

abstraction (_)	: Action	→ Abstraction-Term
_ with the _	: Abstraction-Term, Name	→ Abstraction-Term
_ with all bindings	: Abstraction-Term	→ Abstraction-Term
enact _	: Abstraction-Term	→ Action

C.2. ML-specific Sorts and Operations

Signature

Integer

sum (_, _)	: Integer, Integer	→ Integer?
etc.		

Real

sum (_, _)	: Real, Real	→ Real?
etc.		

String

size (_)	: String	→ Integer
etc.		

Constant

Constant	⊃ Value-Constructor, Boolean, Integer, Real, String	
_ is _	: Constant, Constant	→ Boolean?

Construction

construction (_, _)	: Value-Constructor, Value	→ Construction
constructor of _	: Construction	→ Value-Constructor
component of _	: Construction	→ Value

Record

unity	:	→ Record
single-field (_, _)	: Record-Label, Value	→ Record
_ joined to _	: Record, Record	→ Record?
field _ of _	: Record-Label, Record	→ Value?

Value

Value	⊃ Constant, Construction, Record, Abstraction, Value-Cell

Exception

exception	: Exception-Identifier-Cell	→ Exception
bind-exception	:	→ Exception
match-exception	:	→ Exception

addition-exception	:	→ Exception
etc.		

Packet

packet (_, _)	: Exception, Value	→ Packet
exception of _	: Packet	→ Exception
excepted-value of _	: Packet	→ Value
propagate _	: Packet-Term	→ Action
report _	: Packet-Term	→ Action

Nameable

Nameable ⊃ Value, Exception, Packet, Token, Cell

Bindable

Bindable ⊃ Value, Exception

Token

Value-Token	= Value-Variable	
Exception-Token	= Exception-Identifier	
establish-standard-bindings :		→ Action

Storable

Storable ⊃ Value, Exception-Identifier

Definitions

Construction

constructor of construction(c,v)	= c
component of construction(c,v)	= v

Record

_ joined to _ *is commutative, is associative, and has identity* unity

single-field(l,v) joined to single-field(l,v')	= undefined	
field l of unity	= undefined	
field l of (single-field(l,v) joined to r)	= v	
field l of (single-field(l',v) joined to r)	= field l of r	*if* $l \neq l'$

Packet

exception of packet(e,v)	= e
excepted-value of packet(e,v)	= v

propagate : Packet-Term → Action

propagate T =
obtain a packet from T then escape

report : Packet-Term → Action

report T =
implementation-defined action

Token

establish-standard-bindings : → Action

establish-standard-bindings =
both bind [[`bind`]] to bind-exception
and bind [[`match`]] to match-exception

```
and     bind [[not]] to abstraction (
                    obtain a boolean from the argument value  then
                    obtain an resulting value from not (the boolean) )
and     bind [[+]] to addition-exception
and     bind [[+]] to abstraction (
                    either  obtain a 1st integer from
                                field [[#1]] of the argument value  and
                            obtain a 2nd integer from
                                field [[#2]] of the argument value,  then
                            obtain a resulting value from
                                sum (the 1st integer, the 2nd integer)  else
                            propagate packet (addition-exception, unity)
                    or      obtain a 1st real from
                                field [[#1]] of the argument value  and
                            obtain a 2nd real from
                                field [[#2]] of the argument value,  then
                            obtain a resulting value from
                                sum (the 1st real, the 2nd real)  else
                            propagate packet (addition-exception, unity)
                    )
and     bind [[!]] to abstraction (
                    obtain a value-cell from the argument value  then
                    obtain a resulting value from contents of the value-cell )
and     bind [[:=]] to abstraction (
                    obtain a value-cell from field [[#1]] of the argument value  and
                    obtain a value from field [[#2]] of the argument value,  then
                    store the value in the value-cell  and
                    obtain a resulting value from unity )
etc.
```

(Note: Underlining is used above to distinguish exception-identifiers from value-variables, e.g. bind.)

C.3. ML Semantic Functions

Definitions

Expression

```
                    evaluate : Expression → Action

        evaluate [[Var]] =
                obtain a resulting value from binding of Var

        evaluate [[Con]] =
                obtain a resulting value from Con

        evaluate [[fn M]] =
                obtain a resulting value from abstraction (
                        apply M  else  propagate packet (match-exception, unity)
                        ) with all bindings

        evaluate [[E1 E2]] =
                first   evaluate E1  then  obtain a 1st value from the resulting value
                before  evaluate E2  then  obtain a 2nd value from the resulting value
                then    either  obtain an abstraction from the 1st value  and
                                obtain an argument value from the 2nd value,  then
```

```
                    enact the abstraction with the argument value
              or    preferably
                          check the 1st value is [[ref]] then
                          create a value-cell, and
                          copy the 2nd value; then
                          store the 2nd value in the value-cell and
                          obtain a resulting value from the value-cell
                    else  obtain a value-constructor from the 1st value and
                          copy the 2nd value, then
                          obtain a resulting value from construction (
                            the value-constructor, the 2nd value)

evaluate [[let D in E end]] =
        locally
              first   elaborate D
              before  evaluate E

evaluate [[{ }]] =
        obtain a resulting value from unity

evaluate [[{ RE }]] =
        evaluate-record RE then
        obtain a resulting value from the resulting record

evaluate [[E handle H]] =
        first   evaluate E
        then exceptionally
              handle H else propagate the packet

evaluate [[raise Exn with E]] =
        first   obtain an exception from binding of Exn and
                evaluate E
        then    propagate packet (the exception, the resulting value)
```

Record-Expression

```
                evaluate-record : Record-Expression → Action

evaluate-record [[Lab = E]] =
        first   evaluate E
        then    obtain a resulting record from single-field (Lab, the resulting value)

evaluate-record [[RE1 , RE2 ]] =
        first   evaluate-record RE1 then
                obtain a 1st record from the resulting record
        before  evaluate-record RE2 then
                obtain a 2nd record from the resulting record
        then    obtain a resulting record from the 1st record joined to the 2nd record
```

Handler

```
                handle : Handler → Action

handle [[Exn with M]] =
        first   check exception of the packet is binding of Exn and
                obtain an argument value from excepted-value of the packet
        then    apply M else propagate packet (match-exception, unity)

handle [[? => E]] =
        evaluate E
```

handle [[H_1 | | H_2]] =
handle H_1 else handle H_2

Match

apply : Match → Action

apply [[*P* => *E*]] =
locally
first match *P*
before evaluate *E*

apply [[M_1 | M_2]] =
apply M_1 else apply M_2

Pattern

match : Pattern → Action

match [[_]] =
skip

match [[*Var*]] =
bind *Var* to the argument value

match [[*Con*]] =
obtain a constant from the argument value then
check the constant is *Con*

match [[*Con P*]] =
either obtain a construction from the argument value, then
check constructor of the construction is *Con* and
obtain an argument value from component of the construction, then
match *P*
or obtain a value-cell from the argument value and
check *Con* is [[`ref`]], then
obtain an argument value from contents of the value-cell, then
match *P*

match [[*Var* **as** *P*]] =
both bind *Var* to the argument value
and match *P*

match [[{ }]] =
skip

match [[{ . . . }]] =
skip

match [[{ *RP* }]] =
obtain an argument record from the argument value then
match-record *RP*

match [[{ *RP* , . . . }]] =
obtain an argument record from the argument value then
match-record *RP*

Record-Pattern

match-record : Record-Pattern → Action

match-record [[*Lab* = *P*]] =

obtain an argument value from field *Lab* of the argument record then
match *P*

match-record [[RP_1 , RP_2]] =
both match-record RP_1
and match-record RP_2

Declaration

elaborate : Declaration → Action

elaborate [[**val** *VB*]] =
elaborate *VB*

elaborate [[**exception** *EB*]] =
elaborate *EB*

elaborate [[**type** *TB*]] =
skip

elaborate [[**datatype** *DB*]] =
skip

elaborate [[**abstype** *DB* **with** *D* **end**]] =
elaborate *D*

elaborate [[**local** D_1 **in** D_2 **end**]] =
first copy all bindings before elaborate D_1
within elaborate D_2

elaborate [[*empty*]] =
skip

elaborate [[D_1 D_2]] =
first elaborate D_1
before elaborate D_2

Value-Binding

elaborate : Value-Binding → Action

elaborate [[*P* = *E*]] =
first evaluate *E* then obtain an argument value from the resulting value
then match *P* else propagate packet (bind-exception, unity)

elaborate [[VB_1 **and** VB_2]] =
first elaborate VB_1
then elaborate VB_2

elaborate [[**rec** *VB*]] =
recursively
elaborate *VB*

Exception-Binding

elaborate : Exception-Binding → Action

elaborate [[*Exn*]] =
first create an exception-identifier-cell
then store *Exn* in the exception-identifier-cell and
bind *Exn* to exception (the exception-identifier-cell)

elaborate [[Exn_1 = Exn_2]] =
first obtain an exception from binding of Exn_2
then bind Exn_1 to the exception

elaborate [[EB_1 **and** EB_2]] =
both elaborate EB_1
and elaborate EB_2

Program

run : Program → Action

elaborate [[*empty*]] =
establish-standard-bindings

elaborate [[*Prog D* ;]] =
first run *Prog*
before elaborate *D* then exceptionally report the packet

Appendix D. ML Structural Operational Semantics

Auxiliary Notation

$\{x_1 \rightarrow y_1, \ldots, x_n \rightarrow y_n\}$	Map associating x_1 to y_1, …, and x_n to y_n.
$map_1 \oplus map_2$	Map uniting all associations from map_2 with non-conflicting associations from map_1.
$map\ (x)$	The value associated to x in *map*.
closure (M, *env*, *valenv*)	Function (closure) consisting of the match M, the environment *env* in which the function was defined, and an environment *valenv* containing value bindings to be interpreted recursively when the function is called.
rec (*valenv*)	Environment obtained by recursively unrolling the value bindings in *valenv*.
apply (*f*, *val*)	Result of applying the basic function *f* to the argument value *val*.
construction (*Con*, *val*)	Construction formed from the value-constructor *Con* and the component value *val*.
ass	Standard function to which ":=" is bound.
ref	Standard value-constructor `ref`.

Axioms and Inference Rules

(1) $env \vdash Var \Rightarrow env(Var)$

(2) $env \vdash Con \Rightarrow Con$

(3) $$env \mid\!- \ \mathtt{fn}\ M \Rightarrow \text{closure}\,(M, env, \{\,\})$$

(4) $$\frac{env \mid\!- E_1 \Rightarrow \text{closure}\,(M, env_1, valenv_1) \qquad env \mid\!- E_2 \Rightarrow val_2 \qquad env_1 \oplus \text{rec}\,(valenv_1) \mid\!- M, val_2 \Rightarrow val}{env \mid\!- E_1\, E_2 \Rightarrow val}$$

(5) $$\frac{env \mid\!- E_1 \Rightarrow f \qquad env \mid\!- E_2 \Rightarrow val}{env \mid\!- E_1\, E_2 \Rightarrow \text{apply}\,(f, val)}$$ (where f is a basic function)

(6) $$\frac{env \mid\!- E_1 \Rightarrow Con \qquad env \mid\!- E_2 \Rightarrow val}{env \mid\!- E_1\, E_2 \Rightarrow \text{construction}\,(Con, val)}$$

(7) $$\frac{env \mid\!- \frac{E_1}{store_0} \Rightarrow \frac{\text{ass}}{store_1} \qquad env \mid\!- \frac{E_2}{store_1} \Rightarrow \frac{\{\#1 \rightarrow cell,\ \#2 \rightarrow val\}}{store_2}}{env \mid\!- \frac{E_1\, E_2}{store_0} \Rightarrow \frac{\{\,\}}{store_2 \oplus \{cell \rightarrow val\}}}$$

(8) $$\frac{env \mid\!- \frac{E_1}{store_0} \Rightarrow \frac{\text{ref}}{store_1} \qquad env \mid\!- \frac{E_2}{store_1} \Rightarrow \frac{val}{store_2}}{env \mid\!- \frac{E_1\, E_2}{store_0} \Rightarrow \frac{cell}{store_2 \oplus \{cell \rightarrow val\}}}$$ (where $cell$ is an address not used in $store_2$)

An Algorithmic Approach to p-adic integers

Steven Vickers
Dept of Computing, Imperial College,
London, SW7 2BZ.

Abstract

The ring of p-adic integers can be embedded as the maximal elements in a Scott domain with algebraic structure. We show how definitions and proofs in the mathematical theory of p-adics can be replaced by algorithms on the partial elements and formal programming methods working on the algorithms. Certain types of argument translate naturally into non-deterministic algorithms using the Smyth power domain.

Introduction

The p-adic integers have a rich and well-known mathematical structure, with applications especially in number theory. We do not in this paper add to the known results about p-adics, but rather we show how they can be described and proved using the formal methods of computer programming. The algorithms we present apply ostensibly not to the p-adics themselves (for which the algorithms do not terminate), but to their finite approximations; but continuity principles ensure that they can be used to make statements about p-adics. We have four main aims.

First, the presentation may make p-adics more approachable for computer scientists who find them cropping up in their work – some areas where this is already happening are factorization algorithms (which can use quite deep results from number theory) and the "error-free computation" of Gregory and Krishnamurthy [84].

Second, the richness of the existing theory and the rigour demanded by it provide a stiff test of the formal methods that we wish to apply. We use the standard methods of Scott domains to develop algorithmic constructs better adapted to the mathematical methods.

Third, this approach is patently constructive. Constructivist mathematicians demand that mathematical constructions should in principle be reducible to computer programs; we have a system in which this is not just in principle possible but actually practicable. In particular, some assertions of the form $\phi(x) \rightarrow \exists y.\ \psi(x, y)$ have been shown by proofs of $\phi(x) \rightarrow \psi(x, f(x))$ where f is a domain theoretic construct. This may be seen in programming terms as reducing the problem to a *specification* (of f) $\phi(x) \rightarrow \psi(x, f(x))$, a *definition* of f, and a proof that the definition satisfies the specification.

Fourth, and more speculatively, this system of p-adics brings together a number of notions – profinite algebraic structure, topology, sheaves, domain theory and computing formal methods – that make it look like a good place for testing out grander ideas for unifying categorical logic with computer science.

A curiosity of our approach is that some of our algorithms use non-deterministic calculations to lead to deterministic results. In particular, our "non-deterministic refinement operator" may be thought of as calling for an extra, uninitialized digit to be connected to a register in the computer. In fact, we find it expedient to give a non-deterministic semantics for our constructs, and leave it as part of the programming discipline for the programmers to show that their functions are single-valued just as, in ordinary programming, one ought to show that ones programs terminate.

The present paper is elementary and intended to bring out some computational techniques for handling p-adics. Since these rely heavily on the arithmetic structure, in a companion paper [Vickers 87] we describe how the structures involved may be constructed as fixed points of endofunctors of suitable categories of algebras.

The mathematical background of p-adics is described in many books on algebra or number theory; Cohn [78] gives a very good introduction.

Acknowledgements

I should like to thank Mike Smyth for the original suggestion that the p-adic domain might be worthy of study, Samson Abramsky for the benefit of his encyclopaedic knowledge of domain theory and my other colleagues at Imperial for many stimulating discussions. The work was funded by the Alvey committee.

Notation

Throughout, a *ring* is a commutative ring with a multiplicative identity 1.

$\mathbb{N}$ is the set of natural numbers, including 0.

$N = \mathbb{N} \cup \{\infty\}$, treated as a domain under the usual numerical order ($\sqsubseteq$ is $\leq$).

$\mathbb{Z}$, $\mathbb{Z}/n$ and $\mathbb{Z}_p$ are the rings of integers, integers modulo n and the p-adic integers.

Functions are written on the left and composed in the corresponding order.

For the standard constructions of domain theory, we follow the notation of Abramsky [87].

1. Introduction to the domain of p-adics

1.1 The p-adic integers

The 2-adic integers can be thought of as the values handled by a computer provided with infinitely long registers. Initially, perhaps, these were designed with the idea of eliminating overflow in integer arithmetic by allowing integers that are arbitrarily large, albeit still finite. However, it is found in practice that the same hardware techniques for adding, negating, subtracting and multiplying integers still make good sense when infinitely many of the bits in a register are set: these more general values form a commutative ring $\mathbb{Z}_2$ containing the integers.

The same ideas apply with number bases other than 2, but unlike the case with the ordinary integers, the rings $\mathbb{Z}_n$ of n-adic integers for different n are non-isomorphic. For prime number bases these rings are nicer – they are *integral domains* – and so henceforth we shall assume that the number base is a prime, p. However, the results presented here work without problems in non-prime bases.

From the definition, we can describe a p-adic integer as an infinite sequence of digits

$x = \dots\dots d_i \dots d_2 d_1 d_0$ where $0 \le d_i < p$.

The *truncation* of x to j digits is defined as

$x|j = d_{j-1} \dots d_2\, d_1\, d_0$

x can be described by the infinite sequence $(x|j)_{j\ge 0}$ where $0 \le x|j < p^j$ and $x|(j+1) \equiv x|j \pmod{p^j}$.

Although on the face of it x|j is a natural number, we have in fact got it by pretending that $p^j = 0$ so that all the digits from d_j leftward vanish. Thus x|j is really *an integer modulo* p^j. Note the case x|0 which is always 0 (the unique residue class modulo 1).

The truncation map $x \mapsto x|j : \mathbb{Z}_p \to \mathbb{Z}/p^j$ is a ring homomorphism. We also use the same notation for the unique homomorphism $x \mapsto x|k : \mathbb{Z}/p^j \to \mathbb{Z}/p^k$ if $j \ge k$, so if x is a residue modulo p^j then x|k is x reduced modulo p^k.

1.2 The domain of p-adic integers

We define the *domain* of p-adic integers to be the disjoint union $D_p = \bigcup_{j \in \mathbb{N}} \mathbb{Z}/p^j \cup \mathbb{Z}_p$.

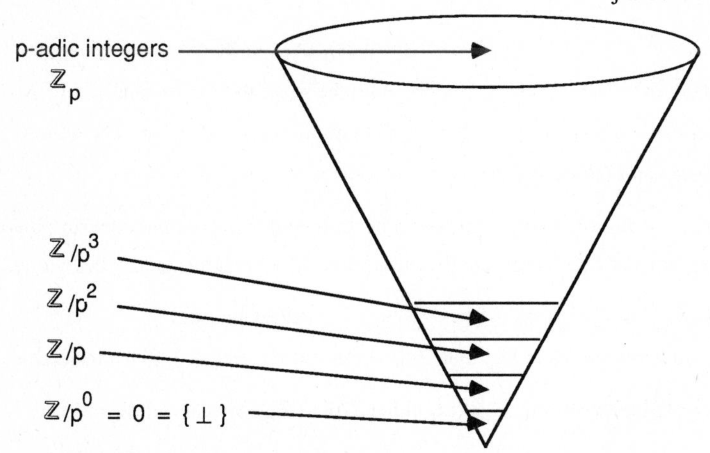

The *extent* $Ex \in N$ of an element x of D_p is defined by

$$Ex = \begin{cases} j & \text{if } x \in \mathbb{Z}/p^j \\ \infty & \text{if } x \in \mathbb{Z}_p \end{cases}$$

This is an extent in the same sense as in Fourman and Scott [79].

We define an approximation ordering $\sqsubseteq$ on D_p by

$x \sqsubseteq y$ iff $x = y|Ex$ (i.e. iff x is a truncation of y)

Proposition 1.2.1 $(D_p, \sqsubseteq)$ is a poset isomorphic to the domain of all finite and infinite lists over an alphabet of p symbols, under the prefix ordering. (This is often known as the *Kahn* domain on p symbols.) It is a Scott domain and its finite elements are those with finite extent.

Proof Identify the p symbols with the base-p digits. A p-adic integer can be represented as an infinite list of digits. An integer modulo p^j can be represented by an integer a with $0 \leq a < p^j$, written to base p and reversed: this is a finite list of j digits (least significant first). Thus every element of D_p corresponds to a finite or infinite list from the Kahn domain. x is a truncation of y (as elements of D_p) iff, as lists, y is x with extra digits at the more significant end, i.e. iff x is a prefix of y.

The rest is now obvious and well-known.]

The Kahn domain has the usual hd, tl and cons operators of list processing. To fit in with the way in which p-adic numbers are written as lists of digits (infinite on the left, finite on the right), we use the notation

$$[x, d] = \text{cons}\ (d, x)$$

We can think of the extent function E as mapping D_p into $N = \omega+1$, the initial fixpoint of the domain equation N = lift N; E is then defined by

$$E\ [x, d] = \text{up}\ Ex = Ex+1$$

1.3 Algebraic structure on D_p

Interpreted correctly in a suitable category (of sheaves), D_p can be given a ring structure. This is developed in Vickers [87]. Here, rather than going into the sheaf theory, we shall show that although as a set D_p is not actually a ring, with a little care it is virtually as good as one. The notions are those of Stoltenberg-Hansen and Tucker [85].

Let $\omega(x, y, z, \dots)$ (finitely many arguments) be an operator in commutative ring theory, i.e. a polynomial in its arguments with integer coefficients. Some of the arguments may be *unused*, as in

$$\omega(w, x, y, z) \quad = -3x^2y + zx + 81 \qquad \text{(with w unused)}$$

If x, y, z, ... are given values in D_p, then let $j = \min(Ex, Ey, Ez, \dots)$. Since the elements of D_p with extent j form a commutative ring $E^{-1}j$ (i.e. either $\mathbb{Z}/p^j$ or $\mathbb{Z}_p$), we can define

$$\omega(x, y, z, \dots) = \omega(x|j, y|j, z|j, \dots)$$

D_p thus has all the ring operations, but it is not itself a ring. For instance, it fails to satisfy the axiom $x-x = 0$. But it does satisfy the axiom $x-x = 0|Ex$, and this shows that if we regard 0 not as a nullary operator but as a unary operator with an unused argument x, then the axiom holds. The moral is that we must take care to interpret the axioms in such a way that an argument appearing on either side also appears, used or unused, on the other.

Proposition 1.3.1 (i) For each m-ary ring operator ω, the corresponding operation

$$\omega : D_p{}^m \rightarrow D_p$$

is continuous.

(ii) Let $\omega_1(x, y, z, \ldots) = \omega_2(x, y, z, \ldots)$ (same m arguments on both sides) be an identity in ring theory. Then the corresponding operations on D_p are equal.

Proof See Stoltenberg-Hansen and Tucker [85] or Vickers [87].]

Definition We extend the notation [x, d] by

$$[x, d] = [x, 0] + d \qquad (x \in D_p, d \in \mathbb{Z})$$

Proposition 1.3.2 If $x, y \in D_p$ and $d, e \in \mathbb{Z}$ then

$$[x, d] + [y, e] = [x+y, d+e]$$

$$[x, d]\cdot[y, e] = [pxy + dy + ex, de]$$

$$[x, d] = [y, e] \Leftrightarrow d \equiv e \pmod{p} \text{ and } y = x + (d-e)/p$$

Proof See Vickers [87].]

Definition We define $x \in D_p$ to be *null* iff $x \sqsubseteq 0$, i.e. $x = 0|Ex$. 0 here means the ring theory 0; in D_p its value is the infinite list of zero digits. Thus the null elements are the finite or infinite lists whose digits are all zero.

nullity $x \in N$ is defined as the greatest j such that $x|j$ is null:

nullity [x, d] = **if** d = 0 **then** (**nullity** x) + 1 **else** 0 **fi** $(0 \le d < p)$

It is the number of initial zero digits in x, and may be a finite natural number or ∞.

1.4 Syntactic constructs

Given a continuous function g_d defined on D_p for each digit d, or integer d with $0 \le d < p$, we can define a strict continuous function f on D_p (with the same codomain as the g_d's) by the notation

$$f([x, d]) = g_d(x) \qquad (0 \le d < p)$$

g_d will commonly also be defined for integers d outside the given range, and it will often be useful to show that $f([x, d]) = g_d(x)$ for all d and x. The mere definition of f says nothing about these other values of d.

We can also describe functions as the least fixpoints of functionals, e.g.

$$F(g)([x, d]) = h_d(g, x) \qquad (0 \le d < p)$$

$$f = Y F \qquad \text{(Y is the fixpoint operator)}$$

For conditionals, we use the Algol 68 **if** … **then** … **elif** … **else** … **fi** notation.

2. "Take an Integer Refinement"

2.1 Introduction

A number of traditional constructions with the p-adics involve taking a more or less arbitrary integer representative of a congruence class and calculating with that representative. We incorporate this into our system by treating it as a non-deterministic operator, and the appropriate domain setting is one that includes the Smyth power domain.

This method leads to the following style of argument: we wish to define some fully determinate function from D_p to itself, but in fact we define it as a non-determinate function from D_p to the power domain $P_S(D_p)$. It is then necessary to prove that the function is actually determinate, i.e. it factors through the singleton embedding $\{| - |\}: D_p \to P_S(D_p)$. We do not claim that these functions can be defined *only* in this way; in fact there are definitions that don't leave the base domain D_p and from the point of view of domain theory these are much less problematical. But the non-deterministic versions do seem to be computationally simpler, and closer in spirit to the standard mathematical constructions.

We define two basic non-deterministic operations on D_p:

refine x is the set of all 1-step refinements of x

nrefine (x, m) is the set of all m-step refinements of x ($m \in N = \omega+1$)

Our main application is to algebraic operations that can be refined to give more digits of a result than would be expected just from the definition of the arithmetic structure. For instance,

$\lambda x.\ px$ can be refined to $\lambda x.\ [x, 0] = \lambda x.\ p$ (**refine** x)

A more complicated example arises from the exponential function $\lambda x.\ x^p$. From the binomial expansion

$$(x + kp^n)^p = x^p + p \cdot x^{p-1} \cdot kp^n + \Sigma_{i=2}^{p} \binom{p}{i} x^{p-i} \cdot (kp^n)^i$$

we see that if x and y are congruent modulo p^n, then their p^{th} powers are congruent modulo p^{n+1} (provided $n \geq 1$). We can deduce that $\lambda x.\ x^p$ can be refined (modulo some adjustments at $\perp$) by the function $\lambda x.$ (**refine** x)^p.

These are particular examples of a more general result (our Theorem 2.6.2) that relates refinability of a polynomial function to the nullity of its formal derivative, and shows how to compute a deterministic refinement using the non-deterministic operation **nrefine**.

2.2 The Smyth power domain

The basic properties of power domains are summarized in Smyth [78]. Given a Scott domain D (although the theory works for more general domains), we write:

- PD for the Smyth power domain. We think of its elements as being all compact (in the Scott topology) subsets of D, modulo a congruence $\equiv$. The congruence classes have canonical representatives, namely the compact, *upper closed* (or *saturated*) subsets.

- set notation with barred curly brackets "$\{|\ \ldots\ |\}$" to mean the congruence class of the corresponding set. This agrees with Abramsky's notation $x \mapsto \{| x |\} : D \to PD$ for the singleton embedding.
- $\sqsubseteq$ for the approximation ordering in PD, and also for the corresponding preorder on subsets of D ($X \sqsubseteq Y$ iff X minorizes Y, i.e. every element of Y refines some element of X, i.e. $Y \subseteq \uparrow X$). Note that if $X \subseteq Y$, then $X \sqsupseteq Y$.
- $\equiv$ for the congruence corresponding to $\sqsubseteq$ ($X \equiv Y$ iff $X \sqsubseteq Y$ and $Y \sqsubseteq X$, i.e. $\uparrow X = \uparrow Y$).
- $\sqcap$ for the meet in PD, represented by set theoretic union. Since this means non-deterministic choice, it corresponds with Dijkstra's $\Box$ notation.
- $\sqcup$ for the join in PD. On the compact upper closed subsets, this is represented by intersection; on the finite elements of PD (represented by finite sets of finite elements of D) it can be computed by

$$X \sqcup Y = \{| x \sqcup y : x \in X, y \in Y, \{x, y\} \text{ bounded above } |\}$$

- **stop** for the top element $\varnothing$ of PD. This is sometimes excluded, but we include it. In a nondeterministic calculation, we interpret **stop** as the value of a possible branch of the computation that has been shown to be a blind alley (a detected error as opposed to bottom, $\{| \bot |\}$, which is divergence).

An element of PD_p is a *singleton* iff it is of the form $\{| x |\}$ for some $x \in D_p$; it is *null* iff it is of the form $\{| x |\}$ for some null $x \in D_p$. A function into PD_p is *single-valued* iff it factors through $\{| - |\}$.

2.3 Extending operations from D_p to PD_p

- Any function into D can be composed with $\{| - |\}$ and considered a map into PD.
- A continuous function $f: D \to PE$ can be extended uniquely to a left exact (**stop**-and-$\sqcap$-preserving) continuous function $f: PD \to PE$.
- A continuous function $f: D_1 \times D_2 \to PE$, in particular one of the algebraic operations on D_p, can be extended to a continuous function $f: PD_1 \times PD_2 \to PE$ by extending to $P(D_1 \times D_2)$ and precomposing with the tensor product map (Hennessy and Plotkin [79])

$$(X, Y) \mapsto X \otimes Y = \{| (x, y) : x \in X, y \in Y |\} : PD_1 \times PD_2 \to P(D_1 \times D_2)$$

e.g. $X+Y = \{| x+y : x \in X, y \in Y |\}$

It is important to realize that PD_p is not a ring under these operations, not even in the redefined way that D_p is. It satisfies some laws, such as associativity $(X+Y)+Z = X+(Y+Z)$, but not laws with repeated arguments on one side. For instance we have $X-X \sqsubseteq 0(X)$ but not full congruence here:

$$X-X = \{| x-y : x, y \in X |\}, \qquad 0(X) = \{| x-x : x \in X |\}$$

We have similar notation for other expressions. Thus $X{\cdot}X \sqsubseteq X^2$:

$$X{\cdot}X = \{| xy : x, y \in X |\}, \qquad X^2 = \{| x^2 : x \in X |\}$$

We regard D_p as a subdomain of PD_p under the natural inclusion $x \mapsto \{| x |\}$, and using the obvious extensions implicitly.

The following lemma is very useful in proving that functions are single-valued:

Lemma 2.3.1 (Algebraic criterion for a singleton)

If $X \subseteq D_p$ is non-empty, then the following three conditions are equivalent:

$X-X \sqsupseteq 0(X)$

$X-X \equiv 0(X)$

X is a singleton

Proof That the first two are equivalent is clear, given that $X-X \sqsubseteq 0(X)$.

If $X \equiv \{| x |\}$, then $X-X \equiv \{| x-x |\} = 0(\{| x |\}) \equiv 0(X)$.

Now suppose that $X-X \sqsupseteq 0(X)$, and choose $x_0 \in X$ of minimal extent. If $y \in X$, then there exists z in X such that $x_0-y \sqsupseteq 0|Ez$, so $Ez \leq Ex_0$. Therefore, $Ez = Ex_0$, $x_0-y = 0|Ex_0$ and so $x_0 \sqsubseteq y$. Thus $X \equiv \{| x_0 |\}$.]

2.4 Polynomial operators on PD_p

If f(t) is a polynomial with coefficients in $\mathbb{Z}$, then f induces a continuous map from D_p to D_p which extends to a left exact continuous map (which we also write as f) from PD_p to PD_p:

$$X \mapsto f(X) = \{| f(x): x \in X |\}.$$

Note that f(X) is *not* necessarily obtained by substituting X in the polynomial expression and using the operations defined above.

Theorem 2.4.1 (versions of **Taylor's Theorem**) Let f(t) be a polynomial of degree n and $X \subseteq D_p$; we write $f^{(i)}(t)$ for the ith formal derivative of f. As is well-known, the coefficients of $f^{(i)}$ are all divisible by i! and so $[f^{(i)}/i!](t)$ is a polynomial over the integers.

(i) The *Weak Taylor's Theorem:*

$$f(X) - f(X) \subseteq \Sigma_{i=1}^{n} \ (X-X)^i \, [f^{(i)}/i!](X)$$

(ii) The *Linearized Taylor's Theorem:* Suppose

$X \equiv X + X - X$ (roughly, X–X is a subgroup and X is a coset of it), and

$(X-X)^2$ is null.

Then

$$f(X) - f(X) \equiv (X-X)f'(X)$$

Proof Taylor's Theorem tells us that

$$f(x+h) = f(x) + \Sigma_{i=1}{}^{n}\ h^i\ [f^{(i)}/i!](x)$$

Although this is normally proved by analytic methods, for polynomials over the integers it is a combinatorial result valid in all commutative rings. For monomial f ($f(t) = t^n$) it follows from the binomial theorem (Birkhoff and MacLane [77], p. 14) and the fact that $[f^{(i)}/i!](t) = ({}^n{}_i)t^{n-i}$, while for general polynomials it follows using the additivity of differentiation.

(i) A typical element of f(X) – f(X) is f(y) – f(x), where $x, y \in X$. Putting $h=y-x \in X-X$ and working at the level of min (Ex, Ey), we get

$$f(y) - f(x) = f(x+h) - f(x) = \Sigma_{i=1}{}^{n}\ h^i\ [f^{(i)}/i!](x) \in \Sigma_{i=1}{}^{n}\ (X-X)^i\ [f^{(i)}/i!](X)$$

(ii) From (i) we get that $f(X) - f(X) \sqsupseteq (X-X)\ f'(X)$.

Suppose $x, y, z \in X$, so $x-y+z \sqsupseteq u$ for some $u \in X$. Then

$$(x-y)f'(z) = ((x-y+z) - z)f'(z) = f(x-y+z) - f(z) \sqsupseteq f(u) - f(z) \in f(X) - f(X)$$]

2.5 Non-deterministic refinement

The essential step in our non-deterministic methods is the use of the functions **refine** and **nrefine**. These could be defined as in Propositions 2.5.2 and 2.5.4 and proved to be continuous, but there is some interest in showing that they can be defined in terms of the constructions of Abramsky [87] and we have done this. We also prove some lemmas about them.

Definition 2.5.1 Using temporary auxiliary functions tl_a and f_a for each digit a ($0 \leq a < p$), define

tl_a [x, d] = **if** d = a **then refine** x **else stop fi** $(0 \leq d < p)$

f_a x = $[tl_a\ x, a]$

refine x = $\sqcap_{a=0}{}^{p-1}\ f_a\ x$

Note that the meet $\sqcap_{a=0}{}^{p-1}$ does not imply any kind of computed iteration; it is a metalinguistic construction. In an implementation p is fixed. If it is – say – 7, then **refine** x is defined as

$$f_0\ x \sqcap f_1\ x \sqcap f_2\ x \sqcap f_3\ x \sqcap f_4\ x \sqcap f_5\ x \sqcap f_6\ x$$

It is worth examining the implicit identifications here. All three functions are defined as functions from D_p to PD_p, so the expression $[tl_a\ x, a]$ uses the extension to PD_p of $\lambda x.\ [x, a]$.

Proposition 2.5.2 $\textbf{refine}\ x \equiv \{|\ y \in D_p: y \sqsupseteq x, Ey = Ex+1\ |\}$

(As an upper closed set, it is $\{|\ y \in D_p: y \sqsupseteq x, Ey \geq Ex+1|.)$

Proof For finite x, we use induction on Ex. For infinite x, the result follows by continuity.

First, $tl_a \perp = \perp$, $f_a \perp = \{|\ [\perp, a]\ |\}$ and so

$$\textbf{refine}\ \perp = \{|\ [\perp, a]: a \in A\ |\} = \{|\ y \in D_p: Ey = 1\ |\}$$

Next,

$$\textbf{refine}\ [x, d] = \sqcap_{a=0}{}^{p-1}\ f_a\ [x, d] = \sqcap_{a=0}{}^{p-1}\ [tl_a\ [x, d], a]$$

$$= \sqcap_{a \neq d}\ [\textbf{stop}, a] \sqcap [\textbf{refine}\ x, d] = [\textbf{refine}\ x, d]$$

$$\equiv [\{|\ y \in D_p: y \sqsupseteq x, Ey = Ex+1\ |\}, d] \quad \text{by induction}$$

$$= \{|\ [y, d]: y \sqsupseteq x, Ey = Ex+1\ |\}$$

$$= \{|\ y': y' \sqsupseteq [x, d], Ey = E[x, d]+1\ |\} \qquad]\!]$$

Then **refine** can be thought of as a function that takes x (stored as n = Ex digits in a computer register) and returns a register with n+1 digits: n of them store x and there is an extra, uninitialized digit. Another interpretation, to explain the definition, sees the elements of D_p as being partially read streams of digits. Then **refine** is essentially a copy function except that at each stage, while waiting for the next digit, it puts forward in parallel p suggestions as to what it might be. When the digit arrives, the wrong suggestions are cancelled by refining them to **stop** and the right one is copied.

Definition 2.5.3 Using a temporary function f, define

$$f\ (x, m+1) = x \sqcup f\ (\textbf{refine}\ x, n) \qquad (x \in D_p \text{ and } m \in N = \omega+1)$$

$$\textbf{nrefine}\ (x, m) = f\ (x, m+1)$$

Because pattern matching definitions are strict, this implies that $f\ (x, 0) = \perp$.

Note the use of the join $\sqcup$ for PD_p in the definition; if this is unavoidable then it indicates that there is an essential element of parallelism in **nrefine**. Said another way, if $\sqcup$ can be defined in terms of **nrefine**, then there is an implicit parallelism in **nrefine**.

Proposition 2.5.4 If m is finite, then $\textbf{nrefine}\ (x, m) = \textbf{refine}^m\ x$

Proof By induction on m. If m = 0 then

$$\textbf{nrefine}\ (x, m) = f\ (x, 1) = x \sqcup f\ (\textbf{refine}\ x, 0) = x \sqcup \perp = x$$

Otherwise, m = m'+1 for some m' and

$$\textbf{nrefine}\ (x, m) = f\ (x, m'+2) = x \sqcup f\ (\textbf{refine}\ x, m'+1)$$

$$= x \sqcup \textbf{nrefine}\ (\textbf{refine}\ x, m')$$

$$= x \sqcup \textbf{refine}^{m'}\ \textbf{refine}\ x \qquad \text{(by induction)}$$

$= x \sqcup \textbf{refine}^m\ x$

$= \textbf{refine}^m\ x \qquad (\text{using } x = \{|\, x\, |\} \sqsubseteq \textbf{refine}\ x)$]

Lemma 2.5.5 (i) $\textbf{refine}\ (x+y) = \textbf{refine}\ x + \textbf{refine}\ y$

(ii) $\textbf{refine}\ (-x) = -\ \textbf{refine}\ x$

(iii) $\textbf{refine}\ (xy) \sqsubseteq \textbf{refine}\ x \cdot \textbf{refine}\ y$

The same three properties hold if we replace **refine** by **nrefine**.

Proof By monotonicity, the sum, negation or product of 1-step refinements is a 1-step refinement of the sum, negation or product. This gives the inequality $\sqsubseteq$ in each case. Let us use the same temporary functions tl_a and f_a as in the definition of **refine**.

(i) $\textbf{refine}\ \bot + \textbf{refine}\ \bot = \sqcap_a [\bot, a] + \sqcap_b [\bot, b] = \sqcap_{a,b} [\bot, a+b] = \sqcap_c [\bot, c] = \textbf{refine}\ \bot$

$\textbf{refine}\ x + \textbf{refine}\ \bot = \textbf{refine}\ \bot + \textbf{refine}\ \bot = \textbf{refine}\ \bot = \textbf{refine}\ (x+\bot)$

$\textbf{refine}\ [x, d] + \textbf{refine}\ [y, e] = [\textbf{refine}\ x, d] + [\textbf{refine}\ y, e]$

$= [\textbf{refine}\ x + \textbf{refine}\ y, d+e]$

$= [\textbf{refine}\ (x+y), d+e]$ by structural induction

$= \textbf{refine}\ [x+y, d+e]$

$= \textbf{refine}\ ([x, d] + [y, e])$ if $d+e < p$

or $= [\textbf{refine}\ (x+y) + 1, d+e-p]$ if $d+e \geq p$

$= [\textbf{refine}\ (x+y+1), d+e-p]$ using induction

$= \textbf{refine}\ [x+y+1, d+e-p] = \textbf{refine}\ ([x, d] + [y, e])$

(ii) is similar. The cases for **nrefine** follow from lemma 2.5.4 and continuity.]

Lemma 2.5.6 $\textbf{nrefine}\ (0|m, n) \cdot \textbf{nrefine}\ (0|n, m) = 0|(m+n)$

(The two factors have respectively m and n factors of p, so their product has m+n factors of p.)

Proof If $m = 0$ then $\textbf{nrefine}\ (\bot, n) \cdot 0|n = 0|n$, and the case $n = 0$ is similar.

For m+1, n+1, we have

$\textbf{nrefine}\ (0|(m+1), n+1) \cdot \textbf{nrefine}\ (0|(n+1), m+1)$

$= [\textbf{refine}^{n+1}\ 0|m, 0] \cdot [\textbf{refine}^{m+1}\ 0|n, 0]$

$= [p \cdot \textbf{refine}^{n+1}\ 0|m \cdot \textbf{refine}^{m+1}\ 0|n, 0]$

$\sqsupseteq [p \cdot \textbf{refine}\ (\textbf{nrefine}\ (0|m, n) \cdot \textbf{nrefine}\ (0|n, m)), 0]$

$= [p \cdot \textbf{refine}\ 0|(m+n), 0]$

$= [[0|(m+n), 0], 0] = 0|(m+n+2)$]

Lemma 2.5.7 If $n \leq Ex$ and $X = \textbf{nrefine}\ (x, n)$, then $(X - X)^2$ is null.

Proof We can take $n = Ex$. The non-trivial direction is

$(X-X)^2 \sqsupseteq (X-X)\cdot(X-X)$

$= \textbf{nrefine}\ (0|n, n) \cdot \textbf{nrefine}\ (0|n, n) = 0|2n$ by Lemmas 2.5.5, 2.5.6]

2.6 Functional refinement

The ring theoretic functions all give a result with extent no greater than that of the arguments, and this will hamper us when we try to define new functions by recursion. However, certain functions are more approximate than they need be. A given extent of argument actually defines more digits of result than the function gives. The refined function, giving as many digits as are defined, is still continuous.

There is a general theory of these refined functions; we show how to use **nrefine** to get good refinements of polynomial functions explicitly.

Proposition 2.6.1 (This is well-known.)

Let $f: D_p \to D_p$ be continuous, and suppose also that it maps maximal elements ($\mathbb{Z}_p$) to maximal elements. Then f has a greatest continuous refinement $f^{\sim}$ defined by

$$f^{\sim}(x) = \sqcap \{f(z) : x \sqsubseteq z, Ez = \infty\}$$

Proof Note that the meet in the definition of $f^{\sim}$ is non-empty and so it exists by the bounded completeness of D_p.

Certainly $f(x) \sqsubseteq f^{\sim}(x)$, and if g is a continuous refinement of f then for every maximal refinement z ($Ez = \infty$) of x we have $g(x) \sqsubseteq g(z) \sqsupseteq f(z)$. But by hypothesis, f(z) is maximal, so $f(z) = g(z)$. Thus $g(x) \sqsubseteq f^{\sim}(x)$.

It remains to show that $f^{\sim}$ is continuous. If $x \sqsubseteq y$ then x has more maximal refinements than y and so $f^{\sim}(x) \sqsubseteq f^{\sim}(y)$. If $x = \sqcup x_\lambda$ is a directed join, then we only need consider the case where x is infinite. Then $f^{\sim}(x) = f(x)$ by maximality of f(x), and $\sqcup f^{\sim}(x_\lambda) \sqsupseteq \sqcup f(x_\lambda) = f(x) = f^{\sim}(x)$.]

Definition We call $f^{\sim}$ the *greatest refinement* of f.

We now concentrate on the case where f is a polynomial function (with integer coefficients) and relate the greatest refinement to the formal derivative.

Rather intuitively, we can compare this with functions on the real line. Think of a real number x as being an approximation to all the numbers $x+\varepsilon$, where ε is infinitesimal. We are interested in the case where refinements $x \to x+\varepsilon$ all have the same effect on f(x), i.e. $f(x+\varepsilon) = f(x) + \varepsilon f'(x)$ is independent of ε, i.e. $f'(x) = 0$. Our proposition says that (within limits) the extent to which f'(x) approximates 0 is the extent to which f(x) is independent of refinements to x, or the extent to which $f^{\sim}(x)$ refines f(x).

Theorem 2.6.2 Let f(t) be a polynomial over the integers, giving a polynomial function f on D_p, and let f'(t) be its formal derivative. Define g(x) by

$$g(x) = f\,(\textbf{nrefine}\,(x, \textbf{nullity}\ f'(x)))$$

Then g is single-valued and $f \sqsubseteq g \sqsubseteq f^{\sim}$. If f'(x) is non-null, then $g(x) = f^{\sim}(x)$.

Proof Clearly $f \sqsubseteq g$. If g is single-valued, then by Proposition 2.6.1, $g \sqsubseteq f^{\sim}$.

Given x, let X = **nrefine** (x, **nullity** f'(x)). We wish to show that g(x) is a singleton, i.e. (by Proposition 2.3.1) that f(X) – f(X) is null. In view of Lemmas 2.5.7 and 2.5.5 (iii) we can use the linearized Taylor's Theorem (2.4.1 (ii)), $f(X) - f(X) \equiv (X-X)f'(X)$.

Let n = Ex, y = f'(x), m = **nullity** y. By Lemma 2.5.5, X–X = **nrefine** (0|n, m). Also, by definition of nullity, $y \in$ **nrefine** (0|m, n–m), so $f'(X) \supseteq$ **nrefine** (0|m, n). Now Lemma 2.5.6 tells us that (X–X)f'(X) is null.

Now suppose that f'(x) is non-null, i.e. that $m < n$; let X' = **nrefine** (x, m+1). We still have

f(X') – f(X') = (X'–X')f'(X') = **nrefine** (0|n, m+1)f'(X')

and f(X') $\subseteq$ **nrefine** (y, m+1).

We show that for any $z \in$ **nrefine** (y, m+1) there is a $u \in X'-X'$ such that zu is non-null. If z' is an integer representative of z, then $z' = kp^m$ for some k not divisible by p. Put $u=p^n|(m+n+1)$. Then zu, equal to $kp^{m+n}|(m+n+1)$, is non-null.]

Note the content of this: $f^{\sim}$ is defined in terms of the maximal, infinite elements of D_p; our theorem allows us to calculate it using purely finite information.

3 Some Examples of Functions

We give two detailed examples of functions defined using the non-deterministic methods, and proofs of some properties they satisfy. We hope to use similar methods to solve polynomial equations by successive approximations to the roots, giving algorithmic point to Cohn's statement that the traditional method is analogous to the Newton-Fourier rule. We hope also to be able to treat power series and other aspects of p-adic analysis.

3.1 The refined exponential $x\uparrow p$

The polynomial x^p has derivative $p\cdot x^{p-1}$ which has nullity at least 1 for all x other than $\perp$. Therefore, using Theorem 2.6.2, we can define a single-valued function $x\uparrow p$ refining x^p by

[x, d]$\uparrow$p = (**refine** [x, d])^p

Note that $\perp\uparrow p = \perp$. This is unavoidable, since the fact that $0^p = 0$ and $1^p = 1$ shows that (**refine** $\perp$)^p is not a singleton.

Proposition 3.1.1 $(xy)\uparrow p = x\uparrow p\cdot y\uparrow p$

Proof This is obvious if x or y is $\perp$. Otherwise,

(xy)$\uparrow$p = (**refine** (xy))^p $\supseteq$ (**refine** x)$^p\cdot$ (**refine** y)^p = $x\uparrow p\cdot y\uparrow p$

and we can now use single-valuedness and compare extents.]

3.2 The exponential $(1+p)^x$

Here is a function for which we need the refined p^{th} power. On p-adics $x = [a, d] = pa+d$ we want

$$(1+p)^x = (1+p)^{pa+d} = ((1+p)^a)^p \cdot (1+p)^d.$$

This shows us the necessary step in the recursion, but to maintain accuracy for finite x we need

$$(1+p)^{[a, d]} = (1+p)^a \uparrow p \cdot (1+p)^d.$$

Define, for $g: D_p \to D_p$,

$$F_1(g)([a, d]) = g(a)\uparrow p \cdot (1+p)^d \qquad (0 \le d < p)$$

$$F(g)(x) = [\text{tl } F_1(g)(x), 1]$$

Bearing in mind that – syntactically – $g(a)\uparrow p$ is an element of PD_p, $F_1(g)$ is an element of $[D_p \to PD_p]$. To use fixpoint theory, we extend the definition to cover $g: D_p \to PD_p$. This works automatically, following the remarks in 2.3. Now, F_1 and F are in $[[D_p \to PD_p] \to [D_p \to PD_p]]$.

Continuity is guaranteed syntactically.

Let $\lambda x.(1+p)^x$ be the least fixpoint YF of F.

Proposition 3.2.1 (i) YF is single-valued

(ii) $(1+p)^{x+y} = (1+p)^x \cdot (1+p)^y$

(iii) $E\,(1+p)^x = Ex + 1$

(iv) If d is a non-negative integer, then

$(1+p)^d$ (as defined here) = $(1+p)^d$ as defined by repeated multiplication.

Proof (i and ii) Let G be the set of $g \in [D_p \to PD_p]$ satisfying

1. g is single-valued
2. $g(\perp) = [\perp, 1]$
3. $g(x+1) = g(x)\cdot(1+p)$ for all $x \in D_p$
4. $g(x+y) = g(x)\cdot g(y)$ for all $x, y \in D_p$

Clearly G is closed under directed joins, and also $F(\lambda x.\perp) = \lambda x.[\perp, 1] \in G$. We show that G is closed under F, from which it will follow that the fixpoint of F is in G.

Note the role of the four conditions for G. Clearly the first and last are natural ones to choose, because they are what we want to prove; but the other two also are not as ad hoc as they might look. The definition of F contains two fiddles, the condition $0 \le d < p$ and the [tl ..., 1] construction, which are only there really to ensure that F(g) is defined for all g. We'd rather restrict ourselves to a domain of g's on which we can define

5. $F(g)(\perp) = [\perp, 1]$

6. $F(g)([a, d]) = g(a)\uparrow p\cdot(1+p)^d$ for *all* integers d

These will make it easier to relate F to the algebraic structure. Condition 5 comes straight from the definition; we show that conditions 2 and 3 on G entail condition 6.

Suppose, then, that $g \in G$. If $x = [a, d] = [a', d'], 0 \le d' < p$, then

$F_1(g)(x) = g(a')\uparrow p\cdot(1+p)^{d'}$ by definition

$\sqsupseteq g(\perp)\uparrow p\cdot(1+p)^{d'} = [\perp, 1]\uparrow p\cdot(1+p)^{d'}$ by condition 2 on g

$\sqsupseteq [\perp, 1]^{p\cdot(1+p)^{d'}} = [\perp, 1]$

so 6: $F(g)(x) = F_1(g)(x)$ as a consequence of the above

$= g(a')\uparrow p\cdot(1+p)^{d'}$ by definition

$= g(a')\uparrow p\cdot(1+p)^{p((d'-d)/p)}\cdot(1+p)^d$

$= g(a')\uparrow p\cdot((1+p)^{(d'-d)/p})\uparrow p\cdot(1+p)^d$

$= (g(a')\cdot(1+p)^{(d'-d)/p})\uparrow p\cdot(1+p)^d$ by Proposition 3.1.1

$= g(a'+(d'-d)/p)\uparrow p\cdot(1+p)^d$ by condition 3 on g

$= g(a)\uparrow p\cdot(1+p)^d$

Now we show if g is in G, so is F(g).

1. F(g) is single-valued because all the functions used in its definition are.

2. $F(g)(\perp) = [\mathrm{tl}\ F_1(g)(\perp), 1] = [\mathrm{tl}\ \perp, 1] = [\perp, 1]$

3. $F(g)(\perp+1) = F(g)(\perp) = [\perp, 1] = [\perp, 1]\cdot(1+p) = F(g)(\perp)\cdot(1+p)$

$F(g)([a, d]+1) = F(g)([a, d+1])$

$= g(a)\uparrow p\cdot(1+p)^{d+1}$ using condition 6 just in case $d = p-1$

$= g(a)\uparrow p\cdot(1+p)^d\cdot(1+p) = F(g)([a, d])\cdot(1+p)$

4. $F(g)([a, d]+[a', d']) = F(g)([a+a', d+d'])$

$= g(a+a')\uparrow p\cdot(1+p)^{d+d'}$ using condition 6 in case $d+d' \ge p$

$= (g(a)\cdot g(a'))\uparrow p\cdot(1+p)^d\cdot(1+p)^{d'}$

$= g(a)\uparrow p\cdot(1+p)^d\cdot g(a')\uparrow p\cdot(1+p)^{d'}$

$= F(g)([a, d])\cdot F(g)([a', d'])$

and clearly $F(g)(x+x') = F(g)(x)\cdot F(g)(x')$ if either x or x' is $\perp$.

We have now shown enough to demonstrate that Y F is in G, and in particular that it satisfies the equation in the statement.

(iii) We prove this by induction on n = Ex for finite x. It holds for $x = \perp$, and if it holds for all x of extent n then for a typical element [x, d] of extent n+1 we have

$$E\,(1+p)^{[x,\,d]} = E\,(1+p)^{x}\!\uparrow\! p\cdot(1+p)^{d} = E\,(1+p)^{x}\!\uparrow\! p = E\,(1+p)^{x} + 1 = n+2$$

Now by continuity, the result also holds for non-finite x, i.e. $(1+p)^x$ maps maximal elements to maximal elements.

(iv) Let us temporarily write f = Y F, and reserve exponential notation $(1+p)^d$ for the repeated multiplication when d is a non-negative integer.

We proved in (i) that $f(x+1) = f(x)\cdot(1+p)$, so $f(d) = f(0+d) = f(0)\cdot(1+p)^d$. It therefore suffices to show that $f(0) = 1$.

We show first that if $x \sqsubseteq 0$ then $f(x) \sqsubseteq 1$, by induction on n = Ex for finite x. It is certainly true for $x = \perp$; suppose then that it is true for $x = 0|n$. Then

$$f(0|(n+1)) = f([0|n,\,0]) = f(0|n)\!\uparrow\! p \sqsubseteq 1\!\uparrow\! p = 1$$

By taking limits, $f(0) \sqsubseteq 1$; and by comparing extents, $f(0) = 1$.]

4 Semantics and Implementation notes

4.1 Semantics

The distinction between D_p and its embedded image in PD_p has been deliberately blurred. This enables us to incorporate **refine** and **nrefine** without a proliferation of the constructs {| - |} (singleton embedding), P (Smyth power domain functor including the morphism part) and ⊗ (from PD×PD to P(D×D)). However, it makes the semantics slightly non-standard.

All functions are to be considered non-deterministic: they are actually functions from a domain to a power domain. For instance,

$+: D_p \times D_p \to PD_p$ (so x+y really denotes {| x+y |})

$0: 1 \to PD_p$

refine: $D_p \to PD_p$

(Of course, the algebraic operations are actually single-valued, and this fact will be used heavily in proofs.) The functions are thus morphisms in the Kleisli category of the monad P_S on the category of Scott domains (see MacLane [71]) and are composed accordingly. For instance, if f, g: $D_p \to PD_p$ then f (g (x)) denotes $\bigcup \{f(y) : y \in g(x)\}$.

4.2 Implementation

Potential implementers should not be put off by the appearance of the Smyth power domains. These are really a device to give a correct semantics for the **refine** function, and certainly do not call for the implementation of set-valued or probabilistically indeterminate functions.

The most natural interpretation of **refine** seems to be that it is a memory allocator. If x is stored as n digits in some register in the computer, then **refine** x uses the old n digits connected to a new one, got either from the computer's spare memory, or, if there's none left, from the chipmongers. On delivery it is uninitialized. The implementer may of course initialize it at this point, but he would be ill-advised to since this has been found to lead to bad programming habits.

5 Bibliography

Abramsky, Samson: Domain Theory in Logical Form, in: Proceedings of LICS '87.

Birkhoff, Gareth, and MacLane, Saunders: *A Survey of Modern Algebra* (4th edition), Macmillan, New York, 1977, 500 pp.

Cohn, P M: *Algebra*, volume 2, Wiley, London, New York, Sydney, Toronto, 1977, 483 pp.

Fourman, M P, and Scott, D S: Sheaves and Logic, in: *Applications of Sheaves,* Lecture Notes in Mathematics **753**, Springer-Verlag, Berlin, Heidelberg, New York, 1979, p. 302-401.

Gregory, R T, and Krishnamurthy, E V: *Methods and Applications of Error-free Computing*, Springer-Verlag, Berlin, Heidelberg, New York, 1984.

Hennessy, M, and Plotkin, G: Full Abstraction for a Simple Parallel Programming Language, in: J. Becvar, ed., *Mathematical Foundations of Computer Science. Proceedings 1979,* Lecture Notes in Computer Science **74**, Springer-Verlag, Berlin, 1979.

MacLane, Saunders: *Categories for the Working Mathematician,* Graduate Texts in Mathematics **5**, Springer-Verlag, New York, Heidelberg, Berlin, 1971, 262 pp.

Smyth, Michael: *Power Domains*, Journal of Computer and System Sciences vol. 16, 1978.

Stoltenberg-Hansen, V, and Tucker, J V: *Complete Local Rings as Domains*, Report 1.85, Centre for Theoretical Computer Science, Leeds University, England, 1985, 31 pp.

Vickers, Steven: A Fixpoint Construction of the p-adic Domain, to appear in: *Summer Conference on Category Theory and Computer Science. Proceedings 1987,* Springer-Verlag, Berlin, Heidelberg, New York, 1988.

The Shuffle Bialgebra

David B. Benson

CS–87–168

May 1987

Abstract

The *shuffle* multiplication and the *cut* comultiplication, a generalized car–cdr pairing, form a bialgebra. The *concatenation* multiplication, sometimes called tensor product, and the *spray* comultiplication form another bialgebra. The concatenation–spray bialgebras are the free bialgebras in the category of precise, graded bialgebras over a semiadditive symmetric monoidal category. The shuffle–cut bialgebras are the cofree bialgebras in the same category of bialgebras. These categories include many of the settings of interest in the theories of formal languages and the theories of distributed, concurrent and parallel computation. We analyze the *marked shuffle*, of interest in theories of distributed computing, in terms of its resolutions into the cofree shuffle–cut bialgebra.

1. Introduction

Shuffling, in the sense of preparing a deck of playing cards for a round of play, is a combinatorial process with uses in theory of bialgebras, [Sweedler]. Shuffling is important in our current understanding of many aspects of formal languages and in theories of distributed and concurrent processes. The shuffle operation is also called the *Hurwitz product* in [Salomaa & Soittola]. The dissertation [Gischer] attributes the shuffle operation on formal languages to S. Ginsburg's 1966 monograph. But [Ginsburg] attributes formal language shuffle to [Ginsburg & Spanier]. Without surprise then, [Spanier], shuffling in vector spaces is related to homology. Perhaps the earliest homological paper mentioning shuffle is [Eilenberg & MacLane]. Also see [MacLane 1963]. While I have not surveyed the older literature, surely there are more ancient mathematical papers on shuffling decks of cards.

V. Pratt has advocated modeling concurrency with partial orders, for which see [Pratt]. Also see Gischer's dissertation and [Gaifman & Pratt] for other works in this development. Here I follow the older tradition of representing a partial order by a distribution of the total orders respecting the given partial order. The term *distribution* has a particular technical meaning in this paper, related to one's knowledge of a particular situation. For the moment we may consider a distribution to be a set. For example, the following labeled Hasse diagram denotes a pomset (partially ordered multiset) in Pratt's terminology.

In this paper we simply consider the distribution consisting of the two words,

$$aabb$$
$$abab$$

which are all and only the total orders respecting this particular pomset.

Another motivating example is the data flow process server called a *merger*, which nondeterministically shuffles together *datons* from two streams of these datons. In the picture

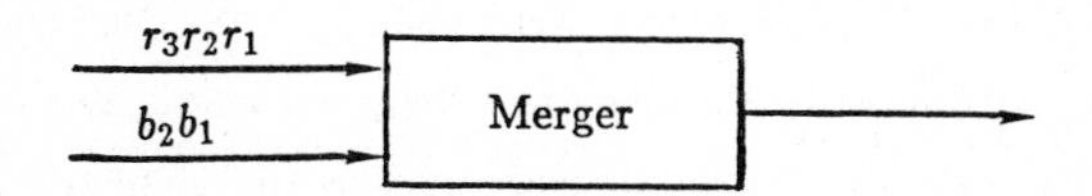

there are three datons on the red (r) input line and two datons on the blue (b) input line. The full distribution of nondeterministic output possibilities from the merger is

$$r_3r_2r_1b_2b_1$$
$$r_3r_2b_2r_1b_1$$
$$\vdots$$
$$r_3b_2r_2r_1b_1$$
$$\vdots$$
$$b_2b_1r_3r_2r_1.$$

That is, the full distribution is all possible shuffles of the red datons with the blue datons.

So long as the streams are finite, this process is isomorphic to shuffling cards. We must, of course, include the most imperfect shuffles in which all cards of one color either precede or else follow the cards of the other color. Further, we must consider strings of cards rather than decks of cards. A string is obtained by selection with replacement. Strictly speaking, selecting a card from a deck is selection without replacement. In this paper, "deck" is simply a synonym for "string."

I shall note here that so–called fairness considerations, for which see [Francez], have no direct role in this paper. An analysis of finite fairness is perhaps possible in this framework.

Whatever the particular application may be, *shuffle* is to be viewed as a binary operation in an appropriate algebra, where here I use the term "algebra" in the sense of universal algebra. At the least, the algebra needs to be an algebra of distributions. While we will become more abstract later on, one should have in mind the primary examples of distributional algebras as abelian monoids, also called commutative monoids, written additively as $(M, +, 0)$. A fundamental computer science example of an abelian monoid is obtained when the carrier is the power set of some given set, "+" is set union, and "0" is the empty set.

The tensor product of abelian monoids always exists by the universal considerations of bilinear maps given, for example, in [MacLane 1971]. We denote the tensor product of $(M, +, 0)$ and $(N, +, 0)$ by $M \otimes N$. When M and N are freely generated by X and Y respectively, one may view $M \otimes N$ as freely generated by the product set $X \times Y$. The collection of all finite subsets of a given set X will be called the *finitepowerset* of X. Thus when M and N are finitepowersets of X and Y respectively, $M \otimes N$ is the finitepowerset of $X \times Y$.

For additional mathematical background in the case of finitely generated abelian monoids, see [Redei]. In general, algebraic categories of abelian monoids are commutative algebraic categories. In such categories a tensor product, universal for bilinear maps, always exists and has the expected properties. See any of [Manes 1976, Eilenberg & Kelly 1966 pp. 548–550, Pfender 1974] for further discussion of this setting, attributed by Eilenberg and Kelly to Linton. Here we only mention that if the sum of every cardinality is defined and homomorphisms are completely additive in the sense of respecting sums of every cardinality, then when M and N are in fact the power sets of X and Y respectively, in this setting $M \otimes N$ is the power set of $X \times Y$.

Thus various notions of infinite nondeterminism may be obtained as well through considerations of infinite addition. One approach to infinite additions to serve as nondeterministic distributions may be found in [Manes 1986a, 1986b and *these proceedings*]. As of this writing there appear to be several different approaches to the questions of infinite nondeterminism. All of these must have the same finite structure, and it is therefore this case of finite nondeterminism which we consider in this paper.

The distributional algebras of abelian monoids with *finite* sum capture every notion of finite nondeterminism known to me. For example, the distribution of the pomset previously noted would be written

$$aabb + abab$$

as an element of the finitepowerset of strings over the alphabet $\{a, b\}$.

All this leads to defining shuffle as a particular *linear* function

$$\sigma\colon T \otimes T \to T$$

when the distributional algebra T is equipped with an appropriate notion of multiplication. Shuffling only makes sense in the context of an appropriately free multiplication and an appropriately cofree comultiplication which together describes total orderings in T.

The multiplication needed is exactly the tensor algebra over a distributional algebra M. In formal language theoretic terms, T_0 is the collection of distributions over the null word, $T_1 \cong M$ is the collection of distributions over words of length one, and in general $T_{n+1} = M \otimes T_n$ is the collection of distributions over words of length $n > 0$. The multiplication in the tensor algebra respects word length,

$$\kappa\colon T_m \otimes T_n \to T_{m+n},$$

with $(\alpha \otimes \beta)\kappa = \alpha\beta$, using the formal language notation for concatenation on the right side of the equation. Notice also that we write function application on the right. Now let $T = \bigsqcup_{n \geq 0} T_n$, where "$\sqcup$" denotes the categorical coproduct in some category of distributional algebras. The foregoing provides a multiplication

$$\kappa\colon T \otimes T \to T$$

which we might as well call *concatenation.*

There is one more operation we need to complete the picture, a comonoid operation, herein called a coalgebra operation. Using formal language terminology, let α be a distribution of strings. We may decompose α into all possible prefixes and corresponding suffixes to form the distribution of tensor pairs,

$$\alpha\Gamma = \sum_{\beta\gamma=\alpha} \beta \otimes \gamma$$

where $\beta\gamma = (\beta \otimes \gamma)\kappa$. Thus we have the linear map

$$\Gamma\colon T \to T \otimes T$$

which we shall call *cut*, although the cut in card play according to Hoyle requires reversing the order of the prefix and suffix. The cut operation used here may be viewed as a generalized car–cdr pair.

Shuffle and cut form a bialgebra with a certain universal embedding property. This property appears to be of value in understanding distributed and concurrent computation. The essence of the argument is obtained by thinking about card shuffling.

Imagine having two decks of cards of any finite size. The deck in the left hand is red backed; the deck in the right hand is blue backed. The goal is to shuffle the two decks together, preserving the original order of each deck separately, as is traditionally done by card players. If the decks are large, shuffling the decks may have to be done in parts. Set down both decks. Take any number of cards from the top of the red deck into the left hand. Similarly, take any number of cards from the blue deck into the right hand. Shuffle together these prefixes of the two decks. Lay the shuffled deck carefully to the side. Now shuffle together the remainder of the two original decks. Place the previously shuffled portion back on top of the shuffled remainders.

Could this shuffle be obtained by just shuffling the two decks, assuming one has big enough hands to hold the cards? Yes, the two means of shuffling—either in parts or all together—can always give identical results. For any shuffled result with a total of n cards, pick any prefix of p cards. This shuffled prefix of p cards can be obtained in exactly one way by, from the original two decks of red and blue cards, selecting a prefix of each deck of the appropriate number of cards and shuffling these in exactly one way to form the desired prefix of p cards.

Irrespective of the technicalities to follow, this is what one means by the shuffle and cut bialgebra. One may always use the intuition of decks of cards to aid in following the technical presentation.

This bialgebra is *graded* by length. Furthermore, cut is *precise*, which just means that single cards remain just exactly the same single cards and a deck of two or more cards has nontrivial cuts. The shuffle and cut bialgebra is couniversal in the sense that any other precise, graded bialgebra maps to the shuffle and cut bialgebra in just one way once the mapping into the individual cards is fixed.

The application in the ultimate section of the paper is *marked shuffle*. Marked shuffle arises in attempting to formalize the properties of three or more distributed processes intercommunicating. Rather than explain the origins of the problem, perhaps first explicated in [Riddle 1972, Riddle 1979], consider again the two decks of cards, one in each hand. This time, however, there are black marker cards at various positions throughout each of the red and blue decks. The marked shuffle rule requires shuffling the red and blue decks up to the first marker, keeping the markers, and repeating on each succeeding segment between markers. The resulting distribution of marked shuffles is a subdistribution of the distribution obtained if one forgot about the marker cards. In effect, but not in detail, this is the couniversal property of the shuffle and cut bialgebra.

As the patient reader will discover, this paper is written to preserve as much generality for as long as possible. This results in repeatedly returning to the various topics. Perhaps the paper should be subtitled "A Spiral Approach."

2. Shuffles as Permutations

A permutation on n letters is a bijection on a set of cardinality n. We take the sets of letters underlying permutations to be sets of positive integers equipped with the usual order on the integers. In the beginning, the permutations of n letters are taken as permuting the set $\{1, 2, \ldots, n\}$. A (k, ℓ)–*shuffle*, π, is a permutation on $k + \ell$ letters in which the first k letters and the last ℓ letters have stable orderings under the permutation:

$$i < j \le k \text{ implies } i\pi < j\pi,$$
$$k < i < j \text{ implies } i\pi < j\pi.$$

Fix natural numbers k, ℓ and a (k, ℓ)–shuffle π to consider the first p letters of the result of shuffling by π for some $p \le n = k + \ell$. For some k_1, ℓ_1 such that $k_1 + \ell_1 = p$, $\{1, 2, \ldots, k_1\} \cup \{k + 1, k +$

$2,\ldots,k+\ell_1\}$ is the set of letters with image $\{1,2,\ldots,p\}$ under π. The values k_1,ℓ_1 may be zero depending upon π and the choice of p. Similarly, for some k_2,ℓ_2 such that $k_2+\ell_2=q$ with $p+q=n$, we have $k_1+k_2=k$, $\ell_1+\ell_2=\ell$ and the last q letters of the result of shuffling by π is the image of the set $\{k_1+1,k_2+2,\ldots,k\}\cup\{k+\ell_1+1,k+\ell_1+2,\ldots,n\}$ where $\{k_1+1,k_1+2,\ldots,k\}$ has cardinality k_2 and $\{k+\ell_1+1,k+\ell_1+2,\ldots,n\}$ has cardinality ℓ_2. Thus for each $p\leq n$ there is a (k_1,ℓ_1)–shuffle π_1, uniquely depending upon p and π, such that the first p letters of the result of shuffling by π is identical to the result of shuffling $\{1,2,\ldots,k_1\}\cup\{k+1,k+2,\ldots,k+\ell_1\}$ by π_1. Similarly, for each q such that $p+q=n$ there is a (k_2,ℓ_2)–shuffle π_2 for which the last q letters of the result of shuffling by π is identical to the result of shuffling $\{k_1+1,k_1+2,\ldots,k\}\cup\{k+\ell_1+1,k+\ell_1+2,\ldots,n\}$ by π_2.

Let $\mathbf{k}_1$ denote the finite sequence $(1,2,\ldots,k_1)$, $\mathbf{k}_2$ denote the finite sequence $(k_1+1,k_1+2,\ldots,k)$, ℓ_1 denote the finite sequence $(k+1,k+2,\ldots,k+\ell_1)$, and ℓ_2 denote the finite sequence $(k+\ell_1+1,k+\ell_2+2,\ldots,n)$. Using a now obvious notation, we have

$$(1,2,\ldots,n)=\mathbf{k}_1\cdot\mathbf{k}_2\cdot\ell_1\cdot\ell_2.$$

There is a composition of elementary transpositions t: $\mathbf{k}_1\cdot\mathbf{k}_2\cdot\ell_1\cdot\ell_2\rightarrow\mathbf{k}_1\cdot\ell_1\cdot\mathbf{k}_2\cdot\ell_2$ which interchanges the positions of the subsequences $\mathbf{k}_2$ and ℓ_1. Write $\pi_1\dot{+}\pi_2$ for the permutation on n letters which acts as π_1 on the first p letters and acts as π_2 on the last q letters. We have

$$\pi=t(\pi_1\dot{+}\pi_2).$$

3. Algebras, Coalgebras and Bialgebras in a Symmetric Monoidal Category

Let $(\mathbf{M},\otimes,e,\alpha,\lambda,\rho)$ be a monoidal category in which e is the unit object with respect to the family of left and right unitary isomorphisms, λ and ρ respectively, and α is the family of associativity isomorphisms. For details see [MacLane 1971]. We further posit that $\mathbf{M}$ is symmetric via the family of *twist* isomorphisms, $\tau_{U,V}$: $U\otimes V\rightarrow V\otimes U$, for all pairs of objects U,V of $\mathbf{M}$. All these isomorphisms are *fixed*, once and for all, as the data of a symmetric monoidal category. While we shall endeavor to suppress this detail whenever possible, one understands that the choice of isomorphisms does make a difference. Thus this data always lies in the background of the presentation. For more about this technicality, see [MacLane 1971].

3.1 Definition. An $\mathbf{M}$–*algebra* is a monoid in $\mathbf{M}$, an object U of $\mathbf{M}$ equipped with two arrows μ: $U\otimes U\rightarrow U$ and η: $e\rightarrow U$ such that the diagrams

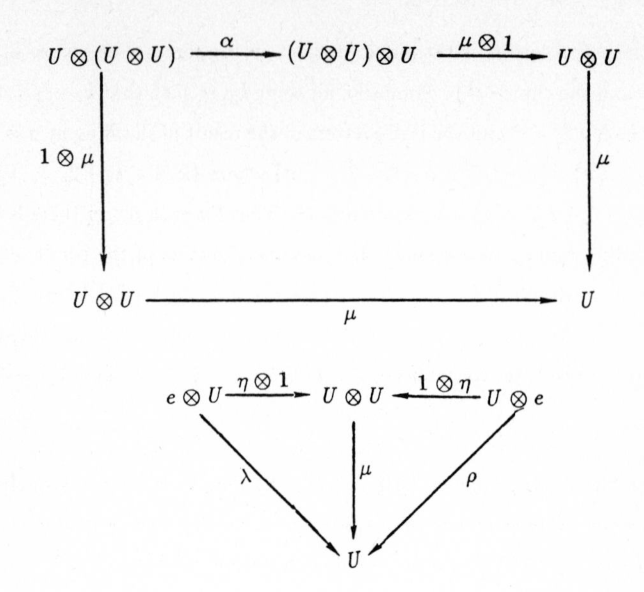

commute, where **1** denotes the appropriate identity arrow.

A morphism of M–algebras is an arrow $h\colon\ U \to V$ such that

$$\mu h = (h \otimes h)\mu'\colon\ U \otimes U \to V$$

and

$$\eta h = \eta'\colon\ e \to V.$$

3.2 Definition. An M–*coalgebra* is a comonoid in M, obtained whenever the previous diagrams commute with the direction of the arrows reversed. We typically write triples such as $(U, \Delta_U, \varepsilon_U)$ for an M–coalgebra.

A morphism of M–coalgebras is an arrow $h\colon\ U \to V$ such that the diagrams

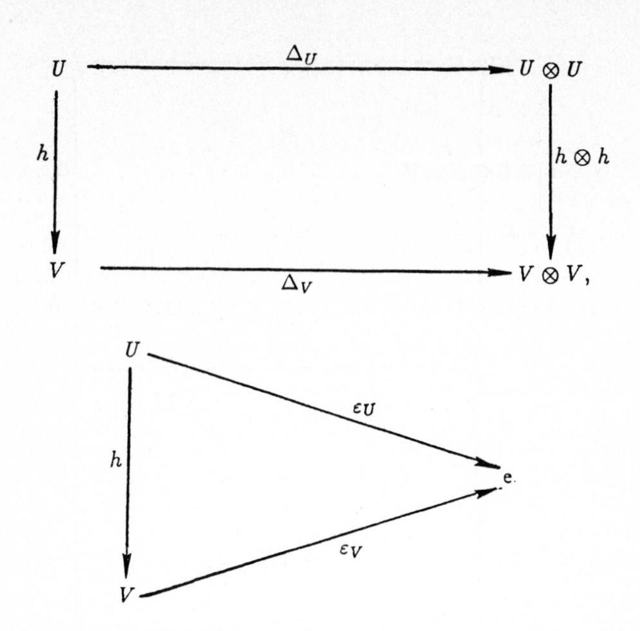

commute.

If U, V are M–coalgebras, the *tensor product* of U and V is the object $U \otimes V$ equipped with structure maps

$$\Delta_{U \otimes V} = (\Delta_U \otimes \Delta_V)(\mathbf{1} \otimes \tau \otimes \mathbf{1}),$$

$$\varepsilon_{U \otimes V} = \varepsilon_U \otimes \varepsilon_V,$$

making $U \otimes V$ an M–coalgebra.

3.3 Example. For each set X write $X \cdot 2$ for the collection of all finite subsets over X. Let $\bullet$ be any singleton set to write 2 for $\bullet \cdot 2 = \{0, 1\}$. These are the objects of the category **FSub**, which is symmetric monoidal with 2 as the unit object. Let $T = \Sigma^* \cdot 2$, with Σ^* the collection of all finite strings over Σ, to observe that κ: $T \otimes T \to T$ as concatenation and $0\eta = \phi, 1\eta$ the singleton set containing the null string, also denoted by 1, provides T with an **FSub**–algebra structure. Further, observe that Γ: $T \to T \otimes T$ and ε: $T \to 2$ given by $1\eta\varepsilon = 1$, $\gamma\varepsilon = 0$ for $\gamma \neq 1\eta$, provides T with an **FSub**–coalgebra structure.

3.4 Definition. An M–*bialgebra* in a symmetric monoidal category **M** is an object U which is both an M–algebra and an M–coalgebra with the property that the monoid structure arrows are comonoid morphisms of the M–coalgebra structure and, equivalently, the comonoid structure arrows are monoid morphism of the M–algebra structure. Explicitly, the following diagrams must all commute for the M–algebra (U, μ, η) and the M–coalgebra (U, Δ, ε) to form an M–bialgebra:

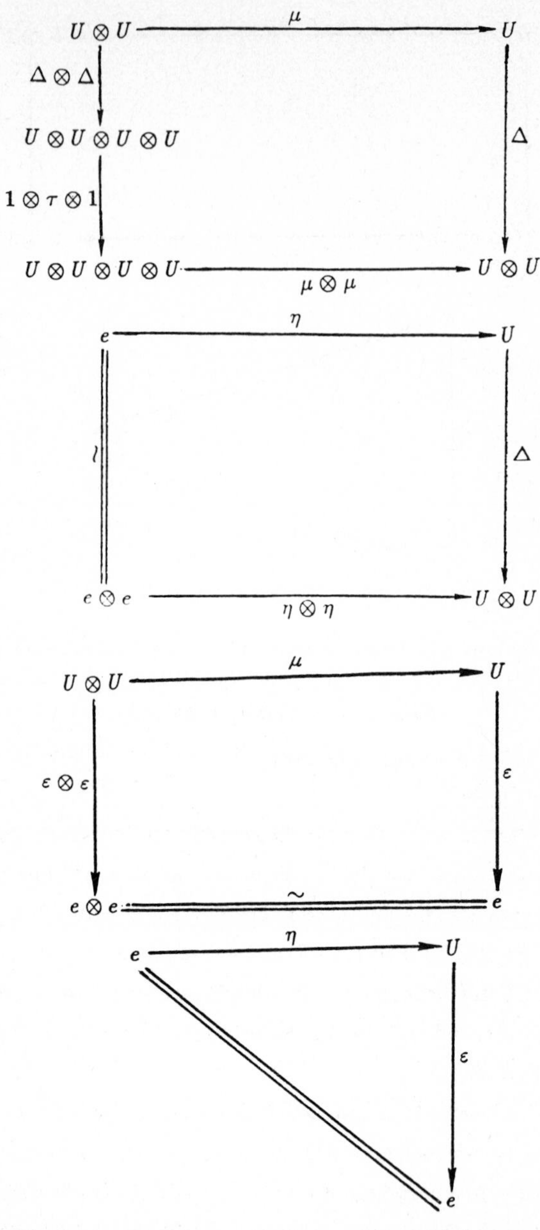

where we have denoted the isomorphism between $e \otimes e$ and e which is guaranteed by the left and right unitary isomorphisms as $e \otimes e \cong e$.

3.5 Definition. An M–*bialgebra morphism*

$$h\colon (B, \mu_B, \Delta_B, \eta_B, \varepsilon_B) \to (C, \mu_C, \Delta_C, \eta_C, \varepsilon_C)$$

is a morphism of M which is simultaneously an M–algebra morphism $h\colon (B, \mu_B, \eta_B) \to (C, \mu_C, \eta_C)$ and an M–coalgebra morphism $h\colon (B, \Delta_B, \varepsilon_B) \to (C, \Delta_C, \varepsilon_C)$. When and only when h is an

M–bialgebra morphism the cube diagram

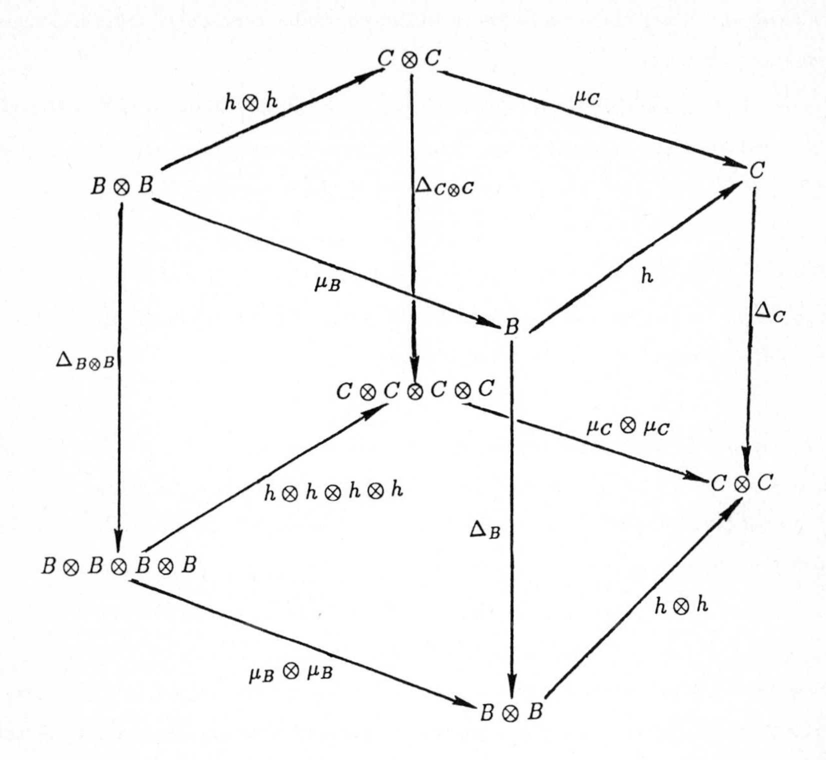

commutes. We refer to this cube diagram later in the paper.

3.6 Definition. Let N denote the natural numbers, $\mathrm{N} = \{0, 1, 2, \ldots\}$. An N–*graded object* over M is a function from N to the objects of M. A *morphism of* N*-graded objects* is a family of M–arrows indexed by N so as to preserve the N–grading. Explicitly, if $\{U_n | n \in \mathrm{N}\}$ and $\{V_n | n \in \mathrm{N}\}$ are N–graded objects over M, a family of M–arrows $\{f_n \colon U_n \to V_n | n \in \mathrm{N}\}$ is a morphism of these N–graded objects.

3.7 Definition. The M–(*tensor algebra*) over object V in M is the free monoid over V in M, when it exists. That is, the M–(tensor algebra) is an M–algebra freely generated by V. The recursion equations

$$T_0 = e,$$

$$T_{n+1} = V \otimes T_n,$$

define the N–*graded tensor* of V as an N–graded object over M. For each $m, n \in \mathrm{N}$ there is a unique associativity isomorphism given by the data of the monoidal categoy M from $T_m \otimes T_n$ to T_{m+n}, unique via the coherence conditions on the associativity isomorphisms. Denote these particular associativity isomorphisms by

$$\kappa_{m,n}\colon T_m \otimes T_n \to T_{m+n}.$$

3.8 Proposition. If the monoidal category M has countable coproducts, there is a tensor algebra over every object in M.

Proof: Let $T = \coprod_{n \geq 0} T_n$ be the countable coproduct of the N–graded tensor of V. Let $\kappa\colon T \otimes T \to T$ be the coproduct of the morphisms $\kappa_{m,n}$. The unit structure map $\eta\colon e \to T$ is simply the injection of $e = T_0$ into T. The proof that T is a free monoid in M is standard. □

When M has countable coproducts, the tensor algebra $T = \coprod_{n \geq 0} T_n$ has graded components T_n which inject into T but cannot, in general, be retrieved. Despite this handicap, we may define the graded components of $T \otimes T$ in the usual fashion as

$$(T \otimes T)_n = \coprod_{p+q=n} (T_p \otimes T_q)$$

which injects into $T \otimes T$ provided tensor ($\otimes$) distributes over coproduct ($\sqcup$). When M has countable coproducts and tensor distributes over these coproducts the M–(tensor algebra) may be viewed as the N–*graded tensor algebra* with components of the algebra structure map the coproducts of the $\kappa_{p,q}$ with $p + q = n$,

$$\kappa_n\colon (T \otimes T)_n \to T_n,$$

$$\kappa_n = \coprod_{p+q=n} \kappa_{p,q}.$$

Now the monoidal category M we are considering is also symmetric. Therefore, for each permutation π on n letters, as π is a composition of transpositions, there is a natural isomorphism on T_n formed solely of identity arrows and twist isomorphisms by composition and tensor,

$$\pi_n\colon T_n \to T_n.$$

For example, $(1 \otimes \tau \otimes 1)\colon T_4 \to T_4$ transposes the two center factors and is induced by the transposition (2 3).

3.9 Definition. A *shuffle* $\sigma_\pi\colon T_p \otimes T_q \to T_{p+q}$ induced by the (p,q)–shuffle π is the composition of natural isomorphisms

$$T_p \otimes T_q \xrightarrow{\kappa_{p,q}} T_{p+q} \xrightarrow{\pi} T_{p+q},$$

$\sigma_\pi = \kappa_{p,q}\pi$. Since we want to form the distributions over all the possible shuffles induced by all the (p,q)–shuffles, we must now specialize slightly by considering only the objects of M which are abelian monoids. Furthermore, we will need an addition of the morphisms between the abelian monoids, and this addition of morphisms will have to have the usual properties.

We leave this ramification to the next section. Instead we conclude this section by noting that we can obtain the sections necessary to construct the cut coalgebra from the family of natural associativities.

The N–graded tensor over object V in M is equipped with a subfamily of the associativity isomorphisms,

$$\Gamma_{m,n}: T_{m+n} \rightarrow T_m \otimes T_n.$$

3.10 Proposition. If the monoidal category **M** has countable products, there is a cut coalgebra over every object in **M**.

Proof: Let $C = \prod_{n\geq 0} T_n$ be the countable product of the **N**–graded tensor over V. The cut structure map Γ: $C \rightarrow C \otimes C$ is the product of the $\Gamma_{m,n}$. The counit structure map ε: $C \rightarrow e$ is the projection to T_0. □

4. Semiadditive Symmetric Monoidal Categories

From [Herrlich & Strecker], a category **A** is semiadditive if for each pair of objects U, V in **A**, the hom set $\mathbf{A}(U, V)$ is an abelian monoid with respect to $+$, morphism composition is left and right distributive over $+$, whenever morphism composition is defined on either side, and **A** has zero morphisms which are the monoid identities with respect to $+$. If **A** also has finite coproducts, then the semiadditive structure is unique and **A** has finite biproducts. This last means that each pair of **A**–objects U, V possess a single object $U \oplus V$, which is both the categorical coproduct and the categorical product of the pair U, V.

Now the category **AbMon** of abelian monoids and morphisms between them is semiadditive as well as symmetric monoidal. Further, tensor ($\otimes$) distributes over biproduct ($\oplus$). The only other examples of semiadditive symmetric monoidal categories I know about are well-known symmetric monoidal subcategories of **AbMon** such as R–**Mod**. In all these examples tensor distributes over biproduct, the category is closed monoidal and the tensor may be formed by universal bilinear considerations. Each of these properties naturally arises in considering nondeterminism. I believe there is a structure theorem to the effect that any category with all these properties is embeddable in **AbMon**, but I am unable to locate this result just now.

For the above reasons, the rest of the work is effectively within **AbMon**. Now **AbMon** is a lovely category which was used in [Lorentz & Benson] to explore certain issues of nondeterminism in programming language semantics. Each object V of **AbMon** has an internal addition which is compatible with the external addition provided by the coproducts. Thus we may freely write finite sums, as in

$$\sum_{i \in I} v_i$$

where while the various v_i may lie in different objects, the sum lies in the coproduct of all finitely many objects. The **AbMon**–(tensor algebra) may be viewed as the **N**–graded **AbMon**–(tensor algebra) and we shall do so as this avoids an annoying technicality which arises: The countable coproduct $T = \coprod_{n\geq 0} T_n$ is in general a different object than the countable product $C = \prod_{n\geq 0} T_n$. If one wishes, one may identify the N-graded **AbMon**-(tensor algebra) object with C. In any case, concatenation is the sequence of **AbMon**-arrows

$$\kappa_n\colon (C \otimes C)_n \to C_n$$

where $(C \otimes C)_n$ denotes $\bigoplus_{p+q=n} (T_p \otimes T_q)$ and $C_n = T_n$.

In general we have the following definitions.

4.1 Definition. Let **A** be a semiadditive symmetric monoidal category with biproducts and in which tensor distributes over biproduct. Such categories will be called *abmon* categories. Only the properties of abmon categories are required for the remainder of the paper. An N-*graded* **A**-*algebra* (A, μ, η) consists of an N-graded object $A = \{A_n | n \in \mathrm{N}\}$ in **A** and N-graded morphisms $\mu = \{\mu_n | n \in \mathrm{N}\}$, $\eta = \{\eta_n | n \in \mathrm{N}\}$ called the *structure maps* on the N-graded **A**-algebra. Each grade n component of η is an **A**-morphism from e to A_n such that η_n is the zero morphism from e to A_n for $n > 0$, $\eta_n = 0_{e,A_n}$. Thus we may view η as consisting solely of its interesting component $\eta_0\colon e \to A_0$. Each grade n component of μ is an **A**-morphism from $(A \otimes A)_n = \bigoplus_{p+q=n} A_p \otimes A_q$ to A_n, $\mu_n\colon (A \otimes A)_n \to A_n$. Clearly, μ is an N-graded morphism from the N-graded object $A \otimes A = \{(A \otimes A)_n | n \in \mathrm{N}\}$ to the N- graded object A. Dually, an N-*graded* **A**-*coalgebra* (C, Δ, ε) is an N-graded object C with N-graded structure maps Δ, ε such that

$$\varepsilon_n\colon C_n \to e = 0_{C_n, e} \text{ for } n > 0,$$

$$\Delta_n\colon C_n \to (C \otimes C)_n,\ n \geq 0.$$

An N-*graded* **A**-*bialgebra* $(B, \mu, \Delta, \eta, \varepsilon)$ consists of an N-graded **A**-algebra (B, μ, η) and an N-graded **A**-coalgebra (B, Δ, ε) such that μ is an N-graded coalgebra morphism, Δ is an N-graded algebra morphism, and similarly for η, ε.

We are now in the position to define the N-graded shuffle operation in abmon category **A**. First, the components

$$\sigma_{p,q}\colon T_p \otimes T_q \to T_{p+q}$$

are given as the sum

$$\sigma_{p,q} = \sum_{\pi \in S_{p,q}} \sigma_\pi = \kappa_{p,q} \sum_{\pi \in S_{p,q}} \pi$$

where $S_{p,q}$ is the set of all (p,q)-shuffles. Now define

$$\sigma_n\colon (C \otimes C)_n \to C_n$$

as the coproduct of all the appropriate $\sigma_{p,q}$,

$$\sigma_n = \coprod_{p+q=n} \sigma_{p,q}.$$

In the usual way we extend the domain of each $\sigma_{p,q}$ to all of $(C \otimes C)_n$ by writing

$$\widehat{\sigma}_{p,q} = \coprod_{i+j=n} \delta_{i,p;j,q} \cdot \sigma_{p,q}$$

where the δ morphisms are the Kronecker deltas: the zero morphism from $C_i \otimes C_j$ to C_n unless $i = p$ and $j = q$. In this latter case, $\delta_{p,p;q,q}$ is the identity morphism. Now we slightly abuse the notation to set $\sigma_{p,q} = \widehat{\sigma}_{p,q}$ and may now write

$$\sigma_n = \sum_{p+q=n} \sigma_{p,q}.$$

Note that $\sigma_n = \kappa_n \cdot \sum_{p+q=n} \sum_{\pi \in S_{p,q}} \sigma_\pi$ by the distributivity of composition over addition. It is both routine and tedious to verify that the family $\sigma = \{\sigma_n | \in \mathbf{N}\}$ is an $\mathbf{N}$–graded $\mathbf{A}$–algebra structure map with the unit of the shuffle algebra given by the identification of $C_0 = T_0$ with the unit e of monoidal category $\mathbf{A}$. In **AbMon** this is the free abelian monoid over one generator, $(\mathbf{N}, +, 0)$. (Of course in R–**Mod**, one uses R for $C_0 = T_0$, and similarly for other subcategories of **AbMon**.) Thus η_0 is the identity while η_n for $n > 0$ is the zero map.

Similarly, we may define the $\mathbf{N}$–graded cut coalgebra in abmon category $\mathbf{A}$ with structure map components

$$\Gamma_n \colon C_n \to (C \otimes C)_n$$

by

$$\Gamma_n = \sum_{p+q=n} \Gamma_{p,q}.$$

The counit component $\varepsilon_0 \colon C_0 \to \mathbf{N}$ is simply the identity, while the remaining counit components, ε_n for $n > 0$, are all zero maps. Let $\Gamma = \{\Gamma_n | n \in \mathbf{N}\}$ and $\varepsilon = \{\varepsilon_n | n \in \mathbf{N}\}$.

4.2 Theorem. With the notation from above, $(C, \sigma, \Gamma, \eta, \varepsilon)$ is an $\mathbf{N}$–graded bialgebra.

Proof: This is a combinatorial exercise to show that every shuffle permutation is accounted for once and only once in the diagram we may abbreviate as

$$\sigma\Gamma = (\Gamma \otimes \Gamma)(1 \otimes \tau \otimes 1)(\sigma \otimes \sigma).$$

Explicitly, we must show that the diagram

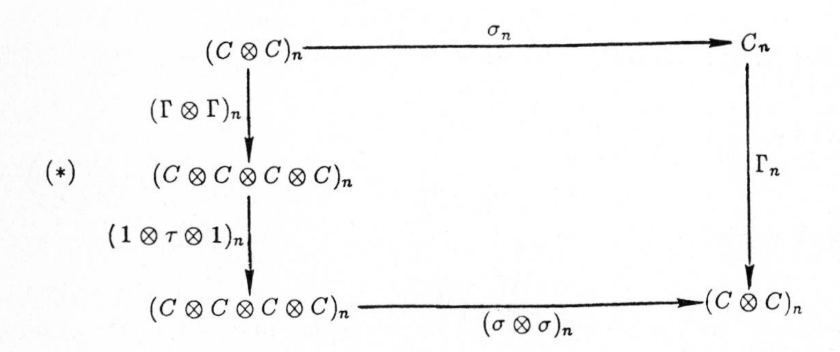

commutes for every $n \in \mathbf{N}$.

Fix some k, ℓ, p, q such that $k + \ell = p + q = n$ to consider a component of the previous diagram,

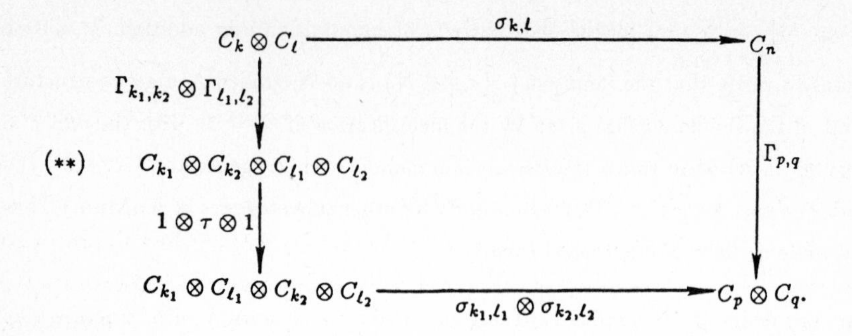

We have $\sigma_{k,\ell}\Gamma_{p,q} = \kappa_{k,\ell} \sum_{\pi \in S_{k,\ell}} \sigma_\pi \Gamma_{p,q} = \sum_{\pi \in S_{k,\ell}} \kappa_{k,\ell}\sigma_\pi\Gamma_{p,q}$. By the earlier section on shuffles as permutations, for each $\pi \in S_{k,\ell}$ and each pair (p,q) there exist unique choices of k_1, k_2, ℓ_1, ℓ_2 with $k_1 + k_2 = k$, $\ell_1 + \ell_2 = \ell$, $k_1 + \ell_1 = p$, $k_2 + \ell_2 = q$, a unique choice of (k_1, ℓ_1)–shuffle π_1 and a unique choice of (k_2, ℓ_2)–shuffle π_2 such that the above diagram commutes for σ_{π_1} in place of σ_{k_1,ℓ_1} and σ_{π_2} in place of σ_{k_2,ℓ_2}. Thus each σ_π determines the component of the down–and–right side of the diagrams with which it commutes.

In the other direction, fix k_1, k_2, ℓ_1, ℓ_2 with $k_1 + k_2 = k$, $\ell_1 + \ell_2 = \ell$. Then $k_1 + \ell_1 = p$ and $k_2 + \ell_2 = q$ is required. Now each pair of shuffles π_1,π_2 determining some $\kappa_{k_1,\ell_1}\sigma_{\pi_1} \otimes \kappa_{k_2,\ell_2}\sigma_{\pi_2}$ uniquely determines a (k,ℓ)–shuffle π and hence uniquely some $\kappa_{k,\ell}\sigma_\pi$.

Since the correspondence is a bijection in both directions, forming all the sums still leaves a bijective correspondence. Therefore, the diagram (**) commutes for each k, ℓ, p, q. A summation argument shows that the diagram (*) commutes for each n. □

The *spray* operation is in a sense dual to shuffle, being determined by the collection of inverses to the shuffle permutations.* Let $R_{p,q} = \{\pi^{-1} | \pi \in S_{p,q}\}$ and let the subcomponents of spray be determined on the N–graded tensor over C_1 by

$$\chi_\pi\colon C_{p+q} \to C_p \otimes C_q = \pi\Gamma_{p,q}\colon C_{p+q} \to C_{p+q} \to C_p \otimes C_q$$

for each $\pi \in R_{p,q}$. Then

$$\chi_{p,q} = \sum_{\pi \in R_{p,q}} \chi_\pi = \sum_{\pi \in R_{p,q}} \pi\Gamma_{p,q}.$$

Finally, the graded component of χ at $n \in \mathbf{N}$,

$$\chi_n\colon C_n \to (C \otimes C)_n$$

is given by the product map

$$\chi_n = \prod_{p+q=n} \chi_{p,q}$$

---

* The term *spray* is from [Pratt].

which by the usual techniques we view as the sum,

$$\chi_n = \sum_{p+q=n} \chi_{p,q}.$$

With $\chi = \{\chi_n | n \in \mathbf{N}\}$ the spray morphism, κ concatenation, and η,ε as before, we obtain another N-graded bialgebra.

4.3 Proposition. $(C, \kappa, \chi, \eta, \varepsilon)$ is an **N**-graded **A**-bialgebra.

Another piece of our apparatus is the observation that the components of concatenation, $\kappa_{p,q}$, and the components of cut, $\Gamma_{p,q}$, are purely structural on an **N**-graded tensor. Specifically,

4.4 Proposition. Let C be the **N**-graded tensor over C_1. For each $p, q \in \mathbf{N}$, $\kappa_{p,q}\Gamma_{p,q}$ is the identity morphism on $C_p \otimes C_q$ and $\Gamma_{p,q}\kappa_{p,q}$ is the identity morphism on C_{p+q}.

If V is an object of **A**, let $V \approx T_1$ and let (T, κ, η) be the **N**-graded tensor algebra over V. For each **N**-graded **A**-algebra (A, μ, η_A) and each **A**-morphism $f\colon V \to A_1$ there is a unique **N**-graded **A**-morphism $f!$ extending f to an **N**-graded **A**-algebra morphism. In detail,

$$f!_0\colon T_0 \to A_0 = \varepsilon\eta_A,$$

$$f!_{n+1}\colon T_{n+1} \to A_{n+1} = \Gamma_{n,1}(f!_n \otimes f)\mu_{n,1}$$

renders commutative

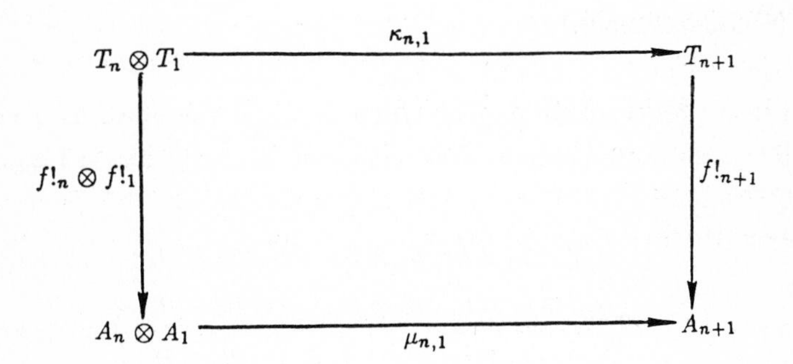

since $f!_1 = \Gamma_{0,1}(\varepsilon\eta_A \otimes f)\mu_{0,1} \approx f$. Simple induction arguments then finish the proof.

Similarly, let (T, Γ, ε) be the **N**-graded cut coalgebra over **A**-object V. For each **N**-graded **A**-coalgebra $(C, \Delta, \varepsilon_C)$ and each **A**-morphism $f\colon C_1 \to V$ there is a unique **N**-graded **A**-morphism $f!\colon C \to T$ extending f to an **N**-graded **A**-coalgebra morphism. Define

$$f!_0\colon C_0 \to T_0 = \varepsilon_C\eta,$$

$$f!_{n+1}\colon C_{n+1} \to T_{n+1} = \Delta_{n,1}(f!_n \otimes f)\kappa_{n,1}$$

to observe that

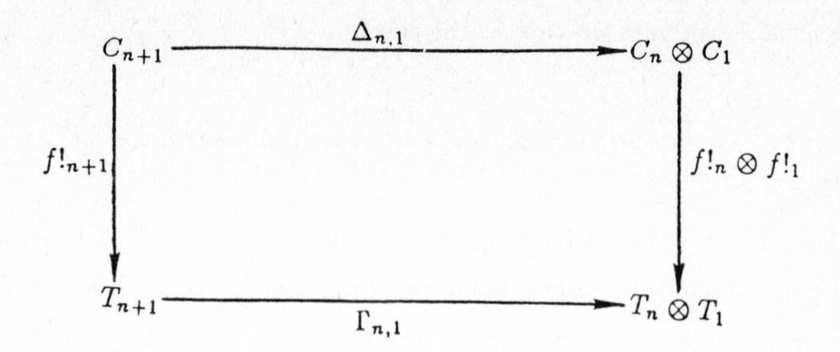

commutes as $f!_1 = \Delta_{0,1}(\varepsilon_C \eta \otimes f)\kappa_{0,1} \approx f$ so that

$$\begin{aligned} f!_{n+1}\Gamma_{n,1} &= \Delta_{n,1}(f!_n \otimes f)\kappa_{n,1}\Gamma_{n,1} \\ &= \Delta_{n,1}(f!_n \otimes f!_1). \end{aligned}$$

Thus (T, κ, η) is the free **N**–graded **A**–algebra over $V \approx T_1$ and (T, Γ, ε) is the cofree **N**–graded **A**–coalgebra over V.

5. Precise N–Graded Bialgebras

5.1 Definition. An **N**–graded **A**–bialgebra $(B, \mu, \Delta, \eta, \varepsilon)$ is *precise* if $B_0 \approx e$ so that $\eta_0\colon e \to B_0$ and $\varepsilon_0\colon B_0 \to e$ are isomorphisms and the components $\mu_{n,0}$, $\mu_{0,n}$, $\Delta_{n,0}$ and $\Delta_{0,n}$ are the specific isomorphisms

$$\mu_{n,0} = \mathbf{1}_n \otimes \varepsilon_0\colon\ B_n \otimes B_0 \to B_n \otimes e \cong B_n,$$

$$\mu_{0,n} = \varepsilon_0 \otimes \mathbf{1}_n\colon\ B_0 \otimes B_n \to e \otimes B_n \cong B_n,$$

$$\Delta_{n,0} = \mathbf{1}_n \otimes \eta_0\colon\ B_n \cong B_n \otimes e \to B_n \otimes B_0,$$

$$\Delta_{0,n} = \eta_0 \otimes \mathbf{1}_n\colon\ B_n \cong e \otimes B_n \to B_0 \otimes B_n,$$

where $\mathbf{1}_n$ denotes the identity on B_n and the left and right unitary isomorphisms specified as part of the monoidality data of **A** are denoted by "$\cong$". We simply say $(B, \mu, \Delta, \eta, \varepsilon)$ is a *precise bialgebra.**

The shuffle–cut bialgebra and the concatenation–spray bialgebra are both precise bialgebras.

5.2 Definitions. When B is an **N**–graded **A**–object, the subobject of $(B \otimes B)_n$, $n > 1$,

$$(B \otimes B)_n^c = \bigoplus_{\substack{i+j=n \\ i,j>0}} B_i \otimes B_j$$

is called the *core* of $(B \otimes B)_n$. The core of $(B \otimes B)_1$ is $(B \otimes B)_1$ itself. If $\mu\colon B \otimes B \to B$ is an **N**–graded **A**–morphism the restriction of μ to the core,

$$\mu_n^c\colon\ (B \otimes B)_n^c \to B_n,$$

---

* [Sweedler] uses a related condition, called *strict*, in vector space categories.

is called the core of μ_n. Dually, if $\Delta\colon B \to B \otimes B$ is an N-graded A-morphism the projection of Δ to the core,

$$\Delta_n^c\colon B_n \to (B \otimes B)_n^c,$$

is called the core of Δ_n. If $h\colon B \to C$ is an N-graded A-morphism between N-graded A-objects, the restriction of $(h \otimes h)_n$ to the core $(B \otimes B)_n^c$ followed by the projection to the core $(C \otimes C)_n^c$ is denoted by $(h \otimes h)_n^c$. Similar notations apply to obtain the core of $B \otimes B \otimes B \otimes B$ and $h \otimes h \otimes h \otimes h$. For example, the core of $(B \otimes B \otimes B \otimes B)_2$ is $(B_1 \otimes B_0 \otimes B_1 \otimes B_0) \oplus (B_0 \otimes B_1 \otimes B_0 \otimes B_1)$.

We say the N-graded A-morphism $\Delta\colon B \to B \otimes B$ is a *monomorphism at the core* if for each $n \in$ N, the core of Δ_n is an A-monomorphism. Similarly, the N-graded A-morphism $\mu\colon B \otimes B \to B$ is an *epimorphism at the core* if for each $n \in$ N, the core of μ_n is an A-epimorphism.

5.3 Lemma. Let $(B, \mu_B, \Delta_B, \eta_B, \varepsilon_B)$ and $(C, \mu_C, \Delta_C, \eta_C, \varepsilon_C)$ be precise bialgebras. Let $h\colon B \to C$ be an N-graded A-morphism. If h is an N-graded A-coalgebra morphism, $h\colon (B, \Delta_B, \varepsilon_B) \to (C, \Delta_C, \varepsilon_C)$, and Δ_C is a monomorphism at the core, then h is an A-bialgebra morphism from $(B, \mu_B, \Delta_B, \eta_B, \varepsilon_B)$ to $(C, \mu_C, \Delta_C, \eta_C, \varepsilon_C)$. Dually, if h is an N-graded A-algebra morphism from (B, μ_B, η_B) to (C, μ_C, η_C) and μ_B is an epimorphism at the core, then h is an A-bialgebra morphism from the given B bialgebra to the given C bialgebra.

Proof: Suppose $h\colon (B, \Delta_B, \varepsilon_B) \to (C, \Delta_C, \varepsilon_C)$ is an N-graded A-coalgebra morphism, where $(B, \mu_B, \Delta_B, \eta_B, \varepsilon_B)$ and $(C, \mu_C, \Delta_C, \eta_C, \varepsilon_C)$ are precise bialgebras. Clearly $\mu_{B,1,0} h_1 = (h_1 \otimes h_0)\mu_{C,1,0}$ as $\mu_{B,1,0}$, h_0, and $\mu_{C,1,0}$ are isomorphisms. Similarly when the 0 and 1 indicies are reversed. Therefore,

$$(\mu_{B,1,0} \otimes \mu_{B,1,0})(h_1 \otimes h_1) = (h_1 \otimes h_0 \otimes h_1 \otimes h_0)(\mu_{C,1,0} \otimes \mu_{C,1,0})$$

and similarly with the 0 and 1 indicies reversed. We now have the basis for an induction using the cube diagram of *3.5*, but with cores on the bottom face of the diagram. All four side faces commute by hypothesis. It remains, therefore, to show by induction that the top face commutes. The proof at each induction step is a diagram chasing argument. For simplicity of notation we have suppressed the induction index n. Chasing around the cube diagram,

$$\begin{aligned}
\mu_B h \Delta_C^c &= \mu_B \Delta_B^c (h \otimes h)^c \\
&= \Delta_{B \otimes B}^c (\mu_B \otimes \mu_B)^c (h \otimes h)^c \\
&= \Delta_{B \otimes B}^c (h \otimes h \otimes h \otimes h)^c (\mu_C \otimes \mu_C)^c \\
&= (h \otimes h) \Delta_{C \otimes C}^c (\mu_C \otimes \mu_C)^c \\
&= (h \otimes h) \mu_C \Delta_C^c,
\end{aligned}$$

show that as Δ_C is a monomorphism at the core we have the commutivity of the top face, $\mu_B h = (h \otimes h)\mu_C$. Since the bialgebras in question are precise, we further deduce that

$$\mu_B^c h = (h \otimes h)^c \mu_C^c$$

and this completes the induction step. The remaining assertion of the lemma is obtained by duality. □

5.4 Corollary. If $(B, \mu, \Delta, \eta, \varepsilon)$ is a precise bialgebra and $(C, \sigma, \Gamma, \eta, \varepsilon)$ is a shuffle–cut bialgebra, an **A**–morphism $f\colon B_1 \to C_1$ extends uniquely to an **N**–graded **A**–bialgebra morphism from the given bialgebra B to the shuffle–cut bialgebra. Dually, if $(A, \kappa, \chi, \eta, \varepsilon)$ is the concatenation–spray bialgebra, an **A**–morphism $g\colon A_1 \to B_1$ extends uniquely to an **N**–graded **A**–bialgebra morphism from the concatenation–spray bialgebra to the given precise bialgebra B.

To obtain any permutation of an ordinary deck of playing cards, first separately obtain the desired ordering among the black cards in one pile and the red cards in another pile. Then shuffle the two piles together. So when we consider the previous result for the unique morphism from the concatenation–spray bialgebra to the shuffle–cut bialgebra over the same grade 1 object, $A_1 \approx C_1$, it is immediate that the unique morphism at component n is the sum of all the permutations in S_n, the symmetric group on n letters. In symbols,

$$h_n\colon A_n \to C_n = \sum_{\pi \in S_n} \pi$$

where A is the concatenation–spray bialgebra and C is the shuffle–cut bialgebra.

Let B be a precise **A**–bialgebra with grade 1 component B_1. Let $A_1 \approx B_1 \approx C_1$ and continue to use the notations of the previous lemma and corollary. There are unique **N**–graded **A**–bialgebra morphisms $f\colon A \to B$ and $g\colon B \to C$ uniquely determined by the isomorphisms $A_1 \approx B_1 \approx C_1$. Furthermore, $fg = h$. One suspects that a pleasant characterization of the precise bialgebra B in terms of f and g is available.

6. Marked Shuffle

The marked shuffle operation shuffles together the strings between distinguished marked entities, but preserves the order at the marks. Let C_1 be some **A**–object which to aid intuition may be thought of as the collection of ordinary, unmarked cards. The shuffle–cut bialgebra $(C, \sigma, \Gamma, \eta, \varepsilon)$ describes the strings and the operations on the ordinary cards. We now view $C \otimes C$ as describing strings containing an intermediate marked card. Any notation for the intermediate marked card is suppressed, as we assume that the matching of the distinguished cards has actually taken place. Similarly, $C \otimes C \otimes C$ describes strings containing two intermediate marked cards and so on. Now each of $C \otimes C$, $C \otimes C \otimes C, \ldots$ inherits a bialgebra structure from $(C, \sigma, \Gamma, \eta, \varepsilon)$ by the very existence of the bialgebra structure on C.

To begin, when (A, μ, η) is an **A**–algebra, $(A \otimes A, \mu_{A\otimes A}, \eta \otimes \eta)$ is also an **A**–algebra when $\mu_{A\otimes A} = (1 \otimes \tau \otimes 1)(\mu \otimes \mu)$. The morphism $h\colon A \to A \otimes A$ is an **A**–algebra morphism from (A, μ, η) to $(A \otimes A, \mu_{A\otimes A}, \eta \otimes \eta)$ when the diagram

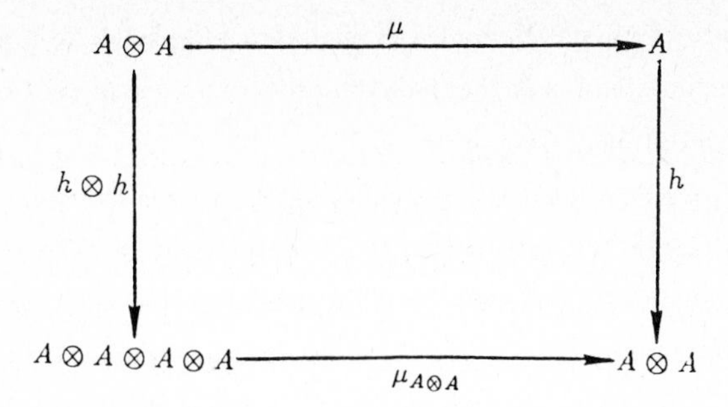

commutes. Since C is equipped with the shuffle–cut bialgebra structure, $\Gamma\colon C \to C \otimes C$ is an $\mathbf{A}$–algebra morphism from (C, σ, η) to $(C \otimes C, \sigma_{C\otimes C}, \eta \otimes \eta)$. Dually, $\sigma\colon C \otimes C \to C$ is an $\mathbf{A}$–coalgebra morphism from $(C \otimes C, \Gamma_{C\otimes C}, \varepsilon \otimes \varepsilon)$ to (C, Γ, ε).

Now $\sigma_{C\otimes C}\colon (C \otimes C) \otimes (C \otimes C) \to C \otimes C$ is the desired marked shuffle of two strings each containing one intermediate mark. Similarly, $\sigma_{C\otimes C\otimes C}$ is the desired marked shuffle of two sequences each containing two intermediate marks.

We now change notation. Let $M_{(0)} = e$, $M_{(n+1)} = M_{(n)} \otimes C$ be the sequence of precise bialgebras alluded to above. Let

$$\begin{aligned} \sigma_{(0)} &= \mathbf{1}_e, \\ \sigma_{(n+1)} &= (\mathbf{1} \otimes \tau \otimes \mathbf{1})(\sigma_{(n)} \otimes \sigma)\colon M_{(n+1)} \otimes M_{(n+1)} \to M_{(n+1)} \end{aligned}$$

be the sequence of marked shuffle algebra maps. Finally, let

$$\begin{aligned} \Gamma_{(0)} &= \mathbf{1}_e, \\ \Gamma_{(n+1)} &= (\Gamma_{(n)} \otimes \Gamma)(1 \otimes \tau \otimes 1)\colon M_{(n+1)} \to M_{(n+1)} \otimes M_{(n+1)} \end{aligned}$$

be the sequence of cut coalgebra maps. As η,ε for each of the $M_{(n)}$ are just the η,ε of C, we have a sequence of bialgebras $M_{(n)}$ with connecting morphisms

$$M_{(0)} \underset{\varepsilon}{\overset{\eta}{\rightleftarrows}} M_{(1)} \underset{\mathbf{1}_0 \otimes \sigma}{\overset{\mathbf{1}_0 \otimes \Gamma}{\rightleftarrows}} M_{(2)} \underset{\mathbf{1}_1 \otimes \sigma}{\overset{\mathbf{1}_1 \otimes \Gamma}{\rightleftarrows}} \cdots$$

where $\mathbf{1}_n$ denotes the identity morphism on $M_{(n)}$. Each $\mathbf{1}_n \otimes \Gamma$ is to be thought of as introducing a new mark while each $\mathbf{1}_n \otimes \sigma$ forgets the last intervening mark to shuffle the last two strings together.

The sequence of morphisms from $M_{(n)}$, $n > 0$, to $M_{(1)}$

$$\begin{aligned} \sigma^{(0)} &= \mathbf{1}_1, \\ \sigma^{(n+1)} &= (\mathbf{1}_n \otimes \sigma)\sigma^{(n)} \end{aligned}$$

provides the sequence of unique **N**-graded **A**-bialgebra morphisms to the shuffle–cut bialgebra. Each $\sigma^{(n)}$ is the unique extension of the isomorphism between the grade 1 component of $M_{(n)}$ and the grade 1 component of $M_{(1)}$, C_1.

Each $\sigma^{(n)}$ forgets all the marks, and shuffles all the strings together. In the context of distributed computing theory, this is a significant loss of information. Therefore, in this setting one may view a string containing a single intermediate marked card as an element of $C \otimes C_1 \otimes C$, with the central factor C_1 denoting the position of the marked card. Forgetting the distinguished status of the marked card is equivalent to forgetting the bialgebra structure necessary to shuffle the first factors and the last factors as described earlier. Forgetting the distinguished status of the central factor may now be viewed as the concatenation

$$(\mathbf{1} \otimes \kappa)\kappa\colon\ C \otimes C_1 \otimes C \to C.$$

Since the central factor is only C_1, this concatenation is not an algebra structure map. This concatenation is not an algebra morphism for shuffle nor is it a coalgebra morphism for cut. Thus further analysis appears desirable.

References

1. M. A. Arbib and E. B. Manes, Adjoint Machines, State–Behavior Machines, and Duality, J. Pure Appl. Alg. **6**(1975), 313–344.
2. M. Broy, Semantics of Communicating Processes, Inform. & Control **61**(1984), 202–241.
3. S. Eilenberg and G. M. Kelly, Closed Categories *in* Proc. Conf. Categorical Alg., La Jolla 1965, Springer–Verlag, New York, 1966, pp. 421–562.
4. S. Eilenberg & S. MacLane, On the Groups $H(\prod, n)$. I., Ann. of Math **58**(1953), 55–106.
5. N. Francez, *Fairness*, Springer–Verlag, New York, 1986.
6. H. Gaifman and V. Pratt, Partial Order Models of Concurrency and the Computation of Functions, Proc. IEEE Symp. Logic in Comput. Sci., Ithaca, NY, 1987.
7. S. Ginsburg, *The Mathematical Theory of Context–Free Languages*, McGraw–Hill, 1966.
8. S. Ginsburg and E. H. Spanier, Mapping of Languages by Two–tape Devices, *J. Assoc. Comput. Mach.* **12**(1965), 423–434.
9. J. L. Gischer, Partial Orders and the Axiomatic Theory of Shuffle, Stanford Univ. Report STAN–CS–84–1033.
10. H. Herrlich & G. Strecker, *Category Theory*, Heldermann–Verlag, Berlin, 1979.
11. R. J. Lorentz & D. B. Benson, Deterministic and Nondeterministic Flowchart Interpretations, J. Comput. Sys. Sci. **27**(1983), 400–433.
12. S. MacLane, *Categories for the Working Mathematician*, Springer–Verlag, New York, 1971.
13. S. MacLane, *Homology*, Academic Press, New York, 1963.